Smart Innovation, Systems and Technologies

Volume 221

The Smart Innovation, Systems and Technologies book series encompasses the topics of knowledge, intelligence, innovation and sustainability. The aim of the series is to make available a platform for the publication of books on all aspects of single and multi-disciplinary research on these themes in order to make the latest results available in a readily-accessible form. Volumes on interdisciplinary research combining two or more of these areas is particularly sought.

The series covers systems and paradigms that employ knowledge and intelligence in a broad sense. Its scope is systems having embedded knowledge and intelligence, which may be applied to the solution of world problems in industry, the environment and the community. It also focusses on the knowledge-transfer methodologies and innovation strategies employed to make this happen effectively. The combination of intelligent systems tools and a broad range of applications introduces a need for a synergy of disciplines from science, technology, business and the humanities. The series will include conference proceedings, edited collections, monographs, handbooks, reference books, and other relevant types of book in areas of science and technology where smart systems and technologies can offer innovative solutions.

High quality content is an essential feature for all book proposals accepted for the series. It is expected that editors of all accepted volumes will ensure that contributions are subjected to an appropriate level of reviewing process and adhere to KES quality principles.

Indexed by SCOPUS, EI Compendex, INSPEC, WTI Frankfurt eG, zbMATH, Japanese Science and Technology Agency (JST), SCImago, DBLP.

All books published in the series are submitted for consideration in Web of Science.

More information about this series at http://www.springer.com/series/8767

Amaresh Chakrabarti · Ravi Poovaiah ·
Prasad Bokil · Vivek Kant
Editors

Design for Tomorrow—Volume 1

Proceedings of ICoRD 2021

Springer

Editors
Amaresh Chakrabarti
Centre for Product Design
and Manufacturing
Indian Institute of Science, Bangalore
Bengaluru, Karnataka, India

Prasad Bokil
IDC School of Design
Indian Institute of Technology Bombay
Mumbai, India

Ravi Poovaiah
IDC School of Design
Indian Institute of Technology Bombay
Mumbai, India

Vivek Kant
IDC School of Design
Indian Institute of Technology Bombay
Mumbai, India

ISSN 2190-3018 ISSN 2190-3026 (electronic)
Smart Innovation, Systems and Technologies
ISBN 978-981-16-0040-1 ISBN 978-981-16-0041-8 (eBook)
https://doi.org/10.1007/978-981-16-0041-8

This Springer imprint is published by the registered company Springer Nature Singapore Pte Ltd.
The registered company address is: 152 Beach Road, #21-01/04 Gateway East, Singapore 189721, Singapore

About the Conference

Design is ubiquitous; it pervades all spheres of life and has been around as long as life has taken up the task of purposefully changing the world around it. Research into design and the emergence of a research community in this area have been relatively new. Its development has been influenced by the multiple facets of design (human, artefact, process, organisation, the micro- and macro-economy and the ecology by which design is shaped) and the associated diversification of the community depending on the facets of focus or that of their applications. Design is complex, balancing the needs of multiple stakeholders and requiring a multitude of areas of knowledge to be utilised, with resources spread across space and time.

ICoRD'21 is the eighth in a series of conferences intended to be held every two years in India to bring together the international community from diverse areas of design practice, education and research. It aims to showcase cutting-edge research about design to the stakeholders; aid the ongoing process of developing and extending the collective vision through emerging research challenges and questions; and provide a platform for interaction, collaboration and development of the community in order for it to take up the challenges to realise the vision. The conference is intended for all stakeholders of design and, in particular for its practitioners, researchers, pupils and educators.

The collection of papers in these two book volumes constitutes the Proceedings of the Eighth International Conference on Research into Design (ICoRD'21) held on virtual platform during 7 to 10 January 2021 at the IDC School of Design, IIT Bombay, Powai, Mumbai.

ICoRD series was initiated by Centre for Product Design and Manufacturing (CPDM) at Indian Institute of Science (IISc), Bangalore, in 2006. Since then it has been hosted in 2009 and 2011 (both at IISc), 2013 (at IIT Madras), 2015 (IISc), 2017 (IIT Guwahati) and 2019 (IISc). CPDM has pioneered design research in India for the last two decades. IISc is one of India's leading science and technology institutions and is one of the Institutes of Eminence decreed by MHRD, Government of India.

ICoRD'21 has been organised jointly by IDC, IIT Bombay, and CPDM, IISc Bangalore. ICoRD'21 has been hosted on virtual platform by IDC School of Design, Indian Institute of Technology Bombay, Mumbai, India. IDC School of Design has overseen the inception of design in a free India for the past 50 years. IIT Bombay is one of India's leading technological institutions and is decreed by the Government of India as an Institute of National Importance. As a city, Mumbai serves as a home to a variety of technological and service sectors.

The theme of ICoRD'21 has been "Design for Tomorrow". The world as we know it in our times is increasingly becoming connected. In this interconnected world, design has to address new challenges of merging the cyber and the physical, the smart and the mundane, and the technology and the human. As a result, there is an increasing need for strategising and thinking about design for a better tomorrow. Our theme for ICoRD'21 serves as a provocation for the design community to think about rapid changes in the near future to usher in a better tomorrow.

The conference contained:

- Keynote presentations from eminent (inter)national experts and practitioners
- Presentations of refereed papers as podium presentations
- Panel discussions to present perspectives on topics of general interest
- A series of workshops on topics of special interest
- Networking sessions for young researchers.

Sponsors

Centre for Product Design and Manufacturing, Indian Institute of Science, Bangalore, India
Indian Institute of Science, Bangalore, India
Springer

Endorsers

The Design Society, UK
Design Research Society, UK

Conference Organisation

Steering Committee

Balan Gurumoorthy, Indian Institute of Science, Bangalore, India
Cees de Bont, Loughborough University, UK
John Gero, University of North Carolina, Charlotte, USA
Kota Harinarayana, National Aeronautical Laboratories, India
Panos Papalambros, University of Michigan, USA
Steve Culley, University of Bath, UK

Advisory Committee

S. Saleem Ahmed, Vellore Institute of Technology, Vellore, India
Tracy Bhamra, Loughborough University, UK
K. S. Bhanumathi, Kovair, India
Henri Christiaans, Ulsan National Institute of Science and Technology, South Korea
Anjan Das, Confederation of Indian Industry, India
U. B. Desai, Indian Institute of Technology Hyderabad, India
Kees Dorst, University of Technology Sydney, Australia
Srinivasan Dwarakanath, Airbus, India
Prem Shankar Goel, National Institute of Advanced Studies, Indian Institute of Science, Bangalore, India
Anil K. Gupta, Honey Bee Network, SRISTI and GIAN, India
Imre Horvath, Delft University of Technology, Netherlands
Jitendra Jadhav, CSIR-National Aerospace Laboratories, Bangalore, India
Arun Jaura, Michelin, India
Deep Kapuria, The Hi-Tech Gears, India
Yong Se Kim, Sungkyunkwan University, South Korea
Ananda Kumar, LPerspective, India
Larry Leifer, Stanford University, USA

Banshildar Majhi, Indian Institute of Information Technology, Design and Manufacturing, Kancheepuram, India
Udayant Malhoutra, Dynamatic Technologies Limited, Bangalore, India
K. Rajalakshmi Menon, Centre for Airborne Systems, Defence Research and Development Organisation, India
Arabinda Mitra, Office of PSA to GOI India
T. S. Mruthyunjaya, Chairman Emeritus, CPDM, Indian Institute of Science, India
M. Muruganant, Indus University, India
Nagahanumaiah,, Central Manufacturing Technology Institute, India
Shailesh Nayak, National Institute of Advanced Studies, Indian Institute of Science, Bangalore, India
Srinivas Nidamarthi, ABB Ltd., Michigan, USA
Sudarsan Rachuri, Department of Energy, USA
P. Radhakrishnan, PSG Institute of Advanced Studies, India
V. Bhujanga Rao, National Institute of Advanced Studies, Indian Institute of Science, Bangalore, India
B. Ravi, Indian Institute of Technology Bombay, India
Sofia Ritzén, KTH Royal Institute of Technology, Sweden
Rajkumar Roy, City, University of London, UK
Al Salour, The Boeing Company, USA
Raman Saxena, USID Foundation, India
Reena Sharma, Defence Research and Development Organisation, India
Vijay Srinivasan, National Institute of Standards and Technology, USA
B. N. Suresh, Indian Institute of Space Science and Technology, India
Kristin L. Wood, Singapore University of Technology and Design, Singapore

Conference Chair

Ravi Poovaiah, Indian Institute of Technology Bombay, India

Programme Chair

Amaresh Chakrabarti, Indian Institute of Science, Bangalore, India

Vice-Chairs

Prasad Bokil, Indian Institute of Technology Bombay, India
Vivek Kant, Indian Institute of Technology Bombay, India

Co-chairs

Lucienne Blessing, Singapore University of Technology and Design, Singapore
Monica Bordegoni, Politecnico Di Milano, Italy
Debkumar Chakrabarti, Indian Institute of Technology Guwahati, India
John Clarkson, University of Cambridge, UK
Benoit Eynard, University of Technology of Compiègne, France
Tim Mcaloone, Technical University of Denmark, Denmark
Yukari Nagai, Japan Advanced Institute of Science and Technology, Japan
Luciana Pereira, Federal University of ABC, Brazil
Raghu Prakash, Indian Institute of Technology Madras, India
Sandro Wartzack, University of Erlangen-Nuremberg, Germany

International Programme Committee

Shakuntala Acharya, Indian Institute of Technology Guwahati, India
Sofiane Achiche, Polytechnique Montréal, Canada
Rajat Agrawal, Indian Institute of Technology Roorkee, India
Janet Allen, University of Oklahoma, USA
Manish Arora, Indian Institute of Science, Bangalore, India
Sandipan Bandyopadhyay, Indian Institute of Technology Madras, India
Niccolò Becattini, Politecnico Di Milano, Italy
Alessandro Bertoni, Blekinge Institute of Technology, Sweden
Marco Bertoni, Blekinge Institute of Technology, Sweden
Mehul Bhatt, Örebro University, Sweden
Ritwij Bhowmik, Indian Institute of Technology Kanpur, India
Braj Bhushan, Indian Institute of Technology Kanpur, India
Abhijit Biswas, Indian Institute of Science, Bangalore, India
Pradipta Biswas, Indian Institute of Science, Bangalore, India
Jeremy Bonvoisin, University of Bath, UK
Yuri Borgianni, Free University of Bozen-Bolzano, Italy
David C. Brown, Worcester Polytechnic Institute, USA
Francesco Buonamici, University of Florence, Italy
Felician Campean, University of Bradford, UK
Gaetano Cascini, Politecnico Di Milano, Italy
Philip Cash, Technical University of Denmark, Denmark
Amaresh Chakrabarti, Indian Institute of Science, Bangalore, India
Jayanta Chatterjee, Indian Institute of Technology Kanpur, India
Yong Chen, Shanghai Jiao Tong University, China
Peter Childs, Imperial College London, UK
François Cluzel, University of Paris-Saclay, France
Nathan Crilly, University of Cambridge, UK

Umberto Cugini, Politecnico Di Milano, Italy
Pranab K. Dan, Indian Institute of Technology Kharagpur, India
Amarendra Kumar Das, Indian Institute of Technology Guwahati, India
Suman Devadula, MIT World Peace University, India
Debayan Dhar, Indian Institute of Technology Guwahati, India
Jay Dhariwal, Indian Institute of Technology Delhi, India
J. E. Diwakar, Indian Institute of Science, Bangalore, India
Alex Duffy, University of Strathclyde, UK
Raghu Echempati, Kettering University, USA
Claudia Eckert, The Pennsylvania State University, USA
Boris Eisenbart, Swinburne University of Technology, Australia
Yvonne Eriksson, Mälardalen University, Sweden
Georges Fadel, Clemson University, USA
Mario Fargnoli, Sapienza University of Rome, Italy
Francesco Ferrise, Politecnico Di Milano, Italy
Susan Finger, Carnegie Mellon University, USA
Kikuo Fujita, Osaka University, Japan
L. S. Ganesh, Indian Institute of Technology Madras, India
Georgi V. Georgiev, University of Oulu, Finland
Detlef Gerhard, Ruhr-University Bochum, Germany
Kilian Gericke, University of Rostock, Germany
Ashitava Ghosal, Indian Institute of Science, Bangalore, India
Philippe Girard, University of Bordeaux, France
Milane G. Goncalves, Delft University of Technology, Netherlands
Ravindra Goonetilleke, The Hong Kong University of Science and Technology, Hong Kong
Lapo Governi, University of Florence, Italy
Serena Graziosi, Politecnico Di Milano, Italy
Matthias Guertler, University of Technology Sydney, Australia
Bala Murali Gunji, Vellore Institute of Technology, Vellore, India
Sean Hanna, University College London, UK
Bisheshwar Haorongbam, National Institute of Design Assam, India
Peter Hehenberger, University of Applied Sciences Upper Austria, Austria
Ben Hicks, University of Bristol, UK
Stephan Husung, :Em Engineering Methods AG, Darmstadt, Germany
David Inkermann, Technical University of Clausthal, Germany
William Ion, University of Strathclyde, UK
Ashok Iyer, Manipal Academy of Higher Education—Dubai Campus, UAE
Santosh Jagtap, Blekinge Institute of Technology, Sweden
R. Jayaganthan, Indian Institute of Technology Madras, India
Camille Jean, Arts et Métiers, France
Shiva Ji, Indian Institute of Technology Hyderabad, India
Anirudha Joshi, Indian Institute of Technology Bombay, India
Purba Joshi, Indian Institute of Technology Bombay, India
Iestyn Jowers, The Open University, UK

Tero Juuti, University of Tampere, Finland
Pratul Kalita, Indian Institute of Technology Guwahati, India
Udo Kannengiesser, Johannes Kepler University Linz, Austria
Sougata Karmakar, Indian Institute of Technology Guwahati, India
Sonal Keshwani, Indian Institute of Science, Bangalore, India
Mohammed Rajik Khan, National Institute of Technology Rourkela, India
Maaike Kleinsmann, Delft University of Technology, Netherlands
Edwin Koh, Singapore University of Technology and Design, Singapore
Christian Köhler, Saarland University of Applied Sciences, Germany
Saptarshi Kolay, Indian Institute of Technology Roorkee, India
Lauri Koskela, University of Huddersfield, UK
Srinivas Kota, Birla Institute of Technology and Science, Pilani, India
Dieter Krause, Hamburg University of Technology, Germany
Vinayak Krishnamurthy, Texas A&M University, USA
Rishikesha Krishnan, Indian Institute of Management Indore, India
Tore Kristensen, Copenhagen Business School, Denmark
Peter Krogh, Aarhus University, Denmark
Petter Krus, Linköping University, Sweden
Jyoti Kumar, Indian Institute of Technology Delhi, India
Udaya Kumar, Indian Institute of Technology Guwahati, India
Senthil Kumaran, Indian Institute of Information Technology, Design and Manufacturing, Kancheepuram, India
Guillaume Lamé, University of Paris-Saclay, France
Tobias Larsson, Blekinge Institute of Technology, Sweden
Sang Leigh, Massachusetts Institute of Technology, USA
Torben Lenau, Technical University of Denmark, Denmark
Rungtai Lin, National Taiwan University of Arts, Taiwan
Ying Liu, Cardiff University, UK
Ying Chieh Liu, Chang Gung University, Taiwan
Joaquim Lloveras, Polytechnic University of Catalonia, Spain
Jianxi Luo, Singapore University of Technology and Design, Singapore
N. Madhusudanan, Corporate Technology, Siemens, India
Johan Malmqvist, Chalmers University of Technology, Sweden
Monto Mani, Indian Institute of Science, Bangalore, India
M. Manivannan, Indian Institute of Technology Madras, India
Nicolas Maranzana, Arts et Métiers ParisTech, France
Dorian Marjanović, University of Zagreb, Croatia
Christian Marxt, ETH Zürich, Switzerland
Yoshiyuki Matsuoka, Keio University, Japan
Deepak John Matthew, Indian Institute of Technology Hyderabad, India
Jason Matthews, University of the West of England, UK
Christopher Mccomb, The Pennsylvania State University, USA
Alison Mckay, University of Leeds, UK
Gavin Melles, Swinburne University of Technology, Australia

Subhas Mondal, Indian Institute of Engineering Science and Technology, Shibpur, India
Charu Monga, Indian Institute of Technology Guwahati, India
Prabir Mukhopadhyay, Indian Institute of Information Technology, Design and Manufacturing, Jabalpur, India
Elena Mulet, Universitat Jaume I, Spain
Raguraman Munusamy, Indian Institute of Information Technology, Design and Manufacturing, Kancheepuram, India
Tamotsu Murakami, University of Tokyo, Japan
Ramanathan Muthuganapathy, Indian Institute of Technology Madras, India
Nanki Nath, National Institute of Design, Ahmedabad, India
Oscar Nespoli, University of Waterloo, Canada
Susanne Nilsson, KTH Royal Institute of Technology, Sweden
Josef Oehmen, Technical University of Denmark, Denmark
Prasad Onkar, Indian Institute of Technology Hyderabad, India
Swati Pal, Indian Institute of Technology Bombay, India
Massimo Panarotto, Chalmers University of Technology, Sweden
Jitesh Panchal, Purdue University, USA
Koumudi Patil, Indian Institute of Technology Kanpur, India
Neven Pavkovic, University of Zagreb, Croatia
Vesna Popovic, Queensland University of Technology, Australia
Sebastien Proulx, The Ohio State University, USA
Ravi Punekar, Indian Institute of Technology Guwahati, India
K. Radhakrishnan, Central Institute of Technology, India
Palaniappan Ramu, Indian Institute of Technology Madras, India
B. S. C. Ranjan, Indian Institute of Science, Bangalore, India
Balkrishna Rao, Indian Institute of Technology Madras, India
N. V. Chalapathi Rao, Indian Institute of Science, Bangalore, India
P. V. M. Rao, Indian Institute of Technology Delhi, India
N. Venkata Reddy, Indian Institute of Technology Hyderabad, India
Caterina Rizzi, University of Bergamo, Italy
Sergio Rizzuti, University of Calabria, Italy
Paul Rodgers, University of Strathclyde, UK
Serge Rohmer, University of Technology of Troyes, France
Anna Öhrwall Rönnbäck, Luleå University of Technology, Sweden
Federico Rotini, University of Florence, Italy
Satyaki Roy, Indian Institute of Technology Kanpur, India
Tomohiko Sakao, Linköping University, Sweden
B. Santhi, National Institute of Technology Goa, India
Gurunathan Saravana Kumar, Indian Institute of Technology Madras, India
Biplab Sarkar, Nokia, India
Prabir Sarkar, Indian Institute of Technology Ropar, India
Peer Sathikh, Nanyang Technological University, Singapore
Keiichi Sato, Illinois Institute of Technology, USA
Michael Schabacker, Otto von Guericke University Magdeburg, Germany

Dirk Schaefer, University of Liverpool, UK
Vinodh Sekar, National Institute of Technology, Tiruchirappalli, India
Harivardhini Selvakumar, PES University, India
Dibakar Sen, Indian Institute of Science, Bangalore, India
Joy Sen, Indian Institute of Technology Kharagpur, India
Anshuman Sharma, TCS, India
Apurbba Kumar Sharma, Indian Institute of Technology Roorkee, India
Arlindo Silva, Singapore University of Technology and Design, Singapore
Inderdeep Singh, Indian Institute of Technology Roorkee, India
Sumer Singh, Indian Institute of Technology Delhi, India
Vishal Singh, Indian Institute of Science, Bangalore, India
Ricardo Sosa, Auckland University of Technology, New Zealand
Mario Storga, University of Zagreb, Croatia
Eswaran Subrahmanian, Carnegie Mellon University, USA
Shankar Subramanian, Indian Institute of Technology Madras, India
Abinash Swain, Indian Institute of Technology Roorkee, India
Puneet Tandon, Indian Institute of Information Technology, Design and Manufacturing, Jabalpur, India
Asokan Thondiyath, Indian Institute of Technology Madras, India
Anshuman Tripathy, Indian Institute of Management Bangalore, India
Yasushi Umeda, University of Tokyo, Japan
Paul Varghese, Trysquare Engineers and Builders, India
Nilesh Vasa, Indian Institute of Technology Madras, India
Gokula Vasantha, Edinburgh Napier University, UK
Srikanth Vedantam, Indian Institute of Technology Madras, India
Srinivasan Venkataraman, Indian Institute of Technology Delhi, India
Alok Verma, Old Dominion University, USA
Nikola Vukasinovic, University of Ljubljana, Slovenia
Fedja Vukic, University of Zagreb, Croatia
Parag Vyas, GrauBär Design Studios, India
Robert Wendrich, University of Twente, Netherlands
Andrew Wodehouse, University of Strathclyde, UK
Pradeep Yammiyavar, Indian Institute of Technology Guwahati, India
Yong Zeng, Concordia University, Canada

Organising Committee

Shakuntala Acharya, Indian Institute of Technology Guwahati, India
Rupa Agarwal, National Institute of Fashion Technology, India
Prasad Bokil, Indian Institute of Technology Bombay, India
Vivek Kant, Indian Institute of Technology Bombay, India
Ajanta Sen, Indian Institute of Technology Bombay, India
Avinash Shende, Indian Institute of Technology Bombay, India

Web Admin

Roopa Narayan Sahoo, IDC, Indian Institute of Technology Bombay, India

Student Organising Committee

Abdul Mohsin Ali, Indian Institute of Technology Guwahati, India
Shivalika Bansal, Indian Institute of Technology Bombay, India
Apoorv Naresh Bhatt, Indian Institute of Science, Bangalore, India
Kaushik Bhattacharya, Indian Institute of Science, Bangalore, India
Debajit Brahma, Indian Institute of Technology Bombay, India
Spandan Das, Indian Institute of Technology Guwahati, India
Deepshikha Dash, Indian Institute of Technology Bombay, India
Arshiya Gahlot, Indian Institute of Technology Bombay, India
Felnunmoi Gangte, Indian Institute of Technology Bombay, India
Anubhav Gangwar, Indian Institute of Technology Bombay, India
Divya Gaur, Indian Institute of Technology Bombay, India
Kiran Ghadge, Indian Institute of Science, Bangalore, India
Bhavana Ghanta, Indian Institute of Technology Bombay, India
Puneeth K. S., Indian Institute of Science, Bangalore, India
Anirvina Karnala, Indian Institute of Technology Bombay, India
Ishaan Kaushal, Indian Institute of Science, Bangalore, India
Shreya Kondvilkar, Indian Institute of Technology Bombay, India
John Kujur, Indian Institute of Technology Bombay, India
Anurag Kumar Singh, Indian Institute of Technology Bombay, India
Siddharth Kumar, Indian Institute of Technology Guwahati, India
Anubhab Majumder, Indian Institute of Science, Bangalore, India
Saikat Mondal, Indian Institute of Technology Bombay, India
Deblina Pandit, Indian Institute of Technology Bombay, India
Arivukkarasu Periyasamy, Indian Institute of Technology Bombay, India
Stuti Prakashkumar, Indian Institute of Technology Bombay, India
Subin Raj, Indian Institute of Science, Bangalore, India
Priyanka Rawal, Indian Institute of Technology Bombay, India
Madhuri S,, Indian Institute of Technology Bombay, India
Nishath Salma, Indian Institute of Science, Bangalore, India
Pratiti Sarkar, Indian Institute of Technology Bombay, India
Rounak Sengupta, Indian Institute of Technology Bombay, India
Namrita Sharma, Indian Institute of Technology Guwahati, India
Shivangi, Indian Institute of Technology Bombay, India
Sanjay Singh, Indian Institute of Science, Bangalore, India
Paridhi Subbaian Kaliamoorthy, Indian Institute of Science, Bangalore, India
JayaSurya Sudakaran, Indian Institute of Technology Bombay, India
Charu Tripathi, Indian Institute of Science, Bangalore, India

Preface

Design is ubiquitous; it pervades all spheres of life and has been around ever since life has been engaged in purposefully changing the world around it. While some designs have transcended time, most designs are in a perpetual process of being evolved. Research into design and the emergence of a research community in this area have been relatively new. Its development has been influenced by the multiple facets of design (human, artefact, process, organisation, ecology, micro- and macro-economy by which design is shaped and which it shapes in turn) and the associated diversification of the community depending on the facets of focus or that of their applications. Design is complex, balancing the needs of multiple stakeholders and requiring a multitude of areas of knowledge to be utilised, with resources spread across space and time.

The collection of papers in these two book volumes constitutes the Proceedings of the Eighth International Conference on Research into Design (ICoRD'21) held at Indian Institute of Technology, Bombay, India, (this time on virtual platform), during 7–10 January 2021. ICoRD'21 is the eight in a series of biennial conferences held in India to bring together the international community from diverse areas of design practice, teaching and research. The goals are to share cutting-edge research about design among its stakeholders; aid the ongoing process of developing a collective vision through emerging research challenges and questions; and provide a platform for interaction, collaboration and development of the community in order for it to address the global and local challenges by forming and realising the collective vision. The conference is intended for all stakeholders of design and, in particular, for its practitioners, researchers, teachers and students.

Five hundred and fifty-four abstracts were submitted to ICoRD'21, from which 529 were accepted for full paper submission. A total of 295 full papers were submitted, which were reviewed by experts from the ICoRD'21 International Programme Committee comprising 238 members from over 149 institutions or organisations from 27 countries spanning five continents. Finally, 234 full papers, authored by 460 researchers (460 unique authors, actually 606 authors' entries in 236 papers) from 21 countries spanning 5 continents, were selected for presentation at the conference and for publication as chapters in this book. ICoRD has steadily grown over the last seven editions, from a humble beginning in 2006 with 30 papers and 60 participants, through 75 papers and 100 participants in ICoRD'09, 100 papers and 150 participants in ICoRD'11, 114 papers and 170 participants in ICoRD'13, 118 papers and 200 participants in ICoRD'15, 178 papers and 230 participants in ICoRD'17 and 169 papers and 352 participants in ICoRD'19.

All papers were presented in ICoRD'21 in the podium mode. It had keynotes from prominent researchers and practitioners from around the world such as Steve Eppinger, MIT, USA; Punya Mishra, Arizona State University, USA; Armand Hatchuel, MinesParisTech, France; Tetsuo Tomiyama, International Professional University, Tokyo, Japan; Paul Hekkert, TU Delft, The Netherlands; Tomas Ramos, Nova University, Lisbon, Portugal; Dibakar Sen, Indian Institute of Science, Bangalore, India, etc.

The chapters in the three book volumes together cover all three major areas of products and processes: functionality, form and human factors. The spectrum of topics ranges from those focusing on early stages such as creativity and synthesis, through those that are primarily considered in later stages of the product life cycle, such as safety, reliability or manufacturability, to those that are relevant across the whole product life cycle, such as collaboration, communication, design management, knowledge management, cost, environment and product life cycle management. Issues of delivery of research into design, in terms of its two major arms: design education and practice, are both highlighted in the chapters of the book volumes. Foundational topics such as the nature of design theory and research methodology are also major areas of focus. It is particularly encouraging to see in the chapters the variety of areas of application of research into design—aerospace, health care, automotive, biomedical and so on.

The theme of ICoRD'21 is 'Design for Tomorrow'. The world as we know it in our times is increasingly becoming connected. In this interconnected world, design has to address new challenges of merging the cyber and the physical, the smart and the mundane, and the technology and the human. As a result, there is an increasing need for strategising and thinking about design for a better tomorrow. Our theme for ICoRD'21 serves as a provocation for the design community to think about rapid changes in the near future to usher in a better tomorrow. ICoRD'21 is organised jointly by IDC, IIT Bombay, and CPDM, IISc Bangalore. IDC School of Design has overseen the inception of design in a free India for the past 50 years. IIT Bombay is one of India's leading technological institutions and is decreed by the Government of India as an Institute of National Importance. As a city, Mumbai serves as a home to a variety of technological and service sectors.

On behalf of the Steering Committee, Advisory Committee, Organising Committees and Co-Chairs, we thank all the authors, delegates, institutions and organisations that participated in the conference. We also thank the members of the International Programme Committee for their support in reviewing the papers for ICoRD'21, which is essential for maintaining the quality of the conference, and for their support in putting this book together.

We are thankful to the Design Society and Design Research Society for their kind endorsement of ICoRD'21. We thank Indian Institute of Technology (IIT) Bombay and its IDC School of Design, and Indian Institute of Science (IISc), Bangalore, and its Centre for Product Design and Manufacturing (CPDM), for their support of this event. We also wish to place on record and acknowledge the enormous support provided by the Student Organising Committee in managing the review process, in preparation of the conference programme and this book and the conference as a whole. In particular, we wish to thank Mr. Apoorv Bhatt, Mr. Roopa Narayan Sahoo and Ms. Nishath Salma for their valuable contributions. We also thank the large and dedicated group of student volunteers of IIT Bombay, IISc Bangalore, IIT Guwahati and other institutions in the organisation of the conference. Finally, we thank Springer, especially its Editor Ms. Swati Meherishi and its editorial support team, for their wonderful support, including their sponsoring of Springer book vouchers for the Winners of Distinguished Paper Awards at ICoRD'21.

Bengaluru/Mumbai, India

Amaresh Chakrabarti
Ravi Poovaiah
Prasad Bokil
Vivek Kant

Contents

Part IV Eco-design, Sustainable Manufacturing, Design for Sustainability

About the Editors

Prof. Amaresh Chakrabarti is a Senior Professor and current Chairman for the Centre for Product Design and Manufacturing, Indian Institute of Science (IISc), Bangalore. He has B.E. in Mechanical Engineering from Indian Institute of Engineering Science and Technology, Shibpur, M.E. in Design of Mechanical Systems from Indian Institute of Science, Bengaluru, and Ph.D. in Engineering Design from the University of Cambridge, UK. He led the Design Synthesis group at the EPSRC funded Engineering Design Centre at the University of Cambridge before joining IISc as an Associate Professor in 2001. His interests are in synthesis, creativity, sustainability, and informatics. He has published 16 books, over 300 peer-reviewed articles, and has 11 patents granted/pending. He has been on the Advisory Board and the Board of Management of the Design Society, UK; member of the CII National Committee on Design India; member of the Jury of India Design Mark; and member of the CII Smart Manufacturing Council India. He founded IDeASLab—India's first Design Observatory. He is the Founding Programme chair for International Conference Series on Research into Design (ICoRD), Conference Chair for the 22nd CIRP Design Conference 2012 and the 3rd International Conference on Design Creativity 2015 (3rd ICDC) and vice-Chair for AI in Design and Design Computing and Cognition Conferences. In 2014, he co-initiated India's first Smart Factory R&D platform.

Prof. Ravi Poovaiah is senior faculty member at the Industrial Design Centre (IDC), Indian Institute of Technology (IIT) Bombay. He has backgrounds in Mechanical Engineering, Product Design and Graphic Arts Education having studied respectively at IIT Madras and Bombay and at the Rhode Island School of Design (RISD), Providence, USA. His current pedagogic as well as research and design interests are in fields related to Interaction Design, New Media Design, Visual Design and Product Design and his research interests are in areas related to Visual Language, Information Visualisation, Visual Narratives, Way Finding Systems, Interaction Devices, Collaborative Social and Learning Environments and Designing for Children. He is involved with building digital resources related to 'Design Learning', 'Folk Tales', 'Interactive Systems', 'Designing for Children', 'Design of Way-finding Systems' and 'Design in India' with free and open access to networked information. He is also the co-director of a research project named COSMIC on experimenting with Social Media at IIT Bombay along with collaborative Institutes NTU and NUS in Singapore. Prof. Poovaiah has worked on projects with major industry leaders such as Microsoft, Yahoo, Google India, Motorola India, Indian Oil Corporation, Bharat Petroleum Corporation, Bharat Electronics Limited, Siemens India, among others.

Dr. Prasad Bokil is a mechanical engineer and communication designer with Master's and Ph.D. in communication design from IDC School of Design, Indian Institute of Technology (IIT) Bombay, India. He teaches courses in typography and graphic design at undergraduate as well as postgraduate levels. His research interests are visual language, visual narratives and game design. He is currently working towards 'design for wellbeing'. Dr. Bokil also worked as a Research Assistant at IDC, IIT Bombay, where he worked with a team of artisans at studio under guidance of Prof. A. G. Rao to enhance the packaging of some of the products. He is involved in several funded projects, including the Kanimuni project funded by MHRD under the Design Innovation Centre at IIT Guwahati. This project aims to develop educational games to build interest and confidence to learn English, Mathematics, Science and 21st century skills in school students. Dr Bokil has (co-)organised many conferences and workshops, including Typography Day 2013, Himalayan Languages Symposium 2016, ICoRD'17 in 2017, Type Design Workshop 2013, Game Design Workshop 2015, Design Research Workshop 2015, Workshop on Software product development process in Agile environment March 2015, Informatica 2016, Kanimuni: DIC Game Design Workshop 2016, etc.

Dr. Vivek Kant is an Assistant Professor at IDC School of Design, Indian Institute of Technology (IIT) Bombay, India. He is an active researcher in the field of systems design, human factors (cognitive ergonomics/cognitive engineering), sociotechncial systems, and history and philosophy of design engineering. Dr. Kant pursued his B.S. in Electrical Engineering from Dayalbagh Educational Institute, India in 2004, M.S. in Cognitive Science from University of Allahabad, India in 2008, M.A. in Psychology from University of Connecticut, USA in 2011, and Ph.D. in Systems Design Engineering from University of Waterloo, Canada in 2015. He was a Postdoctoral Fellow at the Institute of Catastrophe Risk Management, Nanyang Technological University (NTU), Singapore, and the Division of Sociology, School of Humanities and Social Sciences, NTU along with Future Resilient Systems, Singapore-ETH Centre. He has also worked as an Assistant Systems Engineer for Tata Consultancy Services, India.

Part I
Human Factors in Design (Physical and Cognitive Ergonomics; Design for Emotions, etc.)

Chapter 1
The Role of Letter Anatomy in Type Design: An Eye-Tracking Study of Bengali Letters

Subhajit Chandra and D. Udaya Kumar

Abstract The elementary unit of letter construction is the basic anatomical features of letterforms. The letter identification is a process of decoding and encoding of the anatomical information of the letters. They are the fundamental components of letter identification. This paper explains the process of Bengali letter identification and reveals the crucial anatomical features for design purposes. Here, an eye-tracking study has been conducted to identify the involvement of letter features in reading. In order to reveal the letter features, we exposed the letters under low contrast conditions to establish the consequence of anatomical features. The objective of the experiment is to identify the important letters features during the reading process that aids the letter identification task. A total number of six participants had performed the task under low contrast condition, and data was captured using an eye-tracking device. The aim of the paper is to understand the role of anatomical features in the letter identification process, so that such information can be useful for the Type designers. Type designers are designers who design typefaces and develop fonts for everyday use. The saccades, fixation, and heat-signature data from eye-tracking study have provided two major sets of information—(1) a list of letter features (individually) involved in the letter identification process, (2) a list of common and unique letter features across all Bengali letters. The result indicates that there is an active involvement of letter-parts and features in the letter identification process, and the list of common and distinctive letter features has been prepared to guide the type designers.

1.1 Introduction

Letters are meant for reading in the form of a single letter or word or continuous text. One of the major concerns of the reading of any single letter or word or continuous

S. Chandra (✉)
Shiv Nadar University, Delhi (NCR), India
e-mail: subhajit.chandra@snu.edu.in

D. Udaya Kumar
Indian Institute of Technology Guwahati, Guwahati, Assam, India

A. Chakrabarti et al. (eds.), *Design for Tomorrow—Volume 1*, Smart Innovation, Systems and Technologies 221, https://doi.org/10.1007/978-981-16-0041-8_1

text is the 'legibility'. Without it, letters are not readable. Legibility is the quality of a letterform that controls letter perception. In other words, the distinctness of letters makes perception easy to recognize [1, 2]. Letter anatomy is the structural information of letters. A letter of a typeface is confined space or shape that represents the combination of anatomical structure with a design grammar. Letter anatomy contributes to letter legibility by differentiating structural formation of letters. The letter identification is a decoding and encoding method of structural formation or anatomical information [3]. Such anatomical information is the foundation of the letter design process.

1.1.1 The Bengali Script

There are many scripts in India that are used by Indian populace to write their own languages. Bengali is one of them, and it is the official language of two Indian states West Bengal and Tripura. Bengali language is written using Bengali script. The Bengali script has evolved from 'Siddham' script. 'Siddham' is an offspring of 'Brahmi', which is the origin of all Indic scripts. The Modern Bengali is standardized by Pandit Ishwarchandra Vidyasagar (also known as Ishwarchandra Bandyopadhyay) who was one of the pioneers of Bengali literature. Bengali script has twelve vowels and thirty-seven consonant letters [4, 5]. Bengali has few unique characteristics such as Matra or a distinctive horizontal line/stem at top of letters, diacritics or dependent vowel signs and conjuncts or a fused letter of the two or more consonant letters.

1.1.2 Typeface Anatomy and Identity

A script is a collection of letters and characters in crude form. However, a typeface is a collection of characters, letters, or glyphs that share common design features [3, 6]. Typeface anatomy is the grammar of letter construction or design characteristics. According to Pflughaupt (2007), letters are constructed with a certain number of curved, straight, simple, and complex strokes. They are connected in specific ways regardless of their style. Each letter demonstrates the order and direction of combined strokes [7]. The grammar of structural formation and the vocabulary of letter-parts are known as the typeface anatomy.

The term 'typeface' is often confused with another word 'font'. In type design, a font is a set of glyphs in a particular size, weight, and style of a typeface. In the digital era, a font is defined as reproducible characters that are printable or displayable in a specific style and weight.

There are many available typefaces of Bengali that have been designed over the last two centuries: from the eighteenth century to till now. Every typeface has its own unique design characteristics. Yet, most of the letters of any typeface are legible and readable despite variations in their design. Most of all typefaces are designed in certain configurations or based on letter grammar that holds the key to letter identity. According to Pflughaupt (2007), letter strokes fall into two categories, straight lines and curved lines. The arrangement of the straight lines such as horizontals, verticals, and diagonal lines creates different types of angles. The openness of its angles or counter of the angles provides a letter its specific characteristics. The curved lines add more uniqueness to the letters. The shape and location of the combined lines and empty spaces in and around a character play a determining role in constructing this letter 'identity' [7]. Therefore, letter perception refers to the typeface anatomy.

1.1.3 Letter Anatomy and Identification

Anatomy provides the information about structural formation of letters. The anatomical information of the letters is already known to most of the native readers. This information, which is stored in the human brain, helps a reader to take the decision quickly about the 'letter identity' if a letter appears in front of the reader. Thus, 'designing a letter' is a crucial process that creates a letter perception. The anatomy of a designed letter should not deviate too much from the original perception of a letter. A reader always possesses the fundamental anatomical information of a letter [2]. This letter information is crucial to identify the letter without any hesitation.

1.1.4 Objective of the Study

Bengali is one of the most widely used scripts in India. Nevertheless, the legibility of the letterforms has not been a concern in most of the cases. The reason could be the nature of historical development of the letterforms. In the beginning of the digital age, digital type production has been prioritized over the functionality or quality issues. The same was similar when first printing came into light. The production of wooden or metal block letters was more important than quality of letters such as structure and spacing. The foundation of the legibility study of Bengali is merely at the beginning stage with the few literature available in the domain. Here, an attempt has been made to develop the crucial information about letter structure that can contribute to the letter design process for quality Bengali font production. As discussed by Tracy (1986), letter recognition is the term that refers to letter legibility [1]. Hence, anatomical information is crucial for type designers to design letters in legible form. Here, the objective of the research is to reveal the crucial anatomical information of the Bengali letters. The anatomical information will guide the type designers with their letter design process with maintaining the legibility.

1.2 Letter Information in and for Reading

Here, the low contrast technique has been used to make the letter strokes less visible to the normal eye, so that the participants realize the letter information with adequate attention. Pelli et al. (2004) extensively discuss the letter identification under low contrast sensitivity. They concluded that the visual objects such as letters can be studied under threshold contrast for identifying their anatomical features [8]. Low contrast pushes the human brain to understand the letter features and reveal the unique feature of letterforms. The process reduces the chance of getting recognized easily at a glance with less anatomical information as native readers do regularly. The native readers can easily identify any letter without looking at the whole letter structure. Here, it is directed to read the letters at low contrast condition, so that they read the letters with precise attention [9–12].

1.2.1 Eye Movement in Reading

The eye movement explains how the visual system behaves during reading. Human eyes move with stops and jumps, called fixations and saccades. Fixations are the stops where eyes stand still for a short period of time to focus on a letter. Saccades are fast movement and shifting from one fixation to another fixation [13]. The human brain collects information from a fixation after every saccade even though the fixations are very short. In general, reading is a complex cognitive process of decoding letter information in multiple stages [14].

Like saccades, there are backward movements around ten to fifteen percent of saccades, known as regression. Regressions are usually going back due to oculomotor errors. If eyes move far away from the last fixation then the next fixation may hamper the reading process. In such cases, the eyes go back-ward to recognize in-between words for comprehension [15, 16]. The approximate span of a fixation in Latin script is around 225–250 ms but can vary from 50 to 600 ms. The average saccade is 7–9 letters jump but can also range from one single letter to 20 letters or more [16]. The average saccade commences on a 2-degree move and takes around 30 ms to execute (Fig. 1.1).

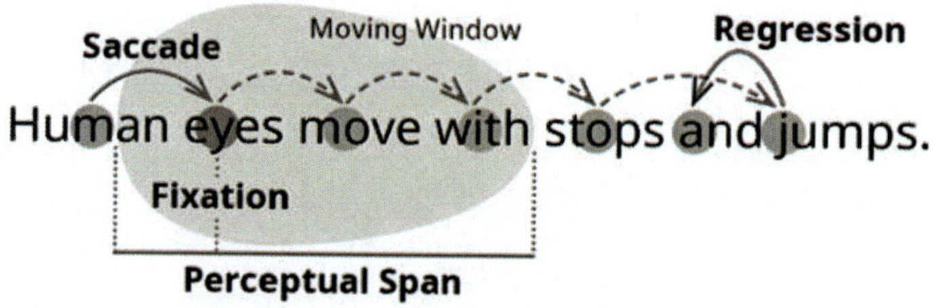

Fig. 1.1 Eye movement in Latin

When we look at a continuous text, we focus on a particular part of a line which is known as perceptual span. This field, perceptual span, actually moves based on the last fixation spot. It extends to three or four characters to the left and fourteen or fifteen characters to the right of a fixation. The perceptual span is asymmetric or uneven and biased toward the direction of moving attention [17].

1.2.2 The Writing System

The writing system is one of the major aspects in reading. For example, the Chinese script is an ideographic writing system where word density is higher than Latin. The average fixation duration and regression rate of Chinese script are similar to Latin. However, average saccade length is different compared to Latin. They move their eyes an average of 2–3 letters, whereas in Latin 7–9 letters [16]. Similarly, in Japanese, saccade length is about 2–5 letters [18]. Bengali is read from left to right, and there is hardly any information available in literature about the same.

1.2.3 The Control of Eye Movement

Eye movement is a moving window that controls the fixations and saccades. The moving window and the stimulus onset can be controlled to understand the reading pattern. Generally, the fixation duration and saccade length vary independently [19]. However, there are cases of regressions and re-fixation that happen due to higher cognitive processes and comprehension problems [20].

The extraction of visual information of text is an independent process from lexical processing that is responsible for the reading of letters, words, or text [21]. The fixations are placed in fovea, whereas the process aids by parafoveal preview benefits [13, 21]. Juhasz et al. (2008) show the benefits of the parafoveal preview of next fixation. The parafoveal preview benefit is the span within the perceptual span but beyond foveal vision. Parafovea is the area beyond two to three letters from a fixation to left and right [22]. It allows and shows improved reading performance in many script systems.

1.2.4 Reading on Screen

In the beginning of the digital era, several studies have presented that reading on screen is slower than reading from paper [23]. Ackerman and Lauterman (2012) suspect that the display screen is a direct source of light, intermittent light, which is difficult for eyes to read. On the other hand, reading from paper is under reflected light which is continuous in nature. This could be the possible reason for slower reading on screen since frequency of light could create interference with the ocular system

[24]. Recently, many researches have concluded that reading is faster on paper but can be achieved as faster as paper on screen [25, 26], if following factors are met:

- The resolution of the display screen needs to be high enough for display. Increasing pixel density can result in better performance in reading in letter identification tasks. Sheedy et al. conclude that if conditions are met in LCD and CRT screens and paper, there is no main effect on letter identification [27, 28].
- Vector fonts should be used to meet the quality of paper in CRT displays.
- The font must use anti-aliased hinting to meet paper quality on screens.

The anatomical information can be revealed at low contrast if exposed letters meet the above conditions. Now, it is hypothesized that the anatomical structure can easily be useful for letter identification tasks, and it contributes to the legibility of the letters. One of the popular methods of measuring legibility is eye-tracking, although there is hardly any literature that studies or measures the legibility of Bengali using eye-tracking.

1.3 Experiment Design

The following experiment is designed to understand the role of anatomical feature in letter identification. The study was conducted using an eye-tracking device. The test was performed with Bengali script, and letters were exposed in low contrast condition. It is presumed that the readers will identify the letters by their key anatomical structure, and the information will be captured by the eye-tracking device.

1.3.1 Preparation of Test Materials

The experiment was set up under a controlled environment. It was equivalent to a reading space scenario with adequate light from 290 to 310 lx.

1.3.2 Apparatus

1. A 20-inch TFT-LCD monitor was used which had a 438 mm diagonal screen with a viewing area of 340 mm horizontally and 276 mm vertically. The screen resolution was 1280 by 1024 pixels. The screen images were refreshed at a rate of 60 Hz.
2. A SMI eye-tracker with BeGaze software.

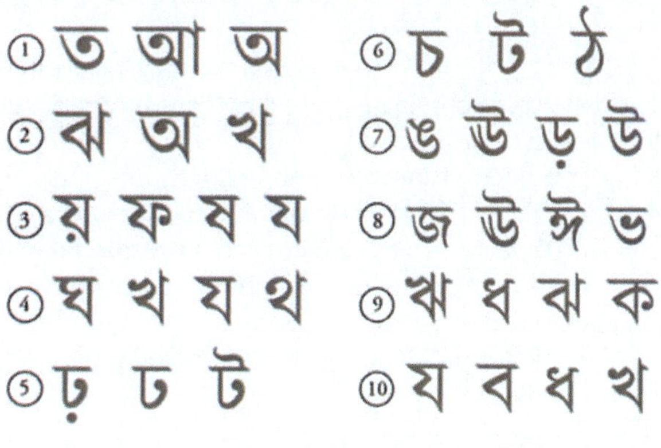

Fig. 1.2 Letter groups

1.3.3 Letters as Sample

The size of the letters was 2.5-inch in height and presented horizontally at the center of a monitor. Three or four letters were exposed from left to right at low contrast. A sufficient gap was maintained among the exposed letters to reduce the parafoveal preview benefit. Each letter was presented in the equal distance of each other. The color of the letters were light gray (RGB—235, 235, 235) on white (RGB—255, 255, 255) background. We placed letters on the screen in two ways:

- Distinctive letters belong to a different structural group and
- Similar letters belong to the same structural group.

There are many letters in Bengali which are very much similar in anatomical structure. Such letters were kept in a group to identify the specific anatomical information. The sample letters were arranged in a manner where letters with higher common anatomical features can be compared and vice versa. A total number of ten groups of letters were exposed as shown in Fig. 1.2. They cover most of all letters in Bengali who are involved in the confusion with other letters.

1.3.4 Experimental Setup

The TFT-LCD monitor was placed on a table 65 cm in height with an inclination angle of 105 degrees with respect to the plane axis of the table [29]. A total number of six participants took part with their consent. They were asked to get in rest position on chin rest. The head-mounted eye-tracker was adjusted accordingly to take all measurements. The eye-tracker was calibrated by using five points calibration modules (four points at corners and one point at center) on the display screen.

1.3.5 Methodology

The test was a combination of 'short exposure' and 'naming' test [3]. The exposure time was 5 s for each letter. After every 5 s, a new letter appears next to the first letter at a fixed distance of 250 pixels from left to right. The eye-tracking device was mounted on the head during the experiment to collect the data. To reduce their unwanted eye movement, a black point was placed in the center of the screen before the beginning of each sequence of letters. They were asked to read aloud each letter as quickly and as accurately as soon as they realized the letter. A total number of thirty-six letters in 10 groups were exposed randomly in front of them. The groups are mentioned in Fig. 1.2.

1.3.6 Result

The data of the exposed letters were recorded using the eye-tracking device. Scan-path and heat signature were determined by using the SMI BeGaze software. The captured data was analyzed and processed with the SMI BeGaze software to determine the eye-tracking information such as fixations and saccades. The reliability statistic was conducted considering fixation time and saccades, and Cronbach alpha values of each participant are given in Table 1.1. The data shows that the association of fixation and saccades is significantly correlated with alpha values more than 0.9. The values signify that the process of reading of exposed letters was hardly with any interruption.

Further, one sample *t*-test and correlation statistics were conducted to validate the results. One sample *t*-test was performed on fixation time with stimulus placement. There is no significant within subject variation observed in fixation time due to stimulus placements with p value $p < 0.001$. Similarly, one sample *t*-test was performed on saccades and stimulus placement. There are no significant variations found in saccades and stimulus placement with p value $p < 0.001$. The eyes follow in a similar pattern in most of the cases to identify the letters with specific fixation locations.

The correlation statistics were performed using Pearson's correlation method on fixations with saccades. It signifies the consistency in reading performance of the

Table 1.1 Cronbach alpha values

Participants	Cronbach alpha's value
1	0.973
2	0.995
3	0.963
4	0.979
5	0.939
6	0.991

Table 1.2 Pearson's r correlation values

Fixation–saccades items	Pearson's *r*
Participants 1 (Fx1–Sa1)	0.947
Participants 2 (Fx2–Sa2)	0.991
Participants 3 (Fx3–Sa3)	0.930
Participants 4 (Fx4–Sa4)	0.959
Participants 5 (Fx5–Sa5)	0.885
Participants 6 (Fx6–Sa6)	0.983

Table 1.3 Correlation data of saccades

	Sa1	Sa2	Sa3	Sa4	Sa5	Sa6
Sa1	–	0.976	0.985	0.993	0.988	0.996
Sa2	0.976	–	0.965	0.967	0.963	0.973
Sa3	0.985	0.965	–	0.994	0.994	0.986
Sa4	0.993	0.967	0.994	–	0.993	0.995
Sa5	0.988	0.963	0.994	0.993	–	0.986
Sa6	0.996	0.973	0.986	0.995	0.986	–

participants. The values of '*r*' are given in Table 1.2. The notation 'Fx' identifies the fixations, and 'Fx1' is the fixations of the first participant and similarly the other fixations 'Fx2', 'Fx3', and so on. Similarly, 'Sa' signifies the saccades, and 'Sa1' is the saccades of the first participant and similarly 'Sa2' to 'Sa6'. The result shows that there are strong correlation values of '*r*' in all cases.

The inter-item correlation of saccades was performed to verify the consistency of the eye-movement pattern among participants. The correlation data of saccades are given in Table 1.3. The data shows a strong correlation with saccades.

1.3.7 Discussion

The reliability of the study has been established using Cronbach alpha test. The significant anatomical information was determined by the location value from the heat signature. To validate the fixation locations, one sample *t*-test (within and between) has been performed. The result shows that there are no significant variations in fixation time and saccades with the stimulus placement. The reading pattern was mostly consistent for all participants. Most of the fixation locations remain in the same position with no difference in mean fixation count.

Later, Pearson correlation has been performed to determine the consistency of fixations and saccades data to validate the fixation location in order to determine the significant anatomical information. This validity has been further extended among inter-item to confirm the consistency in eye gaze. The data shows that a strong

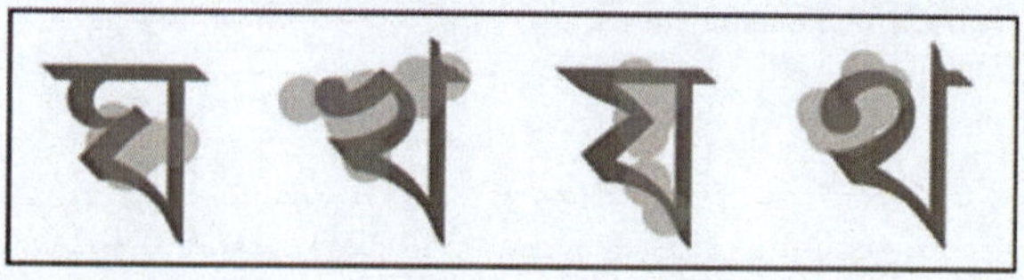

Fig. 1.3 Heat signature data

correlation among participants with the fixation time and saccades. The fixation locations are important to determine the important anatomical information of the letters (Fig. 1.3).

The eye-tracking data has been analyzed qualitatively considering the fixation location on letterforms. The heat signatures are considered to determine the important letter features. It was observed that fixations on a particular letter vary from three (3) to nine (9) fixations. Most of the letter groups were identified correctly except few confusions that were identified in group of letters (ঘ, খ, য, থ), (ঋ, ধ, ঝ, ক), and (য, ব, ধ, খ). More than ninety-seven percent correct pronunciations were noticed during the experiment.

To determine the anatomical features, the heat signature of all participants was superimposed, and the common zone was determined. Afterward, common letter-parts were noted down to determine crucial letter features as given in Table 1.4. The data in Table 1.4 enables us to capture the specific letter-parts that play a significant role in the identification process. The anatomical nomenclatures of the letter-parts are used here as referred by Chandra et al. [30]. The confusion in groups (ঘ, খ, য, থ), (ঋ, ধ, ঝ, ক), and (য, ব, ধ, খ) also suggests that the specific letter features like 'shoot of ঘ' or 'Bud & Arm of খ' needs special attention for correct identification.

The analysis of the superimposed heat signature is further compared and evaluated to determine the significant anatomical features that are important for the recognition task. Thereafter, two anatomical groups are formed—(1) common anatomical features and (2) distinct anatomical features. The common anatomical features the common letter information across all exposed letters. They are responsible for the letter confusion. However, the distinctive letter features are the unique features of the letters that provide unique identity to the letters as a whole. The common and distinctive features are given in Table 1.5.

1.4 Conclusion

The novelty of the study is the methodology to extract anatomical information of the vital letter-parts that are involved in the letter identification process using eye-tracker under low contrast condition. Here, the combined method of short exposure under low contrast condition with eye-tracker is used to identify the crucial letter-parts. The result indicates that there is an active involvement of case-specific letter-parts in the letter identification process. As discussed, the distinctive letter-parts differentiate letters during the letter identification process. However, common letter features are

Table 1.4 Table of letter features

Letters - ত, আ, অ	**Letters -** ঝ, অ, খ
• Counter of Matra and Bowl – ত, আ, অ • Bowl – ত, আ, অ • Initial/ Finial – ত, আ, অ • 'T' Junction – আ, অ • Wedge – আ, অ	• VStem and Wedge – ঝ, অ • Initial – খ • Finial – অ • Nose – খ
Letters - য়, ফ, ষ, য	**Letters -** ঘ, খ, য, থ
• Wedge – ফ, ষ, য • VStem and Arm – ফ • Crossbar – ষ • Initial & Close Counter – ফ, য • '>' Junction – ফ, য • Nose – য়, ষ, য	• Nose – ঘ, খ • Initial – খ, য, থ • Counter of VStem & Nose – খ • Wedge – য • Short Matra – খ
Letters - ঢ়, ঢ, ট	**Letters -** চ, ট, ঠ
• Terminal/ Finial – ঢ়, ঢ, ট • Dot – ঢ় • Counter – ঢ • Matra & Tail Counter – ট • Arm – ঢ়, ঢ	• Junction of HStem & Crossbar – চ • Matra & Tail Junction – ট, ঠ • Upper Tail – ট • Close Counter – চ • Terminal/Finial – ট
Letters - ঙ, উ, ড়, উ	**Letters -** জ, উ, ঈ, ভ
• Nose – ঙ • Loop – ঙ • Matra & Upper Tail Junction – উ • Finial – উ, উ • Dot – ড়	• Nose – জ, উ • Shoot & leg Junction – জ • Finial – উ, ঈ, ভ • Initial – ঈ, ভ • Upper Tail – উ, ঈ • Leg – জ
Letters - ঋ, ধ, ঝ, ক	**Letters -** য, ব, ধ, খ
• Wedge – ঋ, ঝ • VStem – ঋ • Delta & Shoot Junction – ঋ • Arm – ধ • Nose – ঝ • Delta & Matra Junction – ক	• VStem – য • Delta & Matra Junction – ব • Arm – ধ • Initial – খ • '>' Junction – য • Short Matra – খ

responsible for letter confusion. Both together or by combination of common and distinctive letter-parts, a letter becomes unique in anatomical structure. Therefore, the anatomical information is crucial not only in the letter identification process but also in letter design. The confusion caused by the common anatomical parts needs to be designed carefully to maintain overall uniqueness of the letters. Therefore, slight changes in common letter-parts may create enough distinguishable features which can solve letter confusion. Though, overall legibility of the letters needs to

Table 1.5 Important letter features

Common features (detected in more than one letter)	**Distinct features (mostly unique features)**
• Wedge – আ, অ, ঝ, অ, ঋ, ফ, ষ, য • Initial/ Finial – ত, আ, অ, খ, য, থ, ঢ়, ঢ, ট, ঈ, ভ, উ, উ • '<' / Nose – ব, য়, ষ, য, ঘ, জ, উ, ঙ • Short Matra – খ • '>' Junction – ফ, য • Matra & Tail Junction – ট, ঠ	• Bowl – ত, অ • Delta & Shoot Junction – ঋ • Delta & Matra Junction – ক • VStem and arm – ফ, ক • Crossbar – ষ • Arm – ধ • Dot – ঢ়, ড় • 5-point Cross/ Junction – ঠ • Loop – ঙ • Shoot & Leg Junction – জ • Arm – ধ

be maintained. The design of the letters becomes crucial considering the anatomical structure where letter identity and legibility both are important.

This particular experimental setup is designed to determine the important letter-parts. The specific anatomical features for each letter are not identified during this study due to limitation of the instrument and experimental setup. The critical letter-parts further can be evaluated to determine their role in letter legibility.

Acknowledgements The study could not be possible without the experimental facility (the eye-tracking system) that was provided by the Ergonomics Laboratory, Department of Design (DoD), Indian Institute of Technology Guwahati. A heartfelt thanks to Dr. Sougata Karmakar and his team members of Ergonomics Laboratory, DoD, IIT Guwahati for allowing and helping us to do the same.

References

1. Tracy, W.: Letters of Credit: A View of Type Design. Gordon Fraser, London (1986)
2. Beier, S.: Typeface Legibility: Towards Defining Familiarity. PhD Thesis, London (2009).
3. Chahine, N.: Reading Arabic: Legibility Studies for the Arabic Script. PhD Thesis, Leiden, (2012).
4. Banerji, R.D.: The Origin of the Bengali Script. University of Calcutta, Calcutta (1919)
5. Ross, F.G.E.: The Printed Bengali Character and its Evolution, 2nd edn. Shishu Sahitya Samsad Pvt. Ltd., Kolkata (2009)
6. Bringhurst, R.: The Element of Typographic Style, 3rd edn. Hartley and Marks Publishers, US (2004)
7. Pflughaupt, L.: LETTeR by Letter: An Alphabetical Miscellany. N. Bednarek, Princeton Architectural Press, New York (2007)
8. Pelli, D.G., Levi, D.M., Chung, S.T.L.: Using visual noise to characterize amblyopic letter identification. J. Vis. **4**(10), 904–920 (2004)

9. Majaj, N.J., Pelli, D.G., Kurshan, P., Palomares, M.: The role of spatial frequency channels in letter identification. Vision Res. **42**(9), 1165–1184 (2002)
10. Pelli, D.G., Burns, C.W., Farell, B., Moore-Page, D.C.: Feature detection and letter identification. Vision Res. **46**(28), 4646–4674 (2006)
11. Pelli, D.G., Majaj, N.J., Raizman, N., Christian, C.J., Edward, K., Palomares, M.C.: Grouping in object recognition: The role of a Gestalt law in letter identification. Cogn. Neuropsychol. **26**(1), 36–49 (2006)
12. Pelli, D.G., Tillman, K.A.: Parts, wholes, and context in reading: a triple dissociation. PLoS ONE **2**(8), (2007)
13. Rayner, K., Pollatsek, A.: The Psychology of Reading. Prentice-Hall International, Englewood Cliffs, NJ (1989)
14. Schotter, E.R., Angele, B., Rayner, K.: Parafoveal processing in reading. Attention, Percept. Psychophysics **74**(1), 5–35 (2012)
15. Rayner, K., Chace, K.H., Slatt, T.J.: Eye movements as reflections of comprehension processes in reading. Sci. Stud. Read. **10**(3), 241–255 (2006)
16. Rayner, K.: Eye movements and attention in reading, scene perception, and visual search. Q. J. Exp. Psychol. **62**(8), 1457–1506 (2009)
17. Reichle, E.D., Rayner, K., Pollatsek, A.: The E-Z reader model of eye-movement control in reading: comparisons to other models. Behav. Brain Sci. **26**(4), 477–526 (2003)
18. Ikeda, M., Saida, S.: Span of recognition in reading. Vision Res. **18**(1), 83–88 (1978)
19. Rayner, K., Pollatsek, A.: Eye movement control during reading: evidence for direct control. Q. J. Exp. Psychol. Sect. A Human Exp. Psychol. **33**(4), 351–373 (1981)
20. Rayner, K., Kambe, G., Duffy, S.A.: The effect of clause wrap-up on eye movements during reading. Q. J. Exp. Psychol. **53A**(4), 1061–1080 (2000)
21. Inhoff, A., Eiter, B., Radach, R., Juhasz, B.: Distinct subsystems for the parafoveal processing of spatial and linguistic information during eye fixations in reading. Q. J. Exp. Psychol. Sect. A Human Exp. Psychol. **56**(3), 803–827 (2003)
22. Juhasz, B.J., White, S.J., Liversedge, S.P., Rayner, K.: Eye movements and the use of parafoveal word length information in reading. J. Exp. Psychol. Hum. Percept. Perform. **34**(6), 1560–1579 (2008)
23. Gould, J.D., Alfaro, L., Barnes, V., Finn, R., Grischkowsky, N., Minuto, A.: Reading is slower from CRT displays than from paper: attempts to isolate a single-variable explanation. Human Fact.: J. Human Fact. Ergon. Soc. **29**(3), 269–299 (1987)
24. Ackerman, R., Lauterman, T.: Taking reading comprehension exams on screen or on paper? A metacognitive analysis of learning texts under time pressure. Comput. Hum. Behav. **28**(5), 1816–1828 (2012)
25. Stoop, J., Kreutzer, P., Kircz, J.G.: Reading and learning from screens versus print: a study in changing habits: Part 2—comparing different text structures on paper and on screen. New Library World **114**(9), 371–383 (2013)
26. Myrberg, C., Wiberg, N.: Screen versus paper: what is the difference for reading and learning? Insights **28**(2), 49–54 (2015)
27. Sheedy, J.E., Subbaram, M.V., Zimmerman, A.B., Hayes, J.R.: Text legibility and the letter superiority effect. Human Fact.: J. Human Fact. Ergon. Soc. **47**, 797–815 (2005)
28. Sheedy, J.E., Tai, Y., Subbaram, M., Gowrisankaran, S., Hayes, J.: ClearType sub-pixel text rendering: Preference, legibility and reading performance. Displays **29**(2), 138–151 (2008)
29. Horikawa, M.: Effect of visual display terminal height on the trapezius muscle hardness: quantitative evaluation by a newly developed muscle hardness meter. Applied Ergonomics **4**, 473–478 (2001)
30. Chandra, S., Bokil, P., Udaya Kumar, D.: Anatomy of Bengali Letterforms: A Semiotic Study. In: Chakraborty, A. (ed.) ICoRD'15—Research into Design across Boundaries 1. Smart Innovation, vol. 34, pp. 237–247. Springer, New Delhi (2015)

Chapter 2
Factors to Look into for Designing Interiors of Long-Distance Pilgrimage Buses in India: A Survey-Based Study

G. Chinmaya Krishnan and Debkumar Chakrabarti

Abstract There are not many differences in buses used for long journeys for pilgrimage purposes, with a common deluxe bus as regards to interiors. Passengers of different attitudes and comfort perceptions do differ. On inquiry to pilgrims and tour organizers in a preliminary investigation, it was noticed that sufficient attention is required to be given while designing the interiors of such public carriers but not much attention is paid so far, and no specific study was also carried out, and normal buses with need-based modifications are being used. This may be due to the lack of in-depth knowledge about vehicle design, vehicle ergonomics, etc. and the lack of sufficient financial strength of these body building companies to employ knowledgeable persons. This study aims at understanding the critical factors affecting passenger physical comfort as concerned with interior aspects of buses for long journeys. The scope of the study is limited only to the buses operating under pilgrimage activities. The survey was done to find out those factors, and the samples were taken from Vrindavan, Dwaraka and Somanatham. The pilgrims were selected based on purposive and random sampling as the target respondents were mostly senior citizens for whom the concept of comfort and fatigue is pronounced. Questionnaires and meeting observations were used to have their views on physical comfort, accessories and other facilities, seating considerations, the place for movements, etc. The study finds certain issues that appear to be vital for designing long-distance pilgrimage purpose buses.

2.1 Introduction

In a country like India, which holds the distinction of having the most vibrant culture on earth and a philosophy that is extremely inclined towards the spirituality, one of

G. Chinmaya Krishnan (✉) · D. Chakrabarti
Department of Design, Indian Institute of Technology Guwahati, Guwahati, Assam, India
e-mail: k.chinmaya@iitg.ac.in

D. Chakrabarti
e-mail: dc@iitg.ac.in

A. Chakrabarti et al. (eds.), *Design for Tomorrow—Volume 1*, Smart Innovation, Systems and Technologies 221, https://doi.org/10.1007/978-981-16-0041-8_2

the major reason for long journeys is pilgrimages and primarily uses buses and other large transport facilities for very long group travels. Also, it is generally observed that greatest number of pilgrims fall under higher age group. This points towards the importance of proper design of facilities inside the vehicle, where the occupants spend maximum time during a tour period.

If we consider a car, all the manufacturers have proper research wings for the ergonomic comfort of the users/passengers. Hence, their products are designed with proper knowledge background. In some of the foreign and developed countries, the entire bus is made in a factory floor. So if an original automobile manufacturer is making it, there will be proper designing of interiors, facilities like seats, etc., and when we examine the bus building in India, there are some disturbing notes. Only a handful of companies come with factory fitted bodies. All others supply vehicles as cowl and chassis only. This opens a huge market for the bus body builders in the country. But these coach manufactures lack sound knowledge about ergonomical aspects like comfort, or they purposefully neglect it as they are primarily concerned with costs and profits. And their customers too are concerned with price; almost all of the products run in the commercial market with an aim of profit generation alone. It is also pronounced by the fact that in the commercial sector, the customer may not be the user as in the case with non-commercial vehicle market like cars, where the customer who buys the product is the end user and travels in it.

This triggers many health issues for the passengers as continued sitting in a forced or static posture, such as in a bus, causes an individual to long-term static loading of the body which is usually seen as a risk factor for the development of musculoskeletal discomfort [1–5]. It has been also claimed that prolonged sitting opens a potential risk to spinal and paraspinal discomfort and disorders [1, 4]. Confined seating limits physical activity and makes it difficult to leave the seat or do any stretching exercise. It is claimed that this physical inactivity causes physiological problems, such as swell, pain and ischaemia of the lower limbs [6] or any other body parts.

This study aims to bring light on the factors contributing to the physical ergonomic comfort for a traveller inside a bus for a very long journey. Comfort is a very complex and subjective concept that is difficult to understand, measure or quantify and specifically defines due to its cognitive as well as physical nature [7]. There are many external factors to the feeling of comfort or discomfort, such as visual input, smell, noise, temperature, humidity, vibration, pressure/touch, posture and movement. Here, we are concerned only with seat system feedback and amenities needed, neglecting the temperature and humidity as all the buses, in which this study is performed, were air conditioned. It is also to be noted that there are some scales and other methods available for measuring the aspect of comfort. Also, there are some objective tool developed for measuring seat comfort [8]. The objective comfort evaluation is somewhat highly complex for the aspect of comfort and is more subjective in nature [9].

2.2 Methodology

In this study, a group of surveys have been performed to find out the most problematic areas as concerned with long-distance bus travel. The survey particularly aimed at finding out the basic amenities needed by the passengers, feedback on seats, various pain regions after travel to find out the problematic areas. As it is a practice to consider long-term comfort after the first 30 min of travel [10], we focused on very long journeys that may last up to several days. We took pilgrims as the target audience because of two reasons. One, in India, the long-distance bus travel with duration of more than one day is opted mainly by the pilgrims. And two, the target respondents were mostly elderly individuals for whom the concept of wellbeing and fatigue is noticeable, and if you are selecting the pilgrims, it would be easier to include the subjects as needed by the study. The subjects were selected based on purposive and random sampling methods. Questionnaires and meeting observations were used to have their views on various aspects like mentioned above. A total of 87 subjects have been studied travelling in three different buses with three different type seat systems and interiors, in three different trips covering a total of 1474 kms in nine days consecutively. We travelled in the same bus for the entire period and taken feedback just after each day's journey personally. The buses were deemed to be of deluxe type. One bus was having adjustable back rest as well as adjustable hand rest (last one in Fig. 2.1). One was having adjustable back rest only (first one in Fig. 2.1). One was not having any of the features (middle one in Fig. 2.1).

Fig. 2.1 Interior types of selected three buses

Out of these 86 subjects, responses of three were excluded in the seat system feedback and pain regions feedback as they were having surgical or accidents history, so, the pain they are feeling could not attributed to travelling alone. There were 42 men as well as 41 ladies. Regarding the age, the mean age of the group is 57.15 with a standard deviation 12.024. Apart from personal details which included name, sex, age, details about current medications if any, history of any psychological and physiological problems, some thirty open-ended questions were used. The questions were divided into four sections. The section one was concerned with different pain regions after a daylong travel (eight questions), section two was concerned with seat systems feedback (seven questions), section three was concerned with the needed improvements in seat design (six questions), and final section was concerned with needed amenities.

The seat feedback questions were designed for analysing the comfort feeling of passengers. There were five-point comfort level scoring system used, starting with troublesome (score 1), not okay (score 2), neutral (score 3), okay (score 4) and comfortable (score 5). The pain feedback questionnaire was also designed as above with only difference being that the questions were asked to assess the pain rather than comfort. Extreme pain (5 points), moderate pain (4), low pain (3), slight discomfort (2) and absolute comfort (1). There were also questions to assess the needed amenities which was open-ended as well as optional which included questions about fixed or movable leg/foot support, preferred design for hand rest, etc.

2.3 Results and Discussion

As per the seat feedback, the major problematic area was neck support or probably the lack of it. In a comfort scale, it scored the least marks at 1.578, followed by lumbar (1.65), leg (1.759) and back (1.78) (Fig. 2.2).

When asked about the pain regions, again neck area holds the highest pain region at 3.489 on a scale of 5, followed by upper back (3.012), lower back (3), elbow (1.9), ankles (1.3), head (1.22), thigh (1.2) and shoulder (1.07). Also, it is interesting to note that a significant portion of the population has three major pain regions, namely neck (85.54%) followed by upper back (78.3%) and lower back (77.11%) (Figs. 2.3 and 2.4).

We asked about amenities needed, and it was an open-ended survey. After conducting preliminary survey, we zeroed in on 11 items, namely place for water bottle, medicine pouch, coat hanger, foot wear storage, mobile charging points, split curtain, walking ramp, seat belt, floor lights, toilet and trays. As far as the amenities needed are concerned, almost everybody in the target group voted for walking ramp inside the bus as very long duration confined posture is leading to many pain regions. Also, almost all demanded for an attachment by which they can apply leg movements freely. Apart from this, the highest priority was of toilets (Fig. 2.5).

It was also interesting to note that all women voted for toilets inside bus, and also they have given highest mark (5) for toilet. Facilities for keeping water bottles,

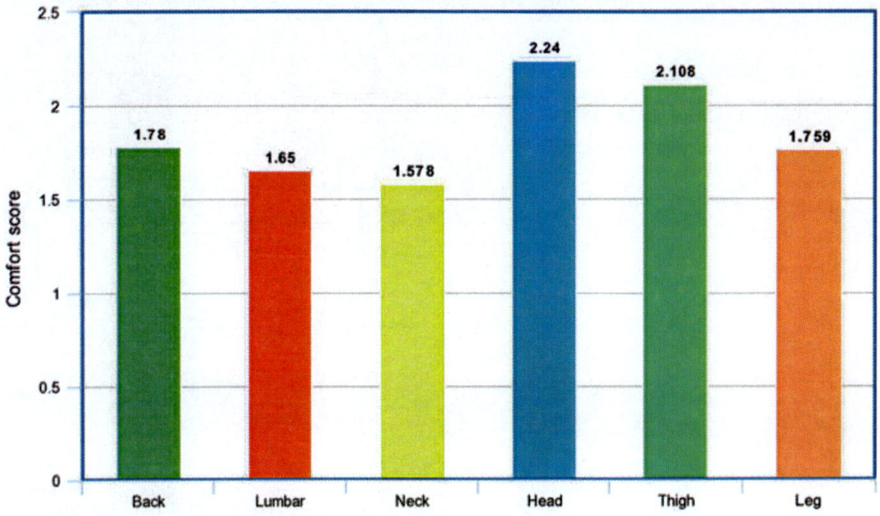

Fig. 2.2 Response about seat system—aspect of comfort

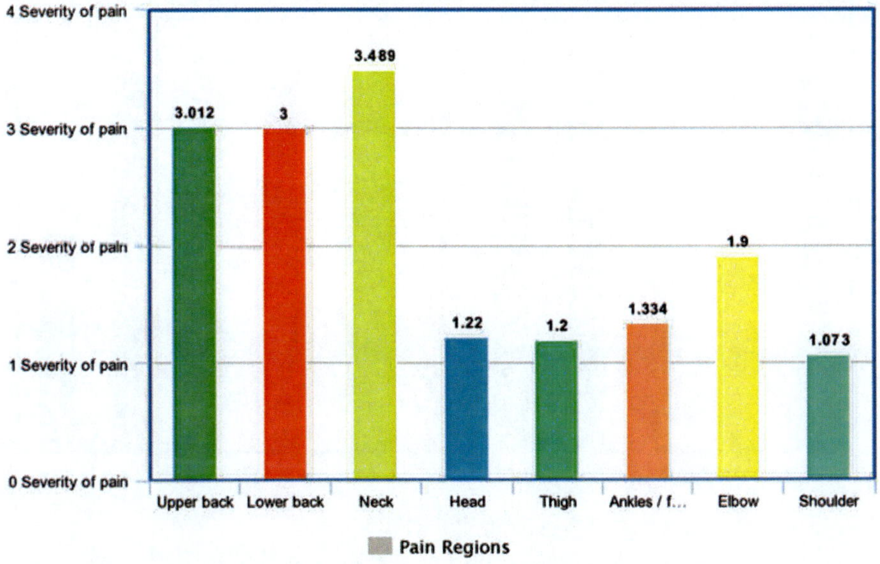

Fig. 2.3 Response about pain regions

medicine pouches, coat/secondary cloth hanger, individual mobile charging points as well as place to keep foot wears without slipping were almost equally important (Fig. 2.6).

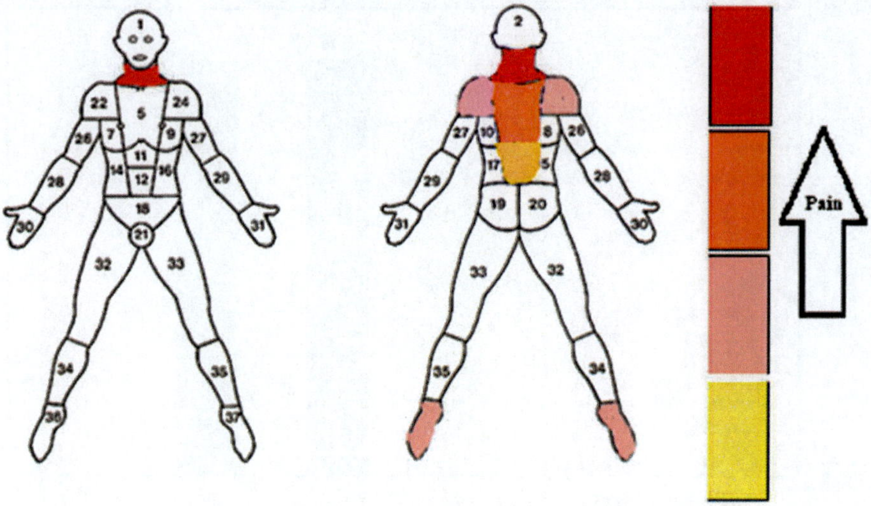

Fig. 2.4 Pain regions with respect to severity of pain

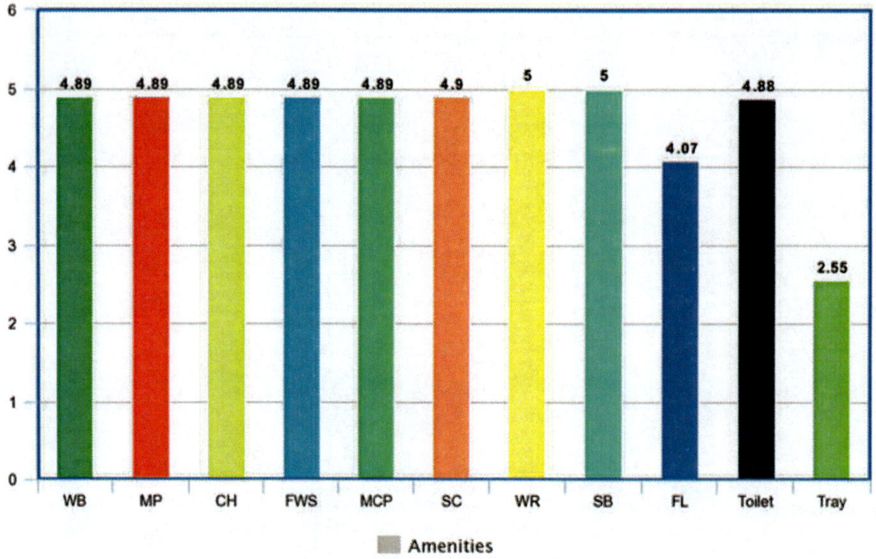

Fig. 2.5 Response about amenities needed. WB—Water bottles; MP—Medicine pouch; CH—Coat hanger; FWS—Footwear storage; MCP–Mobile charging points; SC—Split curtain; WR—Walking ramp; SB—Seat belt; FL—Floor lights

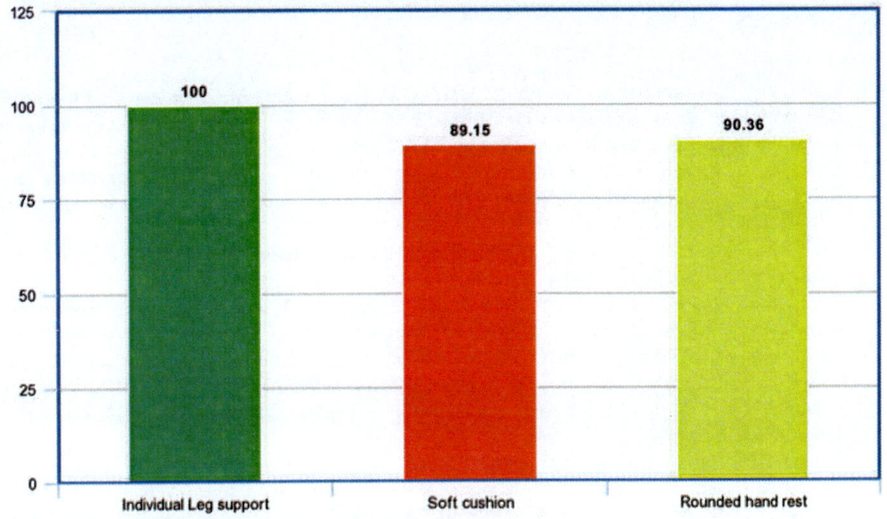

Fig. 2.6 Response about leg rest cushioning and hand rest

Regarding the question about individual leg support, cent per cent voted for it. And 89.15% people voted for soft cushioning together with 90.36% voting for rounded hand rest.

2.4 Conclusion

Ergonomics looks into utility and pleasure comfort issues while ideating and executing a design; it may be a product or a travelling envelop like bus with occupants in it. Majority of people in this study were complaining about various pain regions such as neck, lumbar, lower back, etc. Due to lack of facilities for physical stretching, many are at the risk of musculoskeletal complaints. Hence, it can be safely assumed that there are enough scope to incorporate ergonomic comfort of the occupants in buses. The identified issues thus may be used for designing the bus bodies for long-distance travel, specifically for elderly pilgrim tours. Prototype testing would give more feasibility clarity.

Declaration of Conflicting Interests The author(s) declared no potential conflicts of interest with respect to the research, authorship and/or publication of this article.

References

1. Aldington, S., Pritchard, A., Perrin, K., James, K., Wijesinghe, M., Beasley, R.: Prolonged seated immobility at work is a common risk factor for venous thromboembolism leading to hospital admission. Intern. Med. J. **38**, 133–135 (2008)
2. Cascioli, V., Heusch, A.I., McCarthy, P.W.: Does prolonged sitting with limited legroom affect the flexibility of a healthy subject and their perception of discomfort? Int. J. Ind. Ergon. **41**, 471–480 (2011)
3. Fazlollahtabar, H.: A subjective framework for seat comfort based on a heuristic multi criteria decision making technique and anthropometry. Appl. Ergon. (2010)
4. Healy, B., Levin, E., Perrin, K., Weatherall, M., Beasley, R.: Prolonged work- and computer-related seated immobility and risk of venous thromboembolism. JRSM **103**, 447–454 (2010)
5. Luttmann, A., Schmidt, K.-H.H., Jäger, M.: Working conditions, muscular activity and complaints of office workers. Int. J. Ind. Ergon. **40**, 549–559 (2010)
6. Hinninghofen, H., Enck, P.: Passenger well-being in airplanes. Auton. Neurosci. Basic Clin. **129**, 80–85 (2006)
7. Shen, W., Parsons, K.C.: Validity and reliability of rating scales for seated pressure discomfort. Int. J. Ind. Ergon. **20**, 441–461 (1997)
8. Wegner, M., Martic, R., Frans, R., Vink, P.: A system to measure seat-human interaction parameters which might be comfort relevant. Appl. Ergon. **84**, 1308 (2020)
9. Vink, P.: Advances in social and organizational factors. In: Advances in Human Factors and Ergonomics Series. CRC Press, Hoboken (2012)
10. Mergl, C.: Entwicklung eines Verfahrens zur Optimierung des Sitzkomforts auf Automobilsitzen. Dissertation. Technischen Universität München (2006).

Chapter 3
Effect of Tool Handle Design Parameters on Upper Extremity Muscle Performance in Periodontology

Vibha Bhatia, Jagiit Singh Randhawa, Parveen Kalra, Ashish Jain, and Vishakha Grover

Abstract Periodontists—dental professionals associated with tooth scaling (cleaning) tasks—are often prone to work-related musculoskeletal disorders (WMSDs). Tooth scaling involves the repetitive and forceful grip of scaling tool handle. Trauma related to hectic scaling job leaves periodontists susceptible to upper extremity disorders like cumulative trauma disorders (CTDs) like carpal tunnel syndrome (CTS), etc. Limited studies have been conducted to evaluate effect of tool handle design on upper extremity muscles involved in periodontal task. Optimally designed dental scaling tool design is expected to reduce the prevalence of WMSDs in periodontists. Twelve variations of modified dental scaling tool handle sleeves were fabricated using 3D printing technology. Design parameters like diameter and shape were assessed while performing simulated scaling task. The individual and interactive effects of design parameters on upper extremity flexor/extensor muscles were recorded by Electromyography (EMG). Thumb pinch force readings were recorded simultaneously. Results of the study indicated the reduction in muscle and thumb pinch forces in octagonal handle having 11 mm diameter. Proving taper to the tool handle design was also responsible for reduction in muscle loads. The current study not only aims at guiding the periodontists for adopting optimally designed tool but also engineers and researchers to design ergonomically sound tool handle designs.

V. Bhatia (✉) · J. S. Randhawa · P. Kalra
Department of Production and Industrial Engineering, Punjab Engineering College, Chandigarh, India
e-mail: vibhabhatia.phdidp17@pec.edu.in

J. S. Randhawa
e-mail: jagjitsingh@pec.ac.in

P. Kalra
e-mail: parveenkalra@pec.ac.in

A. Jain · V. Grover
Harvansh Singh Judge Institute of Dental Sciences and Hospitals, Punjab University, Chandigarh, India
e-mail: ashish@pu.ac.in

V. Grover
e-mail: vishakha_grover@rediffmail.com

A. Chakrabarti et al. (eds.), *Design for Tomorrow—Volume 1*, Smart Innovation, Systems and Technologies 221, https://doi.org/10.1007/978-981-16-0041-8_3

3.1 Introduction

Instruments being used by dentists for precise repetitive dental tasks over long duration create the stress in wrists, hands and forearms. This highly repetitive work is related to the occurrence of CTS in dentists [12, 15, 16].

The incidence of CTS in dentists varied 7–8.4% in US [3, 20] and 15.5–25.3% in other places like Isfahan, Saudi Arabia, Pakistan and Malaysia [9, 11, 19, 22] showed the incidence of CTS varying from 15.5–25.3%. Twenty per cent of the dentists were suffering from CTS in Mangalore, India [17]. Variation in percentage might be due to the variety in type of data collected depending upon the demographic variations, age factors, gender, work environment and work load.

Dentistry requires fine movement of hand and fingers repetitively in a limited space that is why dental professionals are among those who suffers higher occurrence of CTS. CTS usually leads to motor skills fatality which may result in slip of instrument while performing dental task. This slip of instrument is not affordable as human oral cavity contains plenty of nerves and serious injury to the patient.

American Dental Association [1] surveyed that 9.2% of dentists being diagnosed with upper extremity MSDs, requiring surgery and reduction in work hours. Due to lengthy duration of periodontal work and high dexterity of manual work, dental hygienists are more prone to upper extremity MSDs than regular dentists [21]. High value of pinch force is needed for periodontal scaling tasks.The value of average pinch force exerted during dental scaling task is 11–20% of the maximum pinch strength [4].

Dental scaling is nothing but scraping off the tartar which gets accumulated in the tooth root accompanied by upward pulling motion along the tooth surface, all starting from the gum to the top of the tooth. The periodontal handle gets pulled up in this process along its longitudinal axis, flexing the fingers and causing the forearm supination. Push motion is only responsible for placing the cutting tool back on the gum line.

The participation of Indian population of dentists in the study is necessary to find the effect of related hand anthropometric variables for tool handle design as anthropometry varies with variation in parameters like gender, race etc. Change in demographics may serve as the factor responsible in inappropriateness of the ergonomics related to any tool [13].

Survey was conducted in target locality prior to experimentation to find the incidence of CTS in dentists who performed dental hand scaling task (at least 6 h a week). Twenty seven dentists (15 males and 12 females) from Regional Dental College, Chandigarh were interviewed, and it was concluded that six (22.2%) of the dentists are already having one or more symptoms related to CTS like numbness or pain in fingers, change in finger bone shape, lack of stability of hand, etc.

The survey clearly showed the high incidence of CTS in target population. With the aim to provide reliable solution to the problem, possible areas of improvisation in system were found. The dental scaling tool handle design was focussed as the possible area of improvement for the current study. The tool with better design may provide better coupling to the fingers and thus may reduce the value of pinch force needed.

Pinch force values can be related to the design of the tool handle. Design of dental tools does hold some qualities that should follow some ergonomic norms [2]. Tool design parameters like weight and diameter has impact on pinch forces and hand muscle loads of hygienists. Lighter weight (<15 g) were related to low value of muscular loads and pinch force [7] and less diameter(<10 mm) tapered round cross-sectional shape required less value of forces than the other customised shapes [8]. Canadian Centre for Occupational Health and Safety recommended 12 mm of diameter of cross section in precision grips, ranging 8–16 mm. Small diameters provide better speed and dexterity of work, whereas larger diameters provide better torque [5].

The diameters available with doctors at regional dental hospital varied between 5–12 mm. In study, the four shapes were assessed for three different diameters, 7, 9 and 11 mm. The shapes selected were conventional round and no taper, conventional round with taper, octagonal and no taper, octagonal with taper.

Tools with round and octagonal cross-sectional shapes were selected for study due to the wider acceptability of these tool handle shapes in the current dental market. Hexagonal shape forces the hand to hold the tool at particular angles and that is why it was not chosen for study. Round cross-sectional shaped pencils provide smooth rotation but cause slippage and can lead to tension in hand musculature. Rounded triangular shape proves most compatible with the three-finger pinch grips giving support to the index finger pad, thumb's distal phalanx and inside of moderately lateral placed middle finger and seems perfect for the tasks which do not need frequent rotation [23], but it may hinder the smooth angular movement needed while doing scaling task. However, round and octagonal cross section of tool handle is intended to provide easy rotation in hand of the dentist for smaller torque movements. The dental scaling task was simulated to evaluate the effects of the different shapes of the grip area of the scaling tool handle on the forearm muscles and pinch force.

3.2 Methods

Dental hygienists and dentists having experience in dental scaling task were enlisted from regional dental hospital in Chandigarh, India. Dentists having any present or preceding injuries in wrist or hand, or suffering from musculoskeletal disorders of upper extremity were ruled out from participation. To maintain the compatibility and standard with the available apparatus, dentists with dominant right hand were only included in the study.

Diameter and shape of cross section of tool handle were chosen as design parameters for the current study which may have an impact on muscle loads and pinch forces. Twelve designs of tool handle sleeves were 3D printed at Centre of Excellence, Punjab Engineering College, Chandigarh, for the customisation of tool handle. The shape of cross section was varied as round (R), round tapered (RT), octagonal

(O) and octagonal tapered (OT) for all 7, 9 and 11 mm diameters as shown in Fig. 3.1. The dimensions for diameter were considered as the maximum distances between parallel surfaces. In the case of tapered handles, the taper of 2.316° was provided at the distance of 45 mm from the end point of the tool handle of the tool tip side. The material of the tool handle was ABS M30i which reduced the overall weight of the tool. The anterior scalars (15/30) tool tip made up of German Stainless Steel of API (Inc., India) brand were press fit into each of the customised 3D printed tool handles.

For experimentation, simulation was carried out using phantom head (CRB Intl., Chandigarh), equipped with human anatomy superior to shoulders (neck, head and mouth) and typodont. As shown in Fig. 3.2, the position and orientation of phantom head was matched with the clinical environment and EMG electrodes along with pressure sensor were placed on muscles of interest. The scaling task was performed on the anterior mandibular group of teeth (incisors and canines) to standardise the task. Mock experiment was done on several materials like P.O.P., nail paint mask and green stick wax by ten dentists to find out best simulating material for calculus deposits. The amount of force required to remove the deposits and the texture of the

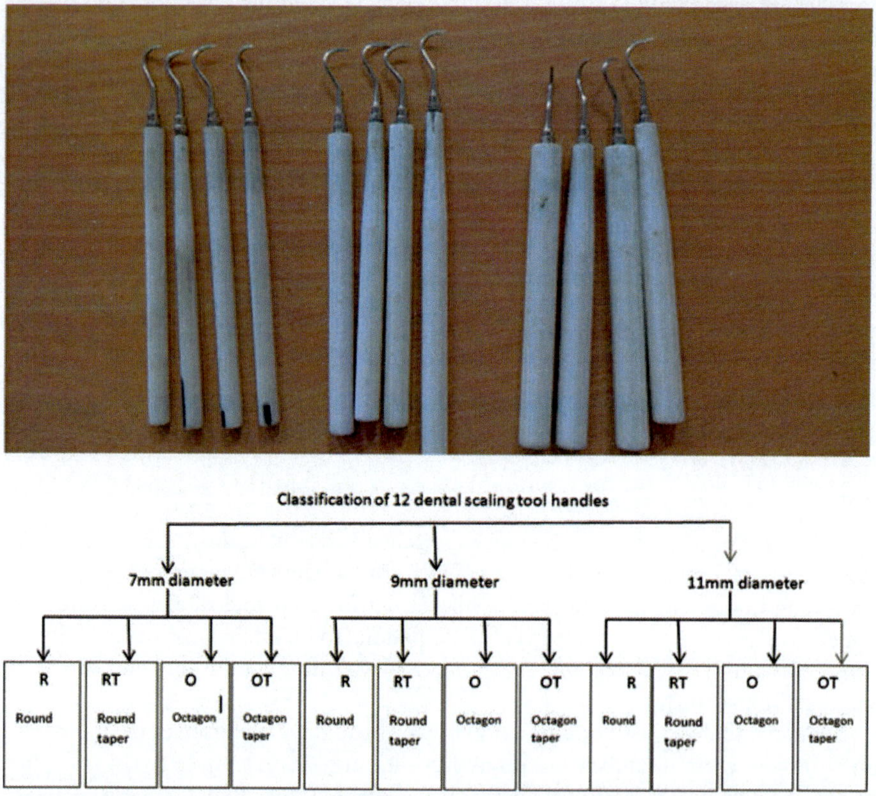

Fig. 3.1 Customised tool handles with 7, 9 and 11 mm diameters and classification table of tools used in the experimentation

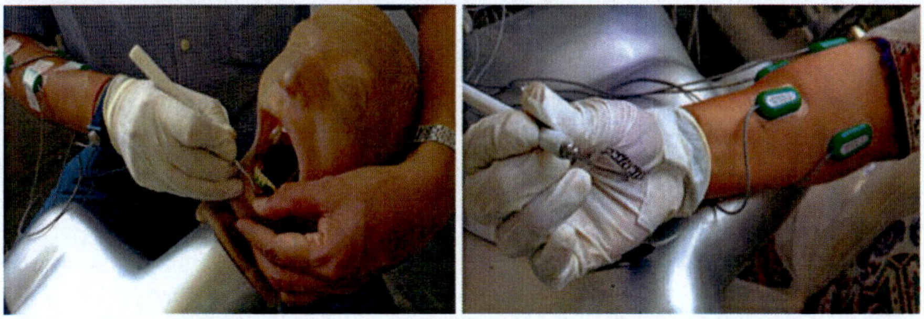

Fig. 3.2 Dental scaling task being performed on typodont and EMG electrodes placed on upper extremity and flexi force sensors on gloves

dry mimicking material was considered as the basis of rating scale. According to the cumulative subjective preferences, green stick was selected to be used as mimicking material for calculus.

Thirteen subjects were selected to perform simulated task. Out of 13 subjects, 9 (69.23%) were females and 4 (30.77%) were males. Three (23.07%) were dentists, 5 (38.46%) were dental hygienists, 3 (23.07%) were interns (final year students of B.D.S.), and 2 (15.38%) were second year post-graduate students.The age varied as: 3 (23.07%) were having age less than 25 years, 2 (15.38%) were between 25 and 30 years, 2 (15.38%) were between 30 and 35 years, 2 (15.38%) were between 35 and 40 years, and 4 (30.77%) were between 40 and 45 years. The average weight of the subjects was 67.5 ± 15.5 kg, and average height was 154 ± 28 cm. Within all thirteen subjects: 7 (53.84%) were having up to 6 years of dental scaling work experience and 6 (46.153%) were having more than 6 years of dental scaling work experience, 3 (23.07%) performed dental scaling task for less than 5 h a week, 5 (38.46%) performed dental scaling task for 5–15 h a week, and 5 (38.46%) performed dental scaling task for more than 20 h a week. Further, it was found out that out of 13 subjects, 7 (53.84%) preferred manual hand scaling tools than the automated ultrasound scaling tools, 3 (23.07%) preferred automated ultrasound scaling tools and 3 (23.07%) gave equal weightage to manual hand scaling and ultrasound scaling tools.

To find out best tool handle design for given population, extensive set of data was collected. Electromyography (EMG) was used to measure muscle activation within various muscles while performing scaling task with different tool designs. Active surface electrodes SX230 by Biometrics Ltd., Ladysmith, USA, were used for obtaining the EMG signals. These are bipolar, differential and reusable EMG sensors with fixed centre-to-centre electrode distance of 20 mm. The data was collected using DataLOG MWX8 (Biometrics Ltd.) at 100 Hz. Biometrics DataLOG software was used to display and analyse raw signals on computer with windows operating system.

Modified pinch grip is usually practised by dental hygienists to perform dental scaling task. Four extrinsic muscles which experienced high muscle activation while pinching the dental scalars with maximum constant force were flexor pollicis longus

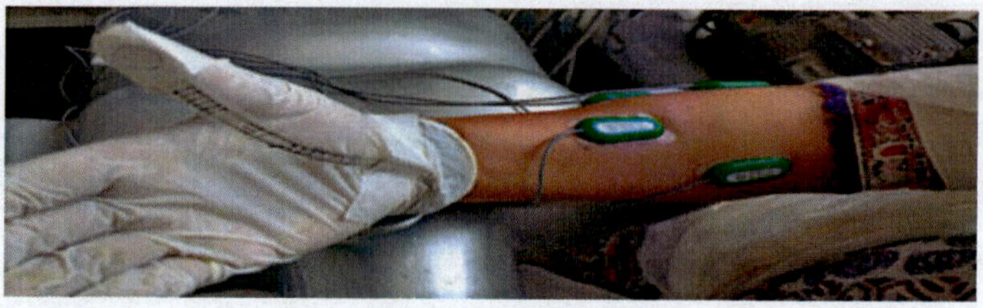

Fig. 3.3 Placement of the flexi force sensor on the thumb tip

(FPL), flexor digitorium superficialis (FDS), extensor carpi radialis (ECR) and extensor digitorium communis (EDC) [6] and were selected for the study. The selected muscles were spotted with the help of physiotherapists (PGIMER, Chandigarh). The reference electrode was placed over the head of the Ulna bone. Peak amplitude normalisation method was used for normalising the EMG data, so that the data of all the subjects can be used for analysis.

For the measurement of thumb pinch force while performing the dental scaling task, flexi force sensors (Biometric Ltd.) were used. The flexi force sensors with sensing area of 9.53 mm diameter gave 0–11 kg force range using DataLog. The sensor was placed on the thumb tip of gloves of the dentist such that the pinch force applied by the thumb tip gets transmitted through the sensor strip to the tool surface. It serves better purpose than the case when sensor is placed on the tool as it allows better manoeuvrability to the hygienist while using the scaling tool. The exact placement of the flexi force sensor is shown in Fig. 3.3.

Experimentation was carried out to record EMG and pinch force data. Time slot for EMG and pinch force data collection of each subject (dentist) was 3.5 h. approximately. Each subject was instructed to use all the twelve modified scaling tools of different sizes and shapes one by one in random order. Actual task was performed only when the subject gets comfortable and accustomed with the standard task procedure. Stress was given to perform the dental scaling task with consistency. For each modified tool, scaling task was performed for 1 min followed by 3–4 min of break to prevent variation in EMG and pinch force values due to muscle fatigue. As instructed, subjects were time bound to scale off approximately same volume of green stick wax to standardise the effort. Extra rest time was provided if in case any of the subjects felt even a bit of tiredness after using few instruments.

All subjects were asked to share their perception and preference about each modified scaling tool with respect to the scaling task performance. The ranks were given to the tools according to the subjective comfort level of usage. The ratings were given on 1–5 scale: 1-Most comfortable, 2-Comfortable, 3-Satisfactory, 4-Uncomfortable and 5-Highly Uncomfortable.

The 30 s window was snipped out of recorded data. Raw signals were processed using low pass filters of first order and rectification. After applying filters, RMS values of EMG signals for four muscles and pinch force looks like a typical curve shown in Fig. 3.4., where high perturbation in force curve shows the pull action of the dentist while performing dental cleaning. Signal values associated with specific

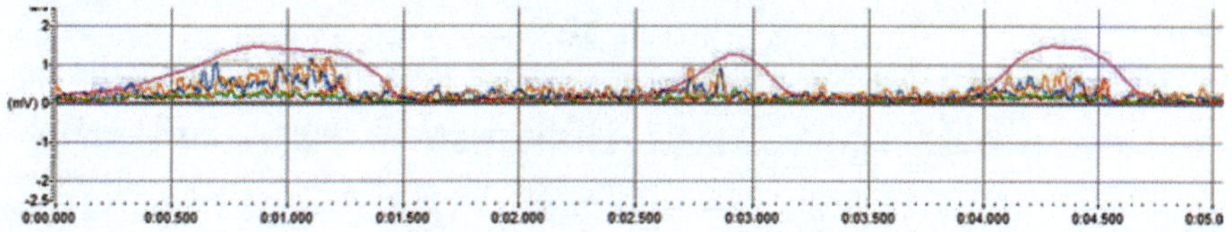

Fig. 3.4 Five second window of EMG and pinch force recordings. Following colours represent respective EMG signals and pinch force from various muscles, (A1) Red-FDS, (A2) Green-FPL, (A3) Blue-EDC, (A4) Orange-ECR, (A5) Pink-Pinch force. Values of pinch force were visible while analysing the pinch force signals separately

muscle and force values can be spotted out with the different colours assigned to each type of signal. The values of force and EMG are in kg and mV, respectively. The approximate synchronisation between EMG signals of all muscles and pinch force can be noticed from the composition of EMG signal window.

EMG and pinch force data was normalised using internal peak value for each trial carried using each modified tool by all subjects. The script in MATLAB was created for normalisation of data and calculation of amplitude probability distribution function (APDFs) at 50% level (equivalent to median values) for the EMG values for all four muscles and thumb pinch force values [18]. EMG data and thumb pinch force values of all the twelve scaling tool handle designs were analysed using statistical software (Minitab 17). The repeated measures analysis of variance F(RMANOVA) was applied at significance level of $\alpha = 0.05$. Subjects, diameter, cross-sectional shape and interaction between diameter and shape were taken as factors.

For further analysis, data was classified into three groups of 7, 9 and 11 mm diameter tools. Pairwise comparisons of multiple modified tools belonging to various groups were done using Tukey method. Cross-sectional shapes and diameter of tool handles were taken as factors and corresponding median values as response. Wilcoxon signed-rank test was used to analyse the subjective perception rating values, and its results were compared with statistical results to find out co-relation.

3.3 Results

3.3.1 Effect of Shape and Diameter Factors on Tool Handle Design

From RMANOVA technique, the combined effect of diameter and shape of the dental scaling tool, was found to be significant ($p < 0.05$) on FPL, FDS and PF as given in Table 3.1.

Table 3.1 Statistical analysis of instrumental handles used in study (*R* = Round, RT = Round taper, O = Octagonal, OT = Octagonal taper)

	E1 (Median)	E2 (Median)	E3 (Median)	E4 (Median)	PF (Median)
D*S	*	*	~	~	*
OT7–O7	*	*	~	~	~
R7–O7	*	~	~	~	~
RT7–O7	~	*	*	~	*
R7–OT7	*	*	~	~	~
RT7–OT7	*	*	*	~	*
RT7–R7	*	*	*	~	*
OT9–O9	*	*	~	~	~
R9–O9	*	~	~	*	*
RT9–O9	~	~	~	~	~
R9–OT9	*	*	*	*	~
RT9–OT9	*	~	~	~	~
RT9–R9	*	~	~	*	*
OT11–O11	*	*	~	~	~
R11–O11	*	~	~	*	~
RT11–O11	*	~	~	~	~
R11–OT11	~	*	~	*	~
RT11–OT11	~	~	~	~	~
RT11–R11	~	~	~	~	~

3.3.2 *Effect of Shape in Different Tool Handle Diameter Groups*

From Tukey method, significant differences for pairs of cross-sectional shapes in 7 mm diameter group were found to be more in number as compared to larger diameter groups (9 and 11 mm). Results of Tukey method are given in Table 3.1 where E1, E2, E3, E4 and PF are median-normalised values corresponding to FPL, FDS, ECR, EDC muscles and pinch force. '*' symbol means significant difference of $p < 0.05$ exists, whereas ' ~ ' symbol means there is no significant difference.

Median normalised values of EMG and pinch forces for different shape and diameter combination for 7, 9 and 11 mm groups of tools were used to extract the findings of the study. From Fig. 3.5, significant differences in normalised median values of EMG and pinch force can be seen for octagonal tapered and round tapered 7 mm tools and octagonal tapered and round larger diameter tools (9, 11 mm). Further it was found that tapered shape in all diameters mostly resulted in lower EMG and pinch force values, whereas round tapered shape in 7 mm tools and round shape in larger diameter (9, 11 mm) result in higher EMG and pinch force values.

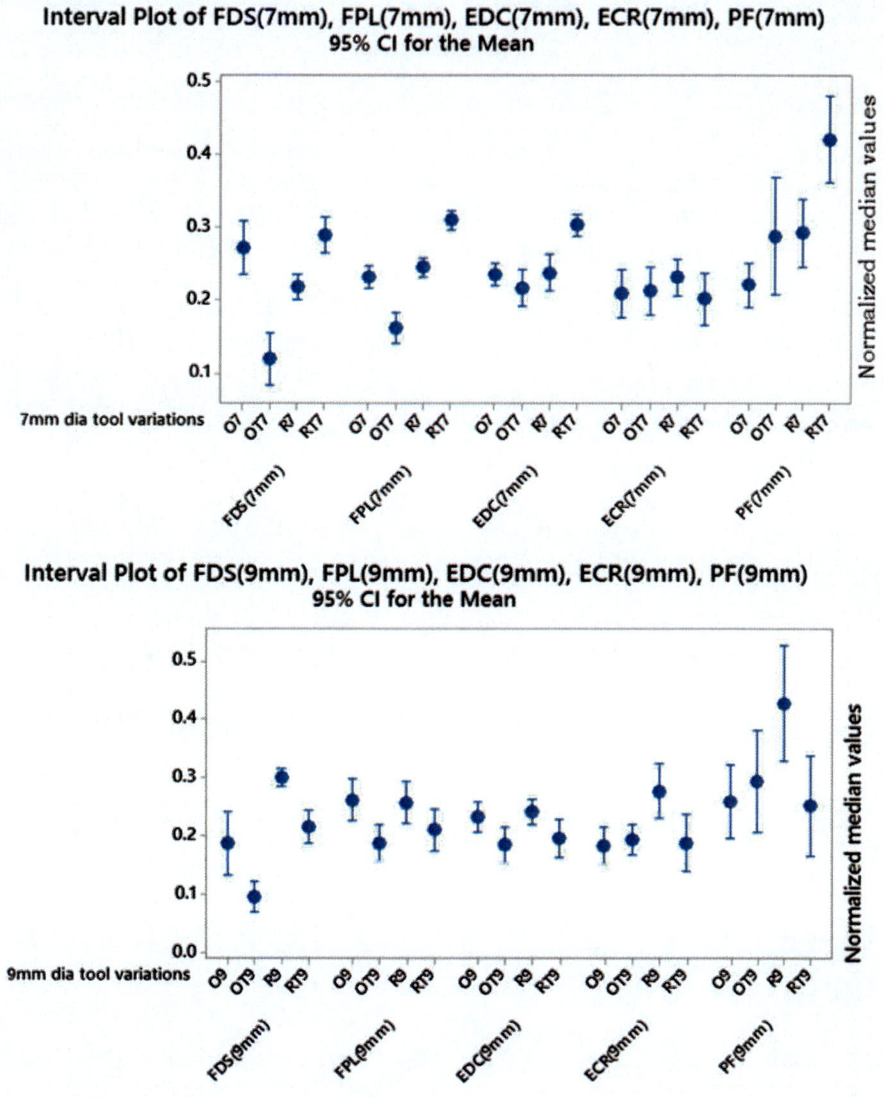

Fig. 3.5 Comparison of different cross-sectional shapes in 9, 7 and 11 mm diameter

From Wilcoxon signed-rank test, octagonal and octagonal tapered shapes in 7 mm diameters and octagonal tapered and round tapered shapes in 11 mm diameters were perceived as comfortable which notably matched with statistical findings. However, in 9 mm tools, no such match was found in ranking analysis.

The line plot (Fig. 3.6) suggests that the smaller diameter dental scaling tool handles (7 mm) offered higher pinch force and EMG values than larger diameter

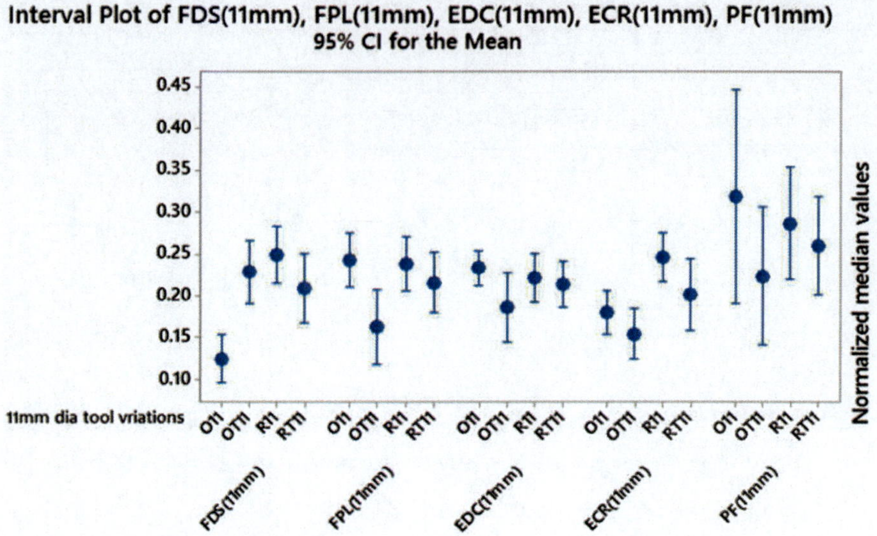

Fig. 3.5 (continued)

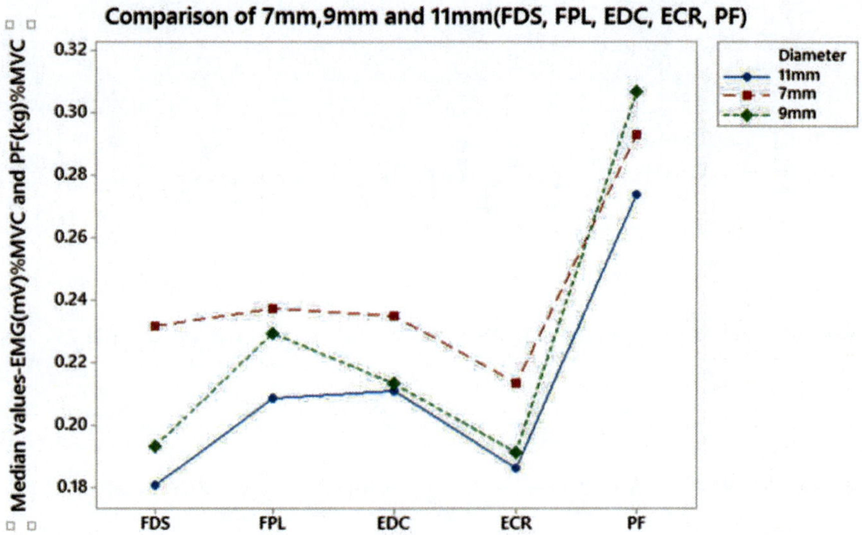

Fig. 3.6 Effect of 7, 9 and 11 mm tool handle diameters on various muscles and pinch force

dental scaling tool handles (9 and 11 mm). Tools with 11 mm diameter resulted in lowest effort of the dentists.

3.4 Discussion

The current preliminary trial was an attempt to improvise scaling tool handle design based on available commercial tools in the market and analyse the effects of tool design in a convenience sample of limited size mixed population of dentists and other dental health workers enrolled from the regional dental institute. The investigation suggested possible effect of diameters and shapes of scaling tool handles on the hand muscle load and pinch forces experienced by dentists while performing manual dental scaling task. Smaller diameter tools resulted in higher loads on hand muscles while performing dental scaling task. This may be due to the smaller dimension of cross section of tool handle which is not in accordance with the hand anthropometry of the dentists. The subject's preference for smaller diameter tools maybe due to the habitual usage and convenenience developed due to its abundant availability in market and continuation in early trends. Most of the dental students are given scaling task training using these smaller diameter tools only due to their cheaper prices and easy availability, they may have developed habit of using smaller diameter tools from there only. In larger diameter tool handles, the modified pinch grip (as in case of smaller diameter tool handles) gets converted to chuck grip. This maybe the possible reason that larger diameter tool handles need less force requirement. Also, the subjects usually preferred octagonal straight tools because of its common usage and availability. The preferences recorded from the subjects should be interpreted keeping the small time span allotted to the task in view in the current study, as the preferences may alter in case the duration of the task at hand is extended for long or the task is repetitive, which is a routine in real-life clinical scenario. ABS material used to 3D print dental tools has excellent mechanical properties like dimensional stability, abrasion resistance, rigidity, surface hardness, superior strength, chemical resistance and is easily machinable. The ABS material also serves wide range applications in medicine, surgery and even in dentistry [3]. Thus, using ABS as a tool handle material may not be a matter of concern.

Tapered handles irrespective of shape proved better than straight tools. For 7 mm diameter tools, round tapered tools required high forces. This may be due to the reason that the value of taper angle provided was at higher side than the diameter of tool which made it difficult for the dentist to hold the tool properly. Octagonal tapered and octagonal shapes in scaling tools provided better grip. Round tools are subjected to rotation while performing scaling task, hence are bit difficult to hold than octagonal ones. In 9 and 11 mm tools, round tapered tools were better in terms of force requirement. It can also be observed from Fig. 3.6 that for most of the times (for different diameters), when the EMG reaches its peak value the pinch grip force also attains the peak. The same trend was observed in the study by Fang et al. [10] where the overall trend for the sEMG and pinch force remained consistent and from the mere observation of the raw data can be indicative of the potential relationship between the two parameters. Taper provided to the tool handle tip increases the surface area of fingers in contact with the tool, which results in increasing the pressure points and reducing the force. Due to the taper provided, component of reaction forces gets minimised, and less effort is required. In the present study, the cross-sectional shape had more effect on muscles in smaller diameters of handles. The consistently repetitive jobs during the course of long time span can give boost to the occurrence of problem of carpal tunnel [14, 22]. Reduction of average median pinch forces were observed when taper was provided. These differences in average median pinch force values may prove to be major factors to cause hand muscle disorders if the task is performed repetitively. Providing taper to straight handles may give phenomenal decline in occurrence of diseases like carpal tunnel syndrome.

3.5 Limitations

In the present study, task was simulated on typodont by maintaining environment identical to clinical environment. The experimentation was carried out only for 1 min using each modified tool, whereas the dental hygienists perform the same task repetitively for hours. The effect of scaling task on muscle load and pinch forces may be different in actual practice. The study was performed only by considering 13 subjects, whereas large number of subjects may be considered. Cross-sectional shapes of dental tool handles other than round and octagonal may be considered for future study. Hand anthropometry can also be considered for designing better dental scaling tool handles.

3.6 Conclusions

Present study suggests that the dental scaling tool handles with larger diameters maybe preferred. Tapered handles in general may prove to be better option. Octagonal tapered tools may give the best results as octagonal shape can provide better gripping and taper may improve the holding of the tool leading to reduction in effort required to perform the dental scaling task.

Acknowledgements This study was funded by Centre of Excellence in Industrial and Product Design, PEC University of Technology, Chandigarh, India, under National Project Implementation Unit scheme with grant number SPFU/(TEQIP-II)/2013/1517. The authors wish to thank the dentists who participated in the study.

References

1. ADA: Survey of Current Issues in Dentistry: Repetitive Motion Injuries, pp. 2–8. American Dental Association, Chicago (1997)
2. Ahern and Rys.: Evaluation of dental scalers. Ind. Syst. Eng. Rev. (ISER) **1**(1) [Online] (2013). Available: https://watsonojs.binghamton.edu/index.php/iser/article/download/6/6

3. Alsandi, Q., Ikeda, M., Nikaido, T., Tsuchida, Y., Sadr, A., Yui, N., Suzuki, T., Tagami, J.: Evaluation of mechanical properties of new elastomer material applicable for dental 3D printer. J. Mech. Behav. Biomed. Mater. **100**, 103390 (2019)
4. Anton, D., Rosecrance, J., Merlino, L., Cook, T.: Prevalence of musculoskeletal symptoms and carpal tunnel syndrome among dental hygienists. Am. J. Ind. Med. **42**(3), 248–257 (2002)
5. Bramson, J.B., Smith, S., Romagnoli, G.: Evaluating dental office ergonomics. Risk factors and hazards. J. Am. Dent. Assoc. **129**(2), 174–183 (1998)
6. Canada, Canadian Centre for Occupational Health and Safety, Government of Canada [Web log post]. Retrieved from https://www.ccohs.ca/oshanswers/ergonomics/handtools/tooldesign.html#wb-info. 4 April 2017
7. Dong, H., Barr, A., Loomer, P., Laroche, C., Young, E., Rempel, D.: The effects of periodontal instrument handle diameter and weight on hand muscle load and pinch force in a simulated dental scaling task. J. Am. Dent. Assoc. **137**(8), 641–658 (2006)
8. Dong, H., Barr, A., Loomer, P., Laroche, C., Young, R., Rempel, D.: The effect of tool handle shape on hand muscle load and pinch force in a simulated dental scaling task. Appl. Ergon. **38**, 525–531 (2007)
9. Ehsan, A., Ehsan, S., Arshad, H.S.: Frequency of Carpal tunnel syndrome in dentists working in government hospitals of Lahore Muhammad. Int. J. Sci. Res. **5**(5), 1672–1675 (2016)
10. Fang, Y., Ju, Z., Zhu, X., Liu, H.: Finger pinch force estimation through muscle activations using a surface EMG sleeve on the forearm. In 2014 IEEE International Conference on Fuzzy Systems (FUZZ-IEEE), pp. 1449–1455. Beijing (2014). https://doi.org/10.1109/FUZZ-IEEE.2014.6891790
11. Haghighat, A., Khosrawi, S., Kelishadi, A., Sajadieh, B.H.: Prevalence of clinical findings of carpal tunnel syndrome in Isfahanian dentists. Adv. Biomed. Res. **1**(13), (2012)
12. Hamann, C., Werner, R.A., Franzblau, A., Rodgers, P.A., Siew, C., Gruninger, S.: Prevalence of carpal tunnel syndrome and median mononeuropathy among dentists. J. Am. Dent. Assoc. **132**, 163–170 (2001)
13. Haslegrave, C.M.: Characterizing the anthropometric extremes of the population. Ergonomics **29**, 281–301 (1986)
14. Jonsson, B.: Measurement and evaluation of local muscular strain in the shoulder during constrained work. J. Hum. Ergon. **11**(1), 73–88 (1982)
15. Lalumandier, J.A., McPhee, S.D., Riddle, S., Shulman, J.D., Daigle, W.W.: Carpal tunnel syndrome: effect on army dental personnel. Mil. Med. **165**(5), 372–378 (2000)
16. Lalumandier, J.A., McPhee, S.D.: Prevalence and risk factors of hand problems and carpal tunnel syndrome among dental hygienists. J. Dent. Hyg. **75**(2), 130–134 (2001)
17. Liss, G.M., Jesin, E., Kusiak, R.A., White, P.: Musculoskeletal problems among Ontario dental hygienists. Am. J. Ind. Med. **28**(4), 521–540 (1995)
18. Mithun Pai, B.H., Shenoy, R., Rajesh, G., Rao, A.: Symptoms of Carpal tunnel syndrome in a dental work force of a developing country. Int. J. Adv. Res. **2**(2), 87–94 (2014)
19. Normasutra, A.R., Munirah, M.A., Aziza, Y., Aziah, D.: Prevalence of probable carpal tunnel syndrome and its associated factors among dentists in Kelantan. Int. J. Collab. Res. Internal Med. Public Health **6**, 247–259 (2014)
20. Osborn, J.B., Newal, K.J., Rudney, J.D., Stoltenberg, J.L.: Carpal tunnel syndrome among Minnesota dental hygienists. J. Dent. Hyg. **64**(2), 79–85 (1990)
21. Rice, V.J., Nindl, B., Pentikis, J.S.: Dental workers, musculoskeletal cumulative trauma, and carpal tunnel syndrome, who is at risk? A pilot study. Int. J. Occup. Saf. Ergon. **2**(3), 218–233 (1996)
22. Roquelaure, Y., Mechali, S., Dano, C., Fanello, C., Benetti, F., Bureau, D., Marial, J., Martin, Y.H., Derriennic, F., Penneau-Fontbonne, D.: Occupational and personal risk factors for carpal tunnel syndrome in industrial workers. Scand. J. Work Environ. Health **23**(5), 364–369 (1997)
23. Strasser, H., Bullinger, H.J.: A Systematic Approach for Analysis and Ergonomic Design of Hand-Held Tools and Control Actuators-Visulaized by some Real Life Examples in Assessment of Ergonomic Quality of Hand-Held Tools and Computer Input Devices, 1st edn. pp. 1–22. IOS Press, Amsterdam (2007)

Chapter 4
An Experimental Investigation on Postural Risks in Floor Mopping

Mohammed Rajik Khan and Gouri Naik

Abstract The present study investigates awkward human body postures adopted in floor mopping by push and figure-of-eight methods. Three mopping professionals each in three different height categories (5th, 50th and 95th percentile person according to Indian anthropometric data) participated in floor mopping experiments conducted in two different laboratory set-ups. In the first set-up, continuous positional data of participant's joints (wrist, elbow and shoulder) were recorded using six dimensional electromagnetic tracking system (ETS) and angular variation of the joints were evaluated for the two mopping techniques. In the second set-up, the three-dimensional positional data of body movements were recorded using optical motion capture cameras in mopping experiments to identify Rapid Upper Limb Assessment (RULA) score at extreme postures. The total RULA score obtained in push (4.6–5) for all height categories is less than figure-of-eight (6.4–6.6) method, indicating less postural load during push mopping technique. This research confirms that mopping professionals are highly susceptible to postural loads/discomfort and the need of ergonomic design solutions are essential to reduce the observed risk.

4.1 Introduction

Millions of people around the world work part-time or full-time as cleaning professionals for their daily earnings. Among them, a high percentage of these workers comprise aged women who are under-educated for other jobs and mostly belong to the lower economic background [1]. Such economic constraints drive these individuals towards professions of cleaning, swabbing, housekeeping, etc. using manual methods as the automated cleaning system is usually expensive. These occupational tasks involve awkward body postures for long duration which leads to severe pain

M. R. Khan (✉) · G. Naik
Industrial Design, National Institute of Technology Rourkela, Rourkela, Odisha, India
e-mail: khanmr@nitrkl.ac.in

G. Naik
e-mail: gourinaik93@gmail.com

A. Chakrabarti et al. (eds.), *Design for Tomorrow—Volume 1*, Smart Innovation, Systems and Technologies 221, https://doi.org/10.1007/978-981-16-0041-8_4

in various body regions [2, 3]. Chang [1] reported that around 90% of the cleaning professionals' experience musculoskeletal disorders (MSDs) in at least one of the major load-carrying joints in the body. The most commonly reported problems in those individuals include discomfort at the wrist, shoulder, lower back and elbow joints.

The working environment, the ergonomic design of the equipment as well as the posture adopted by cleaning personnel's greatly influences the workload [4]. Unhygienic work environment makes the cleaning staff subjected to chronic, respiratory and dermatologic diseases [5]. Keeping the control over working environment while cleaning is challenging. Hence, from the designer's perspective, design modifications in tools/equipment are prioritized. As stated by Bisht and Khan [6, 7] and Dianat et al. [8], shape changes in the tool handles greatly influences the muscle workload of the wrist flexor and extensor. Søgaard et al. [9] mentioned that using a long handle mop decreases the trunk's maximum forward–backward flexion compared to conventional methods of scrub and cloth mopping. It is highly recommended that the mop handle height be adjusted from shoulder to chin level to lessen the shoulder-muscle activity [10].

A review on cleaning equipment recommended modifications in the design of buffing, mopping and vacuum machining system based on the standard ergonomic guidelines [4, 11]. Hagner and Hagberg [12], Khan and Singh [3] and Khan et al. [13] used Rapid Upper Limb Assessment (RULA) and Rapid Entire Body Assessment (REBA) techniques to evaluate the musculoskeletal risks involved in workers working in unorganized sectors. The emphasis was focused in conducting motion and EMG analysis on work activities [2, 10, 12]. Motion capture instrumentations are being used to assist RULA assessment and postural evaluation to predict the ergonomic risks [13–15]. Data on precise range of motion (ROM) of different body joints involved in various cleaning methods are essential for any design modifications.

The work focuses on comparing the ergonomic risks on working professionals involved in the two mopping methods (push and figure-of-eight). The upper limb movement of nine professionals involved in the mopping activity is recorded using six dimensional electromagnetic tracking system (ETS) and optical motion capture cameras. Data captured using ETS were evaluated in MATLab to measure the angular variations of various body joints (wrist, elbow and shoulder). Maximum joint variations for the two mopping methods were identified. Movement data captured through motion cameras are simulated in OpenSim to record extreme positions of awkward postures. RULA scores were calculated to identify the risk factors in the two mopping methods. It was evident that push method involves less risk as compared to figure-of-eight method.

4.2 Methodology

Two sets of experimental procedures have been adopted to analyze the awkward postures adopted during mopping operation. The first procedure captures the continuous positional data of both hand joints using six-dimensional electromagnetic sensors. The second procedure utilizes optical motion cameras to capture the body movement during the tasks.

4.2.1 Participant Selection

In the present study, a group of nine mopping professionals working at the National Institute of Technology Rourkela, India had been selected with a minimum one-year mopping experience. Among them were grouped 3 participants each from three separate height groups' i.e. 5th, 50th and 95th percentile [16, 17]. Proper ethical guidelines as per institute ethical regulations were followed during the study. The research design was explained to the participants prior to the experimental activities, and their signed consent was obtained for their participation. The timing of the study was based on their leisure time, without disturbing their regular work. The demographic and anthropometric (height and weight) data for all the responders were recorded.

4.2.2 Experimental Task

The task has been designed for two most common types of mopping techniques performed by the professionals known as push and figure-of-eight. Figure 4.1 shows the two mopping methods where, the tasks performed by moving the mop forward is push and when the mop makes an eight-shaped curve while retracting is figure-of-eight. A cleaning mop, model Roots EZE clean of make Taara Hygiene Enterprises, India has been used in the tasks. The study is conducted in a highly polished floor. The damping due to dry microfiber of mop is controlled by wetting the mop pad in water before carrying out the experiment. The dominant hand grips the top of the mop handle and the non-dominant hand gives support, as per user comfort, randomly in between the top and middle of the handle. All the participants were instructed to practice cleaning prior to experimental tasks so as to make them comfortable.

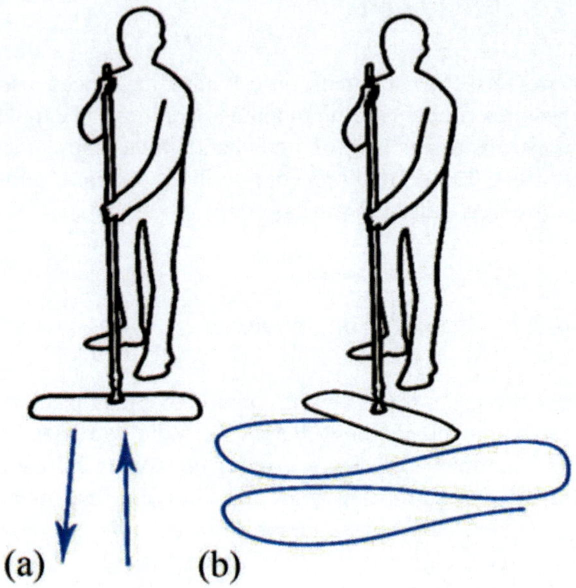

Fig. 4.1 Mopping methods **a** push **b** figure-of-eight

4.2.3 Motion Capture

In the first experimental procedure, a six-dimensional electromagnetic tracking system (ETS), model Liberty of make Polhemus, Chicago having sixteen channel data acquisition (DAQ) with 09 sensors providing 240 updates per second per sensor is used to trace the participant's hand joint trajectory during the tasks. The electromagnetic field is generated (emitted) by the transmitter (source) to recognize three dimensional position and spatial orientation of the sensors. The device has sensitivity within an operating range of 2 m between transmitter and receptor sensor. The experimental tasks were conducted in an open space environment of concrete ground floor where no metal reinforcement was present to avoid metallic interference. The electromagnetic source of the device is positioned on a non-magnetic material based stand (stable support) of height approximately to the mid-height of a 95th percentile person so as to keep all the sensors within the range of magnetic field. The 09 receptor sensors are properly fixed on the skin of centre metacarpal region, wrist joint, trochlea of elbow joint and acromion of shoulder joint of both the hands and C7 vertebra by means of double sided tape and elastic strap. Figure 4.2 depicts the experimental set-up of ETS used during the study. The metallic (Al alloy) rod of mop has been replaced with 03 PVC pipe of same diameter and varying lengths having high stiffness so as to avoid the electromagnetic interference with the ETS system. The length of the three pipes are made at shoulder height of 5th, 50th and 95th percentile person's height according to Indian anthropometric data [16]. The weight of the replaced pipe is made similar as of the metallic mop handle i.e. 0.4 kg by inserting Plaster of Paris (POP). The task of floor mopping for each individual

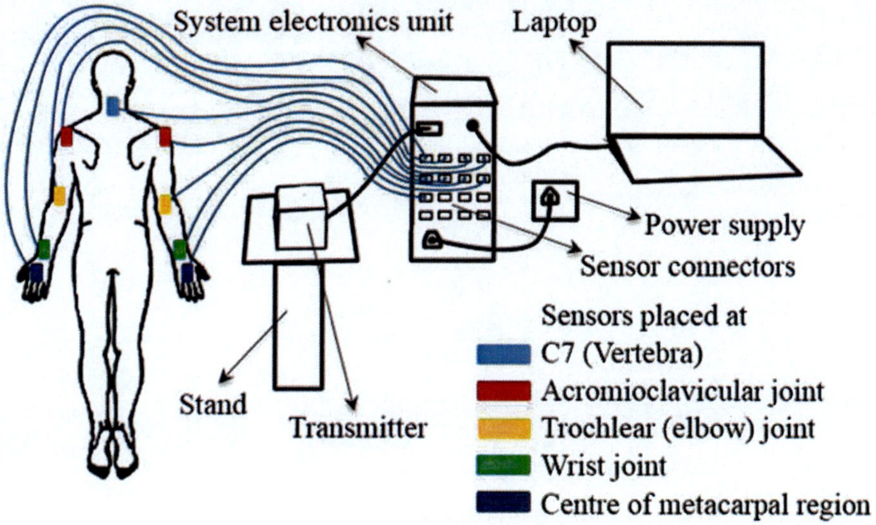

Fig. 4.2 Equipment setup of ETS showing sensor positions

was recorded for 10 trials in each method. The trajectory followed by the 09 sensors placed at various position of the participant's body captured for both the mopping methods during experiments were processed in MATLAB 2016 to determine the angular variation of wrist, elbow and shoulder joints. The mean of the maximum and minimum joint angles calculated for all the trials of each participant is recorded. The mean range of angular variations of various hand joints categorized under three heights were evaluated.

The continuous dynamic movement of mopping professionals was captured in the second experimental procedure in an optical motion capture laboratory environment surrounded by four high-speed cameras (Oqus 5.0, Qualisys, USA). A total of 18 passive retro-reflective markers were mounted on various body anatomical landmarks such as acromial, knuckle, anterior superior iliac spine (ASIS) and lateral elbow, wrist, knee and ankle of both sides of the body and forehead, C7 (vertebra prominens), clavicle (collarbone) and sternum. All the 09 professionals were asked to perform the mopping activity three times for each methods. The marker movements in 3-dimensional space were reported with a data capture frequency (rate) of 100 frames per sec (Hz) in Qualisys Track Manager (QTM). In OpenSim 4.0 environment the 3D gait model with simple arms was scaled to create the model subject specific. The recorded positional data were converted into required motion file (.mot) to simulate the dynamic motions of mopping activity in OpenSim. In addition, the individual mopping simulations were viewed at a rate of 20 fps (frames per second) to assess the frequent change in posture. Therefore, all the postures adopted in each mopping period are then divided into seven prime positions. In both approaches to identify the extreme postures, RULA scores of these seven positions for all participant trials

Table 4.1 Demographic data of mopping professionals

Variables	Number of participants	Percentage (%)
BMI (kg/m^2)		
<18.5	1	11.11
18.5–25	6	66.66
25–30	2	22.22
Literacy level		
Illiterate	1	11.11
Primary school	4	44.44
High school	4	44.44
Work experience		
≤1 year	1	11.11
1–2 years	3	33.33
2–3 years	4	44.44
3–4 years	1	11.11

were determined. The mean of the maximum RULA score of each trails under the three height categories were calculated and compared.

4.3 Results and Discussion

4.3.1 Demographics

Six (06) out of nine (09) selected participants were female. The average age and height of all the mopping professionals were 23.3 years and 1.6 m with a standard deviation of 1.699 and 0.11 respectively. Table 4.1 mentions the demographic data of the mopping professionals involved in this experimental study. During mopping task experiments, 55.5% participants preferred to keep their left hand in the mop handle's upper position as they are right handed.

4.3.2 Postural Assessment

The extreme postures and the ergonomic risks involved in mopping activity under two different experimental setups are discussed here. ETS as used in the 1st setup helps to precisely measure the hand joints angular variations. In the second setup, dynamic motion during the activities were recorded using optical cameras which helps to minimize the potential error arising in photography method.

4.3.3 Variation of Joint Angle

Figure 4.3 shows the range of joint angle variations of wrist, elbow and shoulder for both the hands during mopping with push and figure-of-eight methods. In push method, for participants having 5th percentile height the range of joint angle variations are 153.5–173.08, 91.82–155.3 and 127.72–149.56 for wrist, elbow and shoulder joints respectively where as in figure-of-eight method, it varies from 138.91 to 177.41, 105.95 to 154.87 and 125.34 to 156.58 respectively for the hand positioned at upper portion of mop handle (Fig. 4.3a). Similarly, for the hand on lower position of the mop handle (Fig. 4.3b), these angle ranges from 144.83 to 175.89, 58.23 to 119.76 and 117.34 to 143.58 and 120.46 to 176.03, 59.91 to 109.57 and 118.42 to 141.38 respectively for both the mopping methods.

For the participants having 50th percentile height, the range of joint angles in wrist, elbow and shoulder varies from 126.4 to 171.99, 69.49 to 104.38 and 104.38 to 140.08 and 122.79 to 238.11, 65.77 to 96.97 and 104.3 to 144.9 respectively for push and figure-of-eight methods for the hand positioned at upper portion of mop handle (Fig. 4.3c). Similarly, for the hand on lower position of mop handle (Fig. 4.3d),

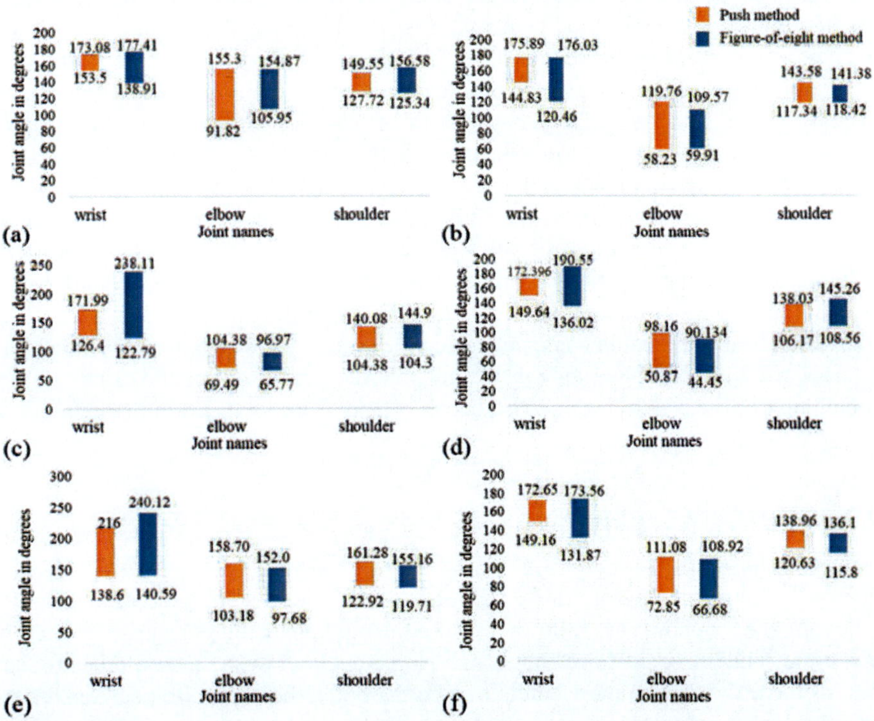

Fig. 4.3 Range of joint angles of wrist, elbow and shoulder during mopping with participant's hand on **a**, **c**, **e** upper position and **b**, **d**, **f** lower position of mop handles respectively for 5th, 50th and 95th percentile heights

these angles vary from 149.64 to 172.39, 50.87 to 98.16 and 106.17 to 138.03 in push method and 136.02 to 190.55, 44.45 to 90.13 and 108.56 to 145.26 in figure-of-eight method.

The 95th percentile height person shows a variation of wrist, elbow and shoulder joints for their hand on upper portion (Fig. 4.3e) of mop handle as 138.6–216, 103.18–158.70 and 122.92–161.28 in push and 140.59–240.12, 97.68–152.0 and 119.71–155.16 in figure-of-eight methods respectively. While the hand is on bottom position (Fig. 4.3f), joint variations in wrist, elbow and shoulder are 149.16–172.65, 72.85–111.08 and 120.63–138.96 and 131.87–173.56, 66.68–138.96 and 115.8–136.1 respectively for push and figure-of-eight mopping methods.

As observed from the graphical representation in Fig. 4.3, the angle variation of wrist and shoulder joints are lower in push mopping technique as compared to figure-of-eight. An increased range of angle of wrist joint in case of figure-of-eight method can be justified with the reason that a larger twisting of wrist happens in the transverse direction while mopping. Similarly, tracing the trajectory of figure-of-eight in the transverse direction of mopping involves larger movement of shoulder joint as compared to the forward movement in push method. In case of elbow, the range of joint variation is higher for push as compared to figure-of-eight method. But, the push method traces the to and fro motions along the mopping direction and the figure-of-eight involves movement across the mopping direction. Considering the height variations of participants as 5th, 50th and 95th percentile, the results of wrist and shoulder joints angle variation of the hand on top position of mop handle increases with increase in height in push method. This is obvious as the person having more height intends to cover a larger distance in the forward direction in push mopping technique. While in figure-of-eight method, joint angle variations of wrist and shoulder is maximum for 50th percentile person. Push mopping method involves linear movement of mop pad whereas a complex curvilinear path is traced in figure-of-eight method which leads to multiple twisting movement of hand in the latter method. This could increase the pain and discomfort in the user's hand. The results can be more reliable while increasing the population size is recommended for future work. It is also suggested that the mop handle can be redesigned to avoid over stretching or excess twisting of wrist, shoulder and trunk joints during mopping.

4.3.4 RULA Analysis

Figure 4.4 depicts a participant's sequential movement (95th percentile height) following the mopping path followed in both the methods with seven prime postures in a cycle. The prime postures shown in a single cycle of push method relate to the start of the task with the mop handle near to the body, the other two postures relate the forward mopping, the fourth relate to the farthest end the mop reaches and the other three postures relate the mop movement in reverse direction before it reaches the body nearest. Similarly, in a single cycle of figure-of-eight method, these seven postures relate to the initial position of user with mop handle, tracing the mopping

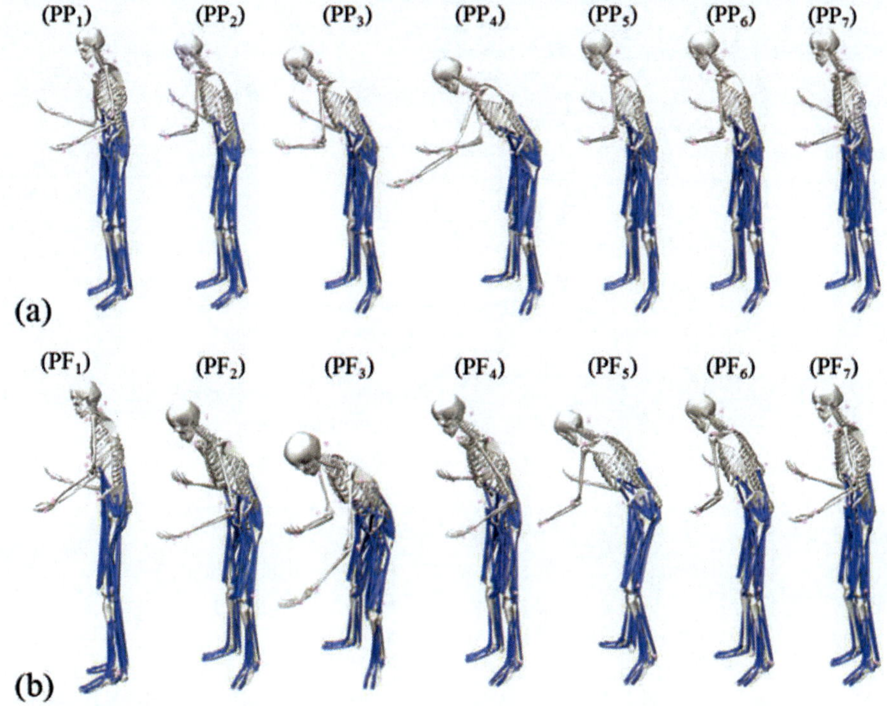

Fig. 4.4 Array of postures showing the sequential movement in mopping activities using **a** push (PP1–PP7) and **b** figure-of-eight (PF1–PF7) methods. (PPi: Posture in push method; Pfi: Posture in figure-of-eight method; i: index from 1 to 7)

trajectory along one of the side, reaching the farthest end on the same side, returning and reaching at the middle of the trajectory, reaching again at the farthest end on the other side, again movement of mop in reverse direction and finally reaching nearest to the body.

Figure 4.5 illustrates RULA score of all the 09 mopping professionals in push and figure-of-eight methods. The RULA score in push method for all the participants falls between 4 and 5. Hence the ergonomics risk falls under low to medium level, whereas RULA score varies from 6 to 7 in figure-of-eight method making the technique a medium to high level risk [2]. Table 4.2 compares the overall mean RULA score and risk index for the participants under three different height categories performing mopping by push and figure-of-eight methods. The overall mean RULA score in mopping with push technique is measured as 4.8, 4.6 and 5 for 5th, 50th and 95th percentile height person respectively whereas the scores are 6.6, 6.4 and 6.5 respectively for figure-of-eight method. The mean RULA is least for 50th percentile person in both the methods. The highest RULA score obtained in push method (5.0) is lower than the lowest score (6.4) in figure-of-eight method. Hence, the risk involved in mopping by figure-of-eight method for all the three height categories are higher

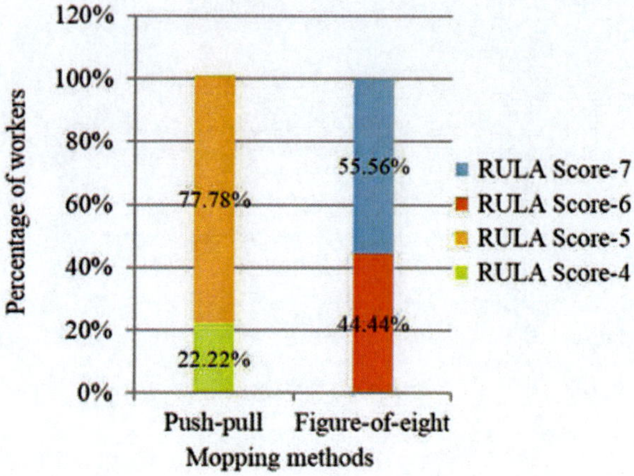

Fig. 4.5 Percentage of mopping professionals showing variations in RULA score in push and figure-of-eight mopping experiments

Table 4.2 Comparison of RULA score and risk index for participants of height 5th, 50th and 95th percentile performing mopping by push and figure-of-eight methods

Height percentile	Push method			Figure-of-eight method		
	Mean RULA Score (SD)	Risk Index (SD)	Risk Level	Mean RULA Score (SD)	Risk Index (SD)	Risk Level
5th	4.8 (±0.5)	1.602 (±0.17)	Low to medium	6.6 (±0.58)	2.198 (±0.19)	High to very high
50th	4.6 (±0.55)	1.54 (±0.19)	Low to medium	6.4 (±0.89)	2.13 (±0.295)	High to very high
95th	5.0 (±0)	1.67 (±0)	Medium	6.5 (±0.58)	2.165 (±0.19)	High to very high

as compared to push. As it can also be seen in the dynamic simulation of mopping activities (Fig. 4.4) that push method involves stretching of arms and trunk in forward direction (PP2 to PP6) but, figure-of-eight method leads to stretching and twisting (PF2, PF3, PF5, PF6) simultaneously. For future investigation, a swiveling head with an optimum bend at the top and provision of an offset at the farthest end of the mop handle with proper ergonomic mediations are suggested to achieve improved posture and lessen risk level in mopping task.

4.4 Conclusion

In the present investigation of mopping activities, six-dimensional ETS and optical systems has been utilized to precisely record the three-dimensional positional coordinates of various body parts. The range of angular variations of various hand joints are evaluated to identify the postures adopted in such occupational tasks. The key findings are (a) angular variation of wrist and shoulder joints in push mopping method is lower as compared to figure-of-eight method, (b) elbow joint shows increased range of angular variation in push method than figure-of-eight, (c) for the hand placed at the top of the mop handle, an increased angular range of wrist and shoulder joints are observed with an increase in height of mopping professionals in push mopping technique and (d) highest RULA score achieved in figure-of-eight method is 7 (high level risk) and in push, it is 5(medium level risk).

References

1. Chang, J.H., Wu, J.D., Liu, C.Y., Hsu, D.J.: Prevalence of musculoskeletal disorders and ergonomic assessments of cleaners. Am. J. Ind. Med. **55**(7), 593–604 (2012). https://doi.org/10.1002/ajim.22064
2. Naik, G., Khan, M.R.: Prevalence of MSDs and postural risk assessment in floor mopping activity through subjective and objective measures. Saf. Health Work **11**(1), 80–87 (2020). https://doi.org/10.1016/j.shaw.2019.12.005
3. Khan, M.R., Singh, N.K.: Prevalence of musculoskeletal disorders among Indian railway sahayaks. Int. J. Occup. Environ. Health **24**(1–2), 27–37 (2018). https://doi.org/10.1080/10773525.2018.1507187
4. Kumar, R., Kumar, S.: Musculoskeletal risk factors in cleaning occupation-A literature review. Int. J. Ind. Ergon. **38**(2), 158–170 (2008). https://doi.org/10.1016/j.ergon.2006.04.004
5. Charles, L.E., Loomis, D., Demissie, Z.: Occupational hazards experienced by cleaning workers and janitors: a review of the epidemiologic literature. Work **34**(1), 105–116 (2009). https://doi.org/10.3233/WOR-2009-0907
6. Bisht, D.S., Khan, M.R.: A novel anatomical woodworking chisel handle. Appl. Ergon. **76**, 38–47 (2019). https://doi.org/10.1016/j.apergo.2018.11.010
7. Bisht D.S., Khan M.R.: Anatomically shaped tool handles designed for power grip. In: Chakrabarti, A., Chakrabarti, D. (eds.) Research into Design for Communities, vol. 1. ICoRD 2017. Smart Innovation, Systems and Technologies, vol. 65. pp. 135–148. Springer, Singapore (2017). https://doi.org/10.1007/978-981-10-3518-0_12
8. Dianat, I., Nedaei, M., Nezami, M.A.M.: The effects of tool handle shape on hand performance, usability and discomfort using masons' trowels. Int. J. Ind. Ergon. **45**, 13–20 (2015). https://doi.org/10.1016/j.ergon.2014.10.006
9. Søgaard, K., Fallentin, N., Nielsen, J.: Work load during floor cleaning The effect of cleaning methods and work technique. Eur. J. Appl. Physiol. Occup. Physiol. **73**(1–2), 73–81 (1996). https://doi.org/10.1007/bf00262812
10. Wallius, M.A., Rissanen, S.M., Bragge, T., Vartiainen, P., Karjalainen, P.A., Rasanen, K., Jarvelin-Pasanen, S.: Effects of mop handle height on shoulder muscle activity and perceived exertion during floor mopping using a figure eight method. Ind. Health **54**(1), 58–67 (2016). https://doi.org/10.2486/indhealth.2015-0108
11. Woods, V., Buckle, P.: An investigation into the design and use of workplace cleaning equipment. Int. J. Ind. Ergon. **35**(3), 247–266 (2005). https://doi.org/10.1016/j.ergon.2004.09.004

12. Hagner, I.M., Hagberg, M.: Evaluation of two floor-mopping work methods by measurement of load. Ergonomics **32**(4), 401–408 (1989). https://doi.org/10.1080/00140138908966106
13. Khan M.R., Singh N.K., Shinde D.: An ergonomic study: bicycle repairer in rural India. In: Chakrabarti, A. (ed.) Research into Design for a Connected World. Smart Innovation, Systems and Technologies, vol. 135, pp. 509–517. Springer, Singapore (2019). https://doi.org/10.1007/978-981-13-5977-4_43
14. Khan M.R., Patnaik B., Patel S.: Design and development of a novel Sit-to-Stand and mobility assistive device for ambulation and elderly. In: Chakrabarti, A., Chakrabarti, D. (eds.) Research into Design for Communities, vol. 1. pp. 801–811. Springer, Singapore (2017). https://doi.org/10.1007/978-981-10-3518-0_69
15. Mahadas, S., Mahadas, K., Hung, G.K.: Biomechanics of the golf swing using OpenSim. Comput. Biol. Med. **105**, 39–45 (2019). https://doi.org/10.1016/j.compbiomed.2018.12.002
16. Chakrabarti, D.: Indian anthropometric dimensions for ergonomic design practice. National Institute of Design (1997).
17. Khan M.R., Giri P., Palarapu P.K.: Redesign and ergonomic analysis of scoop stretcher for full body immobilization during casualties. In: Chakrabarti A. (ed.) ICoRD'15—Research into Design Across Boundaries vol. 1. pp. 411–420. Springer, New Delhi (2015). https://doi.org/10.1007/978-81-322-2232-3_36

Chapter 5
Development of Novel Biodynamic Model of the Seated Occupants

Veeresalingam Guruguntla and Mohit Lal

Abstract Since the beginning of the machine age, vibration has become a significant part of the human being at workplace. In general, the operators of the vehicle/machine expose to vibration due to supporting surface, operating tools directly or indirectly in their daily-life activities. These prolonged vibrations affect the human nervous system, digestive system, muscle function, upper and lower limbs, etc. Head is the most crucial part of the human body; hence, the study of vibration transmissibility between the source and head under dynamic conditions is an important consideration in biodynamic modeling. In order to study the biodynamic response of human beings under a wide range of vibration environments, most of the experimental work has been done until now and limited edition of mathematical models is developed. However, very few mathematical models mimic the structure of the real human being. In this article, a novel 10° of freedom (*dofs*) human model which mimics the real human structure is developed. The stiffness and damping of a human body (bone and muscle) are represented by mass, spring, and damping constants and are optimized through Firefly Algorithm (FA). The conclusion is drawn based on the goodness of fit of experimental data presented in the literature with the tuned parameters of the human body obtained from FA. The accuracy of the developed human model is superior to those reported in the literature. The developed human model helps to monitor the whole body vibration in seated occupants.

5.1 Introduction

People for conveyance prefer vehicles like on/off-road vehicles, trains, ships, etc. that transmit vibration to the human body through its surface. The low-frequency vibration (4–6 Hz) generated in these vehicles makes a human being felt uncomfortable

V. Guruguntla · M. Lal (✉)
Department of Industrial Design, National Institute of Technology Rourkela, Rourkela, Odisha 769008, India
e-mail: lalm@nitrkl.ac.in

V. Guruguntla
e-mail: 518id1001@nitrkl.ac.in

A. Chakrabarti et al. (eds.), *Design for Tomorrow—Volume 1*, Smart Innovation, Systems and Technologies 221, https://doi.org/10.1007/978-981-16-0041-8_5

[1] and leads to tissue damage and lumbar spine problems [2]. Hence, to estimate and improve the ride comfort and safety, vision biodynamic models have been developed [3–6]. Effective suspension systems can reduce these levels of vibration [7, 8].

To analyze the biodynamic responses due the ship shock motion, a 2-*dofs* biodynamic model is proposed [9]. In that whole human body was divided into two segments, namely the head and main body; both the segments were connected with the linear spring–damper system. The same 2-*dofs* model was used by [10] to predict the vehicle ride quality. Liang [1] reviewed the lumped parameter models developed from 1-*dof* to 11-*dofs* and suggested Wan's 4-*dofs* human model [11] as the best model to estimate the biodynamic responses. A full vehicle car model attached with the Wan's human model was analyzed to estimate the vehicle ride quality [8]. The study [12] modified the Wan's model with 23 different configurations and suggested one best model out of the 23 models.

From above review, it may be noticed that the models developed in the past and obtained biodynamic responses do not match with the biodynamic responses of a real human. Hence, a 10-*dofs* biodynamic model is developed in this article to best predict the biodynamic responses of a real human being. The critical parameter to examine the effect of whole body vibration on the human being is biodynamic response that has been estimated and optimized in the present article. In the past, to obtain these biodynamic response, experiments of real human beings have been performed that lead to a human in hazardous and unfavorable situations. To overcome this problem, a numerical investigation has been performed to obtain biodynamic responses. This study will help design engineers to make effective human dummies, automobile seats, suspension systems, etc.

5.2 Human Modeling

A 10-*dofs* human model for the seated occupant is developed as shown in Fig. 5.1. It consists of 10-masses connected with 10-springs and 10-damper. The main body segments are head, thorax, abdomen, and pelvis represented with masses m_1, m_2, m_3, and m_4.

The right upper arm, right forearm, and hand are represented as m_5, m_6, and m_7. Similarly, left-hand parts are indicated as m_8, m_9, and m_{10} as shown in Fig. 5.1. The stiffness and damping between adjacent segments are represented by k_{ij} and c_{ij}, respectively. The suffix '*i*', '*j*' indicates the number of the respective human segments.

5.2.1 Evaluation of Model

Upon application of Newton's second law of motion in each segment in Fig. 5.1, the equation of motion for the whole body can be written as,

Fig. 5.1 A 10-*dofs* biodynamic model with 30 parameters

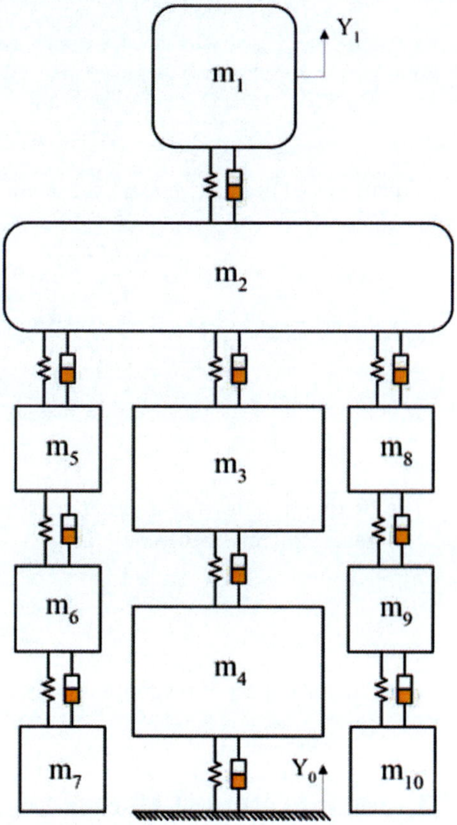

$$[M]\{\ddot{y}\} + [C]\{\dot{y}\} + [K]\{y\} = \{f_y\} \tag{5.1}$$

where $[M]$, $[C]$, and $[K]$ are mass, stiffness, and damping matrices, respectively. The mass matrix is diagonal matrix, whereas stiffness and damping are symmetric matrices. The excitation force is represented by $\{f_y\}$.

$$[M]_{10\times 10} = diag\left[m_1\ m_2\ m_3\ m_4\ m_5\ m_6\ m_7\ m_8\ m_9\ m_{10}\right];$$

$$\{y\}_{10\times 1} = \left[y_1\ y_2\ y_3\ y_4\ y_5\ y_6\ y_7\ y_8\ y_9\ y_{10}\right]^T;$$

The elements of $[K]$, $[C]$, and $\{f_y\}$ may be represented as:

$K_{11} = -K_{12} = k_{12}$, $K_{22} = (k_{12} + k_{25} + k_{23} + k_{28})$, $K_{23} = -k_{23}$, $K_{25} = -k_{25}$, $K_{28} = -k_{28}$, $K_{33} = (k_{23} + k_{34})$, $K_{34} = -k_{34}$, $K_{44} = (k_{34} + k_{40})$, $K_{55} = (k_{25} + k_{56})$, $K_{56} = -k_{56}$, $K_{66} = (k_{56} + k_{67})$, $K_{77} = -K_{67} = k_{67}$, $K_{88} = (k_{28} + k_{89})$, $K_{89} = -k_{89}$, $K_{99} = (k_{89} + k_{910})$, $K_{1010} = -K_{910} = k_{910}$

$C_{11} = -C_{12} = c_{12}, C_{22} = (c_{12} + c_{25} + c_{23} + c_{28}), C_{23} = -c_{23}, C_{25} = -c_{25}, C_{28} = -c_{28}, C_{33} = (c_{23} + c_{34}),$
$C_{34} = -c_{34}, C_{44} = (c_{34} + c_{40}), C_{55} = (c_{25} + c_{56}), C_{56} = -c_{56}, C_{66} = (c_{56} + c_{67}), C_{77} = -C_{67} = c_{67},$
$C_{88} = (c_{28} + c_{89}), C_{89} = -c_{89}, C_{99} = (c_{89} + c_{910}), C_{1010} = -C_{910} = c_{910}$

$$f_{y41} = k_{40}y + c_{40}\dot{y}$$

Equation (5.1) in the frequency domain may be represented as:

$$(-\omega^2 M + j\omega C + K)Y = \{F_Y\} \tag{5.2}$$

where $\{Y\} = [Y_1, Y_2, Y_3, Y_4, Y_5, Y_6, Y_7, Y_8, Y_9, Y_{10}]$.
From Eq. (5.2), the human response at each segment may be evaluated as:

$$Y = A^{-1}F_Y \tag{5.3}$$

where $A = (-\omega^2 M + j\omega C + K)$.
The biodynamic response under seated conditions can be predicted through the seat-to-head transmissibility ratio as:

$$\text{STOH} = \frac{Y_1}{Y_0} \tag{5.4}$$

where Y_0 is the input response at seat–buttock interface.

5.3 Identification of Human Segmental Properties

The present study utilized the objective function that minimizes the error between experimental and analytical value for a seat-to-head transmissibility ratio and represented as,

$$f_{\min} = \sum_{i=1}^{N} (\text{STOH}_e(f_i) - \text{STOH}_p(f_i))^2 \tag{5.5}$$

The experimental reading for a seat-to-head transmissibility is taken from [13], whereas the corresponding analytical readings are obtained from the proposed model. The goodness of fit (ε) is calculated to measure the accuracy of the proposed model as:

$$\varepsilon = 1 - \frac{\sqrt{\Sigma(\beta_e - \beta_p)^2/(N-2)}}{\Sigma\beta_e/N} \tag{5.6}$$

where 'βe' and 'βp' are data points corresponding to the experimental and proposed model, respectively. 'N' is the number of data points in the selected frequency range; for the present analysis, it is (0–20 Hz).

To optimize the mechanical parameters of the human body, Firefly Algorithm is used with appropriate boundary conditions mentioned in [14]. The constraints for the (mass, stiffness, and damping) are chosen from [12] and [15]. From Fig. 5.2, the optimum value of goodness of fit for the proposed model may be seen as 0.952. The tuned mechanical parameters with the help of FA for the best iteration are presented in Table 5.1.

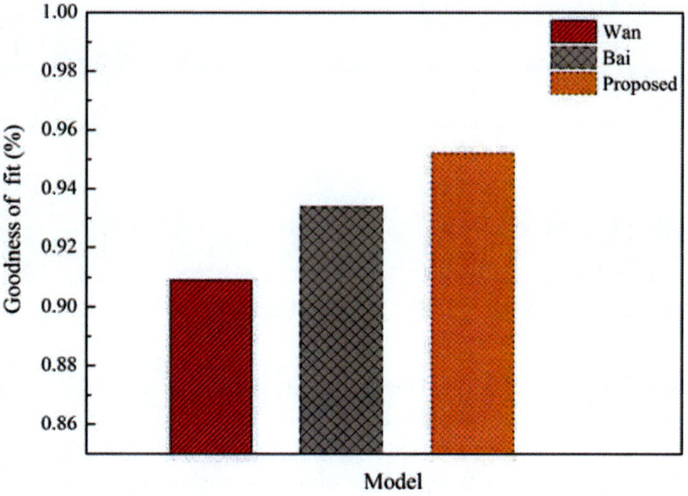

Fig. 5.2 Performance of goodness of fit

Table 5.1 Optimized human segmental properties with FA

Mass (kg)									
m_1	m_2	m_3	m_4	m_5	m_6	m_7	m_8	m_9	m_{10}
7.5	16.7	12.2	10.9	2.2	3.0	1.6	1.1	0.7	0.7
Stiffness parameters (Nm^{-1}, $\times 10^3$)									
k_{12}	k_{23}	k_{34}	k_{40}	k_{25}	k_{56}	k_{67}	k_{28}	k_{89}	k_{910}
201.7	129.5	208.3	77.1	3.0	159.7	83.8	283.8	271.9	117.8
Damping parameters ($N\text{-}sm^{-1}$, $\times 10^2$)									
c_{12}	c_{23}	c_{34}	c_{40}	c_{25}	c_{56}	c_{67}	c_{28}	c_{89}	c_{910}
5.8	28.5	34.3	39.0	6.9	20.7	25.3	29.0	30.1	27.7

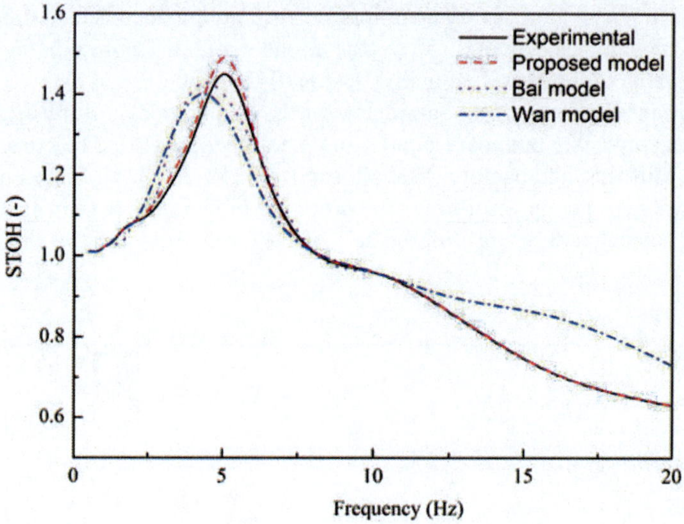

Fig. 5.3 Performance of STOH

5.4 Comparison with the Existing Model

In this section, the seat-to-head transmissibility ratio obtained for the proposed model is compared with the two other models that are widely accepted in the literature, namely Bai's model and Wan's model along with the experimental response. From Fig. 5.3, it can be observed that the response obtained from the proposed model represents the best match with experimental results. The Wan model is particularly matching in the range of 6–12 Hz, whereas Bai's model failed to capture the peak resonance frequency. The goodness of fit for Wan's, Bai's, and proposed model are 0.909, 0.934, and 0.952, respectively. From Fig. 5.2, it can also be proved that the FA can effectively optimize the human segmental properties in comparison with other optimization techniques.

5.5 Conclusion

In this article, a 10-*dofs* biodynamic model is developed with 30 mechanical parameters (mass, stiffness, and damping). The mechanical parameters are optimized by minimizing the error between experimental and analytical values through FA. After obtaining the optimized values, the biodynamic response (STH) is calculated. The biodynamic response obtained from the proposed model shows 95.2% match with the experimental response on the real human being.

References

1. Liang, C.C., Chiang, C.F.: A study on biodynamic models of seated human subjects exposed to vertical vibration. Int. J. Ind. Ergon. **36**(10), 869–890 (2006)
2. Ruoxun, F., Jie, L., Jun, L., Weijun, W.: Presentation of an approach on determination of the natural frequency of human lumbar spine using dynamic finite element analysis. Appl. Bionics. Biomech. **2019**(2), 1–8 (2019)
3. Bhiwapurkar, M.K., Saran, V.H., Harsha, S.P., Goel, V.K., Berg, M.: Influence of mono-axis random vibration on reading activity. Ind. Health **48**(5), 675–681 (2010)
4. Corbridge, C., Griffin, M.J.: Effects of vertical vibration on passenger activities: writing and drinking. Ergonomics **34**(10), 1313–1332 (1991)
5. Khan, M.S., Sundstrom, J.: Effects of vibration on sedentary activities in passenger trains. J. Low Freq. Noise Vib. Active Control **26**(1), 43–55 (2007)
6. Bhiwapurkar, M.K., Saran, V.H., Harsha, S.P.: Quantitative evaluation of distortion in sketching under mono and dual axes whole body vibration. Ind. Health **49**(4), 410–420 (2011)
7. Sekulic, D., Ivkovic, I., Mladenovic, D.: Effects of the seat cushion oscillatory parameters on vibration exposure and dynamic seat comfort in the bus. J. Appl. Sci. Eng. **15**(4), 433–441 (2017)
8. Wagner, J., Liu, X.: An active vibration isolation system for vehicle seats. SAE Trans. **109**, 1076–1087 (2000)
9. Allen, G.: A critical look at biomechanical modeling in relation to specifications for human tolerance of vibration and shock. Agard Conf. Proc. **253**, 6–10 (1978)
10. Cho, Y., Yoon, Y.S.: Biomechanical model of human on seat with backrest for evaluating ride quality. Int. J. Ind. Ergon. **27**(5), 331–345 (2001)
11. Wan, Y., Schimmels, J.M.: A simple model that captures the essential dynamics of a seated human exposed to whole body vibration. Adv. Bioeng. **31**, 333–334 (1995)
12. Bai, X.X., Xu, S.X., Cheng, W., Qian, L.J.: On 4-degree-of-freedom biodynamic models of seated occupants: Lumped-parameter modelling. J. Sound Vib. **402**, 122–141 (2017)
13. Boileau, P.E., Wu, X., Rakheja, S.: Definition of a range of idealized values to characterize seated body biodynamic response under vertical vibration. J. Sound Vib. **215**(4), 841–862 (1998)
14. Yang, X.S.: Firefly algorithms for multimodal optimization. Int. Symp. Stochast. Algorithms **5792**, 169–178 (2009)
15. Plagenhoef, S., Evans, F.G., Abdelnour, T.: Anatomical data for analyzing human motion. Res. Q. Exerc. Sport. **54**(2), 169–178 (1983)

Chapter 6
Urge for Human-Centered Design Intervention for Harvesting Aquatic Food Crops

Jitesh Singh Chauhan and Sougata Karmakar

Abstract Aquatic food crops which have significant economic importance due to their nutritional value are presently being cultivated globally. Farmers in developing countries still follow the traditional cultivation practices for these crops. Aquatic cultivation is entirely different from terrestrial cultivation in terms of the physical environment, field condition, and cultivation technique. Among different stages of agricultural operations for a particular crop, harvesting is considered to be the most time consuming and requires utmost care to collect consumable part of the crop. The current study aims to understand future research avenues in aquatic food crop harvesting. The paper highlights medicinal/nutritional value, challenges in harvesting, and existing scenarios of available tools and techniques used in the harvesting of aquatic as well as terrestrial crops. It will help to identify the future scope of research toward design intervention to reduce the drudgery of aquatic farmers and improvement of productivity of aquatic food crops. For this purpose, relevant literature was searched from the major three electronic databases, namely Web of Science, Scopus, and Google scholar. A prominent research gap was found regarding the design and development of aquatic food crop harvester considering the key constraints of the aquatic environment. Hence, the attention of agricultural engineers/designers, scientists, and researchers is of utmost necessity for designing and manufacturing innovative tools and techniques of aquatic crop growers. The need of the hour is to strengthen the research and devolvement efforts in this overlooked domain for minimizing the sufferings of the underprivileged aquatic agricultural workers.

J. S. Chauhan · S. Karmakar (✉)
Department of Design, IIT Guwahati, Guwahati, Assam 781039, India
e-mail: karmakar.sougata@gmail.com

A. Chakrabarti et al. (eds.), *Design for Tomorrow—Volume 1*, Smart Innovation, Systems and Technologies 221, https://doi.org/10.1007/978-981-16-0041-8_6

6.1 Introduction

The world population is increasing at an alarming rate. The current world population is 7.8 billion, with an average increase of 81 million per year [1]. An increase in population certainly increases the demand for nutritional food and livestock. Agriculture production should be enhanced to fulfill future requirements.

Agriculture is an important, vast, and evergreen sector. It contributes to the country's economy and provides the raw material for the non-agricultural industries like textile and food processing industries, and generates employment among rural people as they mostly depend upon agriculture. This sector does not stop even the whole world struggling with the recession; as we have seen that during the COVID-19 lockdown, the work related to agriculture activities was not stopped during that tough period. Aquaculture is a branch of agriculture commonly known for aquatic plants, fish, algae, and other organisms.

Aquaculture is playing an essential role in food security and economic stability worldwide [2]. It involves cultivating freshwater and saltwater populations under controlled conditions. Aquaculture is considered the most diverse field as it constitutes various species, different growing environments, and farming methods [3]. Aquatic crops are cultivated all over the world, considering their nutritional and economic importance [4]. Some examples of the aquatic crop are lotus (*Nelumbonuciferа Gaertn*), water bamboo (*Zizaniacaduciflora Hand. Mazz.*), taro (*Colocasiaesculenta Schott*), water dropwort (*Oenanthestolonifera DC. C.*), water spinach (*Ipomoea aquatic Forsk.*), water chestnut (*EleocharistuberosaRoem. et Schult*), water caltrop (*Trapa spp.*), Gordon euryale (*Euryale ferox Salisb.*), watercress (*Nasturtium officinale R.Br*), common cattail (*Typhacaduciflora L.*), and water shield (*Braseniaschreberi Gmel*) [4, 5].

Aquatic cultivation involves many challenges comparing with terrestrial crops. Wheat, sugarcane, and potato are among the few examples of the terrestrial crop. These crops/plants are grown in the land and have roots inside the soil while stems in a free atmosphere. The farmer can easily see the growth of the plant. On the contrary, the plant either floats on the water surface or remains immersed inside the water body or grows inside the marshy land in the case of aquatic crops. These conditions make the cultivation more difficult for farmers to perform various practices for these wetland crops.

Agricultural practices are time bond operation, and mechanization plays an essential role in increasing productivity and efficiency. Tools, implements, and machinery used in agricultural land help farmers to accomplish tasks that are difficult to perform. These are considered as part of mechanization [6], and it also reduces the drudgery in farm activities with appropriate design interventions. Among various agricultural operations, gathering of plant's useful parts called harvesting is labor-intensive, costly and requires care for handling [7, 8]. Harvesting and threshing operations are known as the critical processes on quality and production costs [9].

The current study aims to understand future research avenues in aquatic food crop harvesting. The paper highlights medicinal/nutritional value, challenges in

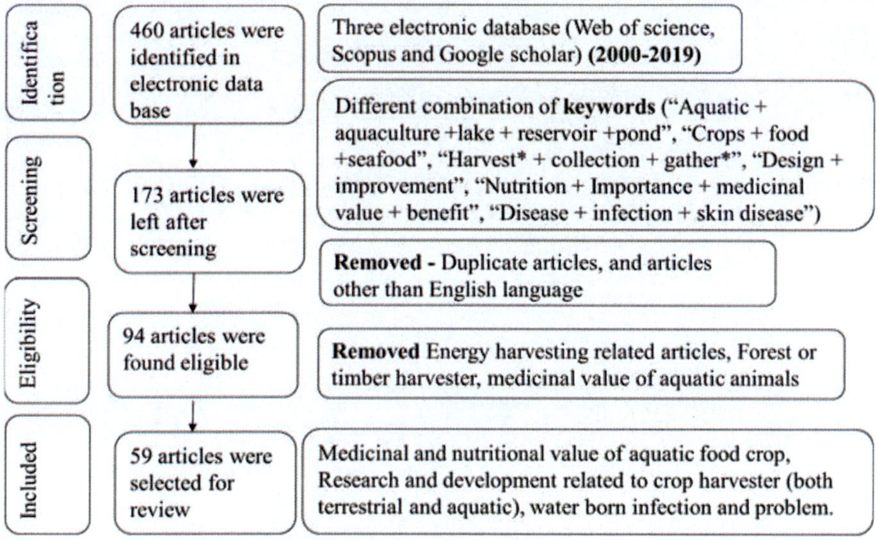

Fig. 6.1 Systematic diagram of the review process

harvesting, and existing scenarios of available tools and techniques used in aquatic as well as terrestrial crop harvesting to identify the future scope of research toward design intervention to reduce the drudgery of aquatic farmers and improvement of productivity of aquatic food crops.

6.2 Review Process

The PRISMA guidelines [10, 11] were followed for the selection of articles as presented in Fig. 6.1. Articles from the year 2000 to 2019 were selected for the review.

6.3 Finding from the Literature

6.3.1 Medicinal and Nutritional Value of the Aquatic Food Crop

Plant of wetland ecosystem not only fulfills food demand but also full of nutrients. Much research has been done on the importance of aquatic plants. Some of the crop nutrients are discussed in Table 6.1.

Table 6.1 Medicinal value of the aquatic food crop

Crop	Nutrients	Benefits	Reference
Water chestnut	Protein (up to 20%), starch (52%), tannins (9.4%), fat (up to 1%), and sugar (3%),	commonly known as cooling food	[12] [13]
Foxnut	9.7% protein, 0.1% fat, 76.9% carbohydrates, 0.5% mineral matters, and 1.45% iron	kidney problems, chronic diarrhea, and excessive leucorrhea	[14] [15] [16]
Lotus stem	100 g fresh of underground stem containing 1 g protein, 19.8 g carbohydrate, 19 mg Ca, 51 mg P, 0.5 mg Fe, and 25 mg Vitamin C	beneficial effects on cardiovascular diseases	[5] [17]
Kelp laminaria japonica	mannitol, alginate, and kelp anti-hypertensive peptide	Help in lower blood pressure, blood sugar, and blood lipids	[18] [19]

Aquatic crops are used for their medicinal value since the ancient time [4]. They have many health benefits; frequent consumption of these crops will be good for a healthy life.

6.3.2 Challenges of Aquatic Food Crop

Long stay inside the water can cause many skins and health problems. As aquatic crops grow in the wetland area, farmers have to remain in contact with water for longer duration without any protective measure. Some of the common diseases and problem are listed below:

- Swimmer's itch is an allergic condition that occurs when the motile and infectious stage of avian schistosomes penetrate the skin of an individual. Flatworm parasites that cause swimmer's itch belong to the family schistosomatidae commonly found in water bodies [20].
- Cercarial dermatitis is a waterborne non-communicable skin condition caused by schistosome cercariae released by aquatic snails. Cercarial dermatitis appears worldwide, but may be caused by different trematode species [21, 22].
- Cyanobacteria, also known as blue-green algae, are commonly found in fresh water, and reservoirs throughout the world. These bacteria can affect the skin [23].
- Water bodies can have harmful insects and reptiles like snake and leech (Jonk) inside the muddy soils [24].
- Hot tub rash (pseudomonas dermatitis) caused in humans due to contact of contaminated water for a long period of time [25].

- The pond environment is unhygienic due to mud, thorns/prickles [26].
- Manual harvesting of aquatic plants using a boat leads to harmful posture [27].

Workers are bound to stay immersed (up to waist or even neck) for prolonged duration in stagnant unhygienic water, which causes swelling of limbs, skin infections, itching in body parts, and injury to body parts. Such valuable crops are slowly on the way of extinction because of the problematic cultivation process.

6.3.3 Existing Scenarios of Available Tools and Techniques for Harvesting

It was found that 44 articles were dealing with the design and development of the agriculture food crop harvester. These articles explained the different designs and working principles of harvesters. We have considered each article as different harvester, as no article discussed the same design; every article showed variation in design. Hence, the categorization of articles was done based on the crop classification [28], which is shown in Table 6.2.

Various crops have different physical properties and methods of harvesting. It requires a different design of harvester to collect the eatable/consumable part successfully. Maze, sorghum, barley, wheat, etc. are examples of cereals as these plants have different geometry, which affects the design of harvester. In the case of root crops,

Table 6.2 Number of harvesters developed based on crop classification

Crop type	Crop classification	No. of harvester designed
Terrestrial crops	Cereals	7
	Vegetables and melons	8
	Fruit and nuts	10
	Oilseed crops	6
	Root/tuber crops	5
	Beverage and spice crops	2
	Leguminous crops	1
	Sugar crops	2
	Total	**41**
Aquatic crops	Root crop (lotus)	1
	Weed (kelp)	1
	Fruit and nuts (makhana)	1
	Total	**3**

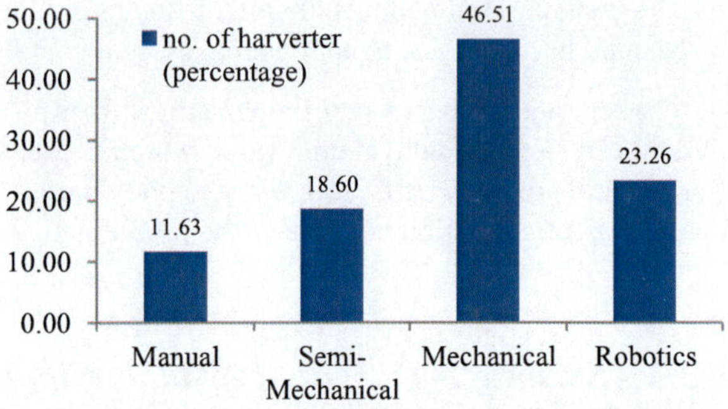

Fig. 6.2 Different harvester found in respective categories are represented in the bar diagram

the eatable part grows inside the soil. Harvester designed for root crops cannot be used for the cereal crop that has an eatable part above the ground. Another factor responsible for the need of different harvesters is the landholding of farmers. If a farmer has vast land to harvest, he will not go with the small harvester.

It was found that only three articles were for the aquatic crop and 41 articles for the terrestrial crops, as shown in Table 6.2. These harvesters were designed to solve majorly three problems. Firstly, it helps increase the harvesting rate because manual harvesting by hand takes high time, reducing the total yield. Harvesting at an appropriate time avoids loss in total production and enhances total productivity [29]. Secondly, it helped in the problem of labor shortage. During the peak time of harvesting crops, demand for labor increases and creates a shortage of labor. This results in a delay in harvesting. Lastly, agriculture activities are at high risk of musculoskeletal disorders [30]. To reduce the drudgery in the harvesting of a crop, harvester plays an important role. Ergonomic intervention in harvesters for terrestrial crops reduced fatigue and enhances performance.

6.3.3.1 Categorization of Harvester Design Based on the Level of Mechanization

The design of harvester depends upon its type or level of mechanization. The harvesters can be categorized into four types: manual, semi-mechanical, mechanical, and robotic. Manual harvesters are simple in design with humans as a source of power. These are generally used when the farmer has small landholding and cannot bear the cost of a highly mechanized machine [31]. Semi-mechanical harvesters have engines less than 10 horsepower. While mechanical harvesters have high power engines up to 135 horsepower, these harvesters can be tractor operated or self-propelled. For saving time and reducing farmers' efforts during harvesting, mechanical harvesters were developed. These harvesters can perform multitask like gathering, cleaning, and transport with the less human intervention [32]. Robotic type harvesters are also gaining popularity considering the shortage of labor in future [33].

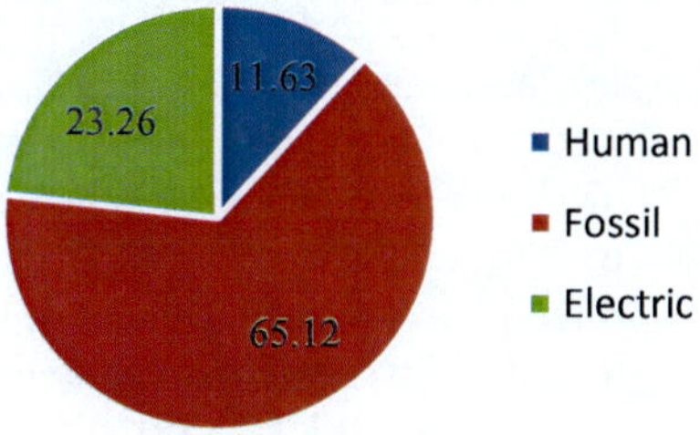

Fig. 6.3 Source of energy used for different harvesters in percentage

The maximum number of harvester developed is for mechanical harvester that is nearly 46%, as shown in Fig. 6.2. Because these harvesters can perform multitask at one time and save much time in a large field.

6.3.3.2 Source of Energy

Energy or power source is one of the crucial factors in the design of harvester. It is found that three types of energy sources are used in the design of the harvester. These are human, fossil (diesel, petrol), and electricity. Fossil fuels are majorly used as the primary energy source for mechanical and semi-mechanical harvesters. It has high calorific value and easily transported to the remote areas. Human energy can be used, where less force is required. Electricity is mostly used in the case of robotic harvesters as it mostly has electric components like motor, camera, and manipulator in design. It was noticed that around 65% of harvesters designed used fossil fuel as an energy source, as shown in Fig. 6.3.

6.3.3.3 Status of Aquatic Food Crop Harvesting

Discussion on the traditional methods and tools used for harvesting of the aquatic crop has been done in this section. Interventions in harvesters designed to facilitate harvesting of the aquatic crop are highlighted in brief. It is observed that the majority of the tool and techniques adopted in aquatic agriculture are very primitive in nature and locally fabricated by the farmers. For the cultivation of Makhna, they use to fix a bamboo pole known as "Kaara." Farmers use this pole as a reference point and go inside the water to pull the Makhna seeds with the help of hands near the bamboo. They clean the collected seed in a locally made container known as "Auka" [14]. The following are some interventions found for aquatic food crops.

- Intervention in the harvesting of Makhna—The improved system consists of a floating platform with 10 L of a compressed air cylinder. The workers use a mini diving kit and mask. In this system, the workers could breathe inside the water, as air is supplied with a 10 m long hose [26]. The system reduced the drudgery of aquatic farmers in comparison to the traditional method.

- Intervention in kelp—Manual harvesting of kelp was labor-intensive and had low production efficiency [34]. Developed kelp harvesting machine reduced the workload and doubled the harvesting efficiency. In the harvester design, they have used two locomotion modules, a kelp collection/transmission module, a power module, and other supporting devices [34].
- Intervention in lotus—A robot using the k-means cluster algorithm was developed to harvest lotus. It can accurately obtain a lotus clustering center for fast recognitions of lotus [35].

6.4 Research Gap and Future Scope

It is found that an enormous amount of research has been carried out by earlier researchers in design and development toward terrestrial crop harvesters. In contrast, there is lack of reported research in the domain of harvesters to be used for the aquatic food crop (Fig. 6.4). A prominent research gap was found regarding the design and development of aquatic food crop harvester considering the key constraints of the aquatic environment. Various research questions, which could be conceived to address the research gap, have been listed below to elaborate the current lacunae in this domain.

- Which type (manual, mechanical, robotic) of harvester would be suitable for aquatic food crops?

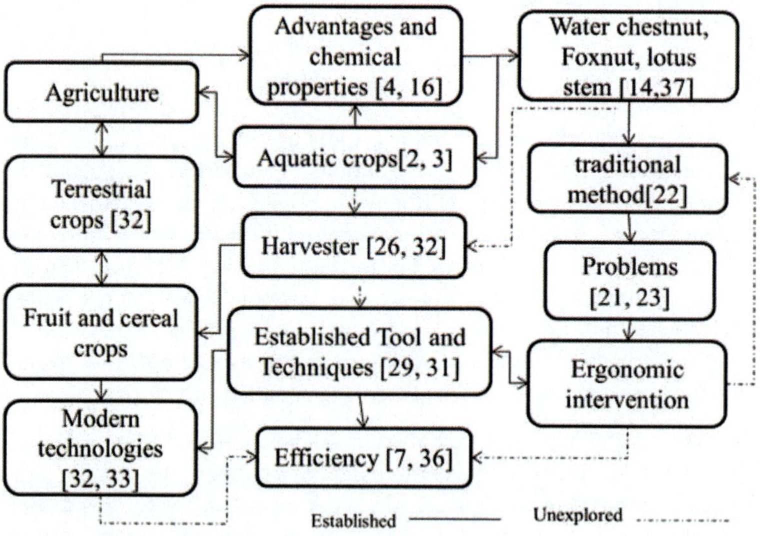

Fig. 6.4 Representation of the research gap through the block diagram

- How to implement effective ergonomic design intervention in line with the traditional method of cultivation of aquatic food crops?
- How is the feasibility for adoption of modern technology (artificial intelligence, automation, etc.) in design and development of aquatic crop harvester?
- What are the various strategies to be adopted by the aquatic farmers for the safe and effective cultivation/harvesting to ensure higher productivity and lesser drudgery?

Farmers of aquatic food crops face numerous challenges in the whole cultivation process. Hence, the attention of agricultural engineers/designers, scientists, and researchers is of utmost necessity for designing and manufacturing innovative tools and techniques of aquatic crop growers. There are ample scopes for carrying out research activities to ensure sustainable livelihood of the aquatic farmers along with ensuring their quality of life. Future scope of research pertaining to aquatic agriculture and more specifically for design and development of suitable tools/techniques includes:

- Study on socio-economic/demographic characteristics of aquatic food crop growers.
- Finding out the various challenges faced by the aquatic farmers in the cultivation process.
- Study on traditional practices and indigenous tools/techniques used by the aquatic farmers.
- Development of ergonomics databases (anthropometric and biomechanical data) of the aquatic farmers for various types of design interventions.
- Conceptualization and prototype development of innovative tools/techniques to facilitate the cultivation process.
- Manufacturing of the innovative tools and techniques of aquatic crop growers and providing training to use these redesigned tools/equipment.
- Design and development of suitable Personal Protective Equipment (PPE) that can be used by aquatic farmers in waterbodies.

6.5 Conclusion

A cost-effective human-centered or ergonomic design intervention can play an essential role for the farmers in different agricultural operations/activities. Aquatic crops have the potential to generate good revenue as they have high nutritional value. This could be a good source of income for the farmers if simpler methods for harvesting techniques/tools are developed for them. Research related to aquatic crops will help farmers as well as motivate other farmers to start the cultivation of these crops.

In the field of aquatic crop cultivation, the attention of designers/engineers, agricultural scientists/researchers, is utmost required to fulfill the need for "design for tomorrow."

References

1. World Population Clock. https://www.worldometers.info/world-population/#ref-1. Last accessed 06 Sept 2020
2. Anderson, J.L., Asche, F., Garlock, T., Chu, J.: Aquaculture: Its role in the future of food. Front. Econ. Glob. **17**, 159–173 (2017). https://doi.org/10.1108/S1574-871520170000017011
3. Metian, M., Troell, M., Christensen, V., Steenbeek, J., Pouil, S.: Mapping diversity of species in global aquaculture. Rev. Aquac. **12**, 1090–1100 (2020). https://doi.org/10.1111/raq.12374
4. Swapna, M.M., Prakashkumar, R., Anoop, K.P., Manju, C.N., Rajith, N.P.: A review on the medicinal and edible aspects of aquatic and wetland plants of India. J. Med. Plant Res. **5**, 7163–7176 (2011). https://doi.org/10.5897/JMPRx11.005
5. Li, L., Teixeira, J.A., Cao, B.: Aquatic vegetable production and research in China. Asian Australas. J. Plant Sci. Biotechnol. (2007)
6. Sims, B., Kenzle, J.: Farm power and mechnization for small farms in sub-Saharan Africa, pp. 1–67. (2006)
7. Joshi, P., Jethi, R., Chandra, N., Roy, M.L., Kharbikar, H.L., Atheequlla, G.A.: Ergonomics assessment of post harvest finger millet threshing for reducing women drudgery. Indian Res. J. Ext. Edu. **15**, 25–30 (2015)
8. Sola-Guirado, R.R., Blanco-Roldan, G.L., Castro-Garcia, S., Castillo-Ruiz, F.J., Gil-Ribes, J.A.: Innovative circular path harvester for mechanical harvesting of irregular and large-canopy olive trees. Int. J. Agric. Biol. Eng. **11**, 86–93 (2018). https://doi.org/10.25165/j.ijabe.20181103.3265
9. Alizadeh, M.R., Allameh, A.: Evaluating rice losses in various harvesting practices. Int. Res. J. Appl. Basic Sci. **4**, 894–901 (2013)
10. Moher, D., Liberati, A., Tetzlaff, J., Altman, D.G., Group P: Preferred reporting items for systematic reviews and meta-analyses: the PRISMA statement. PLoS Med. **6**, e1000097 (2009)
11. Fargnoli, M., Lombardi, M.: Safety vision of agricultural tractors: an engineering perspective based on recent studies (2009–2019). Safety **6**, 1 (2020)
12. Jana, H.: Water caltrop: a potential crop of water bodies. Rashtriya Krishi. **11**, 7–11 (2016)
13. Kundu, M., Joshi, R.: Production technology of water chestnut (Trapa natans L.). https://www.krishisewa.com/articles/production-technology/142-chestnut.html. Last accessed 06 July 2020
14. Mandal, R.N., Saha, G.S., Sarangi, N.: Harvest and processing of Makhana (Euryale ferox Salisb.)—An unique assemblage of traditional knowledge. Indian J. Tradit. Knowl. **9**, 684–688 (2010)
15. Das, S., Der, P., Raychaudhuri, U., Maulik, N., Das, D.K.: The effect of Euryale ferox (makhana), an herb of aquatic origin, on myocardial ischemic reperfusion injury. Mol. Cell. Biochem. **289**, 55–63 (2006). https://doi.org/10.1007/s11010-006-9147-1
16. He, S., Wang, D., Zhang, Y., Yang, S., Li, X., Wei, D., Zhang, M., Qin, J.: Chemical components and biological activities of the essential oil from traditional medicinal food, Euryale ferox Salisb., Seeds. J. Essent. Oil-Bearing Plants **22**, 73–81 (2019). https://doi.org/10.1080/0972060X.2019.1595165
17. Jiang, J.Z., Cao, B.S.: Varieties of aquatic vegetables and their utilization in China. In: Asian Plants with Unique Hort. Potential, pp. 77–82. (2008)
18. Ge, H., Zang, Y., Cao, Z., Ye, X., Chen, J.: Rheological properties, textural and compound preservative of kelp recombination noodles. LWT **118**, 108729 (2020). https://doi.org/10.1016/j.lwt.2019.108729
19. Zhao, X., Guo, F., Hu, J., Zhang, L., Xue, C., Zhang, Z., Li, B.: Antithrombotic activity of oral administered low molecular weight fucoidan from Laminaria Japonica. Thromb. Res. **144**, 46–52 (2016)
20. Rudko, S.P., Turnbull, A., Reimink, R.L., Froelich, K., Hanington, P.C.: Species-specific qPCR assays allow for high-resolution population assessment of four species avian schistosome that cause swimmer's itch in recreational lakes. Int. J. Parasitol. Parasites Wildl. **9**, 122–129 (2019). https://doi.org/10.1016/j.ijppaw.2019.04.006

21. Tracz, E.S., Al-Jubury, A., Buchmann, K., Bygum, A.: Outbreak of swimmer's itch in Denmark. Acta Derm. Venereol. **99**, 1116–1120 (2019). https://doi.org/10.2340/00015555-3309
22. Soldánová, M., Selbach, C., Kalbe, M., Kostadinova, A., Sures, B.: Swimmer's itch: Etiology, impact, and risk factors in Europe. Trends Parasitol. **29**, 65–74 (2013). https://doi.org/10.1016/j.pt.2012.12.002
23. Stewart, I., Robertson, I.M., Webb, P.M., Schluter, P.J., Shaw, G.R.: Cutaneous hypersensitivity reactions to freshwater cyanobacteria—human volunteer studies. 9, 1–9 (2006). https://doi.org/https://doi.org/10.1186/1471-5945-6-6
24. Parmanad Naik, R.K., Sahu, R.: Studies on traditional techniques of harvesting of Lotus studies on traditional techniques of harvesting of Lotus rhizomes in Dhamtari district Of Chhattisgarh, India. Int. J. Agric. Sci. **11**, 8297–8299 (2019)
25. Hot Tub Rash (Pseudomonas Dermatitis/Folliculitis), https://www.health.state.mn.us/diseases/pseudomonis/index.html#:~:text=HotTubRash(PseudomonasDermatitis%2FFolliculitis),withthegermPseudomonasaeruginosa. Last accessed06/08/2020.
26. Khadatkar, A., Gite, L.P., Gupta, V.K.: Interventions to reduce drudgery of workers in the traditional method of harvesting Makhana (Euryale ferox salisb.) seeds from ponds. Curr. Sci. 109, 1332–1337 (2015). https://doi.org/10.18520/v109/i7/1326
27. Novaes, A.L.T., de Andrade, G.J.P.O., dos Santos Alonço, A., Magalhães, A.R.M.: Ergonomics applied to aquaculture: a case study of postural risk analysis in the manual harvesting of cultivated mussels. Aquac. Eng. 77, 112–124 (2017). https://doi.org/10.1016/j.aquaeng.2017.03.005
28. FAO: Classification of crops. https://www.fao.org/fileadmin/templates/ess/documents/world_census_of_agriculture/appendix3_r7.pdf. Last accessed 01 Sept 2020
29. Baneh, N.M., Navid, H., Alizadeh, M.R., Ghasem Zadeh, H.R.: Design and development of a cutting head for portable reaper used in rice harvesting operations. J. Appl. Biol. Sci. **6**, 69–75 (2012)
30. Osborne, A., Blake, C., Fullen, B.M., Meredith, D., Phelan, J., McNamara, J., Cunningham, C.: Prevalence of musculoskeletal disorders among farmers: a systematic review. Am. J. Ind. Med. **55**, 143–158 (2012). https://doi.org/10.1002/ajim.21033
31. Nourani, A., Kaci, F., Garbati Pegna, F., Kadri, A.: Design of a portable dates cluster harvesting machine. Agric. Mech. Asia Africa Lat. Am. **48**, 18–21 (2017)
32. Hachiya, M., Amano, T., Yamagata, M., Kojima, M.: Development and utilization of a new mechanized cabbage harvesting system for large fields. Jpn. Agric. Res. Q. **38**, 97–103 (2004). https://doi.org/10.6090/jarq.38.97
33. De Preter, A., Anthonis, J., De Baerdemaeker, J.: Development of a robot for harvesting strawberries. IFAC-PapersOnLine. **51**, 14–19 (2018). https://doi.org/10.1016/j.ifacol.2018.08.054
34. Zhang, Y., Chang, Z., Zheng, Z., Yang, J.: Harvesting machine for kelp culture in floating raft. Aquat. Eng. **78**, 173–179 (2017). https://doi.org/10.1016/j.aquaeng.2017.07.005
35. Tang, S., Zhao, D., Jia, W., Chen, Y., Ji, W., Ruan, C.: Feature extraction and recognition based on machine vision application in lotus picking robot. In: International Conference on Computer and Computing Technologies in Agriculture, pp. 485–501. (2015)
36. Stankevitz, K., Schoenfisch, A., de Silva, V., Tharindra, H., Stroo, M., Ostbye, T.: Prevalence and risk factors of musculoskeletal disorders among Sri Lankan rubber tappers. Int. J. Occup. Environ. Health. **22**, 91–98 (2016). https://doi.org/10.1080/10773525.2016.1168073
37. Karg, S.: The water chestnut (Trapa natans L.) as a food resource during the 4th to 1st millennia BC at Lake Federsee, Bad Buchau (southern Germany). Environ. Archaeol. **11**, 125–130 (2006). https://doi.org/10.1179/174963106x97106

Chapter 7
Research Design for Simplifying Anthropometric Data Collection Process Using PCA

Amare Wibneh, Ashish Kumar Singh, and Sougata Karmakar

Abstract The current study is aiming at demonstrating how to reduce the numbers of variables using Principal Component Analysis (PCA) to be measured in an anthropometry survey that involves sample data of large number of variables. A case study of analysis of anthropometric data involving PCA method has been reported based on the anthropometric data (32 anthropometric variables) of Ethiopian army personnel collected from 250 male participants. The linear regression models were also constructed using least square method to predict the regression equation of relevant body variables that had correlation coefficients (R) > 0.70. Variables having the lesser factors loading coefficient (<60%), commonality and correlation coefficients (<70%) were counted as independent variables and included in a minimum data set for measurement. The PCA provided six principal component factors. Total 20 regression equations of dependent variables were constructed from the six influential predictors. Therefore, we observed that the total 12 variables (six dominant variables, five variables with less commonality and/or correlation coefficient from their respective predictors, and one targeted variable 'mass') can create a minimum data set that almost accounts for the variability produced by 32 original variables. Hence, during data collection, 12 independent anthropometric variables can extensively represent 32 variables. The current case study would help the researchers to save time and reduce their anthropometric survey to these 12 variables that can predict the remaining ones. It would also guide the researchers to adopt PCA to identify the representative anthropometric variables from large number of variables.

A. Wibneh · S. Karmakar (✉)
Department of Design, Indian Institute of Technology Guwahati, Guwahati 781039, India
e-mail: karmakar.sougata@gmail.com

A. K. Singh
School of Design, Vellore Institute of Technology, Vellore 632014, Tamil Nadu, India

A. Chakrabarti et al. (eds.), *Design for Tomorrow—Volume 1*, Smart Innovation, Systems and Technologies 221, https://doi.org/10.1007/978-981-16-0041-8_7

7.1 Introduction

Designing any facility for the targeted user population requires a large set of anthropometric variables [1]. For developing the anthropometric database, data collection of many anthropometric dimensions becomes quite tedious, time-consuming, costly, and needs special support. Also, considering the large sample size while data collection is constrained by a lack of resources, manpower, and financial support. PCA is one of the widely used methods for the dimensional reduction process [2]. In statistics, variable selection is the method of selecting a subset of predictors for the purpose of dimensional reduction for easy interpretation by the researchers [3]. Predictors from each component factor can represent/predict the measurement of respective inter-correlated variables without conducting all measurements.

Matsunaga [4] noted that PCA is required whenever it is concerned to define the minimum number of variables/features that account for the large set of variables. It helps to remove/reduce the larger data set that contains redundant (strongly correlated) or irrelevant features/variables [5, 6]. During correlation study by Ulgen et al. [7], similar process was followed to reduce into minimum data set. Bermingham et al. [5] also stated that redundant or irrelevant variables would be removed without significant change of relevant information. The words 'redundant' and 'irrelevant' are the two different ideas as it was discussed by Guyon and Elisseeff [6], due to the presence of another relevant variable, one relevant variable may be redundant even though they are strongly correlated. Identifying the most dominant variables from those relevant variables has a vital role in estimating a predictor for the construction of regression equations and prediction of outcomes for redundant or highly correlated variables. Moreover, many redundant but relevant variables may be excluded by a dominant variable which represents them.

This paper targets mainly on constructing and identifying the most dominant variables that are useful to build a good predictor. Therefore, the aim of the current study is to demonstrate how to reduce the numbers of variables involving a large sample based on the results obtained from PCA. The PCA technique was applied to the collected anthropometry data set with a large number of dimensional variables.

7.2 Methodology

A case study of anthropometric data analysis involving the PCA method has been reported in the current research based on the anthropometric data (32 anthropometric variables) of Ethiopian army personnel (collected from a sample of 250 male participants) to achieve the aim.

7.2.1 Anthropometric Measurement Techniques

This is a longitudinal study, and the protocols followed in the previous study [8] were used for sample size determination, anthropometric procedures, and data analysis. Ethiopian army personnel were the focused group for our study. A total of 250 male and 60 female armed personnel were participated in anthropometric data collection, though only male data were used for PCA analysis due to a smaller sample size of female participants. The age from 18 to 52 years (mean = 30.86; SD = 6.7) for the male and 18 t30 years (mean = 24.21; SD = 3.26) for female subjects were included. Table 7.1 shows the statistical database of anthropometric measurements for both (male and female) Ethiopian soldiers. The anthropometric database was presented in terms of mean, SD, 5th and 95th percentile values. The normality test indicated that most of the anthropometric variables follow Gaussian distribution [8].

7.2.2 Statistical Analysis

The measurement data representing body dimensions of Ethiopian male army personnel were analyzed using IBM SPSS version 25 statistical tool for identifying the closely related sets of dimensions (clustering) and regression analysis. PCA was used to cluster anthropometric variables into higher-order category [2, 9]. Then, the highest factor load was used within each cluster to identify the dominant variable, which was followed by regression analysis for predicting other dimensions in the cluster as a function of the dominant variable. The 12 relevant anthropometric variables which are account for 32 measured variables were identified based on their component factor loadings, commonality values, and correlation coefficient of regression analysis. Principal components were categorized based on their eigenvalues greater than 1.0. The higher factor loading coefficient (>60%) variables in each factor category along with commonality (>70%) were considered to be highly inter-correlated variables. The linear regression models were constructed using the least square method. These models were used to predict the regression equation of relevant body variables that had correlation coefficients (R) > 0.70 [10]. Regression coefficients were used to measure the degree of relationship and dependency of most dominant predictor variables to represent all the anthropometric variables. Moreover, the majority of dimensions that would be represented by the most dominant variables or predictor were dropped from the list of relevant anthropometric dimensions. Variables having the lesser factors loading coefficient, commonality, and correlation coefficients were counted as independent variables and included in a minimum data set for measurement.

Table 7.1 Measurements of anthropometry for Ethiopian soldier population (Male (n) = 250; Female (n) = 60)

Anthropometric variables		Male			Female		
		Mean (SD)	5th	95th	Mean (SD)	5th	95th
Standing posture							
1	Stature	170.34 (6.40)*	161.00	179.90	158.46 (5.31)	150.80	169.20
2	Eye height	159.83 (6.40)*	150.60	168.90	143.97 (5.57)	142.30	156.86
3	Acromial height	140.86 (5.60)*	132.50	149.00	123.7 (4.66)	120.80	139.68
4	Chest height	131.22 (4.90)*	117.50	134.50	116.95 (3.86)	110.26	124.98
5	Elbow height	108.22 (4.90)*	101.00	116.00	101.11 (3.10)	96.07	107.51
6	Waist height	96.96 (5.60)*	87.10	105.00	89.64 (4.94)	80.00	99.40
7	Hip height	84.41 (5.10)*	75.50	91.00	80.57 (4.60)	73.81	89.50
8	Buttock height	75.91 (4.20)*	68.48	81.42	72.79 (3.79)	67.07	80.17
9	Arm length	75.93 (3.70)*	70.00	81.90	72.79 (3.79)	60.00	73.38
Sitting posture							
10	Sitting height	85.01 (3.6)*	79.77	90.86	79.05 (2.77)	75.48	83.82
11	Eye height	73.80 (4.0)*	67.48	80.20	67.13 (2.76)	63.44	72.48
12	Acromial height	58.69 (3.4)*	52.71	64.15	53.51 (2.19)	49.54	58.52
13	Elbow rest height	21.74 (2.9)*	17.20	26.31	17.12 (2.37)	13.76	22.10
14	Thigh thickness	15.89 (1.1)*	14.33	17.68	14.68 (1.55)	12.34	17.72
15	Bideltoid breadth	45.31 (1.9)*	42.34	48.58	39.70 (2.01)	36.50	43.68
16	Forearm-forearm breadth	50.19 (2.6)*	46.21	54.55	42.68 (2.68)	38.40	48.00
17	Chest breadth	29.41 (1.7)*	29.00	34.00	27.12 (1.37)	24.60	29.90
18	Chest depth	23.27 (2.6)*	23.38	30.02	20.92 (1.80)	17.60	24.60

(continued)

Table 7.1 (continued)

Anthropometric variables		Male			Female		
		Mean (SD)	5th	95th	Mean (SD)	5th	95th
19	Waist breadth	29.41 (1.7)	27.10	32.90	27.88 (1.65)	25.02	31.30
20	Abdominal depth	23.27 (2.6)	19.53	28.44	20.84 (2.50)	16.49	26.04
21	Hip breadth	37.03 (2.28)	34.50	40.90	36.32 (2.28)	32.43	41.67
22	knee height	54.13 (2.9)*	50.83	57.16	51.88 (1.19)	49.55	54.83
23	Popliteal height	43.56 (2.5)	40.35	47.81	39.37 (1.82)	37.30	42.61
24	Buttock to popliteal length	48.54 (2.1)*	44.99	50.86	46.72 (1.86)	42.83	49.84
25	Buttock to knee length	61.18 (2.4)*	57.35	64.25	57.73 (1.98)	53.64	60.51
26	Forearm length	45.83 (3.3)*	41.10	50.40	44.96 (0.86)	40.41	46.37
27	Forward thumb tip reach	81.25 (3.9)*	75.33	87.27	74.08 (3.55)	64.74	78.70
Standing/Sitting Posture							
28	Hand length	18.45 (1.4)*	16.52	20.70	17.95 (0.53)	16.88	18.68
29	Handbreadth	8.40 (0.70)*	7.41	9.50	8.01 (0.49)	6.74	8.51
30	Foot length	25.21 (0.90)*	23.50	26.50	22.82 (0.54)	21.84	24.12
31	Foot breadth	9.55 (0.40)	8.77	10.00	8.16 (0.40)	7.37	9.04
32	Mass (kg)	66.75 (9.4)	55.00	84.00	52.50 (5.47)	45.00	63.80
	BMI (kg/m^2)	23.00 (2.61)*	21.2	25.4	20.90 (2.21)	19.8	22.3

All measurements are in cm unless specified; SD = standard deviation; BMI = body mass index
*The data having a normal distribution

7.3 Result and Discussion

In this section, PCA analysis for variable reductions from the substantial dataset and regression analysis were carried out. In the following sub-sections, PCA for clustering for higher factor loading variables, regression equation analysis for predicting the missing dependent variable, and correlation coefficient analysis to indicate the extent to which two variables move together have been discussed. Due to the smaller sample size of female participants (60), PCA analysis of the collected male anthropometric data (250) was carried out in this study.

7.3.1 PCA Analysis

To minimize the challenge of data collection, analysis of variances, and other applications from large sets of anthropometric variables, PCA was used for selecting the most relevant variable called predictor [9, 11] from each factor. In the study, the higher inter-correlated anthropometric variables were clustered into six factors with eigenvalue ≥ 1 (Table 7.2). The estimated factor loadings and commonality values could be used to define the characteristics of each factor in PCA analysis. These six factors were developed to achieve minimum good data sets that will account for a large set of variables for further applications.

Table 7.2 shows the PCA results for clustering of the higher inter-correlated variables and identification of predictors in each component factor. At least, two anthropometric variables were considered in each factor with a factor loading (correlation) coefficient greater than 0.60. The variable with highest factor loading and commonality values is clustered under the respective component factor. For instance, even though the correlation coefficient of stature was found to be greater than 0.60 for both factors 1 and 3, it was clustered under Components Factor 01 (Fac01). Moreover, even though the factor loading coefficients of popliteal height and popliteal length were high, they were not considered under Fac01 due to their lower recommended commonality values (<0.70).

One dominant variable called predictor (which represents other variables) within the group was taken from each component factor category [11]. These predictors were identified based on their higher factor loading coefficient and commonality. Within each component factor, they would be considered the essential dimensions for anthropometric characteristics of the Ethiopian soldiers in different applications and analysis of variances under a certain limit of errors. The highest dominant variables in each component factor were foot length, arm length, sitting height, hip breadth, bideltoid breadth, and stature. These highly dominant anthropometric variables are listed in the factor-wise category in Table 7.3 on the basis of inferences drawn from Table 7.2. The higher factor loading coefficient (>60%) variables in each factor category along with commonality (>70%) were highly inter-correlated [10] and the characteristics of variances expected to be nearly the same [8, 12]. Therefore, we

Table 7.2 Factor loadings and commonality results for anthropometry (*Adapted from Amare* et al., *2020)*

Anthropometric variables	Component factors						
	Fac01	Fac02	Fac03	Fac04	Fac05	Fac06	Commonality
Stature	**0.949***	0.540	0.751	0.474	0.588	0.575	0.927#
Acromial height	**0.944***	0.600	0.703	0.491	0.543	0.517	0.923#
Standing eye height	**0.939***	0.572	0.728	0.512	0.610	0.558	0.919#
Buttock height	**0.909***	0.348	0.536	0.366	0.422	0.423	0.882#
Hip height	**0.908***	0.349	0.535	0.366	0.421	0.422	0.881#
Chest height	**0.902***	0.581	0.641	0.492	0.516	0.518	0.852#
Standing elbow height	**0.896***	0.590	0.677	0.488	0.452	0.549	0.878#
Waist height	**0.893***	0.374	0.606	0.333	0.489	0.499	0.816#
Knee height	**0.835***	0.428	0.607	0.248	0.401	0.375	0.714#
Popliteal height	0.805	0.284	0.565	0.229	0.293	0.324	0.690
Buttock to knee length	**0.760***	0.592	0.673	0.074	0.525	0.487	0.756#
Popliteal length	0.727	0.503	0.620	0.021	0.462	0.474	0.686
Bideltoid breadth	0.472	**0.899***	0.710	0.283	0.347	0.354	0.864#
Elbow to elbow breadth	0.471	**0.897***	0.702	0.283	0.346	0.353	0.862#
Chest depth	0.342	**0.878***	0.702	0.392	0.395	0.479	0.837#
Chest breadth	0.334	**0.867***	0.609	0.372	0.252	0.479	0.835#
Foot breadth	0.428	**0.625***	0.597	0.210	0.196	0.300	0.713#
Handbreadth	0.257	**0.607***	0.560	0.140	0.192	0.269	0.702#
Sitting hip breadth	0.465	0.634	**0.964***	0.347	0.349	0.316	0.938#
Abdominal depth	0.481	0.625	**0.963***	0.362	0.354	0.341	0.932#
Waist breadth	0.435	0.646	**0.958***	0.311	0.343	0.315	0.931#
Thigh thickness	0.483	0.611	**0.946***	0.308	0.347	0.362	0.895#
Mass	0.520	0.772	**0.824***	0.516	0.327	0.490	0.824#
Sitting height	0.572	0.465	0.375	**0.892***	0.568	0.403	0.866#
Sitting eye height	0.589	0.403	0.452	**0.877***	0.576	0.437	0.867#
Sitting acromial height	0.535	0.327	0.392	**0.866***	0.517	0.419	0.828#

(continued)

Table 7.2 (continued)

Anthropometric variables	Component factors						
	Fac01	Fac02	Fac03	Fac04	Fac05	Fac06	Commonality
Elbow rest height	0.428	0.182	-0.112	**0.802***	0.454	0.093	0.823#
Arm length	0.608	0.294	0.180	0.539	**0.963***	0.517	0.952#
Forearm length	0.591	0.283	0.171	0.500	**0.956***	0.608	0.945#
Sitting thumb tip reach	0.678	0.372	0.170	0.580	**0.895***	0.451	0.901#
Foot length	0.556	0.322	0.328	0.429	0.484	**0.931***	0.951#
Hand length	0.544	0.280	0.285	0.504	0.445	0.834*	0.938#
Eigenvalue	**16.555**	**4.641**	**2.607**	**1.856**	**1.649**	**1.056**	–
Total variance (%)	**50.167**	**14.062**	**7.900**	**5.626**	**4.996**	**3.199**	–
CV (%)	**50.17**	**64.23**	**72.13**	**77.76**	**82.75**	**85.95**	–

*Factor loading ≥ 0.60
CV = Cumulative of Variance
#Commonalities ≥ 0.70

Table 7.3 List of factor-wise independent variables

Component factors	Dimension
Fac01	Stature*, acromial height, standing eye height, buttock height, hip height, chest height, standing elbow height, waist height, knee height, buttock to knee length,
Fac02	Bideltoid breadth*, elbow to elbow breadth, chest depth, chest breadth, foot breadth, handbreadth
Fac03	Sitting hip breadth*, abdominal depth, waist breadth, thigh thickness, mass,
Fac04	Sitting height*, elbow rest height, sitting eye height, sitting acromial height
Fac05	Arm length*, sitting thumb tip reach, forearm length
Fac06	Foot length*, hand length
V	Popliteal height, popliteal length

V = independent variables having less commonality values (<70%)
*The highest factor loading variables within component factor

proceeded to compute linear regression models with these six dominant independent variables to verify the dependency of other relatively less dominant variables on them. The variables which are grouped in the same component factors were used to formulate the linear regression equation of the anthropometric variables, which were under the respective component factor.

7.3.2 *Regression Analysis and Correlation Coefficient*

The regression equations predict dependent variables from the predictors if the dependent variable and predictors (independent variables) have higher factor loadings and commonality coefficients in the same factor loading category. Based on these criteria, regression equations for the most relevant dependent variables were predicted from the six dominant independent variables, viz. stature, bideltoid breadth, hip breadth, sitting height, arm length, and foot length (Table 7.3). These six independent variables do not limit the regression analysis, but there are multiple chances to formulate several simple regression equations from these six good predictors from Table 7.3. Table 7.4 reports the list of equations derived to estimate the value of one vari-

Table 7.4 Linear regression equations for the estimation of body variables from their predictors

No	Regression equation	F-ratio	SEE	R	R^2
1	Standing acromial height = 0.844 × (Stature) – 2.93	1430*	1.92	0.94	0.88
2	Standing eye height = 0.974 × (Stature) – 6.10	4836*	1.20	0.98	0.96
3	Buttock height = 0.523 × (Stature) – 13.22	340*	2.54	0.80	0.64
4	Hip height = 0.627 × (Stature) – 22.38	339*	3.05	0.82	0.67
5	Chest height = 0.795 × (Stature) – 10.00	955.5*	2.22	0.91	0.83
6	Standing elbow height = 0.6802 × (Stature) – 7.65	690*	2.29	0.88	0.78
7	Waist height = 0.713 × (Stature) – 24.53	393*	3.22	0.82	0.67
8	Knee height = 0.3395 × (Stature) – 10.12	231*	1.89	0.74	0.55
9	Buttock to knee length = 0.2782 × (Stature) + 13.8	261*	1.52	0.76	0.58
10	Elbow to elbow breadth = 1.34 × (Bideltoid breadth) –10.4	2429*	0.72	0.97	0.83
11	Chest depth = 0.843 × (Bideltoid breadth) –11.69	227*	1.45	0.76	0.58
12	Chest breadth = 0.644 × (Bideltoid breadth) + 2.16	135*	1.06	0.74	0.55
13	~~Foot breadth= 0.072×(Bideltoidbreadth)+ 6.27~~	234*	0.54	0.51	0.26
14	~~Handbreadth= 0.0727× (Bideltoidbreadth)+ 5.09~~	219*	0.35	0.47	0.22
15	Waist breadth = 0.759 × (Hip breadth) – 1.31	1517*	0.55	0.94	0.88
16	Thigh Thickness = 0.486 × (Hip breadth) – 2.10	1534*	0.35	0.93	0.86
17	Abdominal depth = 0.697 × (Hip breadth) + 15.2	1804*	0.80	0.95	0.95
18	Sitting eye height = 0.977 × (Sitting height) – 9.21	1318*	1.37	0.94	0.88
19	Sitting acromial height = 0.701 × (Sitting height) – 0.90	321*	2.01	2.01	2.01
20	~~Elbow rest height= 0.375× (Sitting height) –10.172~~	132*	2.44	0.50	0.25
21	Forearm length = 0.696 × (Arm length) – 7.07	601*	1.47	1.47	0.76
22	Thumb tip reach length = 1.026 × (Arm length) + 3.32	3947*	0.84	0.87	0.76
23	Hand length = 1.44 × (foot length) -17.25	507*	0.66	0.66	0.72

All dimensions are in cm; SEE = standard error of the estimate; R = correlation coefficient.
The ~~strikethrough~~ line signifies the excluded equation and its dependent variable having low coefficient of determination ($R^2 < 50\%$).
*$p < 0.05$.

able from the given value of another anthropometric variable. These equations were formulated based on the categories of highly correlated dominant variables, as it was referred from Table 7.3. The regression model for equation #13, #14, and #20 in Table 7.4 was rejected since the dependent variable cannot be predicted from the independent variable with a lower value of the coefficient of determination (denoted by R^2). Therefore, 12 anthropometric measurements from the six measured values of dominant variables using the developed equations shall be applicable for data collection.

Henseler et al. [13] proposed a rule of thumb for acceptable R^2 values of 0.75, 0.50, and 0.25, be perceived as substantial, moderate, and weak, respectively. On the contrary, Hair et al. [14] opined that there could not be a thumb rule applicable to all disciplines. R^2 acceptability is reliant upon the model complexity and the research discipline. Kroemer et al. [15], in their study, also pointed out that the independent variable has a higher correlation coefficient (>0.70) which predicts the dependent variables. The variables having higher factor loadings, commonality, and correlation coefficients could provide a good prediction of regression equations [10]. Altogether from Table 7.4, as the correlation and regression coefficients ranging from 0.71 to 0.98 and 0.50 to 0.96 in all the cases (moderate to substantial), it may be concluded that the mathematical model has good fitness for predicting analysis. However, due to the less correlation coefficient ($R < 70\%$) of elbow rest height and breadth with their respective predictors, the mathematical model has no good fitness for predicting analysis.

The correlation and regression coefficients from Table 7.4 revealed that most of the variables among the same component factor were highly correlated. For example, the dependent variables, viz. sitting acromial height and sitting eye height, were found predictable from their most dominant variable, i.e., sitting height (independent variable) in principle component factor 4 (Fac04). Therefore, it can be concluded that the six predictors nearly account for the characteristics of 21 original variables other than popliteal height, popliteal length, elbow rest length, foot breadth, and handbreadth. Total 21 regression equations of dependent variables were constructed from the six predictors by rejecting variables having less correlation coefficient, and those rejected variables are included in the minimum data set (12 variables) of 32 variables.

Therefore, it is essential to consider independently variables that are not represented by other anthropometric variables in component factor category. Therefore, variables having yet high factor loading coefficients but lower recommended commonality values (<0.70) like popliteal height and popliteal length, and variables having lower correlation coefficients like elbow rest length, foot breadth, and a handbreadth were considered for the representations of those 32 original variables. We have also included mass as one targeted variable [11]. Therefore, the six dominant variables, the two variables having less commonality, the three variables having less correlation coefficient, and the one targeted variable called mass can also be counted as independent variable for representing itself in anthropometric data survey of those the 12 representing variables (Table 7.5).

Table 7.5 Minimum required anthropometric variables accounts for the 32 anthropometric variables

V1	V2	V3	V4
1. Stature	7. Elbow rest height	10. Popliteal height	12. Mass
2. Bideltoid breadth	8. Foot breadth	11. Popliteal length	
3. Sitting hip breadth	9. Handbreadth		
4. Sitting height			
5. Arm length			
6. Foot length			

V1—the six most dominant anthropometric variables in each component factor; V2—variables having less correlation coefficients with their respective predictors (R < 70%); V3—variables having less commonality (<70%); V4—the targeted variable referring many earlier studies.

Hence, during data collection, 12 independent anthropometric variables can represent 32 variables, as shown in Table 7.5. These variables are the six most dominant anthropometric variables, variables having less commonality, the less correlated variables, and the key and targeted variable in many of the study called mass. Moreover, most of the anthropometric variables can be predicted using six influential variables.

7.4 Conclusion

The 12 substantial variables which account for 32 anthropometric variables were identified for simplifying the anthropometric data collection process, particularly in measurement surveys of large sample sizes. Therefore, identifying the representative anthropometric variables/predictors is helpful for the reduction of a large number of variables, while data collection is constrained by a lack of resources, manpower, and financial support. Hence, the variable reduction technique can be helpful while using fewer variables that can be an accurate representation of larger numbers of variables in a data set. The method of reducing variables in the present study can be utilized for any anthropometry survey with a sample size larger than 250 subjects. Therefore, the selection of 12 variables holds good for any larger sample survey. In PCA analysis, the surveyed data should be normal; however, some variables from the female data were not normally distributed. Therefore, we have not considered female data. Being the first army anthropometry study in Ethiopia, the results can be used as a baseline for further studies to create a larger database.

References

1. Pheasant, S., Haslegrave, C.M.: Bodyspace: Anthropometry, Ergonomics and the Design of Work, 3rd edn. CRC Press, Boca Raton (2005)
2. Lu Y, Cohen I, Zhou, XS, Tian Q (2007) Feature selection using principal feature analysis. In *Proceedings of the 15th ACM international conference on Multimedia* 301–304
3. James, G., Witten, D., Hastie, T., Tibshirani, R.: An introduction to statistical learning. New York: Springer **112**, 3–7 (2013)
4. Matsunaga, M.: How to factor-analyze your data right: do's, don'ts, and how-to's. Int. J. Psychol. Res. **3**(1), 97–110 (2010)
5. Bermingham, M.L., Pong-Wong, R., Spiliopoulou, A., Hayward, C., Rudan, I., Campbell, H., Wright, A.F., Wilson, J.F., Agakov, F., Navarro, P., Haley, C.S.: Application of high-dimensional feature selection: evaluation for genomic prediction in man. Sci. Rep. **5**, 10312 (2015)
6. Guyon, I., Elisseeff, A.: An introduction to variable and feature selection. J. Mach. Learn. Res. **3**, 1157–1182 (2003). https://doi.org/10.1162/153244303322753616
7. Ulgen, A., Han, Z., Li, W.: Correlation between quantitative traits and correlation between corresponding LOD scores: detection of pleiotropic effects. BMC Genet. BioMed. Central **4**, S60 (2003)
8. Wibneh, A., Singh, A.K., Karmakar, S.: Anthropometric measurement and comparative analysis of Ethiopian army personnel across age, ethnicity, and nationality. Defence Sci. J. 70(4), 383 (2020). https://doi.org/10.14429/dsj.70.15435
9. Jackson, J.E.: Principal components and factor analysis: part I—principal components. J. Qual. Technol. **12**(4), 201–213 (1980). https://doi.org/10.1080/00224065.1980.11980967
10. Patel, T., Sanjog, J., Chatterjee, A., Karmakar, S.: Statistical interpretation of collected anthropometric data of agricultural workers from Northeast India and comparison with national and international databases. IIE Trans. Occup. Ergon. Human Factors **4**(4), 97–210 (2016). https://doi.org/10.1080/21577323.2016.1194912
11. Lin, H.F., Hsu, C.H., Wang, M.J., Lin, Y.C.: An application of data mining technique in developing sizing system for army soldiers in Taiwan. WSEAS Transactions on Computers **7**(4), 245–252 (2008)
12. Charles O (2019) Crawford. Functions of Independent Variables in Research and Program Planning. Public Health Reports (1896–1970) Sep. 1969. https://www.jstor.org/stable/i409505
13. Henseler, J., Ringle, C.M., Sinkovics, R.R.: The use of partial least squares path modeling in international marketing. In: New Challenges to International Marketing, pp. 277–319. Emerald Group Publishing Limited (2009)
14. Hair, J.F., Ringle, C.M., Sarstedt, M.: PLS-SEM: indeed a silver bullet. J. Mark. Theory Pract. **19**(2), 139–152 (2011). https://doi.org/10.2753/MTP1069-6679190202
15. Kroemer, K.H., Kroemer, H.J., Kroemer-Elbert, K.E.: Engineering Physiology: Bases of Human Factors Engineering/Ergonomics, 4th edn., pp. 362. Springer Science and Business Media (2010)

Chapter 8
An Attempt to Understand Social Relationships Using Facial Expression Electromyography Analysis

Quentin Ehkirch, Saya Kakiuchi, Yuki Motomura, Susumu Matsumae, and Akane Matsumae

Abstract This study served as an attempt to better understand social relationships by using facial expression electromyography (fEMG) analysis. The goal was to be able to visualize social relationships through objective data to broaden understanding of the mechanism of social relationships and social design. This study focuses more on observed daily experiences as a means of comprehending social relationships among the participants and to determine if social relationships can be understood using fEMG analysis. The authors conducted an experiment that compares the results of fEMG analysis with social relationship transitions recorded during a card game called mini-bridge. This is a game played in teams, and it requires that players work in pairs. It can clearly be seen whether the pair is successful in this co-creative relationship by checking the data recorded on video and by seeing what cards are in each player's hand and how they play the game. During the game, the participants wore electrodes on their corrugator supercilii and zygomaticus major, allowing us to collect data using fEMG (objective results). Using a shared behavior and intentions study from the videos, the intersubjective state of the examinees was graded (subjective results) and compared to the objective results interpreted through the Levenshtein method. The research results suggest the possibility of identifying a social relationship as a partnership by using fEMG with data from the corrugator supercilii. This study will contribute to our knowledge of the mechanism of social relationships, which is needed to develop a methodology for social design.

Q. Ehkirch
University of Technology of Belfort-Montbéliard, Sevenans, France
e-mail: quentin.ehkirch@utbm.fr

Q. Ehkirch · S. Kakiuchi
School of Design, Kyushu University, Fukuoka, Japan

Y. Motomura · A. Matsumae (✉)
Faculty of Design, Kyushu University, Fukuoka, Japan
e-mail: matsumae@design.kyushu-u.ac.jp

S. Matsumae
Faculty of Science and Engineering, Saga University, Saga, Japan

A. Chakrabarti et al. (eds.), *Design for Tomorrow—Volume 1*, Smart Innovation, Systems and Technologies 221, https://doi.org/10.1007/978-981-16-0041-8_8

8.1 Introduction

8.1.1 Motivations

What helps bring individuals together? Collaboration is not just the result of sharing individual knowledge and power, but something that creates a dynamic between participants. Personal interaction is at least as important as knowledge sharing and shared understanding during this process, but it is often undervalued in studies of collaboration processes in established teams [1, 2].

Being able to go further in considering such human factors might help us to develop our understanding of co-creativity. This is why there is a need to develop knowledge of intersubjectivity during the collaborative process. However, as it stands now, intersubjectivity is mostly understood via subjective factors that are highly dependent on context. It is essential to identify objective indicators of intersubjectivity to broaden understanding of the interactive processes that support it.

8.1.2 Related Studies

Intersubjectivity. Intersubjectivity is essential as a subjective basis of co-creative collaboration, as many case studies have already emphasized [3–6].

Intersubjectivity was first conceptualized in Husserl's phenomenology as a philosophical idea, but has since been developed in fields such as sociology, psychology and anthropology. The basic idea of intersubjectivity is that subjects do not constitute a world alone, but jointly with other subjects, which domain lies beneath the empathy [7, 8]. Thanks to the development of social cognition research, considerable evidence has been uncovered to support the theory of intersubjectivity from different fields, such as linguistics [9], neuroscience [10] and child development [11]. The formation of intersubjectivity in co-creation was pointed out as a dynamic mechanism [6] that suggested the importance of focusing on interactions between members to emphasize the complex dialogue needed to create this shared world. However, as intersubjectivity is not only conversation, but also the sharing of intentions and affects [12], the need to link it with nonverbal communication and behavior studies was strongly supported.

Facial Electromyography (fEMG). To determine objectively how a person, react to a situation, facial electromyography use the reading from muscle activity by detecting and amplifying the tiny electrical impulses that are generated by muscle fibers when they contract. Many papers demonstrated the ability to read an emotional state using this method [13], focusing on the zygomaticus major and corrugator supercilii muscles [14] by linking them to strong emotional reactions to a situation [15]. Strongly related to the mental state of someone in a social environment and affected by it [16], this methodology confirms the idea of the perception–behavior link [17] giving the fundament of the chameleon effect theory, i.e., that social actors are unconsciously matching postures, mannerisms, facial expressions and other behaviors of one's interaction partners which have given solid tangible argument toward intersubjectivity through cognitive theory. Thus, in this paper, the author used non-intrusive electromyography to measure the activity of some facial muscle in a collaborative situation to see if it could be linked as an objective indicator of intersubjectivity.

8.1.3 Purpose of This Study

The purpose of this study was to determine whether intersubjectivity, in this case meaning the subjectivity of co-creative collaboration between two persons, can be evaluated using fEMG analysis. One particular characteristic of this study was that it used a field research approach (analyzing data coming from an everyday situation) to see if it was possible to apply the theory of intersubjectivity in an actual collaborative process. By capturing objective data along with subjective data during the collaborative process and then comparing them, the authors proposed a methodology of using fEMG oscillation comparisons as an indicator of intersubjectivity.

8.2 Methodology

As the goal of this study was to compare two approaches: objective and subjective data, and to find a correlation between them, the experiment was designed to simplify situational factors. With the idea of examining a simpler sort of cooperative situation, the "mini-bridge card game" was chosen, as it is a game that requires pair-work to win. Using references from earlier case studies (as explained in Sects. 8.1.2 and 8.1.3), the means of evaluation concentrated on comparisons between a subjective evaluation of behavior and intention determined by the intersubjectivity state (explained in Sect. 8.2.2), and the data resulting from an fEMG analysis of two types of muscles: the corrugator supercilii and the zygomaticus major. The methodology used to compare the fEMG data (see Sect. 8.2.4) will be explained along with the results in Sect. 8.5.

8.2.1 Experiment

Participants. Sixteen healthy participants (7 men and 9 women) aged in their twenties participated in the experiment and were divided into eight pairs, from A to H. Each examinee was given a number (from Examinee 1 to Examinee 16) and classified by order into a pair. For example, pair A comprised Examinees 1 and 2, while pair H comprised Examinees 15 and 16. Furthermore, each pair's relationship (i.e., whether the examinees in the pair knew each other beforehand) was recorded by the authors to examine whether it influenced the results.

Procedure. The experiment was carried over four different sessions, as the equipment needed to carry out fEMG was only sufficient to measure four examinees at a time; which is also the number of players needed to play a game of mini-bridge. The opponents in each game were two adjacent pairs. In all, 190 plays were documented. All plays were recorded on cameras to evaluate the intersubjectivity feeling of each pair on each play, as is explained in Sect. 8.2.2. The result was used as subjective data to indicate whether the pair was in an intersubjective state during a given play. This experience was designed as a game to reduce the influence of observation on the examinees, so that they could immerse themselves while not being bothered by the measurements being taken.

Finally, an fEMG analysis was conducted during play, as explained in Sect. 8.2.3. This objective data was then compared to the other dataset to see if they matched.

8.2.2 Evaluation of Intersubjectivity

Evaluation by Examiner. To get a more in-depth view of the intersubjectivity state of a pair in play, the authors followed results from earlier research and decided to assess the behavior of each participant in accordance with their intentions to create a scale of intersubjectivity.

Observations of Shared Behavior. As many previous studies have pointed out, nonverbal communication and conversation are constitutive of shared behavior that can lead to an intersubjective state. Following this research, the authors used standard categorizations of nonverbal communication, through communicative gestures, posture and eye behavior. Using the videos recorded of each play, the researchers studied how each examinee shared his/her behavior with their partner. In addition to studies of these factors, the amount of talk was also taken into consideration as an indicator of shared behavior of a pair during a play.

Intentions. An intersubjective state means not only conversing with one another, but also sharing intentions and finalities. To assess the intentions of the players, the examiner used not only the shared behavior and conversation flow, but also the results of play in addition to the report from the examinee. It was then compared to the other play intentions for use as a reference point for the play studied.

Scale. From the combination of intentions in accordance with the shared behavior, a grading standard of the intersubjective state of the pair in play was created. To have an understanding of the diversity of the play state while minimizing the influence of external factors, a 3-grade scale was chosen going from C to A, C being the lowest intersubjective state and A being the strongest.

8.2.3 *Facial Expression Electromyography Analysis (fEMG)*

Following social cognitive research on the importance of facial muscle activity in forming intersubjectivity in a social environment, the authors decided to record the activity of two major muscles linked to strong emotional states: The corrugator supercilii linked to the smile, hereafter C-muscle and the zygomaticus major linked to the frown, hereafter Z-muscle. They will be referred as, respectively, C-muscle and Z-muscle. Non-intrusive electrodes were used to measure the muscle's activity, minimizing any chance that the examinee might be excessively distracted by them during play. To eliminate white noise, a dummy was placed near the edge of the eye. The muscle activity corresponding to voltage fluctuations resulting from ionic current within the neurons of the brain was recorded on a 0.15 s frequency. These results were first all taken positive as the idea was to capture the muscle movement and then were averaged to a 0.45 s timespan to measure the tendency of the muscle activity.

This data was then divided by play into waves to make it correspond to the standards of the subjective grading system.

8.2.4 *Comparing the Two Waves*

Binary Transformation. The muscle activity wave of an examinee cannot really be used alone, as it is only an indicator of one person in the pair. Thus, there was a need to find a method to compare the waves of both members of the pair. However, to efficiently compare the data, it had to be simplified. We converted the data into a binary series where 1 represented activity from the muscle, and 0 represented an absence of activity. Because each examinee has a different value of voltage fluctuation corresponding to an activity, the most efficient way forward was to use an own value also fluctuating on the play studied. This is why the median value was chosen as a correct binary condition in our study to determine whether there was muscle activity.

Our binary transformation was then processed as in this operation:

$$\text{binary transform} \quad \begin{array}{l} g = (g(1), g(2), \ldots, g(n)) \\ \downarrow \\ g_b = (g_b(1), g_b(2), \ldots, g_b(n)) \end{array}$$

where $g_b(i) \stackrel{\text{def}}{=} \begin{cases} 1 & g(i) > \text{median}(g) \\ 0 & g(i) \le \text{median}(g) \end{cases}$.

$$h = (h(1), h(2), \ldots, h(n))$$

h_b is defined similarly.

Levenshtein Distance. We then used the Levenshtein distance method to compare the binary waves of two examinees, as it eliminates the delay problem inherent with the reaction of the other examinee and also offers reasonable precision while allowing a simplified method of calculation.

To express the Levenshtein distance between two series g and h, a function $L_{\text{dist}}(g_i, h_j)$ is defined whose value is a distance between an i—symbol prefix (initial substring) of series g and a j—symbol prefix of h.

The restricted distance function is defined recursively as:

$$L_{\text{dist}}(g_i, h_j) = \begin{cases} \max(g_i, h_j) & \text{if} \min(g_i, h_j) = 0 \\ \min \begin{cases} L_{\text{dist}}(g_{i-1}, h_j) + 1 \\ L_{\text{dist}}(g_i, h_{j-1}) + 1 \\ L_{\text{dist}}(g_{i-1}, h_{j-1}) + 1(g_i \neq g_{hj}) \end{cases} & \text{Otherwise} \end{cases}$$

where $1_{(g_i \neq h_j)}$ is the indicator function equal to 0 when $g_i = h_j$ and equal to 1 otherwise.

This method computes a value distance that we then divided by the size of the wave to result in a proportional measurement of the difference between the two sets of data. Later in this paper, we will use the Levenshtein distance term to refer to this proportional value, as it is more representative of the behavior.

8.3 Results

8.3.1 Justification of the Levenshtein Distance Methodology

Binary Condition. While measuring the activity of a muscle using electroencephalography does depict all fluctuations, it is too precise to be used as a reference as not all voltage fluctuations recorded correspond to an actual movement of the muscle. Using a simpler binary code not only allowed easier manipulation of the data but allowed the choice of which part of the voltage fluctuations should be studied.

By using the median point of the wave study, the authors decided to focus on the major activities of the muscle, i.e., the ones that were always linked to a movement and reaction of the muscle (Fig. 8.1).

Levenshtein Distance. The Levenshtein distance calculation is often used in various domains to compute the minimum differences between two strings. In this case, it was used on a binary string, providing a fair idea of the divergence between the examinee's muscle activity while eliminating the problem of delay in the answer of the other. After computing the Levenshtein distance for all plays and classifying them by value, the authors visually confirmed the pertinence of this methodology on the study (Fig. 8.2).

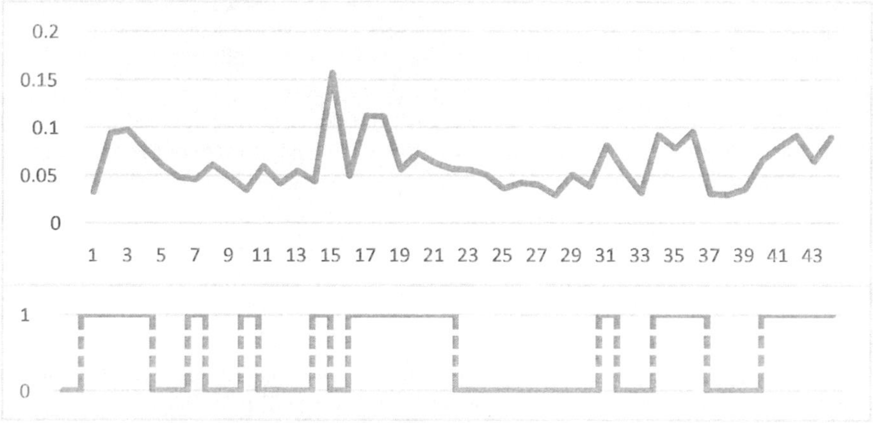

Fig. 8.1 Comparison of waves and their binary series

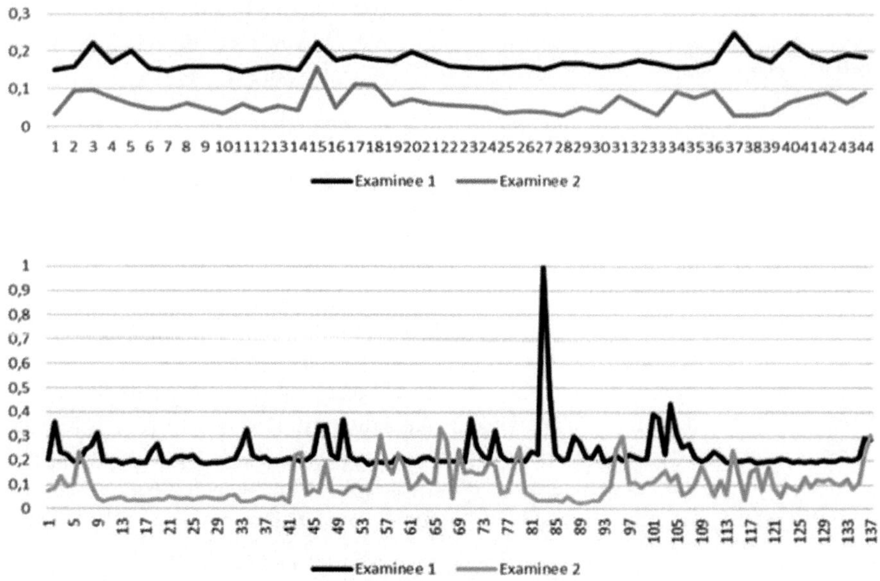

Fig. 8.2 Comparison of fEMG datasets of the pair in the lowest (above) and the highest (bellow) Levenshtein distance play

8.3.2 Probability of Levenshtein Distance Methodology Studied by Comparing Random Functions

The methodology was designed to provide a sense of the differences between two waves, and a standard reference was needed to interpret the results correctly. To do so, we calculated the average size-proportional Levenshtein distance between two random binary series by using probability and an approximation trough simulation at the end.

To conclude, it was determined that by using random set series, the Levenshtein distance methodology provides a distance around 29% of the size of the binary series being compared.

8.3.3 *Selection of the Representative Play*

Mechanical problems interfered with the collection of subjective data from the plays, leaving some data uncollected. To use the data successfully collected, we decided to restrict the study to 72 representative plays coming from four pairs: A, B, G and H. These pairs were chosen because they had different relationships before the experiment; pairs A and B barely knew each other, while pairs G and H were all close friends beforehand. Furthermore, as a representative pool for the play, the 6 lowest, 6 highest and 6 plays from the middle for each pair were selected and ranked by the Levenshtein distance methodology. This pool of plays was confirmed to be representative, as the average Levenshtein distance of these plays was the same as that coming from the average of all the plays collected.

8.3.4 *Results for the C-Muscle*

From All Play. Using the data received from the C-muscle (corrugator supercilii, linked to the frown) from 190 plays, we calculated the differences between the muscle activity of two examinees in a pair using the Levenshtein distance methodology (Sect. 8.2.4). We found that, on average, the pairs did not show a significant difference in their Levenshtein distance, with all clustered around 0.36. It would seem in this case that external factors such as how well the members knew each other beforehand were sufficiently diminished to avoid influencing our results.

Comparing Levenshtein Distance Results and the Subjective Grading System of the Representative Plays. Using the grading system to judge the intersubjective state of play presented under Sect. 8.2.2, the authors classified the 72 plays of the representative pool (see Sect. 8.3.3) by grade and then compare the distribution of the Levenshtein distance of the plays by grade. The results show a correlation between the Levenshtein distance of the C-muscle's activity of two examinees in a pair and their intersubjective state in a play (Fig. 8.3). T-test showed that this result of differences for the variance of the average Levenshtein distance by grade is statistically significant (t = -3.15, df = 15, p = 0.0032).

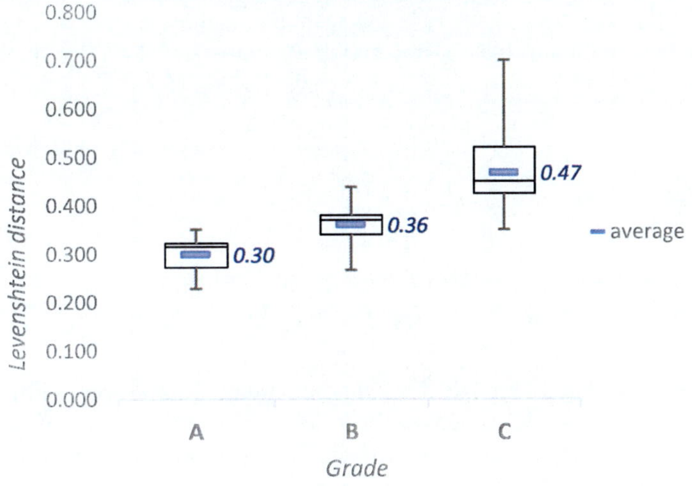

Fig. 8.3 Distribution of the Levenshtein distance by grade for the C-muscle

8.3.5 Comparison to the Z-Muscle Results

The same method was used to compare results coming from the Z-muscle (zygomaticus major, linked to the smile). However, with results showing the same tendency as those from the C-muscle, things become less clear, and there is too much dispersion to draw any conclusions on the correlation between the subjective and objective data (Fig. 8.4). The t-test showed that this result of differences is statistically less significant (t = 1.02, df = 14, p = 0.163). Furthermore, after checking the videos

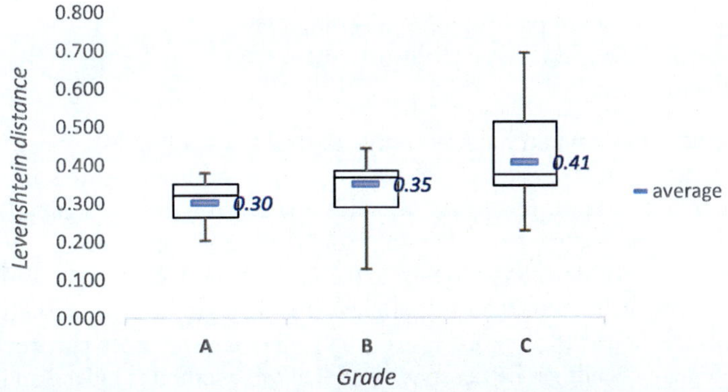

Fig. 8.4 Distribution of the Levenshtein distance by grade for the Z-muscle

taken from the play once again, it would seem that Z-muscle activity is more influenced by the situation of play between all the members of the game (both pairs of examinees).

8.4 Discussions

8.4.1 Proposition of a Methodology to Simplify the Comparison of Electroencephalography Data

Levenshtein Method. Through this study, the authors have proposed a standard methodology (Sect. 8.2.4) to compare the results of an fEMG analysis of two persons in a cooperative environment by first creating a binary simplification of the data and then computing the differences using Levenshtein distance. This method enables an easy and quick way to compute the general behavior of muscle activity in answer to another in a social environment. After trying it on two facial muscles strongly involved in facial expression, it was also pointed out that this method offers a good indication of the conscious and unconscious communicative behaviors between examinees.

Influence of the Method on Results. The Levenshtein method is strongly dependent on the choice of method used to capture the two waves that will be compared. To understand the influence of this factor and to confirm that the focus remains on the intersubjective state between the examinees in a pair, the authors used the Levenshtein method on three separate datasets of 32 plays each (roughly 100 datasets) differently:

- *Basic*: results set randomly using a random function
- *By examinee*: using results recorded during the experiment from examinee but coming from different pairs on non-matching play
- *By pair*: using the results used in the experiment, i.e., coming from the same pair and the same play.

As expected, the random method results tended toward 0.29 (see Sect. 8.3.2 for the development), only 0.31 here because the amount of data used was insufficient. Using this as a reference, there is a clear difference between our standard method giving 0.36, and one using waves coming from examinees of different pairs, which leads to an average Levenshtein distance of 0.39 (Fig. 8.5). The t-test showed that this result of differences is statistically significant ($t = 2.99$, $\mathrm{df} = 31$, $p = 0.0026$). The Levenshtein method then diminishes external influences of other players to allow focus on the pair's feelings. It was concluded that this method is indeed an indicator of the intersubjective state between two persons in a cooperative environment.

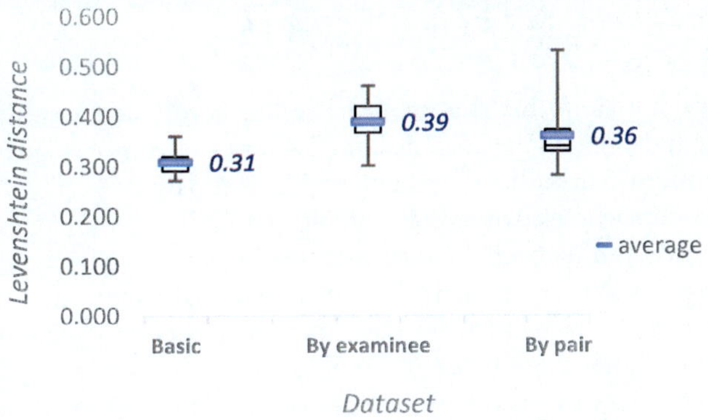

Fig. 8.5 Comparison of the Levenshtein distance average results by method

8.4.2 *Chameleon Effect and Situational Limitations*

Comparing the results from both muscles and the video recorded, the authors learned that the C-muscle seems to be used more unconsciously while the Z-muscle seems to be more often shared between all members playing the game, linked to the situation. It would then explain the differences in results because of the influence of external factors (i.e., not linked to the examinee in a pair) on the zygomaticus major's activity. This explanation is also supported by the chameleon effect theory (Chartrand and Bargh 1999) which explains how social actors tend to mimic the behavior and facial expression of others.

8.4.3 *Findings*

In this study, the authors found:

- The Levenshtein method applied to the reading of facial muscles provides a good indicator of communicative behavior between two persons in a cooperative situation.
- If used on the results of an electroencephalography on the corrugator supercilii, it could be an indicator of the intersubjective state between them.
- Differences in results found between the corrugator supercilii and zygomaticus major might be explained by referring to the chameleon effect theory.

8.4.4 Limitations and Future Research

The main limitations in this study are the influence of situational factors and the problem of achieving simplicity in design. The use of a mini-bridge card game was designed to diminish the influence of external factors. However, it was insufficient to completely eliminate interactions between members of the game (situational factors), which then notably influenced the results on readings coming from the zygomaticus major. During this study, we tried to diminish external factors such as the level of closeness of two examinees before the experience, though it does seem that this is a factor influencing the results of fEMG analysis, and it was observed that the frequency and value of peaks in a wave are stronger in pairs who knew each other beforehand. This could be something to develop in future research with confirmations of the validity of the method.

8.5 Conclusion

Through this study, the authors tried to find a correlation between the intersubjective state of two persons in a collaborative environment and the results of fEMG analysis. The experiment focused on a cooperative card game that emphasized interactions between members while diminishing the influence of external factors. Following the experiment, the Levenshtein method was proposed to help determine communicative behavior in the facial muscles. By using this method on the corrugator supercilii, the authors were able to find a correlation between the objective and subjective data. The Levenshtein distance between the facial muscle activities of two persons in a collaborative environment, computed by this method, could be an indicator of intersubjectivity.

The results fall within the purview of the chameleon effect and will help advance research on intersubjectivity in collaborative social environments from the viewpoint of cognitive science. This study should contribute to a greater understanding of the mechanisms of social relationships, which is needed to develop a methodology for social design.

Acknowledgements This work was made possible by the help of Julien Medici in writing a program to compute the Levenshtein distance. We would also like to thank Hirasawa Hikari and Shouji Keisuke for their participation in the evaluation of each play. This work was supported by JSPS KAKENHI Grant Number JP20K20119.

References

1. Cash, P., Dekoninck, E.A., Ahmed-kristensen, S.: Supporting the development of shared understanding in distributed design teams. J. Eng. Des. **28**, 147–170 (2017). https://doi.org/10.1080/09544828.2016.1274719
2. Kleinsmann, M., Deken, F., Dong, A., Lauche, K.: Development of design collaboration skills. J. Eng. Des. **23**, 485–506 (2012). https://doi.org/10.1080/09544828.2011.619499
3. Publications S: Sage Publications, Ltd. 31:71–85 (2014)
4. Matsumae, A., Matsumae, S., Nagai, Y.: Dynamic relationship design of knowledge co-creating cluster: traditional Japanese architectural industry. SN Appl. Sci. **2**, 1–11 (2020). https://doi.org/10.1007/s42452-020-2209-2
5. Matsumae, A., Burrow, K.: Serviceology for Designing the Future. Serv. Des. Futur. 551–565 (2016). https://doi.org/10.1007/978-4-431-55861-3
6. Matsumae, A., Nagai, Y.: The function of co-creation in dynamic mechanism of intersubjectivity formation among individuals. Proc. Int. Des. Conf. Des. Inn, 1925–1936 (2018). https://doi.org/10.21278/idc.2018.0141
7. Zahavi, D.: Husserl and Transcendental Intersubjectivity. Ohio University Press, Ohio (2001)
8. Bower, M.: Developing open intersubjectivity: On the interpersonal shaping of experience. Phenomenol. Cogn. Sci. **14**, 455–474 (2014). https://doi.org/10.1007/s11097-014-9346-2
9. Linell, P.: Rethinking Language, Mind, and World Dialogically. Information Age Publishing, Incorporated (2009)
10. di Pellegrino, G., Fadiga, L., Fogassi, L., et al.: Understanding motor events: a neurophysiological study. Exp. Brain Res. **91**, 176–180 (1992). https://doi.org/10.1007/BF00230027
11. Stone, L.D., Underwood, C., Hotchkiss, J.: The relational habitus: Intersubjective processes in learning settings. Hum. Dev. **55**, 65–91 (2012). https://doi.org/10.1159/000337150
12. Mundy, P., Kasari, C., Sigman, M.: Nonverbal communication, affective sharing, and intersubjectivity. Infant. Behav. Dev. **15**, 377–381 (1992). https://doi.org/10.1016/0163-6383(92)80006-G
13. Dimberg, U.: For Distinguished early career contribution to psychophysiology: award address, 1988 facial electromyography and emotional reactions. Psychophysiology **27**, 481–494 (1990)
14. Larsen, J.T., Norris, C.J., Cacioppo, J.T.: Effects of positive and negative affect on electromyographic activity over zygomaticus major and corrugator supercilii. Psychophysiology **40**, 776–785 (2003). https://doi.org/10.1111/1469-8986.00078
15. Wolf, K., Mass, R., Ingenbleek, T., et al.: The facial pattern of disgust, appetence, excited joy and relaxed joy: an improved facial EMG study. Scand. J. Psychol. **46**, 403–409 (2005). https://doi.org/10.1111/j.1467-9450.2005.00471.x
16. Hess, U., Banse, R., Kappas, A.: The intensity of facial expression is determined by underlying affective state and social situation. J. Pers. Soc. Psychol. **69**, 280–288 (1995). https://doi.org/10.1037/0022-3514.69.2.280
17. Chartrand, T.L., Bargh, J.A.: The chameleon effect: the perception-behavior link and social interaction. J. Pers. Soc. Psychol. **76**, 893–910 (1999). https://doi.org/10.1037/0022-3514.76.6.893

Chapter 9
Design Using Storytelling and Play to Give Psychological Support to Children and Caregivers Battling Cancer and Other Chronic Illness

Aditi Sit, Aditi Singh, and Parag Anand

Abstract The news of a child being diagnosed with a chronic illness such as cancer is often unexpected and can seem devastating. Many families experience a sense of powerlessness in the beginning at the prospect of dealing with their child's illness. Research indicates that it can be extremely beneficial for the family, including the child, to try and understand as much as they can about the condition and its care. This helps build trust, reduce confusion and anxiety as well as help them feel included in medical exchanges. The child is also more likely to cooperate with the tests and treatments. Children who do not know about their ailment use their thoughts to fill in the knowledge gaps and may start to think that they did something wrong or are at fault. Taking a positive and encouraging approach to educate the child on this matter boosts morale and empowers the child instead of scaring and stressing them. This is exhibited via design of a demonstration kit named "Can Care Pals" that uses storytelling and play to help ease the information transfer for the children and their parents, specifically for children undergoing cancer treatment. The matter is demonstrated through an inspiring story, narrated using smart and interactive dolls, which represent strong and positive characters for treatment, immunity and cancer. "Can Care Pals' also educates and guides the parents or caregiver through the entire process with the help of a supported app. Aided with this app, the dolls are capable of comforting, encouraging, setting a routine, dealing with symptoms of the therapy and recording the child's pain." Can Care Pals' aims at creating a safe, playful and encouraging environment for children fighting cancer.

A. Sit (✉) · A. Singh · P. Anand
Department of Industrial Design, School of Planning and Architecture, New Delhi, India
e-mail: aditisit4@gmail.com

A. Singh
e-mail: aditisingh@spa.ac.in

P. Anand
e-mail: paraganand@spa.ac.in

A. Chakrabarti et al. (eds.), *Design for Tomorrow—Volume 1*, Smart Innovation, Systems and Technologies 221, https://doi.org/10.1007/978-981-16-0041-8_9

9.1 Background

With the advancement in health care, the medical attention and the number of survivors have both increased. The physical effects of a chronic illness are the most obvious aspects to attend to. Health care and medicine is able to manage it by either controlling the condition or retarding the progression [1]. Unfortunately, what gets overlooked or is often taken lightly are the psychological aspects. Childhood chronic illness leads to high risk of psychological and behavioral problems. The increase in life expectancy makes it even more necessary to investigate the psychosocial perspectives of children as they progress as one study by Wilson et al. [2] indicates. So, there needs to be a shift in the focus of health care in enhancing the quality of life of the survivors. To do this, it becomes essential to improve the psychological status of the child while the child is fighting the chronic illness. Attaining a positive and strong mindset in the fight would help strengthen the child's mental and physical state.

The family of a child is an important source of either risk or protection to the child. Also, there are consequent effects of the chronic illness on the functioning of the family. Alderfer et al. [3] studied about the possibility of risks in the manner of psychiatric disorders and social maladjustments on the siblings as well. It is central for the entire family to stay positive and strong which would also motivate and comfort the child. If the child has a safe, protected and positive environment while seeking treatment and coping with the side effects of the disease and procedure, there are higher odds of a smooth transition into adulthood.

9.2 Project Mission

9.2.1 Problem Identification

There are 300,000 childhood cancer cases annually around the world as per World Health Organization [4]. Every day, 43 children are diagnosed with cancer and 12% of them do not survive. Over the past 50 years, there has been a major advancement of the overall survival rate to increase from 10 to almost 90% today [5]. The types of cancer observed in children are very different from what is seen in adults. Some of which are leukemia, brain and spinal cord tumor, neuroblastoma, Wilms tumor, lymphoma, etc. The reason for childhood cancer is unknown, though a small percentage is due to genetic abnormalities. The treatments that the child undergoes are the following or a combination of the following depending on the type and stage of cancer. The treatment could be chemotherapy, radiology, surgery, immunotherapy, and targeted therapy.

Communication of prognosis to children has changed over the past 60 years. Sisk et al. [6] found that traditionally, back in the 1950s, children would be shielded and protected from the bad news. Over the next decade with accumulation of objective evidence, there was a shift to a more open and direct way of communication. There has been persistence in this approach till now. Cooper & Blitz found that with the stress of a severe illness, feelings of anxiety and depression begins which may lead to decreased cooperation, aggression and disruptive behavior. Landreth studied that it becomes extremely crucial to regain their sense of control and freedom. This may be regained by introducing the children to their condition and their forthcoming path to recovery.

As per a research carried out by Children's hospital of Philadelphia, it was concluded that the availability of supportive care tools tailored to the experience of childhood cancer and its treatment is very limited [7]. Sposito et al. attempted to understand the coping strategies used by children of the age 7–12 years with cancer undergoing chemotherapy during hospitalization. They conducted an exploratory study and found that some of the coping methods children use were: understanding the need for chemotherapy, finding relief for the chemotherapy's side effects and pain and engaging in entertaining activities [8]. It was understood that a child who is diagnosed with cancer would cope better by understanding the disease, the need for chemotherapy or other medical procedure.

It is only until the age of 5 and above, that the child starts demonstrating logic, concrete reasoning and organized thinking as per Paiget's stages of cognitive development [9]. This is a critical age for the child to be made aware, so as to be able to cooperate with treatment and also to know what to expect [10]. Children at this age often have many questions. Through this paper's research with pediatric oncologist and healthcare team members from reputed organizations such as AIIMS, New Delhi, and CanKids KidsCan, it was substantiated that it is vital for children at this age to understand and be informed about their cancer and each step of the procedure so that they feel comfortable and involved. This would also allow them to express themselves and feel reassured. Based on a study to examine young people's and parents' accounts of communication about cancer in childhood, Young et al. found that the parents both facilitated and constrained communication. The children used their parents to manage the burden of communication, but also said they tended to feel marginalized during doctor–parent–patient encounters. Through this, it was verified that the amount of knowledge and the knowledge transfer to the child is critical for them to feel in control of their condition [11].

It is natural for a parent to protect their child and not want to disclose the details about the chronic illness to them. They find the task even more complex because as parents it is in their nature to create an identity of safety and guarantee an optimistic reality [11]. However, a parent must prepare the child for their diagnosis and the consequent procedures as it helps them feel safe and included. When a child first hears about the disease, they may think that something they did wrong may have caused it. If the child does not know what to expect, it may cause them anxiety, confusion, stress and fear. It is very important to let them know that they are not at fault. It is necessary to encourage the child to talk about it, openly share their feelings, ask questions and clear any doubts from their mind. The parents need to be open and honest to keep them safe. Being aware and knowledgeable also helps the child to cooperate with the procedures and hence aid in their healing.

Explaining such a complex disease and procedure to a child is an overwhelming duty. It is required to simplify the matter in terms that the child can understand. As parents know their child the best, they know the mannerism to do so for their child. Rendering a positive way of explaining these terms and procedure to the child will boost and encourage them. A review by Rosland and Piette affirmed that in such a situation the family members or caregivers would feel less burdened if they had the right tools, skill and resources to be more effective in helping their child [12].

9.2.2 Existing Solutions

Landreth acknowledged that hospitals have become increasingly aware of the psychosocial, educational and recreational needs of pediatric patients [13]. Through play, children are able to express what they feel, confront their fears, play out reality and feel a sense of mastery of the environment and most importantly themselves [14]. For children with acute illness, the feeling of acceptance and freedom is most significant. Hospitals that include child-centered play therapy create opportunity for children to have control during medical treatment. Through play, children are able to enact the procedures that they fear and can tend to feel familiar or accustomed to the process in turn calming them down during the actual medical course. In a study, Garrot, Caltworthy and Daniel et al. observed that children who underwent play therapy using medical toys, dolls, puppets and art supplies were significantly less anxious during the medical procedure [15–17].

Currently, there are several reading resources for caregivers to help explain cancer to children. Hospitals and organizations have child life specialists who are educated and clinically trained in the developmental impact of illness and injury. They help improve the overall experience and care for the patients and their families [18]. They use various techniques such as books, help create scripts for parents and do science activities to help explain different aspects of the child's cancer [19].

There have been products that help comfort children such as the "Cellie Cancer Coping Kit" developed by behavioral researchers Meghan from the Children's Hospital of Philadelphia and Kentucky Children's Hospital [20]. It helps children and their families cope from stressful situations during the cancer treatment, recovery and medical condition, by providing tips from current research, families and children and oncologists. The Cellie Coping Kit is designed for ages 6–12, which includes a stuffed toy used for engagement, Coping Cards and Caregiver Book [21].

Another product that comforts and entertains children fighting cancer is an innovative social robot named "My Special Aflac Duck" which is developed by Sproutel and is a part of the Aflac Childhood Cancer Campaign [22]. It engages with the child by medical play, nurturing play—feeding and bathing, expressing emotions, soundscapes—music and calming sounds, lifelike motion—singing, nuzzling, heartbeats and breathing and promoting connection by recognizing friends [23].

In a study by Gwilt et al., it was found that young people were better able to convey their pain through visual representations of their own. A workshop was held where young individuals were instructed to use creative techniques to describe their pain experiences. It gave them opportunity to explore and express their unique experience of pain through familiar and contemporary visual language. Study revealed that a representation of their progress was likely to create greater motivation toward self-management and facilitate enhanced pathways to recovery [24].

9.2.3 Design Approach

On studying the above-mentioned solutions, it was determined that there are not too many products and systems that look into the area of explaining the disease to children, the procedure at the initial stages of diagnosis and providing aid for caregivers and parents to do the same. This paper aims at advancing methods, systems or products to support caregivers to explain about the child's condition and the treatment or procedure to them. This will in turn help reduce fear and bring in positivity and motivation to the child fighting against the chronic illness. For ease of research and deriving at a solution, cancer was selected as the chronic disease to solve for, but the design can be applicable to other chronic illnesses specific to their procedures, treatments and their physical and emotional consequences.

9.3 Design Procedure

Story is a powerful approach to convey meaning to others. It is a form of expressing oneself, and there is room for finding meaning, interacting and asking questions for the interpreter. It builds a comfortable space to grasp and create an understanding of the matter. It was considered that storytelling will be a fitting way to address this delicate issue. Using this approach with characters relatable to a child would help dilute the matter, to understand, as well as ease the process of explaining for the caregivers in a safe environment.

The most essential aspect for a child to understand is the importance and role of the treatments such as chemotherapy in fighting cancer and the symptoms that follow. This way, the child understands why they have to go through the procedures and are prepared for the symptoms that follow. Chemotherapy retards or stops the growth of cancer cells which are fast dividing cells. It also harms healthy cells that divide quickly such as white blood cells, cells that line the mouth and intestines or cause hair to grow [25]. When the healthy cells die such as the white blood cells, immunity reduces and the body becomes susceptible to symptoms such as infections, fatigue, pain, fever, anemia, nausea [26]. The patient feels these side effects as a consequence of chemotherapy fighting cancer. It becomes extremely important to help strengthen our immunity by taking care of the body, through coping methods and tracking of symptoms, as advised by the cancer care team.

To help a child comprehend the importance and consequences of the procedure, it was decided to use the medium of demonstration, by developing appropriate characters representing each of the facets of the illness. These characters are in the form of cancer, chemotherapy and immunity. The characters chosen are motivational, inspirational and relevant to the child. An angel or a superhero was chosen to represent chemotherapy, depicting the protector and powerful nature of the treatment. A monster which seems unwanted but not frightening depicts the cancer cells, so that the child is not afraid of the disease but understands that it is unwelcome. As the body's immunity is a fighter but becomes helpless due to cancer cells, it has been shown as an armored soldier, a warrior.

These characters can be understood and are relevant for children of any gender, ethnicity, background and age above 5. Through these strong characters, there is flexibility for different aspects of the story to be adapted based on type of cancer, phase of treatment, treatment type and intensity, etc. Parents generally can gauge how much their children can understand and how involved they can be and want to be in conversations about their health [27]. The story telling method along with the characters give the parents and caregivers the freedom to narrate.

Taking the concept forward of a demonstrational kit for children and their parents, the characters were represented in a tangible and toy-like manner, making it more personal, playful and handy. Cancer brings limitations to the life of the child that can hinder their development. A study by da Silva and Cabral emphasized that addition to promoting children's development, playing is also a source of enjoyment for the child, it is when they, for a moment, forgets the difficulties of the illness and treatment and what they are going through [28].

On exploring the form and materials to use, professionals from cancer care organizations advised that as the child's immunity is low they are sensitive to synthetic material and that the form should be such that it does not allow dust to trap in. Natural material such as rubber, pine or teak wood was selected. For the colors, vegetable dye was selected to ensure that the dolls are safe for children. This was inspired from the traditionally crafted toys of Chanapatna, Karnataka [29].

To bring out compassion and emotions while making the experience engaging, the dolls have been integrated with technology, to enhance the storytelling. By incorporating proximity sensors and OLED screens, the dolls can display emotions. This improves its capabilities and gives it more flexibility to adapt to every child's unique fight. This demonstration kit has been named "Can Care Pals" for its caring and friendly characters which also acts as companions for the child (See Fig. 9.1).

9.4 Working of Design

Each character is represented as a smart doll, which reacts based on proximity with the other smart dolls by expressing emotions. It is operated with an aided app which is used by the parents. It plays an integral role in educating and guiding them through this process. It connects with the smart dolls and helps control their behavior, and it

Fig. 9.1 Can Care Pals Demonstration Kit- Smart dolls, Docking Station and App

also assists the caregivers on how to narrate the story with the smart doll. The "Can Care Pals" doll along with the app aids in three main capacities (Refer Fig. 9.2).

One being the initial intent of helping the child understand their condition and the need for chemotherapy, building trust and being comfortable with their caregivers. Through the interaction between the characters there is visual and audio feedback.

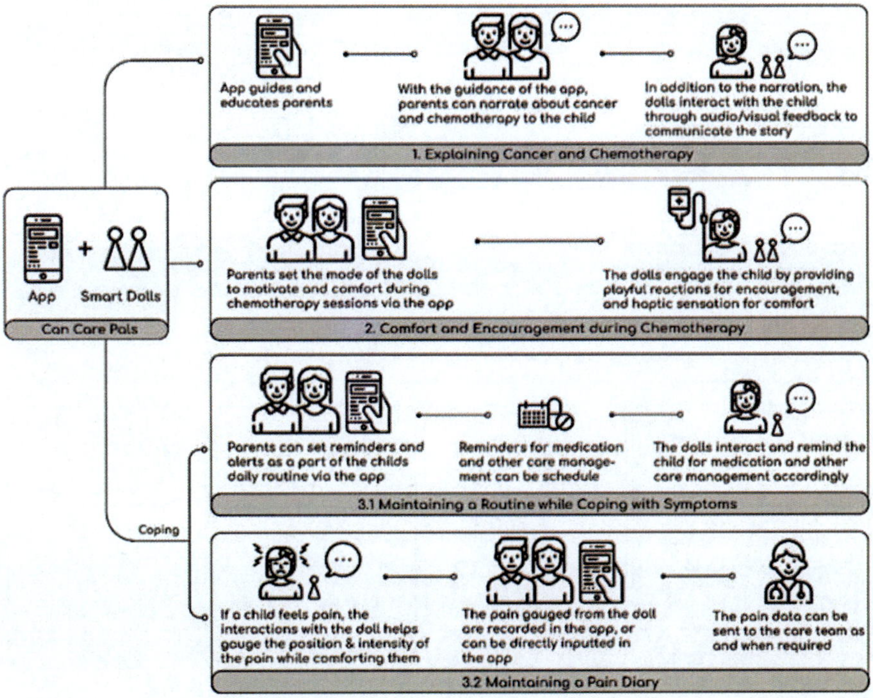

Fig. 9.2 Can Care Pals Demonstration Kit User Experience Process

The colors on the chest of the dolls and horns of the cancer cell along with the changing of expression on the doll's OLED display screen and the audio feedback help explain the process in a more storytelling manner. The parent with the help of the app is able to narrate along with the doll's reactions (See Fig. 9.3).

Secondly, the dolls act as a companion during the chemotherapy sessions. Chemotherapy session are long and stressful. The dolls ease the experience by engaging and encouraging them through play (Refer Fig. 9.4). With completion of every chemotherapy session, the child is a step closer to their improvement. This is conveyed through the smart doll's ability to upgrade by attaching add-ons, which would help boost the child's confidence. Thirdly, it helps while the child is coping with the side effects of chemotherapy. After chemotherapy sessions, the patient undergoes various symptoms such as pain, prone to infections, weakness, and infections. To

Fig. 9.3 Explaining cancer and chemotherapy through story telling using the Can Care Pals

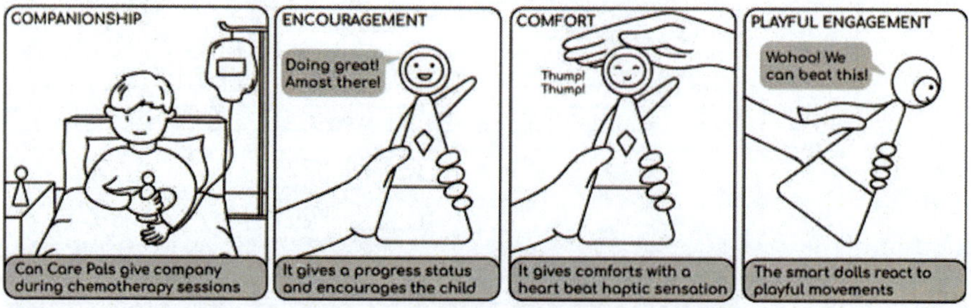

Fig. 9.4 Can Care Pals comforts the child during chemotherapy procedure

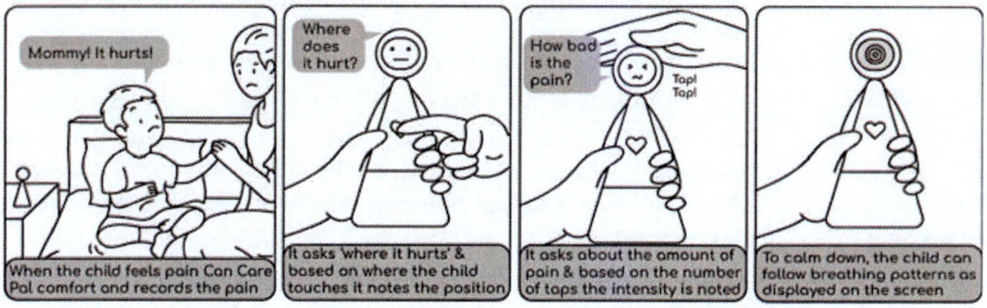

Fig. 9.5 Can Care Pals Kit comforts the child during pain and records the pain

combat these, it is very important for the patient to maintain a proper routine which includes exercising, following a diet, keeping clean and sanitized, sleeping enough and maintaining a record of the pain. This tool includes self-management and medication as a part of the child's daily activities as a way to integrate the illness care into family life and to decrease emphasis on the illness itself. Knafl et al. states that this would have a more positive impact on the family and child [30, 31]. Can Care Pals sets a specific routine for the child and makes the otherwise monotonous and tedious routine into a more interactive and engaging one. The symptoms and side effects of the treatment and medication is tracked, recorded and notified to the cancer care team via the app for faster diagnosis and prevention of potential issues and identification for early intervention. For the child or the caregiver to remember the time, day, position and intensity of pain until the next doctor visit is difficult. This tool helps maintain the pain diary in the app, by recording these parameter of the pain through interactions made with the smart dolls (See Fig. 9.5).

These smart dolls need to be charged for which a docking station has also been designed. It consists of a charging station for the three smart dolls, and it contains a storage space on the top of the dock for placing the child's medicines. Along with its utility as a docking station, it also acts as bedside night lamp. The simplicity in the form and its friendly nature fits in any child's bedroom.

Can Care Pals is designed to emotionally and psychologically strengthen the child and caregivers. The responsive and engaging qualities support play and development of children, while being mindful of the limitations of the ailment.

9.5 Discussion

Through this study, it was substantiated that communication on the condition and the need for therapy is one of the coping strategies for children fighting cancer. Being aware and notified, they feel more in control, reduces anxiety and makes them not feel sidelined. This communication is complex for the parents, especially when overwhelmed with so many emotions. Play gives children a platform to express their feelings, confront their fears and act out different scenarios. For children with chronic illness, being able to access these emotions becomes extremely vital.

Can Care Pals uses storytelling and role-play to portray the need for chemotherapy with motivational characters, and it also acts as a companion during chemotherapy sessions. It seeks to empower and bring confidence for self-management by integrating cancer care into their routine. The theme gives significance to long established methods of play therapy through tangible toys, but integrates technological abilities to express greater emotions, empathy and natural representation. It reflects the tendency for parents to project a more positive reality and hence gives them a tool to narrate the condition in a more courageous and inspirational rendition of the truth. These tense circumstances of the caregivers have also been considered by better educating and supporting them through the convenience of an app.

Previous studies have shown that there is a lack of tools to aid children and their parents to grasp and cope with the condition. Currently, not enough of the existing methods of play therapy and coping strategies use technology as support tools. Can Care Pals leverages the technological advancement and the popularity of mobile applications to create a more interactive and responsive coping experience.

There are some limitations to this study that should be acknowledged. The product currently includes a pain diary on the app which is captured by interactions made through the dolls. Additional research on chronic pain and pain management needs to be done to strengthen this portion of the tool. Similarly, symptom tracking and care management is currently in a conceptual stage and requires more study.

Further research is needed to explore more on the technology of the dolls. Algorithms can be implemented within the software to learn the child's psychological behavior, thereby delivering a customized and personal form of coping. Frank et al. recognized social support and skills as a viable area to target while considering children and their parent's ability to cope [32]. Additional research in this domain is required to expand the tool's scope. Although the kit was deliberated with medical professionals, who showed an inclination to the overall concept, the tool needs to be tested by children and caregivers in the accurate context and setting, to get a better understanding of the product. This will lead to possible improvements and opportunities to positively enhance the child's behavior toward cancer care.

9.6 Conclusion

A child with cancer needs physical, mental and emotional support. Simply putting them on a treatment is not enough. Childhood chronic illness leads to high risk of psychological and behavioral problems. There is now a need for health care to focus on enhancing the quality of life of the children fighting chronic illnesses.

There are limited care tools tailored to the experience of childhood cancer and its treatment, especially in the area of communicating the disease, treatment and its consequences. Children at their concrete operational stage of the Paiget's stages of cognitive development are able to grasp and comprehend. It becomes extremely important for them to understand what is happening, to avoid misconceptions, and help them cooperate with the approaching procedures, while feeling comfortable and safe with their parents and health care team. This project took an approach of storytelling and role-play to help ease the information transfer between the caregivers and child, by using smart responsive dolls that represent strong and positive characters. The tool assists the parents on how to deal with various situations that the child is going through. Aided with the app the dolls are capable of comforting, encouraging, planning a routine, tracking symptoms and recording the child's pain.

With storytelling through play, "Can Care Pals" aims to create a protective and positive environment for children fighting cancer. By aiding them to understand and cope with the treatment and side effects, this will eventually provide them with the scaffolding to move through the treatment phase of their young lives.

References

1. Goh, D.L.M.: The Impact of Chronic Illness on a Child. Child Research Net (1999)
2. Wilson, D.K., Hutson, S.P., Wyatt, T.H.: Exploring the role of digital storytelling in pediatric oncology patients' perspectives regarding diagnosis: a literature review. SAGE Open 1–10, (2015)
3. Alderfer, M.A., Stanley, C., Conroy, R., Long, K.A., Fairclough, D.L., Kazak, A.E., Noll, R.B.: The social functioning of siblings of children with cancer: a multi-informant investigation. J. Pediatr. Psychol. **40**(3), 309–319 (2014)
4. Cancer in Children. 28 Sept 2018 [Online]. Available: https://www.who.int/news-room/fact-sheets/detail/cancer-in-children
5. Childhood Cancer Statistics [Online]. Available: https://curesearch.org/Childhood-Cancer-Statistics
6. Sisk, B.A., Bluebond-Langner, M., Wiener, L., Mack, J., Wolfe, J.: Prognostic disclosures to children. Am. Acad. Pediatr. **138**(3), (2016)
7. Cellie Cancer Coping Kit Research [Online]. Available: https://www.chop.edu/centers-programs/cancer-center/cellie-cancer-coping-kit-research
8. Sposito, A.M.P., Silva-Rodrigues, F.M., Sparapani, V.D.C., Pfeifer, L.I., da Lima, R.A.G., Nascimento, L.C.: coping strategies used by hospitalized children with cancer undergoing chemotherapy. J. Nurs. Sch. **47**(2), 31 (2015)
9. Marcin, A.: What are Piaget's stages of development and how are they used? 29 Mar 2018 [Online]. Available: https://www.healthline.com/health/piaget-stages-of-development
10. NC Institute: Children with Cancer: A Guide for Parents. NIH Publication (2015)
11. Young, B., Dixon-Woods, M., Windridge, K.C., Heney, D.: Managing communication with young people who have a potentially life threatening chronic illness: qualitative study of patients and parents. BMJ (2003)
12. Rosland, A.-M., Piette, J.D.: Emerging models for mobilizing family support for chronic disease management: a structured review. SAGE J. **6**(1), 7–21 (2010)
13. Landreth, G.L.: Innovations in Play Therapy. Brunner-Routledge, New York (2001)
14. Landreth, G.L.: Play Therapy: The Art of the Relationship. Accelerated Development, Muncie (1991)
15. Pa, G.: Therapeutic play: work of both child and nurse. J. Pediatr. Nurs. **1**(2), 111–116 (1986)
16. Clatworthy, S.: Therapeutic play: effects on hospitalized children. Child. Health Care: J. Assoc. Care Child. Health **9**(4), 108–113 (1981)
17. Rae, W.A., Worchel, F.F., Upchurch, J., Sanner, J.H., Daniel, C.A.: Article Navigation. J. Pediatr. Psychol. **14**(4), 617–627 (1989)

18. What is a Certified Child Life Specialist? [Online]. Available: https://www.childlife.org/the-child-life-profession
19. AC Center: How to explain a Child's Cancer Diagnosis. (2017)
20. The Cellie Cancer Coping Kit [Online]. Available: https://www.chop.edu/health-resources/cellie-cancer-coping-kit.
21. Cellie Coping Kit [Online]. Available: https://www.healthcaretoolbox.org/tools-and-resources/cellie-coping-kit.html
22. Play with Purpose [Online]. Available: https://www.sproutel.com/products.html
23. Why: Every Child Deserves Comfort [Online]. Available: https://aflacchildhoodcancer.org/learn.cfm
24. Gwilt, I, Langley, J., Partridge, R., Davis, S., Woodward, L., Spooner, A.: Pain talking: exploring the experience, expression and description of chronic pain through creative processes and visualisation strategies. In: Cumulus Hong Kong 2016. Hong Kong (2016)
25. Chemotherapy [Online]. Available: https://www.medicinenet.com/chemotherapy/article.htm#chemotherapy_facts
26. Chemotherapy in Children [Online]. Available: https://curesearch.org/Chemotherapy-in-Children
27. Talking with Your Child about Cancer [Online]. Available: https://www.seattlecca.org/diseases/non-hodgkins-lymphoma/talking-your-child-about-cancer
28. da Silva, L.F., Cabral, I.E.: Rescuing the pleasure of playing of child with cancer in a hospital setting. Revista brasileira de enfermagem (Brazilian Journal of Nursing) 3(68), (2015)
29. Unique symbols of Karnataka [Online]. Available: https://www.thehindu.com/todays-paper/tp-national/tp-karnataka/unique-symbols-of-karnataka/article3066635.ece
30. Knafl, K.A., Deatrick, J.A., Havill, N.L.: Continued Development of the family management style framework. SAGE J. **18**(1), 11–34 (2012)
31. Beacham, B.L., Deatrick, J.A.: Children with chronic conditions: perspectives on condition management. J. Pediatr. Nurs. **30**(1), 25–35 (2015)
32. Frank, A., Blount, R.L., Brown, R.T.: Attribution, coping, and adjustment in children with cancer. J. Pediatr. Psychol. **22**(4), 563–576 (1997)

Chapter 10
Mental Imagery for Multisensory Designers: Insights for Non-visual Design Cognition

Ingrid Monteiro Maciel, Guilherme Felicio, Edmar Thomaz da Silva, Emília Villani, Petter Krus, and Luciana Pereira

There's plenty of imagery that goes on all the time in blind people. It just isn't visual.
—Paul Gabias, Associate Professor, University of British Columbia, early blind.

Abstract How do people with visual impairment see the world? In this literature review based on cognitive sciences findings, we have analyzed the main concepts used in the human brain's cognitive processes to represent our perception of the surrounding environment. One of these concepts is mental imagery, which resembles perceptual experience without external sensory stimulation. This concept plays a central role in multisensory design cognition. It can help us understand the designer's cognition process, design better systems for people with disabilities, and open opportunities for multisensory design teams.

10.1 Introduction

Designing a new system is a time-consuming and expensive activity. The design of products used in everyday life, such as consumer goods, is an engineering challenge per se. However, the requirements posed by universal design create new complexities

I. M. Maciel · L. Pereira (✉)
Federal University of ABC, Santo André, Brazil
e-mail: luciana.pereira@ufabc.edu.br

G. Felicio
EE Professora Inah de Mello, Santo André, Brazil

E. T. da Silva · E. Villani
Aeronautics Insitute of Technologic, São José dos Campos, Brazil

P. Krus
Linköping University, Linköping, Sweden

A. Chakrabarti et al. (eds.), *Design for Tomorrow—Volume 1*, Smart Innovation, Systems and Technologies 221, https://doi.org/10.1007/978-981-16-0041-8_10

[1]. In addition to dealing with the routines observed in the usual design process, it also must match the demands of people with disabilities [2].

Universal or inclusive design has gained widespread attention as part of the Agenda for Sustainable Development Goals. The United Nations have worked with member countries to promote the Convention on the Rights of Persons with Disabilities. Thus, it is essential to fully understand people with disabilities' demands before attempting to design systems for them.

Considering this matter, we have asked if there are ways where people with visual impairment could have a more proactive design process role. How could people with visual impairment participate in an activity that has been so sighted-centered, where designers "learn to think with drawings" [5]? Designing devices for people with visual impairment is a complicated task for the most innovative designer [6, 7]. At the same time, it can be inspiring since different angles reveal different worlds. Therefore, this chapter aims to deliver a systematic literature review relating it to existing knowledge in cognitive-sensory processes. We aim to understand how people with visual impairment perceive the world around them. This knowledge is critical to define a better design domain. But how?

Our point of departure is the assumption that design is a purpose-driven activity with a limited number of options to act on [8]. Nevertheless, the search for feasible solutions allows learning as well as finding potential alternatives. Said otherwise, designers are part of the design space where their perception of the design problem's circumstances is central to characterize the design requirements' parameters. The other assumption is that design is a depiction of the functional attributes, which transform ideas into design features. In other words, design requirements encode messages in the format of words, numbers, and sketches/graphics [8].

We need to know more about the sensory substitution in non-visual sensory domains [9]. It is an effort to reach out to design requirements based on understanding the person that lack the sense of vision own situation awareness, which only exists in their minds. This knowledge helps describe the design cognition process of a designer with visual impairment. For shedding some light on it, we present a review of the literature's interdisciplinary body to explore how the world's perception happens with and without senses.

We organized the remaining of this paper as follows. In Sect. 2, we discussed the methodological strategy. Finally, in Sects. 3 and 4, we present our findings and conclusion.

10.2 Research Strategy

The research on visual impairment is quite varied in terms of the researcher's profile, goals, and scientific rigor [10, 11]. In general, the literature describes attempts to:

- solve immediate problems with applied solutions,

- advance the understanding regarding the perception of people with visual impaired,
- the participation of people with visual impairment to test hypotheses about vision.

Therefore, the first step would be to define blindness and low vision, which together form a heterogeneous group of disabilities. Blindness and low vision vary in etiology, visual acuity, and the visual field [10]. The visual impairment level differs from no perception of light at all to a slightly blurred vision, with all gradations between the two [12]. The blind may have light perception when they can tell if they are in a dark or illuminated room, they can locate the source of the light, and the field of view can vary in length or include blind spots.

The term legally blind corresponds to blindness criteria and on low acuity or visual field restriction [12]. In the United States, the legal definition of legal blindness is a visual acuity of 20/20 or less or 20° or less field of view.

The term "low vision" describes individuals with severe visual impairment but still have some useful vision [12]. The word "blind" is sometimes used in a restrictive sense to refer to people who, at most, have a perception of light. "Visually impaired" then refers to every person with vision loss that cannot be corrected by usual means, i.e., glasses [10].

The age of onset of blindness is essential information, as it can affect a person's perception of the representation of space and attitude toward disability. A person who has been blind since birth is said to be congenitally blind or premature blind [12]. The term encompasses those who lost their sight in the early years of their life. Acquired blindness can happen later in life.

Despite the continuous progress of medicine and science, people with visual impairment are likely to keep growing. According to the World Health Organization [13], this is due to the aging population.

People with visual impairment and sighted people perceive the world in a different manner [12]. We can only perceive some aspects of the building environment through vision. However, we also can access specific properties of the environment through touch or hearing. It is then natural to ask how visual deprivation affects people's perceptual and cognitive skills.

10.2.1 Managing the Literature Review

This study is part of a larger interdisciplinary project that aims to understand how visually impaired people can have an active role during the design process. Some of the researchers in the group are visually impaired. More than respect, the rights of people with disabilities, "Nothing about us without us" recognizes that diversity drives knowledge advancement.

For this research, we have searched academic databases such as Springer, ScienceDirect, Google Scholar, and the Web sites of the journals relevant in the field to identify the articles, books, and proceedings that fit our research. The selection procedure was determined based on predefined criteria. We started with the question: "how do visually impaired people represent mental image?" After reading the papers, we took notes of the main concepts that shed light on cognitive skills and the sensory system's importance in this process. We then organized a table pointing: authors, type of publication, year, keywords, central concepts, and definitions.

As reported in the references, these papers are mainly part of the field of cognitive sciences, neurosciences, and psychology. However, their findings significantly contribute to innovative applications in engineering, design, and public policy. Thus, there are no biases in the journal search. Even though we are not part of the community of the papers that we read, they are internationally recognized peer-reviewed publications.

10.3 Design in the Eye of the Beholder

Design education emphasizes the idea of putting designers "on the skin" of users [14]. In this way, designers can design from the end-user's perspective to perform a task that works best for the user. Sometimes, in designing some systems, designers and end-users are similar in their skills. It may happen that the activities that the end-user will have to perform are not different from those of the designer. The designer must still consider the user's preferences.

In other cases, the designer and the eventual user are not the same, so naturally, there is a gap in understanding how designers will display a task and how a user will do the same. When designing accessibility systems, the relationship between the task, user, and designer is usually quite different.

It is much more difficult for a designer to imagine himself/herself in the place of a user. That user may have some deficiency that designers can only imagine. A designer without a disability can have a descriptive understanding of the needs of a user with disability. Much more difficult is a designer without a disability to understand the implications of living with these everyday needs [15]. The designer must somehow know about the cognition of users and how they will interact with the system. In this way, the designer can be sure that users will perform a task successfully.

10.3.1 Design with and Without Senses

One of our senses' primary goals is to inform us about the properties of the environment surrounding us, an essential skill for our survival [16]. According to Sternberg and Sternberg [17], perception is a set of processes that recognize, organize, and understand the sensations arising from environmental stimuli. The perceived object is a mental representation of a received stimulus.

Perceptual illusions indicate that what we feel through the senses is not necessarily what the mind perceives. Gibson [18] has provided a framework for understanding perception. The sensation and perception processes are part of a perceptual continuum where distal objects (objects in the environment) offer the structure for the information medium (sound waves, chemical molecules, tactile information, or reflected light). Then, it reaches the sensory receptors to finally reach the object's internal identification (sharp object).

The definition of multisensory design says it is the one that affects the five senses: visual, auditory, gustatory, tactile, and olfactory. However, this notion covers only the most basic concept of multisensory design. There are three other systemic concepts related to the perception that goes into the multisensory design: haptic, kinesthetic, and synesthesia. In essence, these systems merge with the basic five senses, resulting in an active and inclusive system that directly affects users' experience.

The modality that provides more detailed and reliable information guides our sensations about the environment [19]. Therefore, the lack of a sensory modality, such as the vision, results in increased use of the remaining intact sensory systems like tact and auditory senses. The sensory compensation hypothesis says that the remaining senses of the person with visual impairment become more active to compensate for vision loss [20].

10.3.2 Decoding Mental Imagery

Designers play a crucial role in converting new knowledge into systems designed for the real world. Thus, the project's realization occurs through multidisciplinary teams, both for technical projects and those less technical [21].

Within this perspective, the newest modality of team collaboration is related to multisensory design. At first, the multisensory design appeared in the marketing function during the development phase, to anticipate or influence consumer decision making [22]. The challenge now is to form multisensory design teams.

The way forward to understand how multisensory teams work is to focus on studies of cognition. What matters is people's cognitive skills and how they navigate interpersonal communication more effectively. It is necessary to understand how two or more brains interrelate, as this is part of the process that shapes human behavior.

> Before moving to the new building, I asked to have a conversation with the engineer responsible to tactile paving the common areas. In the end, he told me it was good that I had suggested to map my navigation to point him where the tiles should go. He said would have done it differently. Roberta Morita Fernandes da Silva, Financial institution clerk. Lost eyesight during her early 30s

Cognition materializes in an interpersonal space [23]. The emergence of complex behaviors requires the coordination of actions between individuals according to a shared set of rules. For example, sharing other people's emotional states can make it easier to understand their intentions and actions [24].

After studying designers' cognition during the conceptual sketching processes, Sawa et al. [25] categorized the formal methods for design cognition protocol analysis into two types:

a. Process-oriented design

The process-oriented design focuses on representing design processes as a general scientific classification of problem-solving such as problem formulation, plans, objectives, and techniques.

b. Content-oriented design

The content-oriented design retrievers from the designers 'memory while designing the contents of their own experiences, such as what they think, what they see, and with what they have dealt Designer classify these contents into visual information (depicted elements and their perceptual features, and spatial relations), and non-visual information (functional thoughts, and knowledge).

An essential element of both process and content-oriented design is the critical role of mental imagery embedded in the designer's cognition. According to the definition, mental imagery is a mental representation followed by a concomitant awareness of sensory information of any modality that occurs without the need for an exogenous catalyst [26]. The skills of processing and representing information mentally without senses are critical for higher cognitive functions, which involve a set of complex thinking abilities [27].

Mental representations are retrieved from memory (imagery) and lead one to re-experience a version of the first stimulus or some novel mix of the stimulus [28]. More than mere mental representation, imagery is part of a dynamic building block process with connections to episodic memory and future thinking that is the place where complex mental setting up has been formed [29].

One key aspect of the multisensory design is the fact that mental imagery is not only visual. Mental imagery can be described in terms of auditory perception [30], olfactory [31], tactile [29], and motor [32].

10.3.2.1 Visual Mental Imagery

It incorporates reactivating visual representations in long-term memory and utilizing them to construct representations in working memory. These representations can be reinterpreted or changed, with more processing [33]. It also can extract new information not previously encoded, reassembled them in new forms. The ability of visual mental imagery of structuring and blending new information is a critical element in different types of cognitive skills (logic and reasoning, working memory, long-term memory used in engineering, math, and science) [34]. We can find them in people with visual impairment as well:

> I believe one reason why people with visual impairment are good at logic systems, such as programming, is that since we cannot see, we have to develop abstract thinking. Sidney

Tobias, System Analyst and Digital Accessibility Specialist at the Information, and Communication Technology Company of São Paulo City (PRODAM). Lost his eyesight during his teen year.

10.3.2.2 Auditory Mental Imagery

It is a modality of mental imagery characterized by introspective and non-hallucinatory auditory sensory qualities that occur without any external auditory stimulus [30]. It occurs in the form of verbal or musical imagery.

10.3.2.3 Somesthetic Mental Imagery

This modality is composed of tactile sense (touch), proprioception sense (self-movement and body position), and haptic perception (object recognition through touch and proprioception). Some tactile cognitive system examples are pain processing, temperature, body representation, and proprioception [35].

> I had to relearn how to walk. When we lose vision during adulthood, we feel very dizzy. In my case, as I have no light residue, the biological system was significantly altered. I felt hungry and sleepy at the most inappropriate times. The question of balance is very much changed. After the brain's adaptation, I have got used to the sound that the cane brings me when it hits a car, a pole, a tree, and the speed I have to walk. When I decided to walk with the guide dog's help, I had to learn how to communicate with the dog and realign my body with his. In the beginning, I was afraid to walk again because the dog pulls you. It is not like the cane that hits everything .
>
> Cleia Galdino, Speech Therapist, Black Women with Disabilities' intersectionality activist. Leader of the Movement Female Voices (Vozes Femininas).
>
> Lost eyesight during her early 20s.

10.3.2.4 Olfactory Mental Imagery

The definition of olfactory mental imagery is the experience of the sensation of smell when an appropriate stimulus is absent [31]. There is no agreement among scientists about the existence of olfactory mental imagery.

10.3.2.5 Gustatory Mental Imagery

As for olfactory, there are some controversies about the existence of imagery on the sense of taste because of the idea that it may be a combination of complex perception of taste, odor, texture, and trigeminal nerve [36].

10.3.2.6 Motor Mental Imagery

It is defined as a subconscious activity of the motor function that besides being responsible for movements, motor imagery is also in charge of envisioning actions, context awareness, and learning by examining and understanding others' conduct [37, 38]. What is unique about motor imagery is that it must have a motion, which differs from visual imagery that is static [39]. It is the imagination of moving without any physical alteration. For instance, it has critical in medicine (training a surgeon), sports, music, and motor rehabilitation.

> When you offer to conduct someone with visual impairment, please make sure never to grab the cane. Instead, offer your elbow. We know how to read your next steps from the biomechanics movement of your arm. Elton Garrone, Sprint runner, and Paralympic athlete. Lost his eyesight during his late 20s.

10.4 Conclusion

In this paper, we have presented a review of cognitive sciences' concepts to understand how people with visual impairment perceive the world. First of all, we considered that because designers are part of the design domain, this would allow them to grasp the problem's context better to identify the feasible solution. Here, we claim that a designer with a disability is better positioned to establish the design process's purpose-driven activities. It is especially relevant given the nature of visual impairment and the visual dominance in design.

The literature review has expanded our understanding of perception and the role of mental imagery in diverse sensory modalities. This knowledge is particularly relevant for multisensory design because it merges the five senses, affecting users' experience outcomes and universal design possibilities. Multisensory design benefits go beyond the applications in impairments. As future research, the inclusion of people with visual impairment in the design process poses the question about how they could communicate their situational awareness. These concepts could be the foundations to propose specific experiments involving design task protocols and new methods based on fMRI and ECG.

Acknowledgements To Andrea Kamensky and Marcio Aveiro, organizers of the course "Accessibility and Technology", offered by Digital Plural, a social learning network platform, with the financial support from PROEC/UFABC. This research received financial support from the Brazilian National Council for Scientific and Technological Development (CNPq).

Conflict of Interest The authors declare that they have no conflict of interest.

References

1. Persson, H., Åhman, H., Yngling, A.A., Gulliksen, J.: Universal design, inclusive design, accessible design, design for all: different concepts—one goal? On the concept of accessibility—historical, methodological and philosophical aspects. Univ. Access Inf. Soc. **14**(4), 505–526 (2015)
2. Luck, R.: Inclusive design and making in practice: bringing bodily experience into closer contact with making. Des. Stud. **54**, 96–119 (2018)
3. UN Disability Laws and Acts by Country/Area Homepage. https://www.un.org/development/desa/disabilities/disability-laws-and-acts-by-country-area.html. Last Accessed 21 April 2020
4. Manduchi, R., Kurniawan, S. (eds.).: Assistive Technology for Blindness and Low Vision. CRC Press (2012)
5. Gero, J.S.: Computational models of innovative and creative design processes. Technol. Forecast. Soc. Chang. **64**(2–3), 183–196 (2000)
6. Horton, E.L., Renganathan, R., Toth, B.N., Cohen, A.J., Bajcsy, A.V., Bateman, A., Lim, M.K.: A review of principles in design and usability testing of tactile technology for individuals with visual impairments. Assist. Technol. **29**(1), 28–36 (2017)
7. Bhowmick, A., Hazarika, S.M.: An insight into assistive technology for the visually impaired and blind people: state-of-the-art and future trends. J. Multimod. User Interf. **11**(2), 149–172 (2017)
8. Bilda, Z., Gero, J.S., Purcell, T.: To sketch or not to sketch? That is the question. Design Stud. **27**(5), 587–613 (2006)
9. Cattaneo, Z., Vecchi, T. Blind Vision: the Neuroscience of Visual Impairment. MIT Press (2011)
10. Warren, D.H.: Perception by the blind. Percep. Ecol. **10**, 65 (1978)
11. Ackland, P., Resnikoff, S., Bourne, R.: World blindness and visual impairment: despite many successes, the problem is growing. Community Eye Health **30**(100), 71 (2017)
12. Hollins, M.: Understanding Blindness: An Integrative Approach. Lawrence Erlbaum Associates, Inc (1989)
13. World Health Organization. Blindness and vision impairment. Homepage https://www.who.int/health-topics/blindness-and-vision-loss#tab=tab_1. Last Accessed 14 June 2020
14. Nieusma, D.: Alternative design scholarship: Working toward appropriate design. Design Issues **20**(3), 13–24 (2004)
15. Pullin, G.: Design Meets Disability. MIT Press (2009)
16. Schiffman, H.R.: Sensation and Perception: an Integrated Approach. Wiley (1990)
17. Sternberg, R.J., Sternberg, K.:Cognitive psychology. Nelson Education (2016)
18. Gibson, J.J.: The Senses Considered as Perceptual Systems (1966)
19. Bremner, A.J., Lewkowicz, D.J., Spence, C.: The Multisensory Approach to Development (2012)
20. Braun, A.: The sensory compensation hypothesis. In: The Speaker Identification Ability of Blind and Sighted Listeners, pp. 27–29. Springer VS, Wiesb (2016)
21. Love, T.: Design as a social process: bodies, brains, and social aspects of designing. J. Design Res. **3**(1), 45–54 (2003)
22. Spence, C., Gallace, A.: Multisensory design: reaching out to touch the consumer. Psychol. Market. **28**(3), 267–308 (2011)
23. Hasson, U., Ghazanfar, A.A., Galantucci, B., Garrod, S., Keysers, C.: Brain-to-brain coupling: a mechanism for creating and sharing a social world. Trends Cognit. Sci. **16**(2), 114–121 (2012)
24. Nummenmaa, L., Glerean, E., Viinikainen, M., Jääskeläinen, I.P., Hari, R., Sams, M.: Emotions promote social interaction by synchronizing brain activity across individuals. Proc. Natl. Acad. Sci. **109**(24), 9599–9604 (2012)
25. Suwa, M., Gero, J., Purcell, T.: Analysis of cognitive processes of a designer as the foundation for support tools. In: Gero, J.S., Sudweeks, F. (eds.) Artificial Intelligence in Design '98, pp. 229–247. Springer, Dordrecht. (1998)

26. Thomas, N.J.: Mental imagery, philosophical issues about. In: Nadel, L. (ed.) Encyclopedia of Cognitive Science, pp. 1147–1153. Nature Publishing/Macmillan, London (2003)
27. Mellet, E., Petit, L., Mazoyer, B., Denis, M., Tzourio, N.: Reopening the mental imagery debate: lessons from functional anatomy. Neuroimage **8**(2), 129–139 (1998)
28. Pearson, J., Naselaris, T., Holmes, E.A., Kosslyn, S.M.: Mental imagery: functional mechanisms and clinical applications. Trends Cognit. Sci. **19**(10), 590–602 (2015)
29. Schmidt, T.T., Ostwald, D., Blankenburg, F.: Imaging tactile imagery: changes in brain connectivity support perceptual grounding of mental images in primary sensory cortices. Neuroimage **98**, 216–224 (2014)
30. Reisberg, D. (ed.).: Auditory Imagery. Psychology Press (2016)
31. Stevenson, R.J., Case, T.I.: Olfactory imagery: a review. Psychon. Bull. Rev. **12**(2), 244–264 (2005)
32. Guillot, A., Collet, C.: Contribution from neurophysiological and psychological methods to the study of motor imagery. Brain Res. Rev. **50**(2), 387–397 (2005)
33. Kosslyn, S.M., Ganis, G., Thompson, W.L.: Neural foundations of imagery. Nauru Rev. Neurosci. **2**(9), 635–642 (2001)
34. Blazhenkova, O., Kozhevnikov. M.: Visual-object ability: a new dimension of non-verbal intelligence. Cognition **117**, 276–301 (2010)
35. Gallace, A.: Somesthetic mental imagery. In: Multisensory Imagery, pp. 29–50. Springer, New York, NY (2013)
36. Bensafi, M., Tillmann, B., Poncelet, J., Przybylski, L., Rouby, C.: Olfactory and gustatory mental imagery: modulation by sensory experience and comparison to auditory mental imagery. In: Multisensory Imagery, pp. 77–91. Springer, New York, NY (2013)
37. Jeannerod, M., Frak, V.: Mental imaging of motor activity in humans. Curr. Opin. Neurobiol. **9**(6), 735–739 (1999)
38. Moran, A., O'Shea, H.: Motor imagery practice and cognitive processes. Front. Psychol. **11**, 394 (2020)
39. Engelkamp, J., Zimmer, H.D.: Imagery and action: differential encoding of verbs and nouns (1990)

Chapter 11
The Influence of Personal Traits on Indian Millennial's Adoption of 3D Printed Fashion Products

Indranil Saha and Deepak John Mathew

Abstract Additive manufacturing technology, popularly known as 3D printing, is a contemporary method of customizing and manufacturing products. The 3D print market is expected to be worth more than $10 billion by the year 2021. The future of fashion aims to make use of 3D printing technology in the manufacturing process; however, practitioners are still doubtful whether this technology has the potential to revolutionize the fashion industry. Although 3D printing technology offers benefits to the businesses and production processes, whether it can replace the present mass production methods is still questionable. Until now, 3D printed fashion is studied from the perspective of material and technology applications. The perspective and adoption intents of the consumers are not explored, specifically in the Indian context. Hence, this research aims to study possible influences of 3D printed fashion products' adoption intention on Indian millennials. Personal traits are the essential basis of human behaviors that are closely associated with human needs. These personal traits, such as social, functional, hedonic, and cognitive attributes, are also arguably better predictors of consumer behavior than demographic or psychographic factors. Hence, recognizing the importance of studying present-day consumers' adoption preferences relating to 3DP fashion products, the current study explores the role of personal traits in adopting 3D printed fashion products through domain-specific factors such as fashion innovativeness and fashion leadership. The data were collected from Indian millennials via the online survey method. Research findings of this study add to the existing conceptual literature of 3D printing applications in fashion and guide fashion design practitioners by providing influencing factors of 3D printed fashion adoption among Indian millennial fashion consumers.

I. Saha · D. J. Mathew (✉)
Indian Institute of Technology Hyderabad, Hyderabad, India
e-mail: djm@des.iith.ac.in

I. Saha
e-mail: md18resch01001@iith.ac.in

A. Chakrabarti et al. (eds.), *Design for Tomorrow—Volume 1*, Smart Innovation, Systems and Technologies 221, https://doi.org/10.1007/978-981-16-0041-8_11

11.1 Introduction

The market of additive manufacturing, a contemporary way of manufacturing products, also known as 3D printing (3DP) [1], is going to be worth 23 billion dollars globally by 2022 [2]. 3DP technology has transformed various industries like automotive, medical, aviation, fashion, and lifestyle accessories [3]. Particularly, the fashion and lifestyle accessory industry has observed 3DP apparel and accessories in various categories such as haute couture, mass-produced, as well as customized sectors over the last decade [4]. Experts arguably suggest that 3DP can be the future of fashion [5] due to its number of benefits over traditional production methods [4]. 3DP has the potential to revolutionize the fashion industry with user-centric design, economic, and sustainable values by lowering the cost of inventory, production, and distribution [6]. With an increasing involvement of consumers, 3DP can create a delightful product experience [7] due to accuracy in the translation of personalized measurements to the actual products [6]. Indian fashion designer Nikhil Thampi in collaboration with PrintOmake™ has launched 3D printed face mask using sustainable PLA material. Despite having evident instances, the perception of 3DP in mass production are still doubtful [8] because of the limitations of conventional textile-like properties of its available materials [4]. Researchers have explored 3DP materials, machines, societal impact, and their applicability in different industries, including fashion [8, 9]. Simultaneously, millennials, the younger consumer segment born after 1980, are identified as the first tech-savvy generation, who are passionate about technological innovations [10]. Researchers have also mentioned that the millennials have relatively more spending abilities [11], and their consumption preferences are difficult to comprehend [12]. Hence, India, with 400 million millennials, which is the largest in the world [13], makes it relevant to study the perspectives of the potential consumers of 3DP fashion products, which have a considerable potential to conquer the market [4]. In the context of studying consumers' perspective, attitudes and values are the types of social recognition that reflect the fundamental attributes of adaption into a specific environment and process [14]. As personal values are essential to someone's personal traits [8], existing studies have explored the role of values in consumers' characteristics, such as consumer innovativeness [15], which has also been further studied extensively to examine the consumers' inclination toward using new technology relating to consumer products and services, such as electronics and e-commerce portals [16]. The current research aims to investigate Indian millennial consumers' personal traits related to the adoption intention of 3DP fashion products through domain-specific factors, i.e., fashion leadership and fashion innovativeness. The specific research objectives of this study are (i) to identify the factors influencing Indian millennial consumers' adoption of 3DP fashion products, (ii) to determine the relationship between Indian millennial consumers' fashion innovativeness and fashion leadership, and (iii) to examine whether fashion innovativeness and fashion leadership mediate the relationship between personal traits and attitude toward 3DP fashion products.

11.2 Personal Traits and Domain-Specific Innovativeness

Consumer innovativeness is explained as the rate at which a consumer is comparatively earlier in looking for stimulation and innovation from new products than the overall population within their environment [17, 18]. Existing studies have explored the harmony between these early adopters [19] by using consumer innovativeness as an originator of the acceptance of new products [20]. Simultaneously, personal traits are the persistent characteristics, which describe an individual's inclination toward purchasing particular products [18]. Previous studies on consumer innovativeness indicate that each consumer has innovativeness, which is parallel to the personal traits [21]. Therefore, it is relevant to analyze such personal traits to determine the adoption intention of new products [22]. Based on the existing literature, researchers have precisely developed an inclusive four-level scale of measuring personal traits to understand consumers' motivation to accept new products or ideas [24]—(a) social innovativeness, which focuses on the need for social differentiation; (b) hedonic innovativeness, which addresses the anticipated positive expressions after purchasing a new product; (c) cognitive innovativeness, which explains consumers' perception and experience of purchasing new products; and (d) functional innovativeness which describes the perception of the functionality of a new product [23]. Although 3DP technology has been influencing the market since over a decade, 3DP fashion products are relatively a new concept among the consumers [4]. Hence, the current study encompasses present-day consumer innovativeness from a personal trait perspective.

Researchers indicate that even though the levels of personal traits can measure the psychometric parameters of consumers' innovativeness, they alone cannot predict the actual innovative adoption intentions of the consumers [24]. Existing studies have suggested the idea of domain-specific innovativeness, which refers to the affinity toward knowledge acquisition and innovation adoption within a particular product category [25]. The features of domain-specific innovativeness were extensively used in the existing literature of diverse backgrounds, such as products, nations, and industries. [26]. Despite the recognition of the benefits, only a few consumers have bought and experienced 3DP fashion products as yet [8]. Hence, the comprehension of the fundamental motivations at various levels has the potential to enrich the novel context of the study [9]. Opinion leadership, one of the most critical factors of innovation adoption, is defined as the rate at which a person can inspire other people [27]. Existing research indicates that opinion leadership influences the decisions, beliefs, and attitudes of others through interactive communications [28]. Opinion leadership with respect to fashion products, i.e., fashion leadership, is identified to be a means to communicate self-image; for instance, the self-conscious consumers acquire innovative fashion products to amaze others [25]. Fashion leaders look for self-indulgence, and they are prone to experience and relish innovative fashion products. Although there is a lack of evidence that significantly explain the impact of fashion leadership toward 3DP fashion adoption, in the context of this current study, millennial opinion leaders of different fashion product categories are more likely to look for more information about new technologies and products in the domain of fashion [8]. Simultaneously, innovative consumers tend to influence others' buying choices by sharing their experiences and understanding of new products with others [29]. While consumer innovativeness shows consumers' generic personal traits associated with innovation [17], domain-specific innovativeness focuses on a specific area of adoption intention [30]. Researchers indicated that consumers having high fashion innovativeness are inclined to buy the latest trends in fashion [31]. So, consumers having a higher level of general innovativeness are more likely to have a positive perception of 3DP fashion products. Additionally, fashion leadership and fashion innovativeness are interrelated mediating variables between the generic personal traits, and attitude and adoption intention toward 3DP fashion [8]. Since the concept of 3DP is relatively new in fashion and the technology has the potential to transform the industry, consumers' inclination toward accepting or perceiving innovation may describe adoption intention [8]. Attitudes commonly represent a consumer's intentions to adopt technology, which impact their purchasing decision and behavioral consequences. Several studies have established the significant role of attitude in a consumer's adoption intention [8, 32]. Along the lines of the existing studies, the current research theorizes that positive attitude concerning 3DP fashion creates a positive perception in consumers' minds, which ultimately translates into the purchase intention of the product.

Based on the above comprehensive literature review, the following hypotheses are formed to fulfill the research objectives: H_{1a}: Personal traits positively influence fashion leadership; H_{1b}: Personal traits positively influence fashion innovativeness; H_2: Fashion leadership positively influences fashion innovativeness; H_3: Fashion innovativeness positively influences attitude concerning 3DP fashion products; H_4: Fashion leadership positively influences attitude concerning 3DP fashion products; H5: Attitude concerning 3DP fashion products positively influences the adoption intention of 3DP fashion products. Figure 11.1 represents the research model for the current study.

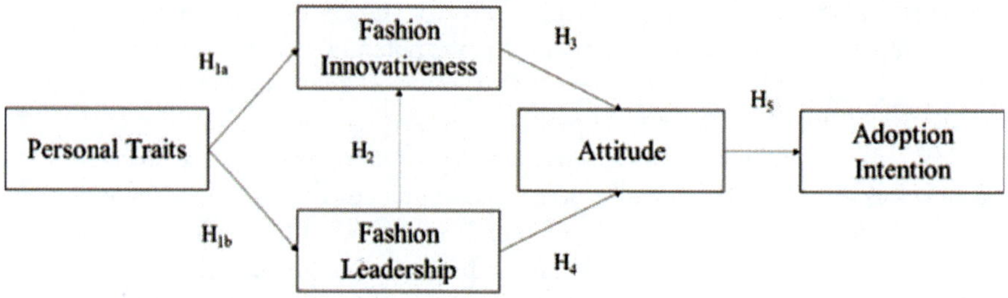

Fig. 11.1 Research model of the study

11.3 Methodology

11.3.1 Approach

This study followed a survey-based quantitative approach to research design. As the existing literature suggested, there is a dearth of studies that focuses on consumer behavior in the context of 3DP fashion products, particularly in the Indian context. Therefore, the study integrates elements from highly accepted frameworks of exploring consumers' drive to execute a specific behavior, i.e., value-attitude-intention hierarchy [14], and diffusion of innovation [8, 27], which explains consumers' new adoption behavior.

11.3.2 Sample and Data Collection

The participants of this study were millennials, both students and working professionals from multidisciplinary technology and fashion education institutes in India. As age is an influential demographic factor in adopting new technology [33], the selection of millennials ranged within the age group of 18–33 years [11], using a convenient sampling method. Data were collected through an online survey, the link of which was emailed to the students. A total of 354 survey responses were gathered, and 339 responses (female: 40.71%; male: 58.11%; prefer not to mention: 1.18%) were filtered for analysis after excluding incomplete and unacceptable entries.

11.3.3 Research Instrument

The questionnaire of the survey included a short introductory description (both visual and textual) of 3DP technology and products, as the topic of this study is relatively new. After the consent statement, the participants were provided with photograph-elicitation references of existing 3D printed apparel and accessories. Measurement items anchored with a seven-point Likert scale using "strongly disagree" and "strongly agree" (1 = Strongly disagree, 7 = Strongly agree) were extracted and adapted from the existing studies. Personal traits were measured using twenty-one items for consumers' general innovativeness [23]. Domain-specific variables, fashion leadership, and fashion innovativeness were measured using six and four items, respectively [8]. Attitude and adoption intention toward 3DP fashion products were measured through five [34] and six items from [35], respectively. Cronbach's alpha values of all items were reported to be more than 0.7, which indicated the reliability of the measurement items (Table 11.1). The questionnaire ended with demographic questions.

Table 11.1 Cross-loadings of all measurement items using CFA

Variables	Items	Loadings
Personal traits (PT)—(CR: 0.936; AVE: 0.566; Cronbach's Alpha: 0.927)		
Social	1. I love to use innovations that impress others	0.683
	2. I like to own a new product that distinguishes me from others who do not own this new product	0.772
	3. I prefer to try new products with which I can present myself to my friends and neighbors	0.720
	4. I like to outdo others, and I prefer to do this by buying new products which my friends do not have	0.571
Functional	1. If a new time-saving product is launched, I will buy it right away	0.681
	2. If innovation is more functional, then I will usually buy it	0.592
	3. If I discover a new product in a more convenient size, I am very inclined to buy this	0.673
	4. If a new product makes my work easier, then this new product is a "must" for me	0.728
Hedonic	1. I am an intellectual thinker who buys new products because they set my brain to work	0.526
	2. It gives me a good feeling to acquire new products	0.692
	3. Innovations make my life exciting and stimulating	0.795
	4. Acquiring an innovation makes me happier	0.780
	5. The discovery of novelties makes me playful and cheerful	0.612
Cognitive	1. I mostly buy those innovations that satisfy my analytical mind	0.684
	2. I find innovations that need a lot of thinking intellectually challenging, and therefore, I buy them instantly	0.761
	3. I often buy new products that make me think logically	0.640
	4. I often buy innovative products that challenge the strengths and weaknesses of my intellectual skills	0.619

(continued)

Table 11.1 (continued)

Variables	Items	Loadings
Fashion leadership (FL)—(CR: 0.943; AVE: 0.847; Cronbach's Alpha: 0.910)		
(Domain-specific variable I)	1. I am aware of fashion trends and want to be one of the first to try them	0.909
	2. I am the first to try new fashion; therefore, many people regard me as being a fashionable leader	0.920
	3. It is important for me to be a fashion leader	0.932
Fashion innovativeness (FI)—(CR: 0.838; AVE: 0.634; Cronbach's Alpha: 0.717)		
(Domain-specific variable II)	1. If I heard that a 3D printed fashion product was available in the store, I would be interested enough to buy it	0.825
	2. I will research 3D printed fashion products even if I have not heard of it before	0.793
	3. I know 3D printed fashion products before most other people in my circle know	0.769
Attitude toward 3DP fashion products (AT) - (CR: 0.803; AVE: 0.577; Cronbach's Alpha: 0.751)		
(7-point semantic scale)	1. Unappealing: Appealing	0.810
	2. Inappropriate: Appropriate	0.757
	3. Impractical: Practical	0.710
Adoption intentions of 3DP fashion products (AD)—(CR: 0.915; AVE: 0.684; Cronbach's Alpha: 0.884)		
(7-point semantic scale)	1. Unlikely: Likely	0.874
	2. Improbable: Probable	0.899
	3. Impossible: Possible	0.811
	4. Uncertain: Certain	0.775
	5. Definitely would not use: Definitely would use	0.767

Chi-square = 1648.006; NFI = 0.921; SRMR = 0.043; $p < 0.05$
Note CR Composite reliability; *AVE* Average variance extracted; *NFI* Normed fit index; *SRMR* Standardized root mean square residual

11.3.4 Data Analysis

To analyze the gathered data, firstly, descriptive statistics were used to summarize respondents' demographic elements (Table 11.1). Secondly, the internal consistency of the measurement items was evaluated through reliability and validity analysis (Table 11.1). Thirdly, confirmatory factor analysis (CFA) and structural equation modeling (SEM) were used to test the influences between the variables for examining the five research hypotheses.

11.4 Results

As given in Table 11.1, the CFA was used to test the construct validity of the proposed research model. The measurement model provided a satisfactory fit to the data (Chi-square = 1648.006; NFI = 0.921; SRMR = 0.043; $p < 0.05$) after eliminating three items with cross-loadings less than 0.40. The composite reliabilities for the five constructs ranged from 0.803 to 0.943. AVE for each construct ranged from 0.566 to 0.847 (Table 11.1). All standardized cross-loadings in factor analysis ranged from 0.526 to 0.932 ($p < 0.001$), indicating convergent validity. For every set of two constructs, Table 11.2 illustrates that the squared correlations between each pair were less than the AVE for each construct [36]. Therefore, the discriminant validity conditions were satisfied, indicating a satisfactory level of validity of the model.

Total effects statistics were computed to determine the directed dependencies and strength, i.e., the total effects of the paths among the set of variables in the model. As per the results given in Table 11.3, it is found that personal traits have the strongest effect on fashion innovativeness. While concluding the support decision of the framed hypotheses, the results specified that the R^2 values of fashion leadership, fashion innovativeness, attitude, and adoption intention are 0.524, 0.563, 0.425, and 0.434, respectively, which indicate an adequate level of the model's predictive accuracy. The values computed through SEM represent t-statistics and p-values, as listed in Table 11.3. As all the t-statistics are above 1.96, and the p-values are less than 0.001, and the effects were concluded to be significant in all cases at a 95% confidence level. At the same time, Table 11.3 also illustrates f-square values, which indicate the effect size. The results show that even though fashion leadership has a statistically significant effect on fashion innovativeness, the effect is not meaningful and large enough, as the f-square value is less than 0.15. Overall, the results suggest that personal traits positively influence fashion innovativeness (H_{1a}) (STDEV = 0.024; p = 0.000) and fashion leadership (H_{1b}) (STDEV = 0.047; p = 0.000); fashion leadership positively influences fashion innovativeness (H_2) (STDEV = 0.054; p = 0.000); fashion innovativeness positively influences attitude toward 3DP fashion products (H_3) (STDEV = 0.061; p = 0.000); fashion leadership positively influences attitude toward 3DP fashion products (H_4) (STDEV = 0.063; p = 0.000); and attitude

Table 11.2 Discriminant validity

	Mean	SD	1	2	3	4	5
Adoption intention	4.49	1.14	*0.827*				
Attitude	5.11	0.92	0.578	*0.760*			
Fashion innovativeness	4.40	1.52	0.698	0.461	*0.796*		
Fashion leadership	3.50	1.80	0.592	0.389	0.663	*0.921*	
Personality traits	4.74	1.22	0.557	0.469	0.723	0.724	*0.752*

Note AVE is shown diagonally in *Italics*

Table 11.3 Significance test results of the structural model

Hypothesis	Constructs	*t*-statistics	Standard deviation	*p*-value	*f*-square	Result	Support decision
H1a	Personal traits → Fashion leadership	30.034	0.024	0.000	1.103	Significant	Yes
H1b	Personal traits → Fashion innovativeness	10.747	0.047	0.000	0.283	Significant	Yes
H2	Fashion leadership → Fashion innovativeness	5.488	0.054	0.000	0.094	Significant but small effect	Yes
H3	Fashion innovativeness → Attitude	5.984	0.061	0.000	0.195	Significant	Yes
H4	Fashion leadership → Attitude	2.374	0.063	0.000	0.160		
H5	Attitude → Adoption intention	17.711	0.033	0.000	0.503	Significant	Yes

toward 3DP fashion products positively influences adoption intention of 3DP fashion products (H5) (STDEV = 0.033; $p = 0.000$).

To examine the mediating role of fashion innovativeness and fashion leadership between personal traits and attitude toward 3DP fashion products, mediation analysis was conducted. The results suggested the indirect effects between personal trait and attitude via (i) fashion leadership was not significant (LLCI = -0.0792; ULCI = 0.0620); (ii) fashion innovativeness was significant (LLCI = -0.0301; ULCI = 0.1417); (iii) via both fashion innovativeness and fashion leadership was significant (LLCI = 0.0122; ULCI = 0.0601). The total indirect effect of personal traits on attitude toward 3DP fashion products was significant (LLCI = 0.0147; ULCI = 0.2042). Hence, there is a significant mediated effect of personal traits on attitude; however, fashion innovativeness played a better role than fashion leadership in mediating between them.

11.5 Discussion and Conclusion

The current study has investigated the factors influencing Indian millennials' attitudes and adoption intention through personal traits and behaviors specific to 3DP fashion products. The findings showed that individuals' personal traits positively influenced fashion specific behaviors, such as fashion innovativeness and fashion leadership, which is different from the findings of previous studies [8]. The possible reason for such a finding could be the increased awareness of the application of 3DP in fashion as 67% of the respondents indicated that they were aware of the 3DP fashion products before the survey description. The probable explanation of such difference in findings may also lie in the different socio-cultural backgrounds and the varied demographic distributions of sample respondents of the current and the previous study; for example, 94% and 41% of respondents were female in the previous and current study, respectively. The difference in time between the previous (2018) and current study (2020) could also be an essential reason behind the difference as the 3DP market was projected to grow sharply in the last two years [37]. Moreover, the strong effect of personal traits on fashion specific variables pointed out that fashion leaders are more open toward innovations and new experiences. Results also suggest that the fashion leadership had a significant influence on fashion innovativeness, which further strengthens the fact that Indian fashion-sensitive millennials are willing to be recognized as the trendsetters by experiencing and acquiring new concepts. Along the lines of previous studies, this study also validates the idea that personal traits based on social, functional, hedonic, and cognitive attributes are positive influencing factors of millennials' innovative behaviors toward 3DP fashion products [16]. Consequently, Indian millennials with high fashion innovativeness were found to have a positive attitude toward 3DP fashion products, which further led them to a positive adoption intention, which is also supported by previous studies [21]. Fashion specific variables showed a significantly positive yet relatively weaker influence on attitude and adoption intention of 3DP fashion products, which is different from the outcomes of a previous study [8]. The nature of participants may explain the finding of mediation analysis that identified fashion innovativeness as a significant mediator between personal traits and attitude—millennials from fashion and technology institutes, who are tech-savvy and well-informed of fashion innovations. Hence, millennials with high social, functional, hedonic, and cognitive personal traits are inclined to have high fashion innovativeness, which ultimately led them to a positive attitude toward 3DP fashion products.

The study contributes to both conceptual and practical insights into the field of fashion design. At a conceptual level, this study contributes to exploring fundamental influencing factors based on multi-level personal traits incorporating fashion specific attributes to provide comprehensive visions to the consumer-centric design discipline. As fashion innovators play a major role in influencing the later adopters to acquire new esthetics, the findings of this study were particularly significant as it explored the relatively new concept of 3DP fashion products in the light of millennial consumers' fashion innovativeness. Since personal traits are significant factors for an innovative fashion practice built on transformative consumption patterns [38], at a practical level, this study can guide the practitioners to explore 3DP fashion products for Indian millennials as they are most likely to adopt them by showing a positive attitude toward the newness of 3DP esthetics in fashion. Although the responses were extracted from the relevant millennial consumers in India, examining the influencing factors specific to different demographics, such as age, gender, and native, will provide more precise insights. This study did not acquire responses relating to specific product experience, future studies may encourage consumers to use and provide their firsthand perspectives about 3DP fashion products. An extension of this paper also requires exploring how the specific personal traits highlighted in the study will drive decision making in the fashion design process.

References

1. Yampolskiy, M., Skjellum, A., Kretzschmar, M., et al.: Using 3D printers as weapons. Int. J. Crit. Infrastruct. Prot. **14**, 58–71 (2016)
2. Razavi Bazaz, S., Kashaninejad, N., Azadi, S., et al.: Rapid Softlithography Using 3D-Printed Molds. Adv. Mater. Technol. **4**, 1900425 (2019)
3. Yap, Y.L., Yeong, W.Y.: Additive manufacture of fashion and jewellery products: a mini review: this paper provides an insight into the future of 3D printing industries for fashion and jewellery products. Virtual Phys. Prototyp. **9**, 195–201 (2014). https://doi.org/10.1080/17452759.2014.938993
4. Vanderploeg, A., Lee, S.-E., Mamp, M.: The application of 3D printing technology in the fashion industry. Int. J. Fash Des. Technol. Educ. **10**, 170–179 (2017). https://doi.org/10.1080/17543266.2016.1223355
5. Tania, R.T.: 3D printing technology : the surface of future fashion. **157**, 48–51 (2017)
6. Thompson, M.K., Moroni, G., Vaneker, T., et al.: Design for additive manufacturing: trends, opportunities, considerations, and constraints. CIRP Ann. Manuf. Technol. **65**, 737–760 (2016). https://doi.org/10.1016/j.cirp.2016.05.004
7. Lim, H.-W., Cassidy, T. D.:. 3D printing technology revolution in future sustainable fashion. In: Sustainability in Textiles and Fashion. Leeds (2014)

8. Lyu, J., Hahn, K., Sadachar, A.: Understanding millennial consumer's adoption of 3D printed fashion products by exploring personal values and innovativeness. Fash Text **5** (2014). https://doi.org/10.1186/s40691-017-0119-8
9. Huang, S.H., Liu, P., Mokasdar, A., Hou, L.: Additive manufacturing and its societal impact: a literature review. Int. J. Adv. Manuf. Technol. **67**, 1191–1203 (2013)
10. Eastman, J.K., Iyer, R., Liao-Troth, S., et al.: The role of involvement on millennials' mobile technology behaviors: the moderating impact of status consumption, innovation, and opinion leadership. J. Mark Theory Pract. **22**, 455–470 (2014). https://doi.org/10.2753/MTP1069-6679220407
11. Muralidharan, S., Rejón-Guardia, F., Xue, F.: Understanding the green buying behavior of younger millennials from India and the United States: a structural equation modeling approach. J. Int. Consum. Mark **28**, 54–72 (2016). https://doi.org/10.1080/08961530.2015.1056328
12. SivaKumar, A.K., Gunasekaran, A.: An Empirical study on the factors affecting online shopping behavior of millennial consumers. J. Internet Commer. **16**, 219–230 (2017). https://doi.org/10.1080/15332861.2017.1317150
13. Chaudhary, R., Bisai, S.: Factors influencing green purchase behavior of millennials in India. Manag. Environ. Q. An. Int. J. (2018)
14. Kahle, L. R.: Social values and social change: adaptation to life in America. Praeger, New York (1983)
15. Burgess, S.M.: Personal values and consumer research: an historical perspective. Res. Mark **11**, 35–79 (1992)
16. Venkatesh, V., Thong, J. Y. L., Xu, X. Consumer acceptance and use of information technology: extending the unified theory of acceptance and use of technology. MIS Q. 157–178 (2012)
17. Hirschman, E.C.: Innovativeness, novelty seeking, and consumer creativity. J. Consum. Res. **7**, 283–295 (1980)
18. Hirunyawipada, T., Paswan, A. K.: Consumer innovativeness and perceived risk: implications for high technology product adoption. J. Consum. Mark (2006)
19. Gatignon, H. Innovative decision process. Handb. Consum. Behav (1991)
20. Chau, P.Y.K., Hui, K.L.: Identifying early adopters of new IT products: a case of Windows 95. Inf. Manag. **33**, 225–230 (1998)
21. Im, S., Bayus, B.L., Mason, C.H.: An empirical study of innate consumer innovativeness, personal characteristics, and new-product adoption behavior. J. Acad. Mark. Sci. **31**, 61–73 (2003). https://doi.org/10.1177/0092070302238602
22. Mowen, J. C.: The 3M Model of Motivation and Personality: Theory and Empirical Applications to Consumer Behavior. Springer Science & Business Media (2000)
23. Vandecasteele, B., Geuens, M.: Motivated consumer innovativeness: concept, measurement, and validation. Int. J. Res. Mark. **27**, 308–318 (2010)
24. Hoffmann, S., Soyez, K.: A cognitive model to predict domain-specific consumer innovativeness. J. Bus. Res. **63**, 778–785 (2010)
25. Goldsmith, R.E., Hofacker, C.F.: Measuring consumer innovativeness. J. Acad. Mark. Sci. **19**, 209–221 (1991)
26. Roehrich, G.: Consumer innovativeness: concepts and measurements. J. Bus. Res. **57**, 671–677 (2004)
27. Rogers, E. M.: Diffusion of Innovations. Simon and Schuster (2010)
28. Tellis, G.J., Yin, E., Bell, S.: Global consumer innovativeness: cross-country differences and demographic commonalities. J. Int. Mark. **17**, 1–22 (2009)
29. Grewal, R., Mehta, R., Kardes, F.R.: The role of the social-identity function of attitudes in consumer innovativeness and opinion leadership. J. Econ. Psychol. **21**, 233–252 (2000)
30. Goldsmith, R.E., Freiden, J.B., Eastman, J.K.: The generality/specificity issue in consumer innovativeness research. Technovation **15**, 601–612 (1995)
31. Choo, H.J., Sim, S.Y., Lee, H.K., Bin, K.H.: The effect of consumers' involvement and innovativeness on the utilization of fashion wardrobe. Int. J. Consum. Stud. **38**, 175–182 (2014)
32. Claudy, M.C., Garcia, R., O'Driscoll, A.: Consumer resistance to innovation—a behavioral reasoning perspective. J. Acad. Mark. Sci. **43**, 528–544 (2015)

33. Blazquez, M., Alexander, B., Fung, K.: Exploring Millennial's perceptions towards luxury fashion wearable technology. J. Fash. Mark. Manag. An. Int. J (2020)
34. Wansink, B.: Advertising's impact on category substitution. J. Mark. Res. **31**, 505–515 (1994)
35. Fishbein, M., Ajzen, I.: Understanding attitudes and predicting social behavior (1980)
36. Fornell, C., Larcker, D.F.: Evaluating structural equation models with unobservable variables and measurement error. J. Mark. Res. **18**, 39–50 (1981)
37. MarketsandMarkets.com.: 3D Printing Market by Source, Type, Application & Geography| COVID-19 Impact Analysis|MarketsandMarkets™. In: marketsandmarkets.com. https://www.marketsandmarkets.com/Market-Reports/3d-printing-market-1276.html. Accessed 14 May 2020 (2019)
38. Reiley, K., DeLong, M.: A consumer vision for sustainable fashion practice. Fash. Pract. **3**, 63–83 (2011)

Chapter 12
Inclusive Design—Designing Barrier-Free Public Spaces

Suhani Arora and Aditi Deshpande

Abstract Interior design as a discipline places humans at the heart of the process and has an underlying commitment to inclusivity. The British Standards Institute (2005) defines inclusive design as: 'The design of mainstream products and/or services that are accessible to, and usable by, as many people as reasonably possible ... without the need for special adaptation or specialised design (Inclusive Design Toolkit).' Inclusive design embraces diversity and enables cohesion among age groups, ethnicities, genders as well as mental and physical abilities. In the context of the Indian scenario, however, inclusive design is not incorporated into mainstream design. Thus, many public spaces remain inaccessible to a large chunk of the population—including the elderly, people with temporary, situational and permanent disabilities among others. Though there are government bodies entrusted to formulate guidelines for making certain buildings barrier free, the ground rules are silent on the implementation of accessible design in public spaces and built environments. This study aims to address the need for inclusivity in designing public spaces and proposes an interior design model that can be accessed by diverse individuals and communities.

12.1 Introduction

Inclusive design, universal design and design for all are terms used to describe an approach of developing products, services and environments, which are usable and attractive for a large number of people regardless of age, gender, language and ability [1]. The development of products and services that offer universal access is a crucial element in facilitating a genuinely inclusive environment in all public spaces. The UK Design Council describes inclusive design as neither a new genre of design, nor a separate specialism, but as a general approach to designing in which designers ensure

S. Arora
sP+a, Mumbai, India
e-mail: suhaniarora188@gmail.com

A. Deshpande (✉)
Symbiosis Institute of Design, Pune, India
e-mail: aditi.deshpande@sid.edu.in

A. Chakrabarti et al. (eds.), *Design for Tomorrow—Volume 1*, Smart Innovation, Systems and Technologies 221, https://doi.org/10.1007/978-981-16-0041-8_12

that their products and services address the needs of the widest possible audience, irrespective of age or ability [3]. The design process is optimised for a specific user or group of users with specific needs to ensure that a wider diversity of people can make use of it. The approach to inclusive design recognises the fact that it is not always possible to cater to the demands of the whole population. Nonetheless, the approach to inclusivity maintains that public spaces must be barrier free in order to be accessible to as many people as possible, keeping the extreme users and their needs in mind. The use of the term "inclusive" rather than "universal" reflects the view that "inclusivity" is a more achievable, and in many situations, appropriate goal than "Universal Design" or "Design for All [10]." Kat Holmes, an advocate for equitable design in the tech world, argues that an inclusive designer "is anyone who recognises and remedies mismatched interactions between people and their world [6]." As design principles continue to evolve, designers must look beyond aesthetic and functional designs and include more complex issues like equality and inclusion.

This paper highlights the design strategies and principles for barrier-free public spaces in India and lays down an interior design model for a bookstore that can be occupied by senior citizens, people with temporary and permanent physical disabilities as well as fully bodied persons. The aim of this paper is to present a methodological design approach that can be used by students, teachers and industrial designers among others to implement inclusive design in order to create spaces for one and all. A summary of the guidelines for designing for users with different capabilities is given along with a description of the model of a bookstore—a public space that caters to people of all ages and backgrounds yet still remains neglected in terms of accessibility and a practical design approach. The model is an interpretation and implementation of inclusivity in design based on a user-centric approach. It can also be used as a template for creating accessible public spaces.

12.2 Designing for All

As we all prepare to live in a world with ageing ailments and disabilities, we need to prepare our infrastructure, systems and technologies to allow inclusion of human diversity as the guiding philosophy for designers. Census 2011 has revealed that over 26.8 million crore people in India suffer from one or the other kind of disability. Several legal regulations like Rights of Persons with Disabilities Act, 2016, passed by the Government of India, ratification of the UNCRPD Act, 2007 and the Accessible India Campaign launched by the Hon'ble Prime Minister in 2015 reflect a growth in the process of including and understanding the needs of the differently abled population in the country. More than fifty cities have been undertaken under the Accessible India Campaign to be made accessible with investments on accessible infrastructures for public use, tourism, mobility, etc. However, it remains a fact that India has only a limited number of professionals competent to deliver accessibility solutions in diverse urban situations [14].

By incorporating universal design principles into plans, designers and professionals can create spaces without discrimination, thereby benefitting a wide range of people with varied abilities, preferences and age groups. Designers, companies and government all have a role to play, by designing, investing and legislating with difference in mind, so that a design process that is inclusive becomes a standard practice [13].

12.2.1 Principles of Inclusive Design

Great places are designed for the inclusion and enjoyment of everyone, including those with differing cognitive, sensory, physical or developmental abilities [11]. Considering how elements can be reached, manipulated and used irrespective of body size, posture or mobility is one of the fundamental principles of inclusive design. A number of design approaches exist for users with impairments, irrespective of whether caused by ageing or congenital conditions. The approaches can broadly be defined by the target user groups and how they aim to enable usability. For instance, most inclusive design approaches define the target population as being older users, functionally impaired users or the whole population [8].

The user pyramid Fig. 12.1 shows how users can be divided into four segments. The average, able-bodied consumer is located at the bottom or base of the pyramid. The next and biggest segment consists of people with temporary or situational disabilities—pregnant women, people who need glasses, are left handed, or have dyslexia as well as those carrying heavy luggage or infants on stroller. Inclusive design aims to reach the population within these two groups to accommodate them into mainstream

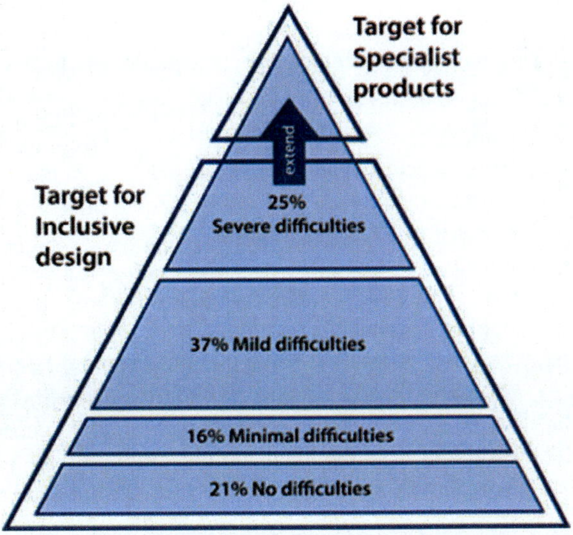

Fig. 12.1 Inclusive design user pyramid presenting a continuum of population diversity (Inclusive Design Toolkit)

products and services in the market. The top two segments are usually not viewed as primary markets for inclusive design, due to the large gap between their needs and the needs of the mainstream user [5].

The UK Commission for Architecture and the Built Environment (2006) published five principles for inclusive design in buildings, places and spaces. The principles are as follows:

1. Inclusive design places people at the heart of the design process.
2. Inclusive design acknowledges diversity and difference.
3. Inclusive design offers choice where a single design solution cannot accommodate all users.
4. Inclusive design provides for flexibility in use.
5. Inclusive design provides buildings and environments that are enjoyable to use for everyone.

12.2.2 *Guidelines for Barrier-Free Built Environments by Government Bodies*

The main objective of the "Persons with Disabilities (Equal Opportunities, Protection of Rights and Full Participation) Act, 1996 enacted by the Government of India is to create barrier-free environment for persons with disabilities and to make special provisions for the integration of persons with disabilities into the social mainstream. In regard to non-discrimination in the built environment, provisions have been made in this Act for ramps in public buildings, adaptation of toilets for wheel chair users, Braille symbols and auditory signals in elevators [2].

The Central Public Works Department of India recommends adequate space for persons using mobility devices like wheelchairs, crutches and walkers as well as those walking with the assistance of other persons. With regard to the special needs of person with disabilities, the design guidelines for built environments are as follows:

1. Clear 900 mm width for entry and exit doors.
2. Compulsory ramp for approach with a minimum slope of 1: 12. Hand rails shall be provided for ramps/slope ways.
3. Entrance landing shall be provided adjacent to ramp with the minimum dimension 1800 × 2000 mm.
4. For stepped approach, size of tread shall not be less than 300 mm and maximum riser shall be 150 mm. Provision of 900 mm high hand rail on both sides of the stepped approach is compulsory.
5. Switches as well as door handles and other fixtures and fittings should be between 900 and 1200 mm height from the finished floor level.
6. Locking and opening controls for window and doors should not be more than 1400 mm from the finished floor usable by one hand.
7. A window should have an unobstructed viewing zone for wheelchair users.

Fig. 12.2 Digital sketches depicting reaches and clear width for wheelchair users

8. Allow a space at least 350 mm deep and 700 mm high under a counter, stand, etc (Fig. 12.2).
9. The minimum size of washroom shall be 1500 × 1750 mm with a clear opening of 900 mm for the door swinging outwards.

The above-mentioned guidelines are a working tool to provide a common point of reference as well as a starting point to design spaces for the disabled users. Design professionals, the building industry and the community as a whole must consider these low-cost guidelines and implement them without sacrificing the aesthetics and feel of the space.

12.3 Bookstore and Inclusive Design

The bookstore is a space that allows retail activity of books and also promotes interaction between people of various backgrounds, culture and age groups. Bookstore is a space that must open its doors to all kinds of people irrespective of their age, disability and needs. In recent years, with the development of the digital publishing technology and online sales platform, physical bookstores that regard traditional printed books face a challenge. However, bookstores provide a more holistic customer experience. Store design, virtual displays and visual merchandising play a huge role in attracting customers and influencing their decision to purchase books in a physical bookstore. Moreover, the elderly population still struggles to cope with technological advancements and thus prefers the traditional idea of visiting a bookstore and spending time in a conducive environment. The disabled population too prefers visiting a bookstore as a leisure activity, although they might be assisted by someone. Thus, the bookstore must be a platform to engage customers and provide an environment that is favourable to all.

12.4 Method

12.4.1 Defining the Target Audience

Inclusive design focuses on choosing a target market for a particular design, and making informed decisions to maximise the usability of the space or product. The definition of inclusivity adopted here also means that inclusive design does not mean designing specifically for people with disabilities—it means considering the needs of able-bodied people as well as people with various levels of capability limitations and designing products and services to meet the needs of all users to the greatest extent possible [12].

For the model of the bookstore, the target audience identified is the elderly population (people above the age of 65) as well as those with temporary, situational and physical disabilities. Wheelchair-bound persons, people with crutches and others who might be temporarily disabled form the focus group for this particular project. It excludes people with visual impairments and mental disabilities.

12.4.2 User Study

User study is important from the aspect of "knowing the users" and designing inclusively for the target audience as well as for able-bodied users. An empathetic approach is important to understand the needs of the users. Empathy is crucial to a human-centred design process such as design thinking and empathy helps design thinkers to set aside his or her own assumptions about the world in order to gain insight into their users and their needs [4].

In order to engage with extreme users (i.e., the elderly and disabled), four old age homes as well as two homes for the disabled were visited in and around Pune. The naturalistic observation method was applied in order to study how the end users interact with products and their surroundings. Researchers using this method of data collection are interested in observing a subject's unaltered behaviour in his normal environment [9]. The insights from the observation pointed the need to consider anthropometric data, keeping in mind reaches and clearances while design for the end users.

12.4.3 Survey

Data collection surveys question a selected sample of individuals to obtain data about their needs, opinions and requirements. It is a standard tool for empirical research that helps in identifying customer preferences. Online questionnaire with multiple choice questions were appointed for the purpose of understanding the needs of customers.

The survey was circulated among 120 adults of various age groups (18–70 years), and the results collected were then analysed to understand consumer propensity.

The results revealed that for most people, the environment and atmosphere of the space are a key factor for opting to visit a bookstore. Moreover, reading spaces within the store enhance customer experience and were selected as the most important aspect in providing a conducive ambience. The survey also brings to light the most preferred and popular genre of books among readers—fiction and novels. This is an important criteria for space allocation within the store for individual categories and their placement within the display arenas.

12.4.4 Task Analysis

One of the most important steps to conduct a meaningful user study is task analysis—a diagram or flow chart explaining the steps taken by a user to complete an activity or a task in the space. Once all the steps are laid out, the designer is in a position to eliminate unnecessary steps, with the objective of minimising the effort required to fulfil a particular activity. The idea is to simply map out the sequence of activities required to achieve a goal. This step is beneficial as it depicts the influence of the physical environment on the users while attempting to meet a goal.

In a bookstore, the goal of the customer is to browse through various genres, using the seating areas to read books and purchasing items after completing payment at the checkout counter. The experience flow of the primary users, i.e., the visitors, is elaborated in Fig. 12.3. The staff and employees form the secondary users in the bookstore. Breaking down the tasks advocates a deeper understanding of the work undertaken by the employees to achieve completion of an activity (see Fig. 12.4. These users occupy the space for a longer duration everyday, thus making this step vital in the design process.

12.4.5 Case Studies

A case study is a research method involving an in-depth study of existing spaces and noting down the observations from the same. Case studies can be used to establish pros and cons in the design of spaces that exist in the market and overcome the limitations to improve the design and customer experience.

Site visits form an integral part of conducting case studies. For the purpose of this project, visiting bookstores and observing the space planning layouts was a crucial step. There were six bookstores visited in Pune, India, including Crossword, Book World, Varada Books, Rasik Sahitya, The Word Book Shop and Pragati Book Centre. The case studies conducted revealed a typical zoning plan followed by bookstores. Most stores attempt to maximise display areas within small spaces to generate

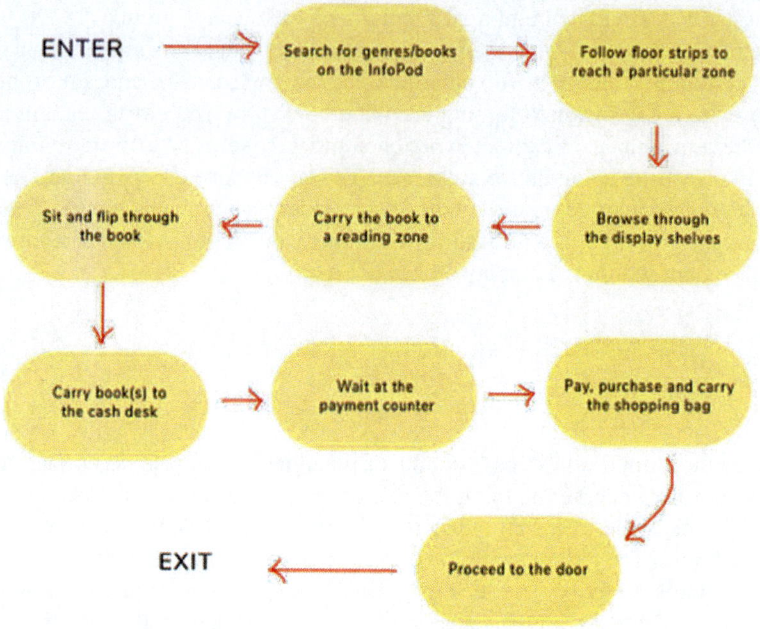

Fig. 12.3 Task flow of customers

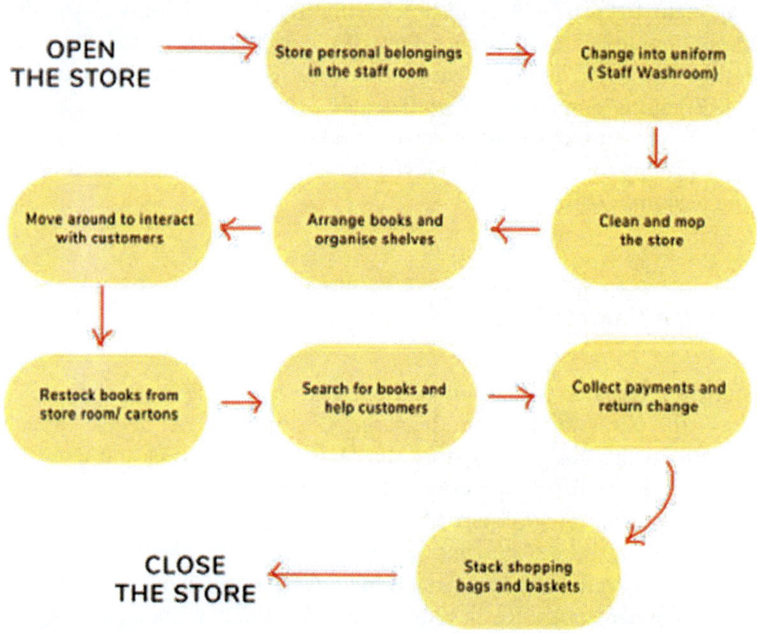

Fig. 12.4 Task flow of employees

Table 12.1 Pros and cons of existing bookstore designs

S. No.	Pros	Cons
1	Maximising display areas	Towering bookshelves
2	Clear identification of various genres	Insufficient seating
3	Space near checkout counter for waiting	Narrow corridors
4	More than one help desk placed with employees to assist	Heavy furniture that cannot be moved
5	–	Stairs for accessing main door
6	–	Absence of handicap washroom

revenue from the sale of books. The following table Table 12.1 shows the takeaways and observations of the case studies conducted.

12.5 Execution

12.5.1 Space Planning

The standalone bookstore has an area of 405 m^2. The rectangular plan has a clear entrance and exit and is divided into various zones. The display area is the most important retail zone in the bookstore and occupies about 40% of the total area. The reception and checkout counter is situated on the right-hand side as one enters (see Fig. 12.5 to allow clear exit after purchasing items. A small cafe attracts customers through the window display. People can pick up a book and enjoy a beverage in the cafe area. A central reading space along with multiple reading zones form about 5% of the plan area. Seating provisions have been provided alongside the display sections

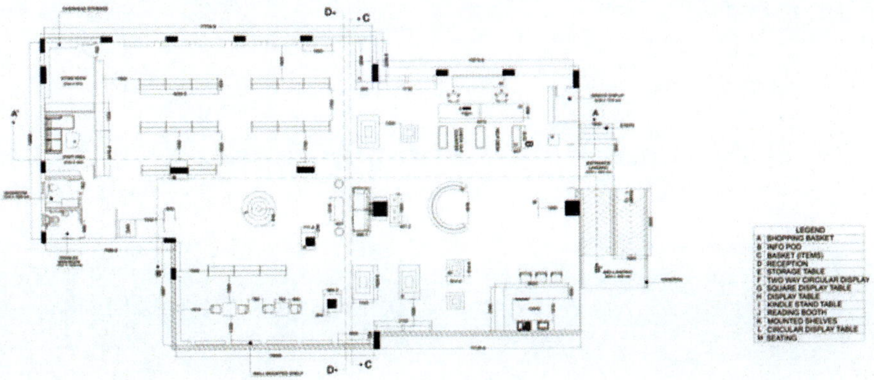

Fig. 12.5 Bookstore furniture plan

so as to allow convenience, especially for senior citizens. Information pods—smart and flexible desks that can be used as information points or computer stations—have been placed at regular intervals in the store. The information pods meet accessibility guidelines due to its adjustable height and can be used by customers to navigate within the bookstore. Other ancillary areas, situated at the back end of the store include specially designed washrooms for the disabled, a staff area as well as a store room for storing inventory. Circulation being an essential part of the bookstore utilises 35% of the space. Wide corridors and clearance along display sections are a key requirement for ease of movement and a smooth customer flow. Thus, the bookstore is designed to have enough room and is built to accommodate users with varying needs.

12.5.2 Design Solutions

One of the most important characteristics of the design process during the conceptual phase is that it is creative, innovative and unpredictable as an activity [15]. The design process allows one to go back to each problem and create practical solutions.

For instance, to solve the existing design problems mentioned in Table 12.1, multiple reading spots have been placed to minimise travel time and distance, thereby reducing the effort put in by customers. Lightweight and movable furniture has been designed along with placement of ottomans near display racks. These stools can be carried and placed wherever required. The wall-mounted shelves are designed to maximise display while keeping the reaches and heights at an accessible level for wheelchair users (Figs. 12.6 and 12.7). Thus, two-way shelves have been employed,

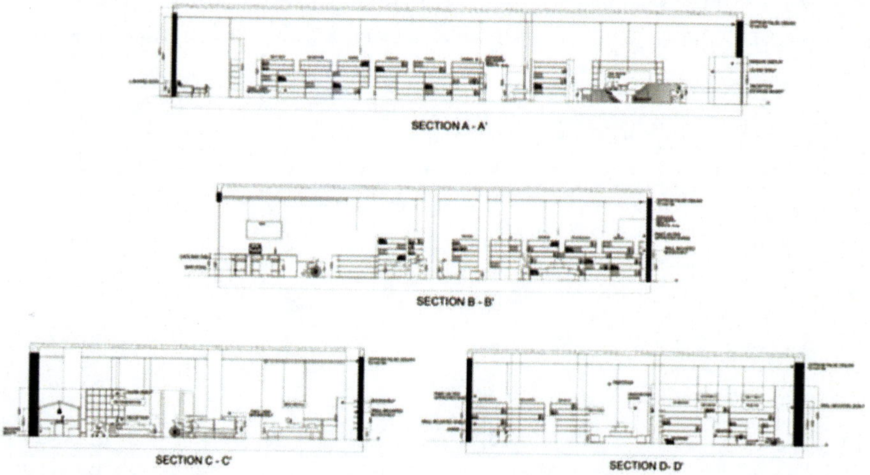

Fig. 12.6 Sectional elevations

Fig. 12.7 Wall-mounted shelves along with reading space

and the distance between consecutive shelves is wide enough for two wheelchairs to pass simultaneously with ease.

12.5.3 Selection of Materials and Finishes

In order to provide traction to easily manoeuvre wheelchairs, linoleum flooring has been employed. The linoleum acoustic tiles also reduce impact sound by 14 dB and have a cushion underfoot that helps in avoiding echo. Ceramic tiles have been opted for the service areas and washrooms as they are durable and slip resistant. For the construction of the shelving units and display fixtures, plywood with a laminate covering provides an ideal and budget friendly solution (Fig. 12.8).

Inclusivity Within the bookstore

The foremost benefit of inclusive design is to welcome everyone to encourage harmonious relations between social groups. In order to create a barrier-free space for the physically disabled and elderly, certain guidelines have been enforced to ameliorate the environment. For instance, an 1800 mm wide ramp leads to an entrance landing to allow obstruction free entry. An automatic sliding door is employed to bring the customers in to the space. The door is clearly distinguished from the glass facade with the help of 75 mm wide high-contrast strips running on all corners of the door. All corridors are 900 mm wide with a turning radius of 1500 mm.

Signages placed throughout the store permit easy navigation and identification of zones as well as genres. Suspended signs at 2000 mm height with a legible font (Helvetica) and size provide clear cut description of the shelving units. In order to

Fig. 12.8 Central display and checkout

make way finding easy for visitors, floor decals and vinyl strips have been colour coded to guide the users to specific areas (see Fig. 12.9. The inclusive design principles meet all accessibility requirements without compromising on the look and feel of the store. The colour scheme of the interiors has a neutral backdrop and flooring surface with pops of colours to highlight certain spaces such as wall mounted shelves and signage boards. The use of multiple colours not only depicts diversity but also serves a functional purpose of colour coding genres and floor strips (Fig. 12.10).

Fig. 12.9 Floor decals to guide customers

Fig. 12.10 Front view of the store

12.6 Conclusion

As a final note, this paper shows a level of inclusion that can be practised and exercised through the application of a methodical user-centred approach. The guidelines can be used to design any barrier-free public space, and the model provides a template highlighting a practical approach to design accessible spaces. Even though accessibility has improved over the last decade, and planning policy has shifted with investment providing new facilities to once-excluded communities, the fact remains that disadvantaged people are far more likely to live in poor quality environments [7]. Inclusion in design is an important step to build a future with equality. Thus, designers must help create places that work for everyone. The principles of inclusive design outlined in this paper help professionals think about subtle yet essential touches that can make a space functional and comfortable without sacrificing the aesthetic value.

Acknowledgements We would like to express our sincere gratitude to Ms. Sanjeevani Ayachit, Dy. Director and Prof. Prasanna Halbe, H.O.D., Industrial Design Department of Symbiosis Institute of Design, Pune, for their constant support and guidance. Special thanks to Prof. Tushar Amin and Dr. Tanmayee Puntambekar, for their kind cooperation to the completion of this project and research paper.

References

1. Bechmann, S.: Inclusive Design: A Perfect Solution?
2. Central Public Works Department Publishings.: https://cpwd.gov.in/Publication/aged&disabled.PDF. Accessed 8th Oct 2020 (1998)
3. Clarkson, P.J., Coleman, R.: History of Inclusive Design (2013)
4. Dam, R.F., Siang, T.Y.: Design thinking: getting started with empathy. In: Interaction Design Foundation, https://www.interaction-design.org/literature/article/design-thinking-getting-started-with-empathy. Accessed 1st Oct 2020 (2020)
5. Eikhaug, O., Gheerawo, R., et al.: Innovating with People—The Business of Inclusive Design (2010)

6. Fletcher, H.: The principles of inclusive design. In: Design Council UK Publications, https://www.designcouncil.org.uk/sites/default/files/asset/document/the-principles-of-inclusive-design.pdf. Accessed 11th Oct 2020 (2006)
7. Inclusive By Design.: In: Design Council UK Publications. https://www.designcouncil.org.uk/sites/default/files/asset/document/inclusion-by-design.pdf. Accessed 12th June 2020 (2008)
8. Keates, S., Clarkson, P.J., et al.: Towards A Practical Design Approach (2000)
9. Naturalistic Observation in Psychology.: Definition and Examples, https://study.com/academy/lesson/naturalistic-observation-in-psychology-definition-examples.html. Accessed 11th Oct 2020 (2015)
10. Newell, A.F., Gregor, P., et al.: User-Sensitive Inclusive Design (2011)
11. Peinhardt, K., Storring, N. Inclusive by design: laying a foundation for diversity in public space. In: Project for Public Spaces. https://www.pps.org/article/inclusive-by-design-laying-a-foundation-for-diversity-in-public-space. Accessed 12th June 2020 (2019)
12. Prasad, U., Clarkson, P.J.: Characterising Users for Inclusive Design—Three Requirements (2005)
13. The Benefit of Designing for Everyone.: In: Centre for Inclusive Design. PwC Australia, https://centreforinclusivedesign.org.au/wp-content/uploads/2020/04/inclusive-design-report-digital-160519.pdf. Accessed 1st June 2020 (2019)
14. Universal Design and Accessibility Planning.: Global Initiative of Academic Networks, https://www.iitr.ac.in/Main/uploads/File/2018/gian12012018.pdf. Accessed 10th June 2020 (2018)
15. Yilmaz, S., Seifert, C.M.: Cognitive Heuristics in Design Ideation (2010)

Chapter 13
Memes that Evoke Emotions: A Neurodesign Strategy for Brand Communication and Experience

Anirban Chowdhury and Prasun Chakraborty

Abstract A meme is a component of a culture or system of behaviour passed from one individual to another by imitation or other non-genetic means. Nowadays, memes are predominately observed in social media like Facebook, WhatsApp, etc. Memes communicate various messages either directly or indirectly in a codified manner. Although the memes are used for sharing social messages, strategies for brand communication using memes are less explored. Memes probably have potential to evoke responses in human brain and activates emotions. Therefore, objectives of this paper are—to analyse the effect of different design elements used in memes and the effect of memes on emotional responses and to demonstrates few strategies for brand communication using memes. A total of 20 memes were analysed to identify design elements such as character, images (human, animal, emoticons, etc.), taglines, dialog cloud, logo, etc., which helps in composition of memes. Later, the level of emotion generated by memes used for brand communication was measured using Self-Assessment Manikin (SAM) scale. It was observed that character image and taglines are effective elements that are evoking different types of emotions among participants. This was a minimalistic neurodesign strategy for brand communication which might be beneficial for effective brand communication.

13.1 Introduction

The first mention about the term "meme" is mentioned in a book "The Selfish Gene" by Richard Dawkins [1, 2]. Commonly emoticons are acting as memes that serve a number of functions in the transmission of information. Memes can be used to frame the composition of the content as positive or negative, serious or joking, or any number of other things [2]. The use of meme in advertising is still a contemporary phenomenon. Few firms are using Internet memes in order to show that they are

A. Chowdhury (✉) · P. Chakraborty
School of Design (SoD), University of Petroleum and Energy Studies (UPES), Dehradun, Uttarakhand 248007, India
e-mail: chowdhuryanirban14@gmail.com

A. Chakrabarti et al. (eds.), *Design for Tomorrow—Volume 1*, Smart Innovation, Systems and Technologies 221, https://doi.org/10.1007/978-981-16-0041-8_13

active participants of the online culture [3]. The advertisers do their utmost to make the adverts memorable so that consumers could buy the product.

This study answers the following research questions:

Q1. How to evoke desired emotions using memes?
Q2. What are the meme elements associated with affective response?
Q3. How to apply memes in brand communication?

13.2 Memes and Types

- *Classic*—most common way of presentation,
- *Dank*—unpleasantly damp and cold kinds of representation,
- *Normie*—these memes are inexperienced mainstream that peasants enjoy. These are characteristically memes that grow very big but become classified as dead very quickly,
- *Wholesome*—these memes are a subgenre of image commands in which creators disrupt audience expectation by taking well-known meme templates and using them expressing supportive, caring sentiments rather than making the jokes usually associated with each template,
- *Surreal*—these memes are not timely expressions or atemporal, a black hole in which sexist celebrities, heart breaking news stories and depressingly relatable real-life scenarios do not exist) (Fig. 13.1).

13.3 Direct and Anonymous Communications

Memes aiding people to communicate social and personal messages in a humorous way [2, 3]. Sometimes, this communication is direct using the image of actual character or person to whom it supposed to convey the message, and sometimes it is indirect maintaining the anonymity of the targeted person. It might be important for memes designer to be aware of the beneficial way which is better than the other ways. In this context, designer should be aware of the kind of reactions and emotions they likely to get because of use of a particular elements and layouts of memes. Understandability of messages conveyed by designers is another important aspect in generation of reaction and emotions of target audience for memes. A quick understanding of communicated message is depending of the past experience of the target audience about the elements used in memes [4–6]. In addition, direct communication of messages in meme might easily be understand by target audience, whereas indirect way of communication or anonymity might take longer time for understandability of message, hence, longer reaction time.

Fig. 13.1 Examples on different types of meme

13.4 Neurodesign, Affective Responses and Memes

Getting a favourable or expected affective or emotional response to memes are very crucial for designers as they always want to convey messages in a humorous and joyful ways. However, getting a vital affective response is not possible to achieve to a particular context of application of memes [3]. For this prospect, designers need to understand how to generate the desired affective response by manipulating different elements and layouts of memes. In this context, neurodesign (neuroscientific basis of acceptance of design deliverables) strategy might be fruitful [3]. Neurodesign strategy suggests that if designers able to evoke the intended emotion by stimulating the amygdala (the brain centre responsible for control of emotion [7–9]) through manipulation of elements of memes, they might get the desired outcome for the designed meme. Literature also suggests that monitoring of emotion is possible through both the psychophysiological and non-psychophysiological methods. Psychophysiological methods for emotion monitoring involve facial EMG, f-MRI, PET, etc. [3, 10, 11]; whereas, non-psychophysiological methods (using psychometric scales) involve SAM scale [12, 13], pleasure scale [14], semantic differential scale [12, 15] and other emotion measurement scales. Among psychometric scales, SAM scale is cost-effective method and very useful for measurement of emotions. Emotions are mental feelings as well as bodily changes [16]. Many times, bodily changes are very prominent and perceived by people as in case of fear, rage and happiness. These emotions are expressed through higher heart rate, sweating, redness of the skin, etc., can externally perceived by humans when at high arousal level. On contrary, bodily changes are not prominent and not externally perceived by people when at low arousal level. Affective or emotional responses might be spontaneous and also might take longer time to occur. For instance, funny or sad images might intuitively generate affective responses; whereas, reading of funny or sad text materials might take comparatively longer time for affective responses.

13.5 Study 1: Observational Study on Existing Memes

Goal of this study is to find out compositions (elements and layouts) of different memes.

13.5.1 Method

13.5.1.1 Participant

A total of five visual designers were volunteered for this study. They have work experience as a visual designer of minimum 10 years in industry (average age = 42.5 years).

13.5.1.2 Procedure

Initially, a group of visual communication designers seat together and randomly searched memes circulated in the Internet to identify different elements and layouts in composition of memes. Then, they randomly shortlist 20 memes (4 memes from each type: classic, dank, normie, wholesome and surreal) from Internet to further analyse the typical characteristics of identified elements through observational study.

13.5.2 Observations

13.5.2.1 Elements

Composition of different memes is involving various elements such as:

- *Photographs:* might be adapted from movies, natures, TV serials, news posters, etc. Most of the images selected for the composition have association with funny scenes or incidents or have some level of sense of humours

- *Characters/Carri catchers:* might be from cartoons, stickers or self-composed
- Text: variations in typeface is observed in text of memes
- *Dialog Cloud:* these are commonly used to compose memes with comics (graphical stories)
- *Emoticons:* these are generally common in social media applications. However, current trends show that meme designers are also using emoji(s) or emoticons for composition of memes
- Icon: use of icons are comparatively recent in meme. It is generally observed in meme composition having mobile or Web interfaces (Please see 2.1).
- Stickers (image/illustrations): various kinds of stickers might be considered for composition of memes.
- UIs of Mobile/Web apps: these elements are recently used by meme designers. Generally, such memes are composed of UIs of social media apps such as WhatsApp, Facebook, etc.
- *Other Graphic elements:* these involve nature illustrations, sceneries, etc.

Although there are various elements used by designers in meme, image along with text (upper and lower) are most common elements in compositions of meme.

13.5.2.2 Layouts

- There are 12 different layouts observed in the images selected for this study.
- Sizes (ratio) varies from 1:1 to 4:5 to 9:16
- Figure 13.2 'a' to 'f' are best suited with size 1:1
- Figure 13.2 'g' to 'i' are best suited with size 9: 16
- Figure 13.2 'j' to 'l' are suited either with size 4:5 or 9:16

13.6 Study 2: Application of Meme for Brand Communication

This part of this study intended to unfold the affective responses generated for the memes designed for brand communication.

13.6.1 Method

13.6.1.1 Participant

A total of 40 adults were randomly assigned for this study (20 participants for Fig. 13.3a and 20 different for participants Fig. 13.3b). Their age ranges from 21 to

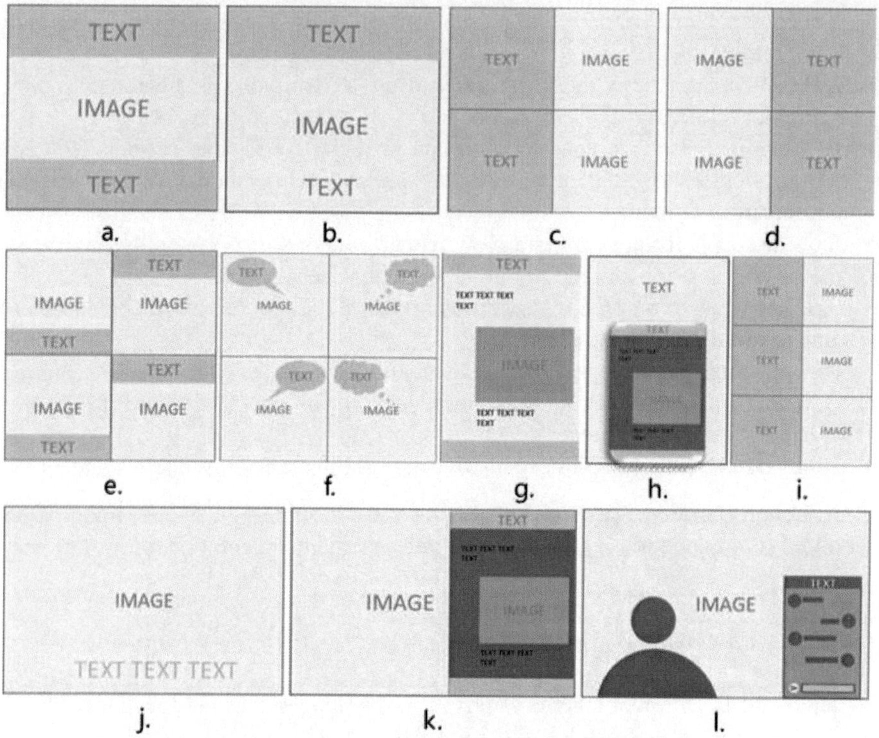

Fig. 13.2 Different layouts of memes commonly observed in Internet

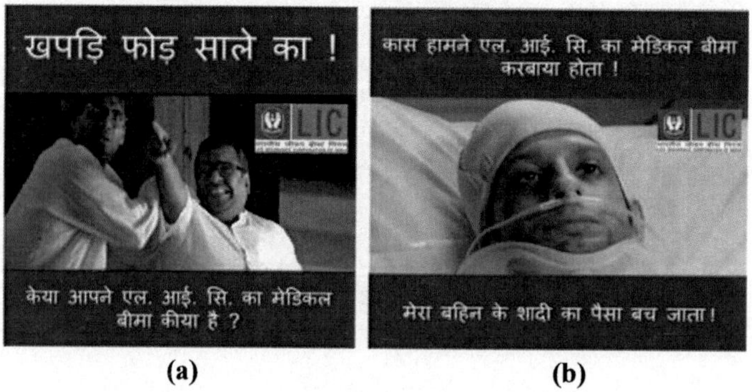

Fig. 13.3 Memes used for study of brand communication: **a** inspired from funny scene of the Hera Pheri; and **b** inspired from sad accident scene of the movie 3 Idiots

25 years and average age of 22.5 years. Among them, 60% are male and 40% are females. They do not have any medical insurance; however, they might be potential customers for any medical insurance company.

13.6.1.2 Procedure

Two memes were designed for this study to promote medical insurance of LIC India Ltd. The both designs have same layout and size. However, texts and images were different. The first design (a) was composed which was inspired from a funny scene from the Hindi film Hera Pheri. The other design (b) was inspired from a sad emotional scene from 3 Idiots. Please see Fig. 13.3 for designed memes. The initial participant group was divided into two group (each had 20 adults). One group was assigned to evaluate first design (a) and another group was assigned for evaluation of second design (b). Memes were presented on computer screen for 30 s and then they were asked to share their affective responses by rating of five-point SAM scale (Please see Fig. 13.4) which was adapted from Bradley and Lang [12] and Lang [15, 17].

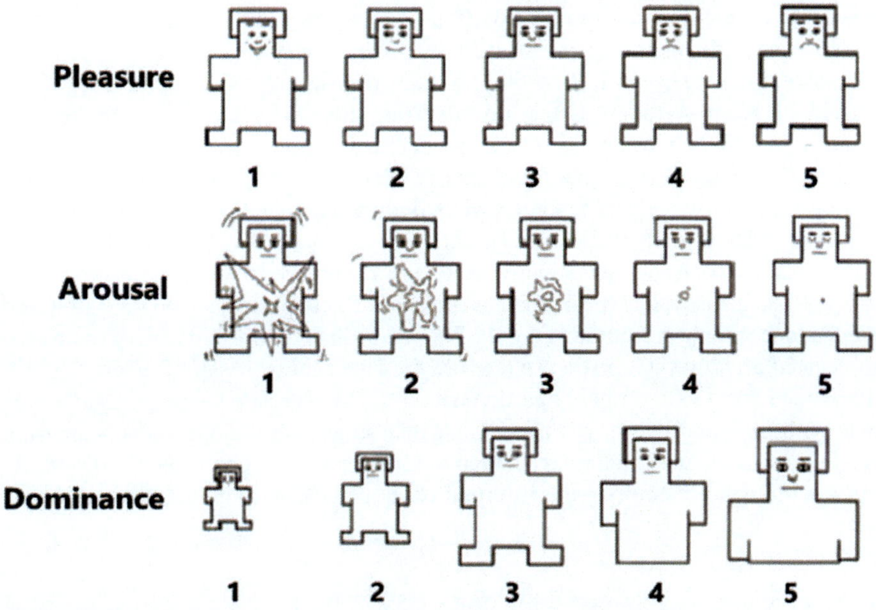

Fig. 13.4 SAM scales for subjective ratings

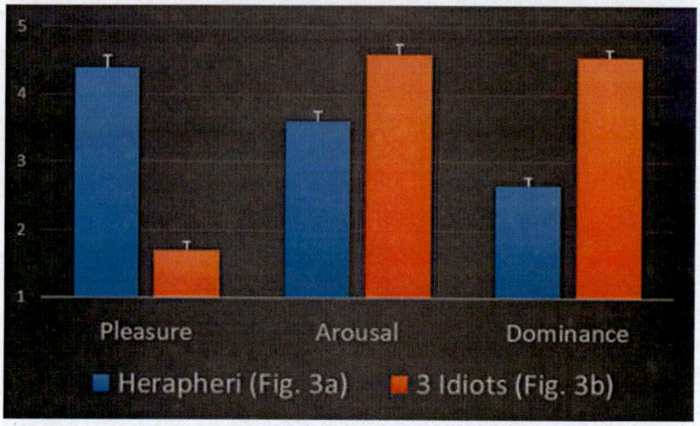

Fig. 13.5 Variation in average score of pleasure (affective state) level, arousal level and dominance

13.6.2 Result and Discussion

Average ratings for pleasure, arousal and dominance for Hera Pheri movie scene-based LIC advertisement were 4.40, 3.60 and 2.65, respectively (Please see Fig. 13.5). On the other side, pleasure, arousal and dominance for 3 Idiots movie scene-based LIC advertisement were 1.70, 4.60 and 4.55, respectively (Please see Fig. 13.5). Results of Mann–Whitney U-test revealed that there were significant differences in mean values of pleasure, arousal and dominance [pleasure Z (39) = −5.529; arousal Z (39) = −3.801; and dominance Z (39) = −5.468; $p < 0.001$]. Therefore, 3 Idiots scene-based advertisement is more dominating than the other advertisement (which was Hera Pheri scene-based); although 3 Idiots scene-based advertisement was associated with negative affective state. It was observed that the meme presented in Fig. 13.3a have evoked the happiness (positive affective and near average activated state) and the meme presented in Fig. 13.3b is associated with afraid feeling (activated unpleasant affective state) when plotted in the circumplex model of affect [18, 19] (Please see Fig. 13.6). Probably participants could better relate the negative affective state with LIC, and hence, got these kinds of results. Literature has also suggested that human mind gives more attention to the images with negative emotion [20, 21] and the brain can memorize the information associated with the negative affective states [22, 23].

13.7 Conclusion

Current study demonstrates the strategy to design memes that might be useful to create brand identity considering affective responses of the people. The used method for evaluation of memes could be used by neurodesign researchers in the future for evaluation of meme-based advertisements. Cartoon-based memes (as presented in

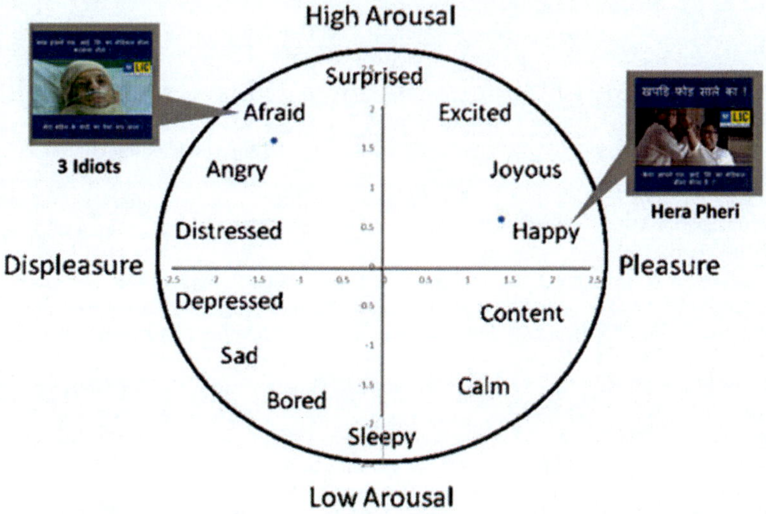

Fig. 13.6 Position of advertisement on circumplex model of affective states

Fig. 13.7) could also be made and evaluated in the future. This is a preliminary study on memes, exploring the possibility of neurodesign strategy applying memes for brand communication. However, detail study applying neurodesign techniques such as EEG and f-MRI could be conducted in the future to see effect of affective memes on human brain.

Fig. 13.7 Meme-based dettol advertisement in COVID-19 scenario

References

1. Dawkins, R.: The Selfish Gene. Oxford University Press, London (2016)
2. Davison, P.: The language of internet memes. The Social Media Reader, pp 120–34 (2012)
3. Chowdhury, A.: Neurodesign and Neuromarketing Startup Guide. Educreation Publishing, India (2019)
4. Simmons, M.P., Adamic, L.A., Adar, E.: Memes online: Extracted, subtracted, injected, and recollected. In: Fifth International AAAI Conference on Weblogs and Social Media, pp 17–21 (2011)
5. Laczniak, R.N., Muehling, D.D.: Toward a better understanding of the role of advertising message involvement in ad processing. Psychol. Market **10**(4), 301–319 (1993)
6. Mizerski, R.W., White, J.D.: Understanding and using emotions in advertising. J. Consum. Mark **3**(4), 57–69 (1986)
7. Davis, M., Whalen, P.J.: The amygdala: vigilance and emotion. Mol. Psychiatr. **6**(1), 13–34 (2001)
8. Murray, E.A.: The amygdala, reward and emotion. Trend Cogn. Sci. **11**(11), 489–497 (2007)
9. Phelps, E.A.: Human emotion and memory: interactions of the amygdala and hippocampal complex. Curr. Opin. Neurobiol. **14**(2), 198–202 (2004)
10. Costafreda, S.G., Brammer, M.J., David, A.S., Fu, C.H.: Predictors of amygdala activation during the processing of emotional stimuli: a meta-analysis of 385 PET and fMRI studies. Brain Res. Rev. **58**(1), 57–70 (2008)
11. Makin, A.D., Wilton, M.M., Pecchinenda, A., Bertamini, M.: Symmetry perception and affective responses: a combined EEG/EMG study. Neuropsychologia **50**(14), 3250–3261 (2012)
12. Bradley, M.M., Lang, P.J.: Measuring emotion: the self-assessment manikin and the semantic differential. J. Behav. Ther. Exp. Psy. **25**(1), 49–59 (1994)
13. Ong, J.C., Cardé, N.B., Gross, J.J., Manber, R.: A two-dimensional approach to assessing affective states in good and poor sleepers. J. Sleep Res. **20**(4), 606–610 (2011)
14. Gard, D.E., Gard, M.G., Kring, A.M., John, O.P.: Anticipatory and consummatory components of the experience of pleasure: a scale development study. J. Res. Pers. **40**(6), 1086–10102 (2006)
15. Hsu, S.H., Chuang, M.C., Chang, C.C.: A semantic differential study of designers' and users' product form perception. Int. J. Ind. Ergonom. **25**(4), 375–391 (2000)
16. Chowdhury, A., Reddy, S.M., Chakrabarti, D., Karmakar, S.: Cognitive theories of product emotion and their applications in emotional product design. In: ICoRD'15–Research into Design Across Boundaries, vol. 1, pp. 329–340. Springer, New Delhi (2015)
17. Lang, P.J.: The emotion probe: studies of motivation and attention. Am. Psychol. **50**(5), 372 (1995)
18. Russell, J.A.: A circumplex model of affect. J. Pers. Soc. Psychol. **39**(6), 1161 (1980)
19. Russell, J.A., Lewicka, M., Niit, T.: A cross-cultural study of a circumplex model of affect. J. Pers. Soc. Psychol. **57**(5), 848 (1989)
20. Sears, C.R., Newman, K.R., Ference, J.D., Thomas, C.L.: Attention to emotional images in previously depressed individuals: An eye-tracking study. Cognitive Ther. Res. **35**(6), 517–528 (2011)
21. Murphy, F.C., Hill, E.L., Ramponi, C., Calder, A.J., Barnard, P.J.: Paying attention to emotional images with impact. Emotion **10**(5), 605 (2010)
22. Kensinger, E.A.: Negative emotion enhances memory accuracy: Behavioral and neuroimaging evidence. Curr. Dir. Psychol. Sci. **16**(4), 213–228 (2007)
23. Kensinger, E.A., Garoff-Eaton, R.J., Schacter, D.L.: How negative emotion enhances the visual specificity of a memory. J. Cognitive Neurosci. **19**(11), 1872–1887 (2007)

Chapter 14
Design of a Low Cost Adjustable Seating System for Comfort of Weaver in 'de Sign Loom: A Semi-Automatic Handloom'

Manohar Mahato and Amarendra Kumar Das

Abstract Seating system of handloom weaver need appropriate height from floor and distance from treadle and sley handle, somehow similar to the tailor working in a sewing machine. Handloom weaver need to work by their hand and leg simultaneously in a regular dynamic pattern for long time. It is difficult to weave with same seating system by weavers having different body dimensions. Currently, weavers are using stool or a fixed bench as the seating system for weaving. Therefore, some weaver weaves in awkward posture which causes occupational health issues over time. Therefore, it is very important to have a proper adjustable seating system with weaver's reachability to handle for beat-up motion and to paddle in the treadle for shedding motion. A field study is carried out in Sualkuchi and mega handloom cluster at Sivasagar in Assam state of India to know the health issues associated with the seating system and to know about existing seating system used by the weavers. This paper is concerned with handloom weavers impacting through design intervention and development of a new seating system considering anthropometry, physical ergonomics, and operational factors. An alpha and beta model with physical model is made with anthropometry data considering man and women. Simulation is carried out with both lower and higher size manikin in CAD tool to get the adjustment range required in the new seating system of 'De sign loom: a semi-automatic handloom' design and developed by Department of Design, IIT Guwahati.

14.1 Introduction

Working by sitting is more comfortable than working by standing. Knees, lower legs, ankles, and especially feet are mostly affected during work by standing. Similarly, discomfort in hips mostly noticed during work by sit/standing [1]. Weavers of handloom are subjected to use hand and leg both during weaving. It may experience them back pain and other musculoskeletal disorders due to prolonged sitting on improperly designed seat [2]. This paper has discussed the design methods to develop a new

M. Mahato (✉) · A. K. Das
Indian Institute of Technology Guwahati, Guwahati, Assam, India
e-mail: manoharmahato@iitg.ac.in

A. Chakrabarti et al. (eds.), *Design for Tomorrow—Volume 1*, Smart Innovation, Systems and Technologies 221, https://doi.org/10.1007/978-981-16-0041-8_14

seating system suitable for parallel engagement of hand and leg. The new design is simple, it has height adjustment through simple mechanical links, which can be placed closer to the loom, and this distance is adjustable both way, can be easily removed from its location for attending to the loom like removal of woven cloth from cloth beam, etc., and is first of its kind in the handloom sector and solved the seating problem of the weavers very efficiently. This paper is targeted for handloom designer, handloom machine manufacturer, weaver, researcher, and innovator in this sector.

14.2 Literature Review

A literature review is done considering the 'adjustable chair design' keyword through keyword search strategy. 561 articles were found, but among all, only 37 were found relevance to this topic for review. Research conducted on seating system earlier have considered following design variables as given in Table 14.1.

Anthropometric body dimensions of India were selected based on the design variables of required seat components for weaver of handloom given in Table 14.2.

Table 14.1 Selected seat components in bold and design variables for seat for weaver of handloom

Seat component	Design variable in the seat
Seat pan	Height, width, length, angle, contour
Lumbar support	Height, depth, contour
Foot rest	Height, angle

Table 14.2 Indian anthropometric data of selected design variables for seat of handloom weaver

Anthropometric variable	Male		Female	
	95 percentile	5 percentile	95 percentile	5 percentile
Sitting height	893	757	809	698
Eye height, sitting	805	678	731	574
Acromion height, sitting	608	484	558	462
Elbow height, sitting	220	158	265	124
Knee height, sitting	567	497	520	462
Popliteal height, sitting	471	380	441	365
Hip breadth, sitting	405	272	429	259
Buttock to knee length, sitting	615	489	585	459
Buttock to popliteal length, sitting	512	399	494	384
Maximum body breadth	482	379	449	319
Abdominal depth	327	227	270	201
Omphalion height, (sitting)	221	154	260	165

Anthropometric variable of seat design are shown in below image showing each variable.

In Fig. 14.1, all the representations that belong to each number are given below in Table 14.3.

A subjective framework for seat comfort based on a heuristic multi-criteria decision-making technique and anthropometry was developed for automobile seat comfort based on feedback by user [3]. Automobile seating is designed for sitting,

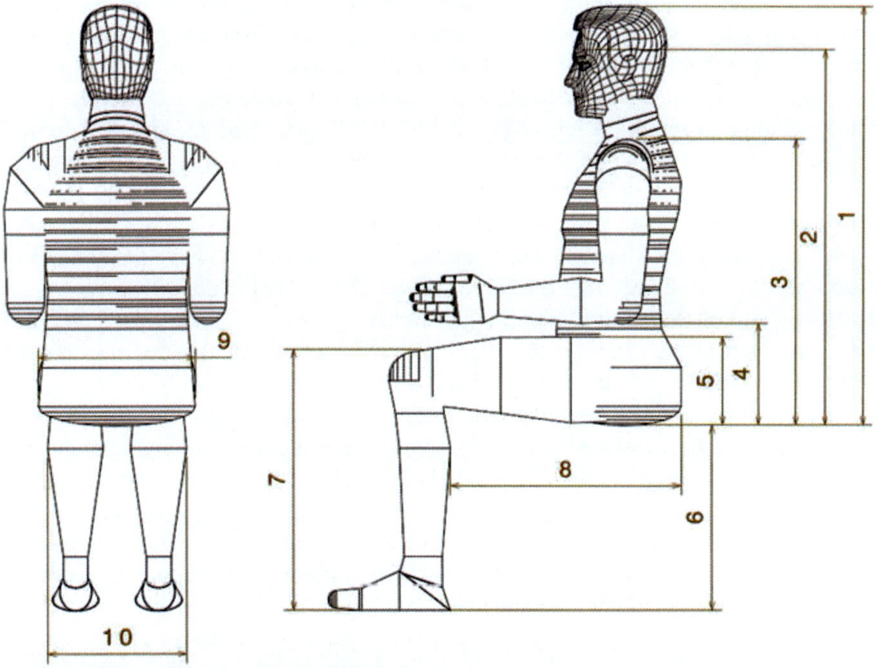

Fig. 14.1 Anthropometric measures for seat design

Table 14.3 Dimensional representation of Fig. 14.1

1	Sitting height
2	Eye height
3	Shoulder height
4	Elbow rest height
5	Thigh clearance
6	Popliteal height (stool height)
7	Knee height
8	Buttock–Popliteal length
9	Hip breadth
10	Knee to knee breadth

putting brake, operating accelerator, and handling steering by driver. Driver uses hand and leg both during driving, but it is not in regular dynamic pattern like weaver does in a handloom. Seat-back angle, seat-bottom angle and foam density, height above floor, and pressure of armrests are the affecting factor in seating posture. Seat length and width should be in between 35 and 40 cm. The seat height should be less than the distance from knee to feet. A seat-bottom posterior backward inclination of 5° is preferable, but it should not go beyond 0–10°. 135° of thigh trunk angle is a neutral position for tension in the thigh muscles. Optimum seat-back inclination is 110° in normal sitting posture [4]. Backrest should have adjustment in both vertical and horizontal plane for proper lumbar support [5]. Seat-back inclination range of 110–120° is better when seat subjected to vibration especially to operators who work for long time. Also, these effects can be minimized by attenuating seat and seat material [6]. Range of seat-back angle 95–120° is found in bus and airplane, but in train seat, it was found till 145°. Seat pan angle is found 11–16° in bus, 5–7° in airplane, and 10° in train [7]. Adjustable height range of seat pan should be 40 to 47 cm from the floor based on Indian anthropometric data [8].

Although chair has the adjustment facility, however, weavers have lack of adjustment knowledge and importance of doing it [9]. Direction of hand movement has the greatest effect on shoulder strength when working at or above shoulder level. The optimum position that maximizes shoulder strength is vertically downward [10]. Optimum and preferred hip joint angle is 79–130°, and knee joint range is 84–147°. Similarly, optimum and preferred elbow, ankle, and shoulder joint range are 80–167°, 77–130°, and 0–63°, respectively. Optimum or preferred neck flexion angle is 33–66°, and the wrist range is 133–206° [11]. Also, it should be noted that pedal position in far under the workstation is preferred for leg comfort [12]. One researcher have checked differences in pressure sensitivity of the body area in contact with seat and found that area of the seat touching the shoulder is more sensitive than lower down the back and the area in between the shoulders [13]. The differing sensitivity in the buttock and thigh areas are very important factors along with anthropometry data of the regional body dimensions during seat design. Height adjustment is an important design variable in case of prolonged use of seat [14]. Concave surface under ischial tuberosities in seat pan area with downward angle in the front 18 cm and maximum 8° forward inclination angle have improved seating comfort than plain seat pain area [15]. Previous studies indicated that flat or rearward sloping seat promote lumbar kyphosis, while forward sloping seats preserve the lumbar lordosis, and therefore, forward sloping seat is popular among researcher to study [16].

14.3 Aim and Objective

Aim of this study is to design and develop an optimum seating system for multiple weaver of handloom considering all affecting factors. Main objective is to check the synchronization of leg and hand with respect to seating system by enhancing comfort for weaver of handloom.

14.4 Methodology

We have an ongoing research on semi-automatic handloom. This research on seating system for weaver during weaving is a supplement to the above research. We have done an extensive literature review using the keyword search strategies. We have not used forward chronological search or backward chronological search strategy due to lack of enough literature on the relevant field. Seating system required for weaver is somehow similar to stitching machine operated by tailor. Also, this will be quite similar to automobile driver. Automobile driver engaged with non-repetitive work in his hand or leg. Tailor also gets some time for rest during re-orientation of cloth. In case of handloom weaver, they have repetitive cyclic work for beat-up and shedding motion.

We have done a field survey through questionnaire to examine the current practices in Sualkuchi, Assam. Weavers are using a stool or bench right now as seating system. These seating system are not adjustable to fit different body dimensions of the weaver. Ideal chair should have proper adjustment of two body dimensions, i.e., popliteal height, buttock–popliteal length. More importantly, dimension between sitting level to shoulder height is required for beat-up motion. Accordingly, beat-up handle position needs to be designed in handloom, because peddle position of the treadle for shedding can be adjusted with the hanging rope according to the position of foot in a comfortable extended leg.

There is extensive review made with all standard seating system available online as well as offline with respect to the ideal condition [2, 6, 7, 12]. Adjustable movable seating system for office was found somehow suitable since many of these have height adjustment and with coasters can move front and back. However, it is not considered for use in conjunction with handloom because, in handloom work, posture requires that the chair is fixed in relative to the loom and due to lack of locking system for movement restriction, these were not considered. Also weaving does not need a back rest like office chair, and cost of these chairs is comparatively high.

There is multiple health issues observed in the literature due to the irregular repetitive posture at occupation. Therefore, a new seating has been conceptualized to overcome from the existing seating system. Popliteal height, buttock–popliteal length, buttock to extended (rested on floor) leg comfortable length are three primary body dimensions considered for shedding motion. Mid-position length–forward arm reach with and without leaning are considered as primary body dimensions for beat-up motion.

14.4.1 Concept Design

An alpha model is conceptualized considering the low cost manufacturing to reduce the cost. Parallel motion linkage or four bar linkage mechanism is considered to adjust popliteal height and buttock–popliteal length. This mechanism was imitated from

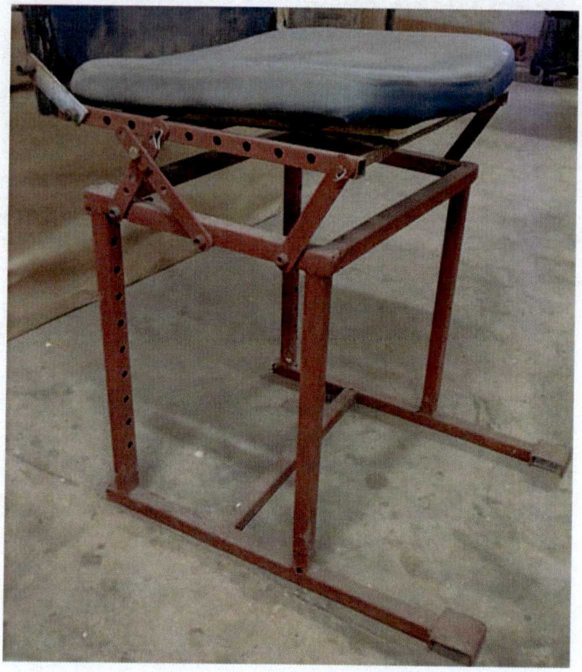

Fig. 14.2 Alpha model of seating system for handloom

pantograph of electric railway engine. Quick fixing stopper is kept for locking the mechanism in desired position. Anthropometric body dimensions were considered with 5th percentile, 50th percentile, and 95th percentile for adjustment to fit weaver with different body dimensions. It has multiple steps to fix to get best fit for any individual weaver. It is designed with 1″ square section steel tube. Welding and drilling operations were used to make this model. Mild steel is used for strength and automation friendly compared to wood. Alpha model is shown in Fig. 14.2.

14.4.2 Beta Model

Alpha model is checked functionally. It has been working well as expected. In beta model, we have eliminated all minor drawback found in alpha model like position shifting of drilled hole due to tubular section. Arc welding was burning out thin sheets during arc welding to build alpha model. Major focus was on low cost simple manufacturing process suitable to make it in mass. Laser cutting and CNC bending manufacturing processes are considered to design major parts of the seating system. 3 mm mild steel sheet metal is used for all laser cut parts except linkage parts, which have made with 5 mm. TIG welding is used to weld wherever applicable for proper welding. This model can be easily assemble and disassemble, along with easy to package. Weavers from Sualkuchi and Artfed organization with variation in weight and body dimensions have operated the handloom and found comfortable in physical loom design by us. Beta model with loom is shown in Fig. 14.3.

Beta model is made physically for trial as shown in Fig. 14.4.

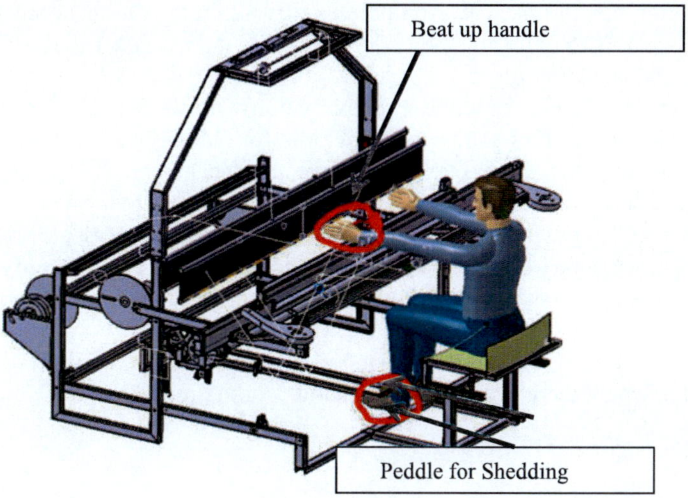

Fig. 14.3 Beta model of seating system with handloom

Fig. 14.4 Physical Beta model of seating system

14.5 Result and discussion

Our finding suggest that minimum 3 mm mild steel sheet thickness is needed for structure of seating system and 5 mm mild steel sheet thickness is needed for linkage parts. 125 mm length link with stop hole in every 25 mm is used to accommodate 5th percentile to 95th percentile body dimensions of the weavers for central position of beat-up and shedding point.

This seating system is suitable in combination with the De sign loom, a semi-automatic handloom. This can be manufactured for mass requirement in short time and can support in handloom cluster development. This research has provided few guidelines to design and manufacture handloom for comfort of weaver.

This seating system is limited to body dimension with a fixed aspect ratio. Also, spare parts may not be available without proper supply chain and inventory management. Adjustment cannot be done below 25 mm in this system, which can be improved further for micro adjustment.

This mechanism of seating system can be imitated by the profession like tailoring, driving, etc., where there is cyclic and dynamic work required by both hand and leg.

14.6 Conclusion

Newly designed seating system is subjected to many variations due to treadle movement, sley movement, and simultaneously leg and hand movement. Also, the variation in body dimensions was considered to accommodate majority weavers. This seating system will help all those weavers who worked for long hours along with De sign loom which works at 60 picks per minute compared to 30 picks per minute of traditional handloom.

References

1. Chester, M.R., Rys, M.J., Konz, S.A: Leg swelling, comfort and fatigue when sitting, standing, and sit/standing. Int. J. Indust. Ergon. **29** (2002)
2. Oyewole, S.A., Haight, J.M., Freivalds, A.: The ergonomic design of classroom furniture/computer work station for first graders in the elementary school. Int. J. Ind. Ergon. **40**(4), 437–447 (2010)
3. Fazlollahtabar, H.: A subjective framework for seat comfort based on a heuristic multi criteria decision making technique and anthropometry, Appl. Ergon. **42**(1), 16–28 (2010)
4. Harrison, D.D., Harrison, S.O., Croft, A.C., Harrison, D.E., Troyanovich, S.J.: Sitting biomechanics Part I. J. Manipulat. Physiol. Therapeut. **22**, 594 (1999)
5. Drury, C.G., Francher, M.: Evaluation of a forward-sloping chair. Appl. Ergono. **16** (1985)
6. Goonetilleke, R.S., Feizhou, S.: A methodology to determine the optimum seat depth. Int. J. Indust. Ergon. **27** (2001)
7. Jung, E.S., Han, S.H., Jung, M., Choe, J.: Coach design for the Korean high-speed train: a systematic approach to passenger seat design and layout. Appl. Ergonom. **29** (1998)

8. Sahu, M., Joseph Vijay, S., Wilson Kumar, M., Gnanaraj Solomon, D., Gnanaraj, D., Kumar, W.: Article ID: IJMET_09_13_006 design and fabrication of an automated seat pan and armrest height adjustable chair. Int. J. Mech. Eng. **9**(13), 45–53 (2018)
9. Underwood, D., Sims, R.: Do office workers adjust their chairs? End-user knowledge, use and barriers to chair adjustment. Appl. Ergono. **77**, 100–106 (2019)
10. Chow, A.Y., Dickerson, C.R.: Shoulder strength of females while sitting and standing as a function of hand location and force direction. Appl. Ergono. **40**(3), 303–308 (2009)
11. Schmidt, S., Amereller, M., Franz, M., Kaiser, R., Schwirtz, A.: A literature review on optimum and preferred joint angles inautomotive sitting posture, Appl. Ergono. **45**(2 PB), 247–260 (2014)
12. Delleman, N.J., Dul, J.: Sewing machine operation: workstation adjustment, working posture, and workers' perceptions. Int. J. Indust. Ergono. **30** (2002)
13. Vink, P., Lips, D.: Sensitivity of the human back and buttocks: the missing link in comfort seat design. Appl. Ergono. **58**, 287–292 (2017)
14. Magnusson, M., Hansson, T., Pope, M.H.: The effect of seat back inclination on spine height changes. Appl. Ergono. **25** (1994)
15. Graf, M., Guggenbiihl, U., Krueger, H.: Investigations on the effects of seat shape and slope on posture, comfort and back muscle activity. Int. J. Indust. Ergono. **12** (1993)
16. Goonetilleke, R.S., Rao, B.G. Forward sloping chair effects on spinal shape in the Hong Kong Chinese and Indian populations. Int. J. Industr. Ergono. **23** (1999)

Chapter 15
Redesigning Walker for Elderly

Nilanjana Bairagi and Prathmesh Mahajani

Abstract The present paper aims to redesign a walking aid for the elderly through a user-centric design research approach. The study involved 12 elderly volunteers. During the research, detailed observation has been carried out in the area of the daily activities of the elderly, in their real environment with assistive devices and interaction with other objects using ethnographic research method. Users' problems and preferences toward assistive devices were also studied. It was found from the pilot study that the existing walkers make the elderly feel psychologically dependent and disabled. Many elderly persons found it shaky while using. This leads to non-use or abandonment of the walker, though medically advised. Therefore, this study focused on redesigning of walker to address the above issues. The designing criteria were derived from the observations and interviews with stakeholders, concepts were developed, prototypes were made and user tested, before arriving at the final design. During prototyping and material exploration, indigenous natural material like rattan cane was used to design the walker with additional features like bottle holder, newspaper holder, and mobile phone holder. User trials indicated the advantages of the rattan walker-like easier handling, faster gait speed due to 16% lighter in weight, and aesthetically pleasing appearance. There was an increase in the gait speed of the users using rattan cane walker as compared to the conventional walking frame. The psychological effect of feeling disabled was significantly reduced with the new design as observed during the study. The study shows the possibility of designing a user-centric walker using a rattan cane walker for the elderly in India.

N. Bairagi · P. Mahajani (✉)
National Institute of Fashion Technology, Bengaluru, Karnataka, India
e-mail: prathmesh.mahajani@nift.ac.in

N. Bairagi
e-mail: nilanjana.bairagi@nift.ac.in

A. Chakrabarti et al. (eds.), *Design for Tomorrow—Volume 1*, Smart Innovation, Systems and Technologies 221, https://doi.org/10.1007/978-981-16-0041-8_15

15.1 Introduction

Aging is the process of becoming older, wherein the person's cells and tissues start to change which leads to greater risk of being diseased or dead [1]. As defined in a research [2], the elderly is the chronological age of 65 years. Mass of the lower limb muscles reduces at an older age. This phenomenon is called sarcopenia [3]. This results in imbalance and falls among the elderly. When the center of mass (COM) exceeds the bounds of the base of support (BOS), falls are likely to occur. The elderly walk slowly and tire more quickly due to loss of strength and mass in leg muscles due to aging. More than 70% of the falls among people aged more than 65 take place at home and 20% of them need hospitalization [4]. The most common consequences of falls are hip fractures, restricted activities, psychological impacts like reduced self-esteem, and reduced quality of life [5]. The majority of falls lead to hip fractures, which lead to either long-term hospitalization or death [6]. To prevent this, assistive devices like walkers are used. Walkers increase the base of support of the elderly, thereby improving their balance and stability [7].

Although the likelihood of disability increases with aging, it is not just associated with a particular age of population but may affect at any age. Some common reasons for disability are chronic diseases like cardiovascular disorders, musculoskeletal disorders and mental health conditions like dementia and depression. Other reasons are injuries due to accidents or mishaps and malnutrition [8].

Assistive devices like canes, crutches, walkers and wheelchairs are advised to meet mobility challenges among the elderly. Different novel solutions have been proposed through a user-centric design approach to meet the mobility challenges of the elderly. For instance, Khan et al. [9] have designed a novel sit-to-stand walking aid that incorporates a pantograph mechanism to aid sit-stand transfer, coupled with a walking frame to assist walking. Another lift-assist walker has been proposed by Bulea and Triolo [10] for individuals suffering from spinal cord injury, which uses gas springs to create powerless lift assist. Integration of ergonomics in product design is found to be relevant and useful in designing products for differently-abled users. This has been shown by Das and Seid in their work with the self-transfer assistive devices for wheelchair users [11].

Fundamentally, a walker is a four-legged frame used by people who need support in balance and mobility. It is also used by older adults to prevent falls. During the pilot study, it was observed that out of the different types of walkers like rollators (wheeled walkers), weight-bearing type (with wheels on rear legs of the walker), and non-weight bearing type of walker (with all four legs of walker without wheels), the most commonly used was the non-weight bearing type. As a part of the pilot study, during interviews with doctors and physiotherapists, it was learned that there is a design opportunity in the area of the walker in the Indian context as many patients are not comfortable with the existing commercially available models. As there is limited research in the Indian context, using a user-centric approach in the designing of an assistive device for the elderly, the aim of this research was to redesign a walker for the elderly. This would help towards the improvement of physical rehabilitation.

15.2 Research Methodology

A user-centered design approach was undertaken for this research as presented in Fig. 15.1. User-centered design is an iterative design process in which the designers focus on the users and their needs in each phase of design [12]. The first step in user-centered designing is to understand the user needs and requirements. This is done through an ethnographic research approach where the users were studied in a real environment in their home setting or home-cum-rehabilitation center. During the ethnographic observations, the activities, environment, interaction, objects and users (AEIOU) principles of design thinking were used. The detailed observation was carried out in the area of their daily activities, their living environment, the interactions in their real environment with the assistive devices and other objects and user preferences were studied. Users were also interviewed to study the ease of using a walker and performing daily tasks. Empathy mapping was carried out to understand these needs from the perspective of the potential users.

A case study approach has been used. Out of a total population of 28 patients from an old age home-cum-rehabilitation center, Kadji Care, Vadodara, Gujarat, interviews of 12 senior citizens using walkers, who volunteered to be a part of the

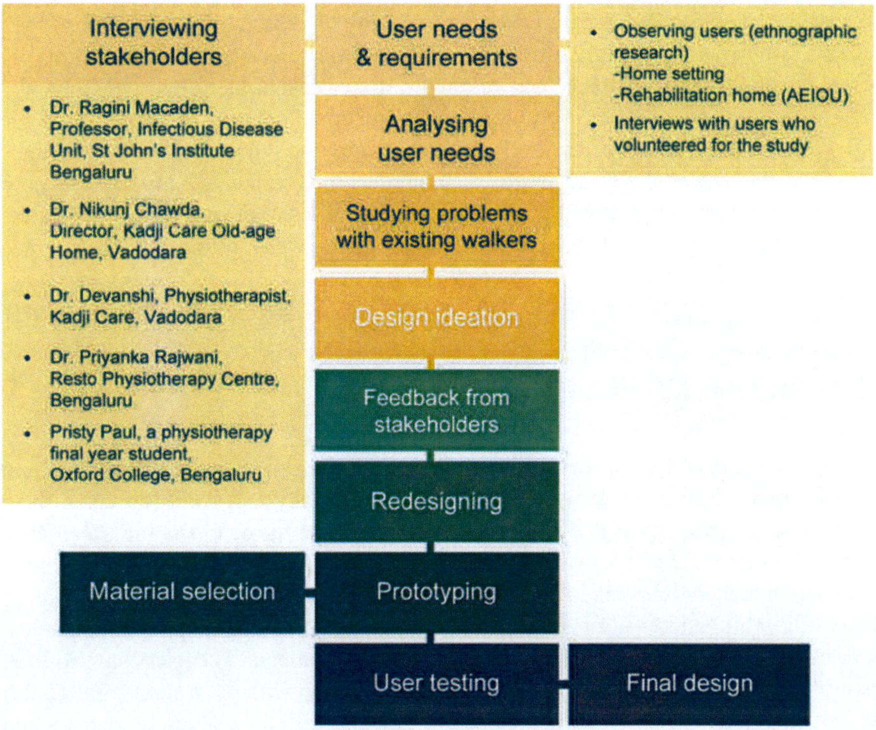

Fig. 15.1 Research methodology

study, were carried out. Informed consent was obtained from all the participants. Contextual interview questions were asked during the ethnographic research like: 1. Have you ever experienced a fall? Where? How? How frequently? 2. What kind of assistive device do you have? How do you find it? How easy it is to use this device? How useful do you think it will be to move around this living environment with this assistive device? How comfortable do you feel when using this device in front of other people? 3. Do you think the device can be improved? How? Answers to these questions have been collected and were noted in the form of insights.

To understand the perspective of healthcare professionals—physiotherapists, orthopedic doctors, and neurologists have been interviewed using convenience sampling. A comparative analysis of the most frequently bought walkers was done with help of 10 surgical shops in Bengaluru, India.

User needs have been assessed at home settings and at rehabilitation center during the research. The main requirements in a walker were identified during the observation and interaction with the users. The emotional feeling associated with the use of a walker was also investigated through care-givers. The activities of the users are observed like—using the walker to stand up, navigate in the home spaces, including accessing the washrooms, sitting on a chair, using mobile phones, carrying small objects like a bottle of water or newspaper, their stability, and convenience of using the walker were also observed.

Designs were conceptualized based on the initial research. User feedback was taken to improve the design and during the material selection process. Prototypes were developed and feedback was taken from the stakeholders. User testing was carried out with two elderly volunteers, with informed consent. To understand the difference in their gait speed, elderly volunteers are asked to travel a 12 m distance using the conventional walker as well as the redesigned walker. A stop timer is used to note the time taken to travel the distance and their average speed was calculated. Based on the feedback and observations, the prototype was modified and redesigned before the development of the final design.

15.3 Research Findings

During the study, it was observed that the main needs of the users are durability and stability of walkers. The conventional walker gets shaky and wobbly over a short period of use. The grip of the walker is too short for the hands of the Indian elderly. Due to smaller base of the walker, the walker's legs are often placed on the user's footwear, which at times leads to falling. The users expressed the need for a pocket to keep utility things like—a bottle of water or essentials like newspaper or spectacles. A foldable seat was also expressed as one of the requirements. The users have to look behind to avoid fall while performing standing-to-sitting with the walker (Fig. 15.2c) while using conventional walkers. Once the user sits with the walker, he gets locked and someone needs to move the walker away (Fig. 15.2b). It was also found that the common reason to prefer a non-weight bearing walker is its stability and sense

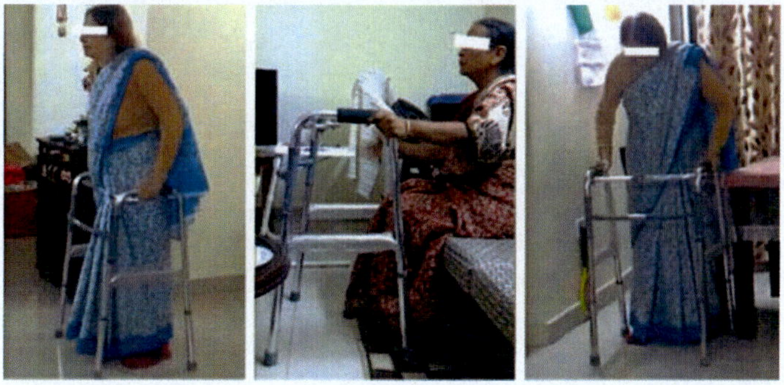

a. b. c.

Fig. 15.2 **a** User navigating inside home using walker, **b** user seated and locked around by walker in front side of body, **c** user performing sit-to-stand and stand-to-sit

of confidence from falls. Also, it has fewer number of moving parts and needs less maintenance. The wheeled walkers were not preferred by most of the users as it is less stable and the users need to walk faster than their comfortable pace. It was also expressed by the users that the walkers make them feel psychologically dependent and weak.

Therefore, the design criteria that were derived from the user studies are:

- To design a more stable and durable walker
- Absence of pinch points and sharp edges.
- To be able to withstand everyday use for 2–3 years
- To be able to fit in 18 × 22 × 31-inch space for transportation [7]
- Must be psychologically and aesthetically pleasing to the user.

15.4 Development of Concept Sketches

Based on the research findings, initial sketches focused on designing the frame that would meet the criteria of better stability as shown in Fig. 15.3. The ergonomics of using a walker were considered while concept generation based on the observations. Figure 15.4 shows the study of grip, sizing, or dimensions and shape for the appropriate handling of walker [13]. Accessories like alarm/ horn, torch, seat and bag were also conceptualized from users' interviews and feedback.

15.5 Prototyping

To meet the basic requirements of stability and durability, different materials were explored like steel, plastic, and organic materials as shown in Fig. 15.5. In terms of

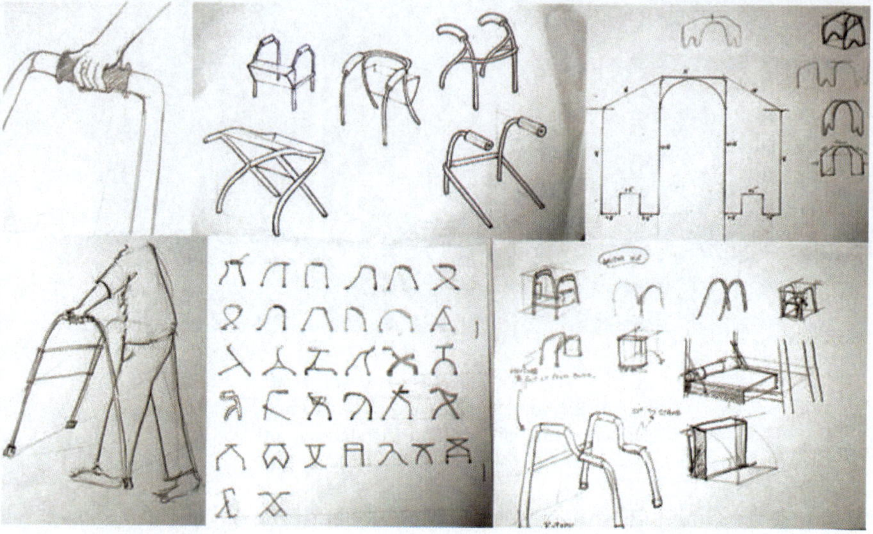

Fig. 15.3 Sketches of concepts of grip design and walker's frame design

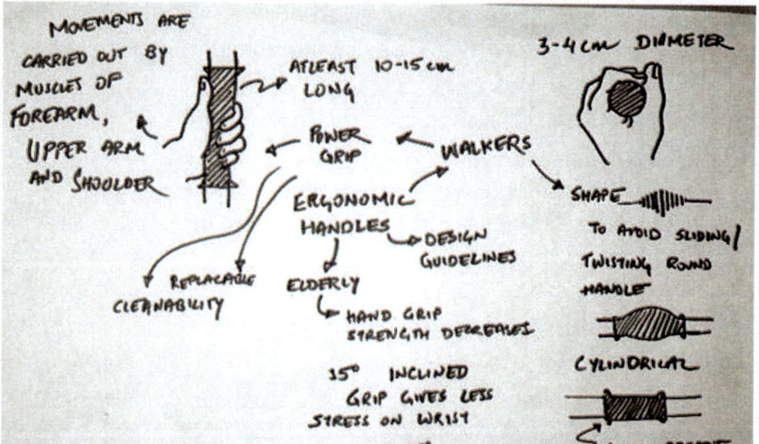

Fig. 15.4 Conceptualization of grips for walker design

steel, low-scale frugal prototypes are made using locally sourced galvanized wire (GI) wire and parts of the Mechanix tool kit. To reduce the weight of the walker, lightweight polyvinyl chloride (PVC) pipes were also explored. It was observed that lightweight PVC structures are prone to bending and twisting.

Walker designs were further developed using natural material rattan cane as presented in Fig. 15.6. The prototype was developed to check the feasibility of

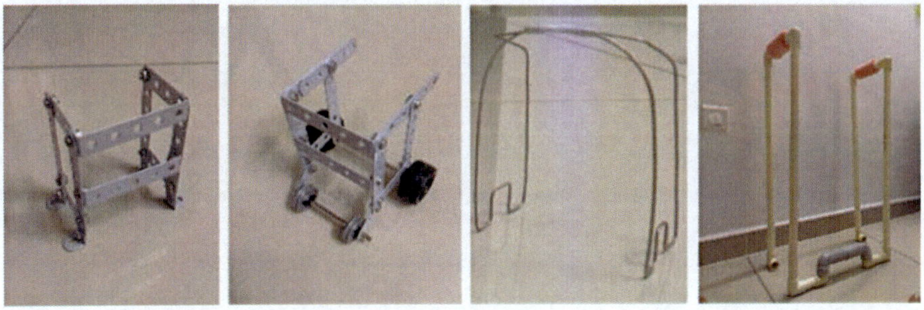

Fig. 15.5 Initial prototyping of walkers using Mechanix tool kit, GI wire and PVC pipes

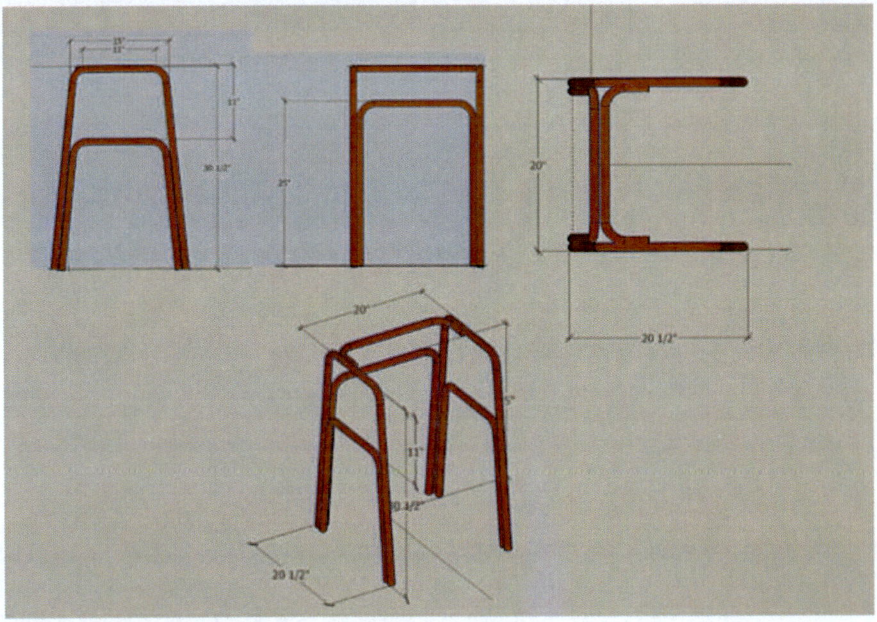

Fig. 15.6 Conceptual design of a walker made out of rattan cane

bending the material, the stability of the structure, and weight assessment as presented in Fig. 15.7. The dimensions of the rattan cane walker were 20 (width) × 20 (depth) × 30 (height) inches.

15.6 User Testing

User testing of the rattan cane walker was carried out in real set up as shown in Fig. 15.8. It was found that the prototype met the required criteria of stability and

Fig. 15.7 Process of walker design using rattan cane

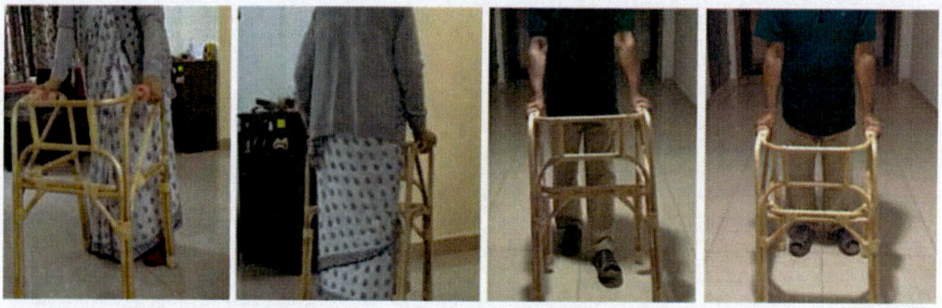

Fig. 15.8 Users trying out the rattan cane walker in home setting

was easy to use. From the user's feedback, it is found that the legs of the walker may be prone to wearing off, thereby leading to instability of the walker. To overcome this problem, rubber stubs for walker's legs were designed. Additionally, gait speed was calculated using the rattan cane walker.

To study the difference in their gait speed, two elderly volunteers are asked to travel 12 m distance using conventional walker v/s rattan cane walker. A stop timer is used to note the time taken to travel the distance and speed was calculated. The results are presented in Tables 15.1 and 15.2. The results of the user testing indicate that the average gait speed of the two elderly users using a rattan cane walker is more

Table 15.1 Gait parameters using aluminum walking frame

	Stride length (cm)	Distance (m)	Time taken (s)	Speed (m/s)
Elderly 1	32	12	222	0.054
Elderly 2	50	12	152	0.078

Table 15.2 Gait parameters using rattan cane walker

	Stride length (cm)	Distance (m)	Time taken (s)	Speed (m/s)
Elderly 1	30	12	120	0.1
Elderly 2	50	12	75	0.16

as compared to the average gait speed while traveling the same distance using a commercially available metallic walking frame. The comparison has been presented in Table 15.3. This may be attributed to the ease of using the rattan cane walker as it is 16% lighter in weight than the metallic frame walker as presented in Table 15.3.

The rattan cane walker was perceived to be more user-friendly and light in weight. The prototype was also viewed to be aesthetically pleasing to the users, thereby reducing the psychological barrier that they have toward the commercially available walkers.

15.7 Conclusion

During the pilot study, it was learned that there is an opportunity in the area of redesigning a walker in the Indian context as many users are not comfortable with the existing commercially available walkers. Additionally, there is limited research in the Indian context; using a user-centric approach in the designing of an assistive device for the elderly. Therefore, the objective of this paper was to redesign a walker for the elderly that would help in physical rehabilitation.

The methodology used was user-centered research, using ethnographic research, where the users were studied in a real environment in their home setting or home-cum-rehabilitation center. 12 senior citizens volunteered for the study. Detailed observation in the area of their daily activities in their actual living environment, the interactions with the assistive devices and other objects were carried out. During the research, various designs are conceptualized, prototypes were developed and natural material-like rattan cane was also explored. The main design needs were the stability of the walker, ease of use while doing the daily activities. Indigenous material like rattan cane was used to design the walker with a larger base area and additional features like bottle holder, newspaper holder, and mobile phone holders. The redesigned walker was preferred by the elderly in the study. The results of the user testing indicated that the average gait speed of the elderly users using a rattan cane walker is more as compared to the average gait speed while traveling the same distance using a commercially available metallic walking frame. This was attributed to the ease of using the rattan cane walker as it is 16% lighter in weight than the metallic frame walker. The psychological effect of feeling disabled was also less among the users with the rattan cane walker. Therefore, rattan cane can be used as an alternative natural material for designing of custom-made walkers for the elderly in the Indian context.

Table 15.3 Parameters and challenges addressed by the redesigned walker

Parameters	Challenges with commercially available metal frame walkers	Design proposed	Features of the developed rattan cane walker
Stability	Walker is wobbly	Need for a more table design/structure	Walker is stable as the weight gets well distributed along the legs of walker
Illumination	Difficulty in seeing obstacles on the ground under poor illumination	Torchlight can be attached to the walker	An attachment may be incorporated
Mobility	Difficulty in sitting-to-standing	For sitting-to-standing, counterweights can be added to the front end of the walker	Weight of walker is naturally more toward the front two legs, making it firm
Ergonomics	Walker's leg might fall on the foot of the user by mistake	Need for bigger base of support	Wide legs give a good base for support and mobility
Ergonomics	Need to adjust the height of the walker for different users	Custom made walkers can be used	The walker may be custom made in the range of 75–95 mm, for a particular height of the user
	This is difficult for the elderly as the studs are not easy to push in the slots for height adjustment		
	Short grip	A longer grip can be used	Grip is long and firm
Maintenance	This also makes the walker noisy and shaky which needs maintenance		Less maintenance, as there are no nuts and bolts
Accessories	No pocket/space to carry essentials like spectacles, newspapers, books	A pocket can be added, even a bottle stand can be added to the walker	Rattan cane basket can be easily attached to the walker
Weight	Around 3 kg weight, heavy	Lightweight material may be used	Rattan cane walker weighs less, 2.5 kg
Temperature	Aluminum gets cold in winter and hot in summer depending upon ambient conditions	Material with poor thermal conductivity may be used	Rattan cane is a poor conductor of heat so the temperature variation in the product is less

(continued)

Table 15.3 (continued)

Parameters	Challenges with commercially available metal frame walkers	Design proposed	Features of the developed rattan cane walker
Bending modulus	68.9 GPa for aluminum [14]	Should meet the requirements	2.6 GPa for rattan cane treated with methyl methacrylate [15]
Compressive strength	30 MPa for aluminum [14]		36.96 MPa for rattan cane treated with methyl methacrylate [15]
Average gait speed to cover distance of 12 m (6 m forward and return to starting point)	Average gait speed for elderly using aluminum walker was 0.066 m/s		Average gait speed for elderly using rattan cane walker is 0.13 m/s
Average cost of walker	INR 1100/-		INR 3000/- (for prototyping as a custom made product)

References

1. Tosato, M., Zamboni, V., Ferrini A., Cesari, M.: The aging process and potential interventions to extend life expectancy (2007)
2. Orimo, H., Ito, H., Suzuki, T., Hosoi, T., Sawabe, M.: Reviewing the definition of elderly. Geriatrics Gerontol. Int. **6** (2006)
3. Spirdus, W.W., Francis, K.L., MacRae, P.G.: Physical dimensions of aging (2005)
4. Berg, R.L., Cassells, J.S.: The Second Fifty Years: Promoting Health and Preventing Disability. National Academy Press, Washington D.C. (1992)
5. Luz, C., Bush, T., Shen, X.: Do Canes or Walkers Make Any Difference? NonUse and Fall Injuries
6. Parkkari, J., Kannus, P., Palvanen, M., Natri, A., Vainio, J., Aho, H., Vuori, I., Jarvinen, M.: Majority of hip fractures occur as a result of a fall and impact on the greater Trochanter of the Femur: a prospective controlled hip fracture study with 206 consecutive patients. Calcif. Tissue Int. **65**, 183–187 (1999)
7. Foley, M., Johnson, K., Kalbach Jr, J.E., McNally, B.: Improved Walker Design: A Major Qualifying Project Report. WORCESTER POLYTECHNIC INSTITUTE (2010)
8. Sarkar, S., Das, A.K.: Design Modularity in the Assistive Devices for the Elderly People. Springer Nature Singapore Pte Ltd., Guwahati (2017)
9. Khan, M.R., Patnaik, B., Patel, S.: Design and Development of a Novel Sit-to-Stand and Mobility Assistive Device for Ambulation and Elderly. Springer Nature Singapore Pte Ltd (2017)
10. Bulea, T.J., Triolo, R.J.: Design and Experimental Evaluation of a Vertical Lift Walker for Sit-to-Stand Transition Assistance. Cleveland, USA (2012)
11. Seid, K., Das, A.K.: Conceptual Design and Evaluation of a Novel Self-transfer-Assistive Device. Springer Nature Singapore Pte Ltd, Guwahati (2019)
12. Abras, C., Maloney-Krichmar, D., Preece, J.: User-centered design. In: Encyclopedia of Human-Computer Interaction (2004)

13. Patkin, M.: A Check-List for Handle Design. Ergonomics Australia On-Line, South Australia (2001)
14. AZoM.: Aluminum—Advantages and Properties of Aluminum, 4 June 2002 [Online], Available: https://www.azom.com/article.aspx?ArticleID=1446.
15. Liu, X., Yang, S., Shang, L., Ma, J.: Properties of Rattan cane and its comprehensive utilization technology. In: International Seminar on Rattan Sustainable Management and Utilization in South-East Asia.

Chapter 16
It is not a Driverless Car!—A Framework for Interacting with the AI in Autonomous Vehicles

Peer M. Sathikh and Guan Yi Tan

Abstract With autonomous vehicles (AV) fast becoming a reality, in the not so distant future, streets are expected to be populated with these intelligent autonomous beings ferrying passengers alongside human-driven vehicles and pedestrians. The bulk of research today seems to focus on setting up the required software and hardware systems, alongside policies and transportation infrastructure. A gap is seen to exist within the realm where the passengers and vehicle AI interface. This calls for research in developing capabilities for the AI in AVs a form of social transactions that are more familiar to human beings, taking cues from natural conversations and body language signals. The reason for this is that autonomous vehicles not only need to perform navigational duties, but also need to communicate, respond, and reciprocate in a social manner. This would give the AI the dignity they deserve when performing their functions, as performing these activities accords with the capabilities of a self-aware, intelligent being. This paper proposes a framework for future communication between autonomous vehicles and humans, focusing on the working relationship between the vehicle and its passengers taking references of the communications theory related to human-to-human interactions, extrapolating to human-to-AI interaction within the AV.

16.1 Introduction

Autonomous vehicles are starting to become a feasible mode of transportation in many countries. In the not so distant future, streets are expected to be populated with intelligent digital beings transporting passengers alongside human-driven vehicles and pedestrians. Progress in realizing this future today appears to focus mainly on setting up the required software and hardware platforms, policies, and infrastructure

---

P. M. Sathikh (✉) · G. Y. Tan
Nanyang Technological University, Singapore, Singapore
e-mail: peersathikh@ntu.edu.sg

G. Y. Tan
e-mail: gytan@ntu.edu.sg

A. Chakrabarti et al. (eds.), *Design for Tomorrow—Volume 1*, Smart Innovation, Systems and Technologies 221, https://doi.org/10.1007/978-981-16-0041-8_16

to make it happen. A shared goal is apparent, in which a network of intelligent beings is capable of making informed choices when performing their work, and seamlessly adapt to the needs of their passengers. However, a gap is seen to exist in the space where the being and its human passengers interact. It is predicted that in future they would be trusted with more sophisticated tasks previously given to humans since it required creative problem solving and emotional awareness. Yet not much research has gone into how an AI in an autonomous vehicle (AV) setting should and could engage in social interactions with its passengers to perform its driving tasks. To perform as a "chauffeur," the being not only needs to perform navigational duties, but also communicate and reciprocate social signals as a human driver would. What is interesting is that this being comes as a new form of intelligence that does not need to restrict its physical form to any iconic appearance, inviting a new approach to enabling natural social interactions. To this end, the authors propose a framework for future communication between this new form of intelligence and its passengers, gaining inspiration from existing communication modes that humans use to synthesize a multimodal human-to-AI interactions.

16.2 Review of Autonomy in Vehicles Today

The end goal for all autonomous vehicle research appears to be a network of intelligent machines with a cognitive ability like human drivers performing its chauffeur services. Human travelers would sit inside without hesitation or surprise as the car navigates to the intended destination, forming a traffic landscape populated with human drivers, pedestrians, small personal mobility devices, large public transport vehicles, etc. While this utopian future of fully autonomous vehicles (AV) has yet to arrive, partially autonomous features in recent models have begun to assimilate into city streets.

The levels of autonomy are defined in five levels [1], from currently available "driver assist" functions to the ideal future of level five fully autonomous vehicles requiring no human intervention. Developments in autonomy levels 2–4 can be seen in places such as Singapore with NuTonomy's taxi fleet, German carmakers producing models such as the 2019 Audi A8 and Mercedes F105 concept, and the Tesla series of electric vehicles in the USA. In this paper, the focus is on levels 4–5 only. Levels from 1 to 3 also known as conditional automation require the human to always pay attention to the vehicle's driving as a backup option. Simulated situations involving level 3 AVs [2] and traffic accident reports involving level 2–3 vehicles [3] show that drivers, if not attentive to the driving task, need more time to respond and transition back to the driver role. Baruch [4] argues that a handover feature should not exist, instead letting either the human or digital being be the driver at all times (Fig. 16.1).

Autonomous vehicles (AVs) eventually are expected to benefit city life by sharing, rather than dominating the layout of a cityscape, decreasing traffic congestion, and contribute to increasing road safety [5]. Streets are projected to enable more shared us

Fig. 16.1 Mercedes F105 concept

as AVs are able to make safer and more efficient use of road area by communicating with each other using vehicle-to-vehicle (V2V) networks and also to the city's traffic infrastructure by updating each other about their whereabouts and routes [6]. Human error in driving also accounts for 93% of traffic accidents in the USA [7], which the introduction of AVs is planned to decrease. Still, fear exists at many levels, starting from governments to the common person that inhibit the projected acceptance of AV use. A 2019 survey [8] has found that 71% of Americans fear riding in an AV, a jump from the 63% in 2017. In a Wizard-of-Oz[1] study, [9] participants when driven by an AV in a simulator had different initial experiences of being in this new mode of transportation which brings different passenger perceptions and mental models of its function to question. Another 2019 survey [10] has also found that between 46 and 48% of its participants attribute fear and anxiety, respectively, to the idea of AVs with 55% of the survey participants also expecting the AVs to make sound decisions during unexpected situations. To meet these expectations of alleviating anxiety and assurance of sound decision-making, the authors believe that an AV needs a capable form of intelligence to be adept at communicating its intent to passengers, graduating into utilizing natural ways of communication that humans are akin to rather than through rigid programmed commands. Kun et al. [11] echo this, highlighting a need to solve the user experience of interactions in AVs.

The question then is, how can design research contribute to bringing about a naturalized interaction between the intelligence in the AV and its human passengers? This paper discusses social interaction aspects of the digital being, making assumptions that the proposed digital being is capable of sensing what is signaled both verbally and non-verbally in a social interaction, while being technically proficient in navigational tasks as part of its core functions package.

[1]A method in which participants are led to believe they are interacting with a working prototype of a system, but in reality, a researcher is acting as the proxy for a system from behind the scenes. Martin B, Hanington B (2012) Universal methods of design. 204.

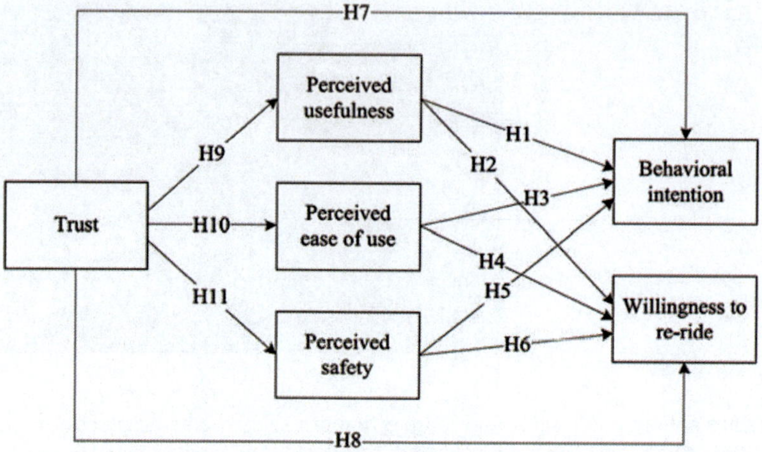

Fig. 16.2 Structural model explaining acceptance of automated vehicle behavior

16.3 Human Expectations and AV Interaction

A 2018 study done in China on the acceptance of AVs as shown in Fig. 16.2 [12] set forth a structural model for explaining acceptance of automated vehicles. Human trust on AVs is dependent on the perceived usefulness, ease of us and safety of the system involved. These three perceptions have a direct influence on behavioral intention and willingness to re-ride on an AV influenced by the trust one has on the automated vehicle. The challenge that arises then is:

1. Is there a need and, and if so, how can the passengers and the vehicle communicate with each other?
2. How can an interaction system be designed that will allow for a familiar communication with the AI in the AV without resorting to dashboards and automated voices?

Building trust, then needs to be designed to bring the AI's ability to fulfill all aspects of a human driver's job scope. This involves the understanding of communication not only in the form of navigation and data input, but also engage itself fully in social interactions with passengers to display evidence of intelligence.

16.4 Theories of Human Interaction

16.4.1 Speech Act Theory

Speech act theory analyzes speech utterances as intent carriers, named due to some forms of speech being able to carry out actions without the need for an external

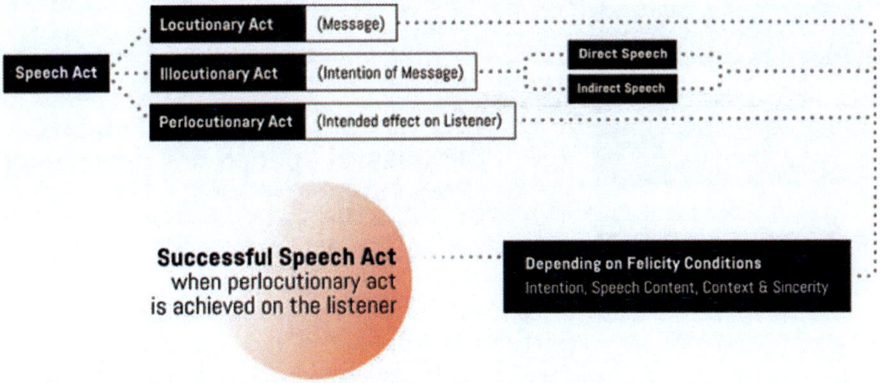

Fig. 16.3 Summary of speech act theory

referent [13]. For example, making a proclamation or a promise, one not only expresses intention or describes a state of affairs, but the performance itself constitutes an act of its own. This makes it a form of rule-based behavior that allows it to succeed. Typically, a human speech performance is composed of three different components, namely locutionary, illocutionary, and perlocutionary acts. A locutionary act is the message that a person is trying to convey, illocutionary act is the intent of the message, and perlocutionary act is the intended effect on the listener. Added together with felicity conditions in which each speech act is performed, a successful and trustworthy speech act is achieved (Fig. 16.3).

16.4.2 Communicative Action Theory

For a trustworthy interaction to happen, there is more than felicity conditions proposed in speech act theory. An utterance alone cannot claim to be sincere or authentic by itself and requires more discerning acts and understanding for illocutionary intentions to be clear. Communicative action theory builds upon this [14], wherein four universal validity claims for mutual agreement are introduced; claim to **truth**, claim to **justice**, claim to **sincerity**, and claim to **power**. While the first three are easy check points for human cognition to understand, claim to power stands out as an exception to the rule where a speaker sends out signals of authority and power of veto which overrides any form of perlocutionary act from the listener that contradicts the illocutionary act of the speaker. The communicative action theory is summarized in Fig. 16.4.

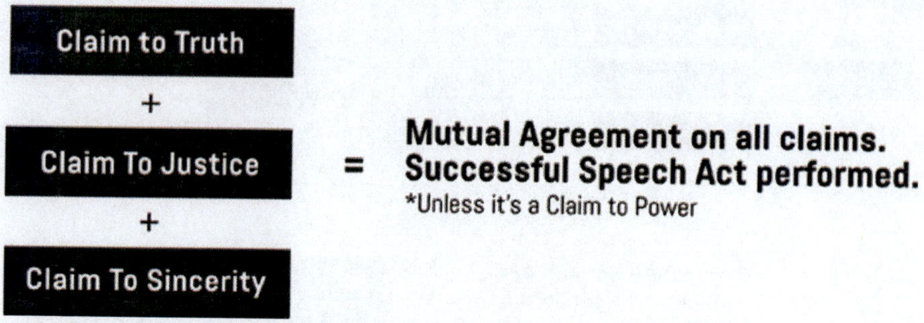

Fig. 16.4 Summary of communicative action theory

16.4.3 Theoretical Model of Social Interaction

In addition to speech, our natural modes of interaction also involve nonverbal signals. Turner's theoretical model [15] for human-to-human interaction provides a basis for all manners of social interaction as shown in Fig. 16.5, bridging processes from both the individual level to the social dyad.

The model is split into two parts, the one in left being the internal self-driven parsing of social signals within an individual, and the right being the outward-facing social signal exchange with the individual's conversation partner.

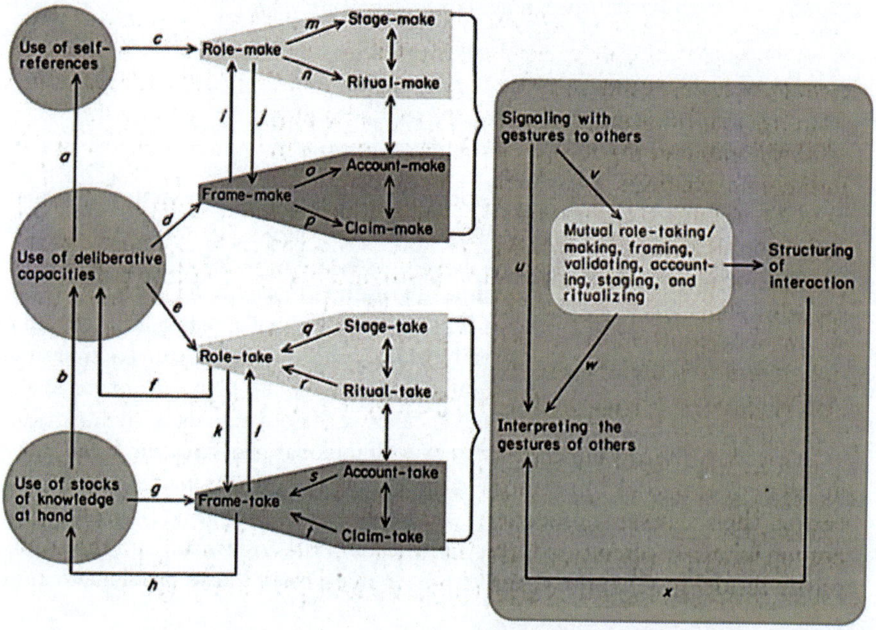

Fig. 16.5 Turner's theoretical model of social interactions

16.4.4 Summary: Human-to-AI in the AV Interaction

Based on the fact that a successful human-to-human interaction is based on trust, verbal communication based on utterances, intentions, claims to truth, justice, sincerity, and the conceptualization of self-image, self-references, and stocks of knowledge at hand, a basis for setting up a framework for human-to-AI in AV interaction may be distilled from the following:

1. Trust or a means for the human travelers and AI to establish a mutual belief in the level of comfort needed to engage in social interactions.
2. A comprehensible mastery of verbal language (e.g., English or the language used) needs to be acquired. For example, an equivalent of a standard (universal) English without use of slangs and colloquialism to build upon.
3. A sense of truth, justice, and sincerity built into the AV as both learned rule-based libraries and acquired in its everyday tasks.
4. A sense of self-established through initial and learning algorithms within the AI
5. A networked knowledge map built using machine learning methods able to access information streams such as other vehicle AIs and urban infrastructure.

16.5 Assumptions and Elements of the Framework

16.5.1 Being Present with the Traveler

In this framework, it is proposed that the AI being would be the "mind" of the AV, a software brain in a vehicle body driving the decision-making processes of the vehicle the traveler is in. When the vehicle is unfit for service, the "mind" can transfer itself from one robot body to another to resume via a networked cloud. It can also be such that the AI is able to transfer to whichever vehicle "body" the traveler happens to be riding in. The AI being then exists without any physical form within the AV, with the possibility of utilizing any part of the vehicle interior of the AV and become a "presence" within the vehicle. This frees up the need to follow existing norms of using fixed display surfaces or heavy anthropomorphism as the "face" of the AI (as in the case of hospitality robots in use today [16]). If one is to consider the social interaction of the AI in AV with the human travelers within its vehicular "body", it is important to introduce concepts such as proprioception [17] in the AI. This allows for self-awareness of physical limits and knowledge of its usable articulations in the "body" of the vehicle interior. This allows for the possibility of natural self-guided expression, as shown in Fig. 16.6. By being in the vehicle physically and in presence together with the traveler, the AI being is ready to perform social interactions on near-equal terms.

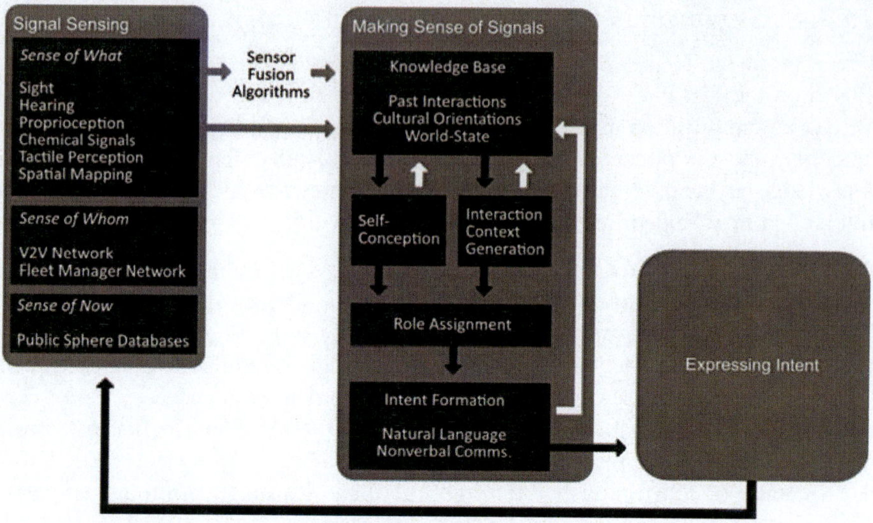

Fig. 16.6 Model of social interaction for AI being in AV

16.5.2 Building a Model of Social Interaction

The need for complex social signal comprehension and expression allows for a revisit to Turner's model of Fig. 16.5 again to develop a similar model for the AI, akin to a specialized neural network built for social interactions. Looking at the AV ecosystem, which could be said to consist of the following:

- Information gathering to build up knowledge bases of the world-state through networked public sphere databases, a "**sense of now.**"
- Machine learned knowledge of AV fleet employers, other AIs in AVs or regular travelers which gives it a "**sense of whom**" supported through external networks.
- Sensor-based feedback of the AV "body," an internalized subjective experience of social context and state of affairs "**sense of what.**"

Social signals received are sent to this knowledge base which can be described as the storage unit for past and new experiences of events that relate to the larger collective experience of a community sharing this understanding, and the AI being builds up its ability to interact socially as its own personality with humans through records of past interactions, cultural orientations, and the world-state. The resultant information is further processed through a subjective self-conception within the interaction context. Based on this, the social roles in the conversation can be identified, building up character personas of the interaction partner to facilitate intent formation for coherent conversations. Based on this, a model of social interaction of AI being in AV is derived as shown in Fig. 16.6.

This combination of a physical "body" to sense incoming social signals, and a non-physical presence or "mind" to make sense of these signals to generate coherent responses prepares the AI being for a framework for social interaction to evolve.

16.6 Framework for Interacting with the AI in AVs

Figure 16.7 below shows the schematic for social interaction between the AI being in an AV and a human traveler. It may be noted that the elements of intent, utterances, trust as well as the networked knowledge in AV are the results of processing through complex technologies based on the model of social interaction mentioned in Fig. 16.7 above. In contrast, these forms of discernment are built-in seamlessly through the cognitive brain network in humans.

Through the schematic of Fig. 16.7, a framework for the AI in AV to perform social interactions with human travelers is proposed in Fig. 16.8.

The framework as shown in Fig. 16.8 depicts the requirements, contents, means, and the AI's sense of being in social contexts that will bring about natural social interaction with humans. Achieving the means in each of the requirement requires two necessary ingredients:

1. Computational power for processes involving AI/computer science, large data management, and high-speed networking capabilities
2. Designed communication involving both audio and visual spectrums perceptible within human sensory ranges.

The authors are aware of the humongous task in front of computer scientists and AI engineers to develop the technology required to build the networked knowledge and the processing required to enable the expression of intent, utterance, and trust

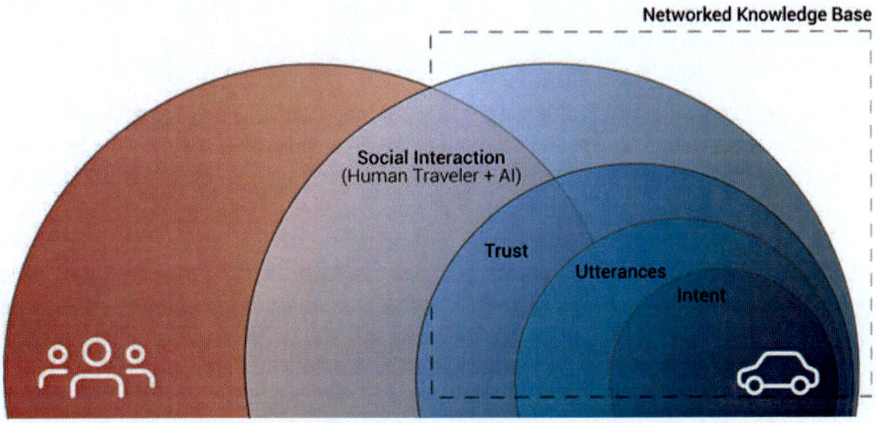

Fig. 16.7 Schematic of social interaction for AI being in AV

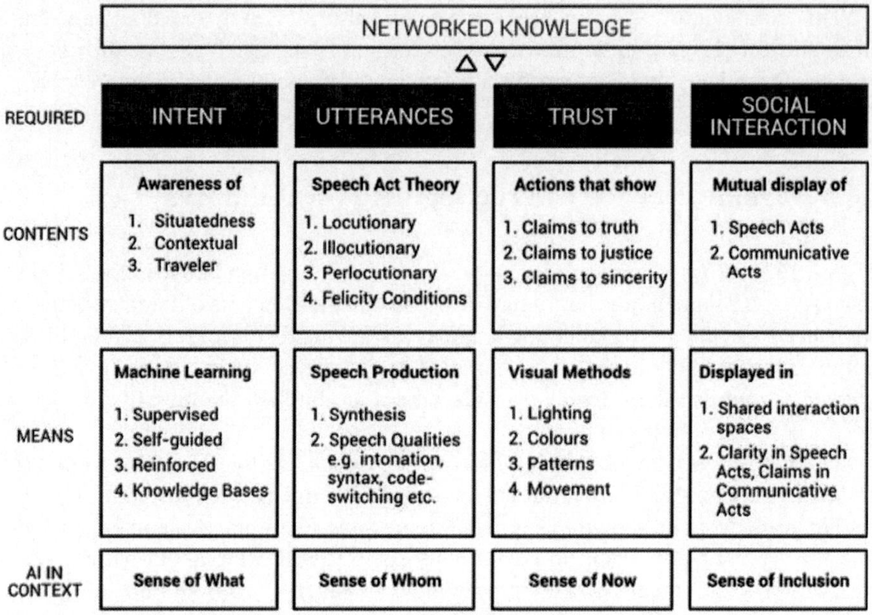

Fig. 16.8 Framework for social interaction of the AI in AV

within an AI being. However, as that future grows nearer design research can explore the manifestations of intent, utterance, and trust in the AI being in an AV setting. With informed research and solutions from such design research, the authors believe it will shed light in the research and development of the necessary computational technology to realize fully capable AI in AVs.

16.7 Conclusion

The advent of AV and deployment of these vehicles seems a reality in the near future. The authors feel that there needs to be paradigm shift in how an AV may be designed for it to be assimilated into society. As a next step to this research, speculative design methods are proposed to realize designed communication solutions for AIs in AVs. This could indicate the possible directions on how a sense of inclusiveness can be derived in future social interactions of the AI in AV and human travelers in their journeys. In this research, a framework is generated concentrating on human factors, cognitive aspects of interaction design and the aesthetics of the AV and the communication elements. The availability of computational technology to achieve the content shown in the framework is assumed to be fully realized. The framework proposed in this paper, the authors believe, not only sets up a direction for design research but also suggests a clearer idea for computational technology development.

Acknowledgements This paper is a result of a National Research Foundation grant given to TUMCREATE, Singapore

References

1. Society of Automotive Engineers.: SAE International Releases Updated Visual Chart for Its "Levels of Driving Automation" Standard for Self-Driving Vehicles (2018)
2. Mok, B., Johns, M., Miller, D., Ju, W.: Tunneled. In: Proceedings of the 2017 CHI Conference on Human Factors in Computing Systems (2017). https://doi.org/10.1145/3025453.3025713
3. National Highway Traffic Safety Administration.: Automatic Vehicle Control Systems. US Department of Transportation, Washington DC (2017)
4. Baruch, J.: Steer driverless cars towards full automation, https://www.nature.com/news/steer-driverless-cars-towards-full-automation-1.20390. Accessed 13 Jun 2020 (2016)
5. O'Toole, R.: Policy implications of autonomous vehicles. 5 (2014)
6. Siripanich, S.: Crossing the road in the world of autonomous cars. In: Medium, https://medium.com/teague-labs/crossing-the-road-in-the-world-of-autonomous-cars-e14827bfa301. Accessed 15 Jun 2020 (2020)
7. National Highway Safety Traffic Administration.: National Motor Vehicle Crash Causation Survey. 26 (2008)
8. Ellen, E.: Three in Four Americans Remain Afraid of Fully Self-Driving Vehicles (2019)
9. Stayton, E., Cefkin, M., Zhang, J.: Autonomous Individuals in Autonomous Vehicles: The Multiple Autonomies of Self-Driving Cars. Ethnographic Praxis in Industry Conference (2018)
10. The Autonomous Car A Consumer Perspective. 5–7(2019)
11. Kun, A., Boll, S., Schmidt, A.: Shifting gears: user interfaces in the age of autonomous driving. IEEE Pervasive Comput. **15**, 32–38 (2016). https://doi.org/10.1109/mprv.2016.14
12. Xu, Z., Zhang, K., Min, H., et al.: What drives people to accept automated vehicles? Findings from a field experiment. Transp. Res. Part C Emerging Technol. **95**, 320–334 (2018). https://doi.org/10.1016/j.trc.2018.07.024
13. Searle, J.: Speech Acts. Cambridge University Press, Cambridge (2012)
14. Eriksen, E., Weigård, J.: Understanding Habermas. Continuum, London (2004)
15. Turner, J.: A Theory of Social Interaction. Stanford University Press, Stanford, Calif (1988)
16. Tussyadiah, I., Park, S.: Consumer evaluation of hotel service robots. Inform. Commun. Technol. Touri. **2018**, 308–320 (2017). https://doi.org/10.1007/978-3-319-72923-7_24
17. Kwiatkowski, R., Lipson, H.: Task-agnostic self-modeling machines. Sci. Robot. **4**, eaau9354 (2019). https://doi.org/10.1126/scirobotics.aau9354

Chapter 17
MONEY TALKS: "Back to the Future"—Challenges in Banknote Design

Rukmini Dahanukar

Abstract As artifacts of popular culture, banknotes signify more than just tokens of payment. Banknotes are at the heart of our exchanges and interactions. They are omnipresent and ubiquitous. The first modern banknotes were printed in Sweden in 1661 (Cribb in Money, Dorling Kindersley, New York, 1990); circulating for more than 350 years now; they bear witness to our everyday lives, our possessions, our stories and our memories—close to our hearts and deep in our pockets. With galloping technological advancement, on-going additions to payment gateways and alternate global currencies; banknotes would seem to be following the path of the dodo. In fact, it is just the contrary. The last decade has seen a growth of 175% of the US Dollar bills as per the data available on the United States Federal Reserve's Web site (https://www.federalreserve.gov/paymentsystems/coin_currcircvolume.htm). The Bank of England's Web site (https://www.bankofengland.co.uk/statistics/banknote) also documents an increase of almost double at 190% of the British pound notes; even the relatively nascent Euro hit a staggering 200% increase in note circulation according to the European Central Bank's Web site (https://www.ecb.europa.eu/stats/policy_and_exchange_rates/banknotes+coins/circulation/html/index.en.html). With counterfeiters relentlessly at their heels; banknotes; substrates adorned with complex and layered artworks integrating constantly updated security technology into production techniques; ceaselessly needing to stay ahead; are befitting to the adage, *back to the future*; perhaps, like no other product design or innovation. The size is restricted, the medium usually paper, recently polymer, two-dimensional and two sided, the method being printing, foiling and special security technologies; not much has changed over the past three centuries considering the influx and constant bombardment of inventions leaving most products and practices obsolete or even redundant, but not banknotes. A clear indication that makes the banknote and its design so extraordinary and challenging. Apart from the usual mundane exchange; our interactions with banknotes run deep. They are experienced and consumed by our senses; even if less apparent and more subliminal. Seeing the magnificent and indigenous artworks, feeling the raised and intaglio impressions, hearing the flutter and folding (the crushing sounds are tests for

R. Dahanukar (✉)
Nirmiti, 21 Khushnuma Apartments, M.L. Dahanukar Marg, Mumbai 400026, India
e-mail: moneytalksonline@gmail.com

A. Chakrabarti et al. (eds.), *Design for Tomorrow—Volume 1*, Smart Innovation, Systems and Technologies 221, https://doi.org/10.1007/978-981-16-0041-8_17

real versus counterfeit notes); they have an extremely personal relationship with each of us on a daily basis. To make their design decisions further complex, they are also handled by machines regularly. Optical and magnetic sensors to assist the multiple interfaces and technology need to embedded in the notes for effective exchange. Banknotes are, thus, part of an elite group of products that are handled both by men and machines. And as far as the quantum of exchange is concerned, they securely lead by a hefty margin. Designed exclusively and uniquely, individually customized, artistically bespoke and yet for a wholesale, across-the-board, extensive exchange; to be used anywhere, anyhow, anytime; a perfect blend of form and function, of information and security, of exchange and trust, of durability and bio-degradability; thereby making the banknote the most coveted of design possibilities, practices and commissions. Not to forget, each note, in its medium and its message integrates *zeitgeist*—the spirit of the time; reflecting the promise of its currency. This research presents; *back to the future*; the on-going journey of banknotes that continue to offer challenges and opportunities; a noteworthy specimen for the design of tomorrow.

17.1 Introduction

The art of banknote printing is a much-guarded secret and rightly so as are most security documents. Various accounts of economic warfare during the Second World War also reinforce the need for its secrecy. However, a dichotomy exists at the heart of its exchange. The premise of modern money is trust and authenticity. The general public has to be made aware of the peculiarities and originalities of the banknote design but only in part. And here lies the challenge in its design. This paper addresses the need to develop an effective framework in the user-interface design with keeping security safely protected.

17.2 Literature Review

Toward the end of the twentieth century, the introduction of the World Wide Web, information accessibility and sharing, the break-down of the Berlin Wall, the start of the European monetary integration, have all contributed to an interwoven political, social, technological and cultural fabric to our world with a global redefinition in our interactions, including the experience and exchange of our money, specifically, banknotes.

However, the counterfeiters have also benefited with the shrinkage of innovations and their timelines owing to the galloping speed of technology. The initial life-cycle of banknotes was determined by usual wear and tear but now determined by counterfeiting. Significantly crashing the time between two successive issues from 10–15 years to sometimes, just 1–2 years [5]. Thereby forcing design and

Image 17.1 Hector Breeze's cartoon published in the *Guardian*; from Williams, et al. Eds. (1998), p. 15

development processes to follow certain parameters and models to crash lead times. A definite argument and the premise of this research.

However, banknotes are a peculiar lot. From most end-user's perspectives, they are looked upon as monopolistic and thrust upon to use by issuing authorities in their usual nonchalant, top-down approach, making looking twice or even once at their images not worth the effort. Perhaps if design production was more interactive; banknotes would be seen in a different light, with more interest and enthusiasm. Add to that, they are produced exclusively and secretively within restricted circles owing to their sensitive nature. Perhaps that is why much discussion, debate or even academic research and models were not considered; let alone pursued and applied. That has now changed (Image 17.1).

On January 1, 2002 at midnight, Europeans rushed to cash dispensing machines to get their first notes; designs being shortlisted and voted for by members of the general public; never before on such a large and diverse scale; long lines were seen outside banks; to get the first peek at the notes.

But they are constantly at risk; making their shelf lives shorter and shorter; replacements or additions have to move in constant tandem calling for a need of using tested design models to be introduced in the banknote production process to keep money ahead in the game.

To better understand the future; it pays to visit the past.

The history of money began when humans learned that they could trade for things they wanted rather than produce them; themselves. The first primitive populations resorted to a simple form of barter, or exchanges in kind. However, one made tools and implements, another cloth and still another collected animal furs. The lack of an acceptable exchange value, need for a standard unit of measurement and a storage of value that would not perish with passage of time; combined with the ingenious human intellect; led to the introduction of modern money. Initially, as commodity money like rice, salt-bricks, tea-cakes; and eventually with the discovery of metals; money changed hands to ingots that were further polished to coins; still prevalent today.

Coins subliminally introduced a novel concept of documenting information on money [2]; the date, issuing authority and in due course denominations; changed the very fabric of this medium of exchange that is so embedded in our everyday lives almost 2500 years later. As money historian Glyn Davies said, "for over 500 years the coins of Rome publicly portrayed the events, hopes, ambitions, lives and lies, of its rulers" [10]. The art on coins was truly splendid. But coins were small, round, hard objects, making it difficult for artists to really experiment on such a limited canvas. The art of money truly took off with the evolution of paper currency-today, money talks.

Historically, commodity and metallic money had an inherent value. Paper, however, has limited intrinsic exchange value unlike its predecessors and thus relies on trust and the philosophical premise on which modern money rests. The modern banknote is a combination of the promissory note's tradition ensuring trust and guarantee and the coin's character of information dissemination. The 1-lb coin issued in Britain in 1970; inscribed the words; *"decus et tutamen"*—meaning an ornament and a safeguard; a phrase from the "Aeneid" of Virgil; originally inscribed on coins from the seventeenth century; referring in this case to the serrated edge of the coin as a protection against clipping for its metal content; capturing best the essence of monetary design with "form" as its ornament and a safeguard as it "function."

Additionally, the design of a banknote also carries the burden of being a national symbol important for many reasons; ranging from philosophic to pragmatic. Since currency is viewed as a symbol of sovereignty and independence; design decisions become emotional and contentious, sometimes, also resulting in long production delays which is not a privilege anymore in today's competitive times. Another objective to be achieved is to establish or restore public confidence even via its design. The selection process, thus, should minimize delays and seek to ensure that the banknote is user-friendly, durable, easily recognizable and reasonably secure against counterfeiting. Surely a need for tried and tested product development process.

One of the earlier academic writings on these lines was by Adams [1]; on the considerations that needed to be taken into account when designing banknotes; how the designs would be selected, the design philosophy, the main themes and texts, the size, the user conditions, security features, durability and more. Perhaps, the need for such guidelines was motivated by the break-up of Soviet Russia; leading to the formation of 17 nascent nations in central Europe while 12 nations in Western Europe chose to shed individual identities toward a unified trans-national identity; all in need of banknote design.

Adams [1] suggested the best option for design selection would be to have the decision-making carried out by a broad-based, high-level committee that includes representatives from banks, government, academia, the arts (designers and commercial artists) and experts on currency printing and handling. This paper is significant as being one of the earliest writings where monetary design was considered as a part of a mainstream design process.

Adams, further added that the highlight of the committee's work would be to organize a national competition for the best design and sometimes an appropriate name for the currency. However, sometimes, competitions are expensive and time-consuming. To combat these issues, an alternative of having the central bank or the committee to make the decisions on its own is also available.

The second step proposed was the different approaches to the design. Certain decisions about the design philosophy should be made before the design process begins, such as whether the notes should be an expression of the culture of the country, or whether it should be a part of the serial design, or whether there should be a relationship between the two sides of the note. Design decisions about the art style, traditional, contemporary, or post-modern have also to be made. The style used also contributes to communicating national image, thereby identity (Image 17.2).

A country's currency traditions should also be considered, since people are more likely to accept notes that are broadly similar to those that they have been used to seeing and using. Design decisions might also be influenced by the desire to have notes that are similar to a selected series of notes similar to another nation.

According to de Heij [4]; design motivations range from advanced, new, upgrades, revisions and emergency issues. However, to continue familiarity, consistency and to not shock users completely; the use of dominant colors are considered. Also, keeping in mind, the entire user population including the illiterate, people with vision disabilities, rely on consistent color linked with denominations (Image 17.3).

For security reasons notes also include different shades of gray, since color-copying machines have difficulties with the color balance if more than one gray color is used.

Countries like the USA prefer to have all their currency notes printed in a single set of colors so that they can be easily identified as the national currency also leading

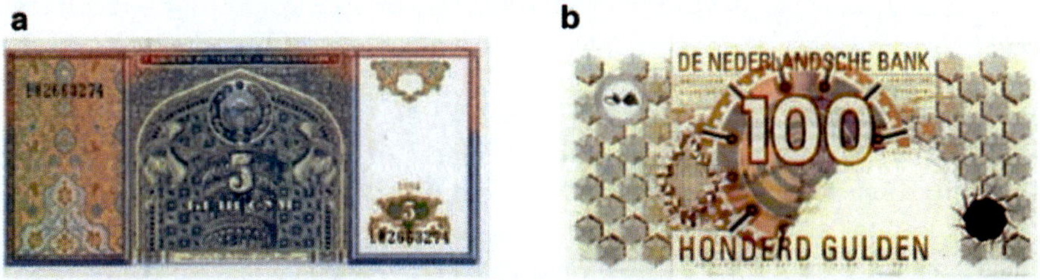

Image 17.2 **a** and **b** On the left is a 5-sum note of Uzbekistan issued in 1994 and on the right is a 100-Dutch Guilder note of the Netherlands issued in 1992. Both notes were issued at around the same time but with completely different design styles indicating the different design philosophies, an outcome of the political ideologies and cultural attitudes of both the nations. (Images from https://banknotes.com)

Image 17.3 **a–d** Four notes of the 10-guilders issued by the Netherlands Bank across six decades are all printed in blue(s). (Images from https://banknotes.com)

to lower production costs. However, most countries prefer to use different colors for each denomination so that they can be easily recognized. This choice is especially important when a new series of currency notes are issued because once a color is chosen it is highly recommended to remain unchanged in the subsequent issues.

The bookshelves of popular culture have been adorned by authors like Monestier [8], Standish [10] and Mudd [9]; discussing themes chosen and followed. Portraits are usually chosen as the main theme or part of the theme. From royalty, heads of state, leaders all the way to the ordinary and everyday citizens. Queen Elizabeth II though has been the exception; perhaps, more for her royal status; all the same; she has adorned across 85 years in 31 portraits from 38 different countries according to researcher Peter Symes (https://www.pjsymes.com.au/QE2/).

Hewitt [7], in her book "Beauty and the Banknote," was one of the first to collate various essays on female representation from allegorical nation-states to their disappearance to the modern heroes; outnumbered by the men's division. However, Australia since 2006 has been the for-runner against gender discrimination with portraits of men on one side and women on the other; on all its notes.

Other themes like plants, animals or buildings are often used because they are less contentious than portraits of individuals. These are also cheaper to produce than portraits, particularly, engravings of widely recognized individuals. Like the office of Lord Chamberlin indicates definitive guidelines on the use of British royal portraits. The Bank of Canada, too, clearly stated the approval of the metallic portrait used of the Queen on its commemorative polymer issue (2015).

However, portraits still emerge as winners; even the faceless Euro was redesigned with Europa finding herself embedded in the holographic foil window; following a study in the field of neuroscience; conducted at Stanford University; proving that the presence of a human face adds value by recognition, a strong sense of familiarity and more importantly, could reduce counterfeiting [6].

All the discussion and confusion about banknote imagery and its role as culture communication is best appreciated and acknowledged in the case of Slovenia; imagery and design playing the special role in communicating the nation's new identity which declared its independence from Yugoslavia in the June of 1991 [11] (Image 17.4).

According to Tomanic and Hartman (1994), the government of Slovenia took full advantage of the opportunities presented by the design of a new currency to demonstrate the country's strength, vitality, commitment to capitalism and its national values. The intent was clear to produce a currency which is at once distinctive and familiar to Western Europeans. The message; "We are like you. We will do business like you." Along with their coherent economic policies and reforms; over a decade later, in May 2004, Slovenia was one of the new members to be inducted in the European Union.

After images, the text is the next decision. The mandatory information like the country name, the issuing agency is written on the front, while the value should be clearly printed in large numerals on both sides, and spelled out of the front. More than one language is sometimes used depending on the culture of circulation. Typography is now efficiently used as a tough to crack security feature. The next concern is the

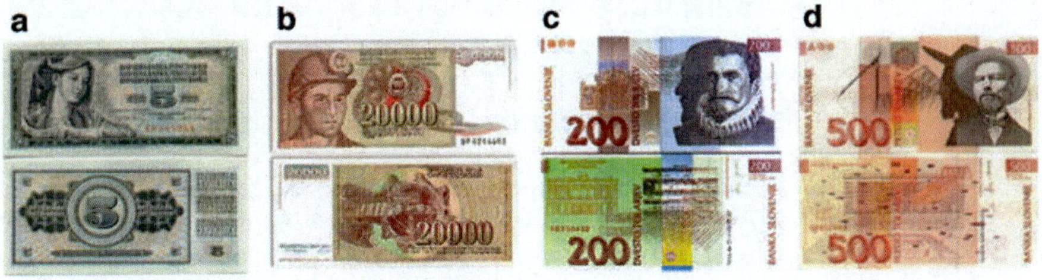

Image 17.4 a–d Faces on the old currency of Yugoslavian banknotes (left) display politically correct anonymous communist icons and suggest the slow ponderous forces of nature and history through its static design. While Slovenia's new currency (right) celebrates the lives and work of its cultural heroes and suggests the extraordinary dynamism of a new nation proud of its past and eagerly embracing the future with its new message to the world using its currency as a medium for communication. (All images from https://banknotes.com)

decision regarding size, small enough to fit in wallets, cash registers and large enough to accommodate all the images and text. Most modern currency notes are between 65 and 80 mm in height. This is an appropriate height for currency handling machines, vending machines, and modern cash registers. The storage and handling are much easier and more efficient when all the denominations are of the same height like the current series of the Indian Rupee. Views on the optimal length of currency notes vary. Different note-lengths help people, especially the visually impaired to discriminate between different denominations. On the other hand, when all denominations have the same size and the same watermark, it is possible to economize on paper ordering, since all the currency notes can be printed using the same lot of paper like in the USA.

All these points deal with the production aspect; switching over; user needs, safety and security are all to be equally considered. The notes should be easy to use, yet secure against counterfeiting. The levels of security features are classified by de Heij [5]; following acceptable guidelines by central banks world over.

Classification of authenticity features		
The features are checked by	Name giving technology-driven	Name giving use-centered
Human senses—subconscious	Level 0	Public trigger features
Human senses—conscious	Level 1	Public features
Hand-held devices	Level 2a	Retail features—human operated
Automatic devices	Level 2b	Retail features—automatic device
Banknote recycling machines	Level 2c	Commercial cash centre features
Sorting machines of central bank	Level 3	Central bank features
Counterfeit deterrence system	Level 4	Features to prevent home-made reproductions

(continued)

(continued)

Classification of authenticity features		
The features are checked by	Name giving technology-driven	Name giving use-centered
Expertise in laboratories	Level 5	Forensic features

In addition to all these; advances in technology have led to other substrates like polymer banknotes as well as hybrid notes; the combination of paper with a special coating to safeguard notes from wear and tear, mutilation but also from easy counterfeiting; thereby contributing to shelf-life. But recent counterfeiting scams in Mexico (2017) and Romania (2020) suggest, otherwise, rendering all substrates equal in the playing field of fakes.

All of the above parameters deal with human interaction. Machine readability is one more factor since banknotes are used by both. The Netherlands (1989) for the first time applied the barcode on their notes proving to be very efficient for machine reading as well as anti-counterfeiting. The recently launched Russian 100 Ruble (2018) also carries a QR code. Identification of fake notes is a job that machines can do more efficiently minus human error. India's recent exercise of demonetization; involved not only the design of new notes but the entire reworking of the back-end to accommodate the machines to recognize new sizes and new features.

A greater understanding and interaction between the front-end (human) and the back-end (machine) usage, experience and interpretation is the need of the hour. Hence, the question; isn't it time to develop a design management model for banknotes?

17.3 Methodology

A study measuring user perceptions and experiences of this nature is extremely difficult in terms of time and resources. However, there have been recent published studies from the European Union; not only studying the Euro but also world banknotes; that do offer a good insight. Also being qualitative in nature, "anecdotal references" provide good indications enabling a fair and sufficient content analyses. Information also on central bank Web sites and issuing authorities is available for certain art and security features and their design motivations.

The launch of the trans-national euro prompted a plethora of social identity research due to the shift in the definition of European identity by member state citizens, thereby promoting and measuring the change in the fabric of Europeanness and banknote imagery's role in the same, which can and is applied to other countries and their currencies too.

The International Banknote Society (IBNS) started in 1960 also has been documenting articles in its quarterly journal about images on notes; their peculiarities and their classification across timelines and regimes.

However, the most unique point of this research is the personal interviews conducted by the researcher with banknote designers, museum curators, banknote manufacturers, central bank design researchers and enthusiastic collectors.

A combination of thus published literature, independent academic, as well as sponsored private and public information, popular and critical writings and testimonials from personal interview transcripts; pave a way toward answering the research questions and proposal provided herein.

17.4 Observations and Analysis

The case of the Euro is interesting. The first-ever design competition on such a large scale is worth appreciating.

> *It is the design coup of the century: a chance to put your work in front of millions of people for years and years to come.*—(Steinmetz 1998).

The historical perspective of the Euro is also important especially while looking at the old currencies of the twelve nations that the Euro replaced. They are imagery revealed the loss and re-creation of Europe's identity. A thorough study of the old design elements and themes along with the new notes; their continuation on the new coins that were allowed to retain their national sides and the spine on which the new designs were based was published. The critical finding that the European Union flag was the at the center of the new designs was unique to this study [3]; reconfirming that the design process does follow standard design management protocols but the acknowledgment and further use of models is not yet in play.

Recently, appreciating the Euro's success; the Bank of England also organized a public opinion poll and vote (2019) to decide who features on the upcoming 50-lb polymer note based on the theme of science; with great success and euphoria. "The theme and design is the soul of the banknote"; Debbie Marriot; Banknote Designer at the Bank of England further quoted in a personal interview with the researcher; "It was better that theme design decisions reverted back to the Bank making them more unique and undiluted; with the fear that they are not just standardized by their printing partner De la Rue."

Hans de Heij from the Central Bank of Netherlands added that "the field even though its super-specialized; the lack of any design management literature is appalling"; he has been working on user-friendly model for cash as a payment instrument as compared to others. He has also proposed for the first time a working model and protocols regarding a banknote design manager who would liaison between banks and printers. According to him; "the Netherlands; not only factored in security with the novel innovations of placing barcodes on notes but also making machine reading ability and tracking the banknote trail very easy."

Inge Madle, self-employed banknote and stamp engraver and designer working with central banks and printers; quotes "the process of modern banknote design can best be summarized as the beauty and the beast; where the beast-security features more or less determine the beauty-form." She concurred; "With a framework in place, it would be easier to take these decisions across time and cost resources. Also, helping to deal with international banks in a professional way."

Steve Gulliford from De la Rue added that; "Business decisions are hard to take for a publicly held company and answerability to stake-holders would mean more efficiency required in the process; and since the lead times have fallen; various frameworks are followed given to meet the time and cost requirements. De la Rue has created standard templates designed to facilitate the process."

Sudhakar Kaza; chief of Bhartiya Reserve Bank Note Mudran; also noted that "the demonisatisation process put great pressure and the designs though conceptualized needed to be moved out at lightning speed. The choice of Mahatma Gandhi to instill trust on the obverse and the series of heritage sites on the reverse; contributed to a balance between the old and the new." He added; "What was more important in India's case was the overhauling of all the ancillary machinery like cash disbursing machines, etc., to accommodate not only the new design but also the new size with width of notes remaining a constant and only lengths varying for the 5 variants of 50, 100, 200, 500 and 2000; a radical change for both human as well as machine interactions." A standard model and protocol would surely have helped.

Kaori Sekiguchi, curator of the Currency Museum Bank of Japan spoke of the upcoming designs of the Japanese yen and the lead time to complete them has been set at 8–10 years; however, their designs were put out for public viewing since 2019 though notes were expected in 2024; a strategy for public recognition and trust; following standard design management protocols. She proudly quotes; "To think that artworks were revealed for public display 5 years in advance to instill familiarity of the banknote; best reflects the Japanese efficient planning systems."

The central theme from all the interviews did point toward a shift in the framework and processes from product development and execution strategies used by industry.

17.5 Findings and Future Research

Banknotes are unique and thus cannot work on *the one size fits all* principle but keeping continuously shrinking lead times and ever-expanding printing and processing technologies; a working model or combination thereof has become the need of the hour. Blending learnings and templates from the art and form department along with machine scanning and reading technologies and can greatly contribute to banknote design implementation and effective exchange.

The design challenge is twofold; the first part being the user interaction with the general public and the second being the machine-interface. The first part can refer to the guidelines by Adams [1] and the classification by de Heij [5]; working as effective starting point for a framework regarding images, colors, sizes and themes. The information dissemination to the public and their involvement will make exchanges more effective; with the use of social media; more involvement can be generated as in the case of the Bank of England. (https://www.bankofengland.co.uk/news/2018/nov

ember/think-science-for-the-next-50-character). The second part that involves the more critical security features will have to be worked upon with firms that specialize in security documents and their implementation. The role of the banknote designer; private or part of a central bank, thus, involves dual interaction from the front-end-users and the back-end manufacturers with keeping resources like time and costs too; in the equation. Various models from the product and brand design literature can thus be combined to enable a faster, more efficient approach to the system by keeping the user and his interaction at the heart of all design decisions. The objectives would be to ensure that the banknote is user-friendly, durable, easily recognizable and reasonably secure against counterfeiting with minimizing delays. With end-users being the starting point of the design process; easy recognition of notes would enable combating counterfeits more easily; and to some extent support an argument toward more complicated counterfeiting; which would be unviable, thereby reducing such cases.

This research concludes by strongly proposing a completely reverse model unlike the presently followed industry-norm; by looking at the origin of banknote design usage and exchange as the starting point; the user interaction being the most important and then working backwards to offer design solutions considering the challenges and opportunities of the banknote; a noteworthy specimen for the design of our tomorrow.

References

1. Adams, R.: The design and printing of Banknotes: considerations when introducing a new currency. IMF Working Paper, WP/95/26. (1995)
2. Cribb, J.: Money. Dorling Kindersley, New York (1990)
3. Dahanukar, R.: Money Talks: Currency Notes as Communication Media. California State University, Fullerton (2004)
4. De Heij, H.: Designing banknote identity. DNB Occasional Studies, 10(3), (2012)
5. De Heij, H.: A Model for User-Centered Design for Payment Instruments Applied to Banknotes: UPID-model. Tilburg University, Netherlands (2017)
6. Eagleman, D.: Money and our minds: can neuroscience stop counterfeiting. Financial Times Magazine, March 2017
7. Hewiit, V.: Beauty and the Banknote. British Museum Press, London (1994)
8. Monestier, M.: The Art of Paper Currency. Quartet, London (1983)
9. Mudd, D.: All the Money in the World. Harper Collins, New York (2006)
10. Standish, D.: The Art of Money. Chronicle Books, San Francisco (2000)
11. Tomanic, M., Linda Hartman. 1994. "Slovenia's Money-A message to the world," Design Management Journal, 5(1), pp. 68(3)

Chapter 18
A Field Investigation of the Average Indoor Thermal Comfort Parameters on the Railway Pantry Car Kitchen at the Different Cooking Period

Md. Sarfaraz Alam, Arunachalam Muthiah, and Urmi Salve

Abstract There are various kinds of literature available on thermal comfort in buildings, hospitals, industries, and others too in developed and developing countries. However, few pieces of literature are available on the kitchen environment like rural homes, restaurants, and hotels, but related to the kitchen environment of the railway pantry car is almost nill. Therefore, this study aimed to investigate the average indoor thermal comfort parameters on the railway pantry car kitchen at a different cooking period. A total of 14 railway pantry cars included were this study in which six and eight pantry cars took in the summer and winter seasons, respectively, in 2018. In this research, with the help of a kestrel anemometer 4500, air temperature, relative humidity, and air velocity were measured. Although globe temperature measured by 6-inch black-globe thermometer. Descriptive analysis and Kruskal–Wallis tests performed were for the investigation using SPSS-16 software. A result of the descriptive analysis indicates that in both seasons, and the value average thermal comfort parameters were found maximum at lunch and snack preparation time. The outcomes of statistical analysis revealed that there is no significant difference between all thermal comfort parameters during the summer season at the cooking period. A significant difference was found between thermal comfort parameters during the cooking period in the winter season except for the parameters of relative humidity and air velocity. Furthermore, this result will help in the design intervention of the ventilation system and estimate the comfort temperature in the pantry car's kitchen.

18.1 Introduction

American Society of Heating, Refrigerating and Air-Conditioning Engineers (ASHRAE) 55 described that thermal comfort is the condition of the mind that the people report contentment with the surrounding environment [1, 2]. Six major factors that influence thermal comfort in which four environmental factors like "air

Md. S. Alam (✉) · A. Muthiah · U. Salve
Department of Design, Indian Institute of Technology Guwahati, Guwahati 781039, India
e-mail: sarfaraz8665@gmail.com

A. Chakrabarti et al. (eds.), *Design for Tomorrow—Volume 1*, Smart Innovation, Systems and Technologies 221, https://doi.org/10.1007/978-981-16-0041-8_18

temperature, radiant temperature (estimated by globe temperature), relative humidity, and air velocity." Similarly, two personal factors like "clothing level and metabolic rate/activity level" [2, 3]. Many thermal comfort studies were carried out on the building environment, bus, car, train, and industries in developed and developing countries [4–7]. Still, some types of research found in the commercial kitchen environments like rural homes, restaurants, and hotels others too, which identified the occupant's comfort level at the time of various cooking activities based on the objective and subjective measurement [8, 9]. But there is no literature found on the kitchen environment of railway pantry cars based on the average cooking period difference between "indoor thermal comfort parameters" during summer and winter seasons. While Alam et al. [10] directed a study on the "railway pantry car kitchen," which compared the indoor and outdoor thermal comfort parameters during the different cooking times and identified the chef's comfort zone. Similarly, another study was conducted by Alam et al. [11] on the Indian Railway pantry car kitchen. In which, air-conditioned and non-air-conditioned railway pantry car kitchen was compared based on subjective and objective evaluation. The results of this study showed that the measured value of thermal comfort parameters in both types of railway pantry car kitchens did not comply with the ASHRAE standard during cooking. Whereas the chef's thermal perception votes during this period indicated high rates of thermal dissatisfaction.

The kitchen of the railway pantry car is usually like the kitchen, which prepares meals for railway passengers. Contribute a "significant role in the catering system" on Indian Railways, which is an integral part of every short and long-distance train throughout the nation. As per the detail of the Indian Railway Board, 338 pairs of pantry cars are running with the trains. There are consists of "3–5 chefs," "40–50 caterers," and "2–3 pantry manager" in a one pantry car coach [12, 13]. Although two different types of pantry cars available at the existing condition, viz., (i) air-conditioned and (ii) non-air-conditioned. In between both pantry cars, it has a considerable difference based on the cooking installation height, air supply system, visual aesthetic in design, space inside the pantry car, etc. [10]. However, as a result of the Alam et al. [13] clearly states that there is no difference between the "air-conditioned and non-air-conditioned" kitchen of the railway pantry car based on the "indoor thermal environmental condition" at the time meal preparation. Whereas in both pantry car using conventional equipment for cooking like "electric heater," "LPG gas stove," "soup warmer," "oven," "kettel," "karahi," and "deep-fryer," etc. These appliances create humidity and heat generated due to which the indoor environment becomes uncomfortable [14, 15]. It will directly affect occupant health and work efficiency.

The above literature distinctly states that thermal comfort is necessary for pantry car chefs. Hence, the present research intended to investigate the average indoor thermal comfort parameters on the railway pantry car kitchen at a different cooking period.

18.2 Method

18.2.1 Sample, Location, and Period

The present research was conducted on 14 railway pantry car kitchens for field investigation. In which, for summer season, six pantry car kitchen was considered, and for winter season, total eight pantry car kitchen was taken. Based on the Koppen climate classification system, two climate zone: "humid-subtropical and tropical wet and dry" has been incorporated in this research. Data from both seasons were collected in 2018. In which, the data of the summer season was taken in August and the data of the winter season was occupied in December. Owning to the limited permission from railway authority board, the current study was conducted in the month of August and December. Moreover, it could be difficult (in terms of cost, energy, time) for anyone to collect/measure data from a range of months with repeated data collection. In addition, the current study was a kind of preliminary study to understand the existing scenarios of thermal comfort in Indian Railway pantry car kitchen. Thus, the study focuses only two months (August and December) for achieving the objectives of the current study.

18.2.2 Objective Measurement

In the present study, four "indoor thermal comfort parameters" recorded were during the field measurement. The current status of the non-air-conditioned railway pantry car kitchen is depicted in Fig. 18.1, which shows the measuring point of thermal comfort parameters. All thermal comfort parameters, viz.; "air temperature Ta (°C), globe temperature Tg (°C), relative humidity RH%, and air velocity va (m/s)" were recorded with the help of kestrel anemometer 4500. However, with the help of the "6-inch black-globe thermometer," globe temperature was measured. The indoor thermal comfort parameters were measured according to ASHRAE 55 and ISHRAE code standard [10, 16]. Where "Ta, RH, va" measured was at 1 foot (0.3 m) close to the cooking zone, and 43 inches (1.1 m) beyond the workstation floor height at the pantry car. Although, in the center area of the cooking zone, "Tg" was measured. Figure 18.2 displays the 3D schematic layout of the pantry car kitchen where the point of measurement is shown. In a day, all the environmental parameters were measured for 4 times (morning, afternoon, evening, and night) using single probed meter. In this study, Ta °C was measured with the "range −29.0–70.0 °C and 1 °C accuracy." RH% measured with the "range 5–95% and ±3% accuracy." Correspondingly, va (m/s) measured with the "0.4–60 m/s and ±0.1 m/s accuracy." And similarly, Tg °C has been measured with the "range −5–95 °C."

The entire study, all "thermal comfort parameters" were measured at the different periods of cooking time, viz.; "breakfast–7:00 am, lunch–11:30 am, snack–4:00 pm, and dinner–6:30 pm" in the pantry car kitchen. Based on the Alam et al. [10] all

Fig. 18.1 Existing condition of the pantry car kitchen

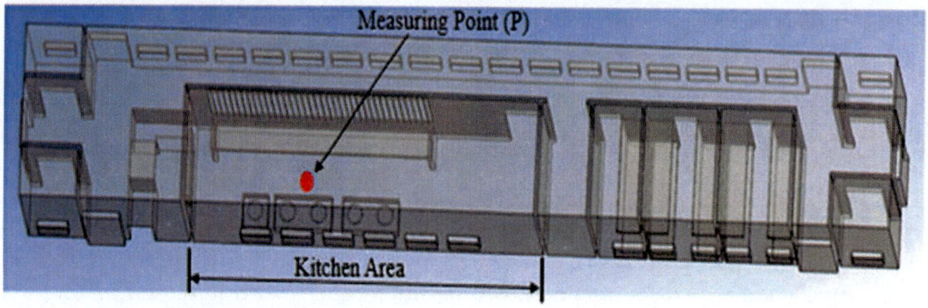

Fig. 18.2 Schematic layout of the pantry car kitchen

indoor thermal comfort parameters recorded were at 10–15 min in the pantry car kitchen with the 5–10 s interval at all cooking periods.

18.2.3 Data Analysis

In this research, to identify the average range of "thermal comfort parameters" at different cooking periods inside the kitchen of the pantry car, descriptive analysis was incorporated in which data are reported based on the histogram of the average value distribution. MS Excel 2016 software was used for this analysis. However, to check the significant difference between all cooking periods, Kruskal–Wallis test was used since data were not normal. While performing the statistical analysis, SPSS 2016 version software was used.

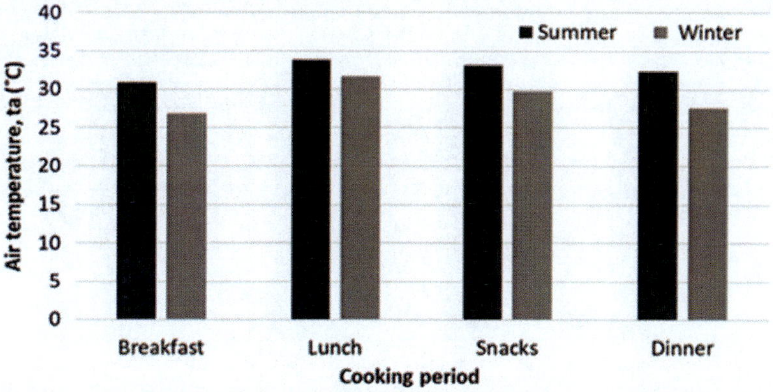

Fig. 18.3 Average air temperature distribution at the different cooking period during summer and winter season

18.3 Result and Discussion

18.3.1 Indoor Environment During the Different Cooking Period

As a previous study suggested that for any thermal environment there are four major thermal comfort parameters, such as "air temperature, globe temperature, relative humidity, and air velocity" [17]. Present research we measured all four thermal comfort parameters at various cooking period, viz., "breakfast, lunch, snacks, and dinner."

Figure 18.3 shows that the average indoor air temperature distributed values at the different cooking periods during the summer and winter seasons. From the graphical result indicating that the maximum temperature was found at the time of lunch (33.8 °C) and snacks (33.2 °C) cooking period, while the minimum temperature was recorded at the time of breakfast (30.9 °C) during the summer season. Similarly, during the winter season, "maximum air temperature" (31.7 °C) was recorded at the time of cooking period at lunch. However, minimum air temperature (26.6 °C) was measured at breakfast time. During both seasonal conditions, maximum air temperatures were measured at the lunch period and minimum at the breakfast cooking period. In "both seasons," the measured indoor "air temperature" data does not comply with the recommended acceptable range "22.6–26 °C" as per ASHRAE 55 standards [18]. Similarly, Fig. 18.4 shows that the average "indoor globe temperature" distributed values at the different cooking periods during the summer and winter seasons. Here also, "during the summer season, the maximum value of globe temperature" was found at lunch (33 °C) and snacks (33.1 °C) preparation time. While the minimum value of globe temperature was found at breakfast (30.4 °C) preparation time. Similarly, "during the winter season, maximum globe temperature (27.9 °C)" was recorded at the time of the cooking period at lunch. However, the minimum globe

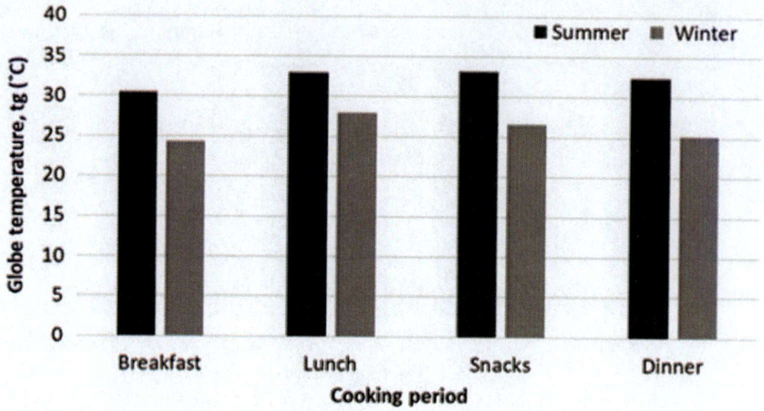

Fig. 18.4 Average globe temperature distribution at the different cooking period during summer and winter season

temperature (24.2 °C) was measured at breakfast time. In both seasons, measured parameters of "indoor globe temperature" were not obey the recommended range of ASHRAE 55 standard.

Figure 18.5 demonstrates that the average indoor relative humidity distributed values at the different cooking periods during the "summer" and "winter" seasons. During the summer season, maximum relative humidity was recorded (77.3%) and (76.1%) at the preparation of lunch and dinner, respectively. Whereas the minimum percentage of relative humidity was recorded at breakfast (74.9%) preparation time. Accordingly, during the winter season, maximum relative humidity was recorded 67.7 and 64.6% at the preparation of breakfast and snacks, respectively. Although

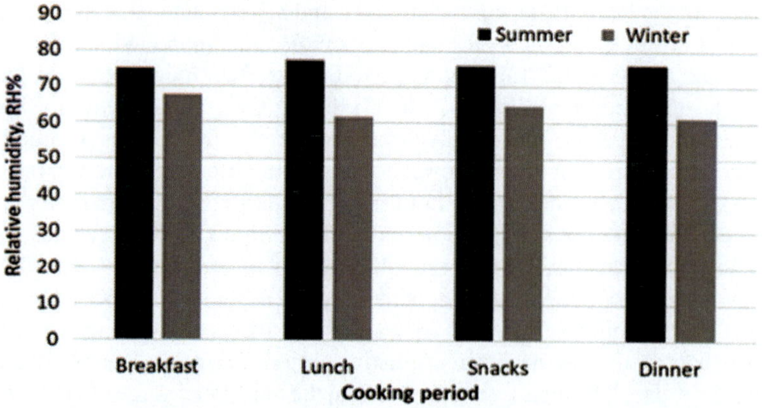

Fig. 18.5 Average relative humidity distribution at the different cooking period during summer and winter season

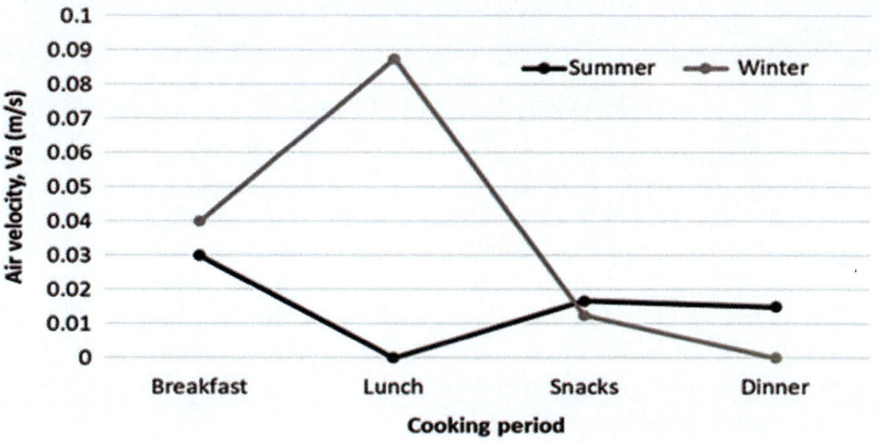

Fig. 18.6 Average air velocity distribution at the different cooking period during summer and winter season

the minimum percentage of relative humidity was recorded at dinner (61.5%) preparation time. A previous study suggested that for an indoor environment recommended comfort range of relative humidity percentage should be between 30 and 60% [10, 18, 19]. In this research, the measured percentage of relative humidity does not follow the recommended comfort range.

Figure 18.6 shows that the average indoor air velocity distributed values at the different cooking periods during the summer and winter seasons. The graphical result indicates that during the summer season the maximum value of "air velocity" was found at breakfast (0.03 m/s) and minimum at lunch (0.00 m/s) preparation time. Similarly, "during winter season," maximum and minimum value of air velocity was recorded at the preparation of lunch (0.09 m/s) and dinner (0.00 m/s), respectively. The previous researchers/ASHRAE standards recommended the air velocity value for summer and winter was <0.25 m/s and <0.15 m/s, respectively [10, 20]. In this study, very less value of air velocity movement was found "inside the pantry car."

In this study, significant differences between all different cooking periods during the winter and summer seasons have been investigated using the Kruskal–Wallis test, which is illustrated in Table 18.1.

Significant differences between different cooking periods during the summer season, the p-value of "Ta (°C), Tg (°C), RH%, va (m/s)" found exceeding the value of the significance level of 0.05. Therefore, in this study, there is "no statistically significant difference" found between all cooking time during the summer season.

Similarly, significant differences between different cooking periods during the winter season, the p-value of "Ta (°C), Tg(°C)" was found less than the significance level of 0.05. It is mean that there is a statistically significant difference found of "Ta (°C), Tg(°C)" between all cooking periods. However, the p-value of the "RH%, va (m/s)" was found more than the significance level of 0.05. It is mean that there is

Table 18.1 Kruskal–Wallis test—summer and winter season

Seasons	Parameters	Mean rank				χ^2	Sig
		Breakfast	Lunch	Snacks	Dinner		
Summer	Ta (°C)	9.5	15.67	14.42	10.42	3.255	0.354
	Tg (°C)	7.5	14.67	14.83	13	4.307	0.23
	RH%	11.67	14.17	13.67	10.5	1.075	0.783
	va (m/s)	13.83	10	11.92	14.25	2.743	0.433
Winter	Ta (°C)	8.56	25.38	19.38	12.69	15.083	0.002**
	Tg (°C)	9.38	25.5	18.56	12.56	14.108	0.003**
	RH%	21.06	14.31	17.44	13.19	3.45	0.327
	va (m/s)	17.31	19.94	15.25	13.5	4.524	0.21

**There is a statistically significant difference

no statistically significant difference found of "RH%, va (m/s)" between all cooking period.

Statistical results showed that "no significant difference" in thermal comfort parameters was found between all cooking periods during the entire summer season. Whereas, significant differences were found between thermal comfort parameters during different periods of cooking throughout the winter season except in the parameters of relative humidity and air velocity. Accordingly, the descriptive analysis shows that the highest thermal comfort values have been measured during lunch and snack periods except for air movement. While in this study, the average values of all measured thermal comfort parameters did not comply with the recommended range of the standard. A previous thermal comfort study also indicated that thermal comfort parameters measured in railway pantry car kitchens during cooking did not follow the recommended limits [10, 11, 13].

The results of the present study will support the design of ventilation systems such as air supply system, exhaust hood, and in the pantry car's kitchen environment to increase the thermal comfort of the chef during different cooking periods.

18.4 Conclusion

This research effort to recognize the present condition of the indoor thermal comfort parameters at different cooking periods in the kitchen of the pantry car on the Indian Railway. Summer and winter seasons were considered for data collection in this study. The analysis of field investigation results is as follows:

- A result of the descriptive analysis indicates that in "both seasons," the value average of "air temperature" and "globe temperature" was found maximum at lunch and snack preparation time. While the "average value of relative humidity" was also found high at lunch-time during the "summer season," but in the "winter season," it was found maximum at "breakfast time." Similarly, the average value of air velocity was found high at breakfast time during the summer season; however, it was found maximum at lunch during the winter season. All parameters of the thermal comfort do not obey the ASHRAE 55 standard.

- Statistical analysis revealed that there is "no significant difference between all thermal comfort parameters" during the summer season at the cooking period. While the "significant difference" was found between thermal comfort parameters during the cooking period in the winter season excluding for the parameters of "relative humidity and air velocity."

In this study, objective measurement is considered. It can be led to better results by identifying the chef's perception rate during the working period. The consequences of this research will help to determine the comfort range of the workers in future, which will make a significant contribution to increasing thermal comfort in the indoor environment of the pantry car kitchen by installing or designing an appropriate ventilation system. Also, vibrations and jerks might be very important for the moving train, which could be included in future research.

References

1. Ebenezer, S.P., Pramanick, A.K., Ramachandran, K.P.: Window air conditioner with orthodox refrigerants. Int. J. Eng. Trans. B **32**(2), 322–327 (2019)
2. Simion, M., Socaciu, L., Unguresan, P.: Factors which influence the thermal comfort inside of vehicles. Energy Procedia **85**, 472–480 (2016)
3. Streinu-Cercel, A., Marza, S.C.M., Streinu-Cercel, A., Marza, M.: Models for the indices of thermal comfort. J. Med. Life **1**(2), 148–156 (2008)
4. Danza, L., Belussi, L., Floreani, F., Meroni, I., Piccinini, A., Salamone, F.: Application of model predictive control for the optimization of thermo-hygrometric comfort and energy consumption of buildings. Instrum. Mesure Métrologie **17**(3), 375–391 (2018)
5. Pala, U., Oz, H.R.: An investigation of thermal comfort inside a bus during heating period within a climatic chamber. Appl. Ergon. **48**, 164–176 (2015)
6. Mezrhab, A., Bouzidi, M.: Computation of thermal comfort inside a passenger car compartment. Appl. Therm. Eng. **26**(14–15), 1697–1704 (2006)
7. Ismail, A.R., Rani, M.R., Makhbul, Z.K.M., Deros, B.M.: Assessment of thermal comfort and optimization of environmental factors at automotive industry. Eur. J. Sci. Res. **31**(3), 409–423 (2009)
8. Simone, A., Olesen, B.W., Stoops, J.L., Watkins, A.W.: Thermal comfort in commercial kitchens (RP-1469): Procedure and physical measurements (Part 1). Hvac&r Res. **19**(8), 1001–1015 (2013)
9. Ravindra, K., Agarwal, N., Kaur-Sidhu, M., Mor, S.: Appraisal of thermal comfort in rural household kitchens of Punjab, India and adaptation strategies for better health. Environ. Int. 124, (2019)
10. Alam, M.S., Muthiah, A., Salve, U.R.: Thermal comfort of the kitchen in pantry cars on Indian railways. Instrum. Mesure Métrologie **18**(5), 465–477 (2019)
11. Alam, M.S., Muthiah, A., Ravindra Salve, U.: Appraisal of thermal comfort in non-air-conditioned and air-conditioned railway pantry car kitchens. Int. J. Integr. Eng. **12**(8), 318–327 (2020)

12. Pethkar, G., Jadhav, S., Dhumal, P., Bhosale, P., Ahir, P.R.: Automated pantry car system in India using RF module and GSM technology. Int. J. Adv. Res. Electron. Commun. Eng. **4**(4), 794–797 (2015)
13. Alam, M.S., Arunachalam, M., Salve, U.R.: A pilot study on thermal comfort in Indian Railway pantry car chefs. J. Phys: Conf. Ser. **1240**(1), 012033 (2019)
14. Matsuzuki, H., Ito, A., Ayabe, M., Haruyama, Y., Tomita, S., Katamoto, S., Muto, T.: The effects of work environments on thermal strain on workers in commercial kitchens. Ind. Health **49**(5), 605–613 (2011)
15. Alam, M.S., Arunachalam, M., Salve, U.: A comparative study of thermal environment between two varieties of pantry car available in Indian Railway. In: 15th International Conference on Humanizing Work and Work Environment. AMU, Aligarh, 8–10 December (2017)
16. Indian Society of Heating, Refrigerating and Air Conditioning Engineers: Indoor Environmental Quality Standard. ISHRAE Standard (2019)
17. Hamzah, B., Ishak, M.T., Beddu, S., Osman, M.Y.: Thermal comfort analyses of naturally ventilated university classrooms. Struct. Survey **34**(4/5), 427–445 (2016)
18. Hussin, M., Ismail, M.R., Ahmad, M.S.: Thermal comfort study of air-conditioned University Laboratories. Int. J. Environ. Technol. Manage. **17**(5), 430–449 (2014)
19. American Society of Heating, Refrigerating, and Air-Conditioning Engineers: Ventilation for Acceptable Indoor Air Quality. ASHRAE Standard-55, Atlanta (2016)
20. American Society of Heating, Refrigerating, and Air-Conditioning Engineers: Thermal Environmental Conditions for Human Occupancy. ASHRAE Standard-55, Atlanta (2004)

Chapter 19
Visual Framework of Color Analysis of Shop Signs in India

Nanki Nath

Abstract Shop signs contribute to the visual representation of business identities through color, form, and materials. Among these, in the context of India specifically, color becomes the foremost significant attribute for business image building in, especially, multi-cultural and competitive commercial street marketplaces. The researcher could observe and integrate certain qualities of colors displayed on shop signs through this methodological research approach of bricolage (Denzin, N., Lincoln, Y.: Handbook of Qualitative Research, 2nd edn. Sage, Thousand Oaks, CA (2000). These qualities are the four quadrants/factors of the illustrated framework in this paper. We have redefined all the four quadrants by following an inductive rationale. This was done by selection of characteristic keywords for each quadrant from the established definitions, etymologies, and findings by researchers and practitioners in the field of design. The classification theories using morphology, thematic basis, groupings, and orderings as empirical data helped to formulate the framework of color analysis. The final picture of the framework illustrates all the four quadrants and their respective classifications along vertically and horizontally aligned domains. This form of the framework is flexible enough to provide multiple perspectives of quadrants as individual qualities as well as relative ones in a collective analytical group in order to analyze colors displayed on shop signs.

19.1 Introduction

19.1.1 Research Question(s) and Research Plan

The present paper brings forth the main contribution as part of the author's doctoral thesis with the aim to formulate a flexible visual framework that would be an analytical aid and conceptual tool for visual communication designers involved in

N. Nath (✉)
Faculty Member, National Institute of Design (NID), Ahmedabad, India
e-mail: nankinath@gmail.com; nanki_n@nid.edu

A. Chakrabarti et al. (eds.), *Design for Tomorrow—Volume 1*, Smart Innovation, Systems and Technologies 221, https://doi.org/10.1007/978-981-16-0041-8_19

research/design projects on signs and signage designs. The two primary research questions in the thesis were:

(1) What sort of visual framework is needed for the analysis of color in shop signs?
(2) How do we use this framework to visually analyze color, its characteristics, and modes of communication on shop signs in India?

A total of 3500 shop signs were collected by the researcher through photography as part of her travels to twelve different cities in India. After elimination of damaged/worn off signs, a population of 2700 shop signs from these cities could be selected. Further, through the method of stratified sampling she obtained a sample size of 450 shop signs belonging to five different cities of India for final color analysis in the thesis. These 450 shop signs were observed in depth for a visual analysis. She could filter out four distinct factors/quadrants: Patterns, trends, tendencies, and conventions in each of the shop signs [1].

19.1.2 Morphology of a Typical Shop Sign in India

The morphology of a typical shop sign in India consists of three component layers of information. In each of the 450 shop sign photographs in this research, these three distinct layers of information were segregated for the final color analysis. Each of the three layers has a different message to communicate about a shop and its business. This is mostly so since each viewer may be interested in a specific part of the information content that a shop sign displays. The layers of main text and secondary text are in the foreground design of a shop's sign face. The third layer is the background sign panel itself (Fig. 19.1).

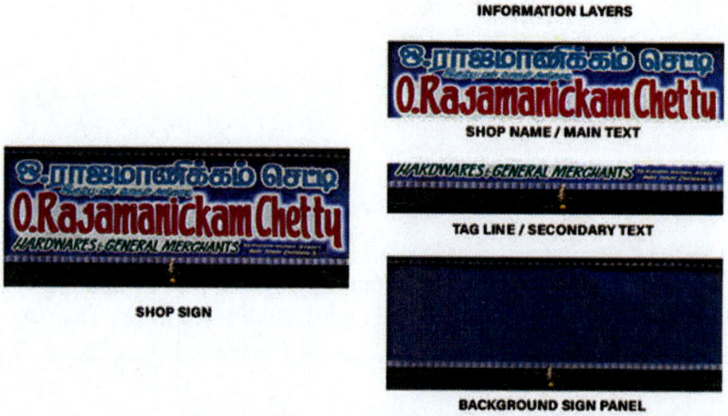

Fig. 19.1 Morphology of a shop sign in India with three layers of information. *Source* Visual framework of color analysis of shop signs in cities of India, Nath (2015)

Fig. 19.2 Color characteristics identified on shop signboards in India. *Source* Visual framework of color analysis of shop signs in cities of India, Nath 2015

19.1.3 Role and Significance: Color as a Sign

The semantics, syntactics, and pragmatic meanings of color semiosis in a typical shop sign in India make the attribute of color 'a sign' itself. In addition, color is significant attribute to identify objects. Color as a visual attribute accompanies the type of the form of the letters that together construct the shop name: main text and the tag line: secondary text (Fig. 19.2). In addition, 'color' as a visual attribute accompanies the material board or the panel form of the shop sign. Color provides the identity to this form acting as the background in a shop sign (Fig. 19.2).

Among the elements of visual design, graphics and text are inseparable from the attribute of color. When applied skillfully and intentionally, color is a valuable communication tool for graphic designers [2].

19.1.4 Color Dimensions and Characteristics

Designers and practitioners have discovered color analysis methods prominently for fashion designers and makeup artists. Among these existing methods, we have considered the *four seasons and their twelve color types* [3], introduced by Carole Jackson in her book *Color me Beautiful* (1987). Following the same method could observe four dimensions of colors for the visual analysis of 450 shop signs in this research. The foremost *description* dimension *describes hue, value, saturation, temperature, and kinds.* In past, the well-known color researcher Johannes Itten saw reactions to color as aesthetic, emotional, or symbolic. According to Johannes Itten, *colors and their combinations* can be particularly evocative when they conjure an atmosphere or a period of history [4]. Such *color relationships* are formed by using two or more *color combinations,* their *hierarchy, contrast, effects,* and the mixing qualities in the form of *color schemes.* The fabrication, make, and production of colors as part of materials are in the dimension that explains their pragmatics. In

Table 19.1 Four dimensions and thirteen color characteristics

Color dimensions	Color characteristics
1 Color description	Hue, value, saturation, temperature, kind
2 Color relationship	Combination, hierarchy, contrast, effect, scheme
3 Color pragmatics	Production, personality
4 Color symbolism	Identity

this dimension, an extra thirteenth characteristic, namely *color functionality* is also considered for final analysis in our research. Symbolism of colors reveals the identity dimension, i.e., the culturally known or arbitrarily implied meanings of colors (Table 19.1).

19.2 Formulation of a Research Methodology

In order to give a functional integration as well as a structure to all the four factors/quadrants of patterns, trends, tendencies, and conventions, following step-by-step formulation of a research methodology for the main analysis was followed:

19.2.1 Methodological Bricolage: Meaning

The term *bricolage* or *making-do* combines elements, components, or concepts into new and original outcomes. The method of combining elements has its roots in the academic studies belonging to social research. Bricolage comes from a traditional French expression that denotes crafts people who creatively use leftover materials from other projects to construct new artifacts. French anthropologist and ethnologist Claude Levi Strauss defines this method in the context of anthropology as a spontaneous creative act that uses whatever is available to reach at a desired outcomes. These two minds adopt two different modes of knowledge acquisition, namely the concrete and the abstract, respectively.

19.2.2 Methodological Bricolage: Selection of the Approach

The basis of methodological bricolage that this research consciously selects is that that approach of 'making-do' or 'tinkering' that helps in charting out a new reference point for the visual semiotics: the semantics, syntactics, and pragmatics of color attribute in the finalized 450 shop signs of the five cities. The bricoleur—the researcher here as a tinkerer brought together the independent factors/quadrants of patterns, trends, tendencies, and conventions—this originality and a deliberate attempt that the approach of methodological bricolage provided the researcher with; in addition, it provided a base of semiotics during the integration of these four factors together. The resulting framework could contextually base firstly, *patterns/arrangements* as semantics (arrangements/expressions as they exist), secondly, *trends/vogues* as syntactics (order/sequence of shop sign vogues showcasing transitions over time periods), and lastly, the third and fourth factors of the *emerging tendencies and standard/ established conventions* as pragmatics (those intuitive aesthetic forces/ underpinnings as tendencies and symbolic rules as conventions that establish the communication of a visual attribute with the viewer).

19.2.3 Methodological Bricolage: Meaning and Salient Features

The term *bricolage* or *making-do* combines elements, components, or concepts into new and original outcomes. Bricolage comes from a traditional French expression that denotes crafts people who creatively use leftover materials from other projects to construct new artifacts.

19.2.4 Salient Features of Bricolage in the Final Analysis of Color

19.2.4.1 Prescribed Affordance in Integrating Elements

The subject of design has the quality to create prescribed affordances[1] for the designed products. According to Yee and Bremner, it becomes a pertinent requisite of the researcher in the context of qualitative research studies such as the present paper for the different theoretical, philosophical, and methodological aspects [5].

19.2.4.2 Flexibility and Openness

A visual analysis involves the role of a meaning-making, and a researcher does not approach knowledge production with concrete plans, methods, tools, or checklists of criterion. In each quadrant, she could utilize different knowledge tools of inquiry during visual analysis of colors. For instance, in order to explain occurrence of tendencies, she could identify social as well as aesthetic underpinnings as different

---

[1]For Gibson (1986) the term *affordance* meant meaning the interactive possibilities of a particular from https://ecologylab.cs.tamu.edu/courses/physicalInterfaces/hostedMaterials/gibsonAffordances.pdf.

causes that have given rise to a parallel tendency of color behavior in the shop signs (see Fig. 19.7).

19.2.4.3 Pragmatic and Self-Reflexive in Nature

Bricolage as a research method is self-reflexive and pragmatic to the extent that it generously appreciates the complexity of the analysis involved. In the context of our visual framework, reflexivity of the four-quadrant framework (see Sect. 19.3) established by the methodological bricolage presents an example of how an inquiry into shop sign as an object of a visual research study can be seen from multiple vantage points-rooted in design, history, technology, production, culture, practice as well as social realities.

19.2.4.4 Inter-Determinacy Leading to Innovative Theories, Methods and Concepts

The adoption of the bricolage method and the questions it raises highlights the value of indeterminacy as a key condition to design research. Dunne and Rabby are well known to employ an alternative approach of using design research as 'a space for designers to reflect upon ideas, theories, logics, and implications of design in and through design practice.'[6]

19.2.4.5 Multiplicity

Kellner describes that the multi-methodological approaches not only provide unique possibilities for knowledge construction, but also create opportunities for informed political action [7]. For Kincheloe (2005b), the criticalisation[2] of inquiry includes an adoption of such research pursuits that appreciate the complexity of the living world that include inquiry processes that do not study objects as detached things-in-themselves but rather as connected objects-in-the-world [8]. Since the visual attributes that create the design of shop signs construct new knowledge with respect to each quadrant grounded within the framework (see Fig. 19.11, Sect. 19.3).

---

[2]In his *criminalization*, Kincheloe (2005b) moves Denzin and Lincoln's (1999) articulation of bricolage onto *the next level* (p. 323), by adopting and extending the five categories of bricoleurs.

19.2.5 Bricolage as a Research Approach

The design of a shop sign, as a visual form envisages expression(s) that are explicit and meaning(s) that are implicit in the shop sign identity. This has helped in integrating the explicit structural and visual syntax-based patterns and trends with the implicit expressions and meaning-based tendencies and conventions by applying the bricolage as a research approach. According to Denzin and Lincoln, 'Visual research depends upon and redistributes social power' [9].

19.2.6 Identifying and Defining Individual Parts of the Visual Framework

The visual display of 450 shop signs in five cities was first put on the display space of the present Shenoy Innovation studio at the IDC School of Design, IIT Bombay by the researcher. As a bricoleur, she identifies four independent factors in the visual design of the shop signs. When observing each of the 450 signboard photographs identifying varied shop business, the first observation was of the component parts and their arrangements in the foreground texts and the background panels. This represented the visual syntactics as part of each and every shop sign in her doctoral thesis [10].

19.2.6.1 Defining Quadrants

The identified factors as the four corners of the visualized framework for color analysis are called *quadrants* in this research. By applying bricolage, the four factors of patterns, trends, tendencies, and conventions are integrated to form the structure of the visual framework ahead (Fig. 19.11). To redefine each quadrant, an inductive rationale has been followed, wherein selective keywords from existing definitions, etymologies, and published findings of each of the four quadrants were collected together for the redefinitions as follows:

Patterns are composed of similar elements of a visual attribute and similar arrangements or visual relationships between those elements that guide the concrete form or the formal syntax in the visual design of a shop sign.

19.2.6.2 Kinds of Patterns

According to Christopher Alexander in his book, 'The Pattern Language,' [11].

> 'Each pattern is connected to certain larger pattern which comes above it in language and to certain smaller pattern that comes below it in language. Hence, the pattern helps to complete

Color value patterns	Pattern examples	Business size
i) High value in white and yellow	 STAR ARUN PROPERTIES परम्परा RAJWADI	Medium size old and big size modern stores
ii) Medium value in green	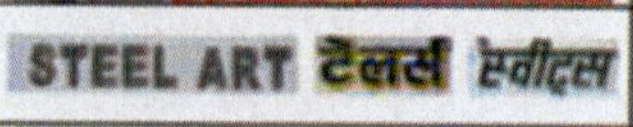 STEEL ART टेलर्स	Small size old stores and a medium size modern stores
iii) Dark value in dark blue, dark green and dark brown	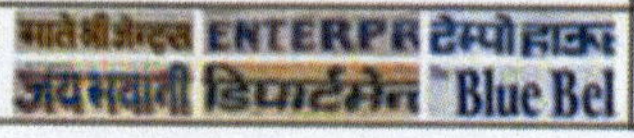 ENTERPR जयभवानी डिपार्टमेन्ट Blue Bel	Small size old shops and one medium size modern store

Fig. 19.3 Small patterns are within large patterns such as color hue, value, etc., for shop signs

the larger patterns that are above it and is itself completed by those smaller patterns that are below it.'

In a nutshell, every small pattern is part of a larger pattern. **Small patterns** are the color characteristics themselves. All these small patterns together within a dimension of large pattern—complete the pattern system of color analysis in this research. For instance, smaller patterns of color value are part of the large pattern of color description (Fig. 19.3).

Trends are the simultaneous movement or change of one or more characteristics (variables) of a visual attribute of design of a shop sign in a certain direction along time as the constant variable. Here, the objective or goal may not be clear for any single factor. But, the direction has some link with respect to the visual, social, economic, or even historical popular themes belonging to a specific time period.

19.2.6.3 Kinds of Trends

Transitory trends are the movement or a change in a single color characteristic that undergoes a transition from 1930–44 time period till 2005 onwards which is a *transitory trend*. For instance, in very old shop signs, only two hues have been used in the main text. With time, both two and three hues came into picture; whereas for modern time periods, there are a sudden drop in the two hue trend and an equivalent upsurge in the single hue (Fig. 19.4).

Period specific trends are those that belong to different color characteristics that are specific to a certain time period. For instance, we have outlined in the temporal scale below the specific trends seen in colors of shop names across marketplaces in the city of Jaipur belonging to specific time periods (Fig. 19.5).

Tendencies are causal behavioral inclinations or prevalence toward a specific line or direction that may or may not vary in the designs of shop signs. They are those innate qualities in a visual attribute that develop because of impelling social, aesthetic, or historical forces.

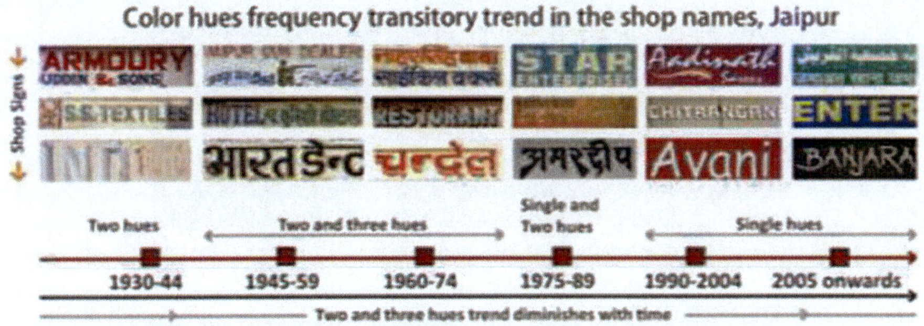

Fig. 19.4 Transitory trend of color hues in the main texts of shop signs, city of Jaipur

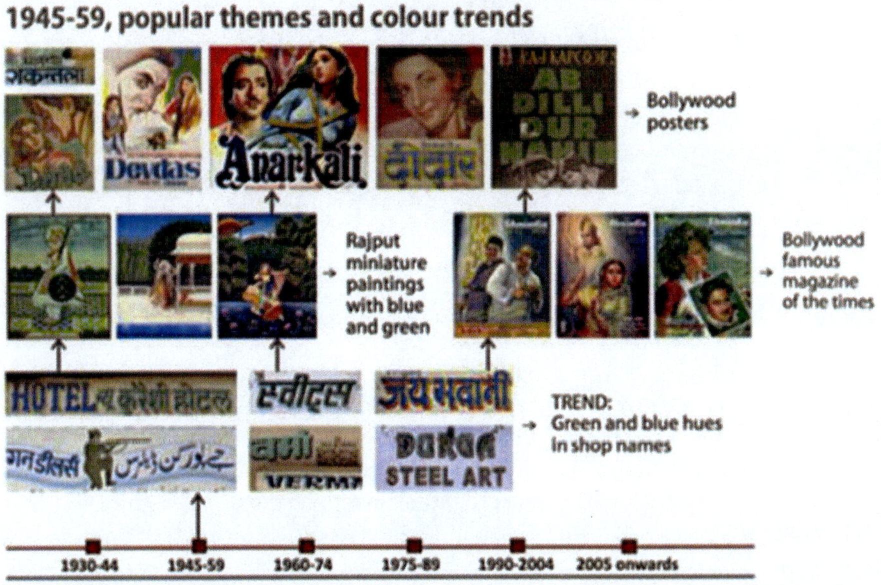

Fig. 19.5 Period specific color hue trend in the main texts of shop signs, city of Jaipur

19.2.6.4 Kinds of Tendencies

General tendencies that are most common inclinations of the color display in the design of shop signs. The word *general* here means that such tendencies are not governed by local influences. For instance, *old shop signs have prominently hand-painted colors in the main text*. Illustrated below is an example of shop signs from the old city of Jaipur (Fig. 19.6).

Distinct tendencies are locally promulgated inclinations of the visual attributes in the key design elements in the shop signs. Such tendencies are the more indigenous

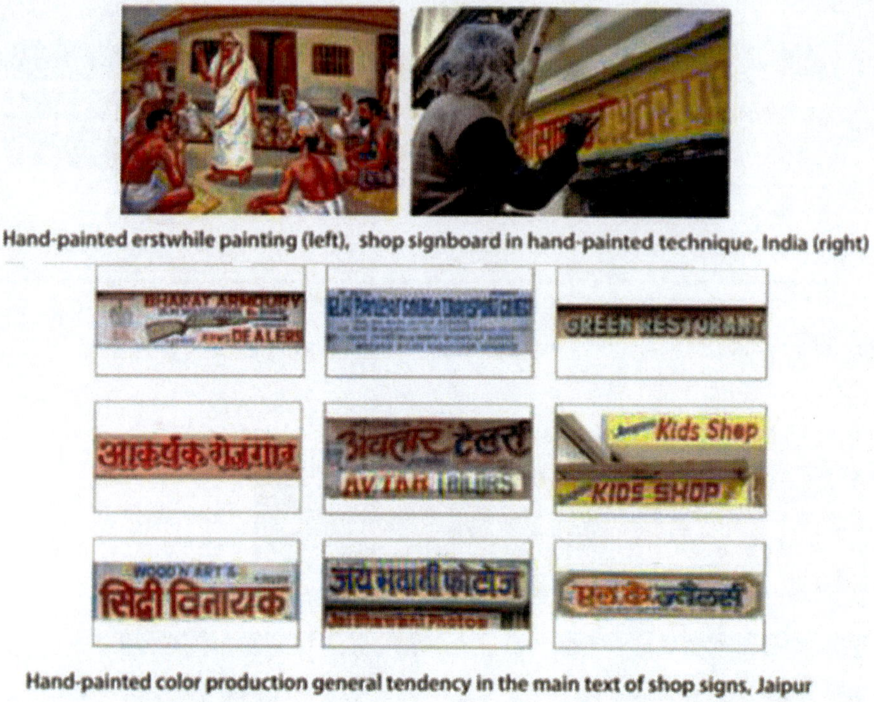

Fig. 19.6 General tendency of color production in the main texts of shop signs, city of Jaipur

that reveal a city identity as a local, social, or cultural byproduct. A local tendency may also indicate city specific inclinations.

For instance, the traditional city of Jaipur sustains a distinct color tendency of using *Jaipur Pink wash* in the background elements of display design of the shop signs (Fig. 19.7).

Conventions are sets of accepted norms, standards, and criteria established since ages. In context of shop signs in our study, most of these conventions are either the color codes applied in context of their business meanings or the historical narratives specific to a particular city or India at large.

19.2.6.5 Kinds of Conventions

Conventions in the historical context are established norms of different colors with respect to specific historical narratives in a city or modes of display in particular colors that imbibe some symbolic meaning in context of India. For instance, we could observe the use of ***red*** or ***orange*** colored *Om* or *Swastika* symbols visually displayed on shop signs. This color and form meaningful relation is an age-old convention practiced in India at the launch of new business or after the purchase

Fig. 19.7 Distinctive tendency of saffron in the main texts of shop signs, city of Jaipur

of a new product (Fig. 19.8). Another instance shows ***Rani Pink*** hue of mystical Rajasthan or the saturated hues of southern India that have their distinct stories that subtly guide their use in shop signs of the specific parent city (Fig. 19.9).

Conventions in the business context are those related to the meanings of different colors with respect to nature of the business/trade that the shop sign signifies: the products/services provided by the shops. The meanings of specific colors are established due to a close association of an old or a modern shop with their respective business identities. For instance, everyday old service shop signs of Moti Doongri road marketplace in Jaipur city have blue shop names (Fig. 19.10).

Fig. 19.8 Historical practice of displaying religious marks on sign facades in context of India

Fig. 19.9 Rani Pink in Sisodia Rani ka Bagh palace *(left)*; in Jaipur's branded shop *(right)*

Fig. 19.10 Rani Pink in Sisodia Rani ka Bagh palace *(left)*; in Jaipur's branded shop *(right)*

19.3 Visual Framework of Color Analysis of Shop Signs

The first part of this section explains the theoretical basis to apply ontological, epistemological, and methodological modes of bricolage in order to filter the appropriate quadrants and their integration along planes within the framework, respectively. In order to understand the functioning of each quadrant, we first need to understand the rationale applied to position them along with these vertical and horizontal planes. Why some quadrants fall in respective functional planes within the framework?

19.3.1 Theoretical Basis of Application of the Bricolage Methodology

In our framework of components based on bricolage methodology, each quadrant in our research takes following modes:

- Ontological mode

The ontological assumptions relate to the knowledge of the way things are, the nature of reality, real existence, or real action [12]. This mode has been applied in the collection of 'the way each quadrant, namely patterns, trends, tendencies, and conventions are in reality' or real action by collecting all known terminologies.

- Epistemological mode

The epistemological assumptions relate to 'how things really are' and 'how things really work' in an assumed reality [12]. This mode has been applied in the case of each quadrant by inductively putting all the definitions together under scrutiny. In this scrutiny, we have redefined quadrants, constructed their classifications as well as understood their significance with respect to shop signs.

- Methodological mode

After setting the essential ontological and epistemological base of each of the quadrants, a methodological assumption of illustrating them together in a framework is required [12]. This assumption is the most crucial of all the modes, since it acts as an assessment tool to know the real world of how things work for the visual design of shop signs.

19.3.2 Formulation of the Visual Framework of Color Analysis of Shop Signs

Application of bricolage methodology as described above has enabled us to formulate the visual framework for color analysis of a shop signboard (Fig. 19.11). There are functional domains along the x- and y-axis of each quadrant. There is a specific rationale that guides positioning of these domains.

For instance, patterns and their analysis represent the domains of *arrangements* and their *spatial configurations* with respect to each other. The rationales per quadrant redefinitions (Sect. 19.2.4.1) are as follows:

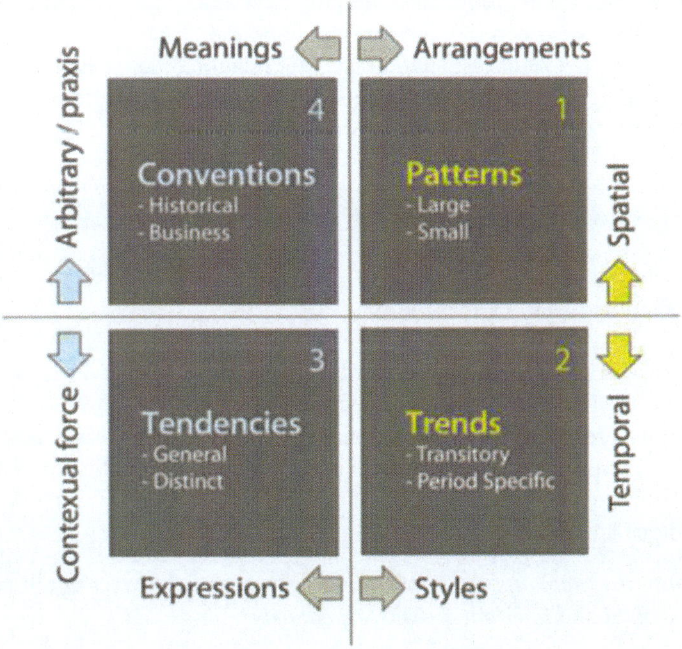

Fig. 19.11 Visual framework of color analysis of shop signs

19.3.2.1 Quadrant I: Patterns

- Patterns are **elemental** or basic units of color composition.
- Patterns are not contextual.
- Patterns are relational due to a **variety of arrangements** in **spatial compositions**. *Therefore, the domains of patterns are* ***spatial-arrangements*** *placed in vertical and horizontal axis, respectively.*

19.3.2.2 Quadrant II: Trends

- Trends are **visual styles** of color that undergo **simultaneous movement**.
- Trends are essentially **temporal** because of this movement across time.
- Trends are hence, not relational or spatial. **Styles** occur freely not based on arrangements or any preordained structure of elements put together.

Therefore, the domains of trends are ***temporal-style-based*** *placed in vertical and horizontal axis, respectively.*

19.3.2.3 Quadrant III: Tendencies

- Tendencies are **behavioral inclinations** of color in a **certain direction**.
- Tendencies are **expressions** that are not temporal. They are **independently occurring prevalence(s)**. So, they are not temporal.
- Tendencies are **prevalence(s)** and hence, not relational nor spatial. **Expressions** imbibe meanings and are **not dependent on** formal visual **arrangements**. Rather, they **are grounded within a contextual force**.

Therefore, the domains of tendencies are ***contextual force-expressions*** *placed in vertical and horizontal axis, respectively.*

19.3.2.4 Quadrant IV: Conventions

- Conventions are **established standards or norms** of color **over a long period of time or applied since ages**. So, they are not temporal.
- Conventions are not spatial or relational or temporal. They are **practice-based symbolic meanings** that are learnt as part of a cultural practice or deep rooted historic values.
- Conventions are **abstract**.

Therefore, the domains of conventions are ***abstract-praxis-based-meanings*** *placed in vertical and horizontal axis, respectively.*

19.4 Insights and Discussions of the Framework

(a) The vertical axis includes the spatial or temporal domains at the right describing patterns and trends, respectively.

- Temporality of trends has a spatial function in terms of historical development of material objects and their visual designs such as those of shop signs (Fig. 19.11). This associates patterns and trends to each other through the function of spatiality.
- Contextual forces that drive tendencies are also intrinsic to the arbitrary nature of conventions. While conventions are standardized practices with a firm traditional praxis since a long time, tendencies are those expressions that are deep rooted cultural inclinations without conscious knowledge (Fig. 19.11).

(b) The horizontal axis includes arrangements domains positioned on the top and bottom peripheries of the framework. This domain is part of patterns and conventions that are otherwise very distinct quadrants without a shared quality. The second occurrence is at the bottom having expressions or styles domains that are part of tendencies and trends.

19.5 Conclusions

19.5.1 Visual Framework of Color Analysis of Shop Signs in India

The four-quadrant open, flexible, and multi-perspective framework is a distinctive contribution of this research (Fig. 19.11). This framework aims to analyze color in the design of shop signs in India. Such a framework provides more flexibility and freedom for imagination to the researchers as well as visual communication designers with whatever knowledge tools they have in their repertoire to deconstruct the meanings of signs through their designs. A designer's role in the present times has changed from the traditional one that applied a unilateral way of thinking and approaching a subject area, to the contemporary need to know:

(1) Multiple points of view and apply them as multi-perspective approach in research
(2) To create designs with multi-functional, versatile, and more open-ended solutions. This would help in continuous evolution of designs, systems, and frameworks for innovative and creative ideas as part of the future work. To address this new role and need in visual designs, the present framework of visual analysis of color in shop signs of India reciprocates this multi-perspective opportunity by providing a range of autonomous quadrants.

The framework of visual analysis of color designed in this research has been envisioned to act as an analytical tool that can aid designers or makers of signboards in the context of promotional design strategy, identity creation, and business image building from the perspective of marketing and branding. This is so because the framework with patterns, trends, tendencies, and conventions views color in this research from multiple perspectives.

19.5.2 Relevance of the Framework for Visual Sign Design Community

Attributes other than color, such as typography, graphics, layout, and material can be analyzed for identifying patterns, trends, tendencies, and conventions of the framework by sign designers in India. To elaborate on these perspectives and influencing factors, for instance, trends of visual display design in shops would give information about the changing fashion and make of designs, related naming trends in case of different business kinds, new techniques of displaying shop names along with tag lines or sometimes tag lines themselves assuming the role of brand identity for a shop's business. The perspective of patterns would help resolve the mechanical coherence of all content on the signboard that at many times is also followed on the printed visiting business card of a shop (Fig. 19.3). These patterns of specific hues and other characteristics would help create a color or any other attribute's distinct grammar for the shop name identity, tag line identity, or even background sign panels. This would give required cues to the sign designer to create a distinct syntax for a shop business in context of information design in these contemporary times. With tendencies and conventions being more locally, nationally, traditionally, historically, and commercially ingrained meanings, the designer would get familiarized with certain unknown qualities and age-old known practices to portray form and content of the business in context of its kind, scale, purpose, history, and future visions (Figs. 19.6, 19.7, 19.8 and 19.9). The suggested visual framework can act as effective analytical tool to decode the communication aspects of different kinds of signs in context of visual design or may be the larger perspective of semiotics and conceptual modes of meaning-making and signification systems with which different signs operate in a range of semiotics. The current framework in the focused context of visual design takes a lead from the mode of semiotic analysis in order to create a design thinking-based visual analysis mode that studies an attribute, its characteristics, and the modes of visual communication.

Acknowledgements I express my utmost gratitude to my guide, Prof. Ravi Poovaiah for motivating, encouraging, supporting the genesis, and successful articulation of this major research contribution as part of my doctoral thesis: 'Visual Framework of color analysis of shop signs in cities of India' (2015), IDC School of Design, IIT Bombay.

References

1. Denzin, N, Lincoln, Y (2000) Handbook of qualitative research (2nd ed.). Thousand Oaks, CA: Sage. https://doi.org/10.1111/j.1365-2648.2001.0472a.x
2. Rider, R.: Color psychology and graphic design applications. In: Color Psychology. Graduation Thesis. Honors Program. Liberty University (2009). https://digitalcommons.liberty.edu/cgi/viewcontent.cgi?article=1118&context=honors. Accessed May 2015
3. Jackson, C.: Color Me Beautiful. Ballantine Books. Revised, Reissue edition. ASIN: B0061C1NHA (2011)
4. Itten, J.: The Art of Color. Van Nostrand Reinhold Company, New York (1961). ISBN: 978-0471289289
5. Yee, J., Bremner, C.: Methodological bricolage: what does it tell us about design? In: Doctoral Design Education Conference. Hong Kong Polytechnic, Hong Kong, 23–25 May 2011
6. Dunne, A., Raby, F.: Speculative Everything: Design, Fiction And Social Dreaming. MIT Press, USA (2013). ISBN: 978-0-262-01984
7. Kellner, D.: Theorizing McDonaldization: A multi-perspectivist approach. In: Smart, B. (ed.) Resisting McDonaldization, pp. 186–206. Sage, London (1999)
8. Kincheloe, J., McLaren, P.: Rethinking critical theory and qualitative research. In: Denzin, N.K., Lincoln Y.S. (eds.) The Sage Handbook of Qualitative Research, pp. 323. Sage Publications, Thousand Oaks, CA (2005).
9. Denzin, N.K., Lincoln, Y.S. (eds.): Collecting and Interpreting Qualitative Materials, 2nd edn. Sage, Thousand Oaks, CA (2003)
10. Nath, N.: Visual Framework of Color Analysis of Shop Signs in Cities of India. IDC School of Design. IIT Bombay, India (2015)
11. Christopher, A.: A pattern language: towns, buildings, construction. USA. In: Volume 2 of Center for Environmental Structure Series. (1978). ISBN: 9780195019193
12. Guba, E., Lincoln, Y.: Competing paradigms in qualitative research. In Denzin, N. Lincoln (eds.) Handbook of Qualitative, pp.105–177. Sage, Thousand Oaks. CA (1994)

Chapter 20
Physical Space Acting as a Catalyst for Improving Mental Health Especially for Teenage Students

Aavrati Kushwaha and Mayur Hajare

Abstract The notion of well-being has been becoming a faraway reality. With evolution, the environment around us has also changed. The need to survive and the inborn nature to compete have eventually led us, humans, to achieve unattainable things which, in turn, have made us susceptible to many of the conditions like depression and anxiety. This project done on mental health works towards creating a space which will indirectly help our senses to enhance our moods emphatically in daily life. This will lead to the so-called empowerment of our mental health even though people are aware of what they should have been doing in order to achieve the state of a healthy mind. They are still incapable of doing so, as it is really hard to do so and it majorly depends on others—spaces and people around us. Time restrictions and financial limitations are one of the important factors contributing to them and also unwanted pressure from parents and society. This project has been done as a part of systems design project through the case study of the students living in Kota, Rajasthan. One of the categories among the research groups includes: age, occupation and family type. The design intervention came through the research will eventually aid in achieving those little yet significant tasks and experiences which amount to our well-being. Space which surrounds us from the moment we wake up till we sleep can affect our mental health immensely. The ambience, which consists of lights, colours, sound textures and aroma, indirectly signals our brain to be at a certain state of mind. The creation of a manipulative yet sympathetic environment can control the space which will include furniture and supportive systems. Hence, designing a solution to effortlessly put our minds at ease.

A. Kushwaha (✉) · M. Hajare
National Institute of Design, Ahmedabad, India
e-mail: aavratikush24@gmail.com

M. Hajare
e-mail: imagini.design21@gmail.com

A. Chakrabarti et al. (eds.), *Design for Tomorrow—Volume 1*, Smart Innovation, Systems and Technologies 221, https://doi.org/10.1007/978-981-16-0041-8_20

20.1 Introduction

This research paper aims to look at the mental health of teenagers in India and how space can help them to stay mentally healthy in day to day life through a systematic study of the physical space around them. Systems thinking when combined with design thinking help to chart a holistic picture of the case. In this research paper, a systems design approach has been followed which is a synthesis of things happening around us and not just a mere analysis. It helps us in understanding underlying issues, which would be primarily hidden in a more linear research approach. With complex problems, the real world is usually multidimensional, dynamic and nonlinear. Every project begins with a prompt which can lead to the initial problem regarding the main topic selected. With further secondary and primary research, the mapping of reality helps in understanding the situation better. Various complex elements are interlinked in a systems design. Here, they are the migration, the education system in India and spatial design, behavioural changes and patterns, urbanization, the rising culture of individuality, social hierarchy in people and objects, and mental health. They were studied, and inferences were linked together to map out the complexity of these elements, their connections and impacts on other subsystems. The situation analysis and comprehension of these elements were done step by step. This particular step allows us to find loopholes and gaps that will later help in devising design solutions [1] for the problem found. Only after the problems and situations are known, we can assess the solution. After this, evaluating the design interventions or solutions to fit into the systems space will determine its end decision for its real-life application. It always has a scope to keep going back and forth between the micro and macro perspectives.

While collecting data about mental health issues in India, we found that a large number of teenagers are victims of it. Student suicides [2] were found to be at the epicentre of the problem. Teenagers, the future generation of our nation, are the most susceptible group to mental illness. To further understand this matter on the ground level, a case study was done on the coaching centre culture in Kota, Rajasthan, which is the hot spot of student suicides in India. We tried to understand the current helping systems by government and coaching institutes and their outcomes and how effective these systems are? We found that no. of suicides [2] have been increasing since the last few years. What are the factors responsible for the increase in the number of suicides? Do we need to rethink these helping systems? Does this flow of the system work? Are students aware of mental health issues and how to talk about them? Are we comfortable enough to talk about mental health issues in India? What kind of systems can we add to tackle the problem? Is it only students who need counselling in mental health-related issues? What kind of tangible and intangible factors are responsible for their state? The initial research done at Kota coaching centres revealed that the surrounding environment around students matter a lot while assessing their mental health.

In our day-to-day lives, as soon as we wake up, we are surrounded by the physical space around us. The physical space (PS) influences our actions both directly and indirectly. What kind of attributes should be added to the space, so that the students' mental being can improve? What effect different elements of space has on users' mental health? Can we increase the cognitive ability of the brain positively through our senses which are responsible for registering most of the information? What effect does colours, materials and finishes have on our cognitive ability? How should we do space planning for a better lifestyle? When it comes to solutions, there are mainly two categories—tangible solutions and intangible solutions. While there are therapists and professionals who are working with the students to address their mental state through intangible means, not much work has been done to improve the tangible means, i.e., the physical space around these students. There is a great scope which is yet to be explored and worked upon which concerns physical spaces and elements around us.

This research paper would like to facilitate the well-being of mental health through physical space, for a single person living, especially for teenage students. One example could be what factors create a sense of our own space? Do we feel the same in a hotel room and our bedroom? How a lot of students who live out of a suitcase, will usually have a pile of clothes on their bed until they go to sleep, which would move to a different spot once they decide to sleep, or worse, they would sleep on the pile itself. Such behaviours are not permanent. Like every once in a while, they would clear up the mess, only to get back to the mess eventually. One could sit and wonder all day as to why our brains function the way it does [3].

Usually, designers and architects only look at the groundwork of creating aesthetic and functional spaces. Unfortunately, not much emphasis has been given to the mental well-being [4] of the inhabitants. Neuro architecture [5] is one such field which caters to designing stimulating physical spaces for its residents. Space that we design eventually shapes the lives of the humans that they have been designed for. Hence, one of the main challenges of this paper would be how physical spaces can act as a catalyst for improving overall mental well-being and make a crucial part of spatial design. Another challenge would be to address the few of many problems like tremendous pressure of competition and family's expectations on students. Hence, the focus has been largely on teenage students. Lastly, with urbanization and rural–urban migration, all of the city-landscapes and urban spaces have to be re-thought while keeping in mind the mental health of their occupants. Can the idea of living spaces move beyond the ages-old definitions of four walls and a roof, to actually consider the effects that our built environment has on our mental health?

20.2 Perspective to Space

For each one of us, the notion of space [3] is presumptive. Space includes elements such as sound, light, colours, nature, objects and textures. The concept of absence and presence of supposing space projects the reality which unconsciously manipulates our actions. Space is not something that only exists as a visual system but also in contexts of what we feel as well. In space we represent to ourselves the objects on which we act, that is "Where we determine their shape, their magnitude and the relationships between them" [6] Mainly there are two kinds of space that exist- the first one is the physical space that surrounds us (PS). Physical Space (PS) is an area or expanse that is free, available, or unoccupied. The entities composing the space i.e. what is seen and what is seized, walked or touched, arranged or transformed in that particular space [5, 6] forms a system that unconsciously regulates our actions. The visual system of the space [6] surrounding us is privileged in the sense that allows us to relate the perception and representation of the projected object, actual approach, virtual form, context and model. Therefore, while planning the architecture of the city landscape and the interior of houses, the intent must be very well thought of. The intent here refers to the actions and movements which the physical space will allow. The second one is the mindspace (MS). Neuroscientists have known for decades that our brains contain representations of the area surrounding us, known as peripersonal space. Each of us is surrounded by a kind of invisible bubble possessing remarkable qualities.

In order to feel comfortable in a space, various elements of space [7] such as natural light and temperature have a changing effect on our moods. As humans have a holistic perception, our senses influence our thinking, feelings and actions and therefore our actions of body and mood. It can either affect us positively or negatively. For instance, in 1971, the red road flats of Glasgow were considered the best high-end estate in the city, but they were soon emptied as the residents found them difficult to warm up in winters. Rather a contradictory notion to how a well-planned space can accelerate our healing processes and keep our mental being in stable condition.

20.3 Physical Space Environment Transformation

20.3.1 Cities Coping Up to Adjust New Population

The Economic Survey of India 2017 estimates that the magnitude of inter-state migration in India was close to 9 million annually between 2011 and 2016. Apart from this, three out of ten Indians are internal migrants. Shifting from one demographic state to another [8] (migration of any sort) will influence human life and the environment around. Due to this, in India, both the cities and villages are experiencing shifts in population, hence leading to an uneven population distribution with a large number of people shifting to urban areas. It will inevitably steer towards rural–urban migration.

As the population keeps increasing, the cities face the predicament of adjusting to the newly shifted numbers. Amidst this, urban city planners tend to neglect to consider the effect of architecture building on the cognitive ability of the city inhabitants. As discussed earlier, space affects our mental well-being. But yet, why is it that many designers and architects do not consider it essential?

A significant change occurs in the physical space environment of the population when shifting to a new place space for better life and education opportunities. There is a notable transition in language, culture and the physical space environment. As per the 2011 census, the level of urbanization in India has increased from 27.81% in 2001 to 31.16% in 2011 [9], reasons being the demographic explosion and poverty induced rural–urban migration. Migration includes processes like making decisions, preparations, going through the procedure, shifting physically to another geographical space, adjusting to the local cultural needs and integrating into the local system. All these tedious complexities can lead to stress and anxiety.

20.3.2 Single Person Living

The prevalence of one-person households in India is low compared to that in other parts of Asia. While the presence of OPH (one-person household) [10] is subjacent, the number of such house spaces is a lot and expected to grow in the next few decades. The results presented revealed significant social, economic and demographic differences between one-person and multi-person households. Elderly females and young migrants [10] who live alone are potentially vulnerable groups [10]. The current emergence of one-person living households or single person living is unprecedented historically. Industrialization in the 1940s, commercialization and the need to become self-reliant have slowly made a single person living possible. In countries like the US, the number of adults [11] who live alone has doubled. Countries like Japan and Sweden are also witnessing this kind of household living.

The emergence of families in society is a collective result based upon fulfilling the basic primal needs and authorizes the appropriate functioning of the social organization in which we live. But with time, these needs change due to social transformation and demands to be fulfilled to live in modern-day society. The current family structures have begun transitioning from joint families to nuclear to a single person living. The rise of the culture of individuality can also be considered one of the many reasons for the same.

Various categories of people who are adopting to a single person living-

Category	Age group
Students (school and college)	15–22 yrs
Young job aspirants	20–26 yrs
Migrant labour workers	22 yrs and above
Independent white collars	24 yrs and above

20.4 How the Physical Space Environment Changes the Human Behaviour Pattern?

Most of us gravitate towards spaces that might not be psychologically favourable for us. We often fail to pay attention to our built environment and appreciate it. When we habituate something [7] whether it is a new office or a pattern of buying, we prefer that pattern even when we are better off not buying something else. Sarah Williams Goldhagen argues [7] that our experiences in the physical environment are nonconscious, but not unconscious, because that would mean it will be inaccessible. Nonconscious refers to the cognitive ability that could access consciously. Our cognitive ability and our mood [12] hold close relations to the built environment around us. For instance, a famous experiment called the London Taxi Drivers study was published [13] in 2009. For one to become a cabbie in London, one must memorize the layout and the street names of the entire city. To acquire such vast knowledge will take around two to four years. So the researchers did FMRI scans of the cabbies in training before they started building these detailed cognitive maps of the city, then scanned their brains again, once the cabbies had passed the test. They discovered that an area of the brain called the hippocampus had grown enormously. Our brain develops and changes when we learn new things. One of the properties of the brain that we have developed is neural plasticity [14], i.e., how our brain changes in response to the physical space environment.

20.4.1 Teenage Students as the Focus Group

We are not meant to survive in isolation, let alone thrive. We are social creatures and require some form of human interaction daily. With urban landscapes gravitating towards a single person living systems: more efforts will have to be made when there will be more physical space interaction and less human interaction (Figs. 20.1 and 20.2).

The pubescent years of an individual (13–19 yrs) is a metamorphic and significant time. Globally, depression [15, 16] is the fourth leading cause of illness and disability for adolescents aged 15–19 years. Teenagers are primarily considered vulnerable to mental health problems, reasons being [15] they go through multiple social, physical and emotional changes, which also includes exposure to poverty, abuse or violence. The parents are the ones who make the majority of life-changing decisions for teenage students. But to be exposed to better education, students tend to migrate to another place for more rewarding academic prospects. The question here arises whether teenagers should experience and endure the complex nuances of living on their own or not? They are most susceptible to mental illness if not provided with a proper support system. Even under the inevitable circumstances of living away from their family and familiar caregivers, what form of support system will be capable of aiding their mental facilities?

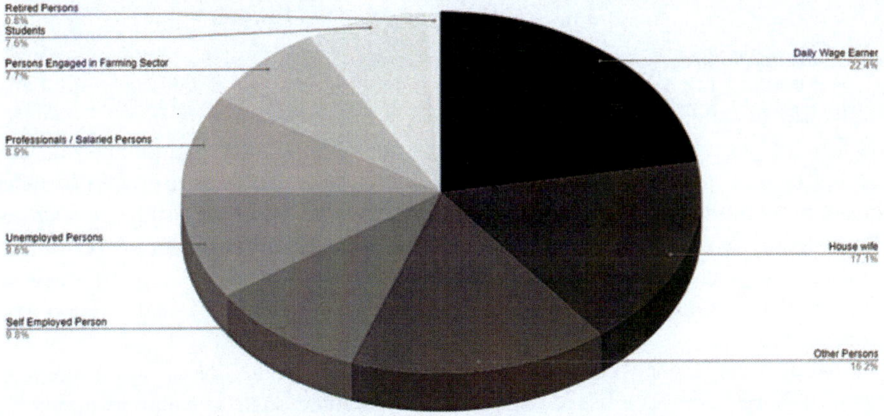

Fig. 20.1 Graph to show the percentage of students committing suicides

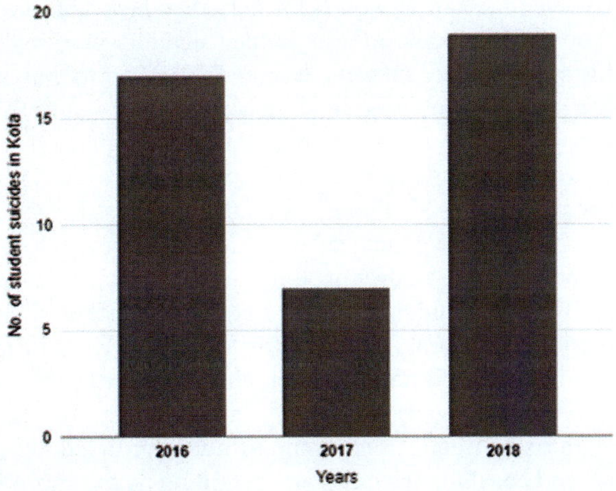

Fig. 20.2 Graph to show the suicide rates in Kota

20.5 Kota Case Study

Over the years, Kota in Rajasthan has become the coaching capital of. Over 1.5 lakh students from all over the country flock every year towards the city for preparation of various exams such as IIT-JEE and NEET-UG. Students live here for two to three years and prepare for the exams. Around five lakh students live in Kota at the same time. Factors like the role of competition, the burden of expectation, education and coaching institute system, hype and false hopes, poor coordination and determination of standards, right direction (career and life), exit policies/refund policies and space exploitation affect the fundamentals on which the student can rely for their well-being on the coaching institute. Moreover, with changes in factors like emotional support, surrounding environment & work consistency can affect motivation over a period of time.

We studied the whole system of Kota coaching classes revolving around the daily life of students. Our objective through this case study is to facilitate the well-being of mental health by developing various design interventions and relevant design methods. Resulting scenarios and interventions can transform the lifestyles of students through a built environment. We attempted to cover subjects such as the schedule of students, current helping systems in Kota, hostels or spaces where the student lives and their status of mental being, with the following approach of design such as qualitative interviews, personas, space scanning, visualization and prototyping.

In order to reduce the increasing no. of suicides and mental illness, various helping systems have been established. Such systems include such as helpline numbers, website portals, counsellors' help, sessions given regularly in classes regarding mental health and lastly organizing events to keep students healthy. It is observed that a kind of similarity exists among all the helping systems. For the most part, the students have to approach for help. But how will the students even recognize that they need help? Are these systems actually working? Looking at the date of how the suicides rates keep increasing, the answer to that is probably no. Student's daily life mainly revolves around their preferences and priorities. First is mostly studies for which students must attend classes regularly and have proper sleep. Students follow two kinds of class timings which are morning classes (8 am–12 pm) and evening classes (3–8 pm). In order to understand the students in-depth, qualitative interviews were conducted on various subjects such as sleep pattern, self-control on emotions, self-satisfaction, time management, concentration levels, decision-making skills, self-confidence, social status, homesickness, aims in life, amount of stress before tests/exams, demands for Internet services or multimedia devices, self-care, self-hygiene and physical strength. It was further observed that these combinations of various subjects are heavily influenced by the functions of education systems. With rising trends of coaching institutes in India, our education systems are failing us, and coaching institutes are just filling up the gap. We require a holistic education policy and infrastructure, where the needs of millions of students in the country can be met.

20.6 Design Intervention

Before people even reach a state of mental illness, these solutions should become a part of our daily life. These design solutions are no outcomes and are no exception to the happening of mental illness. It should become part of the students' day-to-day lives. Here, Fig. 20.3 is explaining the various attributes of a space that should be kept in mind while designing the given space. As per these, two kinds of different design solutions have been formulated. The first one changes the current physical space of the rooms in which the student lives (Fig. 20.5a and b), and the second one is designed

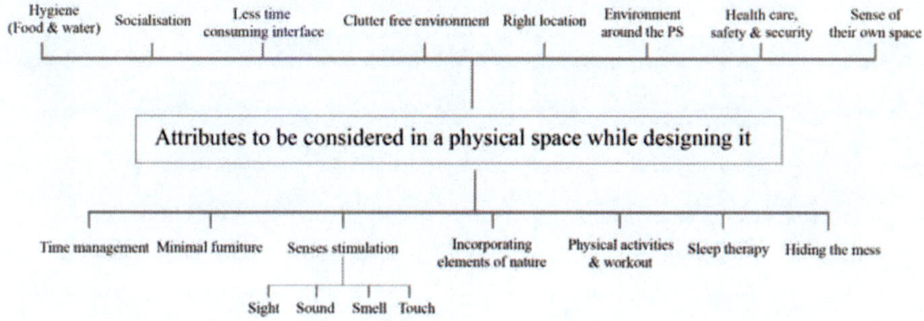

Fig. 20.3 Attributes of space to be considered while designing it

to cater to the future of the physical space that will facilitate mental well-being for the students. (Fig. 20.6a and b). As mentioned in Chapter 3 "Physical space environment transformation" with migration, change of physical space happens. India is a land of diverse cultures and traditions, and the cultures bound to change as you move from one place to another. Kota coaching centres are not really developed as per the culture of the state. But rather has induced a new culture of students which has evolved over a period of time. Hence, the interior design set-ups are generalized concepts. Figures 20.5 and 20.6 have depicted the idea for this particular problem by making the spaces neutral to cultures. The student residing can change it as per his/her need. But even after providing these solutions, who should be responsible for the implementation? The responsibility should be shared among the government, coaching institutions and landlords of housing. The government can form policies stating:—minimum PS to be provided for rent, min to max rent and food per month prices, hygiene and healing security provided revenues and tuition fees generated by coaching institutes. Bare minimum facilities are to be provided. Even after this, the government has to face lots of obstacles as various income groups exist in India. The coaching institutes and landlords will share tangible responsibilities such as finance and space provided where parents will have the intangible responsibility of giving moral support to their students. A pamphlet of some sort must be given to parents, students and landlords alike explaining how various attributes in a physical space can aid in mental health. It can take off the burden of unnecessarily relying on an architect or interior designer for consultation and hence saving money.

Physical space provided at PGs/hostels in Kota (Fig. 20.4a and b) are very compact, cluttered, messy, with no space to move around, books unkept and no attributes of that space contributing to mental wellness in terms of senses stimulation. Currently, the rent per month and food per month in Kota ranges from Rs. 1800–10,000 and Rs. 2000–5000, respectively.

a b

Fig. 20.4 **a** and **b** Images show the current physical space of the rooms in Kota

20.7 Future Scope

This review can prove to be resourceful in providing future directions to designers and architects. Well-designed spaces with naturally inducing attributes of the PS for mental wellness can be beneficial for individuals struggling with anxiety and depression in the long run. Not only for teenagers but for all of us, an empathetic built environment should be made part of our daily lives. It will not only contribute towards designing interiors but also for the upcoming urban landscape for a better lifestyle for the inhabitants. In addition to this, the education system in India should ideally adopt a holistic approach in regard to infrastructure and policies keeping mental health under consideration. Policies should be made by the government, and student accommodation should be certified only after stringent quality checks. Multiple factors such as economic, cultural and availability of space will determine this new framework for designing spaces.

20.8 Conclusion

This paper has focused attention on how the built environment can be conducive to our lifestyle and mental being. How various elements of space such as the right location, the size of space, nature around and physical attributes such as colours, lighting, presence and absence of PS, textures, materials, aroma, sound, fresh air and nature will indirectly affect our moods. PS can act as a tool to put our minds at ease. Especially if a person is living alone irrespective of a hostel or while working, space can become a lifestyle tool. More importantly, both architects and designers can work together to design PS that not only has functional and aesthetic aspects but also caters to mental well-being as well, because sometimes we tend to choose PS or

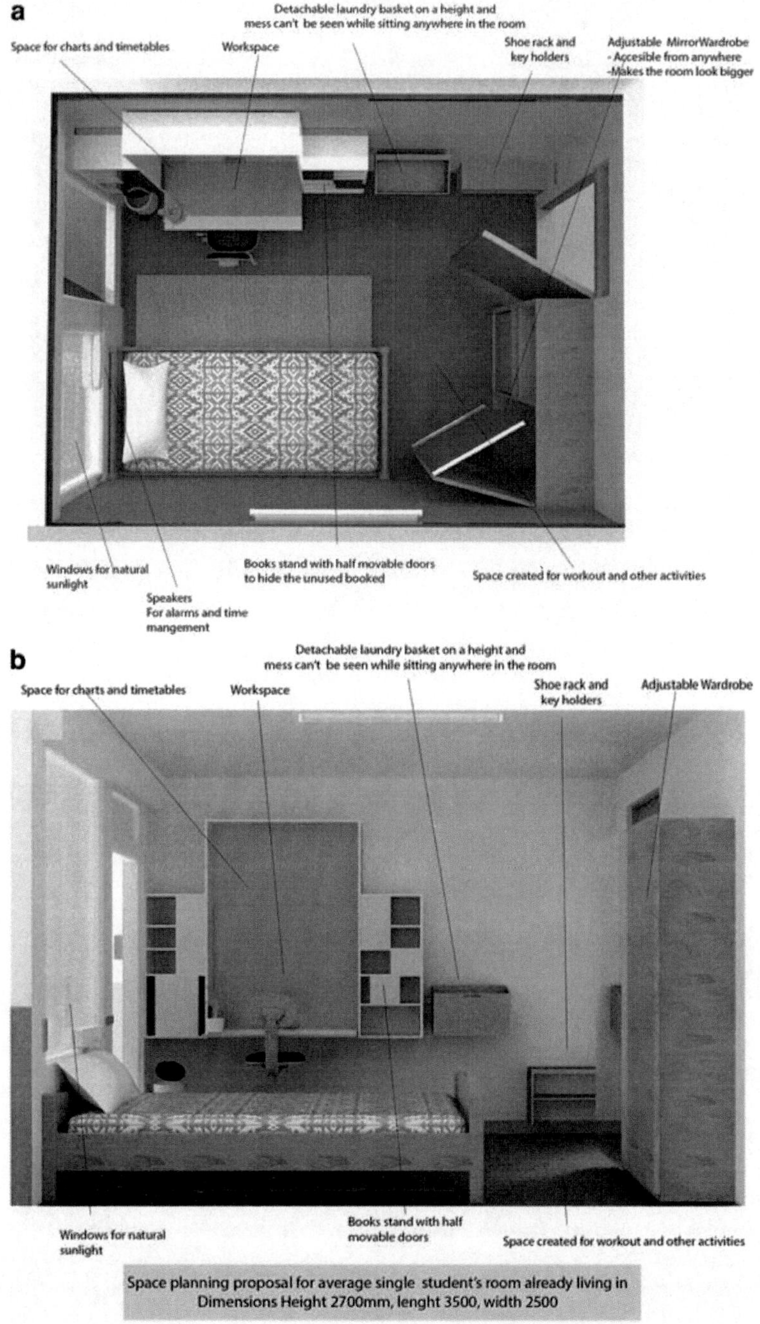

Fig. 20.5 **a** and **b** Design outcomes to alter the current PS of the rooms in Kota

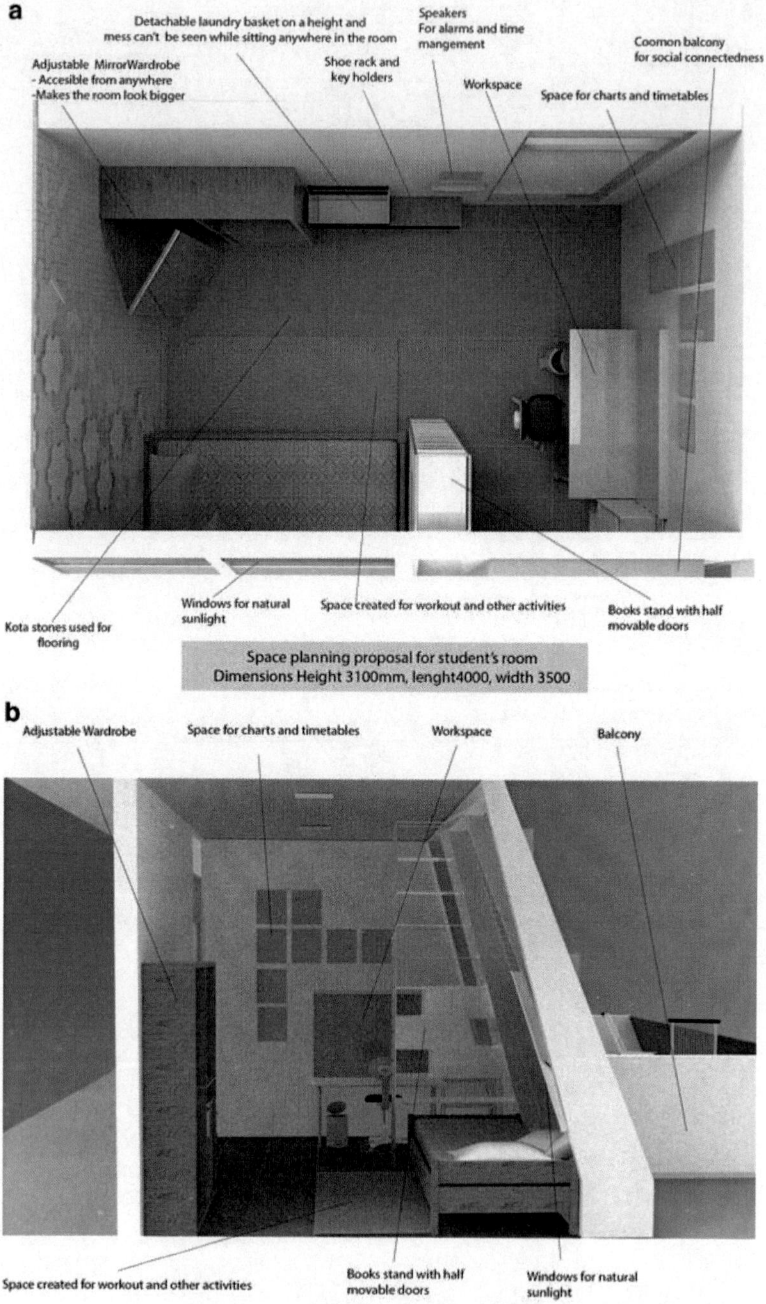

Fig. 20.6 **a** and **b** Design outcomes to design a new PS keeping in mind all the attributes of the space

objects which might not be psychologically sound. Spaces will need personalization depending upon the user as all human beings have different needs. Policymaking by the government is the need of the hour for coaching institutes, hostels, and PGs, with the befitting implementation of rules. Since touching all bases is quite improbable, impact mapping of details on one particular user and then designing accordingly is crucial with not only focusing on the object/space of creation but the intent of the object/ space as well. Whether it is teenage students or any other user group, measures should be taken to modify the built environment to improve the harsh mental mindscapes. Before even depression or anxiety can nip away our mind and take the form of severe mental illnesses. These insights integrated with thoughtful practice have the potential to pave the path to successful solutions and meaningful value propositions for future space inhabitants.

References

1. Lewrick, M., Link, P., Leifer, L.: The Design Thinking Playbook. Wiley, New Jersey (2018)
2. Garai, S.: Student suicides rising, 28 lives lost every day. The Hindu (2020)
3. Ulrich, S.R.: Effects of interior design on wellness: theory and recent scientific research. In: Journal of Healthcare Interior Design: Proceedings from the Annual National Symposium on Health Care Interior Design, pp. 4–7. Researchgate, US (1991)
4. Barnard S.: Design for a healthy mind: interior design and mental health. Medium (2019)
5. Kayan, C.: Neuroarchitecture: Enriching Healthcare Environments for Children, pp. 17–21. MPARC (2011)
6. Pellegrino, P., Jeanneret, E.P.: Meaning of space and architecture of place. In: Semiotica, pp. 2–3. Researchgate, (2009)
7. Goldhagen, S.W.: Welcome to Your World: How the Built Environment Shapes Our Lives. Harper, US (2017)
8. Bhagat, R.B.: Internal migration in India—are the underprivileged migrating more? Asia-Pacific Popul. J. 33–37 (2010)
9. Sharma K.: India has 139 million internal migrants. They must not be forgotten. India Economic Summit (2017)
10. Dommaraju, P.: One person households in India. In: Demographic Research, vol. 32, Article 35, pp. 1239–1266. (2015)
11. Ospina, O.E.: The Rise of Living Alone: How One-Person Households are Becoming Increasingly Common around the World. Oxford Martin School (2019)
12. Pedersen, M.: How architecture affects your brain: the link between neuroscience and the built environment. The Archdaily (2017)
13. Maguire, E.A., Gadian, D.G., Johnsrude, I.S., Good, C.D., Ashburner, J., Frackowiak, R.S.J., Frith, C.D.: Navigation-related structural change in the Hippocampi of taxi drivers. In: Proceedings of National Academy of Sciences of the United States of America. Canada (1999)
14. Lenn, N.: Neuroplasticity: the basis for brain development, learning and recovery from injury. Infants and Young Children **3**(3), 39–48 (1991)
15. Kessler, R.C., Angermeyer, M., Anthony, J.C.: Lifetime prevalence and age-of-onset distributions of mental disorders in the World Health Organization's World Mental Health Survey Initiative. World Psychiatry **6**, 168–176 (2007)
16. Kumar, C.: 10K student suicides in '18, highest in 10 yrs. The Economic Times (2020)

Chapter 21
Exploring the OSH Scenario in Floating Solar PV Projects in India and Opportunities for Ergonomics Design Interventions

Abhijit Sen and Sougata Karmakar

Abstract The renewable energy industry is seeing exponential growth in India. Among the many clean energy sources, floating solar PV (FPV) projects are being preferred because of its many advantages. The installation and maintenance of FPV panels are generally been executed by an informal and inexperienced/semi-skilled workforce who are exposed to the emerging challenges, which may lead to compromising of occupational safety and health (OSH) aspects. To explore the ground scenario, field surveys were carried out to study the OSH scenario which revealed that the workers are highly exposed to different occupational hazards during installation and maintenance of FPV panels due to heavy load handling, awkward postures, inappropriate tools, lack of safety measures against the harsh working environment (hot Sun, electrocution, drowning, etc.), lack of specific training/skill-development, etc. Project managers agreed that immediate attention is required for the safety and better occupational health of the workers and felt the need for protecting the workers from the hazards especially through design interventions. This study is a maiden effort in understanding the OSH aspects in the emerging FPV sector in India and establishes an information gap, which is proposed to be addressed through suitable ergonomics design solutions.

21.1 Introduction

The power of the Sun is virtually infinite. Only 30 min of solar radiation falling on Earth can meet the annual energy demand of the world [1]. To tap this source of clean energy, countries across the world are promoting different kinds of solar PV installations. Such sources of energy are also demand-driven as concerns over climate change and sustainability are taking precedent. The growth of renewable energy in India has

A. Sen · S. Karmakar (✉)
Department of Design, Indian Institute of Technology (IIT) Guwahati, Guwahati, Assam, India
e-mail: karmakar.sougata@gmail.com

A. Sen
e-mail: sen18@iitg.ac.in

A. Chakrabarti et al. (eds.), *Design for Tomorrow—Volume 1*, Smart Innovation, Systems and Technologies 221, https://doi.org/10.1007/978-981-16-0041-8_21

resulted from global and national policy developments (Fig. 21.1), decreasing tariffs of solar PV projects, and the vast availability of solar irradiation and potential in India most of the year [2]. In the last decade, solar PV installations in India have registered higher growth than other renewable sources of energy (Fig. 21.2). Utility-scale ground-mounted solar PV projects require large-scale land acquisition, which is a sensitive socio-economic issue. These circumstances have pushed stakeholders to explore other types of solar PV projects that are land neutral.

Globally, floating solar PV (FPV) is fast emerging as a viable option where water bodies are available. In 2007, the first floating solar project (175 kWp) in the world was installed by SPG Solar of Novato, California, and Thompson Technologies Industries (TTi), at Far Niente Winery in Napa Valley, California [3], and the first FPV project in India was installed in Kolkata, India. A study has estimated that the quantum of water surface available for FPV installation in India is 18,000 km^2, with

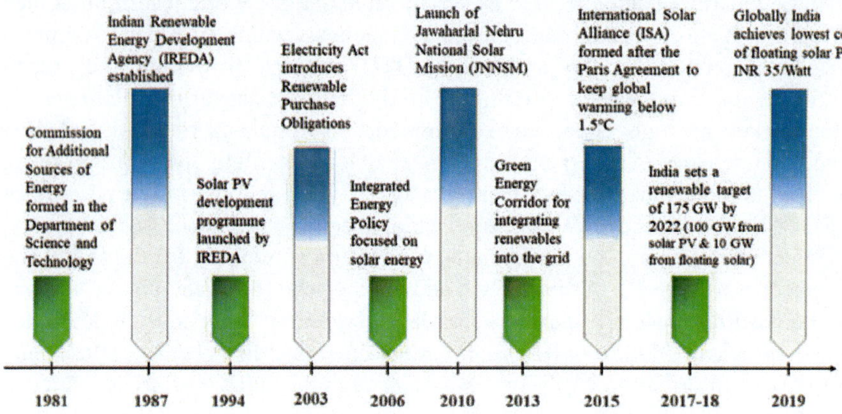

Fig. 21.1 Significant milestones in the growth of renewable energy in India (Author compilation)

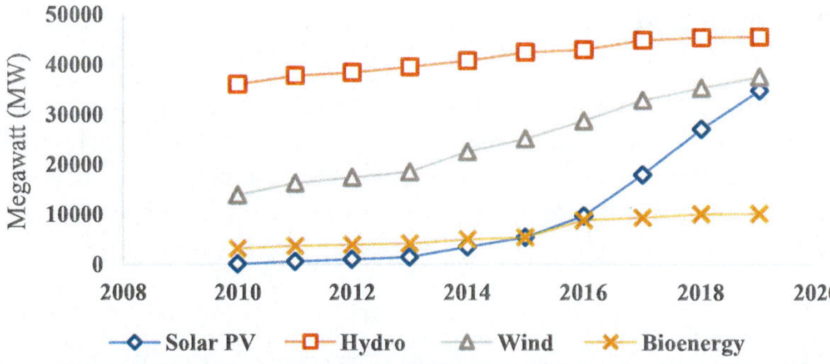

Fig. 21.2 Growth of installed capacities (in MW) of different types of renewable energy sources in India between 2010 and 2019 (Compilation by authors based International Renewable Energy Agency data)

a maximum potential of 280 GW [4] with the state of Maharashtra having the largest number of reservoirs, followed by states of Madhya Pradesh and Gujarat, indicating a vast potential. At the beginning of 2019, the global installed capacity of FPV stood at 1314 MW, and the global potential is estimated at 400 GWp [5].

21.2 Background and Need for the Study

FPV sites are new workplaces. Installation and maintenance involve new technology, new tools, new hazards, and new environments. These circumstances present new challenges to the workers engaged in such projects. Any work that is beyond the capabilities of the worker is likely to lead to serious safety and health hazards. The hazards primarily arise from the various mismatches in the interface at the workplace. The workplace transition and emergence of new hazards are represented in Fig. 21.3. The mismatches may involve physical, cognitive, and organizational factors.

Moreover, if untrained and unprotected workers are exposed to an unknown and uncertain work role or environment, the occurrence of occupational diseases becomes high. The solar PV industry in India is growing fast, and this entails that a large and temporary workforce will be engaged in this sector, exposing a larger worker population to emerging hazards. An estimate of the International Renewable Energy Agency (IRENA) says that globally 11.7 million people will be engaged in the solar PV industry by 2030 with an installed capacity of 2840 GW [6].

As is evident from Fig. 21.3, there are several mismatches between the worker and different components of a typical renewable energy workplace, which are FPV projects in this case. The workers are mostly engaged in a non-standard work arrangement and are likely to be exposed to several psychosocial hazards. These mismatches

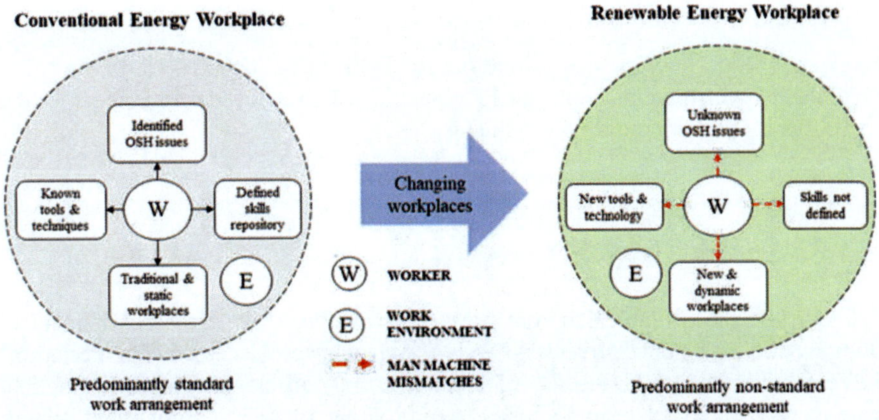

Fig. 21.3 Workplace transition from conventional energy to renewable energy workplaces (Graphic: authors)

are potential situations for the causation of several occupational diseases. An important objective of design is to provide a solution to a problem. Therefore, design solutions can be used to resolve these mismatches by controlling the risk factors and matching the interfaces with the worker to create a safe and comfortable work environment. FPV projects have a bright future in India. Thus, it is essential to address these issues through design for safety approach at the nascent stage, so that present and future workers can be protected.

21.3 Research Questions, Aims, and Objectives

The study began by formulating some research questions. They are (a) what are the occupational hazards that workers are exposed to during the installation and maintenance of floating solar PV projects? and (b) what context-specific tools, equipment, or other workplace aspects can be designed (or redesigned) to control the hazards for which solutions do not exist?

The present study investigates the ground scenario of OSH aspects of the FPV workforce and thereafter checking the possibility of improvement of OSH through innovative design interventions. The objectives are as follows:

- To identify the sources of hazards (occupational, environmental, psychosocial, skill gaps, etc.) regarding the workers engaged in FPV installation and maintenance.
- To study the existing occupational safety and health measures for the FPV sector.
- To identify the possibilities of design interventions from the OSH perspective.

21.4 Methodology

A review of the literature was carried out to identify existing research in the solar PV industry with emphasis on the FPV sector, followed by developing an OSH questionnaire to carry out the field studies.

21.4.1 Literature Review

To understand the available literature in this research area, online databases of Scopus, Google Scholar, ScienceDirect, and Web of Science were accessed. The period of the search was chosen between 1965 and 2020. Only articles published in English were shortlisted. Search terms included "solar/floating solar PV," "ergonomics design interventions," and "occupational safety and health" and their keyword combinations. The existing literature revealed several occupational health and safety hazards, such as chemical (toxic gases), thermal, and physical hazards [7] associated with the manufacturing of photovoltaic energy systems. Studies showed that with large-scale solar PV increasing in number, the focus shifted to studying the hazards related to the installation of PV systems. These include falls, injury from objects, electric shock, traffic accidents, thermal burns, awkward postures, handling heavy loads, and wounds [8]. Such hazards can lead to work-related musculoskeletal disorders (WMSDs), which can cause permanent damage to the workers' health. Fire hazards were identified in grid-connected PV systems [9]. Sources of health effects on workers/employees during PV system installation and maintenance includes heat, non-ionizing radiation, electricity, flora, and hailstorms [10]. The workers are also exposed to hazards from various components of the PV system such as solar panels, batteries, and inverters [11].

To understand the existing design interventions for ensuring the OSH of the workers in the solar PV industry (with emphasis on floating solar), literature was reviewed. Most of the papers concentrated on design considerations related to PV system design in terms of cost and efficiency, ecological design of solar PV system, roof design for PV system efficiency, aesthetic impact, and designing for consumer preference for residential solar panels [10, 12–15]. Only three studies were found which used a design approach for ensuring the safety and health of the workers. One study reports on the floating solar PV system's grounding design, thus ensuring electrical safety [16]. Another research suggests the use of multiple inverters (small) to prevent fire hazard [9]. A comprehensive OSH study on the roof-top solar installations was carried by a group of researchers who have used the prevention through design (PtD) approach to ensure workers' safety by developing a PtD protocol [17]. No studies on training/skill up-gradation, psychosocial aspects, or design interventions specific to the floating solar PV sector were found. This establishes a knowledge gap that this present paper attempts to explore.

Questionnaire development for the study: A literature survey was carried out to explore the existence of OSH related questionnaires specific to the solar PV industry/FPV sector. Only one OSH checklist on roof-top solar was found [18]. Drawing inferences from the OSH checklist on roof-top solar, views of the project managers, and observations during the field studies, a questionnaire was constructed. The questionnaire contains eight sections, namely (a) general information about location, type, and size of the project, type and skill level of workers, etc.; (b) training, safety, and work organization aspects; (c) hazards of working on water, slips, trips, falls, etc.; (d) use of personal-protective equipment (PPE); (e) electricity-related hazards; (f) hazards that may lead to musculoskeletal disorders (MSDs); (g) specific sources of OSH hazards, and (h) suggestions for design/redesign of FPV system components. Photographs were taken of the installation process to understand the workflow and functioning of a floating solar PV system and the workers' interface.

The study included an analysis of the workflow; a survey of the tools and techniques used, and identification of occupational hazards faced by the workers ($n = 22$) using observational methods and unstructured questionnaire seeking views of the project managers ($n = 4$) engaged in the installation and maintenance of two

Table 21.1 Overview of the FPV sites considered for the study

Parameters	Mejia TPS (DVC)	Kawas (NTPC)
Location	The water reservoir of a thermal power station	The water reservoir of a gas-based power station
Capacity	25 kWp	1 MW
Design of FPV	Floating structure type	All in one buoyancy type
Number of workers engaged	06	16
Direct normal irradiation (DNI)*	1187 kWh/m2 per year	1614 kWh/m2 per year
Peak months*	March and April	March, April and May
Power output for 1000 kWp*	1.297 GWh per year	1.478 GWh per year

**Source* Global solar atlas

FPV projects in India. The findings from the field studies were compared with the findings of the literature review to derive opportunities and future scope for design interventions from the OSH perspective.

21.5 Case Studies

To understand the OSH scenario of the FPV projects in India, two sites where the installation was underway were visited. The first site was a 25 kW FPV project at Mejia Thermal Power Station of Damodar Valley Corporation (DVC) in the state of West Bengal, and the second site was a 1 MW FPV installation at the Kawas Gas-based Power Station of National Thermal Power Corporation (NTPC) in the state of Gujarat. An overview of the two locations is given in Table 21.1.

21.6 Results

The project managers completed the questionnaires related to the two sites using online software during the period September–October 2019. The responses are summarized in Table 21.2.

Table 21.2 Summary of responses received from project managers

Query overview	Site 1 (NTPC)	Site 2 (DVC)
Skill level of the workers	Semi-skilled	Semi-skilled
Mode of safety training	Lectures in local language	Demonstration
Cooperation, communication, occupational check-up, and worker feedback for workplace safety	✓	✓
Consideration of needs of workers of different types, gender, and ages	✓	✓
Platforms/scaffolding/rails/barriers/safety nets for work near water	✓	✓
Inspection of all parts of the floating solar PV system	✓	✓
Type of tools used	Tools used in other solar PV projects	General hand tools
PPE (life jackets, etc.) and life preservers	✓	✓
Safety signs, fire extinguishers, first aid, and emergency plan	✓	✓
Precautions to prevent electricity-related hazards such as grounding and dry inverter area	✓	✓
Design of work and layout to reduce manual material handling and ergonomic hazards	×	×
Specific hazards from solar PV panels	Sharp edges, the weight of panels, and lack of proper grips	Sharp edges, lack of grips, electrical hazards, and difficulties in replacement
Specific sources of hazards from floating structures/pontoons	Instability of the floating structure	Falls into water and gaps in the walkways
Specific sources of hazards from tools	Difficulty in use during installation	Difficulties in use during installation and tools may fall into the water
Specific sources of hazards from cables/inverters/batteries	Electrical shocks and carrying cables/wires	Electrical shocks, fire, and chemical hazards from batteries
Environmental hazards	Humidity and hailstorms	Rain and extreme sunlight
Design/redesign required	Scaffolds/walkways/shoes	Solar panels, floaters, personal-protective equipment (PPE), tools, and work methods

21.6.1 Other Inputs from Project Managers and General Observations

- The hazards of workers engaged in FPV projects are different from those of other types of solar PV projects. The additional hazards include unstable floaters working on/near the water. The avoidance of using slippery shoes was emphasized.
- Sometimes workers are casual with the use of PPE, which are monitored and corrected by supervisors. Standardized equipment and items for floating solar projects will help in interchangeability. Need for standards for work procedure, skill sets, etc.
- Site conditions not conducive to the use of mechanized material handling equipment. Instability of floating platforms during strong winds. Possibilities of sunstroke, dehydration, and other heat illnesses.
- A requirement of an OSH checklist and design standards. Training and motivation of workers for skill development, safety consciousness, and alertness during work.
- Workers are exposed to a work environment where both electricity and water are present.

The field surveys revealed that the workers are highly exposed to different occupational hazards (Fig. 21.4) during installation and maintenance of FPV panels. The hazards arise due to heavy load handling, awkward posture, inappropriate tools, lack of safety measures against the harsh working environment (hot Sun, glare, electrocution, drowning, etc.), lack of context-specific training/skill-development, etc. It

Fig. 21.4 **a** A multitude of ergonomic hazards during installation; **b** work near water exposes the workers to a combination of hazards; **c** sharp edges and presence of electricity are occupational hazards, and; **d** work on unstable floaters presents additional hazards

was also noticed that the slippery and unstable working surfaces and unavailability of appropriate tools presented unique challenges for the workers engaged in FPV projects, which are different from other traditional workplaces. Project managers agreed that immediate attention is required for the safety and better occupational health of the workers. They also suggested the need for design/redesign of several aspects from the OSH perspective.

21.7 Discussion and Conclusion

This exploratory study revealed that FPV workers face multifarious OSH hazards on both land and water. The hazards mainly arise due to the mismatches between the workers and FPV system components. The study was designed in such a way so as to collect views of project managers on OSH hazards and their views on the need for addressing them through design interventions in specific areas. The results of the case studies were compared to existing literature available wrt design interventions in the FPV sector (Sect. 21.4.1) from an OSH standpoint, which elicited several research gaps. The identified gaps for design/redesign for ensuring OSH include (a) scaffolds around the walkways: (b) walkway platforms; (c) solar panels; (d) floaters/pontoons; (e) PPE; (f) tools; (g) shoes, and (h) work methods. Apart from this, design interventions toward ensuring training/skilling needs, protecting workers from psychosocial hazards and environmental factors can also be explored.

The protection of the environment is an important human goal, and tomorrow's energy needs are likely to be completely met by clean energy sources. At the same time, the protection of the workers in the renewable energy industries must also be an important priority. Design can serve as a powerful tool to attain this objective. The preliminary field surveys were intended for "problem defining" [19] from the OSH dimension in the FPV sector in the Indian context. The hazards in floating solar PV projects are different from other kinds of solar PV installations. One project manager suggested that "*Floating Solar Technology, still in an evolving stage, needs specific design of equipment, tools, and accessories suitable for Floating installation.*" These findings call for a particular focus on designing different workplace aspects, so that workers have a safe and comfortable work experience. This study has identified some possible design interventions in that direction. Future studies include a detailed analysis of the specific occupational hazards using ergonomic assessments and the development of sustainable design solutions from a human-centered design (HCD) perspective. This paper's novelty is that this is a maiden study to explore the OSH scenario of these new workplaces and identify opportunities for design interventions.

21.8 Limitations of the Study

The present exploratory study was based on a small sample size of workers ($n = 22$) and project managers ($n = 4$) from only two sites limited to the Indian population. This is because very few sites are available for investigation in this emerging sector. The parameters of the work environment and psychosocial factors were not measured.

Acknowledgements The authors are grateful to the officials of NTPC and DVC for providing inputs for the study.

References

1. Kumar, V., Shrivastava, R.L., Untawale, S.P.: Solar energy: review of potential green and clean energy for coastal and offshore applications. Aquat. Procedia **4**, 473–480 (2015). https://doi.org/10.1016/j.aqpro.2015.02.062
2. Shukla, A.K., Sudhakar, K., Baredar, P., Mamat, R.: Solar PV and BIPV system: barrier, challenges and policy recommendation in India. Renew. Sustain. Energy Rev. **82**, 3314–3322 (2018). https://doi.org/10.1016/j.rser.2017.10.013
3. Trapani, K., Millar, D.L., Smith, H.C.M.: Novel offshore application of photovoltaics in comparison to conventional marine renewable energy technologies. Renew. Energy. **50**, 879–888 (2013). https://doi.org/10.1016/j.renene.2012.08.043
4. The Energy and Resources Institute (TERI): Floating solar photovoltaic (FSPV): a third pillar to solar PV Sector? https://www.teriin.org/sites/default/files/2020-01/floating-solar-PV-report.pdf. Last accessed 19 May 2020
5. International Bank for Reconstruction and Development-World Bank: Floating solar handbook-where Sun meets. Water (2019). https://doi.org/10.1596/31880.Lastaccessed19May2020
6. IRENA: Future of solar photovoltaic: deployment, investment, technology, grid integration and socio-economic aspects. (2019). https://www.irena.org/-/media/Files/IRENA/Agency/Publication/2019/Nov/IRENA_Future_of_Solar_PV_2019.pdf. Last accessed 19 May 2020
7. Fthenakis, V.M., Moskowitz, P.D., Lee, J.C.: Manufacture of amorphous silicon and GaAs thin film solar cells: an identification of potential health and safety hazards. Sol. Cells **13**, 43–58 (1984). https://doi.org/10.1016/0379-6787(84)90091-7
8. Bakhiyi, B., Labrèche, F., Zayed, J.: The photovoltaic industry on the path to a sustainable future—environmental and occupational health issues. Environ. Int. **73**, 224–234 (2014). https://doi.org/10.1016/j.envint.2014.07.023
9. Falvo, M.C., Capparella, S.: Safety issues in PV systems: design choices for a secure fault detection and for preventing fire risk. Case Stud. Fire Saf. (2015). https://doi.org/10.1016/j.csfs.2014.11.002
10. Hernández-Callejo, L., Gallardo-Saavedra, S., Alonso-Gómez, V.: A review of photovoltaic systems: design, operation and maintenance. Sol. Energy **188**, 426–440 (2019). https://doi.org/10.1016/j.solener.2019.06.017
11. White, J.R., Doherty, M.: Hazards in the installation and maintenance of solar panels. IEEE IAS Electr. Saf. Work 1–5 (2017). https://doi.org/10.1109/ESW.2017.7914834
12. Scognamiglio, A.: "Photovoltaic landscapes": design and assessment. A critical review for a new transdisciplinary design vision. Renew. Sustain. Energy Rev. **55**, 629–661 (2016). https://doi.org/10.1016/j.rser.2015.10.072
13. Li, H.X., Zhang, Y., Edwards, D., Hosseini, M.R.: Improving the energy production of roof-top solar PV systems through roof design. Build. Simul. **13**, 475–487 (2020). https://doi.org/10.1007/s12273-019-0585-6

14. Bao, Q., Honda, T., El Ferik, S., Shaukat, M.M., Yang, M.C.: Understanding the role of visual appeal in consumer preference for residential solar panels. Renew. Energy **113**, 1569–1579 (2017). https://doi.org/10.1016/j.renene.2017.07.021
15. Sánchez-Pantoja, N., Vidal, R., Pastor, M.C.: Aesthetic impact of solar energy systems. Renew. Sustain. Energy Rev. **98**, 227–238 (2018). https://doi.org/10.1016/j.rser.2018.09.021
16. Bhang, B.G., Kim, G.G., Cha, H.L., Kim, D.K., Choi, J.H., Park, S.Y., Ahn, H.K.: Design methods of underwater grounding electrode array by considering inter-electrode interference for floating PVs. Energies **11**, 1–16 (2018). https://doi.org/10.3390/en11040982
17. Ho, C., Lee, H.W., Gambatese, J.A.: Application of Prevention through Design (PtD) to improve the safety of solar installations on small buildings. Saf. Sci. **125**, 104633 (2020). https://doi.org/10.1016/j.ssci.2020.104633
18. European Agency for Safety and Health at Work: Hazard identification checklist: OSH risks associated with small-scale solar energy applications Part A: Introduction, pp. 1–14. [Online]. Available from https://osha.europa.eu/en/tools-and-publications/publications/e-facts/e-fact-69-hazard-identification-checklist-osh-risks-associated-with-small-scale-solar-energy-applications/view. Last accessed 19 May 2020
19. Norman, D.A.: When you come to a fork in the road, take it: the future of design. She Ji. **2**, 343–348 (2016). https://doi.org/10.1016/j.sheji.2017.07.003

Chapter 22
Musculoskeletal Disorders and Ergonomic Risk Assessment of Postures Adopted by Handcrafted *Kalash* Polishers, Northeast India

Krishna Chaitanya Mallampalli and Swati Pal

Abstract The ergonomic risk assessment of the postural attitude of the workers is essential for evaluating physical workload in the workplace, especially when they interact with workstation elements. Exposure to heavy physical work and repetitive actions are common among informal workers in handicraft work while interacting with non-ergonomic tools. Such working conditions, consequently, lead to the development of musculoskeletal disorders (MSDs). In this context, a study has been conducted on handcrafted *Kalash* polishers in brass metal handicraft production units located in northeast India, where most of the polishers in the polishing task use a traditional polishing tool, and their working postures are mostly constrained by the design of polishing tool. Therefore, the present study aims to determine the occurrence of MSDs and the ergonomic risk of working postures of *Kalash* polishers in brass metal handicraft production units. The study sample comprised of 40 polishers. Nordic Musculoskeletal Questionnaire (NMQ), Rapid Upper Limb Assessment (RULA) tool, non-invasive postural acquisition using cameras, virtual prototyping, digital human modeling, and simulation software were used. The results of this study revealed that 100% of polishers reported musculoskeletal discomfort in at least one body region during the past 12 months. The highest prevalence of MSDs was found particularly in the lower back (75%), followed by upper back (52.5%), elbows (50.0%), and hand/wrist (47.5%) regions. The mean severity of pain using the five-point rating scale indicated moderate to high pain in the lower back, upper back, and shoulder regions. Further, the simulation results indicated that the adopted working postures of polishers were at a very high-risk level with the existing polishing tool. Based on these findings, implications for further research include the design/redesign of the existing polishing tool based on ergonomic principles, which may improve working postures and reduce the risk of MSDs among this occupational group.

K. C. Mallampalli (✉)
Indian Institute of Technoligy Guwahati, Guwahati, Assam, India
e-mail: mkchaitanya555@gmail.com

S. Pal
Indian Institute of Technology Bombay, Powai, Maharashtra, India

A. Chakrabarti et al. (eds.), *Design for Tomorrow—Volume 1*, Smart Innovation, Systems and Technologies 221, https://doi.org/10.1007/978-981-16-0041-8_22

22.1 Introduction

Handicraft work is a long-term traditional activity in many developing countries around the world, including India. Indian handicraft sector is one of the export-oriented businesses, and many people are engaged in this sector. According to one report, approximately, 0.5 million workers are currently engaged in Indian handicraft work [1], and most of them work in cottage industries. In these industries, generally, the number of workers employed is at most 15. These workers often expose to various occupational health hazards due to the lack of various workstation resources. Of which, one of the important problems among handicraft workers is poorly designed tools and work equipment. Several studies carried out in the past demonstrated that the design of tools and work equipment contributes to various occupational problems among handicraft workers [2, 3]. Musculoskeletal disorders (MSDs) are the major occupational health problem in handicraft work. The prevalence rates ranged between 38.5 and 96.5% among various handicraft workers [4–7]. It was also reported that the awkward work postures, repetitive actions, and forceful exertions often exert a heavy physical load on musculoskeletal structures when workers interact with work equipment [2, 8]. In the long run, such poor conditions may lead to serious occupational injuries. Therefore, this research is mainly focused in determining causes for postural risk that are related to the usage of poorly designed tools and work equipment in the workplace.

Kalash (Flower-shaped brass vessel) polishing is a handicraft work and popularly recognized in northeast India. This principle activity is performed at the final stage of *Kalash* manufacturing (i.e., before packing). After polishing, *Kalash* gains golden glow, which attracts customers and also adds value to the product in the market. Usually, this traditional activity is achieved by using a conventional polishing tool (Fig. 22.1). This tool comprises components such as wooden spindle (900 mm length), cotton rope (1.5 m length), and cutting knives. It is typically positioned

Fig. 22.1 Existing *Kalash* polishing tool

100 cm above the floor. The job of the *Kalash* polisher is to peel the outer layers of *Kalash* using cutting knives against the rotating spindle. Usually, polishing operation is performed in a squatting position.

Further, polishers' knees are folded under their bodies when they sit on the floor. The tool position is very nearer to ground level. Therefore, polishers forced to adopt awkward postures such as excessive trunk bending and twisting, shoulders abduction, and neck deviation. All these together impose heavy physical load on the musculoskeletal structure, and this may be attributable to development of MSDs among *Kalash* polishers.

Unfortunately, there is a lack of published literature related to MSD problems and occupational setting of *Kalash* polishers. Therefore, the present study objective was to determine the presence of MSDs and assessing the ergonomic risk associated with postures of *Kalash* polishers.

22.2 Materials and Methods

22.2.1 Study Design, Sample, and Procedure

A cross-sectional study was conducted on *Kalash* polishers at handicraft manufacturing units of Hajo in the Northeastern region of India. From these units, forty *Kalash* polishers were selected randomly. Polishers who were working for more than a year were included in the study. The polishers who had pain history due to musculoskeletal diseases/injuries were excluded. All the respondents were interviewed at their workplace. Participation was voluntary.

22.2.2 Questionnaire Study

A questionnaire was used for the data collection. The questionnaire collected data regarding demographic details (age, height, weight, work experience, and daily work time) and MSDs risk. The MSD risk of *Kalash* polishers was assessed using Nordic musculoskeletal questionnaire [9]. Many studies on handicraft work have also used NMQ, e.g., hand-sewn shoe workers [6], gemstone polishers [10], and potters [7]. A body map was included with nine different body parts (e.g., neck, shoulders, elbows, wrist/hands, upper back, lower back, knees, and ankles/feet) and was also shown to polishers. Latter, polishers were asked, "Have you had any symptoms of MSDs (pain/ache/discomfort) during the past 12 months" (Yes/No). Those who said "yes" were again asked to rate their severity of symptoms on a five-point Likert scale (1—very mild to 5—very severe).

22.2.3 Posture Analysis

Analysis of working postures of *Kalash* polishers was performed via the following steps.

RULA. Rapid Upper Limb Assessment (RULA) tool [11] was used for evaluating the working postures of polishers. The RULA tool provides a final risk score for the working postures, which ranges from 1 to 7 [12]. The risk level is represented with color indicator zones. According to the final RULA score, zone color changes from green to red. The sequence of color indication and corresponding actions is as follows: RULA score 1–2 (green color) indicates low risk level—acceptable postures, score 3–4 (yellow color) indicates medium risk level—need for further investigation and changes may be required, score 5–6 (orange color) indicates high risk-level—investigation and changes are required soon, and score 7 (red color) indicates very high-risk level—investigation and changes required immediately. Figure 22.2 illustrates the RULA color indicator zones in the figure legend.

Posture recording. The polishers' working postures while operating manual polishing tool was recorded using videography and photography techniques. Two cameras were setup to capture the postures adopted by the polishers in non-invasive manner. One camera positioned for side view (perpendicular to sagittal plane) video recording while the second camera recorded video from behind (perpendicular to coronal plane). The analysis of recorded videos performed was using Kinovea software. Kinovea is an open-access online software, which is a reliable and valid tool for postural analysis [14]. With the help of this software, recorded videos were reproduced in the slow motion, and postural analysis was performed frame-by-frame—the most frequent postures were detected (Fig. 22.3). These detected postures are in three different tasks as follows: a: tool fitting, b: polishing the plain surface, and c: polishing the curved surface.

The main objective of the photography technique is to acquire postural angles. In order to acquire these postural angles, photographs comprising critical postures of polishers were processed, and angles were marked using Kinovea software. Figure 22.4 shows an example of acquisition of the postural joint angles. Here,

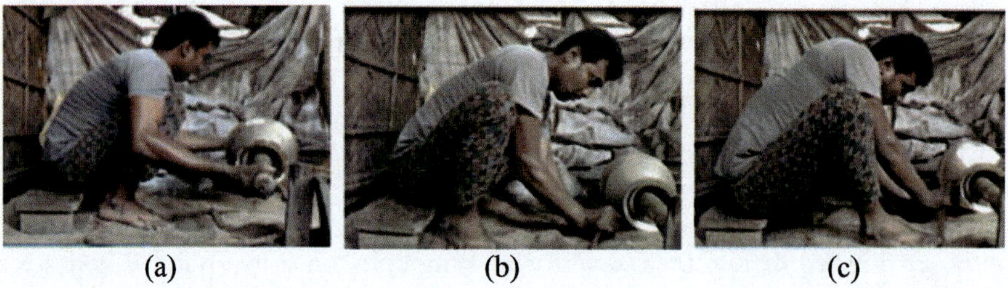

(a) (b) (c)

Fig. 22.2 Figure legend of RULA color indicator zones [13]

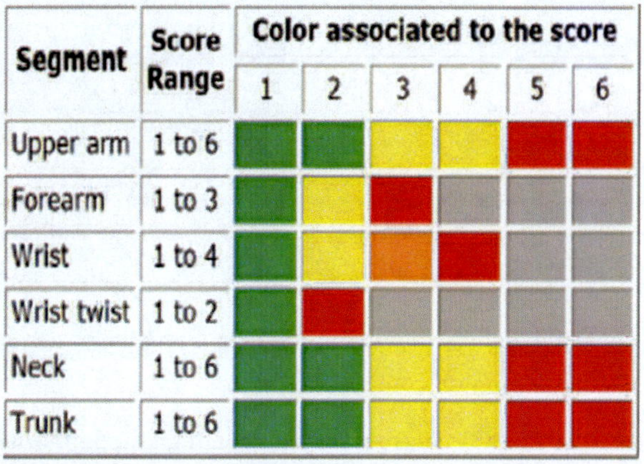

Fig. 22.3 Typical working postures of polishers

Fig. 22.4 Joint angle detection of the polisher

the polishers' trunk, neck, shoulders, elbows, wrists, and lower leg angular positions were represented.

Creation of Digital Human Model. For the creation of a Digital Human Model (DHM), anthropometric data is essential [15]. From the literature, it was observed that non-availability of anthropometric data of polishers who engaged in brass metal handicrafts production. Therefore, during the study, the data for the stature of polishers were collected. The stature ranged between 154 and 169 cm (mean = 161.9 cm, SD = 4.34 cm). Based on this observation, a comparison of stature was carried out with an Indian anthropometric database developed by Chakrabarti in 1997. It was found that polishers' stature was in the acceptable range and matching with the 50th percentile Indian male anthropometric data range (157 to 167 cm). Hence, the

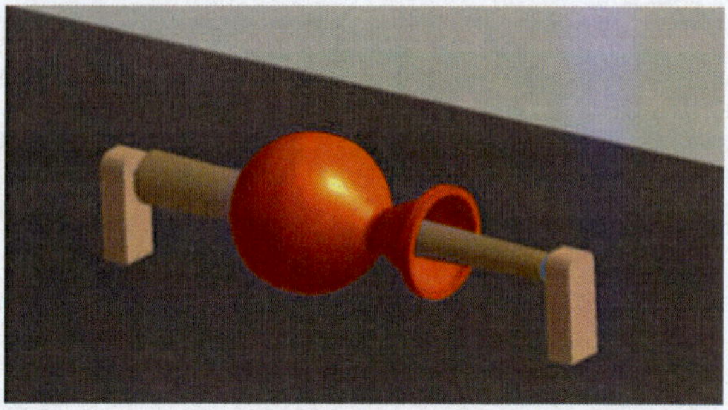

Fig. 22.5 Existing polishing tool—3D model

anthropometric measurements for human modeling were adopted from Chakrabarti 1997 [16]. Using these anthropometric measurements and DELMIA® Digital Human Modeling software, a digital human was created. Further, the created human model has the necessary characteristics (scalability, flexibility, and movements of individual body parts) to simulate real postures.

Existing polishing tool model. The dimensions of the polishing tool were measured from the field using measurement tape. According to the actual measurements, a three-dimensional solid model of the existing polishing tool was developed in CATIA® V5 software. Figure 22.5 presents the manual polishing tool model built in CATIA® V5.

Interfacing. CATIA®V5 software was used for the interfacing of virtual models. The human model interfacing of three-dimensional solid model of polishing tool and human model with anthropometric measurements (50th percentile) was carried out in the virtual environment. To ensure the accuracy of investigation, interfacing of both virtual models carried out in exact proportion (scale 1:1). Human model was positioned in the virtual environment as per the functional requirement of critical working postures of concern. DELMIA® features allowed to incorporate joint angles acquired using Kinovea software and replicated the real postures as closely as possible. A recheck of reconstructed screenshots and photos was also carried out. All working postures were analyzed using the same method. Later, RULA test was performed using DELMIA® software. Figure 22.6 shows the interaction of human model with polishing tool in polishing task. In some cases, recognizing the angles was difficult from the photos because human and tool were relatively positioned. These small differences were adjusted by direct observations of field study before turning to simulation.

22.3 Results and Discussion

A total of 40 male *Kalash* polishers have participated in the study. Table 22.1 presents the demographic characteristics of the polishers. The age of polishers ranged between

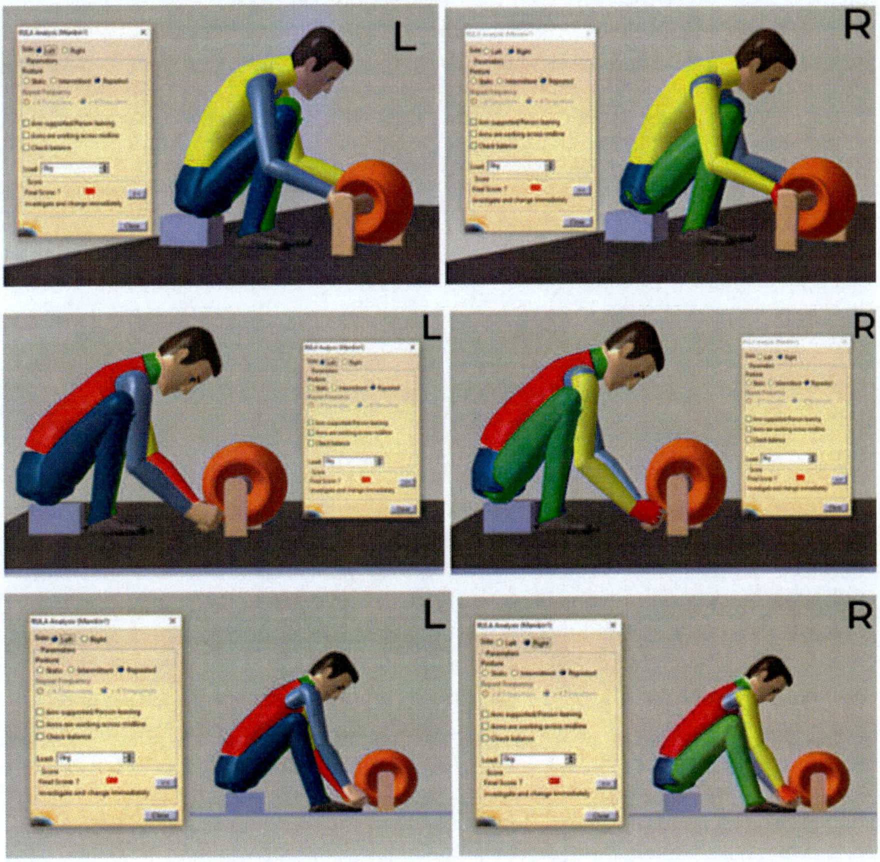

Fig. 22.6 Interaction of human model with polishing tool: a–c working postures (Note: L = left side; R = right side)

Table 22.1 Demographic characteristics of polishers ($n = 40$)

Variable	Units	Mean	Standard deviation	Range
Age	yrs	36.43	11.45	20.0–58.0
Height	cm	161.90	4.34	154.0–169.0
Weight	kg	65.71	5.92	52.2–76.7
BMI	kg/m^2	25.12	2.66	19.6–29.8
Experience	yrs	14.93	9.47	3.0–35.0
Daily working time	hrs	7.30	1.14	5.0–9.0

Table 22.2 Prevalence and severity of musculoskeletal disorders among polishers

Body region	Prevalence in past 12 months (% sample)	Pain severity (0–5 scale)
Neck	32.5	2.2 (1.0)
Shoulders	45.0	3.3 (1.3)
Elbows	50.0	2.9 (1.1)
Wrists/hands	47.5	2.6 (1.3)
Upper back	52.5	3.5 (1.2)
Lower back	75.0	3.9 (1.1)
Hips/thighs/buttocks	25.0	1.7 (0.8)
Knees	20.0	1.5 (0.7)
Ankles/feet	20.0	1.5 (0.6)

20–58 years (mean = 36.4 years; SD = 11.4 years). The mean BMI of polishers was 25.1 kg/m^2 (SD = 2.66 kg/m^2; range = 19.6–29.8 kg/m^2). The polishers had been working in their current job ranged from 3.0 to 35.0 years (mean = 14.9 years; SD = 9.4 years), and their mean daily working time was 7.3 h (SD = 1.14 h; range = 5–9 h). Commonly, all the polishers work six days/week.

One of the important findings of the study is MSDs among polishers. The prevalence of musculoskeletal disorders among handcrafted *Kalash* polishers found to be very high; 100% of workers had experienced some kind of pain in at least one body region during the past 12 months (Table 22.2). This finding confirms that MSDs are significant occupational health problems among polishers. The prevalence rates among polishers were higher when compared with other handicraft workers [2, 6, 7, 10]. With regard to individual body parts, the most commonly affected body part was lower back (75.0%) followed by upper back (52.5%), elbows (50.0%), and wrists/hands (47.5%), respectively. The average severity of pain in the lower back, upper back, elbows, and hands/wrists was generally between moderate to high.

The majority of the polishers had commonly reported lower back pain, and the pain severity of lower back pain was 3.9 (1.1), which is slightly greater than other body parts. This may be attributed to abnormal ergonomic working conditions of polishers. Because polishers were operating the traditional tool at very low level from the ground, and frequent bending of trunk might has increased the lower back pain. This finding is further supported by earlier studies conducted among similar handicraft workers, such as goldsmith workers [17] and gemstone polishers [10]. Further, Choobineh et al. [4] highlighted that tool design plays a significant role in lower back disorders. Similarly, in our study, the lower back pain seems to be due to existing polishing tool design.

The findings of the study provide further evidence that the work of polishers is physically demanding. This was confirmed by the results of RULA method (Table 22.3). The results of postural analysis revealed that all the identified working postures of polishers were at high to very high-risk levels, recommending investigation and immediate changes to postures of polishers to avoid further risk. Perhaps, this result

Table 22.3 Final/grand RULA score and corresponding actions

Task	Posture A	Posture B	Final/grand score	Action
Tool fitting	4	6	7	Investigate and change immediately
Polishing plain surface	5	6	7	Investigate and change immediately
Polishing curved surface	4	6	7	Investigate and change immediately

was not surprising, because most of the working postures were constrained by the polishing tool design. Several studies also found that improper tool design most likely to cause work-related problems and injuries to different body parts [2, 6, 18]. In fact, poorly designed tools are commonly used in handicraft work. Hence, it should be noted that this traditional tools need to be modified/redesigned based on workers' anthropometric data. Also, any improvement in the design of existing polishing tool in terms of raised sitting height may have a considerable impact on the health and safety of polishers in brass metal handicraft production units.

22.4 Conclusion

The present study has demonstrated that handcrafted *Kalash* polishers are at high risk of developing MSDs. The NMQ revealed that the symptoms in the lower back, upper back, elbows, and wrists/hands during the past 12 months are prevalent among *Kalash* polishers. Also, the ergonomic risk assessment of working postures using RULA method also confirmed that postures adopted by the polisher were at very high risk of developing MSDs. Especially, awkward postures—trunk and neck bending, folded knees—need improvement. Further, working on traditional tool caused deviation of body parts from neutral positions. Finally, based on these findings, the study highly recommend that the working conditions of *Kalash* polishers—specifically working postures—can be improved by modifying/redesigning existing polishing tool according to the ergonomics principles and workers' anthropometric data.

References

1. Ministry of Commerce Industry and Industry Report. Department of Commerce. 2017.
2. Motamedzade, M., Choobineh, A., Mououdi, M.A., Arghami, S.: Ergonomic design of carpet weaving hand tools. Int. J. Ind. Ergon. **37**(7), 581–587 (2007)
3. Choobineh, A., Tosian, R., Alhamdi, Z., Davarzanie, M.: Ergonomic intervention in carpet mending operation. Appl. Ergon. **35**(5), 493–496 (2004)
4. Choobineh, A., Lahmi, M., Khani Jazani, R., Hosseini, M., Shahnavaz, H.: Musculoskeletal symptoms as related to ergonomic factors in Iranian hand—woven carpet industry and general guidelines for workstation design. Int. J. Occup. Saf. Ergon. **10**(2), 157–168 (2015)

5. Gangopadhyay, S., Ghosh, T., Das, T., Ghoshal, C., Das, B.B.: Prevalence of upper limb musculo skeletal disorders among brass metal workers in West Bengal. India. Ind. Health **45**(2), 365–370 (2007)
6. Dianat, I., Salimi, A.: Working conditions of Iranian hand-sewn shoe workers and associations with musculoskeletal symptoms. Ergonomics **57**(4), 602–611 (2014)
7. Sahu, S., Moitra, S., Maity, S., Pandit, A.K., Roy, B.: A comparative ergonomics postural assessment of potters and sculptors in the unorganized sector in West Bengal, India. Int. J. Occup. Saf. Ergon. **19**(3), 455–462 (2013)
8. Mukhopadhyay, P., Srivastava, S.: Ergonomic design issues in some craft sectors of Jaipur. Des. J. **13**(1), 99–124 (2010)
9. Kuorinka, I., Jonsson, B., Kilbom, A., Vinterberg, H., Biering-Sørensen, F., Andersson, G.: Standardised Nordic questionnaires for the analysis of musculoskeletal symptoms. Appl. Ergon. **18**(3), 233–237 (1987)
10. Das, D., Kumar, A., Sharma, M.: Risk factors associated with musculoskeletal disorders among gemstone polishers in Jaipur, India. Int. J. Occup. Saf. Ergon. **27**(1), 1–11 (2018)
11. McAtamney, L., Nigel Corlett, E.: RULA: a survey method for the investigation of work-related upper limb disorders. Appl. Ergon. **24**(2), 91–99 (1993)
12. Gnanavel, S.S., Soundararajan, R.: Work posture analysis of human using Catia Ergonomics tool. Appl. Ergon. **57**, 247–269 (2010)
13. Rula color code—Google Search. [cited 2020 May 25] Available from: https://www.google.com/search?q=rula+color+code&tbm=isch&chips=q:rula+color+code,online_chips:musculoskeletal+disorders&hl=en&ved=2ahUKEwjtnPXOgM7pAhWBRSsKHdToBNMQ4lYoAXoECAEQFQ&biw=1226&bih=597#imgrc=CAMecEMpovP2qM
14. Elwardany, S.H., El-Sayed, W.H., Ali, M.F.: Reliability of Kinovea computer program in measuring cervical range of motion in sagittal plane. Open Access Lib. J. **02**(09), 1–10 (2015)
15. Samuel, O.A., Israel, D., Moses, F.: Anthropometry survey of Nigerian occupational bus drivers to facilitate sustainable design of driver's workplace. World J. Eng. Technol. **04**(02), 176–182 (2016)
16. Chakrabarti, D.: Indian anthropometric dimensions for ergonomic design practice, pp. 1–149. National Institute of Design (1997)
17. Ghosh, T., Das, B., Gangopadhyay, S.: Work-related musculoskeletal disorder: An occupational disorder of the Goldsmiths in India. Indian J. Community Med. **35**(2), 321–325 (2010)
18. Sperling, L., Dahlman, S., Wikström, L., Kilbom, Å., Kadefors, R.: A cube model for the classification of work with hand tools and the formulation of functional requirements. Appl. Ergon. **24**(3), 212–220 (1993)

Chapter 23
Calligraphic Typesetting Exploration of Religious Scriptures to Resurrect the Emotional and Spiritual Connection Originally Existed in the Traditional Penmanship

Abhay Verma and Abhijit Padun

Abstract Indian calligraphy has a very noble, decorative, and a unique style which can be seen in the major Indian religious texts composed and handwritten by ancient Indian sages and calligraphers. Technological evolution promoted cheaper mass-produced printed copies of the same handwritten religious scriptures which led to a drastic decrease in the demand for calligraphy artists and handwritten scriptures. It has been observed that available printed scriptures have a monotonous typesetting and printing style. As a result, a hypothesis has been proposed stating that people have lost the emotional connection which they had with the handwritten scriptures. This study addresses this issue by collecting adequate data and conducting a design experiment, to explore the feasibility of offering an absolute alternative to ancient calligraphic scriptures to bring back positive psychological connections through innovative design intervention. Further, the study also tries to explore the future usage and digital implications of the experiment.

23.1 Introduction

Urge for thought and emotional expression is seen in human species throughout all of the ages. For a very long time, humans have tried to discuss values, beliefs, ethical conducts, lifestyle, chants, and mythologies essentially by the verbal and written form of communication [1]. The amount and diversity of Indian religious literature, scripts, and languages are huge. Historically, the literature played a major role in establishing collective laws of society in the form of thoughts and ideas which act as foundational principles upon which modern man exists [2]. Existing literature can

A. Verma (✉) · A. Padun
Department of Multimedia Communication and Design, Central Institute of Technology, BTAD, Kokrajhar, Assam, India
e-mail: vabhay7070@gmail.com

A. Padun
e-mail: a.padun@cit.ac.in

A. Chakrabarti et al. (eds.), *Design for Tomorrow—Volume 1*, Smart Innovation, Systems and Technologies 221, https://doi.org/10.1007/978-981-16-0041-8_23

be found in museums, digital online directories, and personal ownership. Most of the religious literature have been written as manuscripts and are degrading rapidly.

Religious texts evoke certain emotion and respect by the textual information, presentation style, antiquity, and nostalgia. Recently, a shift in the consumption and production of these texts has taken place.

Earlier, all of the religious texts were handwritten by the calligraphy artists, involving intense emotions and feelings of the artisan who is producing them, along with the person who is utilizing these religious scriptures for some purpose. In prehistoric times, the scriptures were only read and recited by the people who were able to access them. The handwritten scriptures were of much significance because every scripture have a distinctive human like feel just like our individual signatures. Conventional printed booklets subjectively have the same information, but it lacks intent and purpose in terms of visual appeal. Production of handwritten texts became extinct due to a lack of demand, calligraphy artists, natural pigments, and handmade writing instruments. Mass production of the printed religious texts is done by directly using conventional typefaces for typesetting. The primary focus is given on objective information with the minuscule effort given to the subjective values in terms of content, style, and antiquity of the printed booklets of the religious texts.

Due to the quality degradation of the ages-old religious manuscripts, future generations would never be able to experience the delight and emotional connection with ancient manuscripts. The utilization of hand-drawn Devanagari letterforms from ancient manuscripts has many scopes for exploration. Experimentation needs to be performed by carefully analysing the letterforms and style details from real samples of these manuscripts. This can be done by the breakdown analysis of aesthetical and functional characteristics of these manuscripts. Conventional technological resources can be used for the mass production of these calligraphic manuscripts, keeping intact the emotional connection antiquity and nostalgia. Furthermore, the adaptability and acceptability of the newly produced scriptures obtained from the performed design experiment can be determined.

23.2 Background Study

As stated by the Author Yuval Noah Harari in his book Sapiens, 'Modern science is not content with creating theories. It uses theories in order to acquire new powers and in particular to develop new technologies' [1]. Basically, by following the path of existing information, new information can be developed and designed which will move us towards a better future.

Two key factors that play a major role for effective absorption of the information from any textual material are visual appeal and distinctive identity. The human brain processes the information more in terms of visuals rather than just plain text and visual images often intrude on verbal thinking [3]. In the study of psychology, a concept of 'Association' refers to mental connections between concepts, events, or mental states that usually stems from specific experiences [4]. So, based on the concept of association, our mind becomes conditioned based on our life experiences and social conditioning.

Hence, a hypothesis has been proposed which states that whenever we see a handwritten scripture, it evokes a stimulus that is very different from the stimuli evoked whenever we see a printed document. It happens because of our past experiences, social conditioning, antiquity, and nostalgia. Handwritten scriptures feel very personal, whereas the printed ones have a monotonous look and feel. A design experiment has been conducted to use technology for exploring mass production possibilities of Indian religious manuscripts for retrieving the lost emotional values originally existed in the ancient handwritten Indian manuscripts. Further, to test the experiment performance, pundits, sages, and regular people have been interviewed for collecting qualitative and quantitative data.

23.3 Literature Review

23.3.1 Significance of Religious Scriptures

For this study, Sanskrit Hindu religious scriptures from north Indian regions were examined and collected which are classified into two different categories: Shruti and Smriti. Shruti consists of religious literature that is recited and is transmitted by sages, whereas Smriti consists of the literature that is remembered, and its central ideas are very influential amongst the traditions and thoughts of the Hindu literature [5]. Different literature pieces fall under Shruti and Smriti but no exact proof of the era in which these texts were written has been confirmed yet.

Diverse Hindu literature includes Vedas, Shakhas, Brahmanas, Itihasa, Upanishad, Mantras, etc. 'Itihasa' is a Hindi word for 'History' which is a chronologically written description of the most socially influential events. The two most significant pieces from Itihasa are: Ramayana and was written by Maharishi Valmiki, and Mahabharata which was written by Vyasa. Both of these texts hold immense spiritual values for millions of people globally [6].

'Mantra' is a group of words in Sanskrit which are tied together and chanted; existence of mantras can be seen in various school of Hinduism, Jainism, Sikhism, and Buddhism. In Hindu culture, these sentences are believed to have religious magical or spiritual powers. Mantras can be short or long depending on the way it is composed by the ancient sages and priests; the shortest mantra is (Aum, Om) [7].

This study has been conducted on the two different historical articles, for examining the calligraphic writing styles and other aesthetical elements.

1. Uttarkand of Ramayana (Part of Itihasa)
2. Mahāmrityunjaya Mantra (Type of mantra).

23.3.2 Penmanship and Significance of Uttarkand of Ramayana

Ramayana is a book of Dharma that is worshipped by millions of people globally because it depicts the duties of relationships by portraying ideal characters and is believed to have solutions to most of the problems and dilemmas human society faces. Ramayana is an epic poem that consists of nearly 24,000 verses and divided into about five volumes (*kāṇḍas*) the ayodhyakāṇḍa, the araṇyakāṇḍa, the kiṣkindakāṇḍa, the sundarākāṇḍa, and the laṅkākāṇḍa. The uttarākāṇḍa and the bālakāṇḍa, although were written later on by unknown sages based on the interpretation of the first five original volumes, written by Maharishi Valmiki [6].

Every page in the collected Ramayana manuscript is beautifully composed by precisely hand drawing the individual letterforms, numbers, decorative patterns, and page margins. Red and black natural pigments and handmade writing instrument with angular tip known as 'Setha Kalam' which has been used for writing these scriptures.

23.3.3 Penmanship and Significance of Mahāmrityunjaya Mantra

Literally meaning 'Great Death—conquering Mantra' which is chanted to gratify the most powerful deity of Hindu religion lord Shiva, to ward of calamities for preventing untimely death. 'In the Vedas, it finds its place in three texts—(a) the Rig Veda VII.59.12, (b) the Yajur Veda III.60, and (c) the Atharva Veda XIV.1.17′. It is said that this powerful mantra re-links the consciousness to one's deeper senses to improve mental, emotional, and physical health when is performed by a technique called Japa which simply means that reciting the mantra again and again for optimal results [8].

Upon interviewing the priest who has written the collected piece of Mahāmrityunjaya mantra (refer to Fig. 23.1) by using a pen made up of pomegranate wood recommended that this to be chanted 108 times on a Rudraksha rosary which

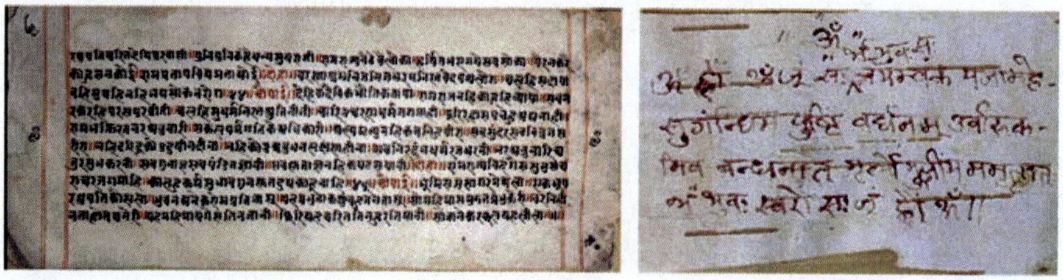

Fig. 23.1 Original manuscript images of Ramayana (left) and Mahāmrityunjaya Mantra (right)

contains exactly 108 beads. This mantra is still handwritten by the priests for the people to keep and recite. Hindu religious texts including this mantra are written on the skin of Himalayan Birch tree also known as 'Bhojpatra' by using sandalwood as a pigment and 'Setha Kalam' as writing instrument to enrich the positive impact.

Compared to the collected Ramayana manuscript, letterform clarity is less in this manuscript. Stroke quality depletion is observed in this manuscript that represents the diminishing artistic skills over a time period. But, according to the priests, the faith and belief in the handwritten scriptures have not been reduced.

23.4 Methodology

To retrieve the emotional values which originally existed in the ancient Indian religious scriptures by performing and testing the performed design experiment, a methodology shown in Fig. 23.2 has been conceptualized to achieve the objective.

23.4.1 Exploration and the Literature Review of Religious Scriptures and Its Styles

Diversified exploration and the literature review of the religious scriptures and its styles have been already discussed in the literature review section (refer Sect. 23.3).

23.4.2 Need Identification

Based on the association theory of psychology, the proposed hypothesis describing the emotional disconnection with the modern printed scriptures has already been discussed in the background study section (refer Sect. 23.2).

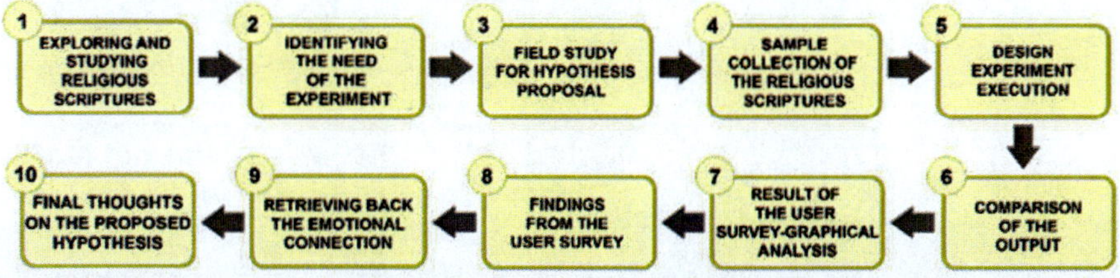

Fig. 23.2 Schematic representation of the proposed methodology

23.4.3 Field Study

To adequately establish the need for a revival design experiment, an initial survey with 30 people was conducted to collect qualitative data for understanding emotional connect, views, and thoughts on the religious handwritten and printed scriptures. In the initial survey, most of the questions were open-ended, and the interviews were recorded with the permission of the interviewees itself. The initial surveys lasted for around 15 min each and were more inclined towards an informal conversation rather than a strictly followed set of questions. The questions of the survey revolved around the following topics:

1. Emotional connection and belief in religious scriptures.
2. The kinds of handwritten and printed scriptures they own.
3. Their personal preference amongst handwritten and printed scriptures.
4. If the scriptures they own were passed on over the generations.
5. Emotional differences in the handwritten and printed scriptures.

From the initial survey, it has been observed that most of the people showed keen interest in the formulated hypothesis. So, the decision to proceed with the design experiment has been made.

23.4.4 Sample Collection of the Scriptures

Ancient handwritten and modern printed manuscripts were explored for comparison in terms of letterforms, layout, colours, texture, and other subjective factors (Fig. 23.3). North Indian Sanskrit manuscripts were collected and narrowed down to primarily two spiritual texts: Ramayana and Mahāmrityunjaya Mantra. Scripture

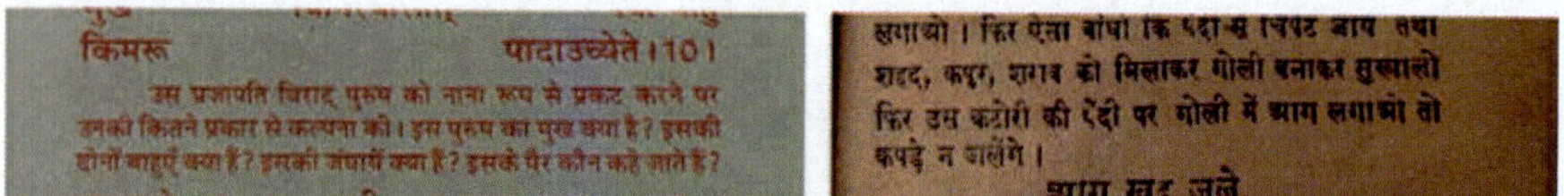

Fig. 23.3 Conventional styles of typographic printed texts from various religious scriptures

selection has been done based on the differences in writing styles from different time periods and usage commonalities.

23.4.5 Design Experiment

23.4.5.1 High-Resolution Digital Scanning for Visual and Technical Analysis

High-resolution digital scanning of the collected manuscripts was done to examine the letterforms, numbers, and decorative patterns individually.

The custom paper of size 319 × 135 mm with horizontal margins of 32 mm and vertical margins of 20 mm was used to write Ramayana manuscript. The space allocated for contents is 255 × 95 mm with page and chapter numbers written outside the red vertical margins. Mahāmrityunjaya Mantra has been handwritten on 'Bhojpatra' of size 125 × 75 mm with small uneven margins from all the sides. Further insights were drawn regarding irregularities of all the visual elements due to pigment behaviour on both manuscripts.

23.4.5.2 Digital Visual Elements Recreation

Both of the Sanskrit manuscripts have a unique style of Devanagari letterforms. Thick and thin calligraphic stokes can be observed in the text of Ramayana manuscript, whereas Mahāmrityunjaya Mantra is written in mono-linear stroke style. Letterforms from both of the manuscripts were digitally converted by tracing and manually refining in different ways because Mahāmrityunjaya Mantra has joint letterforms forming words, whereas letterforms in the Ramayana manuscript are not. To replicate the style of the original text, subtle imperfections were intentionally kept while recreating digital elements (Figs. 23.4 and 23.5).

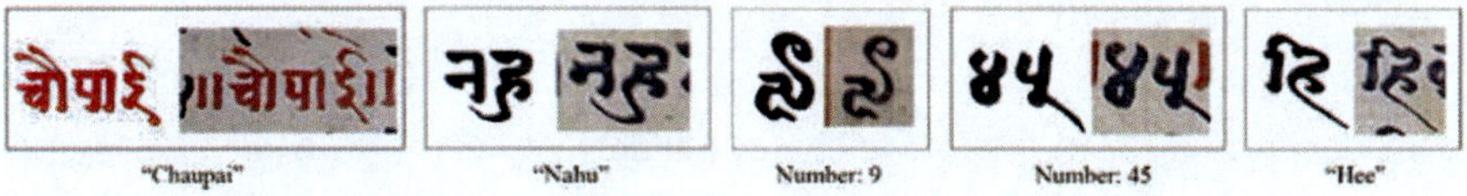

Fig. 23.4 Digitally created letterforms and numbers from original scanned manuscripts

Fig. 23.5 Digitally created decorative patterns from original scanned manuscripts

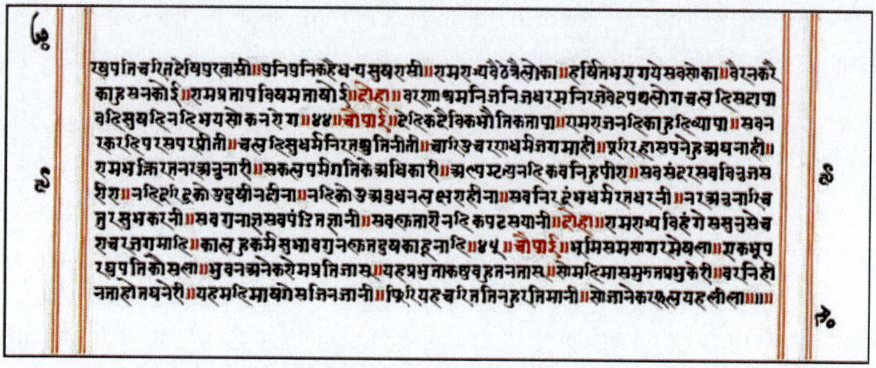

Fig. 23.6 Digitally created manuscript for screen printing on different paper types

23.4.5.3 Digital Manuscript Creation

Character set library of all the numbers, letters, mantras, and patterns was made in a scalable vector format. Then, all of the digital visual elements were placed on the original scanned Ramayana and Mahāmrityunjaya Mantra manuscript proportionally (Fig. 23.6).

23.4.5.4 Cover Page and Decorative Elements Design

Every volume of handwritten Ramayana manuscript has its own handcrafted cover page, which varied depending on the choices made by the artist. Analysis of the worn off Ramayana cover page from the original manuscript has been done. The cover page usually had some decorative block of a pattern containing the title of the volume. The red and yellow coloured block of the detailed pattern has been digitally recreated as shown in Fig. 23.7.

Fig. 23.7 Screen printed sample of the cover page (right) by referring to old manuscript (left)

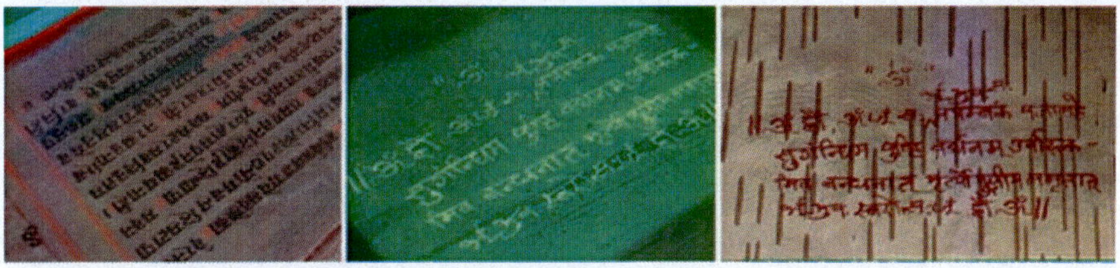

Fig. 23.8 Screen printing process and output

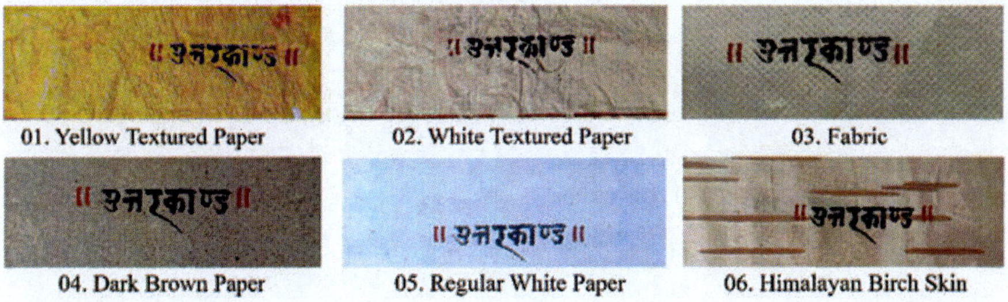

Fig. 23.9 Screen printing on six different kinds of textured surfaces

23.4.5.5 Final Printing

Four different types of paper, one cloth and one skin of the Himalayan Birch tree (Bhojpatra), have been selected for the printing process. Pigment analysis unveiled subtle changes in depth, lustre, and bleed in the handwritten letterforms (Fig. 23.8).

Screen printing has been chosen as printing method because it allows the printing to be done on different textured surfaces; it is easily available, cheap, and capable of replicating handwritten manuscripts. Individual mesh-based screens for every colour were made for transferring the ink on six different writing surfaces for further analysis of the look and feel (Figs. 23.9 and 23.10).

23.4.6 Comparison of Output

The behaviour of the screen printing ink varies on different printing surfaces. Subtle bleed, depth, and lustre in the letters from the screen printing process which gave printed texts human touch and inconsistencies similar to the handwritten texts (Fig. 23.11).

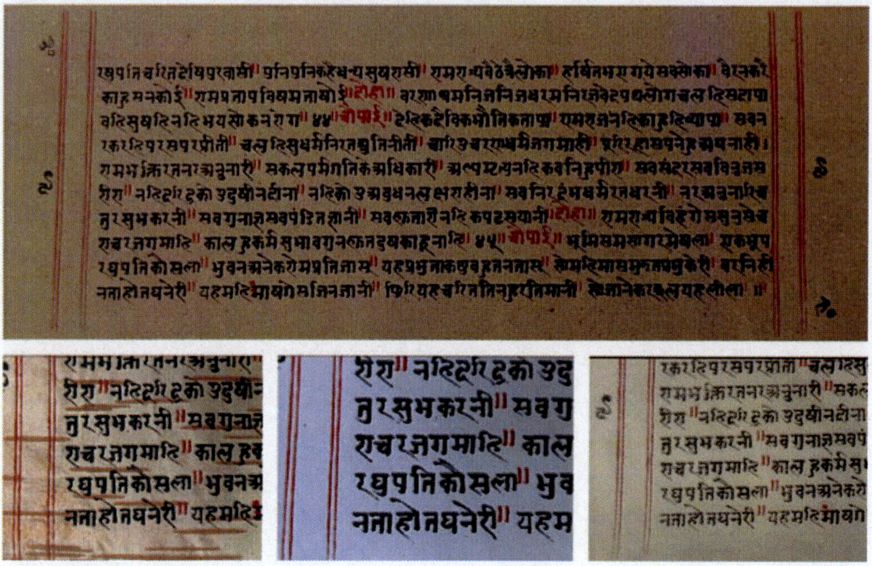

Fig. 23.10 Screen printed page of Ramayana and comparison between different prints

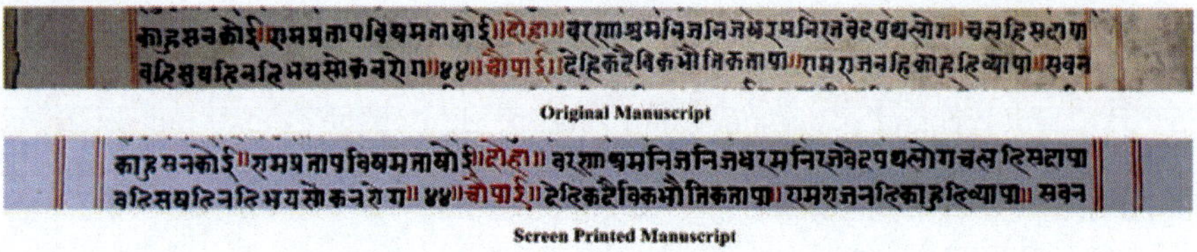

Fig. 23.11 Comparison between original and screen printed manuscripts

23.4.7 Result of User Survey—Graphical Analysis

Final feedback survey was more inclined towards quantitative data collection for the best possible performance measurement of the formulated hypothesis. A total of 45 people were interviewed which include 5 personally known, 15 priests from nearby temples, and 25 people who have faith and are interested in studying religious scriptures.

Around 15 questions were formulated for the questionnaire, and users were requested to provide a response from 1 point being the least to the 10 points being the highest performance rating. Categorization of all 15 questions has been done broadly in 5 different parameters based on the responses and suggestions provided by the users. Figure 23.12 shows a performance bar graph, five performance measurement parameters and rating analysis. On the X-axis, the compiled points given by the users

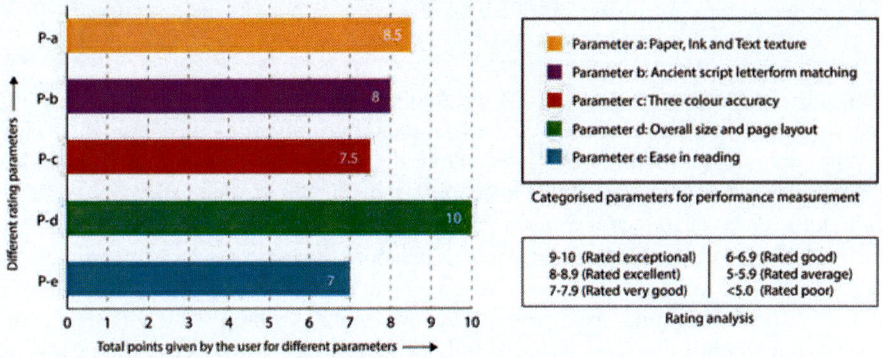

Fig. 23.12 Consolidated results obtained from the final feedback and review sessions

are shown, and along the Y-axis different parameters are shown. On the right of the chart, five chosen parameters and the 1–10 rating analysis is shown.

23.4.8 Findings from User Survey—Emotional Relatability Analysis

Results of the conducted survey show that P-d, P-a, and P-b are rated exceptional and excellent. Colour accuracy (P–c) of the newly printed scripts falls in a very good category. Ease in reading (P-e) is the lowest because the monotonous printed documents use the exact same letterforms throughout the whole scripture, whereas letterforms are slightly inconsistent in the ancient scriptures which are replicated in the newly recreated version.

Performance measurement by combining the quantitative data shows an overall performance rating of 8.2 or 82%, which can be considered as an excellent score to prove the validity of the proposed hypothesis.

23.5 Discussions

Recreation of the whole letter set by tracing and refining the original letterforms gave the multidirectional exploration flexibly with the compositions. Potential users found customization as the most appealing factor of the whole experiment. Preference to utilize the recreated manuscripts over the monotonously printed conventional manuscript booklets has also been found. People and priests appreciated the idea of reviving by revisiting of the old manuscripts and showed an overwhelmingly positive response in terms of emotional and spiritual relatability with the reprinted manuscripts.

23.6 Conclusion

The study found out that conventional printed manuscripts are incapable of capturing the subjective spiritual essence of the ancient handwritten religious documents. This typographic exploration is capable of allowing people besides admiring to actually experience and psychologically connect with the delicate ancient religious art form. Indian handicrafts and art forms are popular globally, so the craftsmen can use this technique to produce craft and artworks to promote the legacy of Indian script globally. Historians, academicians, and designers can take this study as a starting point to further perform similar experiments by exploring a vast number of religious texts and scripts. The prime goal of this study is to encourage the people involved in the print production industry, to redefine the printing process of religious scriptures while taking into account the psychological responses towards religious texts.

References

1. Harari, Y., Noah, S.: A Brief History of Humankind. 1st edition. Penguin Random House (2015).
2. Duhan, R., The Relationship between Literature and Society, Vol. 15(4). (2015). ISSN: 1930-2940
3. The power of picturing thoughts: https://news.harvard.edu/gazette/story/2017/05/visual-images-often-intrude-on-verbal-thinking-study-says/. Accessed on 11 May. 2020
4. Klein, Stephen (2012). Learning: Principles and Applications, 6th edn. SAGE Publications. ISBN: 978-1-4129-8734-9
5. About Hindu Texts: https://en.wikipedia.org/wiki/Hindu_texts. Accessed on 05 Oct 2020
6. About Ramayana: https://en.wikipedia.org/wiki/. Accessed on 05 Oct 2020
7. About Mantra: https://en.wikipedia.org/wiki/Mantra. Accessed on 07 Oct 2020
8. About Mahāmrityunjaya Mantra: https://en.wikipedia.org/wiki/Mahamrityunjaya_Mantra. Accessed on 04 Oct 2020

Chapter 24
Para, *Adda* and Residents of Kolkata: Need for Design of New Neighbourhoods Accommodating Cultural Legacies

Suryendu Dasgupta and Megha Tyagi

Abstract *Para* in Kolkata refers to an informal, organic and urban residential unit, which is the origin point of the significant socio-cultural activity of *adda*—a social informal talk, usually done in Bengali, among friends, colleagues and even family members, tied to intellectual or current affairs topics. Conducted mainly in public spaces within residential areas, *adda* and other social interactions in a *para* are essentially stimulated by their organic design, which in turn, is dictated by the needs of the inhabitants. To accommodate the rising population and in response to the current socio-economic aspirations of the citizens, new neighbourhoods are being developed in Kolkata, which mostly follow the modern design principle of segregation. Due to the segregation of activities, the nature of these neighbourhoods is introverted and exclusionary, reducing social cohesiveness by discouraging social interactions and is detrimental for the mental and physical health of the inhabitants. The study aims to highlight the necessity for extending the socio-cultural legacies of old neighbourhoods of Kolkata into the new ones to address the emotional and social needs of its residents through an alternative approach for neighbourhood design. The study establishes the emotional and social needs of the residents through interviewing the residents of the *paras* and new neighbourhoods and employing content analysis to analyse them. The key objectives of the study are to establish the need for extending the cultural legacies of *paras* to the city's new neighbourhoods and evolve a design philosophy to encourage inclusivity and social interaction within the new neighbourhoods.

24.1 Introduction

The process of urbanization, in the post-independence scenario of India, has transformed the Indian cities to a great extent. The most intense change was perceived after the 1960s due to extensive industrialization. With the onset of the 70s, urbanization in India saw a paradigm shift with the rise in the establishment of satellite

S. Dasgupta (✉) · M. Tyagi
Indian Institute of Technology Roorkee, Roorkee, India
e-mail: suryendu7@gmail.com

A. Chakrabarti et al. (eds.), *Design for Tomorrow—Volume 1*, Smart Innovation, Systems and Technologies 221, https://doi.org/10.1007/978-981-16-0041-8_24

towns around major cities to accommodate middle-class residences or as sites for relocating industries [1]. The change in the urban pattern and ensuing economic changes, the socio-cultural dimension of urban India changed as well. The impact of this change is most visible in the urban neighbourhoods of the country. Neighbourhoods are the smallest units of the urban realm, with an inherent social character than a physio-spatial identity [2]. They are often characterized with permeable boundaries, an ambivalent membership and a contested legal status; but neighbourhoods have a deep influence on the lives of their inhabitants and is their social reality and sites for individual and collective memory [3]. Additionally, everyday practices within a neighbourhood lead to the creation of histories and communities, representing the 'local' context. Including gossip and informal street meetings, these practices may also involve more organized activities of medical care, schooling and political or religious organization. Thus, neighbourhoods are distinctively communal in nature, with their members categorically describing its qualities in terms of association, interdependence and co-operation.

24.1.1 Locating **Para** *and* **Adda** *as Cultural Legacies in the Urban Realm*

In Kolkata (formerly Calcutta), *paras* form the spatial units similar to the neighbourhood, peculiar to the colonial history of the town. They exist at a liminal space, straddling the dimensions of cultural and spatial demarcation. Essentially, rural in its origin, the Bengali word *para* denotes to the localities demarcated by their inhabitants, belonging to a particular caste or profession. This model was replicated in the colonial urban space of Kolkata, and one can still find neighbourhoods with names such as Paik *para*, Ghosh *para* or Bose *para* in the old residential quarters of North Kolkata. This was primarily because the native population of the newly developed city was the inhabitants of the villages constituting and surrounding the boundaries of the city, migrating into Kolkata for providing services to support the colonial state. There are instances of such 'racialized ghettos' in cities all over the world, where a cluster of inhabitants belonging to a certain ethnic, religious or class-based group forms a residential unit. Though during the initial years of British occupation of the town such homogenous neighbourhoods based on caste, clan or occupation came up in Calcutta, the later part of the nineteenth century witnessed the formation of more heterogeneous middle-class neighbourhoods [4]. Unlike the North Indian *mohallas* or the Gujarati *pols*, which have proper physical boundaries and were often gated, the *paras* have a sublime perimeter. The colonial British archives have meagre accounts of the native neighbourhoods with most essays portraying the native part of the city as haphazard, clumsy and devoid of any planning [5]. They did not document the social aspects of the middle-class neighbourhoods, which was the hub of everyday practises and negotiations performed by the inhabitants of the newly developed city in an attempt to transform their alien urban environment into the familiar socio-cultural domain which they previously inhabited. The inhabitants of the *paras* usually maintained an obligatory, amicable relationship with each other, and though occasional feuds and quarrels were common, the sentiments of pride, association and sense of territoriality kept them united.

The *paras* were, hence, essentially territorial in nature with a strong sense of community and attachment ingrained within its inhabitants. This community developed around institutions like libraries, sporting clubs, literary and other cultural associations which were mostly established through the collective voluntary contributions of the inhabitants of the *para* with a substantial donation from the wealthier inhabitants in many cases. Such an environment nurtured discussions and reflections on essential issues of the date, leading to several cultural norms and orders which configured the urban unit. One such practise quintessentially originating out of the cultural customs of meetings and discussions in a para is *adda*. An *adda* is a form of informal social talk whose content is essentially of intellectual significance. It is done [the Bengali verb is 'give'] among friends, neighbours and colleagues, spanning issues of politics, art, literature, music and science and has become synonymous with the urban middle-class Bengali identity and culture [6]. The distinction that *adda* has with casual gossip and informal conversations prevalent in other societies as well as the creative execution of the art, especially responding to the 'cultural hegemony of British imperialism;[7]. It is imperative to note that the salient feature of *adda* is its urban setting and its location in the public sphere, where the young middle-class men of Calcutta engaged in the activity on '*ro'aks*' (elevated platforms outside a dwelling house), markets, parks and playgrounds, canteens and bus stops.

The culture of *adda* soon witnessed the advent of indigenous clubs in various neighbourhoods in response to the elitist British clubs. These clubs concretized the domain of *adda* in the *paras*, providing a defined space in the neighbourhood to partake in the genre. Even at present times, these *para* clubs, run primarily by the youth of the neighbourhood, provide vitality to the middle and lower middle-class society. Many of these local clubs are the strength of the *para*, providing assistance to the residents in times of crisis and serving as the muscle of the community. From providing a platform to position the deliberations and opinions of residents about philosophy, art, literature and politics in colonial times, to organising organizing cultural programmes, community religious festivals (like Durga puja, Kali puja, etc.) and contributing to social causes like blood donation camps, disaster relief, etc., at present, these clubs form an integral part of the urban social dimension. The *para* thus, as the point of origin for such institutions and activities, can be viewed not only as a residential unit but also an urban social unit with strong cohesion in the community. It provides a secure territory in the dynamic urban environment imparting a sense of trust and security among its inhabitants, which still exist in many neighbourhoods of the old quarters of Kolkata.

The situation is, however, rapidly changing in the contemporary megalopolis of Kolkata, where new neighbourhoods are getting constructed to accommodate the growing needs of the city's population. High-rise and mid-rise gated communities are replacing the low-rise neighbourhoods as limited land resource within the city is encouraging vertical urban growth. The peripheral urban areas are experiencing rapid urbanization with the development of new townships of Bidhannagar (also known as Salt Lake) and Rajarhat (also known as New Town). While urbanization has changed the neighbourhood pattern, political and economic changes, with the intercession of globalization, have transmuted the cultural scenario, with the extinction of practices like *adda* and local club activities. With increasing competition and aspirations among the younger generation, there is hardly any time for extensive *adda* sessions which are now more within the confines of cafes and social media applications. This has made the new neighbourhoods socially individualistic and introverted, with weak cohesion within the community. This, in turn, has reduced human interaction within the neighbourhoods, affecting the emotional health of the residents and making the neighbourhood withdrawn and desolate. This has had a detrimental effect on the trust and safety factor of the new neighbourhoods, with more instances of child kidnapping, chain snatching, thefts and burglaries occurring than in the older neighbourhoods. Establishing on this backdrop, the study emphasizes the role of neighbourhoods as a fundamental social institution having critical influence in the lives of its inhabitants. It also compares and analyses the socio-cultural dimension of old *paras* and new neighbourhoods and argues for the infusion of cultural legacies into the new neighbourhoods of Kolkata through an alternative design philosophy for a wholesome social and community existence.

24.2 Methodology

The study employs a narrative-based analysis using the data collected from a comprehensive historical literature study, on field observation, architectural fieldwork and in-depth interviews, based on semi-structured questions, of 16 nuclear families from the new neighbourhoods of New Alipore and Salt Lake in Kolkata. Out of them, ten families have been living in the new neighbourhoods for at least 10 years, after having shifted from older *paras* to new multi-family apartments and gated residential high or mid rises. These residents earlier resided in the older neighbourhoods of Sovabazaar, Baghbazaar, Bhawanipur and Gariahat. The other six respondents comprise of residents, who have been living in the planned neighbourhoods of New Alipore and Salt Lake, since childhood for at least 40 years, in individual houses or mid-rise multi-family apartments. Questions regarding the social cohesiveness in the neighbourhood and the relation with their neighbours, the respondent's perception of the quality of life in the neighbourhood and opinion on necessary improvements, were common for both sets of respondents. Specific questions regarding their reason for shifting to the new neighbourhood and the changes in their social behaviour due to the new environment and the social cohesiveness in the new neighbourhood and how it differed from the *paras* were asked to the respondents who have shifted from the older neighbourhoods. Content analysis of the interviews was done to understand the perception of the residents about their neighbourhood and the issues pertaining to the *paras* as well as new neighbourhoods in the urban realm. The demographic, socio-cultural and physio-spatial needs for the holistic design of a neighbourhood were further revealed for developing a design approach as its response.

24.3 Transforming Neighbourhoods in Kolkata and Its Effects

The new policies of liberalization in India has benefitted the steadily growing middle-class of Kolkata [8, 9], who are aspiring to join the category of elites. They are appealed by the commercialization and commodification of housing and are giving up their old and crumbling family residences for buying residences in the city's 'hot spots' [10]. These new residential developments are mainly group housings comprising of individual houses, mid-rise and high-rise structures in planned and designed residential complexes or standalone mid-rise structures housing several apartments or flats. The former type of residential development is generally located in the adjoining periphery of the city and new townships like Salt Lake and New Town due to the availability of larger tracts of buildable land. The latter, a result of redevelopment of a cluster of low-rise houses in the older neighbourhoods, are found mostly within the city and the core area.

24.3.1 Transformation of the Built Environment

The physio-spatial structure of the older *paras* of the city is almost typical with formal housing stock displaying an amalgamation of hybrid colonial and vernacular architectural styles and 65–100 years old or more. They largely consist of two or three storeyed structures to accommodate joint families or a single family in some cases. Most buildings have a protruding platform jutting into the street forming the *ro'ak*, a space extensively used for *adda* sessions. No construction activities usually take place on the streets, especially the ones which are away from the main road. Several narrower lanes branch off from the main street of the neighbourhood forming pathways and cul-de-sacs in the innermost parts of the neighbourhood. The form of development is organic, and the land use mix is high, with several commercial activities like general stores, essential services like barber and laundry shops, stationary and bookshops, sweet shops, tea stalls and eateries dotting the neighbourhood, providing ample opportunities for casual social interactions.

The new neighbourhoods, however, have polar physiological differences to the *paras*. The development is planned with broad streets enabling unhindered vehicular movements and surgically segregated land use. The housing stock is 10–40 years old, with a few new constructions or modifications of existing structures along the street. The buildings are multi-storeyed structures, most being three storeyed or more with parking space on the ground floor. Majority of the buildings are gated and manned by private security guards, with exclusive access to the residents and their approved guests. Due to the segregation of land use, commercial and recreational activities are clustered at designated zones making the streets in the residential zones almost devoid of people and improbable points of social interactions.

24.3.2 Transformation of the Socio-cultural Dimension

The new neighbourhoods not only stand in stark contrast to the *paras* of Kolkata in their physio-spatial configuration but also in their socio-cultural constructs. One of the biggest sources of discontentment among the ten relocated respondents was the minimal scope of socialization. Though two of the respondent families, both comprising of working, mid-thirties, married couples, with a single child below 10 years each, cited the need of privacy and a quiet neighbourhood as one of the critical factors in their decision to move out of the older neighbourhood, the rest found the new neighbourhoods 'devoid of life'. The rest of the respondents comprised of five elderly couples, whose children have settled elsewhere in the country and three middle-aged couples, whose children are studying in schools or colleges in the city or elsewhere in the country. The elderly respondents stated that their new residences felt depersonalized, with no sense of attachment as opposed to their old houses. Additionally, the quiet neighbourhood and introverted neighbours made their lives solitary and devoid of social interactions. A respondent aged 73, staying in New Alipore with is 65-year-old wife stated that,

> 'We often feel lonely and bored. How long can we keep ourselves busy with TV or books. There is no one to talk [here]. I have a few friends of my age from the neighbourhood, [but] we cannot meet regularly as the park is the only space available for meeting for an *adda,* [and] it is too far for many of us to walk there and back every day'.

The senior residents staying in the new neighbourhoods also lamented on the scarcity of *adda* and the active presence of the local clubs in these neighbourhoods in current times. It was observed that it is mostly the elderly members of the neighbourhood who are now dominating the social space of *adda* in the neighbourhoods, as opposed to the youth, who were the original proponents in the past. With increasing aspirations of the middle-class residents of the neighbourhood and the demands of modern work culture, a significant amount of time is invested in career development with little time to pursue leisurely activities. Socialization among the children was also found to have shifted from playing in the neighbourhood in groups to organized training for art and sport activities, away from the neighbourhood, in the limited time available after their school and co-curricular activities. Many of the youth in the neighbourhood also prefer to spend their leisure time in computer-based activities like online surfing, gaming, chatting and streaming. A housewife in New Alipore stated that she feels anxious and stressed when left alone at home after her husband and child have left for work and school, respectively:

> 'Even a casual exchange of pleasantries from the kitchen window of my old house with the neighbours gave me a sense of security. I have low blood pressure, and I fear that if I feel unwell at any point when I'm alone at my new home, no one would know that I might need help. With my new neighbours being extremely private people, unlike the residents of my older neighbourhood, I would never be comfortable in seeking their help'.

Apart from such anxieties, other residents also feel the lack of social interactions in the new neighbourhood a major safety concern. Respondents who live in individual houses in Salt Lake informed how alarmed they become whenever an unknown person calls at the door, as there have been instances of robbery in broad daylight in the neighbourhood. With empty streets, no neighbours to check on the surroundings and rushing to help in such times of distress, personal safety is a major concern. The respondents who have shifted from the older neighbourhoods asserted that the occasional robberies, which are so common in Salt Lake, are unimaginable in the *paras* of Shovabazaar or Gariahat, where the neighbours know each other personally, and strangers are readily identified and residents, shop owners and even rickshaw pullers rush to the aid of a neighbour in danger.

The lack of open space, at an accessible distance, for children to play is also a serious health concern that many parents of the newer neighbourhoods share. Especially, in the township of Salt Lake, which is designed in sectors with a park in each sector, the parents feel that walking or cycling to the park by children, unaccompanied by an adult, is unsafe due to high speed of vehicular traffic and secluded streets. With a lack of social cohesion, playing in the streets of the new neighbourhoods, in both Salt Lake and New Alipore, is also discouraged by parents out of concern that it may disturb the other neighbours who might complain.

> 'I do not let my son go out of the house unsupervised. The streets are too empty for my comfort, and I hardly know my neighbours personally. I myself do not go out alone for a walk or jog unless I see a substantial number of people in the streets. Chain snatching and eve-teasing are common in such empty streets', said the mother of a 12-year-old boy living in one of the neighbourhoods of Salt Lake.

With the increase of distances between the residences and the amenities, a significant rise in the ownership of private vehicle is observed in the new neighbourhood which is particularly low in the in the old *paras*. This, in turn, has led to an increase in the traffic volume in the new neighbourhoods, which also have wide streets enabling high-speed vehicular movement, making the streets unsafe for pedestrians.

24.4 Need of New Design Approach for Neighbourhoods

The transformation of the *paras* into new neighbourhoods and eventually their gradual eradication can be attributed to an assortment of factors ranging from growing population, governance policies and socio-cultural changes, fuelled by evolving economic and lifestyle conditions. More and more elderly residents of the older neighbourhoods are moving out due to dissolution of the joint family structure and rising cost of maintenance of old houses for a nuclear family, high property taxes, stringent building bylaws and poor infrastructural facilities [11]. The younger generations are moving out due to need to relocate to other cities for better employment and also to fulfil their aspirations of globalized standards of housing configuration than the congested old neighbourhoods, and vehicular accessibility and parking space. This asserts the observation by Bose [12] that 'the modern aspiration of urban living in tune with globalization and the scale of amenity offered by the consumer industries is not fulfilled by living in such historic areas'. However, from the study, it is evident that there are socio-cultural trade-offs involved in the realization of such aspirations which are affecting the quality of living in the new neighbourhoods. The older neighbourhoods are congested spaces with infrastructural issues like flooding during monsoons, narrow streets, lack of natural vegetation and increasing pollution owing to their location in the saturated urban core. These negative aspects have been countered in newer neighbourhoods by knee-jerk design interventions, which valued privacy over sociability and domestic isolation over communication and cooperation among neighbours. They are designed in accordance with the modern concept of mass production, where an ideal house or apartment is created and replicated within a neighbourhood, without due consideration to the plurality of human settlements and the dynamic nature of human society. The study, hence, argues for neighbourhood as an entity formed by the coalescence of the interactions between social and physio-spatial aspects of the urban setting.

The majority of the problems being faced by the residents of the new neighbourhoods are directly associated with weak social cohesion arising out of the lack of interactions. A critical cause for such deficiency is the planning of new neighbourhoods which are designed to enforce segregation of functions such as residential, commercial and recreational activities, public and private domain, and pedestrian and vehicular traffic to name a few. Furthermore, in India, the inconsistent trajectory of neoliberalism has had a deep impact on urban development which has systematically encouraged such exclusivism on the basis of socio-economics as well [13], causing a divide in the demography of a neighbourhood. These stringent rules of segregation are often flouted in older *paras,* where integration and overlapping of such activities and aspects make them successful and vibrant. The points of interaction and social mingling have been meticulously erased in the design of built form in new neighbourhoods. The most evident example is the extinction of the *ro'ak*. The exorbitant rise in land prices made this architectural feature of Bengali houses obsolete and multi-storeyed apartments and housing complexes with globalized design configuration further eliminated its existence. The *ro'ak* formed a space overlapping the domain of private and public appropriated and substantiated by leisure and sociability [5, 14, 15]. Such architectural elements encouraging uninhibited social intercourse are lacking in new neighbourhoods of the city.

With more segregation of spaces for different activities, the density of the residents gets diffused in the neighbourhood, resulting in the loss of activities and vitality in the streets. The commercial activities, essential services, schools and play areas for the children are often located in isolation of the residential areas, encouraging use of personal vehicles, thus reducing pedestrian movement and encouraging high speed traffic, further making the streets unsafe [16]. On the contrary, the older paras have developed historically in an organic manner, shaped by the social and economic agencies, allowing integration of different activities and land uses. This has made the *paras* mostly self-sufficient in their daily needs with all destinations almost within walking distance served by an intricate network of streets acting as critical spaces for public interaction and socialization, devoid of the threats of high speed traffic and improving the quality of life [17]. Additionally, the streets under the dominion of pedestrians in older neighbourhoods reduce the speed of traffic flow and decrease the chances of conflicts and accidents.

The social interactions result in better social cohesion among the residents due to an improvement in trust among neighbours [18], thus contributing to an overall sense of security in the neighbourhood. This also develops a spirit of camaraderie among the residents, making the place more 'human', lacking in the new neighbourhoods. Their impersonal spaces fail to evoke a sense of ownership among the inhabitants and result in dissatisfaction, poor social relations, reduced helpfulness and depression, especially among the elderly and residents spending a major portion of the day in their house or apartment [19, 20]. The absence of social interactions and activity resulting from diverse land use, which attract more people on the streets [21], has also reduced the 'eyes upon the street' [22] making them unsafe [23].

The study also inferred that the new neighbourhoods are designed in response to the globalized needs and aspirations of nuclear families of the young urban populace, within the age bracket of late 20s to early 40s. Both the spouses are working in most such families, spending a significant portion of the day at their workplace. This populace prefers this typology of housing, due to the economic benefits of organized, shared system of maintenance and management, and 24 h security with easy vehicular access to amenities and services, owing to better roads. In these neighbourhoods, the children and elderly are generally given less consideration in the distribution of infrastructure with superficial interventions, like a solitary, windswept open space in the garb of a park, without any relation to the spatial, social and climatic context of the neighbourhood. The natural and social dimension required for a holistic physical and cognitive development of a child living in the neighbourhood [24] or the social needs of an elderly resident are generally pushed down the list of priorities. Though many neighbourhoods attempted to bridge this gap through the establishment of local cultural clubs, organizing socio-cultural events like art and dance competitions, celebration of festivals and blood donation camps, the social cohesion is still weak due to homogeneity in the demographics of such neighbourhoods.

24.4.1 The Way Forward

It can be hence inferred that there is a need for a paradigm shift in the design philosophy for the planning of newer neighbourhoods by taking in experiences from the older ones. The study derives that while the old neighbourhoods developed due to a force essentially 'human' in its needs and scale, the newer ones are developed keeping 'mechanical efficiency' as the crux of the design philosophy. While people dominated the urban space through a dynamic socio-cultural order in the older neighbourhoods, motorized vehicles have taken control of the urban realm in newer neighbourhoods, through direct or indirect design interventions enabling segregation of land use and reduced significance of the socio-cultural dimension. The principles of New Urbanism [25] are based on such argument that traditional people-centric design of neighbourhoods is the solution to the urban issues that planners and designers are striving to achieve.

The study proposes a people-centric design principle for new neighbourhoods of not only Kolkata, but India as a whole, based on a network of interactions between the urban space, inhabitants and activities. The suggested design principle recognizes five crucial dimensions of a neighbourhood, whose interaction in the urban setting establishes the physio-spatial and socio-cultural identity, constituting the *genius loci* of older *paras,* for their integration in the design of new neighbourhoods and are listed below:

i. **Natural environment**—It is imperative that nature and natural elements be respected while new neighbourhoods are developed. Trees, water bodies and other geo-morphological features need to be integrated with new developments to reduce human–nature conflict and enable interaction of residents with nature.
ii. **Demographics**—Ensuring a compact design of neighbourhood with an adequate population density to efficiently utilize and sustain the amenities and services provided. The diversity of the population in terms of age, economic level, culture and profession will also ensure a dynamic social environment with rich cultural values.
iii. **Services**—The compact design of the neighbourhood is to be augmented with provision of amenities and services catering to the daily needs of the residents. This will encourage walkability in the neighbourhood and create points of interaction among neighbours. This, in turn, will ensure familiarity among residents and create a sense of community.
iv. **Connectivity**—The streets of the neighbourhood should ensure connectivity within the neighbourhood and encourage the concepts of 'shared streets' by harmonizing motorized, non-motorized and pedestrian movement in the same space with pedestrians having the right of way. This will ensure pedestrian safety through slow moving traffic and promote walking and cycling among the residents.
v. **Built environment**—The built environment of the neighbourhood should be designed to respond to the needs of the residents and not superfluous. Public spaces, including the streets, should be at a human scale, inclusive, comfortable to use and encourage interaction among residents. Active public spaces and streets make the neighbourhood safe with elements such as street furniture acting as triggers to initiate social interaction.

The synergy between these dimensions has been established organically in the older *paras* over the years, with each layer created in response to the needs of the residents. While each of the dimensions is crucial in the composition of a neighbourhood, the exact concoction for the formation of individual *paras* varies, imparting them their unique character. The study, thus, argues that there is no panacea for an ideal neighbourhood, which can be replicated as is, addressing both the aspirations of the twenty-first century populace and the dynamism of human society. Unlike modern neighbourhood designs, which are based on the cloning of a pre-determined and static module, the suggested design principle needs to be adapted according to the specific socio-cultural context of the area. The study does recognize the need for upgradation of the infrastructural facilities like broad streets and civic services in the older *paras*, but it also emphasizes the establishment of a consonance between well-equipped urban infrastructure of the new neighbourhoods and the physio-spatial and social structure of the *paras*.

24.5 Conclusion

The study locates the position of neighbourhoods in both spatial, as well the socio-cultural dimension within the urban setting. Though the 'wants' of the residents, from an urban neighbourhood, vary with the dynamic changes and trends of the national and global market, the 'needs' are essentially rooted in the security of their socio-cultural constructs, established by their location in the community. The *paras*, which have a dominating social dimension, have transformed with the changing socio-economic conditions of Kolkata, and the design of new neighbourhoods is yet to respond to the social and human needs of their inhabitants. The study concludes that a neighbourhood is created by the intimacy and reciprocating relations of inhabitants and physical elements that coexist within a defined space through sensorial interactions, involving the sights and sounds, and face-to-face encounters among the inhabitants. The study calls for urban designers and planners to adapt and incorporate elements to encourage these interactions in the design of new neighbourhoods which will celebrate the innate 'humanness' and the eclecticism of a quintessential neighbourhood.

References

1. Donner, H., Neve, G.D. (eds.): Space, place and globalisation: revisiting the urban neighbourhood in India. In: The Meaning of the Local: Politics of Place in Urban India, pp. 13–32. UCL Press (2007)
2. Schubert, D.: The neighbourhood paradigm: from garden cities to gated communities. In: Urban Planning in a Changing World, pp. 128–148. Routledge (2012)
3. Donner, H.: The politics of gender, class and community in a central Calcutta neighbourhood. In: Donner, H., Neve, G.D. (eds.) The Meaning of the Local: Politics of Place in Urban India, pp. 153–170. UCL Press, (2007)

4. Ray, A.: Calcutta, Towns and Suburbs: Part 1: Short History of Calcutta (India Census. v. 8. 1901). Bengal Secretariat Press, Calcutta (1902).
5. Sengupta, K.M.: Community and neighbourhood in a colonial city: Calcutta's para. South Asia Res. **38**(1), 40–56 (2018)
6. Fruzzetti, L., Östör, Á.: Calcutta Conversations. Orient Blackswan (2003)
7. Sen, D.: Speech genres and identity: The place of Adda in Bengali Cultural Discourse. J. Emerg. Knowl. Emerg. Markets **3**(1), 30 (2011)
8. D'Costa, A.: The Long March to Capitalism: Embourgeoisment, Internationalization and Industrial Transformation in India. Springer (2005)
9. Fernandes, L.J.U.: The politics of forgetting: class politics, state power and the restructuring of urban space in India. Urban Stud. **41**(12), 2415–2430 (2004)
10. Sengupta, U.: Housing reform in Kolkata: changes and challenges. Hous. Stud. **22**, 965–979 (2007)
11. Shaw, A.: Inner-city and outer-city neighbourhoods in Kolkata: Their changing dynamics post liberalization. Environ. Urbanization Asia **6**(2), 139–153 (2015)
12. Bose, S.: Historic buildings at Chitpur in Kolkata: Problems and prospects through urban conservation and planning. Planning Rev. **48**(1), 68–82 (2012)
13. Banerjee-Guha, S.: New urbanism, neoliberalism, and urban restructuring in Mumbai. In India's New Economic Policy, pp. 88–108. Routledge (2010)
14. Chattopadhyay, S., White, J.: The Routledge Companion to Critical Approaches to Contemporary Architecture. Routledge (2019).
15. Chakrabarty, D.: Provincializing Europe: Postcolonial Thought and Historical Difference. Princeton University Press (2008).
16. Appleyard, D., Lintell, M.: The environmental quality of city streets: the residents' viewpoint. J. Am. Inst. Planners **38**(2), 84–101 (1972)
17. Bardhan, R., Kurisu, K., Hanaki, K.: Does compact urban forms relate to good quality of life in high density cities of India? Case of Kolkata. Cities **48**, 55–65 (2015)
18. Sampson, R.J., Raudenbush, S.W., Earls, F.: Neighborhoods and violent crime: a multilevel study of collective efficacy. Science **277**(5328), 918–924 (1997)
19. Dasgupta, S., Bhattacharyya, S., Asaduzzaman, M.: The impact of tall buildings on elderly residents. Bangladesh J. Psychol. **13**, 7–15 (1992)
20. Chatterjee, M., Perception of housing environment among high rise dwellers. J. Indian Acad. Appl. Psychol. **35**(8), (2009)
21. Kurtz, E.M., Koons, B.A., Taylor, R.B.: Land use, physical deterioration, resident-based control, and calls for service on urban streetblocks. Justice Q. **15**(1), 121–149 (1998)
22. Jacobs, J.: The death and life of great American cities. Vintage (1961)
23. Peterson, R.D., Krivo, L.J., Harris, M.A.: Disadvantage and neighborhood violent crime: Do local institutions matter? J Res. Crime Delinquency **37**(1), 31–63 (2000)
24. Lin, E.-Y., et al.: Social and built-environment factors related to children's independent mobility: the importance of neighbourhood cohesion and connectedness. Health Place **46**, 107–113 (2017)
25. Talen, E.: Charter of the New Urbanism. McGraw-Hill Professional (2013)

Chapter 25
A Comparative Study of Design Methodologies to Integrate Ergonomics Requirements into Design

Krishna Chaitanya Mallampalli, Debayan Dhar, and Swati Pal

Abstract During recent decades, due to demand for customer-oriented product design, designers and manufacturers have been more attentive to define customer requirements during initial stage of design process. As such, importance of integrating ergonomics, particularly for capturing customer requirements, has become an essential activity. The literature provides evidence of numerous studies focusing on various design methodologies that are helpful for incorporating customer requirements at initial design stage. However, existence of multiple design methodologies makes it difficult for designers to choose one over the other. Further, ambiguity among designers while selecting any one of these methodologies may lead to an unnecessary increase in lead time for product design process. Therefore, in order to address this specific issue, a comparative review study across design methodologies with an objective to integrate ergonomics requirements has been conducted and reported here. The focus of the study is on integrating ergonomics requirements in design in initial design stages, specifically to include conceptual design, embodiment design, and detailed design. This paper reports a critical analysis of the applicability of these methodologies. The outcome of the study would be beneficial for designers (with an intention to contribute toward ergonomic led innovation) to choose design methodologies judiciously and importance of customer requirements capturing at the initial stage of the design process.

25.1 Introduction

The design process is an integration of several multi-disciplinary factors such as human, technical, organizational, social, and economical. The challenging activity in this process is to connect all these elements in the best possible way in order

K. C. Mallampalli · D. Dhar (✉)
Indian Institute of Technology Guwahati, Guwahati, Assam, India
e-mail: debayan@iitg.ac.in

S. Pal
Indian Institute of Technology Bombay, Powai, Maharashtra, India

A. Chakrabarti et al. (eds.), *Design for Tomorrow—Volume 1*, Smart Innovation, Systems and Technologies 221, https://doi.org/10.1007/978-981-16-0041-8_25

to achieve outcomes that justify quality, performance, and cost. In this context, user needs/requirements play a significant role in the design process. Often, these user requirements are primarily categorized into two types: functional and non-functional types. Functional requirements are related to technical characteristics of a product/system [1], whereas non-functional requirements refer to how the product should work and how users fulfill their aims [2].

Ergonomics, as a scientific discipline, always plays a central role in the design process. Most importantly, it incorporates the human factor perspective into the design process and improves interaction between humans and products [3]. Moreover, the crucial perspective of ergonomics in design is to resolve specific problems of non-functional requirements. Further, benefits of ergonomics in man–machine interaction include reduction of errors, improved performance, increase flexibility, and human safety [4]. So far, many studies on ergonomics broadly focused on three different specializations, namely physical, cognitive, and organizational. Physical ergonomics mainly concern with physical tasks and related characteristics, such as physiological, anthropometric, and biomechanics [5–7]. Cognitive ergonomics focuses on the method of information processing, memory, perception, reasoning, and decision-making when human interacts with product/system [6, 8]. Lastly, organizational ergonomics aims to optimize the socio-technical system(s) of the organizations, which include structure, policy, and processes [6, 9].

In recent decades, the integration of ergonomics requirements in the initial design stages has gained much attention in the literature. Especially with regard to product design, identifying different user expectations and including these in the initial design process have improved interaction between product and user. On the contrary, the lack of ergonomics information at the design stage results in the poor design of the product [10]. For example, insufficient ergonomics requirements collected at initial design stage may cause modifications at late design stages and ultimately increase the total product cost. Conversely, adequate ergonomics requirements at the early design stage reduce the number of design iterations and limit the design solution space [11].

So far, a wide variety of topics were covered under the integration of ergonomics [12–14]. In this context, previous reviews on this topic provide useful insights and help to understand the current trend of this field. Kaulio [15] review focused on customer involvement in product development in different design stages (specification, conceptual, and prototyping phases). He analyzed specifically seven methods, and the analysis revealed that these methods allow customers to involve at any stage of the design process. Parsons [16] summarized the principles, methods, and models that are commonly applied in environmental ergonomics and their effects such as heat, cold, vibration, lighting, and noise on the user. Wei and Salvendy [17] presented a review on methodologies related to cognitive task analysis and task design for the job. They highlighted the pros and cons of cognitive task analysis in capturing user needs. Van Eerd et al. [18] reviewed participatory ergonomic interventions systematically in order to check the feasibility of ergonomic interventions. They suggested that for addressing the key barriers and facilitators, communication and program support is crucial. Sadeghi et al. [19] provided a systematic review of design methods, tools, and techniques for integrating human safety into the design process. They argued that most of the studies include human safety at the last stage of design process. From these studies, it is clear that most of them focused on a particular specific stage of the design process or existing ergonomics category.

Nevertheless, based on the variability and major components (users, product, environment, and interaction) involved in this field, the topic integration of ergonomics into design is complex. Moreover, a wide range of topics has already by many studies; therefore, providing an overview is difficult in this area. In this context, the present study, in particular, concentrated on identifying and summarizing current literature on design methodologies that integrate ergonomics requirements into different stages of the design process. One important objective of this article is to analyze: "Which studies have used these design methodologies?" and to provide a summary of applications of these methodologies for future researchers. For this purpose, there is a need to explain specific design stages proposed by Pahl et al. [20]. These are:

1. Conceptual design focuses at identification of essential problems, defining a functional structure and developing suitable working principles in order to establish specifications or concepts.
2. Embodiment design aims to formulate concepts and determine overall layout design with respect to technical and economic criteria.
3. Detailed design targets the final product to be completed along with technical details (including dimensions, forms, shapes, and surface properties of corresponding parts).

In this paper, current literature related to design methodologies is summarized and interpreted in order to provide a comprehensive summary of application of design methodologies to designers and researchers. The remainder of this paper is organized as follows: Sect. 25.2 describes the research methodology of this study. Section 25.3 presents a review of six selected design methodologies. Section 25.4 presents the discussion on research findings followed by gaps identified and suggestions for future directions. Finally, the last section provides the conclusion of this study.

25.2 Research Methodology

The literature search was performed in three electronic databases: science direct, Web of science, and Google scholar. Additionally, conference papers, books, and some relevant articles from reference lists of selected articles were are also included.

As mentioned in the introduction section, the aim of this article is to summarize and provide an overview of design methodologies focusing on the integration of ergonomics rather than reviewing all the published literature related to design methodologies. To define the scope of this study, consideration of articles was based on relevance to implementation of methodologies in ergonomics design. An example of a publication within the scope of this literature review is that Marsot and Claudon [14] in which authors proposed a multi-disciplinary model by combining different design methodologies to integrate ergonomics perspective into design process. This strict distinction is necessary to include relevant articles in this study. Therefore, firstly, the design methodologies [4, 21] were studied and selected six methodologies: quality function deployment, axiomatic design, user-centered design, TRIZ, computer-aided design, and function analysis.

Initially, the resulted articles from the literature search were first screened by reading the title of the article, abstract, and keywords. The number of articles obtained after inclusion/exclusion criteria (articles published in English language, articles published during 1999–2019, relevance to study, and exclusion of duplicates, etc.) was 1257 (combing three databases). After the screening, 225 articles were included, and finally, 26 articles were selected for review based on the above-mentioned criteria. Although the maximum possible efforts were made, however, there is no claim from the authors that this search is completely exhaustive.

25.3 Literature Review of Six Design Methodologies

25.3.1 TRIZ

Teorija Rezhenija Izobretatelskih Zadach (TRIZ), a knowledge-based systematic methodology, was developed by Altshuller [22]. In English, it is called as the theory of inventive problem-solving technique [23], which assists in finding several ways of solving a design problem by a systematic approach. TRIZ is a knowledge-based method for eliciting new ideas, and it was formulated after studying over 40,000 technology patents. However, TRIZ does not provide direct solutions to a problem, but it helps to determine several ways in which a new product/system can be improved or innovated [24]. TRIZ employs different tools and methods, to resolve any design problem based on contradiction matrix, inventive principles, scientific effects, and standard solutions.

Marsot and Claudon [14] had applied TRIZ methodology along with two other methodologies (function analysis and quality function deployment) to integrate ergonomics into product design. In this study, TRIZ was used in the conceptual design stage to solve contradictions between ergonomic criteria and functional characteristics. Houssin and Coulibaly [13] aimed at integrating safety into design of a product during the early design stage in order to improve both user performance and productivity. They used TRIZ tool to resolve the contradictions among safety parameters and productivity. Zhang et al. [25] have proposed an integrated model focused at innovating and evaluating product design in the early design stage. Their approach followed four steps: (1) identifying customer satisfaction needs, (2) correlating critical ergonomic parameters using QFD, (3) resolving problems of contradictions and generating innovative alternatives using the TRIZ tool, and (4) evaluating best designs using fuzzy decision-making method. Most importantly, they demonstrated the applicability of TRIZ combining with other methods.

25.3.2 Quality Function Deployment

QFD is a popular method which focuses on introducing quality in design to achieve user satisfaction. Quality in this method is ensured by transforming user requirements into design specifications. Usually, the gathered user requirements are qualitative in nature because these are mostly collected from customer interviews and market surveys. QFD process converts this qualitative data into engineering characteristics for the product to be designed [26]. Later, designers determine the association between user requirements and engineering characteristics using the correlation table. This indicates existing relationships and levels of interaction between them. Generally, this entire QFD process is represented in four different forms, namely house of quality, part characteristics, process planning matrix, and quality control matrix. Among these, house of quality (HoQ) is most commonly used form.

QFD is the most commonly used method for analyzing qualitative user requirements. Marsot and Claudon [14] used QFD method in two steps for correlating ergonomics requirements with design parameters. Firstly, they established user requirements list (Whats list) and secondly, design parameters list (Hows list). Thereby, QFD method helped to correlate these two lists to identify the most important relationships. Marsot [27] extracted feedback from user experience, and by applying QFD, he demonstrated a correlation between ergonomic and technical requirements of product design. Bas [28] proposed an integrated three-step QFD approach for preventing occupational hazards in the workplace. In his study, he determined the correlation between tasks and hazards (using HoQ 1), hazards and events (using HoQ 2), and events and preventive measures (using HoQ 3).

25.3.3 Computer-Aided Design

In the beginning, computer-aided design (CAD) systems were commonly used for modeling, modifying, analyzing, or optimizing the designs. Nowadays, rapid developments in computer systems have expanded the application areas of CAD systems. For example, some of those include virtual reality (VR), digital human models (DHMs), and simulations, respectively. These CAD tools enabled with ergonomics data can solve complex ergonomics problems [29]. Some researchers mainly focused on the applicability of CAD in ergonomics analysis in workstation design. For example, Marcos et al. [30] evaluated the redesigned laparoscopic operating room using DHM simulation. Tian and Duffy [31] proposed a virtual interactive design methodology, which allows dynamic ergonomics analysis of specific body parts in a virtual interactive platform. In 2009, Shahrokhi and Bernard [32] developed a multi-agent/multi-scale human modeling approach, which makes an easy way to model and analyze human performance in 3D virtual environment. A performance analyzer evaluates the requirements for analyzing human performance during the design process. Coulibaly et al. [33] proposed an approach for assessing the risk of products associated with product utilization. In this, a CAD model with pre-installed behavioral semantic data critically analyzes the product and predicts risk at the initial design stage. Similarly, Magistris et al. [34] developed a dynamic DHM model that can mechanically manipulate force and acceleration. These models examine multiple design situations without the need of physical prototypes.

25.3.4 Axiomatic Design

In 1990, Suh introduced the AD theory, which is a theoretical foundation for improving various design activities while designing products [1, 35]. AD theory is supported by independence and information axioms; these axioms will assist in avoiding mistakes while designing a product [1]. According to Suh [1], there are four major domains of AD (customer, functional, physical, and process). The function of each domain in design process is: (1) collection of customer requirements, (2) conversion of requirements into functional characteristics, (3) selection of necessary design parameters in physical form, and (4) formulating process variables.

Karwowski [36] argued that the needs of customers should meet the human–system compatibility needs simultaneously. To address this, he applied AD theory for the ergonomic design process in four steps: (1) defining ergonomic requirements in terms of user and system performance needs, (2) functional characteristics based on user capabilities, (3) physical parameters according to the interaction between human and system, and (4) process parameters based on managing compatibilities. Helander and Lin [37] emphasized the usage of independence axiom in reducing time duration and optimizing iterative design solutions based on the work situation. On the other hand, the information axiom recommends a method for computing anthropometric data. These two combined provide promising applications of AD in ergonomics. Another author, Lo and Helander [38], presented a methodology based on AD theory, which eliminates the couple between user goals and user behavior. Also, this method proposes four domains of AD, such as goals, functions, physical, and processes, into human–machine system.

25.3.5 User-Centered Design

User-centered design (UCD) approach emphasizes on user needs and gives the highest importance for user needs at each design stage [39], because the product performance is improved by fulfilling the needs, wants, or requirements of users, perhaps not to compel the user to alter their behavior. This can be achieved by UCD principles as follows: (1) focusing on user and user tasks at an early stage: first, understand the characteristics of users such as behavior, cognition, attitude, and anthropometric. In addition, identify the work that the user willing to participate. Second, the product design should comply with users' needs and assist users in accomplishing their aims. This will ensure pleasant user experience, and designers need not do any major modifications in the later stages of design. (2) Experimental measures: provide a prototype to the user for testing and identify the functional and usability issues. (3) Iterative design: some user requirements are vague and too ambiguous; hence, refinement of requirements is needed. This principle suggests that this kind of product development may require repeated steps of design, change, and test.

Courage [40] categorized stakeholders of the product into three categories (e.g., primary, secondary, and tertiary) because the collection of user requirements commonly involves diverse groups. In particular, including a wider range of stakeholders' perspectives is vital for successful design. However, it is difficult to collect all stakeholders' needs into the design due to some constraints such as time, energy, etc. Therefore, the effect of the product (i.e., user experience) must consider by the design team [41]. The user experience in the form of subjective satisfaction (effectiveness, safety, efficiency, learnability, and usability) can be recognized and measured with the help of some techniques in the early stage design [41], such as ethnography, participatory design, contextual design, and coherence. Another author, Vredenburg et al. [42] described various methods that are widely used in UCD, such as usability testing, task identification, heuristic evaluation, mockups, etc.

25.3.6 Function Analysis

Functions analysis (FA) describes the functionality of the system. FA is a systematic process for classifying and evaluating the product/system functions and their associations. The conventional approaches of FA are: (1) functional/external needs analysis—deals with the needs of users that involve constraints of product usage and (2) technical/internal needs analysis—describes the technical characteristics and properties of the product.

Jouffroy et al. [43] proposed an approach combining the ergonomics method and FA. By this approach, authors identified and analyzed hazards related to the man–machine system and emphasized on integrating safety principles into design. Marsot and Claudon [14] proposed a multi-disciplinary approach for the integration of ergonomics into design. In their paper, the author examined the role of FA in formulating product specifications at early stage of design. Wu et al. [44] proposed a function combined design methodology based on the Kano model, FA, and QFD to incorporate user requirements into design of a product. Their approach defined the functional structure of the product using FA in order to streamline the user expectations.

25.4 Discussion

This section summarizes the findings of the literature review on design methodologies for integrating ergonomics requirements into design. Based on the literature, the research works that apply design methodologies, specifically including ergonomics aspects, were categorized in respective stages of design process (i.e., conceptual design, embodiment design, and detailed design) (Table 25.1).

25.4.1 Research Findings

Overall, the findings of the literature review indicate that ergonomics integration in different stages of design has a considerable share in positive outcomes of the design process. For instance, including users' ergonomics requirements at early stages of design helps to prevent modifications in later stages and improve the satisfaction of users in terms of performance and operational aspects. Normally, ergonomics requirements mainly comprise non-functional requirements. These non-functional requirements combined with functional requirements will result in improved product performance and greater user experience. However, defining non-functional requirements is easy, but transforming all these requirements into technical parameters is

Table 25.1 Comparison of design methodologies

Design methodologies	Stages of design process		
	Conceptual design	Embodiment design	Detailed design
TRIZ	Marsot and Claudon [14]	Zhang et al. [25] Houssin and Coulibaly [13]	
QFD	Marsot [27] Marsot and Claudon [14]		Bas [28]
CAD	Magistris et al. [34]		Marcos et al. [30] Tian and Duffy [31] Coulibaly et al. [33]
AD		Helander and Lin [37] Lo and Helander [38]	Karwowski [36]
UCD	Courage [40] Preece et al. [41] Vredenburg et al. [42]		
FA	Marsot and Claudon [14] Wu et al. [44]		Jouffroy et al. [43]

not always possible in some cases. In such conditions, design methodologies are promising tools for addressing such aspects in design.

Further, these methodologies permit the designers to perform embodiment design and also include ergonomics information at the same time. This step is time-consuming because the number of iterations is performed to refine and evaluate the design proposal in order to fulfill ergonomics requirements. Moreover, ergonomics integrations often regarded as a scheme for evaluating developed solutions rather than generating solutions. During early design stages, implementing changes are easy and less costly. At prototyping and testing stage, integrating ergonomics requirements become more complex and may extend the design process. Hence, it is better to formalize all residual issues in the detailed design stage. However, few studies also claim integration of ergonomics in the detailed design stage which is not too late.

25.4.2 Research Gaps and Future Directions

There are a large number of studies already conducted on integrating ergonomics into design, which are beneficial for designing a pleasant product (system). In this paper, 26 research studies have been studied on this topic. Nevertheless, to the best of our knowledge, three research gaps have been identified and discussed below.

Based on the literature review presented here, it can be concluded that integration of ergonomics into design is complicated due to involvement of several factors (user, product, environment, etc.). Moreover, while collecting the user requirements (functional and non-functional) from diverse groups (user, manufacturer, and marketing), the definition of functional and non-functional requirements is not clear. This confusion may be cleared by avoiding information loss and overlapping of information. The literature shows that for most of the design methodologies, designers' experience is the pivotal point for design solutions. At this condition, to support designers, a new method for generating design ideas using computer-based tools is indeed required. Lastly, most of the designers use several tools and techniques focusing on designing a product on how to use but hardly consider the conditions of utilization. So, it is important to focus on design, which considers product use as well as human behavior in a particular work situation, simultaneously.

25.5 Conclusion

Over the years, the number of research articles and range of research topics on integration of ergonomics into design has increased. Therefore, it makes very difficult for a researcher to provide an overview of the field. However, this paper reviewed 26 research articles related to the integration of ergonomics and categorized into different stages of design (conceptual, embodiment, and detailed design). The analysis of the literature was performed based on the title of the article, abstract, keywords, and publication year. The findings suggest that most of the studies on integration of ergonomics in design offered solutions in late design phases, which increase complexity of design modifications and number of design iterations over the period. Another significant part of this paper is identifying gaps and recommending future directions. At this stage, it is believed that the designers' expertise and judgments help generate new design solutions and improve man–machine interaction design. Finally, the literature also suggests the need for continuous improvements in this field, and we hope to see new contributions in future.

References

1. Suh, N.: Axiomatic Design: Advances and Applications. (2001)
2. Glinz, M.: On non-functional requirements. In: 15th IEEE International Requirements Engineering Conference. (2007)
3. Rouse, W.B., Cody, W.J.: On the design of man-machine systems principles, practices and prospects. Automatica **24**(2), 227–238 (1988)
4. Salvendy, G.: Handbook of Human Factors and Ergonomics. 4th edn. (2012)
5. Marras, W.S.: Fundamentals and Assessment Tools for Occupational Ergonomics. (2006)
6. Karwowski, W.: The discipline of human factors and ergonomics. In: Handbook of Human Factors and Ergonomics. 4th edn. pp. 1–37. (2012)
7. Pheasant, S.: Bodyspace: Anthropometry, Ergonomics and the Design of the Work, 2nd edn. p. 244. Health (San Francisco) (1996)
8. Hollnagel, E.: Handbook of Cognitive Task Design. (2003)
9. Nemeth, C.P.: Human Factors Methods for Design. (2004)
10. Redström, J.: Towards user design? On the shift from object to user as the subject of design. Des Stud. **27**(2), 123–139 (2006)
11. Eppinger, S.D., Whitney, D.E., Smith, R.P., Gebala, D.A.: A model-based method for organizing tasks in product development. Res. Eng. Des. **6**(1), 1–13 (1994)
12. Hasan, R., Bernard, A., Ciccotelli, J., Martin, P.: Integrating safety into the design process: elements and concepts relative to the working situation. Saf. Sci. **41**(2–3), 155–179 (2003)
13. Houssin, R., Coulibaly, A.: An approach to solve contradiction problems for the safety integration in innovative design process. Comput. Ind. **62**(4), 398–406 (2011)
14. Marsot, J., Claudon, L.: Design and ergonomics. Methods for integrating ergonomics at hand tool design stage. Int. J. Occup. Saf. Ergon. **10**(1), 13–23 (2004)
15. Kaulio, M.A.: Customer, consumer and user involvement in product development: A framework and a review of selected methods. Total Qual. Manage. **9**(1), 141–149 (1998)
16. Parsons, K.C.: Environmental ergonomics: A review of principles, methods and models. Appl. Ergon. **31**(6), 581–594 (2000)
17. Wei, J., Salvendy, G.: The cognitive task analysis methods for job and task design: review and reappraisal. Behav. Inf. Technol. **23**, 273–299 (2004)
18. Van Eerd, D., Cole, D., Irvin, E., Mahood, Q., Keown, K., Theberge, N., Village, J., St Vincent, M., Cullen, K.: Process and implementation of participatory ergonomic interventions: a systematic review. Ergonomics **53**(10), 1153–1166 (2010)
19. Sadeghi, L., Dantan, J.Y., Siadat, A., Marsot, J.: Design for human safety in manufacturing systems: applications of design theories, methodologies, tools and techniques. J. Eng. Des. **27**, 844–877 (2016)
20. Pahl, G.W., Beitz, J.: Feldhusen and KHG. Engineering Design: A Systematic Approach. (2007)
21. Tomiyama, T., Gu, P., Jin, Y., Lutters, D., Kind, C., Kimura, F.: Design methodologies: industrial and educational applications. CIRP Ann. Manuf. Technol. **58**(2), 543–565 (2009)

22. Altshuller: Creativity as an Exact Science. (1984).
23. Ilevbare, I.M., Probert, D., Phaal, R.: A review of TRIZ, and its benefits and challenges in practice. Technovation **33**, 30–37 (2003)
24. Savransky, S.D.: Engineering of Creativity: Introduction to TRIZ Methodology of Inventive Problem Solving, pp. 1–394. (2000)
25. Zhang, F., Yang, M., Liu, W.: Using integrated quality function deployment and theory of innovation problem solving approach for ergonomic product design. Comput. Ind. Eng. **76**(1), 60–74 (2014)
26. Franceschini, F.: Advanced Quality Function Deployment. (2001).
27. Marsot, J.: QFD: a methodological tool for integration of ergonomics at the design stage. Appl. Ergon. 185–192 (2005)
28. Bas, E.: An integrated quality function deployment and capital budgeting methodology for occupational safety and health as a systems thinking approach: the case of the construction industry. Accid. Anal. Prev. **68**, 42–56 (2014)
29. Feyen, R., Liu, Y., Chaffin, D., Jimmerson, G., Joseph, B.: Computer-aided ergonomics: A case study of incorporating ergonomics analyses into workplace design. Appl. Ergon. **31**(3), 291–300 (2000)
30. Marcos, P., Seitz, T., Bubb, H., Wichert, A., Feussner, H.: Computer simulation for ergonomic improvements in laparoscopic surgery. Appl. Ergon. **37**(3), 251–258 (2006)
31. Tian, R., Duffy, V.G.: Computerized task risk assessment using digital human modeling based Job Risk Classification Model. Comput. Ind. Eng. **61**(4), 1044–1052 (2011)
32. Shahrokhi, M., Bernard, A.: A framework to develop an analysis agent for evaluating human performance in manufacturing systems. CIRP J. Manuf. Sci. Technol. **2**(1), 55–60 (2009)
33. Coulibaly, A., Houssin, R., Mutel, B.: Maintainability and safety indicators at design stage for mechanical products. Comput. Ind. **59**(5), 438–449 (2008)
34. De Magistris, G., Micaelli, A., Savin, J., Gaudez, C., Marsot, J.: Dynamic digital human models for ergonomic analysis based on humanoid robotics techniques. Int. J. Dig. Hum. **1**(1), 81 (2005)
35. Suh, N.P.: Axiomatic Design Theory for Systems. Res. Eng. Des.—Theory, Appl. Concurr. Eng. **10**(4), 189–209 (1998)
36. Karwowski, W.: Ergonomics and human factors: The paradigms for science, engineering, design, technology and management of human-compatible systems. Ergonomics **48**(5), 436–463 (2005)
37. Helander, M.G., Lin, L.: Axiomatc design in ergonomics and an extension of the information axiom. J. Eng. Des. **13**(4), 321–339 (2002)
38. Lo, S., Helander, M.G.: Use of axiomatic design principles for analysing the complexity of human–machine systems. Theor. Issues Ergon. Sci. **8**(2), 147–169 (2007)
39. Kraft, C.: User experience innovation: User centered design that works, vol. 9781430241, pp. 1–207. (2012)
40. Courage, C., Baxter, K.: Understanding your users: a practical guide to user requirements: methods, tools, and techniques. Des. Issues **223**(1), 92–92 (2005)
41. Preece, J., Rogers, Y., Sharp, H.: User-centered approaches to interaction design. BT—Interaction design: beyond human-computer interaction. In: Interaction Design: Beyond Human-Computer Interaction, pp. 279–315. Wiley (2002)
42. Vredenburg, K., Mao, J.Y., Smith, P.W., Carey, T.: A survey of user-centered design practice. In: Conference on Human Factors in Computing Systems—Proceedings, pp. 471–478. New York (2002)
43. Jouffroy, D., Demor, S., Ciccotelli, J., Martin, P.: An approach to integrate safety at the design stage of numerically controlled woodworking machines. In: Integrated Design and Manufacturing in Mechanical Engineering '98, pp. 643–650. Springer Netherlands (1999)
44. Wu, X., Hong, Z., Li, Y., Zhou, F., Niu, Y., Xue, C.: A function combined baby stroller design method developed by fusing Kano, QFD and FAST methodologies. Int. J. Ind. Ergon. **75**, 102867 (2020)

Chapter 26
The Relationship Between the Pose of Virtual Character and Virtual Character's Personality

The Animation Software Preset Posture

Chun Yang Su and Chun Heng Ho

Abstract In the psychologic area, the researchers discovered that the realisticness of a character can be improved by the posture, facial expression, and voice. Since the posture of a virtual character is a key element for animation when a dramatic story is being produced, this research focuses on the effect of different virtual character postures. This research uses the big five model, the BFI-44 questionnaire, for participants to evaluate the personality of five virtual character postures. These five postures were built based on common preset postures from major 3D model software. The experimental results evidenced that a participant's view of the personality of the virtual character is affected by the whole body posture. There were significant differences among postures in terms of openness, extroversion, and neuroticism traits, but these were not the case for postures suggesting conscientiousness and agreeableness traits. Thus, the virtual character design could be improved by crafting the personality of the character through different postures as suggested in this paper.

26.1 Introduction

In order to make virtual characters looked like a real person, character designers would imitate actual human's behavior or looks for the designated situations. For the past two decades, neuroscience researchers have concerned with actual human facial expressions [1]. In recent years, researchers have also focused on emotional body language [2]. Among the various types of body language, posture affects viewers greatly so that character designers would study different postures to improve the realisticness of character designs. The quality of character design is one of the main

C. Y. Su · C. H. Ho (✉)
Department of Industrial Design, National Cheng Kung University, Tainan, Taiwan
e-mail: hoch@mail.ncku.edu.tw

C. Y. Su
e-mail: p38041172@gs.ncku.edu.tw

A. Chakrabarti et al. (eds.), *Design for Tomorrow—Volume 1*, Smart Innovation, Systems and Technologies 221, https://doi.org/10.1007/978-981-16-0041-8_26

factors affecting the quality of digital media products. For example, many organizations offer more dramatic characters to enhance the value of the digital product. To make characters communicate with viewers, most designers attempt to make them resemble a Homo sapien as much as possible [3]. While posture research is mostly conducted in the field of neuroscience and psychology, this research is intended to study body expression from an animation designer's point of view. In other words, the body expression might be viewed differently from neuroscience and psychology researchers because of the different purposes and postural issues involved in animation.

There are existing projects developing virtual characters, some of which include animated characters [4–7]. In these projects, short-acting animations (a few seconds) such as a single action are produced with part of the virtual characters created by an animator. Researchers were inspired by these projects to investigate into the factors affecting people's emotions based on the personalities of the virtual characters from the perspective of viewers.

For virtual characters to be natural, believable, and effective, real human features, including personality, emotions, and behavior, may be simulated [4, 8–10]. Previous studies have created methods by which to discuss the emotions of virtual characters. In these studies, the researchers cooperated with animators to create 3D animation character models and used these models as the stimuli in the experiment. To make the visual characters more like Homo sapiens, the factors animators must consider include the character's body shape, posture, voice, and motion. Among these factors, posture is an important factor contributing to making a character looked more like a real person [11].

In addition, posture is a factor contributing to people's attitude toward communication [12]. When people see posture exhibiting an emotion such as fear, brain activity increases in response to what is being expressed by the body language [13]. Whole body expression of emotions can be measured using the Bodily Expressive Action Stimulus Test (BEAST). The BEAST method can be used to detect four types of participant's emotions including anger, fear, happiness, and sadness [14]. In studies on this topic, whole body posture is based on body language, and the emotions include angry, disgusted, fearful, happy, sad, surprised, and neutral (e.g., the Tilburg University image set) [1]. In these studies, facial expression is included as a part of the body posture, rather than masking out the faces. The stimulus sets follow the six basic emotions originally inspired by Ekman's research [15].

In addition to the character's emotions, recent studies on virtual characters also focused on the personality since a visual character with a greater degree of personification increases the characteristics of the persona and made the product more attractive. The big five model was used in these studies to describe the effects of a posture on the personality of a character [16–18]. Although the big five model is usually used for self-report investigations in the field of psychology, researchers have used this model as a tool to discuss the personality of virtual characters for the past fifteen years. These researchers used this model to describe specific elements of virtual characters, which were sometimes called "agents" in their research.

26.2 Method

In this study, 41 participants were included. The age of most participants ranged between 18 and 25 (61.9%), and most of the participants use multimedia products. The participants comprised 55% students, 19% from the media design industry, 7.1% who were public servants, and "others." Since they were college students or above, they were deemed capable of understanding the meaning of the adjectives in the questionnaire.

As discussed above, this study asked participants to evaluate the virtual character's emotions with the OCEAN model questionnaire. The questionnaire has a basic background part (age, education, and occupation) and seven parts with seven different virtual character's posture pictures. Each posture picture as the stimulant comes with 44 questions of the big five questionnaire (BFI-44) with five-scale Likert scale.

26.2.1 The Big Five Model

More than half century, the big five model has been used in personal psychology widely. In virtual digital character evaluate term, the researcher using this model to discuss the personality over fifteen years.

The big five model includes 60 questions and scored by 5 points Likert scale for the participant to self-appraisal [19]. It was based on neuroscience and has been applied in the field of psychology for decades. Improvements in the big five model have made the theory more reliable as a measure of personality [20]. The improved model includes five main personality traits: openness, conscientiousness, extroversion, agreeableness, and neuroticism. This model is called the OCEAN model [21, 22] and is divided into two indicators: a high indicator and a low indicator (see Table 26.1). The positive score will be the high indicator, and negative will be the low indicator in the result. In this research, the OCEAN personality indicator was used

Table 26.1 Example of the adjectives of OCEAN personality traits

	OCEAN personality indicator	
	Low indicator	High indicator
Openness	Bored, intolerant, routine-oriented, uninterested	Creative, curious, insightful, intellectual
Conscientiousness	Disorganized, easily discouraged, unpredictable, unreliable	Dependable, organized, persevering, punctual
Extroversion	Apprehensive, dull, shy, timid	Active, assertive, excitable, sociable
Agreeableness	Aloof, contrary, suspicious, unfriendly	Amiable, cooperative, flexible, trusting
Neuroticism	Anxious, depressed, insecure, susceptible to stress	Calm, resistant to stress, secure, stable

Table 26.2 Questionnaire for analysis

	OCEAN personality: relationship of question
Openness	5, 10, 15, 20, 25, 30, 35Ra, 40, 41R, 44
Conscientiousness	3, 8R, 13, 18R, 23R, 28, 33, 38, 43R
Extroversion	1, 6R, 11, 16, 21R, 26, 31R, 36
Agreeableness	2R, 7, 12R, 17, 22, 27R, 32, 37R, 42
Neuroticism	4, 9R, 14, 19, 24R, 29, 34R, 39

to evaluate the virtual model. The BFI-44 questionnaire [20] is a modified version of the big five model which narrows down the original questionnaire to 44 questions (see Table 26.2). In this research, the BFI-44 edition questionnaire was chosen, and it has been translated into Chinese for the test.

26.2.2 Stimuli

There are seven different standing postures in this study (see Fig. 26.1) including 1 nature standing posture as the control group (the first picture on the left in Fig. 26.1). The models had no texture and facial expression to decrease the factors of the model [24, 25]. The seven standing postures were the standard postures to be setup by MakeHuman 1.1.1 [26] which were already rigging in the default function. The seven stimulates came with the polygon model with transparency and skeleton system be put on the black grid without the shadow [27]. The posture was captured and put into the questionnaire as still image in the JPEG file format.

In Fig. 26.1, the first posture is a nature posture which is a posture symmetrical to each part of the body. The second posture is a relax standing posture. The posture can commonly been seen in different digital media products to show up the virtual character.

- The main issue needed to discuss was comparing more dramatic postures with the nature standing posture (posture 1) and the relax standing posture (posture 2).

Fig. 26.1 Seven standing postures as stimuli

- The postures came with larger arm and hand angle; unsymmetrical posture and stretch legs (posture 3, 4, 5, and 7) were compared with other postures.
- The running posture (posture 6) was compared with other postures.

The seven postures were all default postures from software and could be applied by designers during the design process. At first, three different views (front, side, and top view) of the models were tested in a focus group discussion. The participants of the focus group chose the front view as the best view to reveal the characteristic of the models. This result suggested the way to display the front view of the stimuli in the questionnaire.

26.3 Results

26.3.1 Questionnaire Results

The results of the BFI-44 questionnaire are shown in Fig. 26.2 that the average scores of the big five traits for each posture was plotted according within the following charts.

It can be noted from Fig. 26.2 that:

- The openness indicator in posture 2, 3, and 4 was higher than the control group (posture 1) which means that higher personality indicators like creative, curious, insightful, and intellectual could be detected by the participants in those postures.
- The extroversion indicator in all the experiment groups (posture 2–7) was higher than the control group (posture 1). High extroversion indicator score means that the personality indicators like active, assertive, excitable, and sociable could be detected by the participants.
- Comparing with the posture 1, the neuroticism indicator was lower in posture 3, 4, and 5 but higher in posture 2, 6, and 7 which means that the indicator like calm, resistant to stress, secure, and stable in posture 3, 4, and 5 could be detected by the participants.

In this section, more dramatic postures in these models like posture 2, 3, 4, 5, and 7 are showing higher score in openness and extroversion indicators. The posture 3, 4, and 5 are also showing lower score in neuroticism. However, there are no obvious differences among conscientiousness and agreeableness indicators. To understand the meaningful differences between each indicator in these posture scores, one-way ANOVA analysis was applied.

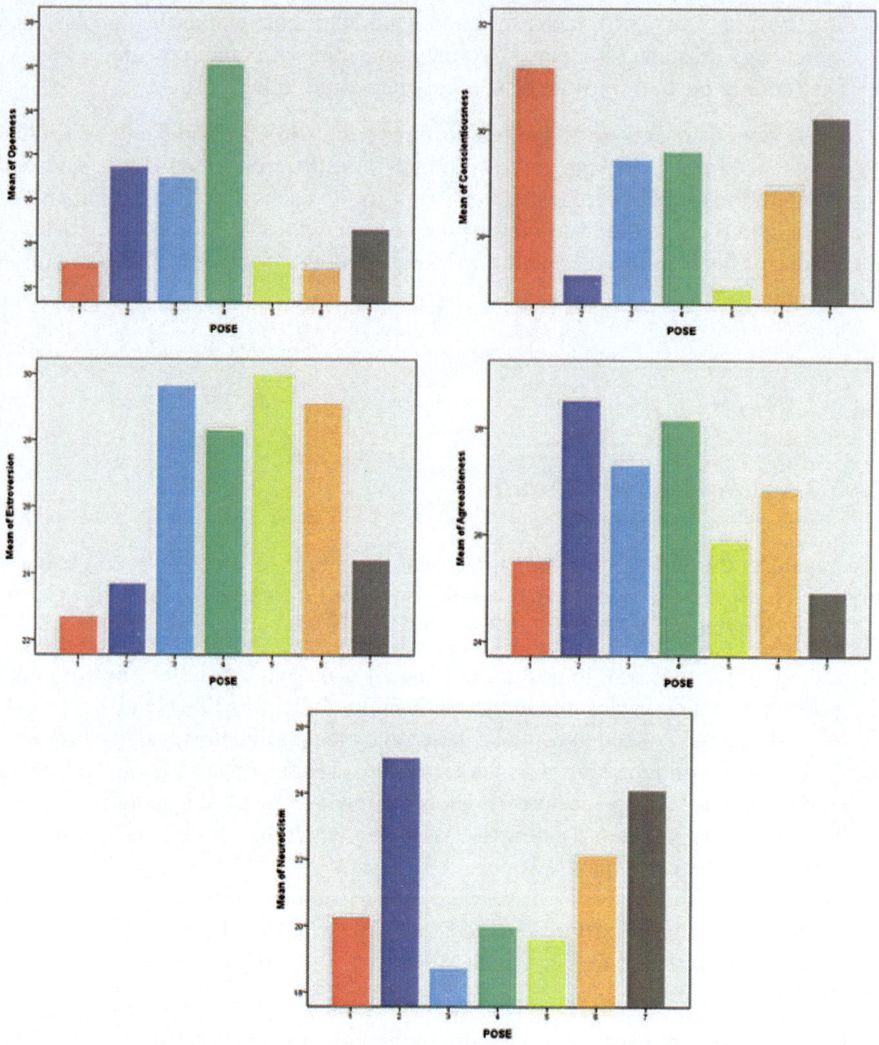

Fig. 26.2 Average scores of the big five traits for the seven standing postures

26.3.2 ANOVA Analysis

The collected data was analyzed using IBM SPSS 20.0. Through a one-way ANOVA analysis, the seven different postures could be compared with the BFI-44 questionnaire, where posture was the only factor. In this research, Tukey tests were used for the post-hoc analyses because of differences in the number of the participants.

The analysis of variance showed a main effect of openness on preferences for each posture at $p < 0.005$ level, F (6, 179) = 5.34, and $p < 0.001$. Post-hoc analyses using the Tukey test indicated that the mean scores for posture 1 ($M = 9.05$, SD = 1.9), 5 ($M = 8.91$, SD = 2.16), 6 ($M = 9.28$, SD = 2.16), and 7 ($M = 7.46$, SD = 2.16) were significantly different than for posture 4. However, there were no significant between group differences in the conscientiousness trait.

Also, the analysis of variance showed a main effect of the extroversion trait on preferences for each posture at $p < 0.005$ level, F (6, 179) = 8.98, and $p < 0.001$. Posture 1 was significantly different from posture 3 ($M = - 6.97$, SD = 1.41), 4 ($M = -5.63$, SD = 1.47), 5 ($M = -7.3$, SD = 1.45), and 6 ($M = -6.43$, SD = 1.45). Posture 2 was significantly different from posture 3 ($M = -5.96$, SD = 1.54), 5 ($M = -6.29$, SD = 1.58), and posture 6 ($M = -5.42$, SD = 1.58) in terms of the extroversion trait.

The analysis of variance showed that the main effect of neuroticism on preferences for each posture was significantly different: F (6, 179) = 6.1, $p < 0.001$. In the post-hoc results, the mean for the posture 2 score was significantly different from posture 1 ($M = 4.82$, SD = 1.21), 3 ($M = 6.33$, SD = 1.37), 4 ($M = 5.08$, SD = 1.42), and 5 ($M = 5.45$, SD = 1.4). However, posture 4 was not significantly different from posture 5.

26.4 Discussion

According to the ANOVA result, participants detected different emotions from different postures. The result show significant differences in openness, extroversion, and neuroticism between each posture. That might be because these indicators were the outer elements that for people to understand the emotion of the model. Even though the big five model questionnaire was originally designed to be self-evaluation, the result of this study suggested that these three indicators can still be detected by participants to evaluate the emotion of a given posture through its external features.

However, the conscientiousness and agreeableness did not have significant differences between each posture. The reason might be because these questions were more likely to evaluate the inner elements of personality which are difficult for people to judge from a still posture only.

Understanding the personality traits is meaningful for design the characteristic and attractive virtual model. Therefore, this research could help us to setup some standard postures to create virtual characters and to enhance the dramatic elements for digital products. This experiment shows the insight into how character postures affecting people's perception of personality. In this research, the stimuli were still images of seven postures.

The seven postures set was discussed in the focus group that posture 2, 3, 4, and 5 are the standing postures with more confident because the body stretches dramatically with higher angle on the upper body (include upper arm, trunk, and head position). This should make our participants feel the emotion right away and reflected in the result. The results show that the score of posture 2, 3, 4, and 5 are higher in openness and extroversion indicators (except for posture 2 got a low score in extroversion indicator). The same situation happened in lower body that posture 3, 4, and 5 with the higher body angle stretch had higher openness and extroversion indicators score. However, the lower body in posture 2 was set to be the pigeon-toed posture as a relax standing posture. As a result, the relax standing posture (posture 2) has a similar score in extroversion indicator compared with the nature standing posture (posture 1).

In this research, the stimuli were full-body model without animation, and the results of the virtual character postures can be applied for digital product design. However, there were studies using only a part of the body posture [16] or using animation [17] in the experiments. Moreover, there are many potentially important factors, like body size and shape, facial expression, or color, which should be discussed in the digital virtual characters design. These factors could be included in future studies to generate more knowledge about how to make a virtual character having a vivid personality.

References

1. Schindler, K., Van Gool, L., de Gelder, B.: Recognizing emotions expressed by body pose: a biologically inspired neural model. Neural Netw. **21**(9), 1238–1246 (2008)
2. Peelen, M.V., Downing, P.E.: The neural basis of visual body perception. Nat. Rev. Neurosci. **8**(8), 636–648 (2007)
3. Courty, N., Gibet, S.: Why is the creation of a virtual signer challenging computer animation? In: Lecture Notes in Computer Science (including Subseries Lecture Notes in Artificial Intelligence and Lecture Notes in Bioinformatics, vol. 6459, pp. 290–300 (2010)
4. Hartmann, B., Mancini, M., Pelachaud, C.: Formational parameters and adaptive prototype instantiation for MPEG-4 compliant gesture synthesis. In: Proceedings of the Computer Animation, pp. 111 (2002)
5. Hartmann, B., Mancini, M., Pelachaud, C.: Implementing expressive gesture synthesis for embodied conversational agents. In: Proceedings of the 6th International Conference on Gesture in Human-Computer Interaction and Simulation, pp. 188–199 (2006)
6. Kopp, S., Wachsmuth, I.: Synthesizing multimodal utterances for conversational agents: research articles. Comput. Animat. Virtual Worlds **15**(1), 39–52 (2004)
7. Thiebaux, M., Marsella, S., Marshall, A.N., Kallmann, M.: SmartBody: behavior realization for embodied conversational agents. In: Proceedings of the 7th International Joint Conference on Autonomous Agents and Multiagent Systems, vol. 1, pp. 151–158 (2008)
8. Mcquiggan, S.W., Mott, B.W., Lester, J.C.: Modeling self-efficacy in intelligent tutoring systems: an inductive approach. User Model. User-Adapt. Interact. **18**(1–2), 81–123 (2008)
9. Piwek, P.: A flexible pragmatics-driven language generator for animated agents. In: Proceedings of the Tenth Conference on European Chapter of the Association for Computational Linguistics, vol. 2, pp. 151–154 (2003)
10. Wang, N., Johnson, W.L., Mayer, R.E., Rizzo, P., Shaw, E., Collins, H.: The politeness effect: pedagogical agents and learning gains. In: Proceedings of the 2005 Conference on Artificial Intelligence in Education: Supporting Learning Through Intelligent and Socially Informed Technology, pp. 686–693 (2005)
11. Ho, E.S.L., Komura, T., Tai, C.-L.: Spatial relationship preserving character motion adaptation. In: ACM SIGGRAPH 2010 Paper—SIGGRAPH '10, p. 1 (2010)
12. Mehrabian, A.: Significance of posture and position in the communication of attitude and status relationships. Psychol. Bull. **71**(5), 359–372 (1969)

13. Gelder, B.: Towards the neurobiology of emotional body language. Nat. Rev. Neurosci. **7**(3), 242–249 (2006)
14. de Gelder, B., Van den Stock, J.: The bodily expressive action stimulus test (BEAST). Construction and validation of a stimulus basis for measuring perception of whole body expression of emotions. Front. Psychol. **2**, 1–6 (2011)
15. Ekman, P., Keltner, D.: Universal-Facial-Expressions-of-Emotions1-2015_03_12-21_10_38-UTC.pdf, pp. 151–158 (1970)
16. Wang, Y., Tree, J.E.F., Walker, M., Neff, M.: Assessing the impact of hand motion on virtual character personality. ACM Trans. Appl. Percept. **13**(2), 1–23 (2016)
17. Neff, M., Toothman, N., Bowmani, R., Fox, J.E.: Don't scratch! Self-adaptors reflect emotional stability, vol. 2792, no. May 2014 (2003)
18. Smith, H.J., Neff, M.: Understanding the impact of animated gesture performance on personality perceptions. ACM Trans. Graph. **36**(4), (2017)
19. Goldberg, L.R.: An alternative description of personality—the Big-5 factor structure. J. Pers. Soc. Psychol. **59**(6), 1216–1229 (1990)
20. John, O.P., Naumann, L.P., Soto, C.P.: Paradigm shift to the integrative Big Five trait taxonomy: history, measurement, and conceptual issues. In: Handbook of Personality: Theory and Research, 3rd edn., pp. 114–158. Guilford Press, New York, NY, US (2008)
21. Costa, P., McCrae, R.R.: A five-factor theory of personality. Five-Factor Model Pers. Theor. Perspect. **2**, 51–87 (1999)
22. Galea, S., Lindell, A.K.: Do the big five personality traits predict individual differences in the left cheek bias for emotion perception? Laterality **21**(3), 200–214 (2016)
23. Mairesse, F., Walker, M.A.: Towards personality-based user adaptation: psychologically informed stylistic language generation. User Model. User-Adapt. Interact. **20**(3), 227–278 (2010)
24. Hsu, F.-C., Hsiang, T.-W.: Factors affecting color discrepancies of animated film characters. J. Interdiscip. Math. **21**(2), 279–286 (2018)
25. Zibrek, K., McDonnell, R.: Does render style affect perception of personality in virtual humans? In: Proceedings of the ACM Symposium on Application Perception—SAP '14, pp. 111–115, August 2014
26. MakeHuman Community: MakeHuman 1.1.1 (2018)
27. Bastioni, M., Re, S., Misra, S.: Ideas and methods for modeling 3D human figures. The principal algorithms used by MakeHuman and their implementation in a new approach to parametric modeling. In: 1st Bangalore Annual Computational Conference, pp. 1–6 (2008)

Part II
Design of/for User Interfaces, User Experience and Human Computer Interaction

Chapter 27
Eye Tracking to Understand Impact of Aging on Mobile Phone Applications

Antony William Joseph, D. V. Jeevitha Shree, Kamal Preet Singh Saluja, Abhishek Mukhopadhyay, Ramaswami Murugesh, and Pradipta Biswas

Abstract Usage of smartphones and tablets has been increasing rapidly with multi-touch interaction and powerful configurations. Performing tasks on mobile phones become more complex as people age, thereby increasing their cognitive workload. In this context, we conducted an eye tracking study with 50 participants between the age of 20 to 60 years and above, living in Bangalore, India. This paper focuses on visual nature of interaction with mobile user interfaces. The study aims to investigate how aging affects user experience on mobile phones while performing complex tasks and estimate cognitive workload using eye tracking metrics. The study consisted of five tasks that were performed on an android mobile phone under naturalistic scenarios using eye tracking glasses. We recorded ocular parameters like fixation rate, saccadic rate, average fixation duration, maximum fixation duration, and standard deviation of pupil dilation for left and right eyes, respectively, for each participant. Results from our study show that aging has a bigger effect on performance of using mobile phones irrespective of any complex task given to them. We noted that participants aged between 50 and 60+ years had difficulties in completing tasks and showed increased cognitive workload. They took longer fixation duration to complete tasks

A. W. Joseph (✉)
IT-Integrated Design, National Institute of Design, Bengaluru, Karnataka, India
e-mail: antony.william@gmail.com

D. V. Jeevitha Shree · K. P. S. Saluja · A. Mukhopadhyay · P. Biswas
I3D Lab, CPDM, Indian Institute of Science, Bengaluru, Karnataka, India
e-mail: jeevithas@iisc.ac.in

K. P. S. Saluja
e-mail: kamalpreets@iisc.ac.in

A. Mukhopadhyay
e-mail: abhishekmukh@iisc.ac.in

P. Biswas
e-mail: pradipta@iisc.ac.in

R. Murugesh
Department of Computer Application, Madurai Kamaraj University, Madurai, India
e-mail: mrswami123@gmail.com

A. Chakrabarti et al. (eds.), *Design for Tomorrow—Volume 1*, Smart Innovation, Systems and Technologies 221, https://doi.org/10.1007/978-981-16-0041-8_27

which involved copy-paste operations. Further, we identified design implications and provided design recommendations for designers and manufacturers.

27.1 Introduction

Population aging is an important emerging demographic phenomenon in the world today. Due to advancement in public health and medicine, life expectancy has improved to a great extent. The 2001 census of India has shown that elderly population in India accounted for 77 million, and number will rise to nearly 140 million by 2021 [1]. As society is progressively moving toward digital technology, aging population do not want to exclude themselves. We may notice that social support system is going through a transformation, and mobile phones have become an enabling factor to remain connected with family and friends. Introduction of smart technology has impacted aging population [2] in terms of accessibility and utilization. In context of aging population and technology adaptation, it is essential to understand aging user behavior on mobile phones.

Visual sense is an important factor to explore the world around us by moving the eyes. Eye movements are vital information for understanding cognitive process. They uncover cognitive activities and indicate the target of our visual interest [3]. In recent years, eye tracking technology has become a promising tool [4] in human computer interaction (HCI).

Eye tracking has been used for more than a century in psychology for recording eye movements while reading. As personal computers were flourishing, in 1980s, researchers started applying eye tracking techniques into issues of HCI. Later, due to technological advances, eye tracking methods were employed to answer many usability issues [5]. Eye tracking is the process of measuring either point of gaze or motion of an eye relative to head with a device called eye tracker. The technique gives numerous possibilities to understand users' mental model and problems faced by them while performing tasks. Eye movement is assumed to indicate the thought in relation to cognitive process [6], thereby suggesting to provide dynamic evidence to users' attention in relation to a visual display. In our study, we used a wearable eye tracking glass to analyze aging user behavior while performing simple to complex tasks on a mobile phone. Main contributions of the paper include:

1. Design studies to understand aging user behavior based on essential requirements for fulfilling daily activities on an android mobile phone
2. Investigate how aging affects user experience on mobile phones while performing complex tasks
3. Identify how aging affects a users' cognitive load while performing tasks on a mobile phone.

27.2 Related Work

From literature, we identified few studies that have evaluated effect of age on usability of mobile devices. Rogers et al. [7] conducted a study to evaluate how task demands and users' age influenced task performance on touch and non-touch screen devices. Their study involved 40 younger (18–28 years) and 40 middle-aged to older (51–65 years) participants. They used control tasks such as scrolling, up/down buttons, list boxes, and text boxes. They noted that older adults were slower than younger adults on pointing and scrolling tasks on a touch screen. Moreover, they noted that small button sizes were particularly problematic for the older adults.

On the other hand, Sakdulyatham et al. [8] conducted a study to investigate how user interface of LINE application affects elderly when using smartphones. Their study involved 38 elderly participants (60–69 years) who performed 8 different tasks on a mobile phone. Results showed that due to decline in physical health, elderly group had difficulty in viewing content on screen leading to more errors. The study noted the user interface, with Arial Unicode MS font with size 12 to 16 pts, and screen brightness at 75% is an appropriate integration for the elderly.

Moumane et al. [9] conducted an empirical study to evaluate influence of screen size of mobile devices for stability of apps using ISO 25062 and ISO 9241 standards for objective measures. They used recordable.mobi for eye tracking. Their study identified a set of challenges when using apps such as screen size, display resolution, and capacity of memory. The study suggested that a smartphone with large screen size, facilitated ease of use.

Majrashi et al. [10] conducted an exploratory study to investigate relationship between eye tracking metrics and cross-platform usability problems, using an online shopping application. Their study involved 31 participants aged between 18 and 60 years. Results from the study noted that usability issues were associated with prior experience and knowledge.

Al-Showarah et al. [11] conducted a study to examine eye movements for young and elderly participants to find dissimilarities in browsing on different smartphone and tablet applications. Their results found that elderly participants have high dissimilarity than younger ages. In other words, elderly participants were less efficient in browsing smartphone applications than younger participants.

Al-Showarah et al. [12] conducted another study to investigate effect of age on smartphones using eye tracking technology. They evaluated usability of smartphone interfaces for three different age groups: elderly (60+ years), middle age (40–59 years), and younger age (20–39 years). They extracted eye metrics like fixation duration, scan-path duration, and saccades amplitude for their analysis. The authors noted that elderly participants were less efficient having lower cognitive ability in browsing smartphone interfaces. They found that elderly participants exhibited greater difficulties in processing information on smartphones across all screen sizes than users of middle and younger age groups. In general, they noted that there exists a possibility of positive relationship between getting older and less experience in using smartphones.

In summary, we may note that a number of studies measuring cognitive workload are often limited to a smaller group or only elderly age groups and not adequately focused across different age groups at large. Thus, we aim to investigate how aging in users of different age groups can affect their experiences on using mobile applications while they perform simple to complex tasks.

27.3 Research Methodology

This study was designed to understand aging user behavior based on essential requirements for fulfilling daily activities on a mobile phone.

Participants: Initially, 180 participants showed interest for the study, but we could select only 50 participants due to various sociocultural factors and selection criteria followed to recruit participants. Short portable mental status questionnaire (SPMSQ) was used to select and assess mental status of participants [13]. Participants with score 8 and above out of 10 were selected. In addition to considering SPMSQ, eye information was collected to identify participants who are comfortable with near vision and could read mobile screen with eye tracking glasses without any difficulty. Participants were grouped into five different age groups: Group-A (20–29 years), Group-B (30–39 years), Group-C (40–49 years), Group-D (50–59 years), and Group-E (60 year and above). Each group consisted of 10 participants (5 male and 5 female), the oldest being 77 years of age. However, other inclusion criteria considered were literacy, usage of android phone, and residing in Bengaluru City.

Material: Samsung Galaxy S7 Edge smart phone was used, with octa core (2.3 GHz, Quad core, M1 Mongoose + 1.6 GHz, Quad core, Cortex A53) processor. It runs on Samsung Exynos 8 Octa 8890 Chipset with 4 GB RAM and 32 GB internal storage. It has 5.5-inch screen and 2560 × 1440 resolution with 535 pixels per inch. For eye tracking, Tobii Pro Glasses 2 with a sampling rate of 100 Hz was used. Tobii Pro Lab software was used to extract raw data and analyze visual information.

Design: We used a non-invasive wearable eye tracking glasses. The language used to communicate with participants was English, and they were approached in their residence with prior appointment. Bengaluru city was chosen for our study because of its multi-cultural nature and home to large number of people migrated from other Indian states. Prior to designing the five tasks, we conducted a survey on technology prior experience; commonly used mobile applications; usage of mobile phones; frequency of usage of various digital technologies; familiarity and awareness of mobile control buttons in general for all 50 participants. Thus, by considering results of above-mentioned survey, we designed five tasks which are essential for fulfilling daily activities on a mobile phone. For each task, scenario and description were given to contextualize the experiment. It may be noted that participants who failed to complete given five tasks were also considered in the study.

Further, each participant was asked to perform a set of five tasks on the mobile phone. The five tasks are- Task-1: Adding a phone number to contact; Task-2: Sending birthday SMS greeting; Task-3: Google search and save information in Memo app; Task-4: Online shopping using Amazon app and Task-5: Sending WhatsApp message. The first two tasks were simple, and following three tasks were complex. This was designed to motivate participants and to understand user experience and usability aspect of mobile applications.

Procedure: Eye tracking system was calibrated for every participant before beginning tasks. An observer was present throughout the experiment. We noted that most participants were unable to complete at least one task due to complex and counterintuitive mobile haptic interface. Among participants of Group-A, 1 participant failed to complete Task-5; in Group-B, 1 participant failed to complete Task-2 and 4; 2 participants each failed to complete Task-3 and Task-4 in Group-C; in Group-D, 6, 4, and 1 participants failed to complete Task-3, 4, and 5, respectively, and among Group-E users, 8 and 6 participants failed to complete Task-3 and 4, respectively. We found that Task-3 was most difficult as many participants were unable to complete the same. Participants of Group-E were unable to complete a few tasks when compared to other user groups. This may be due to lack of precision of finger movements to perform task on a touch screen.

27.4 Results

We recorded ocular parameters like fixation count, saccadic count, average fixation duration, maximum fixation duration, and standard deviation of pupil dilation for left and right eyes for each participant. By analyzing these parameters (also called variables), we aim to identify if aging affects user experience on mobile phones while performing complex tasks and estimate their cognitive workload while performing different tasks on a mobile phone. We extracted ocular parameters like recording timestamp (in milliseconds), pupil diameter left (in millimeter), pupil diameter right (in millimeter), eye movement type (fixation, saccade, eyes not found, unknown eye movement), and eye movement type index (number) from eye tracking glasses. Number of gaze fixations and saccades were extracted using IV-T fixation [14] filtering algorithm of Tobii Studio software with velocity-based threshold set at 30°/s.

We observed that average time taken to complete five tasks increased as age of participant increased. Group-E aged over 60+ years took longest time (161.09 s) to complete five tasks when compared to participants of other age groups. Participants of all age groups took longest time to complete Task-5 (203.94 s).

We conducted ANOVA to identify the main effect and interaction effect of different ocular parameters for participants of different age groups while they performed five tasks. Assumption criteria like independence of samples, experimental errors of data are normally distributed, equality of variance were met, before we conducted ANOVA test. We considered fixation rate (FC), saccade rate (SC) maximum fixation duration (MFD), average fixation duration (AFD), standard deviation pupil dilation left (SPL), and standard deviation pupil dilation right (SPR) as the six dependent variables. Here, FC and SC are nothing but fixation count per second and saccade count per second, respectively. We found that, main effect of task on FC, SC, AFD, SPL, and SPR was not significant ($p > 0.05$). On the other hand, main effect of age group for FC, SC, AFD, MFD, SPL, and SPR was found to be significant with $p < 0.05$. This indicated that aging has a positive effect with task performance on mobile phones.

Additionally, we noted that there was no interaction effect between task and age group on FC, SC, AFD, SPL, and SPR, since $p > 0.05$. But we noted a significance ($p < 0.05$) in the main effect of both task and age group on maximum fixation duration ($F(16.4) = 5.203$, $P < 0.001$, $\eta^2 = 0.085$; $F(16.4) = 6.364$, $P < 0.00$, $\eta^2 = 0.102$) and interaction effect between task and age group ($F(16.4) = 2.291$, $P < 0.05$, $\eta^2 = 0.140$). This indicates that irrespective of age, most participants took longer time to pay attention in completing the five tasks.

We considered post hoc comparison using LSD test for all ocular parameters and noted statistically significant difference between two means for within task and within age group elements. Based on number of ocular parameters that were found to be significantly different, we assigned color-coding scheme as shown in Fig. 27.1. We found more green and brown areas within age group when compared to within tasks. This indicated that most ocular parameters were found to be significantly different for within age group elements. Ocular parameters FC, AFD, SPL, and SPR were found to be significantly different for within age group elements but not for within task elements. This indicates that age has bigger effect on performance irrespective of task the participants performed. Figure 27.1 shows color-coding scheme for indicating significance of ocular parameters considered for our study. In Fig. 27.1, each color is allocated to each of the boxes based on number of variables significant. Color green is assigned when a single variable out of six variables is significant, color orange is assigned when two out of six variables are significant, color brown when three out of six variables are significant, color pink when five out of six variables are significant,

Task

	T1	T2	T3	T4	T5
T1			MFD	MFD	MFD
T2			MFD	MFD	MFD
T3	MFD	MFD			MFD
T4	MFD	MFD			
T5	MFD	MFD	MFD		

Group

	A	B	C	D	E
A		MFD	MFD	FC, SC, AFD, MFD, SPL, SPR	FC, SC, AFD, SPL, SPR
B	MFD			FC, SC, AFD, MFD, SPL, SPR	FC, SC, AFD, SPL, SPR
C	MFD			FC, SC, MFD, SPL, SPR	SPL, SPR
D	FC, SC, AFD, MFD, SPL, SPR	FC, SC, AFD, MFD, SPL, SPR	FC, SC, AFD, MFD, SPL, SPR		AFD, MFD
E	FC, SC, AFD, SPL, SPR	FC, SC, AFD, SPL, SPR	FD, SPL, SPR	AFD, MFD	

Fig. 27.1 Ocular parameters that were found significant with LSD test for task-wise (top) and group-wise (bottom)

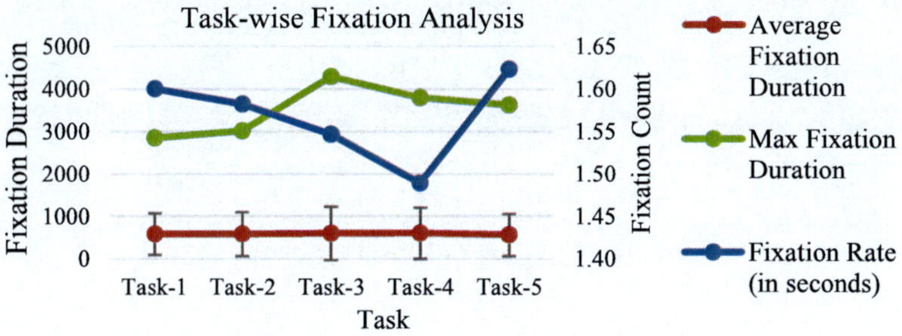

Fig. 27.2 Task-wise fixation analysis

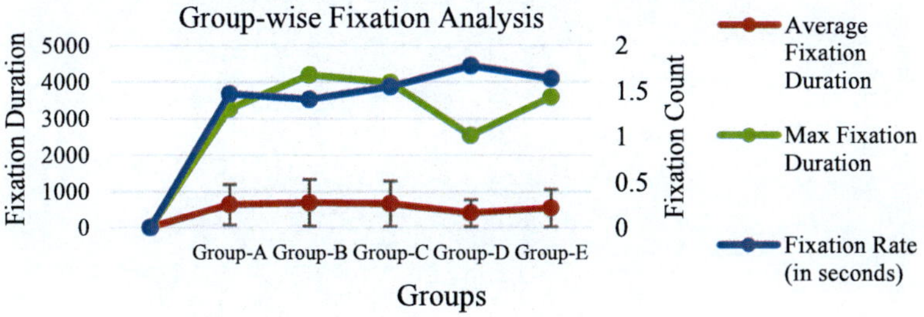

Fig. 27.3 Group-wise fixation analysis

and color blue when all six variables are significant. Color yellow is assigned when all six variables are not significant.

Further, we performed fixation analysis for each task (Fig. 27.2) and age group (Fig. 27.3), respectively. Results state that Task-5 recorded higher fixation rate (Fig. 27.2) requiring longer duration of visual search. Task-3 where participant had to Google search and save information in Memo app took longest fixation duration when compared to other tasks indicating that participants took longest time to pay attention to complete task. Participants of Group-D and Group-E had higher rate of fixation (Fig. 27.2) when compared to participants of other age groups. This means that the above two groups had difficulty in performing and completing five tasks given to them. We further noted that Group-B and Group-C had longer fixation duration values when compared to the rest.

From Fig. 27.2, we may note that average time taken to complete task is increasing linearly from Task-1 to Task-5, respectively, and participants of all age groups took longest time to complete Task-5 (203.94 s) when compared to other tasks. We noted a similar trend in age group wise analysis. Figures 27.4 and 27.5 indicate pupil dilation of tasks and age groups, respectively. We noted that standard deviation of pupil dilation was high while participants performed Task-3 and Task-4. Participants of age Group-D and Group-E showed higher values of standard pupil dilation.

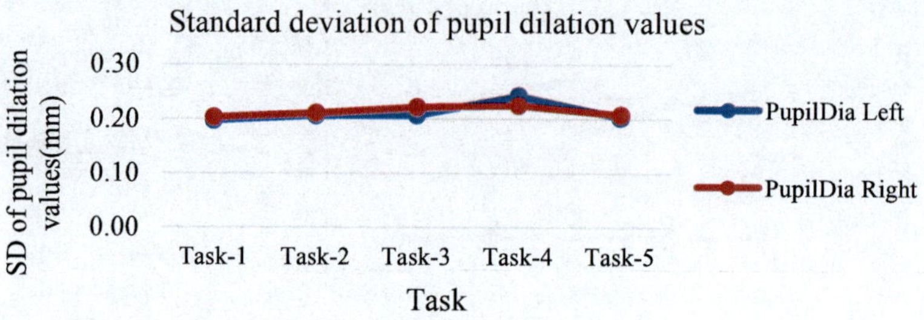

Fig. 27.4 Standard deviation of pupil dilation for five tasks

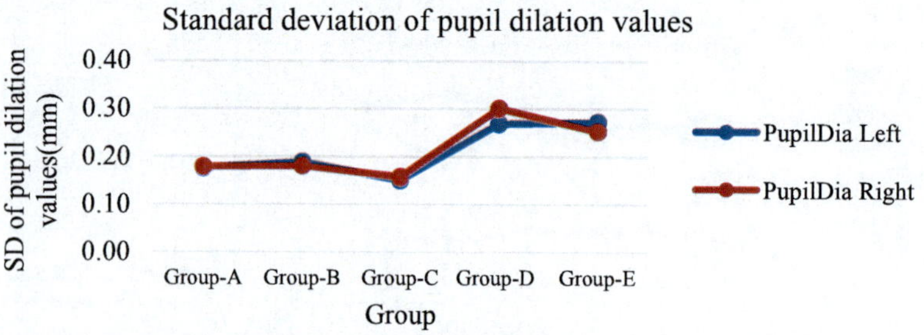

Fig. 27.5 Standard deviation of pupil dilation for five tasks

27.5 Estimating Cognitive Workload

We identified how aging affects users' cognitive workload of each participant in two ways. First, we analyzed ocular parameters like fixation frequency, fixation rate, saccade rate, fixation duration, and standard deviation of pupil dilation for left and right eye values [15]. Study from Coral [16] noted that by analyzing above ocular parameters, we can estimate cognitive workload of participants of different age groups for different tasks. Coral [16] presented a summary of eye-related measurements and their relationship to increased cognitive workload. From summary, it can be noted that increase in fixation frequency, fixation rate, saccade rate, fixation duration, and pupil dilation values indicate increase in cognitive workload. From above results, we can see that participants of Group-D and Group-E were estimated to have higher cognitive workload when compared to rest as they showed increased values in above-mentioned ocular parameters. All participants showed highest estimation of cognitive workload while performing Task-3 and Task-4.

Further, as a second way to estimate cognitive workload for each participant, we analyzed pupil dilation of both left and right eyes for each participant. We developed an algorithm based on pupil dilation using low pass filter (LPF). We divided pupil diameter signal into sections of 100 samples (time buffer of 1 s as sampling frequency of device is 100 Hz). We removed DC offset from original signal by subtracting the mean of original signal. Then, we have applied Butterworth low-pass

Fig. 27.6 Algorithm to estimate cognitive workload

filter to filter signal within 0 to 5 Hz and summed up magnitude of filtered signal in a running window of 1 s with 70% overlap between adjoining sections. The algorithm is described in Fig. 27.6.

Each participants' pupil diameter values were given as input to our algorithm. The algorithm gives a single value as output which is nothing but average LPF for both eyes which can be used to estimate cognitive workload. This procedure is performed for both left and right pupil diameter values for participants of all age groups and for all tasks. Figures 27.7 and 27.8 show average LPF for left and right eyes for group and task, respectively. This can be used to estimate average cognitive workload for participants of all age groups while performing all tasks. It is evident that participants of Group-D and Group-E were estimated to have higher cognitive workload (higher average LPF) when compared to the rest. We can see from graph (Fig. 27.8) that Task-3 was estimated to have higher cognitive workload (higher average LPF).

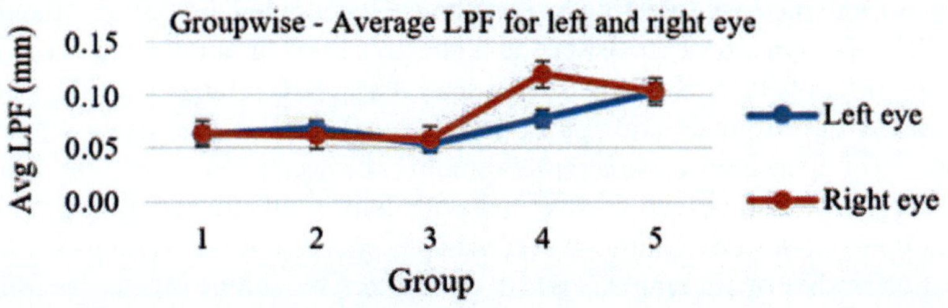

Fig. 27.7 Group-wise average LPF

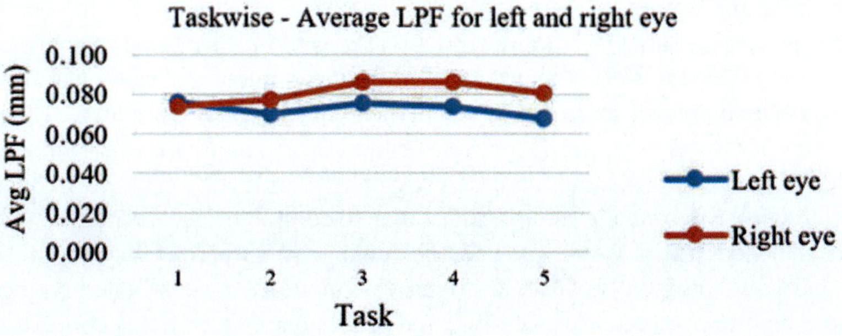

Fig. 27.8 Task-wise average LPF

27.6 Discussion

Taken together, above statistical results show that all ocular parameters considered were found to be significant across all age groups. This indicates that aging affects user experiences on mobile phones while performing complex tasks. Participants of Group-D and Group-E aged between 50–59 years and 60+ years took longer time to complete each task and showed to have higher cognitive workload while performing tasks. Task-1 and Task-2 were perceived as easy task, and Task-3 was perceived as the most difficult task of all. Additionally, our results noted that participants of all age groups took longest time to complete Task-3. It was found that 2 participants from Group-C, 6 participants from Group-D, and 8 participants from Group-E were unable to complete the same. Task-3 showed maximum fixation duration indicating that participants possessed greater attention to complete task. This was due to task involving copying a text from browser and saving the same in Memo app. Though copy-paste operation is familiar to participants, they struggled due to touchscreen interface that was not intuitive. They have problem with soft keys on full touchscreen phones and fail to map physical keys and soft keys. For example, Lee et al. [17] reported that minimal target size for tapping with good speed and accuracy appears to be 8 mm for all age groups. Another important issue discovered is that older participants could not differentiate between short press and long press on touch screen. As a result, controls were not released on time and options distracted their attention. It was noticed that a few participants from Group-C and Group-D tried to pinch and zoom page for selecting and copying text but failed due to unavailability of function. Task-4 was recorded as second highest fixation duration and was perceived as second difficult task of all. It was noticed that 2 participants from Group-C, 3 participants from Group-D, and 6 participants from Group-E were unable to complete the same. This was due to poor representation of interface elements and usability issues that did not serve natural user behavior like ‘product search,’ ‘add to cart,’ and ‘shopping cart icon.’ Many pressed ‘shopping cart icon’ on right top corner of application instead of clicking the ‘add to cart button’ for adding selected product but failed in completion of task. Therefore, we can infer that there is a positive correlation on how aging affects users’ cognitive workload while performing complex tasks on a mobile phone. However, there is no significant difference in interaction effect between task and group.

Participants were able to comprehend all tasks well but struggled to achieve their goal due to touchscreen-based interface. One major problem identified was usability issue of ineffective representation of visual elements for user interaction. There was a mismatch between user interface elements and users’ mental model which led to poor performance of tasks and increased cognitive workload. Task performance was correlated with different age groups, time taken to complete task and number of tasks completed successfully. As people age, they change in numerous ways both biologically and psychologically. Critical cognitive functions most affected by age are attention, memory, speed of processing, and problem solving. In this context, above results provide facts that aging affects users’ cognitive workload while performing tasks on smartphone. Mobile interface was perceived as complex due to usability issues in terms of finding right option to complete task and achieve goal. It is found that familiarity with a feature allows user to use interface quickly and intuitively. Unfamiliar words, symbols, and less well-known functions add complexity to task and increases cognitive load. Lack of clarity and ambiguity observed throughout Tasks 3 to 5 led participants to exhibit a sense of frustration and anxiety. Czaja et al. [18] said that effect of technology use needs to be addressed by developers to accommodate all potential users. Though touch screen interfaces are typically designed to be operated by users' hand, alternate designs appropriate for tactile and motor systems are required to be seriously investigated for elderly users.

Based on our eye tracking experiment study findings, we propose few design guidelines for designers and developers to design intuitive and adaptable interfaces for different age groups. Involving elderly people in early design process and investigating possible steps of users' path will enable designers to address interaction issues that may occur. One good approach to designing adaptable intuitive user interface is applying familiar metaphors to user interface elements. Basically, the design needs to represent users' mental model which is highly recommended. For example, in Amazon online shopping application, elderly users repeatedly pressed the 'shopping cart icon' for adding products but in reality, user has to press the 'add to cart' button which is placed just below the product description. Elderly users typically make mistakes when interacting with small targets on mobile screen. In our study, we noticed that touch area was very small and older users struggled to perform actions. Thus, interactive objects should be larger than 8 mm, facilitating better user experience, and this agrees with previous research [11]. Another important issue which needs greater attention is that the elderly users find difficult to differentiate between short press and long press on touch screen, and they often miss intended targets due to large size of their fingers. Providing intuitive interaction through natural-language interface would benefit them.

27.7 Conclusion

This study aims to investigate how aging affects user experience on mobile phones while performing simple to complex tasks and identify how aging affects a users' cognitive workload while performing complex tasks on a mobile phone. We designed an experiment and collected data from 50 participants aging from 20 to 60+ years. Participants were grouped into five different groups based on their age. Five tasks were designed, and each participant was asked to undertake all the five tasks. From results, we conclude that, as users age, their experience on using mobile phones to perform everyday activities reduces. This can lead to having higher cognitive workload as they age and when the complexity of task is increased. We observed a mismatch between user interface elements and users' mental model which led to poor performance of tasks and increased cognitive workload. As a part of future work, we can develop a prediction model which can accurately estimate cognitive workload of each participant while performing various tasks on a mobile phone and help designers improve mobile user interface.

Acknowledgements The authors are thankful to the Department of Computer Application, Madurai Kamaraj University; I3D Laboratory, CPDM, IISc Bangalore and National Institute of Design, Bengaluru Campus for their encouragement, motivation, and relentless support in carrying out our study.

References

1. Jeyalakshmi, S., Chakrabarti, S., Gupta, N.: Situation Analysis of the Elderly in India. Central Statistics Office, Ministry of Statistics and Programme Implementation, Government of India document (2011)
2. Dowiasch, S., Marx, S., Einhäuser, W., Bremmer, F.: Effects of aging on eye movements in the real world. Front. Hum. Neurosci. **9**, 46 (2015)
3. Kleinke, C.L.: Gaze and eye contact: a research review. Psychol. Bull. **100**(1), 78–100 (1986)
4. Jacob, R.J., Karn, K.S.: Eye tracking in human-computer interaction and usability research: ready to deliver the promises. In: The Mind's Eye, pp. 573–605. North-Holland (2003)
5. Benel, D.C.R., Ottens, D., Horst, R.: Use of an eye tracking system in the usability laboratory. In: Proceedings of the Human Factors Society 35th Annual Meeting, pp. 461–465. Santa Monica, Human Factors and Ergonomics Society (1991)
6. Just, M.A., Carpenter, P.A.: Behavior research methods & instrumentation **82**, 139–143 (1976)
7. Rogers, W.A., Fisk, A.D., McLaughlin, A.C., Pak, R.: Touch a screen or turn a knob: Choosing the best device for the job. Hum. Fact.: J. Hum. Fact. Ergon. Soc. **47**(2), 271–288 (2005)
8. Sakdulyatham, R., Preeyanont, S., Lipikorn, R., Watakakosol, R.: User interface on smartphone for elderly users. Int. J. Autom. Smart Technol. **7**(4), 147–155 (2017)
9. Moumane, K., Idri, A., Abran, A.: Usability evaluation of mobile applications using ISO 9241 and ISO 25062 standards. SpringerPlus **5**(1), 548 (2016)
10. Majrashi, K., Hamilton, M., Uitdenbogerd, A.L.: Correlating cross-platform usability problems with eye tracking patterns. In: Proceedings of the 30th International BCS Human Computer Interaction Conference 30, pp. 1–11 (2016)
11. Al-Showarah, S., Al-Jawad, N., Sellahewa, H.: Examining eye-tracker scan paths for elderly people using smart phones. In: York Doctoral Symbosium (YDS), the paper has been published as a York Computer Science technical report. University of York, UK (2013)
12. Al-Showarah, S., Naseer, A.J., Sellahewa, H.: Effects of user age on smartphone and tablet use, measured with an eye-tracker via fixation duration, scan-path duration, and saccades proportion. In: International Conference on Universal Access in Human-Computer Interaction, pp. 3–14 (2014)
13. Pfeiffer, E.: A short portable mental status questionnaire for the assessment of organic brain deficit in elderly patients. J. Am. Geriatr. Soc. **23**(10), 433–441 (1975)
14. Olsen, A.: The Tobii I-VT fixation filter. Tobii Technology, pp. 1–21 (2012)
15. Babu, M. D., JeevithaShree, D. V., Prabhakar, G., Saluja, K. P. S., Pashilkar, A., Biswas, P.: Estimating pilots' cognitive load from ocular parameters through simulation and in-flight studies. J. Eye Mov. Res. **12**(3) (2019)
16. Coral, M.P.: Analyzing cognitive workload through eye-related measurements: a meta-analysis (2016)
17. Lee, S., Zhai, S.: The performance of touch screen soft buttons. In: Proceedings of the 27th International Conference on Human Factors in Computing Systems (CHI'09), pp. 309–318 (2009)
18. Czaja, S.J., Charness, N., Fisk, A.D., Hertzog, C., Nair, S.N., Rogers, W.A., Sharit, J.: Factors predicting the use of technology: findings from the center for research and education on aging and technology enhancement (CREATE). Psychol. Aging **12**(2), 333–352 (2006)

Chapter 28
Analyzing Ocular Parameters for Web Browsing and Graph Visualization

Somnath Arjun, Kamal Preet Saluja, and Pradipta Biswas

Abstract This paper proposes a set of techniques to investigate eye gaze and fixation patterns while users interact with electronic user interfaces. In particular, two case studies are presented—one on analyzing eye gaze while interacting with deceptive materials in web pages and another on analyzing graphs in standard computer monitor and virtual reality displays. We analyzed spatial and temporal distributions of eye gaze fixations and sequence of eye gaze movements. We used this information to propose new design guidelines to avoid deceptive materials in web and user-friendly representation of data in 2D graphs. In 2D graph study, we identified that area graph has the lowest number of clusters for user's gaze fixations and lowest average response time. The results of 2D graph study were implemented in virtual and mixed reality environment. Along with this, it was observed that the duration while interacting with deceptive materials in web pages is independent of the number of fixations. Furthermore, web-based data visualization tool for analyzing eye tracking data from single and multiple users was developed.

28.1 Introduction

Information visualization is one of the major approaches to analyze data. A great deal of research on information visualization have been implemented in last two decades. With ever-increasing amount of available computing resources and sensing devices, ability to collect and generate a wide variety of large, complex datasets continues to grow. As a result, visualizing and analyzing those complex datasets become challenging. Although an ample amount of visualization techniques has

S. Arjun (✉) · K. P. Saluja · P. Biswas
Centre for Product Design and Manufacturing, Indian Institute of Science, Bengaluru, India
e-mail: somnatharjun@iisc.ac.in

K. P. Saluja
e-mail: kamalpreets@iisc.ac.in

P. Biswas
e-mail: pradipta@iisc.ac.in

A. Chakrabarti et al. (eds.), *Design for Tomorrow—Volume 1*, Smart Innovation, Systems and Technologies 221, https://doi.org/10.1007/978-981-16-0041-8_28

been devised for a wide variety of applications, they tend to follow one-size-fits-all paradigm with respect to users. The designs are rarely personalized to a specific user characteristic or type of dataset. Data-adaptive visualization and personalization of user interfaces are our long-term research goal. Considering our research goal, eye gaze and fixation patterns while users interact with electronic user interfaces were investigated. Eye tracker is a device which can capture eye gaze movements and viewer's gaze on a stimulus. There are different kinds of eye movements like fixations, saccades, pupil dilation and scan paths [1]. The eye movements provide unique insight into visual search task and can be analyzed to derive user's attention patterns. We presented two case studies.

28.1.1 Analyzing Graph Visualization

One of the fundamental and orthodox ways of visualizing data is with graphs and charts. Till today, static graphs and charts are used for visualizing data in substantial amount. Appropriate graph for dataset is difficult to choose as it depends on features of the data and problem statement. With data visualization, issues like huge volumes of data and limitations of algorithms can be confronted easily. Visualization techniques and tools have been developed in abundance to mitigate the above-mentioned scenarios. Despite these tools and techniques, optimizing visualization and interaction techniques still poses research challenges. Static graphs in 2D were considered for the study and results of this study was implemented in 2D and virtual reality environment which is described in Sect. 4.

28.1.2 Analyzing Deceptive Materials in Web Pages

It is in practice to design deceptive user interfaces to manipulate users by exploiting human psychology. Web pages have these deceptive materials which tricks user to do tasks that they did not intend to. There are no easy solutions or alternatives to these deceptive patterns. Nir Eyal, in his book [2], explains how a good understanding of cognitive science can add value to user understanding. Despite these possible solutions, there is still space for significant improvement. With the aim of providing solution to this issue, a study was undertaken for analyzing eye gaze and fixation patterns while user interact with web pages having deceptive materials.

The above two case studies exhibit two different aspect of helping designer through eye gaze data analysis, but still getting access to an eye gaze tracker is not always possible. For example, cheaper eye gaze tracker like Tobii Eye-X model has restrictions in terms of research purpose; hence, we propose webcam-based analysis tool in Sect. 5. The tool is independent of the implementation of webcam-based eye gaze tracker and can generate similar visualization graph as reported in the study.

28.2 Related Work

Information visualization research has maintained one-size-fits-all approach, ignoring an individual user's need, types of data and domain of applications. Ziemkiewicz et.al. [3], as well as Green and Fisher [4] have shown that the personality trait of locus of control can impact relative performance for visualizations. These results indicate that there is an opportunity to apply adaptation and personalization to improve usability. One of the attempts to adapt to individual user differences in visualization is presented in [5]. A user's visualization expertise and preferences are dynamically inferred through monitoring visualization selection (e.g., how long it takes a user to decide on which visualization to choose). Steichen et al. [6] analyzed sequential nature of user eye gaze patterns and found several gaze behaviors differences between different user/task groups during information visualization usage. These results could be leveraged by adaptive information visualization systems in order to automatically identify different user and task characteristics. In this paper, similar kind of study was conducted as Stienchen's [6] for system generated data with the goal of developing optimized visualization techniques for smart manufacturing in both 2D and 3D. Existing visualization techniques for smart manufacturing explored representing relationship among data through establishing ontologies and visualizing network diagram among different items [7]. Sackett [8] presented a review on existing visualization techniques but did not provide detail on visualizing both temporal and spatial information simultaneously. Deceptive user interfaces are interfaces that tricks user to do tasks that they did not intend to. The effect these patterns have on users varies from offensive to subtle or no influence. Fogg [9] described deceptive patterns as techniques that are used to obtain unintended outcome. Gray's taxonomy [10] of deceptive materials is based on the strategic motivator behind patterns. Interface interference that Gray suggested broadly refers to visual and language manipulation. Visual manipulation is a technique where the image, components in the image or visual cue is maneuvered to puzzle user for completing their intended task. Language manipulation refers to writing confusing statement in user interface copy or guilt tripping the user. Rosis [11] stated that these deceptive patterns are both unintentional and on purpose from designers of user interfaces. An eye tracking study was conducted earlier to measure how including text and pictures affects online reading [12]. Pan et al. [13] undertook study to explore the determinants of ocular behavior on web pages. In this paper, a similar study was undertaken to explore severity of impact on users while executing task on web pages with deceptive patterns.

28.3 User Study

In order to investigate eye gaze and fixation patterns while users interact with electronic user interfaces, a user eye tracking study was designed for 2D graphs and deceptive materials on web pages.

28.3.1 Study on 2D Graph Visualization

Aim of the study: A study was undertaken to investigate how users interpret information from graphs. New technique for analyzing gaze data was developed based on soft clustering. In particular, expectation maximization (EM) algorithm was investigated. XB cluster validation index [14] was used for validating optimum number of clusters. Using these soft clustering techniques, the number and locations of areas of interest in a visual display was automatically identified. This study will be useful to point anomalies in the current visualization techniques.

Expectation–Maximization (EM) is an iterative method to find maximum likelihood or maximum a posteriori (MAP) estimates of parameters in statistical models, where the model depends on unobserved latent variables. The EM iteration alternates between performing an expectation (E) step, which creates a function for the expectation of the log-likelihood evaluated using the current estimate for parameters, and a maximization (M) step, which computes parameters maximizing the expected log-likelihood found on the *E* step. These parameter estimates are then used to determine the distribution of the latent variables in the next E step.

XB cluster validation index: A cluster validity function proposed by Xie and Beni [14] is used to evaluate the fitness of partitions produced by clustering algorithms. It is defined as the ratio of the compactness measure and separation measure, i.e., lower index value indicates fitter partitions.

Participants: The user study was conducted with 9 participants, among them 6 males and 3 females, everyone between 20 and 35 years.

Materials: A Tobii Eye-X tracker was used for recording eye gaze, 29-inch display monitor with 1366 × 768 screen resolution and Lenovo yoga laptop with i5 processor for conducting the user study.

Design: A software which consisted four basic visualization techniques and five set of questions with multiple choice answers was developed. An eye gaze tracker was placed at the bottom of the screen. Participants were asked to seat at 75 cm away from the screen. They were instructed to answer a set of questions by investigating the graph. Figure 28.1 shows a sample interface of the system. Bar graph, line graph, radar graph and area graph were considered for the study.

For each graph, the following set of questions was displayed one at a time.

Q1: How many sensors have lesser average value than average of all low values?

Q2: Average of which sensor is approximately same as the average of all sensor's average value?

Q3: Sensors having high value greater than 50 and less than 100.

Q4: Two sensors reading showing nearly equal low values with minimum difference.

Q5: What is the approximate average of all high values of sensors?

Fig. 28.1 Software interface showing graphical and question/answer portion

Procedure: Participants were briefed about the experiment. For each participant, the eye tracker was calibrated using the nine-point calibration routine. After calibration, participants were asked to undertake the study. The *x*–*y* coordinates of the gaze location and response to each question with timestamp were logged in a text file.

Results: We analyzed number of correct answers, average time for correct answers, total time taken for all answers and the optimal (lowest) number of clusters for user's gaze fixation. We found that bar graph had highest number of correct answers with 27 correct answers and radar graph had lowest with 19 correct answers out of 45 questions, all users cumulatively. Area graph had lowest average response time for individual questions (24.6 s) and total time for all questions (155.86 s). We undertook one-way ANOVA for all dependent variables and did not find significant differences for any of the dependent variables, [$p > 0.05$].

28.3.2 Study on Deceptive Materials of Web Pages

Aim of the study: An eye tracking based study was undertaken to investigate deceptive materials on web pages. Three types of deceptive patterns were considered, visual manipulation, language manipulation and combination of visual and language manipulation.

Participants: 11 participants were involved in the experiment: 8 males and 3 females, with the average age of 24.

Materials: The study was undertaken with infrared camera-based Tobii Pro X3-120 eye tracker [15] system with 120 Hz sampling frequency. Tobii studio software

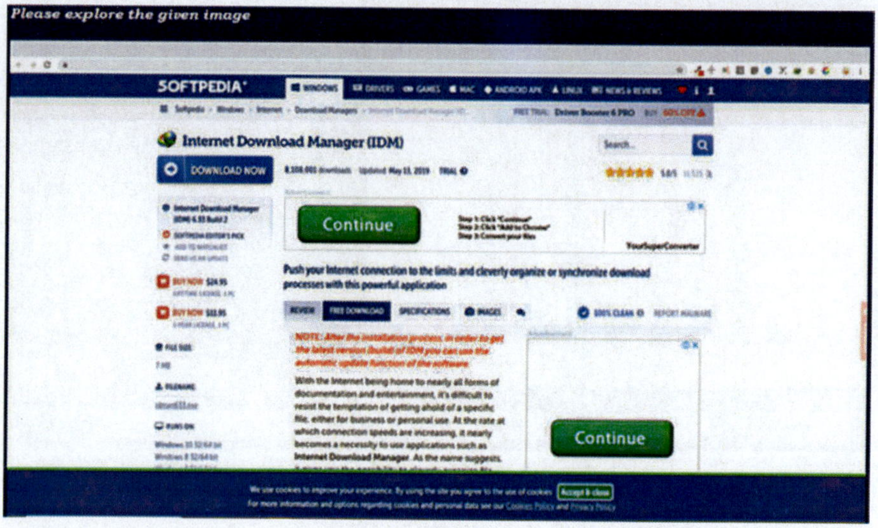

Fig. 28.2 Example of Web site with visual manipulation

working under MS Windows10 (×64) was used along Lenovo yoga laptop with i5 processor.

Design: A software which consisted twelve screenshots of different web pages with deceptive techniques was developed. It also contained a set of questions with multiple choice answers. Out of the twelve images, three images were based on type-1(visual manipulation), three on type-2(language manipulation) and six on type-3(combination of both). Participants were instructed to answer the questions after exploring the images. The questions were presented to participants in randomized order to eliminate the order bias for improving responses. Figure 28.2 shows image with visual manipulation technique.

Procedure: Participants were briefed about the experiment. For each participant, the eye tracker was calibrated using the Tobii calibration routine [15]. Participants were then asked to explore the images before answering the questions. The x–y coordinates of the gaze location and response to each question with timestamp were logged in a text file.

Results: For analysis, the images were divided into two regions. First region is the portion where deceptive patterns were present, and second region was without the deceptive element. Two dependent variables were considered—average time required to complete the task for each image and ratio of fixations in first region to total fixations in image. We undertook repeated measure ANOVA for dependent variables and found significant difference. We found significant main effect for total duration $F\,(11{,}110) = 1.252, p < 0.05, \eta^2 = 0.111$, and ratio of fixations $F\,(11{,}110) = 98.251$, $p < 0.05$, $\eta^2 = 0.908$.

Fig. 28.3 Area graph showing values from light sensors. The graph changes color after it exceeds a threshold value

28.4 Applications

28.4.1 Digital Twin Sensor Dashboard

A digital twin of smart factory layout inside virtual reality (VR) environment was modeled. Visualizations based on results of 2D graph study was used for analyzing values of real-time sensors in VR environment. Visualization graphics was set up at locations of IoT nodes to embed real-time sensor readings on the virtual layout.

We integrated ambient light sensor (BH1750), temperature and humidity sensor (DHT22) to show real-time visualization of data streams in VR setup. Both sensors provide digital output. The BH1750 sensor has a built-in 16-bit A2D converter and output unit is lux. The DHT22 sensor provides temperature in Celsius and humidity as relative percentage. Sensors were interfaced to the VR machine through their respective wireless modules [1]. After establishing a peer-to-peer connection, individual wireless module communicates with VR machine using UDP protocol at a frequency of 1 Hz.

The data stream taken from light sensor is visualized as an area graph. The graph shows change in light intensity over time and color of the graph changes if the value exceeds a threshold as shown in Fig. 28.3. Data obtained from temperature and humidity sensor is shown as separate circular bar. Similar to area graph, color of the circular bar also changes if value exceeds a threshold. Instantaneous values were converted to time-series values when user dwells using his/her eye gaze, providing a detailed view. Any abrupt change in sensor readings was reflected instantly via both visual and/or haptic feedback.

28.5 Eye Tracking Data Analysis Tool

As discussed, cheaper eye tracker like Tobii Eye-X has restrictions in terms of research studies, and high accuracy eye tracker is not always available. As a consequence, researcher uses webcam eye tracker for studies in various domains [16, 17]. In this paper also, we propose a web-based visualization tool to analyze single-user eye tracking data. The tool is developed with the aim to provide an option for analyzing eye gaze data to everyday user. A webcam-based eye tracking system is integrated with web-based interactive visualization techniques for analyzing spatial–temporal aspects of the data. The tool is developed entirely on JavaScript. Webgazer.js library is used for the eye tracker, and visualizations are developed with the help of D3.js library. The tool is hosted on python server and tested for Edge, Chrome, and Mozilla.

Firefox browser on Windows. User can undertake an eye tracking task and then immediately analyze the recorded data visually. Another option is to undertake a task and analyze the data later. Before starting eye tracking task, user has to undergo a calibration routine. The predefined calibration routine of Webgazer.js library is used in the tool. On completion of the task, user should be able to download the recorded data.

The downloaded data can then be analyzed with web-based interactive visualization techniques. The tool allows K-means clustering on fixation position of the data. Fixation positions are shown in a scatter plot visualization. This technique would allow user to automatically identify relevant AOIs in the stimuli. K-means clustering of eye gaze fixation is shown in Fig. 28.4. Heatmaps provide a quick glance on the data distribution over picture observed during an experiment or task. We developed an interactive heat map for this tool to analyze the spatial attributes of the data. The kind of heatmap included in the tool is a fixation count. This means that each fixation of the user adds to the color on that position as shown in Fig. 28.5. Red color indicates high fixation count, and green color indicates low fixation count. The tool also allows to change color gradients of the heatmap with text boxes in the control panel. Interactive gaze plot and scatter plot are also developed for analyzing spatial–temporal aspects of the data. In general, scatter plots are used to show the spatial aspect of the data. This tool uses scatter plot with color-coding technique and interactivity to study

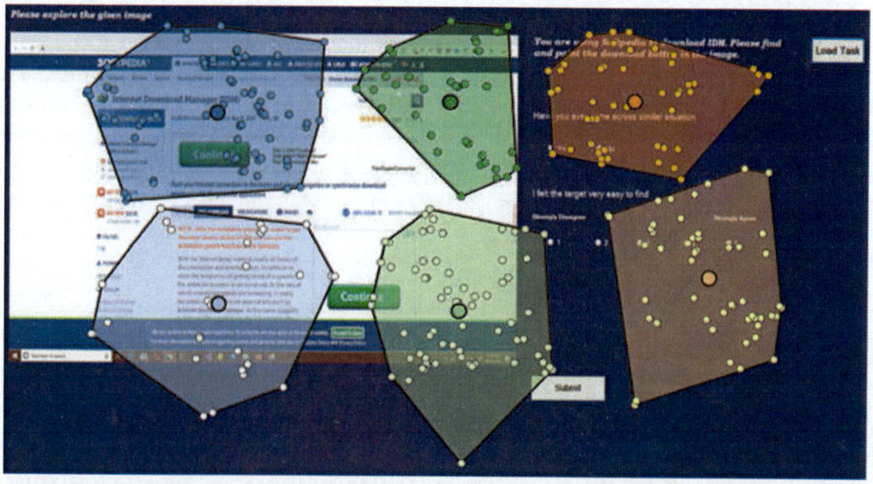

Fig. 28.4 Clustering of fixation points on the stimulus

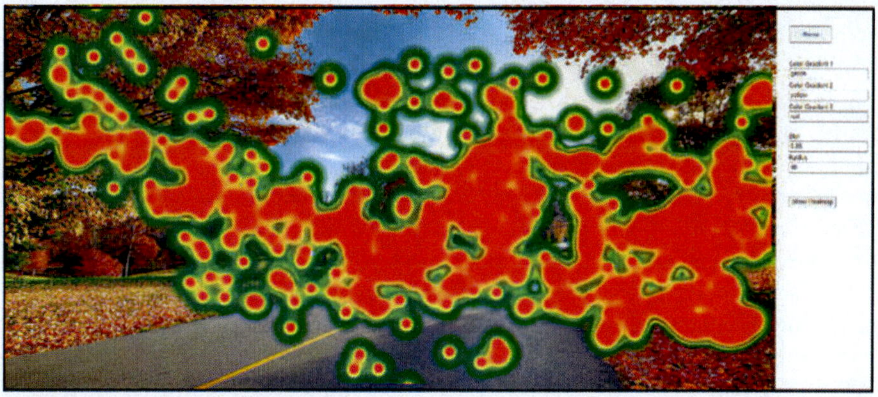

Fig. 28.5 Heatmap of fixation count

the temporal aspect of the data. Gaze plots are similar to scatter plots but reveal more detailed knowledge about the data. Gaze plots can get cluttered and unorganized if a number of fixations are large at a particular AOI. Interaction techniques are used with standard gaze plots to overcome this drawback. An interactive brush tool is used for displaying selected fixation points on both scatter plot and gaze plot.

28.6 Discussion

Our analysis on first case study found significant difference in gaze and response time for different graphs while performing similar tasks. Moreover, we analyzed users' gaze fixation using expectation maximization algorithm. This analysis suggests that more fixations signifies more eye gaze movement, requiring longer duration to analyze data. We noticed that the average response time of bar graph is high and number of correct answers for area graph is low. This speed-accuracy tradeoff may lead us to future research questions. Our initial analysis consisted of dividing the screen in nine regions and analyzing first saccade positions and subsequent gaze movements from initial position. We noted users eye gaze first fixated on the central part and moving to bottom-center of the graph subsequently for all types of graph. Furthermore, we noticed that initial patterns for eye gaze fixations were similar for every graph; however, subsequent gaze movements were distinct for each graph. We found after analyzing those movements that the sequence of bottom-center to middle-center and top-right to middle-center is most frequent among all two-region sequences. For analysis of our second study, we investigated twelve web pages with three different types of deceptive patterns. A user study was undertaken on web pages that contained one of the three types of deceptive pattern, and we found that fixations are independent of deceptive patterns used in the images. We also found that there is no significant correlation between fixations and task duration of the image.

28.7 Conclusion

This paper investigated visualization techniques and to discover suitable techniques for handling datasets. As demonstrated by results of the analysis, bar graph had the highest number of correct response, and area graph had lowest response time. In our second study, we investigated users' eye gaze behavior while interacting with deceptive materials in web pages. We analyzed average task duration and ratio of fixations in deceptive region with entire region. We found that most of the images are significantly different from each other and fixations are independent of the deceptive patterns. We did not find any significant correlation between fixations and task duration of the image. We also developed a webcam-based visualization tool for analyzing eye tracking data. The tool is independent of the implementation of webcam-based eye gaze tracker and can generate similar visualization graph as reported in the study.

References

1. Sereno, A.B., Babin, S.L., Hood, A.J., Jeter, C.B.: Executive functions: eye movements and neuropsychiatric disorders (2009)
2. Eyal, N.: Hooked: How to Build Habit-Forming Products. Penguin (2014)
3. Ziemkiewicz, C., Crouser, R.J., Yauilla, A.R., Su, S.L., Ribarsky, W., Chang, R.: How locus of control influences compatibility with visualization style. In: 2011 IEEE Conference on Visual Analytics Science and Technology (VAST), pp. 81–90. IEEE, New York (2011, October)
4. Green, T.M., Fisher, B.: Towards the personal equation of interaction: the impact of personality factors on visual analytics interface interaction. In: 2010 IEEE Symposium on Visual Analytics Science and Technology, pp. 203–210. IEEE, New York (2010, October)
5. Grawemeyer, B.: Evaluation of ERST–an external representation selection tutor. In: International Conference on Theory and Application of Diagrams, pp. 154–167. Springer, Berlin, Heidelberg (2006, June)
6. Steichen, B., Wu, M.M., Toker, D., Conati, C., Carenini, G.: Te, Te, Hi, Hi: eye gaze sequence analysis for informing user-adaptive information visualizations. In: International Conference on User Modeling, Adaptation, and Personalization, pp. 183–194. Springer, Cham (2014, July)
7. Vrba, P., Kadera, P., Jirkovský, V., Obitko, M., Mařík, V.: New trends of visualization in smart production control systems. In: International Conference on Industrial Applications of Holonic and Multi-Agent Systems, pp. 72–83. Springer, Berlin, Heidelberg (2011, August)
8. Sackett, P.J., Al-Gaylani, M.F., Tiwari, A., Williams, D.: A review of data visualization: opportunities in manufacturing sequence management. Int. J. Comput. Integrated Manuf. **19**(7), 689–704 (2006)
9. Fogg, B.J.: Persuasive technology: using computers to change what we think and do. Ubiquity, 2002(December), 2 (2002)
10. Gray, C.M., Kou, Y., Battles, B., Hoggatt, J., Toombs, A.L.: The dark (patterns) side of UX design. In: Proceedings of the 2018 CHI Conference on Human Factors in Computing Systems, pp. 1–14 (2018, April)
11. De Rosis, F., Carofiglio, V., Grassano, G., Castelfranchi, C.: Can computers deliberately deceive? A simulation tool and its application to Turing's imitation game. Comput. Intell. **19**(3), 235–263 (2003)
12. Beymer, D., Orton, P.Z., Russell, D.M.: An eye tracking study of how pictures influence online reading. In: IFIP Conference on Human-Computer Interaction, pp. 456–460. Springer, Berlin, Heidelberg (2007, September)

13. Pan, B., Hembrooke, H.A., Gay, G.K., Granka, L.A., Feusner, M.K., Newman, J.K.: The determinants of web page viewing behavior: an eye-tracking study. In: Proceedings of the 2004 Symposium on Eye Tracking Research & Applications, pp. 147–154 (2004, March)
14. Xie, X.L., Beni, G.: A validity measure for fuzzy clustering. IEEE Trans. Pattern Anal. Mach. Intell. **13**(8), 841–847 (1991)
15. Tobii Pro X3 120, Available at https://www.tobiipro.com/product-listing/tobii-pro-x3-120. Accessed on 24 June 2020
16. Sharma, V.K., Murthy, L.R.D., Singh Saluja, K., Mollyn, V., Sharma, G., Biswas, P.: Webcam controlled robotic arm for persons with SSMI. *Technology and Disability*, (Preprint), pp. 1–19 (2020)
17. Agarwal, A., JeevithaShree, D.V., Saluja, K.S., Sahay, A., Mounika, P., Sahu, A., Biswas, P.: Comparing two webcam-based eye gaze trackers for users with severe speech and motor impairment. In: Research into Design for a Connected World, pp. 641–652. Springer, Singapore (2019)

Chapter 29
Defining the Non-user: A Classification of Reasons for Non-use

Laura Augustin, Björn Kokoschko, Andrea Wolffram, and Michael Schabacker

Abstract With human-centered design becoming more widely implemented within various industries, the user's needs are being taken into account more than ever. The user of a product or service has been very well defined over the last decades. What is often overlooked is the user's counterpart: the non-user. Integrating non-users into modern development projects provides great additional value, since addressing their needs can turn a previously inactive group into new users. First, current definitions of non-users need to be compiled in order to analyze and evaluate them within the context of product development. A contribution is to be made toward creating a comprehensive definition of the non-user that can be applied to product development. Current literature mixes definitions of non-use with reasons for non-use. An analytical differentiation between the two needs to be made in order to comprehensively define the non-user and reasons for non-use. On the basis of the "non-user map" (Augustin et al. in Proceedings of the Design Society: DESIGN Conference 1:1853–1862, 2020 [1]), which classifies different types of non-users based on frequency of use and degree of self-determination, this paper seeks to define and classify the different reasons for non-use. These reasons can be divided into interconnected groups that show the relationships in between each category. With non-users and their reasons for non-use comprehensively defined, proposals for their integration into product development can be made. With a new concept of the widely used design tool "persona" (Cooper, The inmates are running the Asylum. why high-tech products drive us crazy and how to restore the sanity, Sams, 2004 [2]), the "non-persona," now the non-user of a product or service can be included in the development process as

L. Augustin (✉) · B. Kokoschko · A. Wolffram · M. Schabacker
Otto-von-Guericke University, Magdeburg, Germany
e-mail: laura.augustin@ovgu.de

B. Kokoschko
e-mail: bjoern.kokoschko@ovgu.de

A. Wolffram
e-mail: andrea.wolffram@ovgu.de

M. Schabacker
e-mail: michael.schabacker@ovgu.de

A. Chakrabarti et al. (eds.), *Design for Tomorrow—Volume 1*, Smart Innovation, Systems and Technologies 221, https://doi.org/10.1007/978-981-16-0041-8_29

well as the user in order to give a more comprehensive understanding of possible requirements for a product or service.

29.1 Introduction

The user has been defined very well [3, 4], while the concept of the non-user has only been explored partially [5–10].

> Resistance is a means by which user communicate their discomfort with a system that might be flawed (MARAKAS AND HORNIK, 1996, Paper 31).

Types of non-use and reasons for non-use are often intertwined [10], while they are inherently different concepts [1]. The non-user map [1] compiles all current definitions of non-use, while this contribution is seeking to compile, analyze, and evaluate the different definitions and characterizations of reasons for non-use and give a synoptic view of current literature. The overall objective is to facilitate the incorporation of the non-user in product development practice in order to expand the target audience and improve the product.

29.2 Literature Research Approach

This chapter gives a general overview over the reviewing process before discussing the detailed findings in Chap. 3. The tool used for the search was Google Scholar, and search terms used were centered around the reasons for non-use and how users make decisions, i.e.: "reasons for non-use product," "reasons non purchase," "consumer decision making," consumer decision-making stages," "user resistance theory," and "consumer resistance to product."

The three main requirements for literature to be deemed relevant were:

- Literature cannot only contain case-specific descriptions of non-use (i.e., "How many people are not using xyz").
- Literature cannot only be discussing how to combat non-use while not defining the reasons for it.
- Authors have to address or define categories of reasons for non-use.

29.3 Overview of Current Literature

In order to comprehensively define non-use, the different categories of non-use need to be discussed to then closely examine the reasons for non-use. In a recent paper. a set of definitions for the non-user was defined [1], as the following segment will shortly discuss. In Sect. 3.2 the reasons for non-use found in literature are compiled and analyzed. This differentiation needs to be made since they are often treated as one [10]. Section 3.3 focuses on the current literature gap and how to address it in further research.

29.3.1 Categories of Non-use: The Non-user Map

The non-user map was developed from previous research and combines the definitions found in literature with an array of new definitions [1]. The map in Fig. 29.1 is described via two axes, the diagonal describing the level of willingness to use a product from "do not want to use" to "want to use." The second axis defines non-use in four areas as described on the right-hand side and describes the level of experience with a product. High experience being "currently using" and "have used" and low experience being "might use" and "have never used."

The map then arranges current definitions according to the framework. The different shades of gray represent how frequently the product is used, with the upper right corner signifying the user. This map visualizes the often fluent transition from use to non-use and vise versa.

New non-user types have been added to the map in order to complete the concept: the unconnected (users of a competitor's product, therefore using the product in

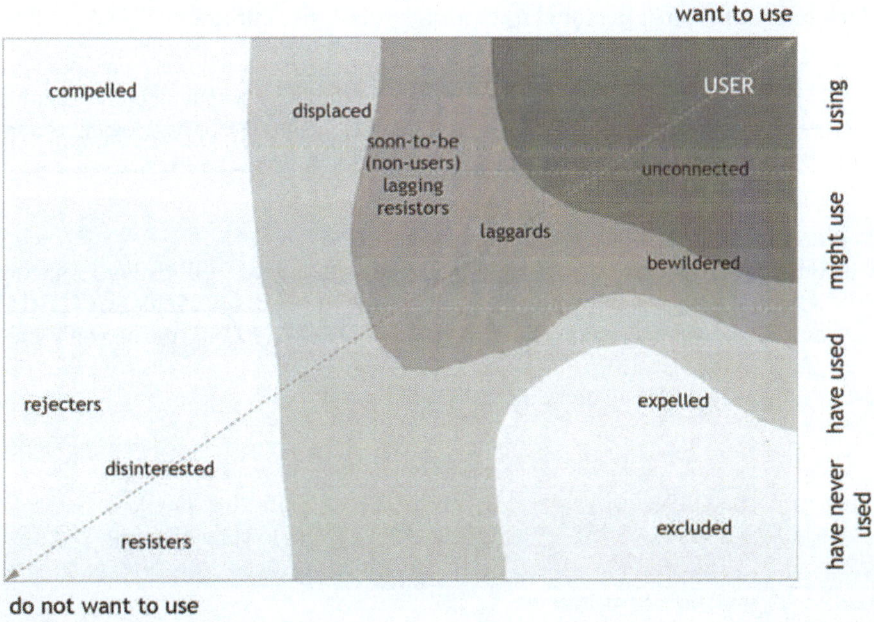

Fig. 29.1 Non-user map [1]

general but not the one in question), the bewildered (confused by choice and therefore still non-users), and the compelled (using a product out of necessity).

The term of the unconnected visualizes that a user of a competitor's product can still be a non-user to the company and therefore a potential user of their product. Therefore, both users and non-users can be viewed as potential users of another product.

29.3.2 Reasons for Non-use

Three main categories can be discerned from the reviewed literature, which will be discussed in the following segments. The reasons for non-use are divided into the psychological (personal), societal, and environmental circumstances. These categories are then divided into subcategories, and different definitions found in the literature review are analyzed. These categories are often intertwined with each other, but for the sake of this overview, a differentiation needed to be made, as shown in Fig. 29.2.

- Personal framework (personal data and psychological profile)
- Marketplace (the market situation: the product and its contenders)
- Social and societal framework (surrounding the people)

Personal framework: personal data and psychological profile

The personal framework of the non-user is the often discussed in current literature [11, 12]. This segment will list the most significant literature with their corresponding sources, highlighting where literature overlaps or dissents. The psychological framework consists of personal information, economic status, personality traits, decision making, and intention.

1. Personal information and economic status
   Personal properties like age and gender are often cited in literature as having influence on decision making in regard to the use of new technologies [11–14] as well as economic status (financial capabilities) [11, 12] and the person's level of education [12, 13].
2. Personality traits and decision making

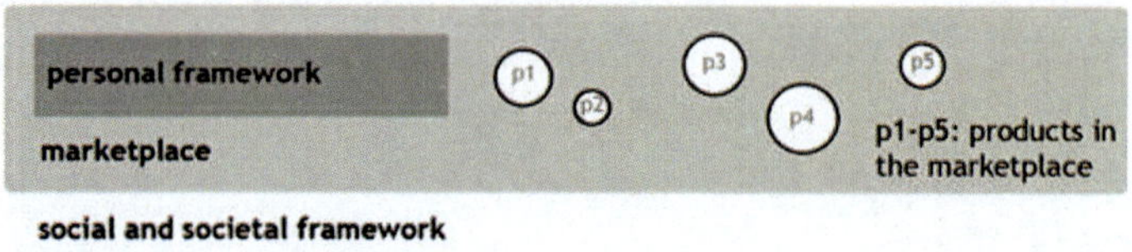

Fig. 29.2 Overview of non-use frameworks

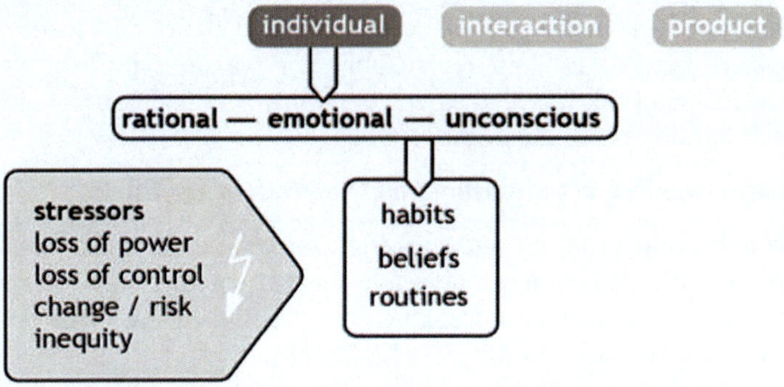

Fig. 29.3 Personal framework

In current literature [11, 15–21], there are several different classifications made in order to give an overview over consumer behavior and what contributes to decision making.

In most literature concerning IT implementation, the three categories discussed are people-oriented, systems-oriented, and interaction theories [15, 16], referencing the person and the product as well as the interaction between the two. The three perspectives of consumer behavior within the field of business management are divided into by Mowen: decision making (rational), experiential (emotional), and behavioral (unconscious) [17], both make insightful differentiations that can be combined for future application as shown in Fig. 29.3.

Beliefs and attitudes are also referenced as influencing factors [18], but are not often discussed in more detail. In the same realm as beliefs and attitudes are habits [11, 19, 20], which are arguably very important for consumer resistance, "Habit is the single most powerful determinant in generating resistance" [21]. Closely linked to habits is routine and its importance to consumers [20]. These heavily influence the individual's cognitive rigidity [18, 20] and inertia [22], which in turn determine how likely a product will be resisted or adopted.

Next to these personal characteristics are stressors that influence the individual and the decision-making process. These stressors are mainly loss of power [1], loss of control [12, 14], lack of trust [11], indication of change [20], perceived risk [23], and fear of inequity [24]. How influential each of them is of course depends on each individual's personality-specific traits.

One of the most written about inhibitors is status-quo bias [18, 25, 26], describing the preference for the current state rather than looking for change.

Limited literature has explored the correlation of stressors and personality traits. Kleijnen et al. for example, define two groups of antecedents for resistance/rejection: degree of change required and conflicts with user's belief structure [27], connecting beliefs and the stressor change. But this study asked consumers why they did not buy a certain product, and it can be challenged whether consumers are always aware of their decisions. People are generally unaware of their true reasons for adoption or resistance, and it is better to only focus on observed choices. Research has found that consumers generate random explanations when asked why they made or did not make a decision [18]. It is also agreed upon that personality traits in relation to resistance need to be examined further [18].

Marketplace: The Market Situation and the Product Itself.

Within the fields of psychology and sociology, decision making has been a field of interest for decades [28], with research focusing on how the consumer deals with different product options and how decisions change based on how many options there are or what characteristics they possess [17, 28, 29].

The marketplace has an undeniable impact on consumer behavior and decision making. Every decision is a trade-off between different factors [18]. If there are too many choices, people tend to not choose a product at all [28], because they are overwhelmed by the options and cannot weigh them all. When two choices are equally attractive, consumers find the more important dimension to them and then choose the alternative that is superior in that singular dimension [28]. Meaning that it might not matter how perfect a product is in one dimension, if that is not the one most relevant to the customer. A choice is likely postponed, when the options are too similar and no clear differentiation can be made [28]. If options are too similar, a third option can help with a decision and influence the choices made, depending on what dimensions it adds [28]. If features are considered unnecessary or wasted, then consumers can reject the whole product all together [28].

The product itself of course has an influence over the decision as well since it might not be chosen due to poor quality [30], price [25] and switching cost [26], maintenance [29], accessibility [12], safety [29], perceived usefulness and ease of use [14] or if the product is disappointing or does not meet expectations [29].

Resistance was also shown to grow according with how aggressively a product was being marketed [19]. Consumers can also be critical and doubt information, especially when they do not understand it [12].

Social and Societal Framework

The social environment undoubtedly has great influence on consumer decisions [18] in groups like families, work, institutions, religions and many more. Group resistance can occur in both negative and positive directions, with in-group favoritism occurring when a certain group supports the use of a product and out-group hostility occurring when all products used outside of the accepted framework are rejected [19].

In a work context, co-worker behavior can also influence opinion formation [26], for example, sabotaging employee acceptance during a new software implementation. This can manifest in employee grumbling [25], describing resistance through conversations between co-workers.

29.3.3 Conclusion of Literature Review and Discussion of Literature Gap

What all literature seems to neglect is the fact that adoption is also rejection. When users adopt one product, they simultaneously reject all others within the same category. Thus, one person might be an adopter to one company's product, but will be a rejector to all others. This has not been discussed in much detail in current literature regarding non-users, but should be explored as much as all other themes in the non-user literature.

Most literature found was located within the information technology field, investigating why software implementation fails in a professional work context [15, 16, 25, 26, 31]. There are of course many factors at play in a work environment, that are not in a personal one, i.e., relationships with co-workers and management [19]. Most importantly, almost all IT related implementations in the workplace are mandatory and therefore met with more resistance. Thus, not all findings can be easily applied in most other non-mandatory contexts.

29.4 Toward a Comprehensive Overview of Reasons for Non-use

This chapter discusses a comprehensive overview of non-use and how that can be applied to a product development context.

29.4.1 Categories of Reasons for Non-use

In Fig. 29.4, all influencing factors for non-use are summarized in one image in

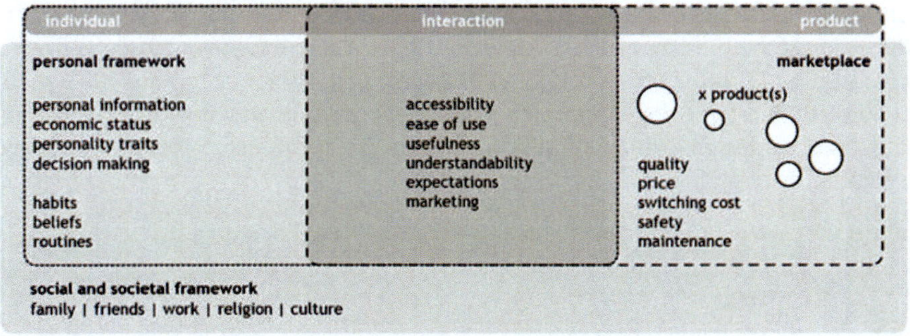

Fig. 29.4 Influencing factors for non-use

order to give a comprehensive overview. It is divided into the three columns individual, interaction, and product [15, 16] in combination with the three frameworks of influence: personal, marketplace, as well as social and societal.

Product development or management can of course mostly influence the product and the interaction to a certain extent, but not the personal or social framework of the individual. Nevertheless, it is imperative to be aware of all frameworks in order to make successful products.

29.4.2 Application of the Non-user: The Non-Persona

Personas [2] are used to define and visualize users and their needs, wants, and wishes. Often, the personas also give insight into the user's frustrations and suggestions, giving developers clues on what to improve or change about their experience.

In order to define the user more clearly, this contribution's suggestion is to give personas a useful counterpart: the non-persona. This tool can help to give a more comprehensive overview of the landscape and therefore adding valuable information to the development process.

When analyzing current literature to define which aspects shall be included in a persona, it is noticeable that different authors apply similar aspects in their templates while there are some which are mentioned less. These are: demographic data, personality (traits), hobbies and interests, technology-related aspects, insight into the daily life, product-related problems, and product goals [32]. These aspects are relevant to both a persona and a non-persona. One aspect that has only been mentioned once is disabilities [32], which can also be very relevant to non-users, especially in terms of the group "excluded."

Furthermore, additional aspects from the previous non-user map can be added: rejection level, existing intent of use, and the categorization of use into currently using, might use, have used, and have never used. These aspects can be represented graphically in order to enable a quick recognition of the type of non-user and are additionally represented in text fields with further information about where possible reasons for non-usage can lie, e.g., in the reasons for resistance, possible reservations against the product group in general or the reflected negative attitudes from a previous use. Figure 29.5 shows a possible persona template integrating all aspects mentioned. This template is currently being developed further, with the possibility of an interactive design displaying the connections and relationship between different aspects of the template.

29.5 Summary

The non-user and the reasons for non-use have been researched in different disciplines over the last decades; however, there is still a gap to fill regarding comprehensive

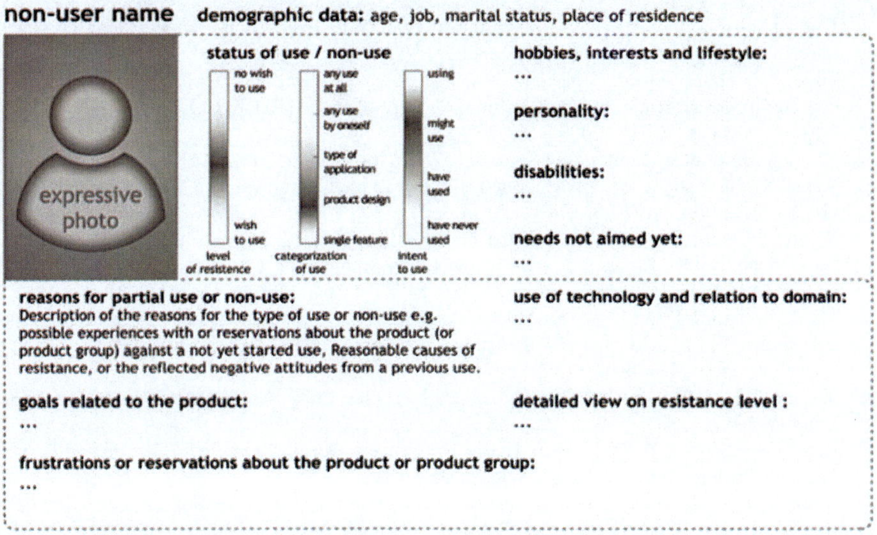

Fig. 29.5 Non-persona

definitions for the two. This contribution has compiled, analyzed, and summarized the most important insights from product development, marketing and social sciences, and translates the non-user to applications in product development by using the non-user map and a categorization of reasons for non-use. Lastly, the non-persona translates the research into an applicable concept for product development.

29.6 Outlook

The non-user map gives an overview over the different types of non-use and the reasons for non-use have been discussed in this contribution. Future research will focus on combining the two in order to give a more concise guide for an integration of the non-user in product development. Further research needs to be done concerning the reasons for non-use and quantify their weight in relation to one another. In order to verify the concept of the non-persona, it needs to be applied in future development projects and adjusted accordingly.

References

1. Augustin, L., Kokoschko, B., Wiesner, M., Schabacker, M.: Toward a comprehensive definition of the non-user. In: Proceedings of the Design Society: DESIGN Conference, vol. 1, pp. 1853–1862 (2020). https://doi.org/10.1017/dsd.2020.124

2. Cooper, A.: The Inmates Are Running the Asylum. Why High-Tech Products Drive Us Crazy and How to Restore the Sanity, 2nd edn. Sams (2004)
3. Gardan, J.: Definition of users' requirements in the customized product design through a user-centered translation method. Int. J. Interact. Des. Manuf. (IJIDeM) **11**(4), 813–821 (2017). https://doi.org/10.1007/s12008-015-0275-2
4. Kurosu, M.: Human Centered Design, vol. 6776. Springer, Berlin (2011)
5. Ribak, R., Rosenthal, M.: Smartphone resistance as media ambivalence. First Monday **20**(11) (2015). https://doi.org/10.5210/fm.v20i11.6307
6. Wyatt, S.: Non-users also matter: the construction of users and non-users of the Internet. In: Oudshoorn, N., Pinch, T.J. (eds.) How Users Matter. The Co-construction of Users and Technologies. Inside Technology, pp. 67–80. MIT Press, Cambridge, MA (2003)
7. Miles, I., Thomas, G.: User resistance to new interactive media: participants, processes and paradigms. In: Bauer, M. (ed.) Resistance to New Technology, pp. 255–275. Cambridge University Press, Cambridge (1995)
8. Rogers, E.M.: Diffusion of Innovations, 3rd edn. Free Press; Collier Macmillan, New York, London (1983)
9. Barsch, T., Heupel, T., Trautmann, H. (eds.): Die Blue-Ocean-Strategie in Theorie und Praxis. Springer Fachmedien Wiesbaden, Wiesbaden (2019)
10. Satchell, C., Dourish, P.: Beyond the user: use and non-use in HCI. In: Foth, M. (ed.) Proceedings of the 21st Annual Conference of the Australian Computer-Human Interaction Special Interest Group (CHISIG) of the Human Factors and Ergonomics Society of Australia (HFESA). The 21st Annual Conference of the Australian Computer-Human Interaction Special Interest Group, Melbourne, Australia, 11/23/2009 - 11/27/2009, p. 9. ACM, New York (2009). https://doi.org/10.1145/1738826.1738829
11. Kahma, N., Matschoss, K.: The rejection of innovations? Rethinking technology diffusion and the non-use of smart energy services in Finland. Energy Res. Soc. Sci. **34**, 27–36 (2017). https://doi.org/10.1016/j.erss.2017.05.024
12. Bazhan, M., Mirghotbi, M., Amiri, Z.: Food labels: an analysis of the consumers' reasons for non-use. J. Paramedical Sci. **6**(1), 2–10 (2015)
13. Helsper, E.J., Reisdorf, B.C.: A quantitative examination of explanations for reasons for internet nonuse. Cyberpsychol. Behav. Soc. Network. **16**(2), 94–99 (2013). https://doi.org/10.1089/cyber.2012.0257
14. Laumer, S., Maier, C., Eckhardt, A., Weitzel, T.: Why are they grumbling about my new system? Theoretical foundation and empirical evidence of employee grumbling as a user resistance behavior. In: Thirty Fifth International Conference on Information Systems
15. Jiang, J.J., Muhanna, W.A., Klein, G.: User resistance and strategies for promoting acceptance across system types. Inform. Manage. **37**(1), 25–36 (2000). https://doi.org/10.1016/S0378-7206(99)00032-4
16. Ali, M., Zhou, L., Miller, L., Ieromonachou, P.: User resistance in IT: a literature review. Int. J. Inf. Manage. **36**(1), 35–43 (2016). https://doi.org/10.1016/j.ijinfomgt.2015.09.007
17. Mowen, J.C.: Beyond consumer decision making. J. Consumer Market. **5**(1), 15–25 (1988). https://doi.org/10.1108/eb008214
18. Laumer, S., Eckhardt, A.: Why do people reject technologies: a review of user resistance theories. In: Dwivedi, Y.K., Wade, M.R., Schneberger, S.L. (eds.) Information Systems Theory, vol. 28. Integrated Series in Information Systems, pp. 63–86. Springer, New York, London (2011)
19. Bagozzi, R.P., Lee, K.H.: How Can Marketers Overcome Consumer Resistance to Innovations? J. Global Acad. Market. Sci. **15**(3), 211–231 (2005). https://doi.org/10.1080/12297119.2005.9707500
20. Oreg, S.: Resistance to change: developing an individual differences measure. J. Appl. Psychol. **88**(4), 680–693 (2003). https://doi.org/10.1037/0021-9010.88.4.680
21. Sheth, J.N.: Psychology of innovation resistance: the less developed concept (LDC) in diffusion research. In: Research in Marketing, vol. 4, pp. 273–282

22. Kim, H.-J., Lee, J.-M., Rha, J.-Y.: Understanding the role of user resistance on mobile learning usage among university students. Comput. Educ. **113**, 108–118 (2017). https://doi.org/10.1016/j.compedu.2017.05.015
23. Laukkanen, T.: Consumer adoption versus rejection decisions in seemingly similar service innovations: the case of the Internet and mobile banking. J. Bus. Res. **69**(7), 2432–2439 (2016). https://doi.org/10.1016/j.jbusres.2016.01.013
24. Joshi, K.: A model of users' perspective on change: the case of information systems technology implementation. MIS Quarterly **15**(2), 229 (1991). https://doi.org/10.2307/249384
25. Mahmud, I., Ramayah, T., Kurnia, S.: To use or not to use: modelling end user grumbling as user resistance in pre-implementation stage of enterprise resource planning system. Inform. Syst. **69**, 164–179 (2017). https://doi.org/10.1016/j.is.2017.05.005
26. Kim, K.: Investigating User Resistance to Information Systems Implementation: A Status Quo Bias Perspective. MIS Quarterly **33**(3), 567 (2009). https://doi.org/10.2307/20650309
27. Kleijnen, M., Lee, N., Wetzels, M.: An exploration of consumer resistance to innovation and its antecedents. J. Econ. Psychol. **30**(3), 344–357 (2009). https://doi.org/10.1016/j.joep.2009.02.004
28. Shafir, E., Simonson, I., Tversky, A.: Reason-based choice. Cognition **49**(1-2), 11–36 (1993). https://doi.org/10.1016/0010-0277(93)90034-S
29. Trocchia, P.J., Janda, S.: An investigation of product purchase and subsequent non-consumption. J. Consumer Market. **19**(3), 188–204 (2002). https://doi.org/10.1108/07363760210426030
30. Hirschheim, R., Newman, M.: Information systems and user resistance: theory and practice. Comput. J. **31**(5), 398–408 (1988). https://doi.org/10.1093/comjnl/31.5.398
31. Chervany, N.L., Lending, D.: CASE tools. SIGCPR Comput. Pers. **19**(2), 13–26 (1998). https://doi.org/10.1145/292349.292353
32. Nielsen, L., Hansen, K.S., Stage, J., Billestrup, J.: A template for design personas. Int. J. Sociotechnol. Knowl. Development **7**(1), 45–61 (2015). https://doi.org/10.4018/ijskd.2015010104

Chapter 30
Usability Analysis of Warning Icons in Passenger Car Dashboards in India Using Modified System Usability Scale (SUS)

Sourav Bhattacharya and Dhananjay Singh Bisht

Abstract In the past few years, automobile dashboard and its instrument panel are becoming increasingly complicated as they offer a variety of digitized functionalities such as warning icons. Past research shows that certain automotive warning icon designs could be hard to interpret by drivers. A study was undertaken for determining the usability of certain standard warning icons in automobile instrument clusters using system usability scale (SUS) and also to investigate the "guessability" of these icons. Here, a set of 20 standard warning icons which are commonly used in new models of Indian cars were selected from ISO 2575:2010, and 50 participants were employed to determine the usability and guessability of these icons. Standard designs of two car dashboard warning icons, namely "Brake Warning Light" and "Tire Pressure Monitoring System" were found to be problematic. A significant and strong positive correlation was found between the mean guessability score and the mean SUS score for all the 20 warning icons.

30.1 Introduction

In the past, only a limited amount of information had to be displayed on automotive dashboards through mechanical gages with little need to provide intricate details about different automotive functionalities and features [1]. However, today mechanical gages have mostly given way to electronic/digital screens which use graphical icons to warn the drivers about status of different automobile operations and functions. This is due to the upgradation of in-vehicle technology utilizing modern equipment, and the requirements for information display as per the guidelines framed by various agencies and standards such as international standard organization (ISO), society of automotive engineers (SAE), federal motor vehicle safety standards (FMVSS), manual of uniform traffic control devices (MUTCD). Today, through automobile dashboards, a lot of situational information could be communicated to the

S. Bhattacharya (✉) · D. S. Bisht
Industrial Design Department, National Institute of Technology, Rourkela 769008, India
e-mail: souravzilla@gmail.com

A. Chakrabarti et al. (eds.), *Design for Tomorrow—Volume 1*, Smart Innovation, Systems and Technologies 221, https://doi.org/10.1007/978-981-16-0041-8_30

vehicle occupants [2]. It is a big challenge for designers to design dashboards that provide an access to multiple vehicular functions and information while keeping the display simple, safe, and efficient–which is typically suggested for better cognitive experience of a driver [3]. Automotive manufacturers often follow certain principles of user-centered design while designing automotive dashboard elements. They also have to set comparative benchmarks in order to create competitive designs [2]. Today, the advancements of vehicular technology offer huge opportunities to enhance the user experience (UX) of the interiors which might include among others, the use of digitized screens for user interfaces (UI) [4]. Therefore, it is important to understand the expertise and expectations of drivers in specific situations. It is also important to ascertain the confusions which might be caused due to such interfaces and the design decisions [3].

A recent study by Jung and Choe [5] illustrates how drivers receive visual notifications about the vehicular state and operational conditions which is effective in preventing accidents. However, in certain situations, certain warning icons may not convey proper meaning to the occupants appropriately due to poor/inappropriate visual design [5]. Such situations could be checked by validating the usability and comprehensibility of these icons during the product design process. For example, in an Indian context, a survey on the appropriateness of different functional icons for a head up display (HUD) system has been recently reported [6]. Inability of the occupant to interpret the meaning could lead to higher risks of road accidents. In a past research, it was observed that 42% of drivers could not accurately comprehend TPMS light icon [5]. In addition to these concerns, distracted driving due to simultaneous performing of secondary tasks is a major cause of automobile accidents [7] and it seems important to consider human factors in designing automobile dashboards to reduce driver distraction. Provision of warning icons/safety symbols in automobile dashboards helps in reducing driving risks. Some of these icons represent "safety precautions," and some signal "caution," indicating a definite course of action required at the driver's end or other actionable directions to reduce risk factors.

The extent of risks and hazards are especially influenced by the nature of colors used in warning icons. "Red" indicates highest risk and used as emergency icons that give warning of an emergency/danger. "Yellow" or "Amber" is used for low risk and stands for caution that typically indicates that a definite course of action is required by the driver. "Green" or "Blue" is used when indicating operational status of the automobile and used in "safe" class of icons (which provide information about safety conditions of the vehicle, e.g., indicating a non-dangerous state) [5]. Safety symbols are designed to identify malfunctioning in cars and to warn against specific hazardous situations with or without use of text [7]. Various studies have indicated that if safety symbols are not easily detectable, then their value in transferring message is rendered ineffective which in turn diminishes their usability [9].

The aim of this research is to evaluate the visual designs of 20 automotive warning icons used in Indian passenger car dashboards in the context of their guessability and usability. The study should serve as a valuable reference for interface designers in the context of usability testing of warning icons and other graphical symbols.

30.2 Methodology

A questionnaire-based survey was conducted to assess guessability and usability for 20 automotive dashboard warning icons.

30.2.1 Subjects

A convenience user sample was employed for the survey with the only selection criterion being possession of at least 1 year driving experience. It was only at the time of data analysis that within this group of participants further sub-groups were realized such as male and female drivers, professional drivers and others, etc. (described in Table 30.4). A total of 50 individuals (31 males and 19 females), aged between 22 and 55 years (Mean = 28.54, S.D. = 5.02) participated. Here, 40% of the participants were found to be professional drivers, 24% were other job holders (professionals), 20% were students from the higher certificate program, and the remaining 16% were housewives.

30.2.2 Materials

Microsoft Excel (2016) was used as a tool for documenting raw survey data, analyzing data, and to perform basic data analysis [10]. Statistical package for the social sciences (SPSS 24) was used as a statistical tool for detailed statistical analysis [11].

30.2.3 Selection of Warning Icons

64 symbols recommended by ISO 2575:2010 [5] were analyzed, and a set of 20 warning icons which are most commonly used in contemporary Indian cars were selected for testing (Table 30.1). The icons were divided into three categories according to their color specific semantics—(i) red-colored emergency icons that give warning of an emergency/danger, (ii) yellow- or amber-colored icons of caution indicated a definite course required by alerting the driver, and (iii) green- or blue-colored safe icons which provide information about safety conditions of the vehicle (Table 30.1).

Table 30.1 List of warning icons used in the experiment

Ref. No.	Image	Standard Label	Ref. No.	Image	Standard Label
S1		Turn Signal	S4		Engine Start Indicator
C1		Low Fuel Indicator	E4		Open Door Warning Light
E1		Air Bag Warning Light	E5		Steering Failure Indicator Light
C2		ABS Warning Light	C5		Wind Shield Water Level
E2		Low Oil Pressure Remainder	E6		Seat Belt Remainder
E3		Brake warning light	E7		Battery Warning Light
S2		Low Beam Light	C6		Trans Axel Warning Light
S3		High Beam Light	C7		Tire Pressure Monitoring System
C3		Fog Beam Light	E8		Coolant Temperature Light
C4		Malfunction Indicator	E9		Service Required Indicator Light

C - Caution icons; E – emergency icons; S – safe icons

30.2.4 Guessability Score Evaluation

A self-administrated questionnaire was provided to the participants. The first section of the questionnaire was about participant's demographic information. Participants were asked to complete several questions like name, age, gender, contact information, profession, license details, driving experience, and time taken to learn driving. Finally, they had to rate themselves on their 4-wheeler driving skill and the extent of "familiarity" with car's dashboard instrument cluster on a Likert scale (ratings from 1 to 5; 1 = minimum familiarity, 5 = maximum familiarity).

The second section of this questionnaire was about assessing "guessability" of the warning icons. Guessability refers to the accuracy of a participant in correctly guessing the meaning/title of an icon. All the warning icons were displayed in colors and in rectangular boxes of 2.4 cm × 1.6 cm without boundary in print. Here, oral responses (guesses) of participants were recorded and compared with the standard label of icon recommended by ISO 2575:2010 to gauge the level of precision. In this exercise,

(a) A precise response was one where participants specifically and appropriately guessed the name of icon.
(b) A partially precise response was one where participants could only guess the title of icon which did not match exactly with the standard label but they had a rough idea about the icon.
(c) A response was considered to be imprecise if it did not fulfill requirements of being whether a precise or a partially precise response.

Precise response fetched 2 points, partially precise responses were awarded 1 point, and for imprecise response, no points were awarded (0 points). Guessability performance for an individual participant was calculated using the following formula:

$$\text{Guessabiliy performance} = \frac{\text{Sum of guessing score of all icons for a certain participant} \times 100}{\text{Maximum score at each response} \times \text{Total no. of icons}}. \tag{30.1}$$

Guessability score for an individual icon was calculated using the following formula:

$$\text{Guessability score} = \frac{\text{Sum of guessing score of an icon across all participants} \times 100}{\text{Maximum score at each response} \times \text{Total no. of participants}} \tag{30.2}$$

30.2.5 SUS Score Evaluation

The third section of this questionnaire consisted of recording usability ratings of warning icons. In this study, usability ratings of warning icons were captured using a

Table 30.2 Modified SUS statements

S. No.	Modified SUS statements
S1	I would like to pay attention to this icon frequently whenever I drive my car
S2	I feel like this icon is unnecessarily complex and does not convey a simple meaning
S3	I think that this icon conveys a simple and a clear meaning, thus easy to understand
S4	I need somebody's assistance to explain me the meaning of this icon without any supplementary text
S5	I think that this icon is well integrated and gives a compact feeling
S6	I think there is too much confusion created in interpreting this icon
S7	I think most of the people can understand this icon only after having suitable training
S8	I feel that this icon is very cumbersome, i.e., difficult to interpret
S9	I feel very confident using this icon
S10	I needed to have a detailed explanation before I could get going with this icon

modified SUS instrument (Table 30.2), originally developed by Brooke [8]. Modification was made only to make the questions more specific to the context of warning icons assessment, whereas in its original form, the questions are framed in a systems-design perspective. Participants were asked to give subjective ratings between 1 and 5 points for each of 10 usability related statements (S1-S10) for every icon. (1 = strongly disagree, 5 = strongly agree). Calculation of the SUS score for an individual icon was adopted from [8]. For each odd numbered statement (S1, S3, S5, S7, S9) 1 was subtracted from the score obtained, and subsequently, for each even numbered statement (S2, S4, S6, S8, S10), the score obtained was subtracted from 5. After addition of scores for all the statements, the total score was obtained. Finally, the total score was multiplied by 2.5 which altogether led to the value of sign usability, known as the SUS score. SUS score ranges between 0 and 100. In accordance with the analysis rules used in [12], products with SUS scores less than 50 are to be judged as inadmissible, score ranging between 50 and 70 are somewhat admissible, and score above 70 are tolerable.

30.3 Results

30.3.1 Guessability Performance

The overall mean and standard deviation of guessability performance for all 50 participants were 87.35% and 15.25%, respectively (using Eq. 30.1). The maximum guessability performance of any participant was 100%, and the minimum guessability performance was 60%.

30.3.2 Guessability Score

The overall mean and standard deviation of guessability score for all 20 warning icons were 87.35% and 9.77%, respectively (using Eq. 30.2). Here in this experiment, guessbility scores 0, 1, and 2 (as mentioned in Sect. 30.2.4) have been converted into percentages in order to map these scores with the usability scores. The icons with maximum guessability score of 100% were *Turn Signal, Low Fuel Indicator, Open Door Warning, Seat Belt Remainder,* and *Battery Warning Light* whereas with minimum guessability score of 68% was for *Tire Pressure Monitoring System* as shown in Table 30.3. It is recommended that an icon must reach a criterion of at least 67% guessability score in a guessability test to be treated as "tolerable." [13]

Table 30.3 Warning icons and their mean SUS scores and guessability scores

Categories	Ref. No. (see Table 30.1)	Mean SUS Score	Mean guessability score (%)
Passable/Tolerable	S1	89.85	100
	E4	87.05	100
	E6	86.75	100
	C1	83.85	100
	E7	83.45	100
	E1	82.3	89
	E9	80.85	95
	E8	77.7	93
	E5	76.8	88
	E2	75.05	86
	C4	70.08	84
Marginally admissible	S3	69.8	90
	C6	69.45	78
	C2	69.4	83
	S2	63.45	87
	S4	61.7	76
	C3	53	82
	C5	52.55	75
Inadmissible	E3	37.95	73
	C7	26.4	68

Table 30.4 Central tendencies in guessability and SUS scores (*N=50*)

Demographic factor	Response	Sample size	Guessability score (%)		SUS score	
			Mean	S.D.	Mean	S.D.
Gender	Male	31	95.64	4.921	69.10	17.74
	Female	19	73.82	18.80	71.22	17.03
License ownership	Have	40	91.44	7.01	69.38	17.58
	Don't have	10	71	22	72.02	17.05
Profession	Driver	20	99.5	1.69	69.41	17.56
	Professional	12	91.25	9.763	67.38	17.94
	Student	10	78.75	18.83	71.57	17.31
	Housewife	8	61.87	27.16	72.86	16.40
Driving skill	Expert	29	93.28	5.970	69.44	17.57
	Average skill	14	81.96	14.83	69.57	17.63
	Learner	7	73.57	20.84	72.5	16.82
Cluster familiarity	Most familiar	21	96.79	3.785	69.42	17.61
	Average familiar	18	81.81	14.34	69.42	17.88
	Non-familiar	11	78.41	16.19	71.64	16.59

30.3.3 SUS Scores

SUS score was calculated using the methodology as described in Sect. 30.2.5. The overall mean SUS score for the entire population was found to be 69.91 (std. dev. = 17.49) which indicated that the warning icons were generally anticipated to be marginally admissible [12]. The maximum SUS score was 92.5, and minimum score was 15. The mean SUS score for 21 participants ranged between 50 and 70, and for the remaining 29 participants was above 70.

The mean SUS score was maximum for *Turn Signal* (89.85) icon and minimum for *Tire Pressure Monitoring System* (26.4) icon as shown in Table 30.3. The icons that were found to be "inadmissible" [12] were *Brake Warning Light* (37.95) and *Tire Pressure Monitoring System* (26.4).

30.3.4 Relationship between Guessability and SUS Scores

Table 30.4 shows sub-groups among the participants, their sample sizes, guessability scores, and SUS scores. Male subjects in driving profession exhibited the highest guessability score than in any other professions, whereas female subjects who were housewives had the least guessability scores. There was relatively small deviation in usability scores among the 50 participants. Overall, the guessability scores were generally greater than usability scores for the entire data sample. Participants those who were moderately familiar with car's instrument cluster and had average driving skills were found to perform average as well in the guessability test. The average familiarity of car's dashboard instrument cluster for the entire sample of 50 participants was found to be 3.85 (std. dev. = 1.22, measured on scale 1–5).

30.3.5 Statistical Analysis

Kolmogorov–Smirnov test and Shapiro-Wilk's test [13] were conducted across all demographic categories, and it was found that the SUS and guessability scores were not normally distributed ($p < 0.05$). Therefore, the effects of user sub-groups (kind of users) on SUS scores were analyzed by conducting two non-parametric tests. For the categories of gender and license ownership (2 levels each), the analysis was done by using Mann-Whitney U test [8], but whereas for the factors like cluster familiarity, profession and driving skill (>2 levels each), analysis was made by using Kruskal-Wallis test [13].

Mann-Whitney U test showed that gender had no significant effect on SUS score (i.e., $p > 0.05$), but license ownership had a significant effect on SUS score ($p < 0.05$). The results of Kruskal-Wallis test showed a significant effect on SUS score by profession ($p < 0.05$), but factors like driving skill and cluster familiarity did not affect the SUS score significantly (i.e., $p > 0.05$). Further analysis was done using Mann-Whitney U test to examine the effects on SUS by professions which included comparisons between "driver and housewife," "professional and housewife," "professional and student," "driver and student," "driver and professional," and "student and housewife." Test results showed that there were significant differences for the first three groups, i.e., between "driver and housewife," "professional and housewife," and "professional and student" ($p < 0.05$), but there were no significant differences for the other 3 groups, i.e., between "driver and student," "driver and professional," and "student and housewife" ($p > 0.05$).

To inspect the interrelationship between mean SUS and mean guessability scores for all the 20 warning icons on the entire data sample, Spearman correlational analysis [8] was conducted. It was found that the mean SUS score was significantly related and had a strong positive correlation with the mean guessability score ($r_s = 0.942$, $N = 20$, $p < 0.01$). Scatter plot (Fig. 30.1) shows that relationship between the mean SUS score and mean guessability score appears linear. The relationship between mean SUS score and mean guessability score for the "drivers" group was less significant and had a weak positive correlation ($r_s = 0.4$, $N = 20$, $p > 0.05$). Data from the Spearman correlation analysis between the mean SUS score and mean guessability score are shown in Table 30.5.

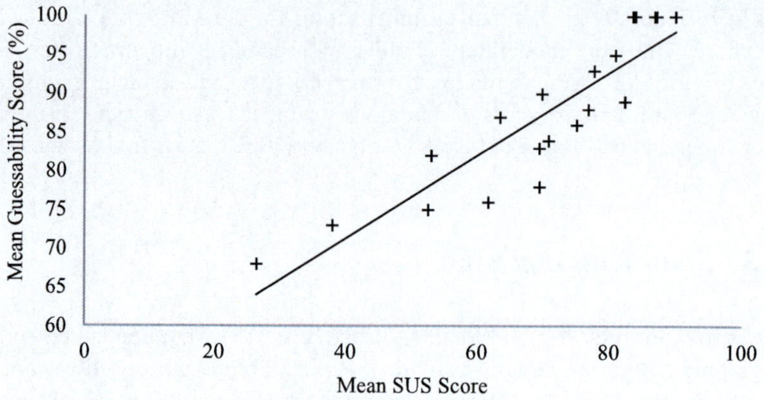

Fig. 30.1. Scatter plot of mean SUS and mean guessability scores for 20 warning icons

Table 30.5 Spearman correlation between mean SUS and mean guessability scores ($N = 20$)

Demographic factor	Response	Correlation coefficients (r_s)
Gender	Male	0.80[a]
	Female	0.92[a]
License ownership	Have	0.91[a]
	Don't have	0.92[a]
Profession	Driver	0.40
	Professional	0.81[a]
	Student	0.85[a]
	Housewife	0.85[a]
Driving skill	Expert	0.85[a]
	Average skill	0.82[a]
	Learner	0.89[a]
Cluster familiarity	Most familiar	0.79[a]
	Average familiar	0.88[a]
	Non familiar	0.90[a]

[a]Correlation is significant at 0.01 level (2-tailed)

30.4 Discussion

Results from survey indicate that male subjects working in driving profession exhibited the highest guessability score than in any other professions. This is in line with the observation in [13] that cultural background is an important determinant affecting sign-guessing performances. Factors like gender, driving skill, and cluster familiarity showed no significant effect on SUS scores but license ownership and profession had a significant effect on SUS score. This might be because male professionals drive cars for longer durations and more frequently and therefore get more accustomed with the warning icons. A significant and strong positive correlation was found between the mean guessability score and mean SUS score for all the 20 warning icons for the entire data. The correlation between mean SUS and mean guessability scores for drivers was not significant at $p < 0.01$ (weak positive correlation). Standard deviation for housewives was maximum in case of guessability scores, and it was minimum for the drivers. This indicated that housewives who drove cars frequently performed better in the guessability test as they were perhaps more familiar with the instrument cluster, whereas those housewives who drove cars only rarely performed poorer. Thus, there was greater spread of performance data for housewives but not so much in case of drivers who generally could be considered skilled at driving.

Also, on an overall basis, the usability scores for all participants were lower than guessability scores. It can probably be inferred that by utilizing modified SUS instrument to measure warning icons usability, more conservative opinions were expressed by the participants. This might be because of greater thoroughness of the SUS analysis instrument in measuring usability than the process used to measure guessability. Two warning icons, namely "Brake Warning Light" and "Tire Pressure Monitoring System" were rated poorly for usability and were regarded as "inadmissible" because of low usability ratings. This is probably due to the nature of these icons, as well as the specific visual design details used in these icons which affect their comprehensibility and usability in general.

30.5 Conclusion

In this study, 50 people from different social and professional backgrounds assessed the guessability and SUS-based usability of 20 common warning icons in Indian cars. The results indicated that guessability score was highest for the following warning icons-*Turn Signal, Low Fuel Indicator, Open Door Warning, Seat Belt Remainder, Battery Warning Light,* and lowest for *Tire Pressure Monitoring System. Turn Signal* icon received the highest usability score on the SUS instrument whereas *Tire Pressure Monitoring System* received the lowest. A significant and strong positive correlation was found between mean guessability score and mean SUS scores for all the 20 warning icons. Two warning icons—*Brake Warning Light* and *Tire Pressure Monitoring System* were found to be of high-risk factor due to low usability scores.

The research approach and results of this study should serve as a valuable reference for interface designers involved in usability testing of warning icons as well as other graphical symbols. However, the scope of this study was limited to 20 warning icons. In the future, other warning icons available in automobile instrument clusters could also be evaluated and examined using a methodology inspired by this research.

Also, this research was constrained by employing a convenience user sample (50 participants). It did not focus on employing specific traits in order to select relatively homogeneous survey participant groups. This work could be considered as an exploratory research in that sense. In future research, participant selection could be more deliberate to ensure among the participants, homogeneities of certain traits, e.g., gender, age, or profession. Such research could be employed to validate the group-related dependencies of performance results (ref. Table 30.4). Also, cultural differences could play an important role in determining use of automotive warning icons, and future research could also focus on the understanding the role of such factors in warning icons' usability.

References

1. Carvalho, R., Soares, M.: Ergonomic and usability analysis on a sample of automobile dashboards. Work **41**(Supplement 1), 1507–1514 (2012)
2. Marcus, A. (2015). The next revolution: vehicle user interfaces. In: *HCI and User-Experience Design,* pp. 91–100.
3. Gibson, Z., Butterfield, J., Marzano, A.: User-centred design criteria in next generation vehicle consoles. Proc. CIRP **55**, 260–265 (2016)
4. Walker, G.H., Stanton, N.A., Young, M.S.: Where is computing driving cars? Int. J. Hum.-Comput. Interact. **13**(2), 203–229 (2001)
5. Jung, S., Choe, J.: Representation method of warning icons on automobile instrument panel for improving driver's cognitive experience. Int. J. Control Autom. **11**(1), 143–152 (2018)
6. Prabhakar, G., Ramakrishnan, A., Madan, M., Murthy, L.R.D., Sharma, V.K., Deshmukh, S., Biswas, P.: Interactive gaze and finger controlled HUD for cars. J. Multimodal User Interfaces **14**(1), 101–121 (2020)
7. Klauer, S.G., Guo, F., Simons-Morton, B.G., Ouimet, M.C., Lee, S.E., Dingus, T.A.: Distracted driving and risk of road crashes among novice and experienced drivers. New England J. Med. **370**(1), 54–59 (2014)
8. Ng, A.W.Y., Lo, H.W.C., Chan, A.H.S.: Measuring the usability of safety signs: a use of system usability scale (SUS). In: Proceedings of the International Multi-conference of Engineers and Computer Scientists, vol. 2 (2011)
9. Caldwell, J.: Safety icons and usability: a Peircean reanalysis. In: Proceedings of the IEEE International Professional Communication Conference, pp. 1–8 (2009)
10. https://www.microsoft.com/en-us/microsoft-365/excel. Retrieved 23 Dec 2019
11. https://www.ibm.com/analytics/spss-statistics-software. Retrieved 28 Dec 2019
12. Bangor, A., Kortum, P.T., Miller, J.T.: An empirical evaluation of the system usability scale. Int. J. Hum.-Comput. Interact. **24**(6), 574–594 (2008)
13. Ng, A.W.Y., Chan, A.H.S.: The guessability of traffic signs: effects of prospective-user factors and sign design features. Accident Anal. Prevent. **39**(6), 1245–1257 (2007)

Chapter 31
What Does the HCI Design Industry Expect from an Entrant? Findings from Interviews with Indian UX Design Team Leads

Surbhi Pratap and Jyoti Kumar

Abstract India is the world's largest exporter of HCI products. However, limited research is available on the HCI design processes adopted in the Indian industry. This paper reports findings from in-depth interviews with 25 Indian HCI design team leads on the design processes followed in the industry and the competencies expected from design professionals. One of the key findings from the interviews suggests that the industry expects a graduate to have an understanding of the business value of the design processes adopted for a given problem. It is also expected that entrants have knowledge of different types of HCI design processes relevant to specific organizations like agile sprints, lean processes, etc. Further, for each of the different design processes adopted by the industry, the specific core skills are expected to be present in the new professional in order to work seamlessly in the industry. Skills like stakeholder engagement ability, sprint usability testing, time management, user research techniques, usage of appropriate terminologies in design communication, co-creation ability with stakeholders of different expertise, etc., are what the industry expects from new HCI professionals. The findings of this study have curricular, pedagogical, and strategic relevance for both HCI academics and industry.

31.1 Introduction

Human–computer interaction (HCI) as a subject of design and computing has been well established globally [1]. It is gaining more prominence in recent times with the advent of digitization in our daily lives [2] and has seen a steady rise in the number of enrolled students who aspire to be HCI design professionals [3]. As a subject, HCI is taught with the intent of making the students explore, understand, and improve the usability and user experience of interactive systems and products [2].

S. Pratap (✉) · J. Kumar
Indian Institute of Technology Delhi, New Delhi, India
e-mail: surbhi.pratap@gmail.com

J. Kumar
e-mail: meetjyoti@gmail.com

A. Chakrabarti et al. (eds.), *Design for Tomorrow—Volume 1*, Smart Innovation, Systems and Technologies 221, https://doi.org/10.1007/978-981-16-0041-8_31

However, HCI design academics in India is relatively new when compared with global HCI design education [4]. Ironically, India has been the world's largest exporter of HCI products [5, 6] which results in a high demand for competent HCI design professionals. This results in an opportunity for the HCI design education in India to modify such as to produce more competent and 'industry-ready' graduates. This paper reports findings from in-depth interviews with 25 HCI design team leads of UX design industry in India working in domains of Web site and mobile application designs. The paper reports the expectations of the design team leads from the entrants, gaps between the HCI design process taught in academics versus the ones being followed in the industry, and the challenges faced by new graduates as they embark on their professional journey. The findings have curricular, pedagogical, and strategic relevance for both HCI academics as well as the HCI design industry in India.

31.2 Related Work

31.2.1 HCI Education in India

There is literature available on HCI education in Western countries [1], however, not much is available on the details of HCI education in India. It has been reported that HCI education has given more importance to design tools and task analysis in New Zealand [7]; to design principles, processes, and cognitive psychology in Sweden [8]; to graphical design and heuristic evaluations in Costa Rica [10]; semiotics and social inclusion in Brazil [9]. These studies show that even though largely the concepts and tools taught in HCI education worldwide remain similar, the aspects which are emphasized by teachers tend to change as a consequence of local academic and professional cultures [1]. Literature reports that the academics in HCI design has not kept pace with the growth of the software design industry in India [4]. Hardly any central Indian university offers HCI design as a separate undergraduate course, and a very limited number offer graduate specialization or Ph.D. in Interaction design (IIT Guwahati, IIT Bombay, IIT Delhi, NID Ahmedabad). However, eminent institutes like IIT Bombay offer courses like interface design, user studies, visual language syntactics, usability evaluation, human factors in interaction design, semantic and communication theory, interactive data visualization, pattern language for interaction design, service design, design management and professional practice, industry project, etc., as part of the interaction design curriculum at graduate level [11].

31.2.2 HCI Industry in India

India has been undergoing the transition from being an outsourcing hub to a center of in-house software product development. This has resulted in a rise and demand of

HCI professionals in the country [12]. This has also meant that the field is witnessing a growth spurt. A decent number of HCI conferences with high international industrial and academic participation are held every year in India now (UX Now, Design Up, etc.) as compared to the first HCI conference which was organized in 2004 by the HCI Professional Association of India, with about 30 attendees [4]. HCI and UX jobs are growing exponentially, and the HCI industry now demands well-prepared entrants in domains of UX research, UX/UI design, development, and visual design to take on the challenges. This study aims to bridge this gap by identifying the challenges posed on an HCI professional in the Indian industry and the expectations and suggestions from industry veterans on how new entrants can be well prepared to take on those.

31.3 Methodology

Conversational analysis [13, 14] was done on the verbal reports collected from the participants during semi-structured interviews conducted one on one on video calls. The participants were team leaders of the UX design industry in India working in companies that were designing and developing web sites and mobile applications. The planned questions and their related probes for the interviews were as follows:

1. Can you please share the UX design process you followed in your last two projects? (Probes: How was the design brief generated? How was the design process selected? How were the users of the design identified? How were the user's needs understood ?)
2. What will be your suggestions to entrants to the industry? (What should they learn during academics, so that they are better prepared for the industry? What courses do you think are required for preparing the UX professionals? Do you think industry internships during education help? What kind of internships students should do during their education?)

While it was ensured that topics planned as above were discussed, however, the interview style was kept conversational rather than question–answer, and therefore, the flow of the interview did not necessarily follow this same sequence of questions and probes as above, rather it developed upon the responses of the interviewee. Each interview lasted 35–45 min, and recording was done with the consent of the participants.

31.3.1 Participants

Purposive quota sampling [15] was used to select the participants for this study. This allowed us to focus on people who would be most likely to experience, know about, or have insights into the research topic of HCI design in India. The contacts were contacted through personal references as well as LinkedIn profiles. Designers were requested for a suitable time for a Skype conversation after introducing to them the intent and the objective behind this interview. The interviews were recorded after seeking their consent. For most of the interviews, the time slot was after their office hours, where they would take their laptops at home or a cafe. The interviews were semi-structured, with the designers explaining the HCI design processes they have followed initially and then responding to probes on their specific experiences in the UX design industry. The probes were developed further upon the conversations of the design practitioners.

The average age of the design professionals interviewed was 33.8 years with Std. Dev. of 3.97 years. The average professional experience of participants was 8.72 years with Std. Dev. of 2.93 years. Out of the 25 participants, 20 were males and 5 were females. An overview of participants' profiles is displayed in Table 31.1.

31.3.2 Study Protocol

The interviews were designed to capture major challenges and experiences of the design professionals with respect to the projects they work on and the design processes that they follow. The interviews started with an introduction of the interviewers and the purpose of the interview. An informed consent was obtained on the Skype call itself and was recorded along with the rest of the interview. The interviewer then asked about the personal details and prior work experiences. The sequence of conversations initiated by the interviewer was planned as: Can you please narrate the design process followed by you in your last two design projects? How was the design brief generated? What challenges do you find in the HCI design industry, especially in India? What suggestions would you have for a new entrant for the industry, academically and professionally?

The qualitative responses from interviews were recorded and then transcribed.

31.3.3 Data Analysis

A 'thematic analysis' was conducted on the interview data post transcription [16, 17]. From the transcript of interviews of each participant, for both the questions, 'themes' in responses were identified and counted. By 'themes,' we mean a consistent set of arguments used in explaining an experience that the respondent had. It is a unified semantic unit of the conversation. A total of six themes were tabulated, and frequencies were counted namely process, brief, stakeholders, guidelines, industry expectations, and academic suggestions. Once the conversational sentences related to a theme were identified and isolated from the rest of the transcript, for each theme, a word cloud was created to identify the most recurring terms. Then, the most frequently used words from these word clouds were used as nodes and subnodes in the NVivo software to make a word-tree map. These word-tree maps indicated how

Table 31.1 Overview of HCI professionals' profiles who participated in the study

Sl. no	Gender	Exp. (y)	Age (y)	Designation	Type of organization	Qualifications
P1	M	7	33	Sr. Lead UX Designer	Product company	M.Des (PD), B.Tech
P2	F	9	34	Sr. Lead UX Designer	Product company	M.Des (Int. D), B.Tech
P3	M	12	37	Product Designer	Service company	M.Des (Int. D), B.Tech
P4	M	14	40	Design Manager	Product enterprise	M.Des (Int. D), B.Tech
P5	F	8	33	Sr. UX Associate	Product enterprise	M.Des (PD), B.Tech
P6	M	5	27	CEO, Founder	Design consultancy	M.Des (Int. D), B.Tech
P7	F	7	31	Sr. UX Design specialist	Product enterprise	M.Des (PD), B.Tech
P8	M	6	30	Sr. Lead UX Designer	Design consultancy	M.Des (PD), B.Tech
P9	M	12	39	Tech Expert, UX	Product enterprise	M.Des (PD), B.Tech
P10	M	9	32	Sr. UX Design specialist	Product enterprise	M.Des (PD), B.Tech
P11	M	9	33	Sr. UX Designer	Design consultancy	M.Des (PD), B.Tech
P12	M	9	34	CEO, Founder	Design consultancy	M.Des (PD), B.Tech
P13	M	7	28	Sr. Lead UX Designer	Product company	B.Tech
P14	M	6	35	Lead UX Designer	Product enterprise	M.Des (PD), B.Tech
P15	M	6	41	Sr. UX Design specialist	Product enterprise	M.Des (PD), B. Arch
P16	F	10	35	Lead designer, research	Product enterprise	M.Des (Comm), BFD
P17	M	5	27	Sr. UX Designer	Design consultancy	M.Des (Int. D), B.Tech
P18	M	9	33	Lead UX Designer	Product enterprise	M.Des (PD), B.Tech
P19	M	5	32	Design Manager	Product enterprise	M.Des (PD), B.Tech
P20	M	5	29	Lead UX Designer	Product company	M.Des (PD), B.Tech

(continued)

Table 31.1 (continued)

Sl. no	Gender	Exp. (y)	Age (y)	Designation	Type of organization	Qualifications
P21	F	9	35	Design consultant	Design consultancy	M.Des (PD), B. Arch
P22	M	12	35	Head of UX	Design consultancy	HFI CUA, GD Animation
P23	M	15	40	Design Head	Design consultancy	M.Des (PD), B. Arch
P24	M	13	39	Associate Design Director	Product enterprise	M.Des (PD), B. Arch
P25	F	9	33	CEO, Founder	Design consultancy	M.Des (Int. D), GDPD

the most frequently used words were used in the conversations. Figure 31.1 depicts one such word cloud and related word-tree map.

31.4 Findings

23/25 design team leads reported that the design process followed in the industry was not singular and changed significantly depending on the project and especially with the type of organization that they were working for. For example, product companies provide time and resources for research, whereas for most service-based companies and design consultancies, user research is compromised because of time and budget constraints. It was also reported (21/25) that the brief mostly came from project managers, marketing teams, and clients directly and was not co-created by the designer after inputs from different stakeholders as is taught in academics. 20/25 of the leading HCI designers did not have a graduate degree in interaction design, rather in product design, they were not insistent on a particular specialization for the entrants but rather felt that design academics needed to focus on the business value of design, time management, soft skills and be updated with the recent processes being followed in the industry, so that the graduating students are well prepared to face the industry.

31.4.1 Challenges in the HCI Design Industry

Unclear design briefs with minimum inputs by design teams. One of the major challenges for the designers was that the briefs would come from marketing teams, who though did their research, lacked the design perspective. This would result in

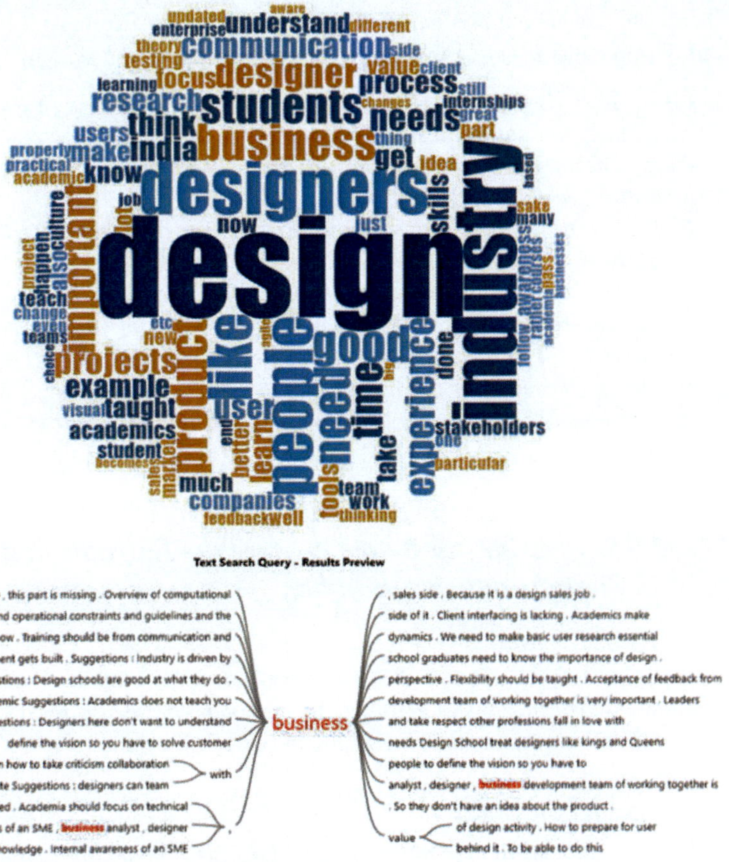

Fig. 31.1 A word cloud for the theme 'industry expectations' and the associated word-tree map for the word 'business' with nodes and subnodes

confusion from the beginning of the project itself. The verbal reports are coded as per Table 31.1.

> *P13: "Brief comes from the manager only. Sometimes the CEO."*
> *P14: "There is no brief, no information architecture and no wireframes. Direct visual design is made. We make 30 versions till the client is happy and also stretch it to a longer time."*
> *P1: "Communication of briefs is not done well."*
> *P6:" initial research done by the client for the brief. Our input was to make actionable items- make personas, scenarios, so we build stories around them."*

Time and budget constraints Designers are quite aware of the classical steps in ideal design processes as prescribed by various D-schools across the globe. However, unless it is a huge product enterprise, the projects have very short timelines. And the clients are not very appreciative of the time spent in research, so it takes a backseat and design concepts are generated with very limited user research.

P25: "They don't see the value in user study. Major part of the time is seen by wireframes and visual design."
P12: "There was no time for user research. Entire flow was charted out on paper. Animation was shown from dribbble."
P23: "In reality, parts of the process are skimmed down to match time and budget constraints."
P10: "Decisions in the mobile world are taken *very fast. By the time these come to us from headquarters we have no time so what happens is, there is no user research. Field study is done by the marketing team (mostly in Delhi), so they do queries, questionnaires* etc.*"*
P3: "clients are result oriented so we skipped a lot of research while in service company."

31.4.2 *Suggestions to Better Prepare the Entrants to the HCI Design Industry in India*

Co-creation ability with all stakeholders It was reported that new entrants have very limited idea and understanding of other stakeholders in the product development team, and this creates a problem during the design process.

P3: "Design School treat designers like kings and Queens and they think engineers should follow them".
P12: "It can also help to understand that people who are not designers contribute *to the design. Like even a client can explain the product better."*
P13: "We need to learn and document failures which is very important."

Understanding the business value of design Most designers spoke about the need to understand the business value of the designs. The entrants are usually unaware of the quantifiable value that design brings to business such as the monetary gains (profits) as well as better user conversion rate.

P4: "designer is a part of a business."
P25: "Academics does not teach one the business side of it."
P2: "process is like doing a college project, but the problem definition is not in your hand you think of the best ROI".
P18:" Students don't understand operational constraints and guidelines and the business value behind it."

Soft skills Students need to be taught how to communicate their design to various stakeholders and also learn soft skills for doing user research.

P25: "Client interfacing is lacking."

P24: "Training should be from a communication and business perspective."
P14: "Students need to be taught communication skills, rationales, office politics, dynamics."
P23:" How to present, communication aspect is missing a lot in designers."
P11: "Indirect qualities as a designer are not brought out for example interpersonal skills. Because of that we do not have good design speakers in India."

Time management Most of the designers emphasized the need for design students to do industry training for a significant time period, so that they can become aware of the time constraints of the industry.

P25: "Academics make projects for *six months plus. In industry they expect you to do research in one week. Ratio of time estimation in academics and industry is skewed."*
P23: "Internship cycles are usually very short. It gives an exposure to practical projects."
P20: "Students should get to companies when they get full-fledged projects and there should be at least 6 to 7 months of internships."
P13: "Getting students involved in a real project is very important."

Being updated with the latest industry trends There are a lot of industry processes and jargons like lean design, agile processes, design sprints, etc., which are not a part of the classic design curriculum. Entrants need to be aware of these before joining the industry, so that the actual industry processes do not pose any shock value to them.

P16: "People should not get scared of design thinking methodologies."
P23: "People focus on software skills rather than why they are doing what."
P1: "designers have platform limitations."
P4: "most institutes focus on theory and not tools which is good because tools become obsolete over time but still designers need to have the mastery over tools. Faculty members sometimes are not well versed with the contemporary scenario, so they need to be aligned with the world of design presently."
P6: "designers need to have a design sprint with all the stakeholders because people are reluctant to understand the value of design."
P10: "Students should be aware and well-versed with agile, sprints and other similar industry trending processes."

31.5 Discussion

The aim of this study was to understand the challenges of the HCI design industry in India and the design processes which the industry leaders follow to overcome those.

Another objective was to find out from the senior HCI professionals about the expectations that they have from fresh graduates. These expectations were communicated in the form of suggestions to the design academics, so that it reaches the maturity of producing graduates who are well prepared to take on these challenges of the industry and can become assets to the HCI design community of India.

One main challenge of the industry as reported by the designers are that the design briefs are not co-created by them as is taught in academics, rather they are given by experts who, more often than not, do not belong to the design domain, and this creates a confusion as to which process should be followed and how. The other challenges are the very tight time and budget constraints because of which they are not able to give enough time and resources to do user research, usability testing and user validation and that results in less than satisfactory design process for them.

The entrants to the HCI design industry would face the above challenges, besides expectations from them of knowing the current methodologies, industry jargon, and the quantifiable business value of their designs, so they can convince clients of the time a proper design process takes and justify it. Time constraints also ensure that the entrants have to be on their toes with design outputs and do not have the luxury of stretched timelines of academic projects. Besides these, the entrants also need to have good communication skills to present their ideas as well as to conduct user research and to co-create with other stakeholders in the team without hassles. The HCI design industry leaders suggested that design academics can tweak its curriculum such that they are taught all the above skills while studying and these expectations are met in fresh graduates. The findings of these interviews are revelatory and have curricular, pedagogical and strategic relevance for both HCI academics and industry.

References

1. Abdelnour-Nocera, J., Austin, A., Michaelides, M., Modi, S.: A cross-cultural evaluation of HCI student performance–reflections for the curriculum. In: International Conference of Design, User Experience, and Usability 2013 July 21, pp. 161–170. Springer, Berlin
2. Fallman, D.: The interaction design research triangle of design practice, design studies, and design exploration. Design Issues **24**(3), 4–18 (2008)
3. Boscarioli, C., Bim, S.A., Silveira, M.S., Barbosa, S.D.: How to join theoretical concepts, industry needs and innovative technologies in HCI courses? The big challenge of teaching HCI. In: International Conference on Human-Computer Interaction 2015 Aug 2, pp. 27–36. Springer, Cham.
4. Thakkar, D., Sambasivan, N., Kulkarni, P., Kalenahalli Sudarshan, P., Toyama, K.: The unexpected entry and exodus of women in computing and HCI in India. In: Proceedings of the 2018 CHI Conference on Human Factors in Computing Systems 2018 Apr 21, pp. 1–12.
5. Thomas, P.N.: Digital India: Understanding Information, Communication and Social Change. SAGE Publications, India (2012)
6. Chandler, C., Zainulbhai, A., (eds) Reimagining India: Unlocking the Potential of Asia's Next Superpower. Simon and Schuster (2013)
7. Sharkey, E., Paynter, J.: CHI education in New Zealand. Bull. Appl. Comput. Inform. Technol. **2**(3) (2004). https://citrenz.ac.nz/bacit/0203/2004Sharkey_CHINZ.htm. Retrieved 28 Feb 2013
8. Gulliksen, J., Oestreicher, L.: HCI education in Sweden. ACM SIGCHI Bull. **31**(2), 4–7 (1999)

9. de Souza, C.S., Baranauskas, M.C., Prates, R.O., Pimenta, M.S.: HCI in Brazil: lessons learned and new perspectives. In: Proceedings of the VIII Brazilian Symposium on Human Factors in Computing Systems 2008 Oct 21, pp. 358–359.
10. Calderon, M.: Teaching human computer interaction: first experiences. CLEI Electron. J. **12**(1), 1–9 (2009)
11. https://www.idc.iitb.ac.in/academic/mdes-programs/interaction-design/course-credits. Last accessed 22 Oct 2020
12. Yammiyavar, P.: Status of HCI and usability research in Indian educational institutions. In: IFIP Working Conference on Human Work Interaction Design 2009 Oct 7, pp. 21–27. Springer, Berlin, Heidelberg
13. Nielsen, M.F., Wagner, J.: Diversity and continuity in conversation analysis. J. Pragmat. **39**, 441–444 (2007)
14. Oak, A.: What can talk tell us about design?: Analyzing conversation to understand practice. Des. Stud. **32**(3), 211–234 (2011)
15. Robinson, O.C.: Sampling in interview-based qualitative research: a theoretical and practical guide. Qualitat. Res. Psychol. **11**(1), 25–41 (2014). https://doi.org/10.1080/14780887.2013.801543
16. Romesburg, C.: Cluster analysis for researchers. Lulu.com (2004)
17. Kaufman, L., Rousseeuw, P.J.: Finding Groups in Data: An Introduction to Cluster Analysis. Wiley (2009)

Chapter 32
Development and Evaluation of Usability Heuristics for Voice User Interfaces

Lokesh Fulfagar, Anupriya Gupta, Arpit Mathur, and Abhishek Shrivastava

Abstract In recent years, voice user interfaces have evolved substantially, enabling seamless and efficient human–machine interaction through spoken language. In spite of the increasing research, there is an absence of explicit evaluation methods for voice user interfaces (VUIs) for enabling their improvement. Presently, the evaluation criteria are primarily based on subjective metrics such as user reviews, ratings, or likelihood of future use. While these metrics have utility, they are often subjective to the users and offer hurdles to research such as cost of recruitment, time, and resources. Other alternatives are performance based such as response time and error rates, offering value to developers, but little insight into the design of these systems. There is a need for a usability-based evaluation method for VUI, analogous to the heuristic evaluation metrics proposed by Nielsen and Molich for screen-based interfaces. To address the same, our study presents a set of heuristics for usability evaluation of VUIs. Initially, existing literature in the domain of VUI was analysed to identify prevalent themes of usability issues. These themes were then categorised to define 11 usability heuristics. The set of heuristics will enable designers to rapidly evaluate and investigate areas of improvement in VUIs. A between-subjects study with 12 HCI professionals, involving usability evaluation of a VUI application, was conducted to test this hypothesis. The study reveals a statistically significant increase in the number and diversity of usability issues identified using the presented heuristics.

32.1 Introduction

Voice user interfaces (VUIs) have recently gained interest as an alternative to graphical user interfaces (GUIs) in many contexts. VUIs offer the unique advantages of mobile and hands-free interaction. In spite of their potential, they are not as prevalent as GUIs [21]. Moreover, VUIs are often synonymous with usability issues that provide negative user-experiences [5, 21, 23]. Why do these issues go undetected?

L. Fulfagar (✉) · A. Gupta · A. Mathur · A. Shrivastava (✉)
Indian Institute of Technology, Guwahati, Assam 781039, India
e-mail: lokeshfulfagar@gmail.com

A. Chakrabarti et al. (eds.), *Design for Tomorrow—Volume 1*, Smart Innovation, Systems and Technologies 221, https://doi.org/10.1007/978-981-16-0041-8_32

Several sets of established heuristics have guided the design of GUI for decades—some of the most notable ones being from Nielsen [18], Norman [19] and Shneiderman [27]. However, an equivalent set of heuristics is missing for VUIs. For VUIs to gain popularity, and to truly leverage their advantages, there is a need to formalise the role of an HCI specialist, by providing a method of usability evaluation of VUIs. In this paper, we present a set of heuristics for practitioners to base the usability evaluation of VUIs upon. These heuristics are designed to identify usability issues with VUIs and were drafted through a literature review of 22 studies conducted in this domain following the methodology suggested by Rusu [24]. We conducted a between-subjects study with 12 HCI professionals, as part of which they evaluated an existing VUI application with and without our heuristics. We collected the number and diversity of the issues identified using both the methods, along with other quantitative and qualitative feedback. With the use of presented heuristics, a statistically significant increase in the number and diversity of issues was observed in the study.

32.2 Related Work

Graphical user interfaces have been extensively studied and hence provide an established set of guidelines for expert evaluation, presented by Nielsen [18], Norman [19] and Shneiderman [27]. However, multiple studies have revealed that these are not sufficient to conduct the analysis of VUIs. Recently, several non-traditional interfaces have also been studied in this regard. Sutcliffe et al. [28] studied the usability issues identified in the domain of virtual reality (VR) and presented a set of guidelines to keep in mind while designing VR interfaces. Similar studies have been done for augmented and mixed realities [9]. Within the context of voice-based interactions, Mathur et al. [12] studied user preferences in interactive voice response (IVR) calls and presented a set of guidelines for outbound automated messages. However, the interaction modality of IVR affords a different set of usability issues when compared with traditional VUIs. Few studies have made efforts to provide a structure to the design and evaluation of VUIs. A recent study aimed at providing usability guidelines to help design the interaction for VUIs [15]. The study, however, gives guidelines that can be used while designing VUI, but cannot essentially be used for evaluation. Two other studies presented sets of speech-based heuristics for VUIs [11, 20]. The former is more centred at ergonomics, and its approach while useful is incomplete, as it does not involve the evaluation of the presented heuristics which are critical to assess their usefulness. The latter, although detailed and complete in its approach, concludes that the presented heuristics were not as effective when used independently and needed to be used in combination with Neilsen's heuristics for a thorough evaluation. Utilising a speech-based set of heuristics in conjunction with another set of visual guidelines is sub-optimal and presents the need to generate a comprehensive list of voice usability heuristics.

32.3 Development of Voice Usability Heuristics

In this section, we describe our analysis of existing literature pertaining to the usability of voice-based interactions. We then propose a set of 11 usability heuristics that designers and voice usability experts can use to conduct usability evaluation of VUIs.

32.3.1 Identifying Usability Themes from Literature Review

In order to form the new set of usability heuristics for VUIs, we followed a method defined by Rusu [24]. We studied 22 relevant papers composed of full and short papers published in the leading HCI conferences and journals. Emerging patterns for VUIs were analysed through these papers, and similar findings and observations were grouped in order to formulate a set of 11 themes. These themes and references are presented in Column 1 and Column 2, respectively, of Table 32.1. Further, prior to defining the exact heuristics, we analysed the existing foundational GUI guidelines and heuristics by Nielsen [18], Norman [19] and Shneiderman [27]. These were then mapped to our identified themes to ensure that the formulated themes did not leave any aspect of the existing literature unaddressed. As shown in the Column 3 of Table 32.1, all the proposed dimensions in existing GUI literature fit well within the identified themes, although none of those dimensions addressed the themes of 'easy to understand system voice' and 'secondary medium of interaction'.

32.3.2 Defining Voice Usability Heuristics

Based on the themes identified in Table 32.1, a set of 11 voice usability heuristics were defined as presented in Table 32.2. Existing terminology was used carefully wherever possible for easier recall, while avoiding potential incorrect associations.

32.4 Testing and Evaluation

An evaluation study was conducted with 12 HCI professionals to assess the formulated usability heuristics for VUI. The participants were asked to do a usability evaluation of the voice aspects of Alexa [1] mobile application. Half of the participants were provided with the proposed heuristics while the other conducted the test without them. The differences in the outcomes of each participant group highlighted the efficiency and usefulness of the proposed heuristics. The study was conducted remotely between April and May 2020 due to the COVID-19 situation.

Table 32.1 Identified usability themes based on existing literature study

Themes	Existing relevant literature	Existing GUI foundations
Easy to understand system voice	Papers [5, 26] talk about the importance of varied characteristics of system voice like vocabulary, expressions, intonation, pauses and pronunciation. Another pattern identifies the subjective of output voice quality with factors like speech, noise and attention [6, 26]	–
Degree of freedom and user control	Papers [29, 31] discuss the ability to interrupt the system and exit a flow anytime. Further [11] talks about enabling user empowerment and higher satisfaction through user control and freedom	User control and freedom [18]; Support internal locus of control, Permit easy reversal of actions [27]
Recognisable cues and less cognitive load	References [2, 4, 13, 29, 31] Papers focus on how presenting large amounts of information at once tends to increase memory load [6, 11]. Knutsen [8] also reports that recalling information given by voice is difficult	Recognition rather than recall, Aesthetic and minimalist design [18]; Affordance [19]; Reduce short-term memory load [27]
Timely feedback and status update	References [3, 6, 10, 11, 16, 30] Papers establish the need for confirmation and conveying appropriate feedback to the users' statements. Since speech is invisible, the absence of feedback leads to errors of all kinds [26]	Visibility of system status [18]; Feedback [19]; Offer informative feedback, design dialogue to yield closure [27]
Matching the perception and conversation models	Papers [6, 17] highlight the need for natural dialogue structure and use of generic vocabulary. Further [7, 29] found that if the users are already familiar with the interactions there is an increase in efficiency and ease in usability [3]	Match between system and the real world [18]; Mapping [19]

(continued)

Table 32.1 (continued)

Themes	Existing relevant literature	Existing GUI foundations
Flexibility and efficiency	References [3, 11, 26] Papers identify the need for maximising flexibility and efficiency by enabling the use of shortcuts and user customised responses	Flexibility and efficiency of use [18]; Enable frequent users to use shortcuts [27]
Preventing and minimising errors	Papers [6, 11, 30] talk about building trust by minimising the errors through dynamic error prevention and conscious design considerations [25]	Error prevention [18]
Recognising and resolving error	References [6, 11, 22, 29, 30] Papers talk about helping the user to identify and recover from errors. System needs to identify the cause of the error as well, since each error needs to be handled differently for recovery, failing which gets the user stuck in error spiral [25]	Help users recognise, diagnose and recover from errors [18]; Offer simple error handling [27]
Secondary medium of interaction	Papers [3, 7, 14] discuss how using speech interfaces at public spaces is a privacy issue. Another paper [26] talks about how alternative mediums like touch can provide faster means for entering data	–
Striving for consistency	Papers [11, 26] talk about striving for consistency and standards across the entire application and also in line with the existing norms	Consistency and standards [18]; Consistency [19]; Strive for consistency [27]
Providing help and documentation	References [5, 6] Papers discuss the importance of providing help and hints either progressively [30] or contextually when needed [5, 31]	Help and documentation [18]

32.4.1 Participants

HCI professionals with industry experience were approached for the evaluation study. Emphasis was given on involving a mixed set of participants with and without expertise in VUI to take into consideration the influence of prior experience. 12 participants were recruited, comprising 6 VUI experts (E1–E6) and 6 non-experts (N1–N6). The participants belonged to the age range of 21 to 29 years, all of whom had formal

Table 32.2 Set of 11 voice usability heuristics

Voice usability heuristics
(1) Clear and Natural Voice. The language and the voice used in a VUI should be easy to understand and interpret for the user. Pronunciation should be clear with appropriate speed and pauses, preferably interacting with users in their natural language and dialect
(2) User Control and Freedom. VUI should empower the user to stay in control of the conversation, giving them the freedom to take contextual actions, undo their mistakes, change course, or interrupt or even opt out of the conversation
(3) Reduce Cognitive Load. VUI should employ minimum cognitive load throughout the conversation. System dialogues should be short, straightforward and intuitive, incorporating familiar words and phrases. System should not expect users to recall details from earlier parts of the conversation
(4) Visibility and Appropriate Feedback. The user should not feel lost while interacting with a VUI. The system should inform the user about the actions that are being carried out, their running status and outcome, all through timely feedback. VUI should also keep them aware of the possible actions that can be taken at any point of time
(5) Mapping Existing Mental Models. A user follows their existing real-life conversation models even when interacting with a VUI, and they expect the system to abide by it, too. VUI should be able to interpret the context of the ongoing conversation and answer in line with the user's existing conversation models
(6) Optimise for Efficiency. Conversation should be crafted for efficiency and speed. VUI should facilitate the user to take shortcuts to quickly navigate through a conversation. It should also accommodate the additional details that a user might give in a single response to bypass the intermediate steps to complete a task
(7) Error Prevention. System should be accurate in recognising the user's natural speech to prevent input-related errors. The conversation should be structured in such a way as to prevent other probable errors
(8) Error Recognition and Resolution. VUI should be able to understand an error and the cause behind it, in order to reduce similar errors in future. It should be able to communicate the same to the user in a comprehensible way and assist them to recover from it easily
(9) Reliable Alternate Medium of Interaction. There might be scenarios based on limitations around accessibility or the user's environment, where voice-based interaction may not be preferred. It should not render the VUI unusable. It is important for VUI to offer an equally usable alternate interaction medium for overall improved usability of the system
(10) Consistency and Standards. All the components of the VUI system including tone, language, terminologies, interactions and the overall persona should be consistent throughout the experience. It should also be in line with existing standards and conventions prevalent in the domain
(11) Help and Documentation. VUI should offer proactive help through guided on-boarding and contextual assistance. It should enable the user to easily access help whenever they need. The help provided should be explicit and actionable

education in design and professional working experience. Participation was based on personal interest and did not involve monetary compensation. The participants were divided into two equivalent groups G1 and G2, each consisting of three experts and three non-experts. The participants in G1 (E1–E3 and N1–N3) were provided with formulated heuristics (Sect. 3.2) to assist in the usability evaluation, while participants in G2 (E4–E6 and N4–N6) were asked to perform the usability evaluation without those heuristics.

32.4.2 Procedure

The testing was conducted remotely through a video call. Prior to the testing, recruited participants were briefed about the project and the requirement of a mobile device and computer for the test. For participants doing the task with proposed heuristics, a separate adequate time was given to them for reading and understanding the heuristics clearly. All the participants were also briefed about the method to record their identified usability issues as part of the task. Two sets of templatised Google sheets for each of the groups—G1 and G2—were shared to record their identified issues. They were asked to focus only on the usability aspect of voice-based interactions concerning the five native capabilities of Alexa excluding third-party 'skills'. These native capabilities included accessing information, music, productivity, shopping and Alexa talents. For the 30 min of task, both the participant and the facilitator continued to stay on the video call with audio and video muted. The participants were suggested to ask their doubts or queries at any point of time by simply unmuting their microphone. This was done to ensure that the participants do not become conscious of the facilitator's presence while still giving them freedom to ask any doubts. At the end of the session, feedback was taken through a short survey and a semi-structured interview to understand certain qualitative aspects of their testing experience.

32.4.3 Results

The usability evaluation sheets of the participants and their feedback from the form and discussion were analysed to draw quantitative and qualitative results. The number of usability issues found by the participants with heuristics (G1) ranged from 7 to 17 with a mean of 11.17, while for the participants without heuristics (G2) it ranged from 5 to 10 with a mean of 6.83, showing a considerable high number of issues discovered by the former with a statistically significant p-value of 0.018. Even most of the non-experts with heuristics (N1–N3) were found to have discovered a greater number of usability issues as compared to experts without heuristics (E4–E6). But interestingly, participants in G1 were less satisfied with the number of issues that they found within the time limit with an average rating of 4.33 of 7 as compared to 4.83 by participants in G2. Also, for participants in G1, the diversity in types of issues ranged from 5 to 9 different types while for G2, it was limited between 2 to 5 with an overall statistically significant p-value of 0.017. Also, G1 rated the clarity and confidence in their articulation as 6.67 of 7 compared to 3.67 by G2. All participants liked the overall task experience with an average experience score of 5.59 out of 7.

Table 32.3 Results (Mean ± Standard Deviation) for G1 and G2

Group	Usability issues	Types of issues	Task satisfaction	Articulation	Task difficulty
G1	11.17 ± 3.87	7.00 ± 2.05	4.33 ± 0.82	6.67 ± 0.75	5.83 ± 0.98
G2	6.83 ± 1.94	4.00 ± 1.10	4.83 ± 1.72	3.67 ± 0.94	4.50 ± 1.64

However, participants in G1 found the task easier in comparison with G2 with the mean ratings being 5.83 and 4.5 of 7, respectively (Table 32.3).

Both, expert and non-expert participants in G1 responded positively to the heuristics and found them helpful during the evaluation. Non-experts found it more useful with a mean score of 6 of 7 as compared to 4.67 by experts. Also, they found the language of heuristics clear and easy to understand. Non-experts rated the same with a mean of 6.33 of 7, while the experts felt that it could be made easier to understand for novice and rated the same as 5.

32.5 Discussion

As part of our research, we have formulated a set of 11 voice usability heuristics for identification and evaluation of usability gaps in voice-based applications. An evaluation study was conducted with 12 HCI professionals including 6 voice experts, the results and feedback from which show the usefulness and efficiency of those heuristics. We elaborate the discussion points from the study in the following sections.

Efficient Discovery of Usability Issues. Heuristics were found to engender increased discovery of usability issues enabling the participants using them to identify considerably higher number of usability issues. They also felt that they could easily identify even more usability gaps given more time. On the other hand, participants without heuristics felt that they had found sufficient issues. This was also reflected in the former's lower satisfaction with the usability evaluation as compared to their counterparts. Also, it was observed that the participants without heuristics included issues that were not essentially usability based. They slightly deviated from the primary objective by including issues like technical glitches and add-on feature requirements.

Diverse Themes. Heuristics were found to help the participants in identifying usability issues in more diverse themes. It was observed that participants without heuristics tend to fixate on the type of issues that they discover in the first few minutes of the task. Also, these types were generally around themes like user control, visibility and feedback, etc., which have their parallels in GUI heuristics. On contrary, the other group found the heuristics helpful, not only for uncovering hidden voice specific issues but also as a trigger whenever they found themselves stuck while discovering issues.

Methodical Articulation. Heuristics were found to enable clear and objective articulation of the identified issues. Participants with heuristics felt that the presence of a defined literature in the form of heuristics helped them in documenting and rationalising their issues. Moreover, they found their issues to be well documented, categorised and actionable as compared to the other group who felt their reported issues needed a level of analysis before use. Apart from identifying issues, heuristics were also found to be useful in acknowledging and validating existing good design decisions in the app.

Structured Approach. Heuristics were found to provide a logical direction and potential avenues for spotting usability issues while conducting evaluation. Participants with heuristics found themselves in control throughout the process and referred to heuristics frequently to seek ideas for identifying possible issues. On the contrary, participants without heuristics found themselves randomly evaluating the application and experienced instances when they hit a roadblock being unable to think of scenarios for finding more issues. Overall, the participants with heuristics found the usability evaluation task easier as compared to the participants without heuristics.

Feedback on Heuristics. Participants with heuristics found the set of heuristics clear and easy to comprehend. They also suggested having examples within the description of the heuristics to improve understanding. Some participants found a list of 11 heuristics too long to refer multiple times. Since the participants were using heuristics for the first time, the finding reveals a scope to study how this comfort changes with familiarity with the heuristics. Further, some participants had apprehensions when they found an issue that could be mapped to multiple heuristics. These participants were encouraged to map such issues to all relevant heuristics in the light of the different perspectives that could emerge, helping them in solving those issues through multiple lenses.

32.6 Conclusion

We present a set of 11 usability heuristics for VUIs based on an extensive study of existing literature in this domain. To evaluate these heuristics, we conducted a between-subjects study with 12 HCI professionals performing an analysis of a VUI with and without the proposed heuristics. The study revealed a statistically significant increase in the number and variety of usability issues, identified with the use of the presented heuristics, with a scope of finding even more provided more time. Heuristics were also found to provide a structured approach to the overall evaluation process. Moreover, they were found to enable better articulation and categorization of the identified issues proving helpful for further design steps.

However, the study had certain limitations. These include reliance on self-reporting data and feedback taken remotely (due to COVID-19 situation). Currently, participants conducted the evaluation by being introduced to the heuristics for the first time. In the future, we would like to further explore the ease of use of heuristics and its effects on performance based on familiarity with the heuristics. Additionally, we aim to build upon the presented heuristics by including examples of heuristics in action, making it more objective and easier to understand. Continuing in the same direction, we also intend to build a complete framework to assist both the experts and budding designers in carrying out efficient usability evaluation of VUIs.

References

1. Alexa Mobile Application (2020) Amazon.com, Inc. https://www.amazon.com/Amazon-com-Amazon-Alexa/dp/B00P03D4D2. Accessed 15 Apr 2020
2. Aylett, M., Kristensson, P., Whittaker, S., Vazquez, Y.: None of a CHInd: relationship counselling for HCI and speech technology. In: CHI '14 Extended Abstracts on Human Factors in Computing Systems, pp. 749–760 (2014). https://doi.org/10.1145/2559206.2578868
3. Begany, G., Sa, N., Yuan, X.: Factors affecting user perception of a spoken language vs. textual search interface. Interact. Comput. **28**, 170–180 (2015). https://doi.org/10.1093/iwc/iwv029
4. Bigot, L., Caroux, L., Ros, C., Lacroix, A., Botherel, V.: Investigating memory constraints on recall of options in interactive voice response system messages. Behav. Inform. Technol. **32**, 106–116 (2013). https://doi.org/10.1080/0144929X.2011.563800
5. Corbett, E., Weber, A.: What can i say? Addressing user experience challenges of a mobile voice user interface for accessibility. In: Proceedings of the 18th International Conference on Human-Computer Interaction with Mobile Devices and Services, pp. 72–82 (2016). https://doi.org/10.1145/2935334.2935386
6. Farinazzo, V., Salvador, M., Kawamoto, A., Soares, J.: An empirical approach for the evaluation of voice user interfaces (2010). https://doi.org/10.5772/9490
7. Howell, M., Love, S., Turner, M.: The impact of interface metaphor and context of use on the usability of a speech-based mobile city guide service. Behav. Inform. Technol. **24**, 67–78 (2005). https://doi.org/10.1080/01449290512331319058
8. Knutsen, D., Bigot, L., Ros, C.: Explicit feedback from users attenuates memory biases in human-system dialogue. Int. J. Hum. Comput. Stud. **97**, 77–87 (2017). https://doi.org/10.1016/j.ijhcs.2016.09.004
9. Ko, S., Chang, W., Ji, Y.: Usability principles for augmented reality applications in a smartphone environment. Int. J. Hum.-Comput. Interact. **29**, 501–515 (2013). https://doi.org/10.1080/10447318.2012.722466
10. Luger, E., Sellen, A.: Like having a really bad PA: the Gulf between user expectation and experience of conversational agents. In: CHI '16: Proceedings of the 2016 CHI Conference on Human Factors in Computing Systems, pp. 5286–5297 (2016). https://doi.org/10.1145/2858036.2858288.
11. Maguire, M., Marcus, A., Wang, W.: Development of a heuristic evaluation tool for voice user interfaces. In: HCII 2019: Design, User Experience, and Usability. Practice and Case Studies, pp. 212–225 (2019). https://doi.org/10.1007/978-3-030-23535-2_16
12. Mathur, A., Bhowmick, S., Sorathia, K.: A study of outbound automated call preferences for DOTS adherence in rural India. In: IFIP Conference on Human-Computer Interaction, pp. 24–33 (2019). https://doi.org/10.1007/978-3-030-29387-1_2.
13. Molnar, K., Kletke, M.: The impacts on user performance and satisfaction of a voice-based front-end interface. Int. J. Hum. Comput. Stud. **45**, 287–303 (1996). https://doi.org/10.1006/ijhc.1996.0053
14. Moorthy, A., Vu, K.: Privacy concerns for use of voice activated personal assistant in the public space. Int. J. Hum.-Comput. Interact. **31**, 307–335 (2015). https://doi.org/10.1080/10447318.2014.986642
15. Murad, C., Munteanu, C., Clark, L., Cowan, B.: Design guidelines for hands-free speech interaction. In: Proceedings of the 20th International Conference on Human-Computer Interaction with Mobile Devices and Services Adjunct, pp. 269–276. https://doi.org/10.1145/3236112.3236149(2018)

16. Myers, C., Nebolsky, J., Caro, K.: Patterns for how users overcome obstacles in voice user interfaces. In: Proceedings of the 2018 CHI Conference on Human Factors in Computing Systems, pp. 1–7 (2018). https://doi.org/10.1145/3173574.3173580
17. Neto, E., Campos, F.: Evaluating the usability on multimodal interfaces: a case study on tablets applications. In: Design, User Experience, and Usability. Theories, Methods, and Tools for Designing the User Experience (2014). https://doi.org/10.1007/978-3-319-07668-3_47
18. Nielsen, J., Molich, R.: Heuristic evaluation of user interfaces. In: Proceedings of the SIGCHI Conference on Human Factors in Computing Systems, pp. 249–256 (1990). https://doi.org/10.1145/97243.97281
19. Norman, D.: The Design of Everyday Things: Revised and Expanded Edition. Basic books (2013)
20. Nowacki, C., Gordeeva, A., Lizé, A.: Improving the usability of voice user interfaces: a new set of ergonomic criteria. SSRN Electron. J. (2020). https://doi.org/10.2139/ssrn.3584206
21. Pearl, C.: Designing Voice User Interfaces: Principles of Conversational Experiences. O'Reilly Media, Inc. (2016)
22. Porcheron, M., Fischer, J., Sharples, S.: Do animals have accents? Talking with agents in multi-party conversation. In: Proceedings of the 2017 ACM Conference on Computer Supported Cooperative Work and Social Computing, pp. 207–219 (2017)
23. Pyae, A., Joelsson, T.: Investigating the usability and user experiences of voice user interface: a case of Google home smart speaker. In: Proceedings of the 20th International Conference on Human-Computer Interaction with Mobile Devices and Services Adjunct, pp. 127–131 (2018). https://doi.org/10.1145/3236112.3236130
24. Rusu, C., Roncagliolo, S., Rusu, V., Collazos, C.: A methodology to establish usability heuristics. In: ACHI 2011: The Fourth International Conference on Advances in Computer-Human Interactions, pp. 59–62 (2011)
25. Schnelle, D.: A pattern language for error management in voice user interfaces. In: Proceedings of the 15th European Conference on Pattern Languages of Programs (2010). https://doi.org/10.1145/2328909.2328920.
26. Schnelle, D., Lyardet, F.: Voice user interface design patterns. In: EuroPLoP, pp. 287–316 (2006)
27. Shneiderman, B., Plaisant, C.: Designing the User Interface: Strategies for Effective Human-Computer Interaction. Pearson (2010)
28. Sutcliffe, A., Brian, G.: Heuristic evaluation of virtual reality applications. Interact. Comput. **16**, 831–849 (2004). https://doi.org/10.1016/j.intcom.2004.05.001
29. Wolters, M., Georgila, K., Moore, J., Logie, R., MacPherson, S., Watson, M.: Reducing working memory load in spoken dialogue systems. Interact. Comput. **21**, 276–287 (2009). https://doi.org/10.1016/j.intcom.2009.05.009
30. Yankelovich, N., Levow, G., Marx, M.: Designing SpeechActs: issues in speech user interfaces. In: Proceedings of the SIGCHI Conference on Human Factors in Computing Systems, pp. 369–376 (1995). https://doi.org/10.1145/223904.223952.
31. Zajicek, M., Wales, R., Lee, A.: Speech interaction for older adults. Univ. Access Inf. Soc. **3**, 122–130 (2004). https://doi.org/10.1007/s10209-004-0091-0

Chapter 33
Liquid Interactions: A Conceptual Interaction Framework for the Future Now

Shweta Chaudhary, Archana Balaji, and Sunil Ganesh

Abstract Gartner predicted that by 2020, customers would manage 85% of their relationship with an enterprise, without human interactions. Here we are—more than halfway past 2020 and in times where this prediction is becoming a reality. To minimize human interactions with customers, the common (mis)perception and low-hanging fruit is the adoption of conversational technologies such as Chatbots and AI via a multitude of channels. Brands are already cognizant that customer journeys have become dynamic, fluid, and unpredictable, as they consider usefulness, stay mindful and design for rapid changes in consumer desires. Customers tend to act on their needs of the moment—their expectations high and patience low. This new-age customer behavior alludes to the relevance and usefulness of making interactions flow like liquids through swift, seamless, and sticky micro-moments for a unified Omnichannel and multimodal ecosystem encompassing online and offline. The framework attempts to evolve a modern layer over known popular Interaction Design Principles to support liquidity in user experience.

33.1 Evolving User Experiences

"User Experience" encompasses all aspects of the end-user's interaction with the company, its services, and its products—Don Norman and Jakob Nielsen.

With digital products going beyond their traditional forms and functions for incorporation with every day human experiences, the line between digital and analog blurs. There is a need for interacting naturally regardless of the device, medium, and in a mode not dictated by predesigned flows. Just as our experiences change, so should our

S. Chaudhary · A. Balaji · S. Ganesh (✉)
Cognizant, Chennai, India
e-mail: mavericksunil@gmail.com

S. Chaudhary
e-mail: shweta.black@gmail.com

A. Balaji
e-mail: archana.balaji@gmail.com

A. Chakrabarti et al. (eds.), *Design for Tomorrow—Volume 1*, Smart Innovation, Systems and Technologies 221, https://doi.org/10.1007/978-981-16-0041-8_33

tools. In this context, the Omnichannel approach to user experiences and marketing has been a clear winner because of its ability to present a personalized message through the unification of traditional and emerging interactive channels.[1]

Despite becoming mainstream, Omnichannel is yet to be leveraged for its deeper aspects—task-based channel recommendation, personalized interaction choices based on behavior, channel preferences, etc. Omnichannel can help object-based uniformity in Enterprise scenarios. Any transaction happening at one place can trigger effortless domino effect to reflect at multiple places in the entire system. A key factor is kinship with technology to build micro-moments supported by a cohesive channel strategy that is reassuring, efficient, and eases task completion.

33.1.1 Understanding Multimodal in the Omnichannel Setting

Traditionally user experience is exclusively visual or gestural design. Contemplating multimodal design is essential for designing experiences that connect our sensory abilities. Take the example of a live concert experience, which is a general combination of listening to music, watching the performance, eating a snack, and clapping hands during an encore. Remove one sense (e.g., hearing), and the experience takes on an entirely different context. Similarly, the full extent of digital experiences can be realized only with seamless multimodal interactions between users and channels supported by recommending the right channels at the right scenarios.

The Merriam-Webster Dictionary defines Multimodal as *"having or involving several modes, modalities, or maxima."*[2] In Human–Computer Interaction, multimodality can refer to *"interaction with the virtual and physical environment through natural modes of communication."*[3] In the same vein, a more colloquial usage of the term would refer to the ability to use two or more channels simultaneously in a single interaction. Therefore, for purposes of the paper, we would like to define multimodal as concurrent channels working together to support a single user interaction or task.

Curiously, for customers on the laggard spectrum of technology adoption, multimodal and Omnichannel are used interchangeably. While multimodality in Omnichannel increases accessibility and usability, our scope was limited to leveraging multimodal experiences as part of the Omnichannel delivery. Multimodal experiences help customers through complicated processes by escorting them through a single experience, which leverages simultaneous and synchronous channels. By this simple definition, a customer can begin an experience through input via desktop (or a preferred channel), expand the experience through a synchronized interaction at a

[1]https://emarsys.com/learn/blog/multi-channel-marketing-omnichannel/#multichannel_omnichannel.

[2]https://www.merriam-webster.com/dictionary/multimodal.

[3]Bourguet, M. L. (2003). "Designing and Prototyping Multimodal Commands". Proceedings of Human–Computer Interaction (INTERACT'03), pp. 717–720.

physical location (store/branch, etc.) and finally complete the experiential transaction through a third different channel such a VUI. A few initial thoughts of multimodal are:

- Disparate experiences working harmoniously
- Experiences that close the "customer preference versus customer usage" gap
  - *Despite investments in a website and/or mobile app, our research indicates that a sizable portion of customers are still utilizing the phone channel for assistance*[4]
- Unwilling participants due to ineffective implementation
  - When an application/product/service is designed incorrectly or inefficiently, customers become unwilling participants as their efforts at self-service are increased. Even more unfortunate are Organizations being averse to allocating resources for process improvement when there are newer ideas or markets to be had.

To frame our hypothesis that an evolution of multimodal interactions through Omnichannel is necessary for enterprises to stay relevant, we needed a research deep dive.

33.2 Research—The Need for an Evolution of Interaction Design Frameworks

The paper's premise and subsequent debate sparked from a customer's requirement of designing a multimodal voice assistant to enable task completion. We conducted primary research in qualitative and quantitative formats, supported by secondary desktop research. This helped to understand the validity of traditional interaction design principles applied in futuristic experiences across multi-device and multi-generational contexts. Our research was exploratory (see Fig. 33.1) and certainly time bound.

The research was conducted through our company's network of friends and family in the USA. Thus, we were able to gain a suitably authentic perspective. Two hundred respondents across ages, genders, and ethnicities in the USA, participated in a survey regarding "Bill Payment." We received 170 responses. The survey gauged user's knowledge of bill payment task-flow, methods of task completion, and enthusiasm for futuristic interactions. Moderated focus group discussions were conducted with twenty percent of the respondents. The outcome of our research demonstrated:

- The CX journey of our client's customers
- Signature moments of these customers
- An uncovering of customers' current understanding of Business processes

---

[4]https://www.jacada.com/blog/what-s-a-multimodal-customer-experience/.

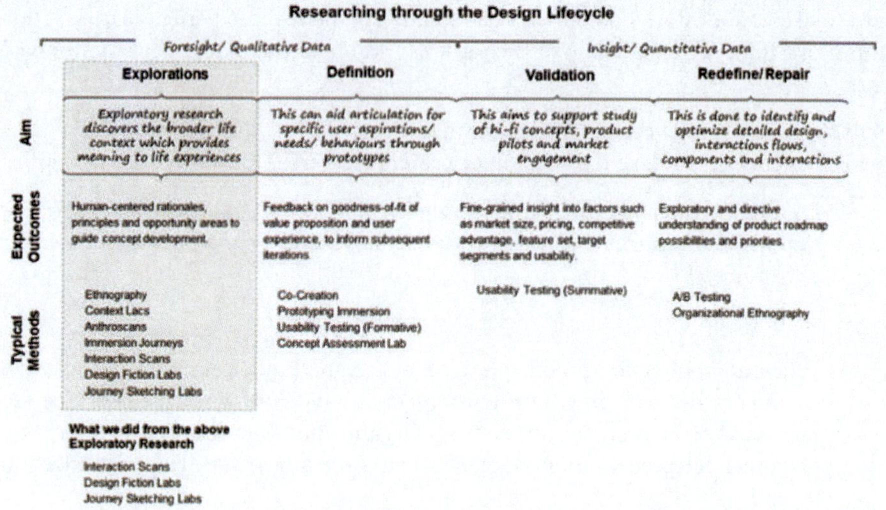

Fig. 33.1 Research methods for every stage of the cycle. Our scope of research was limited to the gray box (Research for Exploration) marked in the figure

- Validation of hypotheses on customers' channel and interaction preferences.

As design practitioners, we are familiar with Dieter Rams' 10 Commandments of Good Design[5,6], Nielsen and Molich's 10 Heuristics[7] and Ben Shneiderman's 8 Golden Rules[8]. These universal interaction design principles encompass the concepts of Visibility, Feedback, Consistency, Affordance, Cognition, Perception, Learnability, and Error handling. Though generic and valid in the era of Chatbots and voice-based interactions, our challenge lay in the direct applicability of these principles from the multimodal experience perspective to our customer's requirement. Thus, clarity was needed to resolve the following themes:

- Understanding how technology affects multi-generational (Baby boomer, Millennial, Gen Z) interactions with a Brand/Products/Services
- Understanding the "customer preference vs. customer usage" gap
- Tapping interaction points that underscore the need for Human Interaction in the end-user journeys.

We created a map, based on research insights, of how different generations have adopted and/or adapted to technology. The construct of a purchase journey was identified because of its universal applicability be it opening a bank account, buying

[5] https://web.archive.org/web/20120904120034/http://www.sfmoma.org/about/press/press_exhibitions/releases/880.

[6] https://www.vitsoe.com/gb/about/good-design.

[7] https://www.nngroup.com/articles/ten-usability-heuristics/.

[8] https://www.cs.umd.edu/~ben/goldenrules.html.

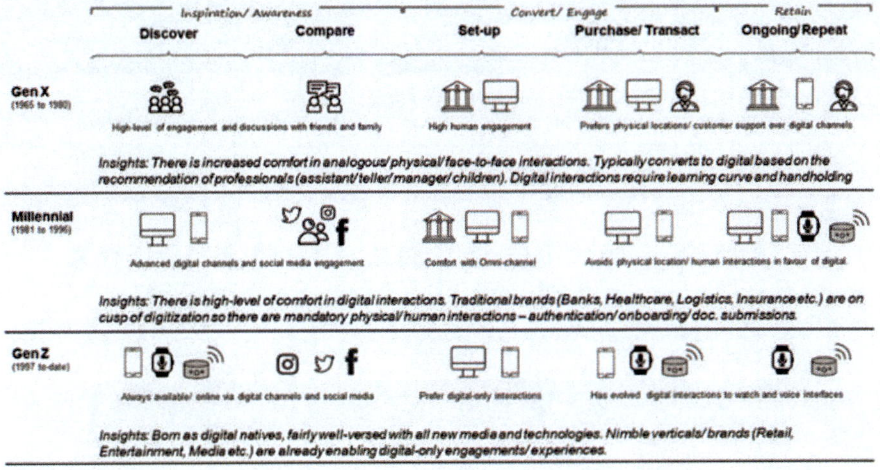

Fig. 33.2 User to channel mapping across a generic purchase journey depicted by generation and technology adoption

a vacation package, finding the right courier/supply partner, purchase on Amazon, etc. (see Fig. 33.2).

In addition to channel mapping by generation, the above figure also reveals customer preferences for interactions (offline vs. online) across generations. In general, it helps us understand how marketers and brands are always attempting to woo the "current generation" or the generation perceived to have the power of influence rather than the power of spend. Not surprisingly the early adopter generation of technology is where much of the digital Omnichannel spend is.

Equally important to note now is that all three generations' point of convergence (in terms of tech. adoption) lies in the "Transaction" phase. Our research suggests that this convergence is because:

- Baby boomers are educated (by their Millennial children) about the convenience of technology to support ongoing/recurring transactions in the digital world.
- Millennials have gone online or digital due to environmental factors such as work requirements, rapid technological advancements, and overall innovations in GUI.
- Gen Z is digital natives or born in the world of digital.

This snapshot of time, therefore, requires Designers, Marketers, and Brands to ensure that their work meets the requirement of generational overlaps. Furthermore, it reemphasizes the need to ensure a liquidity in interactions between the physical, digital, and virtual in a manner that increases accessibility, improves usability, and reduces contact-center support. These were notions were affirmed when representative CX journey through the current state was created (see Fig. 33.3).

Over the course of our conversations with the above groups, several common motivations, aims, and pain-points that surfaced were categorized under three main headers namely Change, Confidence, and Context.

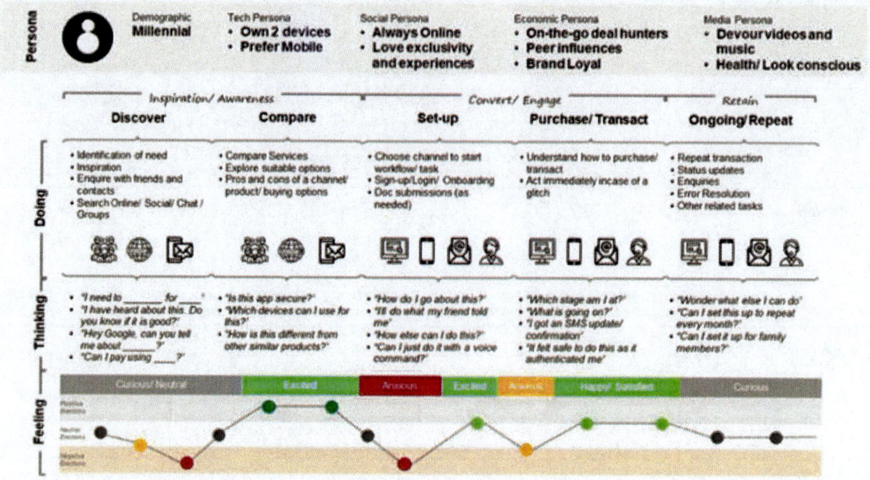

Fig. 33.3 Representative CX journey of current state

Change. Unlike the saying, "Change is the only constant," we observed that users are resistant to change in their status quo. They are happier to stay within the boundaries of what they know. This state persists until they are empowered due a transition that enhanced current knowledge with minimal extra effort. The outcome of this observation meant that it was very pertinent to retain common interaction principles and evolve over it rather than reinventing the wheel.

Confidence. We uncovered that human interactions were still quite significant on account of the reassurance factor. These human interactions offer fewer opportunities for confusion as the interactions automatically bring in the dimension of contextual awareness, intuition, and empathy. Despite its usage as the first point of contact for end-users, new-age AI/Bots redirect complicated requests to human counterparts for resolution. Our takeaway was that Brands/Products/Services need to adopt an Interaction Design strategy which exudes confidence during customer interactions irrespective of channel and/or across multiple channels.

Context. Much like human interactions bringing a level of confidence to end-users, we learned that Brands/Products/Services work to create context as part of their external strategies. Another direction that Brands/Products/Services are exploring is the contextual use of channel. With reduced user attention span and their time spread thin, users are particularly impatient when tasked with perusing dense material or working through complicated workflows. To bolster intuitive and error-free task completion, Brands/Products/Services are designing interactions that put the most appropriate channel and contextual actions at the helm of experience. An example of this is MS Outlook's segregation of emails under the self-explanatory labels of "Focused" and "Other." Another example is the way Uber has ensured its mobile app is a one-stop intuitive interface, while its desktop site subtly encourages users to switch back to the mobile app.

In summary, the above observations necessitated a conceptual framework to be liquid-like in its flexibility, to move seamlessly between various senses while using multiplicity of devices by mixed generations. We envision that this will add layers of maturity to existing Interaction Design principles in a way that is fluid.

33.3 Liquid Interactions

Our human experience demonstrates that interactions are all about connecting devices, interfaces, contexts, environments, and people. Unlike Design Languages such as Material Design by Google, Fluent Design by Microsoft, Living Language by IBM, etc., our proposal is not a new Design Language, but a proprietary interaction framework evolved specifically for Omnichannel and multimodal experiences.

The grand challenge and potential success factor of the Liquid Interaction Framework creation lies in building reliable systems that can analyze and understand multiple forms of communication in real time. On the one hand, multimodal interfaces target a more "human" way of interacting with computers, by means of speech, gestures, or other modalities; on the other hand, Omnichannel delivery interfaces have demonstrated the offer of improved flexibility and reliability over other human–machine interactions.

33.3.1 Properties of the Liquid Interactions Framework

In combining the properties of liquids and the previously noted general principles of design, we can build an initial list of properties that would be acceptable for creating a framework outline. The diagram below depicts few properties of liquids, which inspire the Liquid Interaction Framework.

Liquid is one of four fundamental states of matter whereas fluid is a condition specific to certain substances or a subset of matter. All liquids are fluids but not all fluids are liquids. Liquids can assume the shape of any container or vessel and are incompressible. To explain this further, the following Table 33.1 gives an overview this outline.

33.3.2 Visualizing the Framework

Taking inspiration from what is used to solve current problems; we look to formulate a chain of interactions between humans, devices, and their environment. This conceptual framework explores the need to interact naturally regardless of the device,

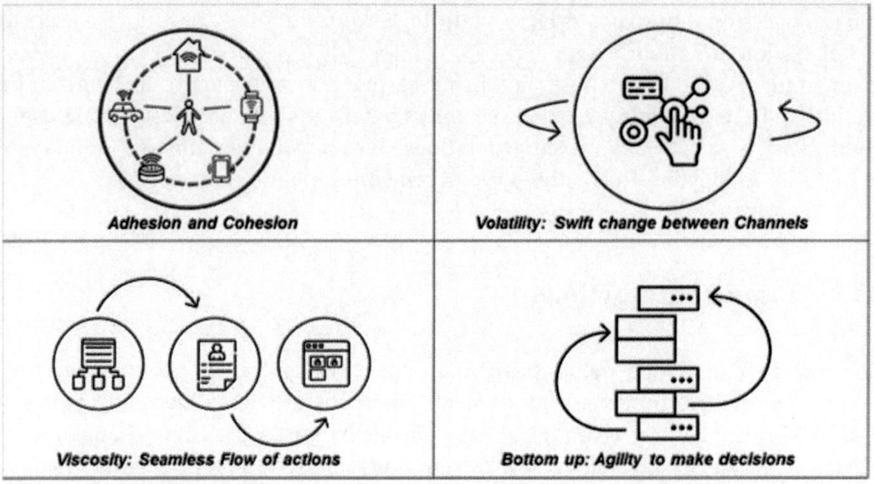

Fig. 33.4 Properties of liquids, which inspire the liquid interaction framework

medium, and via a mode not dictated by predefined flows. We hope that our proposed framework will take shape by revealing new patterns—i.e., voice, multimodal, GUI, etc. For instance, in today's VUI context, building the dialog of a conversational system is an art rather than engineering/science. Patterns and pattern languages offer an approach to design with much potential.

Functions of the Liquid Interactions Framework:

Our framework, like any ecosystem, can have the following functions:

1. It regulates essential transactional processes, supports systems, and renders uniformity through identifiable/purposeful object-based interactions. For example: Generation gap can be filled by multimodal interactions if the medium is flexible.
2. It crosses both scale and context (social science, marketing, applied science, art, etc.) through seamless enveloping.
3. It is gracefully interactive and responsible for balancing supply–demand.
4. It is humane in its identification of a highly agile exchange and its forgiveness of anomalies.

33.4 Future of Liquid Interactions

Automotive designers have cleverly utilized simultaneous multimodality by converging VUI, GUI, and Mobile technologies. For example, users can leverage the VUI to set a destination, the dashboard GUI to interact with maps, and the Bluetooth to continue either a conversation on mobile or the car speakers. Despite three

Table 33.1 Properties of liquids and Interaction principles in support of interaction framework evolution

Existing property/principle	Corresponding framework definition
Adhesion and cohesion (Property of liquid)	Liquid Interactions embed the principles of adhesion and cohesion by being sticky to whichever channel/mode the interactions happen through
Volatility (Property of liquid)	Liquid Interactions are characterized by swift transitions between channels and interaction modes that provide a seamless experience
Viscosity (Property of Liquid)	Liquid Interaction Framework is envisioned as being able to adapt easily for different contexts, user profiles, and application needs
Ripple effect (Property of liquid)	Interactions are tightly coupled with user personalization, channel preference, and device capabilities. A change made in the primary channel of use will see the affects on user choice, personalization, and preferences in other channels
Flow (Property of liquid)	Liquids flow and fill the lowest part of a container first before filling up to the top. Correspondingly Liquid Interactions also take a bottom-up approach thereby giving enough room and options to make agile decisions at any point in the journey
Constraints or internal locus of control (Principle of interaction design)	The framework is designed for a broad range of users but stays aware of environmental factors e.g., the user may benefit from speech input in a car, but text input in a noisy environment
Secured (Principle of interaction design)	There is a particular care to address privacy and security issues when creating multimodal systems, e.g., speech should not be used as a modality to convey private or personal information in public contexts
Error prevention and handling (Principle of interaction design)	Error prevention and handling is a major advantage of liquid interface design, for both user- and system centered reasons. Specific guidelines can support integration of complementary modalities to improve system robustness and give users improved control over modality selection so errors are avoided
Robust (Principle of interaction design)	Enhanced robustness due to combining different partial information sources; flexible personalization based on user and context

generations (Baby boomers, Millennials, Gen Z) having a significant overlap, the key difference lies in their comfort with technology and therein lies the opportunity for Designers and Brands.

Omnichannel multimodal interactions in the retail industry are already complementary to brick-and-mortar because of the BOPIS (Buy-Online-Pickup-In-Store), ROPIS (Reserve-Online-Pickup-In-Store), and BORIS (Buy-Online-Return-In-Store) paradigms, which reduce friction in the retail journey. We have read about

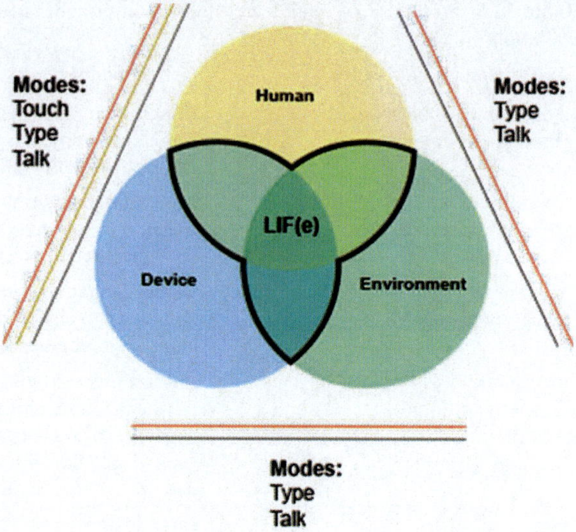

Fig. 33.5 Liquid interaction framework

the "Death of Omnichannel" and heard about the "Revenge of Brick and Mortar," and yet from where we stand, we believe that they are at an impasse where one cannot exist without the other.

To understand possibilities of Liquid Interactions applied in different contexts or domains, Designers and Brands would require a little more imagination. According to Harvard Business Review,[9] "*Customers only know what they have experienced. They cannot imagine what they don't know about emergent technologies, new materials, and the like. What customer, for example, would have asked for the microwave oven, Velcro or Post-It Notes? At the time, the transistor was being developed, radio and television manufacturers were still requesting improved vacuum tubes.*"

Sights in a magic crystal ball

Evolution and future proofing are uncomfortable experiences and seen as daunting for execution and implementation. However, some ideas that are already baby steps in this direction include:

- *"Remind me to pick up bagels at the grocery"* Support conscious contextual decision making
- *"Alexa, ask Uber for a ride"*: Identify the optimal channel for an interaction/ task
- Find the points of overlap to support experience hand-offs between channels
- *Think Netflix*: Recognize the points of break that can enable experience continuity
- *Less is more*: Guide users to logical interaction end points for task completion (without forcing stickiness or overwhelming engagement)
- *Complement, not compliment*: Develop voice and visual languages that can flow with symphonic elegance and precision.

[9]https://hbr.org/2002/01/turn-customer-input-into-innovation.

- *Point that finger*: Think beyond images and video to Interactions using expressions, gestures, sensors, etc.

If correctly done, particularly in a post-pandemic world, the liquidity of interactions across multimodal Omnichannel can result in higher end-user satisfaction and delight. The ideas discussed in the paper are to be followed up with explorations in creating a modular or atomic-design approach toward designing an adaptive, multimodal, and interactive experience that brings together the diverse worlds of brick, glass, and air.

References

1. Suchman, L.: Plans and situated actions: the problem of human-machine communication. Xerox, Palo Alto (1985)
2. Moore, R., Arar, R.: Conversational UX Design: A Practitioner's Guide to the Natural Conversation Framework. Association for Computing Machinery, New York. https://doi.org/10.1145/3304087
3. Schnelle-Walka, D., Lyardet, F.: Voice User Interface Design Patterns, pp. 287–316 (2006)
4. Brown, K., Walia, A., Joshi, P. et al.: Omnichannel natural language. In: Shaked, N., Winter, U. (eds.) Design for Multimodal Mobile Interfaces (2016)
5. Kim, S.W., Jo, H.K., Ha, D.Y.: Different UI, same UX: a design concept for implementing a locally-optimized and globally-unified user experience. In: Marcus, A. (eds.) Design, User Experience, and Usability. Theory, Methods, Tools and Practice. DUXU 2011. Lecture Notes in Computer Science, vol. 6770. Springer, Berlin, Heidelberg (2011). https://doi.org/10.1007/978-3-642-21708-1_50
6. Hayes, P., Reddy, R: Graceful interaction in man-machine communication. In: 6th International Joint Conference on Artificial Intelligence ICJAI-79 Proceedings 9th International Proceedings, pp. 372–374. Morgan Kaufmann Publishers Inc, Tokyo (1979)
7. Rossi, G., Schwabe, D. and Lyardet, F.: User interface patterns for hypermedia applications. In: AVI '00: Proceedings of the Working Conference on Advanced Visual Interfaces, pp. 136–142. ACM, New York. https://doi.org/10.1145/345513.345288
8. Thalmann, D.: The virtual human as a multimodal interface. In: AVI '00: Proceedings of the Working Conference on Advanced Visual Interfaces, pp. 14–20. ACM, New York. https://doi.org/10.1145/345513.345232
9. Business Insider: The Omnichannel Report, https://store.businessinsider.com/products/the-omnichannel-fulfillment-report
10. Deloitte Perspectives: CES 2019: Voice assistants speak to consumers, https://www2.deloitte.com/us/en/pages/technology-media-and-telecommunications/articles/voice-assistant-adoption-and-consumer-behavior.html
11. Word Stream, https://www.wordstream.com/blog/ws/2018/04/10/voice-search-statistics-2018
12. UX Collective: Design patterns in voice interfaces, https://uxdesign.cc/design-patterns-in-voice-interfaces-628a169294ec
13. Zickuhr, K.: Pew Research: Generations 2010, https://www.pewresearch.org/internet/2010/12/16/generations-2010/
14. Conceptual Framework, https://en.wikipedia.org/wiki/Conceptual_framework
15. Business Research Methodology: Exploratory Research, https://research-methodology.net/research-methodology/research-design/exploratory-research/
16. Exploratory Research, https://en.wikipedia.org/wiki/Exploratory_research

Chapter 34
Designing a Donation Portal to Help Underprivileged Indians

Arushi Jindal and Anirban Chowdhury

Abstract Underprivileged people do not get enough donations in the form of cash, food, clothes, or electronic items, as people are not much aware of the problems faced by underprivileged. Below Poverty Line is a benchmark used by the government of India to indicate very low economic status. Requirement of donation for underprivileged as perception and influence from others are the variable that influences people to donate online. Further people seek for the impact they have made on organization after the donation. Preferably, fund raising, need assistance, notifications regarding the impact donors have made, go live, etc., are some of the features of this web application. The objective of the study is to design user interfaces (UIs) of the web application and human centric validation of it. In the current study, two design concepts of a web application were developed considering the user problems identified through user interviews and survey. Then its prototype was created on Adobe XD and usability testing was conducted. It has been observed that the user interface version 2 of the web application was quite favorable to solve the problems faced by donation seeking organizations as well as donors as the flow of the application is acceptable and users can make donation easily to any organization on a single platform. Therefore, it has been concluded that this donation web portal might be useful to provide a good platform to people willing to make donations online.

34.1 Introduction

According to the recent survey, it has been found that 4% (or 20 million) of India's child population are orphans. Moreover, a recent study by Help Age India concluded that Kerala has the maximum number of old-age homes in India. In 1961, 5.8% of the national population rose to 6.8% in 1991. In 2016, the figure is expected to be 8.9% and almost two-thirds of people in India live in poverty. So, decided to connect the donor and non-profit organizations through a web application and expected that the application can bring benefits to both parties and make the donation activity easy to be

A. Jindal · A. Chowdhury (✉)
School of Design (SoD), University of Petroleum and Energy Studies (UPES), Dehradun 248007, Uttarakhand, India

A. Chakrabarti et al. (eds.), *Design for Tomorrow—Volume 1*, Smart Innovation, Systems and Technologies 221, https://doi.org/10.1007/978-981-16-0041-8_34

done as these organizations face a lot of problem with donations. Henceforth, donors can help orphans in different halfway houses by giving them food to eat and fundamental garments to the specific orphanage they wish to, through an online donation portal or by taking assistance if they need by calling in that particular organization. [1]. The findings show that there exists huge scope of intervention that can be taken up for the welfare of the elderly people and donations they need for their living as they are apprehensive and desolate. According to findings, their principle issues are absence of food, garments, and drugs for their everyday use [2]. Today development of online culture and the emergence of smart phones have brought many changes to the donation culture as it encourages people to make donation online however interaction between donors and people has been limited [3]. Also searching for a qualified non-profit organization is one of the major problems that prevents people from making the donation online. So, this application helps to solve the problem by providing a list of a non-profit organization in India. So presented a web application which will comprehend every one of these issues in a simple way just as fast without investing more energy and help to ease the process of donation providing the feature to use message, email, and push-notification in the application that can help the non-profit organization, feature was built so that the donor can provide more detail on the donation and ask questions to the non-profit organizations [4]. In addition, features of existing donation portals and crowd funding when it comes to contents and structure are to be talked about so as to recommend reasonable UI structure [5]. Method and system for sharing visual content across a plurality of mobile or computing devices to generate interest, support, and/or funding for philanthropic or social causes is disclosed. The method enables the generation of interest, support, and/or funding for philanthropic or social causes from a web application [6]. Therefore, this study aims to connect the donor and non-profit organizations through a web application and expected that the application can bring benefits to both parties and make the donation activity easy to be done as these organizations face a lot of problem with donations. Also, people who wish to adopt child can utilize this application to check the details of a child or by arranging a meeting with child by making an appointment with association and they can pick the kid they like and different conventions should be possible by visiting that specific organization.

34.2 Literature Review

34.2.1 Usability of Dynamic Websites

Ease of use testing of web applications created is a significant zone that faces an assortment of difficulties because of interesting highlights of utilization, restricted transfer speed, obligation of remote systems, just as changing plan setting have built up a nonexclusive structure for leading ease of use tests for work area applications through talking about exploratory questions, strategies, and usability attributes

[7]. Usability testing is a broadly utilized strategy to assess user performance and acceptance of products and system.

The significant patterns in ease of usability testing include: web application and Website testing in progressively naturalistic conditions. In the twenty-first century, "fast and clean" usability testing is expected to give legitimate and dependable information on how well individuals use products and system, and how they like utilizing them [8].

34.2.2 Aesthetics and Emotional Aspects of Websites

Web visual style assume significant roles in deciding the accomplishment of the Website. It is hard to choose which perspective should be given more consideration with regard to structuring a Website as style is concentrating on the "look" and "feel" while ease of use is worrying on its functionalities. There is no ideal answer which parts of the web configuration are a higher priority than the others. This research centers around how to adjust the ease of use and aesthetics of web design. A guideline to the Website designer was given to build up a Website which is well equalization in ease of use just as its style perspective [9]. In another exploration work, it has been observed that, both perceived usefulness and perceived ease of use were decidedly associated with passionate valence and negatively corresponded with enthusiastic commitment. Anyway, aesthetics greatly affect valence than saw convenience. In contrast to valence, draw intent could be more affected by apparent ease of use than by apparent style. These discoveries can be utilized as bases for applying customers' enthusiastic responses in every estimation to the items use conditions in the chain of perceptions, emotions, and practices [10].

34.2.3 Psychosocial Development Status of Orphanages

There are an expected 8,000,000 children dwelling in shelters, or private consideration offices, all around and huge numbers of these children are from dismissed homes and their instructive status is low—majority have never gone farther than fifth grade [11]. There are more than 100 million vagrants around the world. Among them, the commonness pace of gloom was 21%, uneasiness was 45%, low confidence was 23% and formative issue was 61%. Conclusion: This investigation inferred that there is high pace of enthusiastic and formative issue among halfway houses youngsters and unequivocally between related with socio-demo-realistic qualities [12].

34.2.4 Importance of User-Centered Design (UCD)

There is a risky connection between new media planners and focused structure (UCD) procedures. To all the more probable grasp and the noteworthiness of the customer inside the innovative antiquated irregularity, these designers normally make "personas"—prototypical customers with names, faces, interests, and tendencies. The assessment suggests that the utilization of personas is pushed as much by political real factors inside new media associations, everything considered by the hankering to understand customer needs [13]. The initial phase in the UCD procedure is deciding the objective client portions of the experience being structured, for instance, the "most-esteemed clients." This progression is regularly guided by the statistical surveying forms that characterize the absolute pool of potential customers, fragment the clients dependent on their needs and needs, and assess each market section's engaging quality. Subsequently client focused plan improves the planners' as it becomes easy to understand the clients, consequently improving the consequences of the design process and acknowledge that personas must be engaged to "become animated" and helps to create UI and UX of design [14].

34.2.5 About Existing Donation Portals and New Work

Neediness, missing laws, ignorance on the dangers of halfway house, travel industry, debasement and missing observing component were the primary driver [15]. Throughout the years the development of the charitable part, expanded commercialization and industrialization, just as diminished government subsidizing in numerous nations, has upgraded the level of com-appeal. Initially, the expanding importance of not-for-profit promoting by illustrating the evolution of advertising all in all is illustrated. Second, a definition that highlights key parts of not-for-profit advertising is proposed. At long last this segment distinguishes and basically surveys different strands of examination specifically social promoting, cause-related advertising, and raising support in order to decide probably the most squeezing worries around philanthropic showcasing today [16].

Numerous donation portals exist in India, each portal has any reason to satisfy and concentrate on the requirements of individuals. In any case, what was executed while dealing with this web-based interface was that to make a bond between the contributor's and individuals who really need it, so the donators become more acquainted with what sway their gift have made on the life of these penniless individuals and getting the persistent criticisms of occasions occurring in these associations. Since this sort of work has never been done and executed one element of "need help" with the goal that giver can contact these associations and they can gather from their homes.

34.3 Methodology

34.3.1 User Need Analysis

34.3.1.1 Observational Study

First consumer behaviors were studied through observational study in different organization with teachers at underprivileged schools, students, caretaker at an orphanage, children, caretaker of old-age home, old people located in Ludhiana and Dehradun to know about the problem they are facing and their daily routine. Based on the information got from them, designed a questionnaire to gather quantitative information. With goal to know how much donation they get, in what form they get donation and what are the problems faced by donors [17].

34.3.1.2 User Interview

The study engaged 50 participants in which 59% users were beneath 24 years old, 24% participants were in 24 to 35 years old and the remaining 17% were from 35 to 65 years old. Around 75% participants said that they are comfortable with smartphones, 15% were more comfortable with desktop and the remaining 10% said they are not fun loving with gadgets. Around 80% participants said they are likely to donate in person (may be due to trust issue) and 20% like to donate through online. Mostly participants prefer to donate old clothes than food, money, electronic items, and electronics in the similar priority as mentioned.

Interviewed caretaker at orphanage, children, caretaker of old-age home at various organizations of old-age homes, orphanages, and underprivileged located in Ludhiana and Dehradun as shown in Fig. 34.1. and asked them few questions to get to know the problems they are facing.

34.3.1.3 User Survey

After the user survey, get to know about the preference of people about items they would more likely to donate with a proper questionnaire and performed chi-square test to check if the collected data is significant or not. Then, performed the affinity analysis on data gathered by user survey and user interviews to make it easy to make a choice regarding the functionalities that could be given in the web application to tackle the issues of these organizations and donors.

34.3.2 Designing Interface

In initial study, started with user interview by talking to people at various orphanages, unprivileged schools, and old-age homes and to get to know about the needs of people in these organizations, also talked to the donors about how often do they donate and what kind of items do they prefer to donate and gathered all the information. Then started with ideation on how to implement features in the application that could solve the problems of these users. Created use case diagrams, class diagram, sequence diagram, flow diagram, information architecture, decision tables, and ER-diagrams as shown in Fig. 34.2 [18]. Then started with paper prototyping of the application by converting it into high-fidelity prototype on Adobe XD [19]. At this point, performed heuristics for surveying the attractiveness of web user interface as proposed based on aesthetic design, corporate personality and brand, and the perceived utility matched to users' prerequisites. The heuristics are tested by evaluating four donation portal web sites to exhibit how unique engaging quality and customary convenience exchange offs add to in general viability [20].

34.3.2.1 UI Design and Prototyping

The UI design of the web application for version 1 and version 2 was designed according to use case diagram and user flow diagrams on Adobe Photoshop CC and Adobe XD as shown in Fig. 34.3. The interface includes consistently assume an essential job in UI acknowledgment by target clients. Besides, ease of use is the key indicator for client acknowledgment. As writing additionally proposes that style is a significant factor for UI or its model plan and may add to the ease of use impression of the product interface.

Fig. 34.1 Visit at underprivileged schools, orphanage, and old-age home

34.3.3 *Evaluation of Interface of Donation Portal*

As there were around 45 number of participants and created the 2 versions of the web application where independent variable was navigation within the system. So, performed 3 usability tests—perceived ease of use (PEoU), perceived usefulness (PU), and willingness to use (WU) and then compared the mean of each response to get to know the usability of web application [21]. It is significant that designers the advantages and constraints of various usability review techniques. This is on the grounds that the nature of the usability testing is subject to the technique used. Two of the standard usability evaluation procedures are ease of use testing and heuristic examination. The principal goal of this examination was to look at the proficiency and viability between 2 diverse web applications designed [22]. Investigation by seriousness of issues found and decreasing profit examination model for the connection between the quantity of new issues found with clients and evaluators utilized demonstrated that the two strategies are similarly productive and successful in tending to various classes of ease of use issues. All together for better assessment results, both usability testing and chi-square testing are up "til now required" [23].

34.4 Result and Discussion

34.4.1 *Interview*

Presently, individuals have built up a solid enjoyment and fascination toward web-based or technology-enabled life. It has been observed that most of the participants need help/assistance to donate items to these organizations. And also, maximum number of participants need to know what impact their donation is making on these organization and to know about the growth of the organization. Later, Ho and Vogel (2014) stated that trust between user and client can be achieved by social networking sites through an interactive and user-centered environment [24]. The Indian customers like to post about the donation they have made on various social platforms with their friends, family-members, relatives, and colleagues. It was also observed that people face issue of trust hence do not prefer to donate online and prefer to donate in-person [25].

Performed affinity analysis on all the data gathered by user interview to understand the needs of these organizations and found that most of the participants want feature so that they could get a tax invoice after making online donation so to keep the track and record of donation and receive an email of certification of making a donation and want to know what impact they have made on life of people in these non-profit organizations and could be one of the reasons to use proposed online system. Significantly more participants said that they want a feature of live video option so that they could interact with the people of these organizations and get updates related to them, so they get to know more about them. According to a survey, it was found that in old-age homes they do not get much donation in form of medicines for old people so considering that provided the option of donate money for medicine [26].

34.4.2 User Survey

After the user survey, get to know about the preference of people about daily-life items they donate (as shown in Fig. 34.4). The sequence of preference was—Clothes > food > books > stationary > medicine > electronic items.

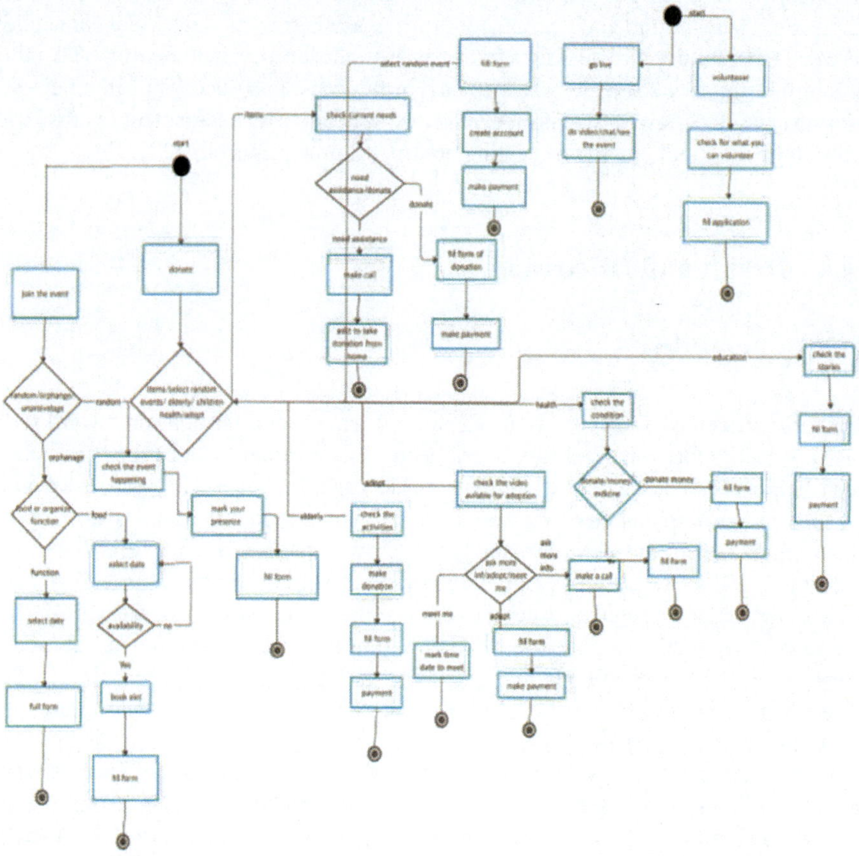

(a)

Fig. 34.2 **a** Flow diagram and **b** use case diagram

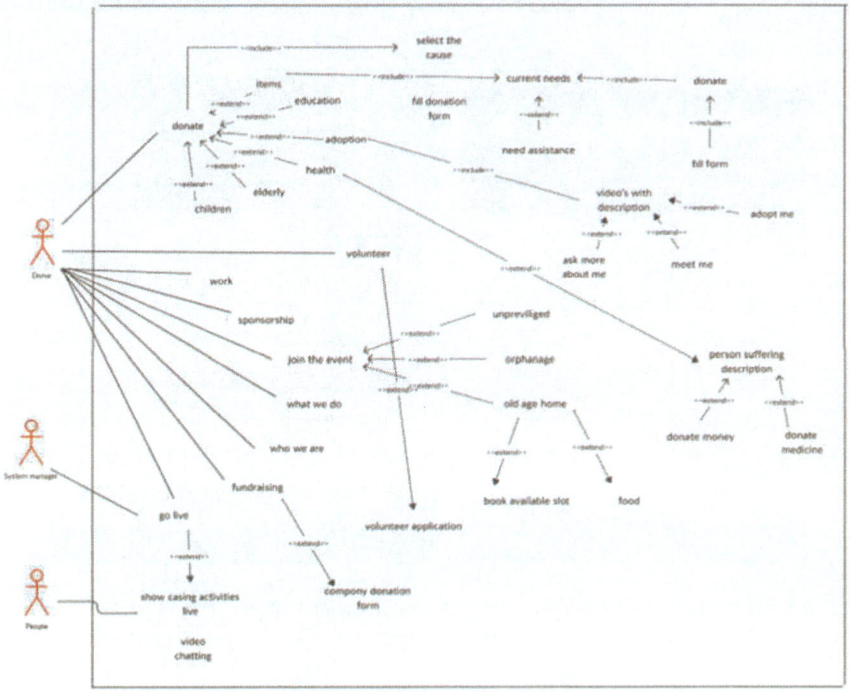

(b)

Fig. 34.2 (continued)

According to data 49.85 wants to know if the charity is genuine, whereas 8.5% people wants to make sure if the organization is financially secure. Also, more percentage of people prefer to donate items rather than donating cash. Performed affinity analysis of data collected by user survey to make it easy to make decision regarding the functionalities that could be provided in the web application to solve the problems of these organizations and donors.

34.4.3 Design Evaluation

About 94% users said they are eager to use the donation portal system with these features and 6% users are not willing to use donation portal to make donations online. However, significantly a greater number of participants were willing to used proposed system it was accepted. Because of progression in the field of "Human Computer Interaction (HCI)" in the most recent couple of decades, consideration has been attracted to ease of use investigations of versatile applications including user need

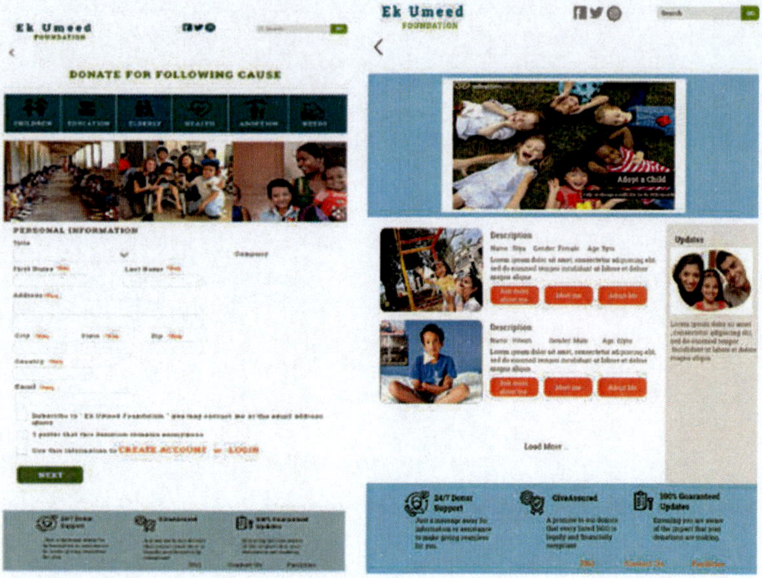

Fig. 34.3 User Interface of donation portal version 2 (left) and version 1(right)

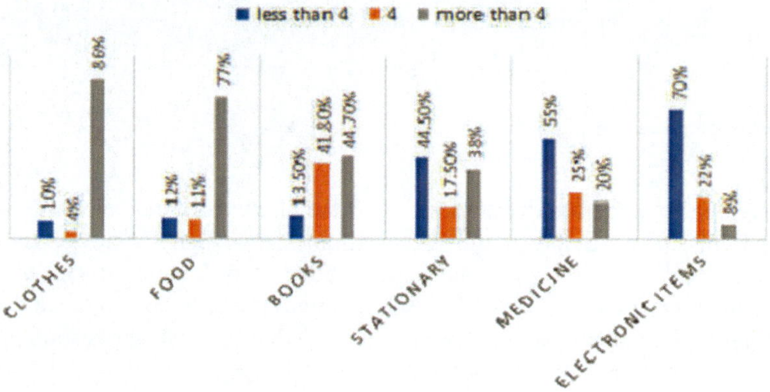

Fig. 34.4 Data of items people prefer to donate more often

recognizable proof, user need examination, usability testing, and so on [27]. Two dependent variables (perceived usefulness and perceived ease of use) were chosen to measure usability of the donation portal and willingness to use was measured for understanding the UI of donation portal. Recorded the data of the perceived ease of use (PEoU) and perceived usefulness (PU) on the nine-point Likert scales (where "1" represents strongly disagree and "9" represents strongly agree). Similarly, willingness to use (WU) was also measured on nine-point Likert scale. As expected, there is significance difference in perceived ease of use, perceived usefulness and

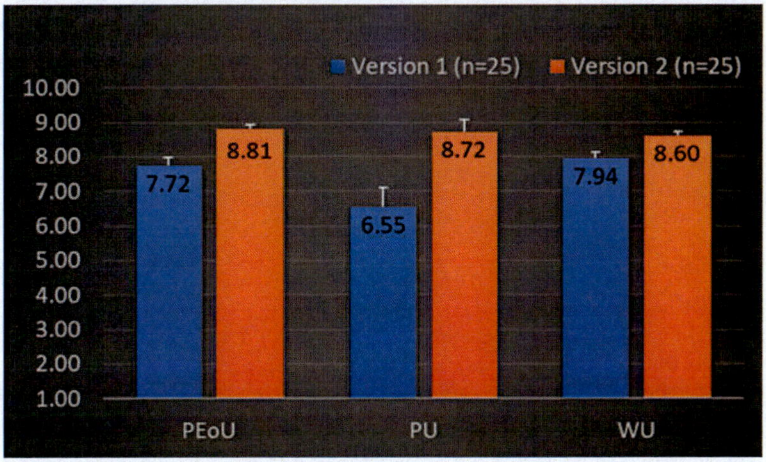

Fig. 34.5 Mean comparisons of perceived ease of use (PEoU), perceived usefulness (PU), and willingness to use (WU) between interface design version 1 and version 2

willingness to use due to change in design found after Wilcoxon test ($p < 0.05$) as after plotting the histogram, it was found out that the distribution came out to be non-normal distribution. It was also observed that the mean values of perceived ease of use, perceived usefulness, and willingness to use were higher in case of version 2 of the design (Fig. 34.5). Hence, the null hypothesis was rejected and it was proved that version 2 design is better than version 1.

34.5 Conclusion

The proposed donation portal will be very helpful to fascinate more people to make donation through online donation portal, will also built trust between user and organization as they will get to know what impact they have made on life of these needy people. This study was performed to provide good user experience to users so that they can make donation to any of these organizations like old-age homes, orphanages, and unprivileged schools according to their preference. Result of the present study suggests that user like to donate clothes and food items more often than by donating cash as shown in Fig. 34.4. This examination was exploratory, researching the variables that impact individuals' willingness to make online donation, including the socio-segment qualities of donors. Additionally, it has been seen that gender, impression of the association, and impact from others are factors that impact the probability of making donation online.

Designed web application is accepted by donors for providing various supports to them by making their task easier through online donation portal. Hence, this web portal could be implemented by government or NGOs in near future to support underprivileged people. Usability testing of design 2 of donation portal has not been with users yet. web portal must be planned in future, high-constancy model of application with a database must be tried before making the application working with the amendment of heuristics evaluation-related issue.

References

1. Elapanti, S.S., Pinthepu, N.S.: Helping Hands—An Android Based Donation System
2. Gruber, H., Chen, J., Jan, Y.Y., Feig, E.: Method and system for soliciting charitable donation during electronic commerce. United States patent application US 09/782,722. 2002 Aug 15
3. Raven, M.E., Flanders, A.: Using contextual inquiry to learn about your audiences. ACM SIGDOC Asterisk J. Comput. Doc. **20**(1), 1–3 (1996)
4. De Boeck, E., Jacxsens, L., Goubert, H., Uyttendaele, M.: Ensuring food safety in food donations: case study of the Belgian donation/acceptation chain. Food Res. Int. **1**(100), 137–149 (2017 Oct)
5. Vredenburg, K., Mao, J.Y., Smith, P.W., Carey, T.: A survey of user-centered design practice. In: Proceedings of the SIGCHI Conference on Human Factors in Computing Systems 20 Apr 2002, pp. 471–478
6. Straka, J.W., Wayne, S.J.: Enhancing charity portal website and methods to help people support charity by making their life activities and every charity donation go farther. United States patent application US 13/092,879. 27 Oct 2011
7. Zhang, D., Adipat, B.: Challenges, methodologies, and issues in the usability testing of mobile applications. Int. J. Human Comput. Interact. **18**(3), 293–308 (2005)
8. Wichansky, A.M.: Usability testing in 2000 and beyond. Ergonomics **43**(7), 998–1006 (2000)
9. Yusof, U.K., Khaw, L.K., Ch'ng, H.Y., Neow, B.J.: Balancing between usability and aesthetics of web design. In: 2010 International Symposium on Information Technology 15 June 2010, vol. 1, pp. 1–6. IEEE
10. Seo, K.K., Lee, S., Chung, B.D., Park, C.: Users' emotional valence, arousal, and engagement based on perceived usability and aesthetics for web sites. Int. J. Human Comput. Interact. **31**(1), 72–87 (2015)
11. Van Leeuwen, M.H., Wiepking, P.: National campaigns for charitable causes: a literature review. Nonprofit Voluntary Sect. Quart. **42**(2), 219–240 (2013 Apr)
12. Fawzy, N.M., Fouad, A.A.: Psychosocial and Developmental Status of Orphanage Children: Epidemiological Study (2010)
13. Hanel, J.: Child exploitation and tourism at the example of orphanage tourism: a situation analysis on the current situation of orphanage tourism in Myanmar including a comparison of Myanmar and Cambodia
14. Massanari, A.L.: Designing for imaginary friends: information architecture, personas and the politics of user-centered design. New Media Soc. **12**(3), 401–416 (2010)
15. Kantola, V., Tiitta, S., Mehto, K., Kankainen, T.: Using dramaturgical methods to gain more dynamic user understanding in user-centered design. In: Proceedings of the 6th ACM SIGCHI Conference on Creativity & Cognition 13 June 2007, pp. 173–182
16. Helmig, B., Thaler, J.: Nonprofit marketing. In: Third Sector Research, pp. 151–169. Springer, New York, NY (2010)
17. Gupta, M., Goel, S.: Case study of old age homes in Delhi: suggesting viable interventions for future. J. Community Mob. Sustain. Devel. **9**(2), 140–144 (2014)
18. Kohler, H.J., Nickel, U., Niere, J., Zundorf, A.: Integrating UML diagrams for production control systems. In: Proceedings of the 2000 International Conference on Software Engineering. ICSE 2000 the New Millennium, 9 June 2000, pp. 241–251. IEEE

19. Križanac, A.: Designing, prototyping and testing a company travel management application (Doctoral dissertation, University of Rijeka. Department of Informatics)
20. Sutcliffe, A.: Assessing the reliability of heuristic evaluation for Web site attractiveness and usability. In: Proceedings of the 35th Annual Hawaii International Conference on System Sciences 10 Jan 2002, pp. 1838–1847. IEEE
21. Islam, M.N., Zavin, A., Srabanti, S., Ferdous, C.N., Suha, S.A., Afroze, L., Shawon, N., Refath, N.S.: GiveMed: a webportal for medicine distribution among poverty-stricken people. In: 2017 IEEE Region 10 Humanitarian Technology Conference (R10-HTC) 21 Dec 2017, pp. 294–299. IEEE
22. Tan, W.S., Liu, D., Bishu, R.: Web evaluation: heuristic evaluation versus user testing. Int. J. Ind. Ergon. **39**(4), 621–627 (2009)
23. Nielsen, J.: Finding usability problems through heuristic evaluation. In: Proceedings of the SIGCHI Conference on Human Factors in Computing Systems 1 June 1992, pp. 373–380
24. Davis, F.D.: Perceived usefulness, perceived ease of use, and user acceptance of information technology. MIS Quart. **1**, 319–340 (1989)
25. Fox, D., Sillito, J., Maurer, F.: Agile methods and user-centered design: How these two methodologies are being successfully integrated in industry. In: Agile 2008 Conference, pp. 63–72. IEEE (2008); Bertram, D.: Likert scales (2007); 2 (2013)
26. Chowdhury, A., Gebretsadik, H., Kumar, S., & Punekar, R. M. (2013). Feasibility of Development of an Immunization Management Tool for Mobile Phones by UCD Approach. International Journal of Computer Applications, 61(18).
27. Koli, A., Chowdhury, A., Dhar, D.: Requirement of new media features for enhancing online shopping experience of smartphone users. In:Intelligent Systems Technologies and Applications, pp. 423–435. Springer, Cham (2016)

Chapter 35
Evaluation and Redesign of SAP Portal for University Students

Yukta Sharma and Anirban Chowdhury

Abstract ERP system, i.e., enterprise resource planning system is a tool that has been used by organizations to integrate and manage all of its different units. In the educational sector, it has been used as a student portal which is providing various services, from providing database to keep student's records to helping them with accessing studies from any part of the world via online classes. In this study, it was focused on one such portal, the SAP portal which is currently used by X University. The aim of this paper is to solve the issues the students are facing while using the SAP portal by applying interface design intervention. Current study was initiated with interviews, collecting and analyzing reviews, and problems faced by students followed by came up with solutions; more of features that could be added to the new design of the website. Results suggest that lots of student from X University is not satisfied with existing student portal. Students of this university have stated that they resist from using the portal as it is not at all easy to use and students often get confused about how to proceed further in it. So, it was decided to resolve these issues by redesigning the portal. Then after redesigning the Web site, it has been tested. It was observed that students find the new design more acceptable, easier, and better than the existing SAP portal inter- face and they are willing to use the newly designed interface. But still there are a few limitations in new design, like the new design has not been developed technically yet.

35.1 Introduction

Today, the time has changed a lot, when we look around us, we see that we have come so far in terms of development with the advancements in technology. Almost all the sectors are using various services and technologies to increase their productivity. One such sector is educational sector [1]. With the advancements in communication technology, now education is not just limited to one place or among some group

Y. Sharma · A. Chowdhury (✉)
School of Design (SoD), University of Petroleum and Energy Studies (UPES), Dehradun 248007, Uttarakhand, India
e-mail: chowdhuryanirban14@gmail.com

A. Chakrabarti et al. (eds.), *Design for Tomorrow—Volume 1*, Smart Innovation, Systems and Technologies 221, https://doi.org/10.1007/978-981-16-0041-8_35

of people, it is spreading throughout states and countries. Students can learn from anywhere, anytime with the help of online tutorial and distant learning services. With the increasing complexity, now we need a system which will help us to manage and integrate all the units, and regulate them as one system. This system is called enterprise resource planning, i.e., ERP system [2].Enterprise resource planning system (ERP) is a tool which allows an organization to manage, integrate, and automate the multiple functions related to business, technology, services and human resources including, product planning and development, manufacturing, sales and marketing in a single database, application (or Web site), and user interface [3]. Educational institutes are now buying the services of ERP known as a student portal which they are using to store students' data and to collect other records for their entire college years like their semester wise attendance, exam scores, timetable, fee payments, etc., and also helping students to continue their lecture through online class via over the Internet. There are various companies that provide ERP services to various organizations. While some are successful, some are still working on improving their system, making it more user friendly [4]. Even besides the success of ERP in the educational sector, there are still some factors which are causing problems in acceptance of student portals [5]. The major factors, along with other factors, which are responsible for making any service more user friendly are its ease of use and usefulness [6, 7]. In this paper, we are talking about one such university X, which is currently using the ERP service of SAP as SAP student portal. It has been seen that a lot of students are not willing to use the SAP portal, rather they get their work done manually. This paper aims to identify the causes for the bad experience of the students with SAP student portal and trying to resolve these issues by using interface design intervention.

35.2 Literature Review

ERP system was first introduced in the early 1960s and since then, it has been developed so far. Followings are the few researches done by various researchers in this field and how it has revolutionized the education system.

35.2.1 Factors Affecting ERP Implementation

Before talking about how it has affected the educational system, let us study the factors which lead to the success of ERP in today's world. The crucial factors on which the user satisfaction and willingness to use the Web site are dependent are explained by following researchers in their papers. Davis in his paper in which he has presented two specific variables, Perceived ease of use (PeU) and perceived usefulness (PU), which are hypothesized to be fundamental determinants of user acceptance [7]. Basoglu et al. have also presented the framework for organizational adoption of ERP systems in their paper. Their modal consist of Technology Acceptance Model (TAM) variables, i.e., PeU and PU of ERP systems, satisfaction, and common actors of an ERP project [8].

Other than these two factors, i.e., PU and PeU, researchers have also explored and studied other factors which would be helpful for successful implementation of ERP system. These factors are called *critical success factors (CSFs).* Abugabah and Sanzogni have discussed in their paper how to improve the ERP systems and make it more accessible for users [9]. Bhatti has talked about the critical factors that lead to success of ERP modal for any organization. He developed certain models and scales which will help in the growth of ERP system [10]. Ehie and Madsen have discussed about the critical factors that drive successful implementation of ERP systems. He identified few factors which have correlation with successfully implementing ERP [11]. Somers and Nelson have also talked about the impact of critical success factors across the stages of ERP implementations. Their results provide advice to management on how to best utilizing their limited resources to choose those critical success factors that are most likely to have an impact upon the implementation of ERP system [12]. Ngai et al. have presented paper on the literature review of the critical success factors for implementation of ERP system. In result, they identified 18 CSFs with more than 80 sub-factors, that are responsible for successfully implementing an ERP system [13]. Dezdar and Sulaiman have also discussed about the CSFs based on current literature for successful ERP implementation [14]. Andreas and Nicolaou have presented their paper on post-implementation reviews of ERP system. In this paper, they talked about the quality of reviews they get post implementation of ERPs, they are known as Post-implementation reviews (PIR). By studying these PIRs, they identified the factors that contribute to high PIRs and what are the potential factors that lead to successful ERP system implementation [15].

35.2.2 ERP Systems as Student Portals

In introduction section, we have discussed a little about the importance of ERP in the educational sector. Following are the papers written by researchers in the development of student portal, making it more useful and reliable for educational purposes:

Secreto and Pamulaklakin have shown how the student portals can act as a support system in an open and distance e-learning environment (ODeL) [16]. Ellison and Arora have presented an amazing idea in his paper on improving the learning experience and to let students engage in social learning along with academic study by using cloud-based suits of collaboration tools, i.e., Microsoft 365 [17]. Gavrilova and Jin have presented an approach in their paper using ontology engineering for building knowledge portal [18]. Boranbayev et al. have discussed about the software architecture of the in-house development of student portal that they have developed for higher education institution in Kazakhstan [19]. Ryan, et al. have developed a portal for an engineering economy class which can improve students' problem-solving skills with IT by creating an Internet-based problem-solving environment [20]. Jones, Provost, and Pascale have developed a university research Web-based knowledge portal in which they developed a system which will help in developing a connection between researchers via the portal so that researchers get aware of the research activities within their own universities [21]. El Said has presented a study on the mobile student portal. It is still under research but his findings suggest design recommendations for the mobile student portal [22]. Thulasi Krishna has written a paper on the online student portal which will act as a platform for students and teachers to exchange study material and information in different formats like videos, images, pdf, etc. [23]. Shord and Moore presented a design for ChemPaths student portal which is not only an online textbook but also it incorporates linked problems, multimedia, and interactive 3D structures [24].

35.3 Methodology (Procedure)

35.3.1 Research

The X University has the existing SAP portal which is redesigned and tested. For this, the bottom-up research has been done.

Target user profile.

Our target audience is all Indian students of bachelors and master's program of X university, having an age group of 15–30, in which 74% are male and 26% are female. These students are from a family background where annual family income varies from 8 to 20 lakhs rupees per annum. A total of 15 student users were participated for interview during problem identification and a total of 30 users were participated in the usability testing.

Data collection procedure.

For data collection, random sampling technique has been used.

User interviews and survey have conducted to know the student experience related to the SAP portal. For this, a questionnaire and a survey form were created.

In user research, we collected data on how frequency of use of the SAP portal, purpose of use of SAP portal, problems while using the portal, time taken in completing the task and willingness of using the portal if it improves.

For *interview questionnaire*, the questions were divided as follows:

1. The user's demographic information, hobbies, and routine
2. The user's computer playfulness
3. The user understanding of portals
4. The user's experience on using their university's SAP portal
5. Comparison of portals (a task which is performed by users).

While recording the interviews of the students, they were also given a task to do. They were asked to open their SAP portal and go and check their attendance. The time and user's reactions were recorded.

In the interview, a *competitive analysis* has been done between three portals. The users were given a Web kiosk-an online student portal of Y University, and asked them to perform the same task, i.e., checking student attendance and time and reaction were recorded. Then users were given one more portal, an online MOOC course Web site, Z.org, and asked them to analyze the Web site. Then reactions were recorded. All the feedbacks for these three portals are compared and analyzed.

For the *usability testing*, a google form has been created which consists of Likert scales, multiple choice questions, multiple select questions, and subjective-type question in the same category as questionnaire was divided. Through survey, we collected frequency of the data.

A total of 30 responses on all the six forms. After analyzing and taking an average of the responses, we get a positive response for new design than the old one. Users find the new design easier to use, more useful, and they are willing to use the new design.

35.3.2 Modeling

After collecting data and results, now modeling has been done.

For modeling, a brainstorming has been done. All the problems were summarized and all the probable features that could be added in new design to solve these problems were chosen. For this, first, an information architecture (Fig. 35.1) was created using an open card sorting technique and then a layout was created using Adobe XD.

After layout, the UML diagrams were created which include use case diagram, activity diagram, sequence diagram, and class diagram. After that, three wireframes (or low fidelity prototypes) were created, out of which one was selected. Then based upon that wireframe, a medium fidelity prototype was created. Please see Fig. 35.2 for both existing SAP interface and newly designed interface of student portal.

35.3.3 Usability Testing

The task that was given to students to check the student attendance in SAP portal; maximum of them took 7–15 min to complete the task. Then, the usability of the new SAP portal was compared with the existing SAP portal on the basis of perceived ease of use, perceived usefulness, and willingness to use. A new survey form was created which consists of questions on PU and PeU and willingness to use on a 9-points Likert scale. The questions on PU and PeU were benchmarked questions which were presented by Fred D. Davis in his paper perceived usefulness, perceived ease of use, and user acceptance of information technology [7], and three questions

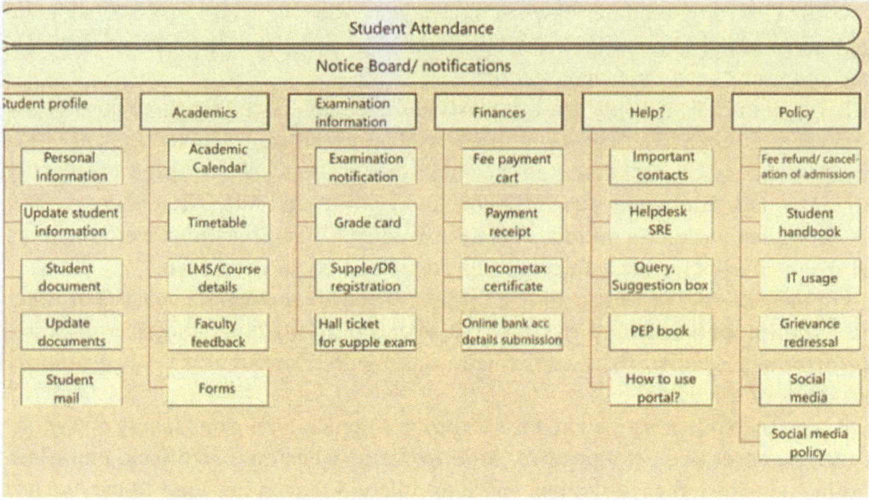

Fig. 35.1 Layout of information architecture of new design of SAP portal

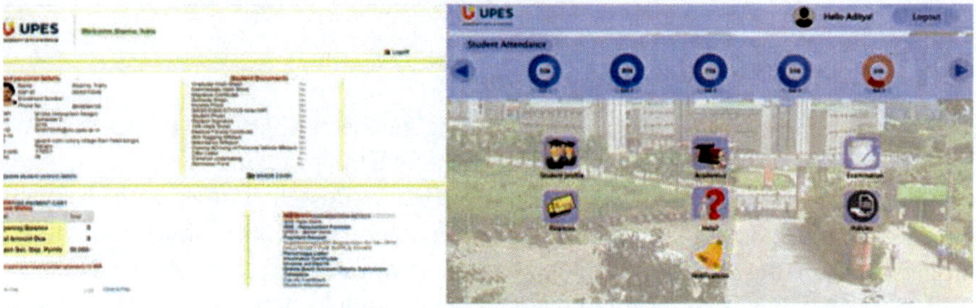

Fig. 35.2 Interfaces of existing and newly designed student portal

on willingness to use were asked. It is reasonable to hypothesize that new interface design is better than the interface of existing student portal in terms of PU, PeU, and willingness to use.

After collecting the data, histograms were plotted for responses for PU, PeU, and willingness to use. The curves in histograms were not following normal distribution patterns. Hence, *Wilcoxon signed rank tests* was conducted for hypotheses testing that had been done on *SPSS* software.

35.4 Results and Discussions

35.4.1 Results of Competitive Analysis and User Interviews

From the results of competitive analysis (Please see Fig. 35.3), we summarized that out of all three portals, Y University's Web kiosk is primary competitor and Z.org is a secondary competitor of X University's SAP portal. Out of all three, our target audience liked the Z.org better than other two. The own X University's SAP portal is least liked by our users.

From the interviews, it was observed that most of the participants are using SAP portal for—(1) checking attendance (100%), (2) checking of exam schedule and other exam related notices (96%), (3) checking personal information and documents (92%), (4) checking timetable (84%), (5) giving faculty feedback (76%), (6) checking fees payment related information (76%).

When asked for problems that are faced by students, they said they are facing problems related to (1) organization and management of data (100%), (2) delayed access to information (96%), and (3) non-interactive GUI (92%).

When asked for the willingness to make SAP portal to be more organized, more accessible, more informative and easier to use, all the participants responded with agree and strongly agreed response.

Then, we find the chi2 value for all the responses and for all the data, we get the value of probability, i.e., $p < 0.05$, which means that the chances of acceptance of the null hypothesis is less than 5%, where the chance of acceptance of alternated hypothesis is more than 95%.

Competitive analysis

	Z.org (Secondary competitor)	Y university webkiosk (Primary competitor)	X university SAP portal
Aesthetics	✓	✓	✗
Ease of use	✓	✓	✗
Informative	✓	✓	✗
Usefulness	✓	✓	✗
User-friendly	✓	✓	✗
Data organisation	✓	✓	✗
MOOC and other study material provided	✓	✗	✗
Visual and navigational flow and coordination	✓	✓	✗
Privacy and security	✓	✓	✓

Fig. 35.3 Results of competitive analysis summary

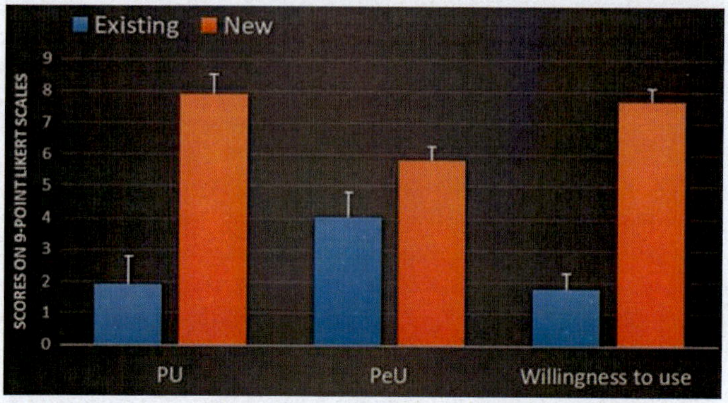

Fig. 35.4 Variations of mean values for PU, PeU, and willingness to use due to two different interfaces

35.4.2 Features that Are Added to Solve the Problems

Along with making the interface more interactive and organized, by adding the following features, it has been made more useful and easier to use:

1. A Notification section where upcoming events, course-related notices, highlights are displayed.
2. Yearly course-related details, requirements of the course, faculty details, faculty info, etc., are provided.
3. Other important helpline numbers like infirmary, wardens contact details, etc., are added.
4. A FAQ/suggestion/complaint e-mail box is added.
5. Cloud storage from where students can store, retrieve, and update their documents anytime.

35.4.3 New Design Was Observed Better in Usability Testing

After new design has been made, testing has been done. The same participant group has shown the new design and asked them to fill the survey form which consists of questions on perceived usefulness (PU) and perceived ease of use (PeU) and a form consists of questions on user acceptance and willingness to use the design. Please see Fig. 35.4 for detailed results.

After taking the average, then comes hypothesis testing. For hypothesis testing, first, we need to verify which test is needed to be done. For verifying that, we plot the histograms of PU and PeU for old and new designs. In all the four histograms, we did not find them following normal distribution curve or bell curve, so we could not go for t-test. Hence, we go for alternative statistics for student t-test, i.e., Wilcoxon signed rank test. We chose Wilcoxon test because it is an alternative for a paired t-test when the graph is not following the distribution curve. After doing the test in SPSS, the resulting z-value for the both PU and PeU is −4.783 and that of willingness to use is 4.800. For all three factors, the df (degrees of freedom) is 30 and the level of significance, i.e., 'p' value is 0.001 which is less than 0.05 for both the tests, which means the chance of acceptance of our alternative hypothesis is more than 95%.

This graph (Fig. 35.4) is showing the descriptive statistics, in which the mean value of the new design, i.e., 5.8263, 7.9047 and 7.6887 is more than that of old design, i.e., 4.0453, 1.926, and 1.7993 for PeU, PU, and willingness to use resp. The values of standard deviation for new design are 0.43823, 0.62329, and 0.41987 and for old design are 0.75983, 0.86560, and 0.49992 for PeU, PU, and willingness to use. This means that users find the new design to be more useful and easier to use; hence, the new design has been accepted by the users and they are willing to use it.

35.5 Conclusion

Presently, many business sectors are spreading all around which lead to increase in the workforce and management sector. To manage all at ones and also to increase the efficiency and profitability of an organization, we need a very effective and efficient ERP system modal that works best for the organization, meeting their needs and expectations. Researchers are working upon upgrading ERP systems, studying their critical success factors, making it more and more efficient. The X University's SAP portal is lagging behind in a lot of factors. To overcome these problems, a new design has been created and tested. It is anticipated that the newly designed SAP portal might be helpful for university students for timely and easy access to important information and fee payment. In the new design, few limitations are there like the number of participants taken were less so it might have given less appropriate results, more exploration could be done in terms of visual design, the design has not been technically developed, i.e., encoded, etc. As the redesigning of the Web site will be continued, few more features would be added like making and app for it or making it a responsive Web site so that it could work on mobile phones as well. We will keep on re-testing and redesigning it until it would become efficient, effective, and user friendly.

References

1. Christensen, R., Knezek, G.: Instruments for assessing the impact of technology in education. Comput. Schools **18**(2–3), 5–25 (2001)
2. Dechow, N., Mouritsen, J.: Enterprise resource planning systems, management control and the quest for integration. Acc. Organ. Soc. **30**(7–8), 691–733 (2005)
3. https://www.webopedia.com/TERM/E/ERP.html

4. Bradford, M., Florin, J.: Examining the role of innovation diffusion factors on the implementation success of enterprise resource planning systems. Int. J. Account. Inform. Syst. **4**(3), 205–225 (2003)
5. Fathema, N., Ross, M., Witte, M.M.: Student acceptance of university web portals: a quantitative study. Int. J. Web Portals (IJWP) **6**(2), 42–58 (2014)
6. Presley, A., Presley, T.: Factors influencing student acceptance and use of aca- demic portals. Journal of Computing in Higher Education **21**(3), 167 (2009)
7. Davis, F.D.: Perceived usefulness, perceived ease of use, and user acceptance of in- formation technology. MIS Quart. **13**(3), 319–340 (1989)
8. Basoglu, N., Daim, T., Kerimoglu, O.: Organizational adoption of enterprise resource planning systems: a conceptual framework. J. High Technol. Manage. Res. **18**(1), 73–97 (2007)
9. Abugabah, A., Sanzogni, L.: Enterprise resource planning systems (ERP) and user performance: a literature review. In: 20th Australasian Conference on Information Systems, pp. 1–10 (2009)
10. Bhatti, T.R.: Critical success factors for the implementation of enterprise resource planning (ERP): empirical validation. In: The Second International Conference on Innovation in Information Technology, vol. 110, pp. 1–10 (2005)
11. Ehie, I.C., Madsen, M.: Identifying critical issues in enterprise resource planning (ERP) implementation. Comput. Ind. **56**(6), 545–557 (2005)
12. Somers, T.M., Nelson, K.: The impact of critical success factors across the stages of enterprise resource planning implementations. In: Proceedings of the 34th Annual Hawaii International Conference on System Sciences, pp. 10. IEEE (2001)
13. Ngai, E.W., Law, C.C., Wat, F.K.: Examining the critical success factors in the adoption of enterprise resource planning. Comput. Ind. **59**(6), 548–564 (2008)
14. Dezdar, S., Sulaiman, A.: Successful enterprise resource planning implementation: taxonomy of critical factors. Ind. Manage. Data Syst. **109**(8), 1037–1052 (2009)
15. Nicolaou, A.I.: Quality of postimplementation review for enterprise resource planning systems. Int. J. Account. Inform. Syst. **5**(1), 25–49 (2004)
16. Percia, V., Pamulaklakin, R.L.: lLarners' satisfaction level with online student portal as a support system in an open and distance elearning environment (ODeL). Turkish Online J. Dist. Educ. **16**(3), 33–47 (2015)
17. Ellison, A., Arora, M.: Harnessing the power of Office 365 to provide a social learning environment through a new Student Portal. In: EUNIS 2013 Congress Proc. **1**(1) (2013)
18. Gavrilova, T., Jin, H.: One approach to knowledge mapping for international student portal (2008)
19. Boranbayev, A., Baidyussenov, R., Mazhitov, M.: Software architecture for in- house development of a student web portal for higher education institution in Kazakhstan. In: Information Technology-New Generations, pp. 759–760. Springer, Cham (2018)
20. Ryan, S.M., Jackman, J.K., Peters, F.E., Ólafsson, S., Huba, M.E.: The engineering learning portal for problem solving: experience in a large engineering economy class. Eng. Econ. **49**(1), 1–19 (2004)
21. Jones, N.B., Provost, D.M., Pascale, D.: Developing a university research web-based knowledge portal. Int. J. Knowl. Learn. **2**(1–2), 106–118 (2006)
22. Said, G.R.: The intention to use mobile student portal: a mobile human computer interaction study in a university context in Egypt. Mobile Information Systems, 1512602:1–1512602:8 (2018)
23. Thulasi Krishna, N.P.: Online student portal—a learning portal for every student. Int. J. Eng. Devel. Res. (IJEDR) **4**(3), 319–321 (2016). ISSN:2321-9939
24. Shorb, J.M., Moore, J.W.: The chempaths student portal: making an online textbook more than a book online. In: Enhancing Learning with Online Resources, Social Networking, and Digital Libraries, pp. 283–308. American Chemical Society (2010)

Chapter 36
LinkedIn Students: An Extension of LinkedIn Designed for College Students to Enhance Their Job-Hunting Experience

Paromita Loha and Anirban Chowdhury

Abstract LinkedIn is a job-hunting application used by people across all age groups and has extensive scope of networking and job postings. As per Forbes maximum users of this application are college students. However, graduating students and freshers get intimidated approaching company HRs. These hindrances limit students to make the best use of this professional platform. The aim of this research was to (1) design a user-friendly application which is an extension of the current LinkedIn specific to graduate and undergraduate students and (2) aid students to create contacts, hunt jobs. Online surveys and interviews lead to concept generation followed by prototypes. Usability testing and Nelson's heuristic evaluation were applied on the working prototype of LinkedIn Students. The prototype satisfied the heuristics and the interface perceived as more useful and easier to use than the existing LinkedIn. Hence, it would indeed be beneficial for college students aiding them to create a professional network, accelerate their job-hunting process and interact with professionals.

36.1 Introduction

Expanding professional network before stepping into the industry is a crucial step for college students. Students are aware of the need to network and create a professional circle but are often held back due to intimidation to approach a professional, profile building, tracking multiple interview schedules and understanding the graph of profile viewership. In the current application job, hunting gets tedious especially for freshers due to lack of specific job filters. Moreover, features for tracking application and interview status are yet not available. After going through the statistics of LinkedIn by 99 firms.com [1], the following factors will influence students to optimize their LinkedIn profile for maximum benefits:

P. Loha · A. Chowdhury (✉)
School of Design (SoD), University of Petroleum and Energy Studies (UPES), Dehradun, Uttarakhand 248007, India
e-mail: chowdhuryanirban14@gmail.com

A. Chakrabarti et al. (eds.), *Design for Tomorrow—Volume 1*, Smart Innovation, Systems and Technologies 221, https://doi.org/10.1007/978-981-16-0041-8_36

(1) 77% of recruiters use LinkedIn to hire potential candidates.
(2) 38% of 25–34-year-olds use LinkedIn.
(3) For 76% of LinkedIn members, Who's Viewed Your Profile is the top feature on the site.
(4) 57% of LinkedIn members use mobile phones to access LinkedIn.

As recruiters majorly use LinkedIn as their professional platform to recruit potential employees, college students and freshers prioritize building a professional online presence [2, 3]. Graduating students are not familiar with features which will boost their profile visibility on LinkedIn which undermines student's employment prospects after graduation. Being in the digital era where we see college students tapping into different social media platform for various self-purposes. However, the online presence seems to be less when it comes to LinkedIn even after knowing the potential advantages of creating a strong virtual identity on this platform [4, 5]. The success of LinkedIn as an application can also be determined when we see students able to bridge the gap between them and the professional world before graduating and with minimal help [5, 6]. Colleges and universities have realized the virtue that comes with this application and thus conduct many workshops to aid students to build a profile on this platform. However, few loopholes exist which shackles the students to escalate their progress. Studies and researches have been carried out to understand the gap between this professional application and its users majorly students; however, no clear-cut solution has yet been provided specifically for students. Lot of scope is visible in the design and development of the current application to cover up all the loopholes and aid them in using the application with minimal assistance. The loopholes are as follows:

- Lack of guidance for students to build a clear and concise profile.
- Endorsements can be done by anyone on your profile which lacks reliability.
- LinkedIn has limited ability to verify recommendations and endorsements.
- There is no mention of estimated stipend or salary.
- No platform for members to publish their own articles and lacks the sense of community building.

The aim of the study was to design of interface for an extension of the current LinkedIn called LinkedIn Students which first and foremost ensures user satisfaction. Freshers and graduating college students would be the main target audience helping them build a strong profile and an online presence. Nielsen's ten heuristic principles were thoroughly applied to understand aspects like satisfaction, usability and functionality of LinkedIn Students. Tracking views, interview schedules and interaction with the HRs would be made easier and everything can be operated from one single platform.

36.2 Literature Review

36.2.1 Social Versus Professional Networking Sites

The inception of LinkedIn happened in 2003 and has become the largest professional networking application, making it an easy access to college students to build their professional circle. It caters to students by enabling them to explore for job opportunities, inquire about corporates, start-ups, provide synopsis of resume and receive recommendations [7]. Also, you can see the number of profile viewers in a day and see the profile of your viewers however keeping a track of all these details gets taxing after a point. The online presence of users on social networking sites is increasing day by day where an average user spends around 22–24 h weekly on all these virtual platforms [8]. Bella Florenthal in her paper talks about the factors motivating college students to use this app and points out the reasons to adopt this platform [9]. Uses and gratification framework was applied to identify the motives. One distinction is that LinkedIn profile displays professional information which is more like an online formatted CV. The use is very specific and the information displayed is based on educational background, work experience, contacts. Its solely for professional purpose and language palette is very different and formal as compared to the social networking sites [10, 11]. The other social media platforms are more eloquent due to display of particulars like pursuits, hobbies, leisure activities mainly used for socializing purposes. Facebook, Instagram and Snapchat mainly satisfy the cognitive needs compelling people to spend hours scrolling down their screens and seems very effortless. The case differs when it comes to professional sites. LinkedIn's user group is majorly working professionals and their aim is to escalate their networking skills. Students need to learn to use LinkedIn for developing on professional grounds and build their virtual network, as articulated by Slone and Gaffney [12]. For graduating students their followers and contacts are alumni and students prefer to contact them on social media platforms and not any professional sites due to the formal code that is usually followed [13]. Unlike other social media sites where you can view other profiles anonymously the same does not apply for LinkedIn where you are instantly notified. This does makes the process more transparent and urges members to opt for a one month free trial of LinkedIn Premium giving access to their viewers. Members owning a premium version of LinkedIn are capable of anonymously viewing profile. Seemingly, graduating students are obtuse to adopt to professional networking apps decreasing their online presence and at the same times recruiters are engaging more and more constantly looking out for potential candidates [14].

36.2.2 Corporate Analysis of LinkedIn

The online activities of recruiters are dramatically escalating contributing to 55% of LinkedIn's revenue. Nowadays job listings have been minimal on print media and even advertisements. Most of it happens in LinkedIn as it is the most trusted network and chances of pretence are lesser due to transparency. The online presence of students is a key factor to gain the attention of any recruiter [2, 3, 13, 14]. Assessment of online activities, profile strength, networking skills, endorsements; all of that can be tracked aiding them to assess the character of any potential employee. Different magnitude of involvement is present amongst university students who have a LinkedIn account. Features such as networking, exchanging professional news, searching for like-minded professionals and interacting with co-workers are the major contributing factors towards the success on LinkedIn [13, 14]. To escalate their values on a professional platform, universities are urging students to activate their social skills on LinkedIn. Students barely tap into the social features which reduces their engagement level as compared to other social networking platforms. According to Brett Cooper and Mary Kate Naatus, LinkedIn can be used as a learning tool in business education. The paper encapsulates the current research on social media as a learning apparatus for higher education.

36.3 Methodology

36.3.1 User Studies

36.3.1.1 Alternative Hypothesis

It can be reasonable to assume that there is a change of response with the change in the existing LinkedIn Application with the addition of features targeted specifically to college going students as they are the ones who are well equipped with the existing application but face difficulty in approaching HR's being a fresher.

36.3.1.2 User Interview

A total of 20 students were targeted for this study in which 60% were postgraduate students and 40% were undergraduate students. It was noted that around 20% of the participants fell under the age group of 17–20 years, 30% fell under 21–25 years of age and the remaining 50% were under the age group of 26–29. A questionnaire was designed for a face-to-face interview wherein students were asked questions to assess the following criteria:

(1) Educational background.
(2) Frequency of using LinkedIn.
(3) Objective of using the application.
(4) Hindrances in the current LinkedIn.
(5) Ease of expanding professional network and connecting with HRs.
(6) Day-to-day online engagement.

Table 36.1 This table depicts the average usage of LinkedIn on a weekly basis

Average usage of LinkedIn	Participants (%)
Active (7 h/week)	15
Moderately (4 h/week)	50
Hardly active (1 h/week)	35

(7) Willingness to use a premium version targeted to college students using LinkedIn.

The interviews were recorded in terms of audio, video and written notes with a prior notice to the participants to keep the process transparent. While recording, participants were asked to use the current LinkedIn app and given small tasks such as profile building, job searches, applying suitable filters while job hunting and connecting to a potential professional of their field. The average usage of LinkedIn was also noted (Table 36.1).

36.3.1.3 Usability Testing for LinkedIn

An online survey was conducted on 30 students who are currently pursuing undergraduate and postgraduate students from fields like Design, Architecture, Computer Science, etc. (Fig. 36.1). Fourth-year undergraduate and postgraduate students were absolute for this method as they are determined to seek career opportunities and utilize skilled applications to plug themselves as compared to initial year graduate students. A Google form was sent to the above participants consisting of 15 multiple choice questions based on a nine-point Likert scale (where 1 indicates strongly disagree and 9 indicates strongly agree). This study was conducted to measure two independent variables (Perceived Ease of use and Perceived Usefulness) and one dependent variable (willingness to use). The independent variables measured the usability of

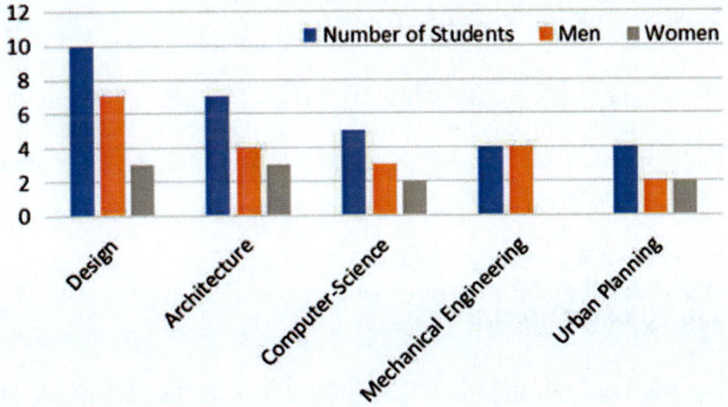

Fig. 36.1 Depicts the demographic details of 30 students chosen for online survey

existing LinkedIn and the dependent variable measured the acceptance of the existing interface.

36.3.2 *Evaluation of LinkedIn Students*

36.3.2.1 Heuristic Evaluation

To further understand the user satisfaction and the usability factors of the application, Nielsen's ten heuristic principles were thoroughly applied (Table 36.2). Minimalistic design approach was adopted for the UI of the mobile application. Consistency was maintained throughout the application to familiarize users with the icons and commands. This will strengthen the navigation system leading to reduction in time taken for task completion. The notification system has been well categorized on the basis of job alerts, interview schedules and connection requests. This will avoid any form of confusion and the notification will always keep the students alert, thus strengthening the heuristic principle of recognition rather than recall.

36.3.2.2 Usability Testing for LinkedIn Students

The same 30 participants were summoned once again to conduct usability testing for a varied version of LinkedIn which is LinkedIn Students. This time the participants were asked to perform two tasks on LinkedIn Students, (1) profile building and (2) job hunting using filters. Post that a Google form consisting of 15 MCQ type questions were sent to the participants. This study was conducted to measure two independent variables.

(Perceived ease of use, PEoU and Perceived Usefulness, PU) and one dependent variable (willingness to use) pertaining to LinkedIn Students. The questions were based on a nine-point Likert scale to measure the usability of the newly designed application. The questionnaire on PU and PEoU were formulated by Fred D Davis in his paper named Perceived Usefulness, Perceived Ease of Use and user acceptance of information technology, published in 1989 [12]. The questions were based on a nine-point Likert scale to measure the usability of the newly designed application. Post the data collection histograms were plotted from the responses received for PEoU, PU and willingness to use.

36.3.3 *Designing Interface*

Post the extensive research and interviews, results were collected followed by brainstorming concepts. All the problem were categorized using Open card sorting technique and feasible features were assigned pertaining to each problem. Use case

Table 36.2 Depicts the checklist designed for heuristic analysis

Heuristic principle	Yes	No	NA
1. Visibility of system status			
- Is there a header on every display?	✓	×	×
- Do the icons follow a consistent design?	✓	×	×
- Is there some sort of system feedback for each operation?	✓	×	×
- Is the current status of the icon highlighted?	✓	×	×
2. Match between system and real world			
- Are the icons familiar?	✓	×	×
- Are the tasks described using familiar technology?	✓	×	×
- Do the shapes correspond to common expectation?	✓	×	×
- Are the function keys labelled clearly?	✓	×	×
3. User control and freedom			
- Can users easily reverse actions?	✓	×	×
- Can the profile details be edited till the end of task completion?		×	×
- Can the users set their own screen defaults and feeds?		×	×
- Can users cancel out operations in progress?	✓	×	×
- Can the data entry time be reduced by copy-pasting data?	✓	×	×
4. Consistency and standards			
- Are icons labelled?	✓	×	
- Does menu structure maintain task structure?	✓	×	×
- Do the commands mean the same thing throughout?	✓	×	×
- Are the back and forward commands left aligned on top?	✓	×	×
- Is the structure of data entry consistent throughout?	×	✓	×
5. Error prevention			
- Is navigation between screen simple and visible within multiple pages?	✓	×	×
- Does the system prevent users from making errors whenever possible?	✓	×	×
- Does the system notify users before task completion?	✓	×	×
- Are the menu choices logical and distinctive?	✓	×	×
6. Help users recognise, diagnose and recover from errors			
- Is sound used to signal error?	×	✓	×
- Are prompts brief and unambiguous?	×	×	✓
- Do alerts place users in control of the system?	×	×	✓
- Do message alerts suggest the cause of problems?	×	✓	×
7. Recognition rather than recall			
- Do multiple filters have breathing space?	✓	×	×
- Is there good colour contrast between image and background?	✓	×	×
- Have the same colours been used to group related elements?	✓	×	×

(continued)

Table 36.2 (continued)

Heuristic principle	Yes	No	NA
- Are long column of information broken into groups or separated by lines?	✓	×	×
8. Aesthetic and minimal design			
- Is the primary purpose clear?	✓	×	×
- Are the texts legible?	✓	×	×
- Are the icons universal?	✓	×	×
- Are long column of information broken into groups or separated by lines?	×	✓	×

diagrams, class diagrams, sequence diagrams were created followed by information architecture. Later low-fidelity and medium-fidelity wireframes were created on Adobe Illustrator CC. The high-fidelity prototype was designed on Adobe XD (Fig. 36.2) and the name LinkedIn Students was finalized targeting the specific user group.

36.3.3.1 New Features and Functions

LinkedIn is the largest professional platform catering to a vast target audience in terms of age, professional backgrounds, type of organization and many more (Fig. 36.3).

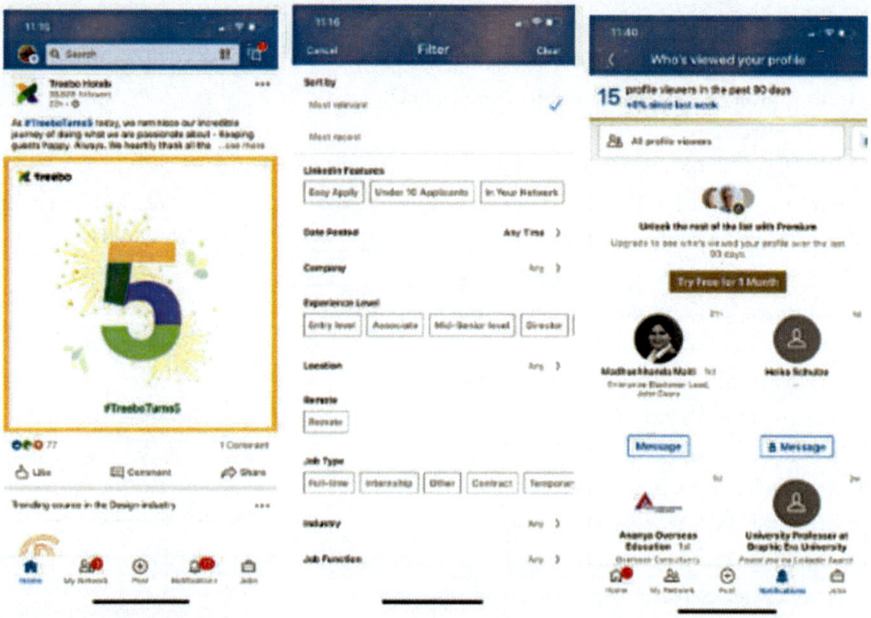

Fig. 36.2 Depicts the homepage, job-hunting filters, profile views of existing LinkedIn

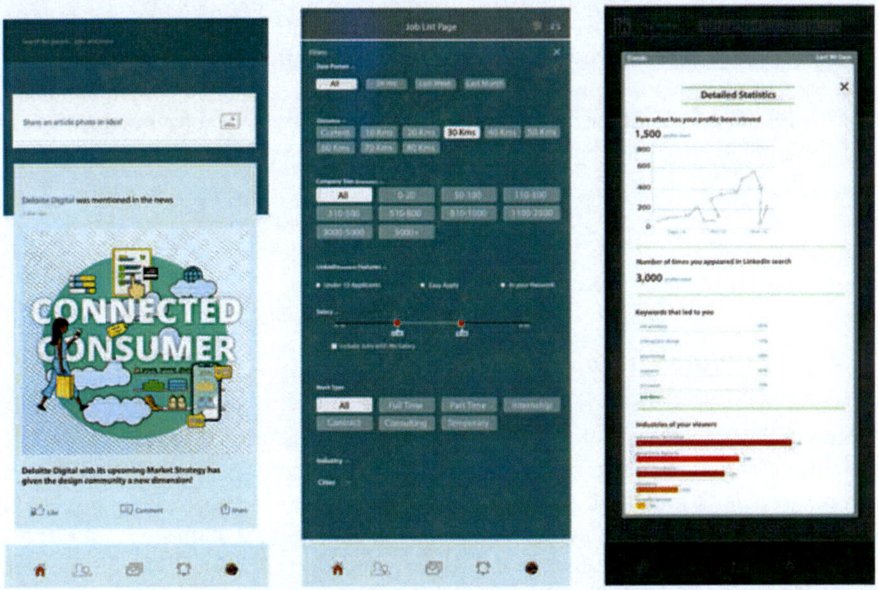

Fig. 36.3 Depicts homepage, job filters, profile statistics on LinkedIn students

However, LinkedIn Students encapsulates a whole new experience specifically for college students and fresh graduates to improve their job-hunting experience and increase the level of engagement on a professional application. New features and functions in LinkedIn Students have helped escalate the usefulness and the experience of this application.

(1) **Job filters**: This is a common and an important feature on job-hunting applications, however, we tried to work on the microelement of this feature by adding filters like industry, cities, distance and salary range. This will help students in getting a real picture of the professional world thus making practical choices.
(2) **One platform**: We tried to integrate multiple platform one on LinkedIn Students to omit maximum confusions. Usually job-hunting portals only help in procuring company information and applying. However, a lot of information is passed to the candidates via their emails, company sites and texts thus making it a taxing experience for candidates to juggle between various platforms to keep a track of information and company HR. On LinkedIn Students all the above procedure take place on one single platform; from job application to interview confirmations.
(3) **Tracking Interview schedules**: Tracking multiple interview schedules and stages of each interview can be tiresome. Thus, LinkedIn Students solves this issue by tracking each interview schedule by categorizing them into upcoming and previous interviews

(4) **Profile Statistics**: This refers to an in-depth analysis of your profile in terms of number of viewers, keywords that led to your profile, industry of your viewers and the number of times you appeared on search.
(5) **GPS**: This is another added feature which helps you reach your location to minimize confusion before heading for the interview.

36.4 Results and Discussions

36.4.1 Findings from User Interviews

From interview responses, it was witnessed that students use LinkedIn to 1. expand their professional circle (60%), 2. connect with industry professionals (75%), 3. receive job alerts (50%), 4. stay at par with the current trends in their respective fields (25%), 5. obtain freelance projects (20%) and 6. reveal their work experience in the form of verified certificates and recommendations (20%). It can be deduced that the majority of the students face hurdles while procuring jobs specific to recent graduates, approaching HRs or professionals, tracking profile views and building profile.

36.4.2 Findings from Heuristics

A total of two usability factors experts participated in the heuristic evaluation of the designed prototype of the app LinkedIn Students. They tried to identify the usability problems in it by using Nielsen's ten heuristics principles. In the beginning activity and task flows were thoroughly checked and analysed. Ensuring the functionalities of the application, the experts went through the UI and the prototype of the application using Nielsen's heuristics. As per the heuristics conducted by both the experts; the following interpretations were agreed upon:

(1) User control, system visibility, flexibility and recognition of UI elements are at par with user satisfaction.
(2) The newly designed interface merges different platforms (while performing tasks) on a single platform. Starting from deciding interview slots, application status, emails, constant interaction and feedback from the recruiters; all of these can very well function on LinkedIn Students itself.
(3) The design is minimalist and thus the user will not have to spend excessive time performing one task. It is time efficient.
(4) The semantic of the newly designed interface can be worked upon as per the guidelines of existing LinkedIn to avoid confusion amongst users and keeping the formal tone and identity of the existing application intact.

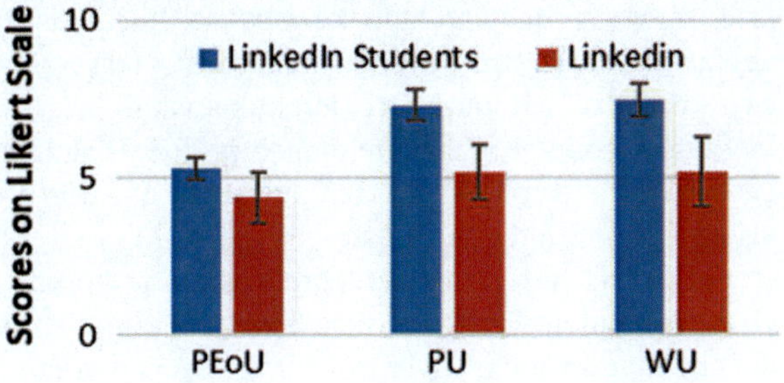

Fig. 36.4 Variations of mean values for perceived ease of use (PEoU), perceived usefulness (PU) and willingness to use (WU) for both versions

(5) The icons and profile templates could be further worked upon in terms of legibility. Colour schemes could be altered as per the target audience yet maintaining the aesthetics of existing LinkedIn.

36.4.3 Findings from Usability Testing

The premium version of the current application i.e. LinkedIn Students went through usability testing wherein the same participants filled an online survey and performed tasks on the new design. Mean, Variance and Standard deviation were calculated to understand the spread out of a set of data. Later error bars were calculated and plotted on the graph. Hypothesis testing was done prior to which verification regarding the type of test to be conducted was decided. Histograms for all the three variables (perceived ease of use (PEoU), perceived usefulness (PU), willingness to use)) were plotted. It was noticed that all the six histograms lacked a normal distribution curve which denied us to conduct t-test. Thus, Wilcoxon Signed-Rank Test was conducted as an alternative to paired t-test as the distribution curve was lacking. After running the test on the software SPSS, value of '*df*' (degree of freedom) was 21 for all the three variables. The resulting '*Z*' values for PEoU, PU and willingness to use are −3.182, −4.015 and −4.025, respectively. The level of significance i.e. '*p*' value is less than 0.001 for PU, willingness to use and PEoU, suggesting that the chances of our alternative hypothesis getting accepted are more than 95% as the '*p*' value is less than 0.005.

Furthermore, the descriptive statistics strengthens the fact that our alternative hypothesis will definitely be accepted by users (Fig. 36.4). This indicates the efficiency, usefulness and the willingness to accept the new version of the existing design.

36.5 Conclusion

LinkedIn Students is a promising application that aids graduating students to accelerate their job-hunting process and create a professional network. The application will also engage students to help them create a strong virtual presence. Students usually neglect building their profile which leads to lesser profile views. Profile building can easily be done without any external help and thus inbuilt templates will guide them in giving a concrete structure to their profile. Inbuilt templates within the application aids students to create and strengthen their existing profile. Similar requirements are also proposed in the recent studies [13, 14].

LinkedIn Students is just a preliminary prototype, the application requires more UI improvements in terms of usability factors like; recognize and recall, help and documentation and recover from error. There is a scope of design improvement as per the brand guidelines of LinkedIn. LinkedIn Students will be a premium version of LinkedIn wherein students buy subscription plans to have a full-fledged job-hunting experience pertaining to their level of educational and work experience [15–18]. The sample size will be increased for both User Survey and User Interview to validate the existing data to enhance micro-interactions within the application and to validate the data chi-square test can also be done for concrete results

References

1. Davis, F.D.: Perceived usefulness, perceived ease of use, and user acceptance of information technology. MIS Quart. **1**, 319–340 (1989)
2. Conlon, J., Taylor, A., Ashruff, P., Bird, L.: LinkedIn groups: building a community to create real connections to benefit students and alumni in fashion and textiles
3. Rahman, N.: Students 'understanding of information revealed in the profile of social networking sites-reviewing and examining facebook, twitter, LinkedIn, google plus and myspace. In: Proceedings of the Australasian Conference on Business and Social Sciences, pp. 1017–1025 (2015)
4. Case, T.L., Gardiner, A., Rutner, P., Dyer, J.N.: A LinkedIn analysis of career paths of information systems alumni. J. South. Assoc. Inf. Syst. **1**(1), 1 (2012)
5. Verschuren, A., Ranganath, K.: Relying on LinkedIn profiles for personality impressions. Master's thesis, Tilburg University, The Netherlands (2012)
6. Zhang, S., Vucetic, S.: Sampling bias in LinkedIn: a case study. In: Proceedings of the 25th International Conference Companion on World Wide Web 11 Apr 2016, pp. 145–146
7. Chang, S.E., Liu, A.Y., Shen, W.C.: User trust in social networking services: a comparison of Facebook and LinkedIn. Comput. Hum. Behav. **1**(69), 207–217 (2017)
8. Parez, M.E.: Linked into a job? The ethical considerations of recruiting through LinkedIn
9. Florenthal, B.: Applying uses and gratifications theory to students' LinkedIn usage. Young Consumers. 20 Apr 2015
10. Archambault, A., Grudin, J.: A longitudinal study of Facebook, LinkedIn, & Twitter use. In: Proceedings of the SIGCHI Conference on Human Factors in Computing Systems 5 May 2012, pp. 2741–2750
11. Oslund, C.: Building a professional ethos on LinkedIn. In: Online credibility and digital ethos: Evaluating Computer-Mediated Communication, pp. 252–268. IGI Global (2013)
12. Slone, A.R., Gaffney, A.L.: Assessing students' use of LinkedIn in a business and professional communication course. Communication Teacher. **30**(4), 206–214 (2016)
13. Papacharissi, Z.: The virtual geographies of social networks: a comparative analysis of Facebook, LinkedIn and ASmallWorld. New Media Soc. **11**(1–2), 199–220 (2009)
14. Auradkar, A., Botev, C., Das, S., De Maagd, D., Feinberg, A., Ganti, P., Gao, L., Ghosh, B., Gopalakrishna, K., Harris, B., Koshy, J.: Data infrastructure at LinkedIn. In: 2012 IEEE 28th International Conference on Data Engineering, 1 Apr 2012, pp. 1370–1381. IEEE

15. Noghabi, S.A., Paramasivam, K., Pan, Y., Ramesh, N., Bringhurst, J., Gupta, I., Campbell, R.H.: Samza: stateful scalable stream processing at LinkedIn. Proc. VLDB Endowment **10**(12), 1634–1645 (2017)
16. Florenthal, B.: Applying uses and gratifications theory to students' LinkedIn usage. Young Consumers (2015)
17. Cooper, B., Naatus, M.K.: LinkedIn as a learning tool in business education. Am. J. Business Educ. (AJBE) **7**(4), 299–306 (2014)
18. https://99firms.com/blog/linkedin-statistics/#gref

Chapter 37
Mobile User Experience (UX) Design Guidelines Considering the Need for Accessibility

Poornima Kapoor

Abstract A growing share of the population is using mobile devices for communication, content snacking, and to an extent computing. These extremely portable devices have become almost ubiquitous. They are everywhere. A rapid rise in cross and multi-device ownership is being seen. With this advent of smartphones and other handheld devices, it has become essential for us as designers to make our websites and applications available and usable to as many people as possible. We traditionally however choose to neglect people who are challenged from our primary user groups until it is an explicit requirement. This research aims to build accessibility design thinking into the UX process, guided by a model of 'Inclusive Design.' This design research helps understand and establish the nitigrities of the environment, pain points, needs, and requirements of a challenged user. It aims to make designing applications easier by eliminating the need to re-visit challenges by proposing a set of guidelines that can be used to improve user experience altogether. These guidelines have been established by using traditional research methods including 'expert interviews' with professionals, a study of literature, data analysis, interviews with challenged users, as well as usability testing along with modern-day design sprint technique to solve problems and test new ideas in a short time. Guidelines proposed at the end of this research by understanding the need for accessibility in UX design tied together with visual design will be of interest to designers, students, and academicians, all alike.

37.1 Introduction

According to the World Bank Group, there are 40–80 million people with disabilities in India [1], that is the same number as twice the population of Australia in 2017 (24.6 million). This number is so large, a fact sheet on persons with disabilities published by the United Nations lists them (persons with disabilities) as the world's largest minority [2].

P. Kapoor (✉)
IndiaMART, Sector 135, Noida 201305, India
e-mail: poornima.kapoor08@gmail.com

A. Chakrabarti et al. (eds.), *Design for Tomorrow—Volume 1*, Smart Innovation, Systems and Technologies 221, https://doi.org/10.1007/978-981-16-0041-8_37

Persons with disabilities, as a user group, are one of the largest in the world. When it comes to designing applications and websites, how often do we think about people who are challenged?

37.1.1 User Group: People with Cognitive Impairment

The user group chosen for this research is that with cognitive impairment. 'Cognitive impairment is when a person has trouble remembering, learning new things, concentrating, or making decisions that affect their everyday life.' [3] It ranges from mild to severe. For example, a 45-year-old female immigrant who settled in the USA 2–3 years ago—English is not her first language, and she needs extra time to read or write in English.

37.1.2 Why is there a need for Mobile UX Design Guidelines?

Interestingly many conventions govern and guide user experience design for the web. The UN Convention on the Rights of Persons with Disabilities [4] signed into effect in 2007, lists the types of appropriate measures which should be taken under Article 9 of the convention promoting access to new ICTs (Information and Communications Technology), such as the internet. Following the adoption of the Convention on the Rights of Persons with Disabilities, the World Wide Web Consortium (W3C) implemented standards and guidelines for the accessibility of facilities and services. It recognizes access to Information and Communications Technology (ICT) as a basic human right. As an effort to improve accessibility, the W3C launched the Web Accessibility Initiatives (WAI) in 1997. The WAI has created three core guidelines for the web: Web Content Accessibility Guidelines (WCAG), Authoring Tool Accessibility Guidelines (ATAG), and User Agent Accessibility Guidelines (UAAG).

In fact, there are well-established tools to check accessibility compliance for the web. But the same does not apply for mobile devices. And with the shift that is happening to more handheld devices being sold and used every day, it is time we look into the specifics for Mobile UX Design. Interestingly, when I began this research project back in 2018, the W3C's (World Wide Web Consortium) Web Accessibility Initiative (WAI) [5] was also developing specific guidelines on mobile accessibility [6]. To this day, June 15, 2020, the W3C has not shared these guidelines but has recognized the need for it, making this research even more significant.

37.1.3 Establishing the Gap from Literature

A study of the literature (between 2009 and 2016 [10–19]) on accessibility concerning UX design was done, to begin with this research. While three studies including Park et al. [7], Serra et al. [8], Fioravanti et al. [9] explicitly state the need for mobile accessibility related guidelines as follows: 'Aiming to bridge this gap, as future work we intend to investigate and propose a set of accessibility guidelines to help in the process of developing mobile learning applications for the elderly. In this context, the results gathered from this systematic mapping will serve as an important input to our research.' [9], 'More research is needed to define more specific sets of guidelines for mobile accessibility and procedures to help practitioners perform effective audits to improve the accessibility of mobile applications.' [8], and 'The proposed UX design elements and UX design principles could be used to create a systematic methodology to generate detailed UX design guidelines for the disabled.' [7], other studies hint the same.

37.2 A Case Study

This research has been brought about by taking up the redesign of BlackBuck (India's largest trucking platform) BOSS application as a case study. The majority of its user group is subjected to temporary cognitive impairment. BlackBuck BOSS has a user group of over 60,000 on its platform spread over more than 2000 locations across India. With such widespread users across culturally different geographies, one major challenge is to ensure application usability since the app is primarily developed in English, providing limited multi-lingual support. Also, the majority of BOSS users are new or challenged smartphone users, creating a case of temporary cognitive impairment, leading to trouble remembering and learning new things making it a good case study to understand usability and accessibility issues. The author of this paper, also the researcher, developed a set of guidelines based on the redesign of this application. The application was redesigned by the researcher with the help of three other designers working closely with the company.

37.3 Research Methodology

The methods used to collect information and data to make design decision during the research included the following:

1. Competitor Analysis
2. Expert Interviews
3. Data Analytics
4. Think-Aloud Usability Testing
5. A/B Testing

Table 37.1 Competitor analysis—sample table

Key factors	Weighting	Rivigo rating	Rivigo weighted	Paytm rating	Paytm weighted
1—Feature importance	0.4	6	2.4	3	1.2
2—Customer feedback	0.3	4	1.2	5	1.5
3—Reach/scale	0.2	3	0.6	3	0.6
4—Level of service	0.1	7	0.7	4	0.4
Totals	1.0	20	4.9	15	3.7

The first three methods of data collection were used before the design process while the last two methods, 'Think Aloud Usability Testing' and 'A/B Testing' are included within step 6, also, the last step of the design process (Sect. 37.5).

37.3.1 Competitor Analysis

It is a strategic research method to find out what competitors are offering and how successful they are at doing so. Qualitative outcome was obtained by thoroughly studying the User Flow/Journey and Information Architecture of the existing competitors. Quantitative data were extracted from user reviews. Rivigo and Paytm were identified as two competitors offering one or more similar services on their respective platforms. BlackBuck users using both or either service were identified and asked for reviews. Each application was reviewed based on four key features, namely, feature importance, customer feedback, reach/ sale, and level of service. Each feature was weighted based on its importance to the study. Users were asked to rate each feature on a ten-point scale. A sum of all rating scores multiplied by weight was calculated to obtain a final review score. Similar scores were collected from 10 users to get an average value.

For example (Table 37.1):

37.3.2 Expert Interviews

To evaluate the context of use, a set of questions were prepared, and an interview (Appendix A) was conducted with the Director of Products and Head of Product Management. Each was an active listening session of about forty-five minutes. These interviews facilitated in creating a mental model as a map to follow users footsteps (if needed at any point) in our mind as researchers. One task as a researcher after these interviews was to transcribe 'How Might We' questions. These are discussed in detail in Sect. 37.5 *Design Process*.

General information obtained from interviewees includes a summary of what BlackBuck is, what the company does, its design process, and challenges.

An excerpt from the transcript:

> 'Majority of our audience is not polished like users of shopping apps, and it is difficult to pitch the idea of digital touch screens or a delete dustbin icon. We need to break from the status quo and create simple products that impact our user's life.'

37.3.3 Data Analytics

With the help of the software development team, we enabled data collection from the mobile app by tracking user actions for ≈24 h including simple text entries, button clicks, navigation, etc. This tracking was brought about by pushing existing users for a forced update (software update that must be installed before the user is allowed to use the app again). This collection of data summed up into tables was facilitated by a business analyst to further analyze and make actionable (Table 37.2).

This registration funnel facilitated a better understanding of user drop off at each step. A drop off indicates that there is a leakage point through which users are dropping off/exiting due to difficulty in understanding an action that must be performed or the user being unable to provide correct required information to proceed, or both. Drop off data were also collected for several other user flows in the application like FASTag purchase and recharge, Fuel card recharge, Tyre purchase, etc. before the redesign.

For example:

37.4 Design Challenges

Based on the research, twenty-nine unique problems or challenges were identified in this study. Some information on each of these is listed below:

Table 37.2 User registration funnel

Onboarding stages	# of users	% users	% drop
App installation	1559	100	–
App launch	1339	85.89	14.11
Login screen	1239	79.47	6.41
Login button click	1117	71.65	7.83
OTP entered	1071	68.70	2.95
Profile screen	987	63.31	5.39
Home screen	698	44.47	18.54

1. *Language Selection*: Users hesitated to select any language other than English at the time of onboarding while an agent is present. They felt inferior.
2. *OTP Field*: While an OTP (One Time Password) is being auto-detected, users try to enter information in blocked input fields. When this does not work, they quit the application.
3. *Referrals*: Some users do not have the required code; they try to fill-in random information to complete the form; leading to failed sign-ups.
4. *Messaging*: Unlike general users who scan information, they tend to read everything on the screen. Too much unnecessary information increases the cognitive load.
5. *Adding an Address*: Users do not understand the concept of landmark and enter chunks of unnecessary data, sometimes repeating the same address information.
6. *Free FASTags*: FASTag (An RFID tag which enables making toll payments on highways directly from a linked bank account) is a free product on the BlackBuck platform. Users place orders but do not activate these cards upon delivery.
7. *Language Translation*: All users do not understand the primary language, English, of this application, creating accessibility issues.
8. *Smartphone Unavailability*: Many users wish to avail of services but do not have phones.
9. *Icons without Support Text*: Users are not familiar with standard setting/delete icons.
10. *Notifications and Feedback*: Users are unable to read quick notifications.
11. *Pure Decoration*: Decorative elements create confusion and add to cognitive load.
12. *Audio Feedback/Alerts*: Multiple feedbacks at the same time create confusion.
13. *Audio Commands*: The application interacts one way via audio. There is no audio output.
14. *Translation Sequence*: While converting text into regional languages Google's API does not always provide the right translation.
15. *Visual Feedback*: Shape, color, size of feedback symbols/icons are not consistent.
16. *Input Field Labels*: Labels are not consistent and clear.
17. *Contrast/ Text Ratio*: Application is used at odd times, early in the morning or late in the night. The visual contrast and text ratio must be apt for clear visibility.
18. *Background Sounds*: The background sounds have an equal volume or loudness to the foreground speech content. These sounds also last longer than one or two seconds.
19. *Design Pattern*: A design pattern is not fixed in the application. Thus, a button with the same action in different flows is placed differently. A user learning to work with the application faces cognitive challenges as each flow requires separate education.

20. *Screen Reusability*: The same screen in different flows has different user interface.
21. *Text Spacing*: It is difficult to distinguish between essential and non-essential information because of improper line height and text spacing.
22. *Shifting Focus*: Once the user does an action, and feedback is to be provided; the focus does not shift, making it difficult to understand if the action has been successfully completed, or is under processing, or has failed partially or altogether.
23. *Keyboard States*: The keyboard covers up important parts of a screen. For example: Hides a call-to-action underneath sometimes.
24. *Keyboard Trap*: Some keyboard states are not customized leading to a trap. A proceed or exit key is missing.
25. *Preserving Data on Timeouts*: If the application closes while onboarding or in the mid of a purchase process, the user must start again from the beginning each time.
26. *Flashy Images*: Application has flashy images and animations that are not necessary, serve no specific purpose, are easily replicable but increase cognitive load.
27. *Motion Animation*: Auto-animation triggered by interaction cannot be disabled.
28. Title and Headings: Page titles and headings missing. Users find it hard to keep track of their path.
29. *UX Copy*: UX copy is unclear, unable to provide the correct and required message in the simplest way.

37.5 Design Process

Each user flow in the application was taken up as a separate entity, one at a time, making tasks manageable. The following design process; summed up accurately by the subsequent six steps, was used based on the research findings for each of these user flows to solve problems: mapping problems, sketching solutions, deciding on an idea, detailing solution(s), building a usable prototype and usability testing [20]. This process (Step 1–5) was ultimately used to generate two variants that were evaluated during the 'Usability Testing' phase (Step 6).

Before beginning with Step 1, an insight into the product and its challenges is a preliminary requirement. This was brought about by conducting and/or participating in additional interviews and discussions (other than ones in Sect. 37.3.2) with a mixed team of people which included (not all at once) product manager(s), data analyst, sales team, in-house designers, and marketing team. Each session duration was between 30 and 45 min. During each of these sessions, all problem statements were reframed into questions to find solutions for. These statements were not too narrow—with one obvious answer or way to go forward; neither too broad or vague—with multiple answers. For example: How might we prevent customer dropouts during signup? A question that points at a substantial problem to be addressed. User personas and transcripts of previously conducted user interviews by the organization were also studied to deeply understand the context of use.

Step 1: Mapping Problems

Each of these questions was sorted under categories based on obvious patterns. For example: User Experience, Marketing, Logistics. These were then prioritized to find challenges that will have the most impact to solve. This presented a chance to align our (designers and other team members) understanding of general challenges. Design thinking was used in the process of prioritizing each of these challenges. Each of these challenges' focus on an area in the product was identified by drawing a map. A simple map was drawn, with actors (users or people that interact with the product) on the left and objectives on right. Each flow within the map connecting its respective actors to their objectives with steps that must be taken to complete their journey using the product/service.

Each of these flows was subdivided into three stages—discovery, learning, and use. Discovery is how users discover our product/service—via referrals, advertisements on social media, etc. Learning is how users learn more about our product/service—via information pages. Use is the steps involved with using the product/service. All statements identified before are put on to this map in respective stages they fit in. If there are two spots on the map where a single problem could fit in, preference is given to the one toward the left as solving problems earlier in the user journey fixes issues later in the journey. This map helped us identify primary problem areas.

Before we moved to step *two-sketching solutions*, we looked at other applications that have had a similar problem and their solution. This helps in coming up with multiple ideas and concepts on how we might be able to solve our problems. For example: We looked at other companies and products that have to coordinate availability for service delivery. These could be from other industries that are solving a similar problem.

Step 2: Sketching Solutions

Solution to each problem statement and user flow on the Map from Step 1 was translated into a visual sketch of the app. To exploit all possibilities, each screen was drawn multiple times. We divided a piece of paper into eight artboards and took almost a minute to draw in each of these boxes. All these boxes contained different versions of the same idea. Once we gathered ideas on each possibility we began with sketching solutions for each flow. The important learning here was however to not use lorem ipsum or wiggles to represent something but to use proper content.

Step 3: Deciding on an idea

Each idea was evaluated based on previous experience with users, inputs from design professionals and its feasibility, in a group discussion to generate a heat map of opinions on the potential and likeliness of solving a problem. A cluster of dots helps identify areas where opinions are really falling. Any questions or clarifications on any idea were addressed. This aligns the understanding of everyone on the team. We decided over variations at this step.

Step 4: Detailing Solutions

We created solutions with correct details and prepared a flow of steps for the prototype. Our aim here was to validate an idea with a simple prototype and not to build a complete product. Storyboarding is an essential part of this step. We wrote six simple steps in the process that a user might go through, starting with a realistic entry point, and an ideal ending. Flows that were not in the scope of this prototype were simply assumed to be there. Rubber Duck Debugging was done by reading out aloud all steps in these interfaces and rectifying any shortcomings.

It is to be noted that there are certain well-known user interface issues like limited area, on-screen interruptions and many others which were also identified during the course of this redesign. These were addressed while detailing solutions based on intuition and previous experiences of the designer.

Step 5: Building a Usable Prototype

The aim of this step was to build something very realistic that can be put in front of people. These are not wireframes. Every detail should be correct—the content, copy, and images.

Step 6: Usability Testing

The aim here is to get feedback from real users and see whether our goals have been fulfilled and if we are moving in the right direction. The target in this case was a small group of five existing customers or our most frequent users. We scheduled tests starting early in the day at 9 AM. The gap between the first and second test was the largest to make quick tweaks for issues that might be noticed by the first user. For example: A button that was not working in the prototype or typos in the content.

The other four testers were scheduled as close together as possible so that we have time at the end of each day to summarize and understand all the feedback we received during the day. Each interview (Appendix B) was timed between 30–45 min and 45–60 min (for more complex prototypes). Ideally, a team of two people would run testing—one speaking and interacting with the tester and the other taking notes. A feedback wall was set up for each test with the names of people and scheduled time on top. On the left side of the board, general topics were listed that we would be talking about in each test. These could be the name of a screen or a feature in the app or other general questions. This was set up before each interview and used as a guide through the test.

Each test was conducted remotely with a run-through of the prototype on call. The person taking notes would use two different colored post-its-green for affirming, confirming and positive comments and orange for misunderstandings, confusions, or negative responses. These were then placed on the feedback wall against each user and a broad list of topics. A/B Testing was also conducted by running two variants. A tracker was used to collect data on each interaction users had with the application—at each step in the flow. A quantitative summary of performance helped determine which fragment of either of the variants performed better.

Both A/B test and user interviews were used to collectively measure usability of the designs keeping in mind the standard usability metrics: success rate, time required per task, error rate, and user's subjective satisfaction.

Iterations

Results from usability testing were iterated on. Starting with a retrospective exercise to identify what worked well in the test and what did not. We begin with sketching solutions and then build a prototype for usability testing. The storyboard need not be updated as the improvements are on small feature changes or simple updates—adding a new drop-down or new copy on a page. The point of this iteration is not to test completely new ideas but iterate on the existing concept.

37.6 Design Guidelines

The end result of any UX design should be usable and defining experiences that meet the expectations of its users. Below is a list of general considerations or guidelines for UX Designer working on mobile platforms for users that might have different needs and experiences, even different attitudes toward technologies created during the design process. These guidelines were developed with repetitive design iterations much like the development of the C language and Unix operating system. The growth of C led to the development of Unix and vice versa. Similarly, here also these guidelines helped build a better UI, and feedback on visual design helped improve on guidelines. The development of guidelines was a later step in the process. Once all design challenges were identified, we began structurizing their solutions with the help of the design process (*Step 1 to Step 6* as mentioned above). During the design or UI iterations, we began using these pointers as guidelines and kept re-shaping and fine tuning them as our interfaces advanced.

It is to be noted that some of these guidelines are generic and have been presented in many other researches over the time (A few of these studies were even a source of reference for this particular research). However, they are a significant part of this paper as they have been re-found during the course of this design research.

37.6.1 Design Considerations for UX Designers

37.6.1.1 Design to Prevent and Minimize Error

1. Design to reduce error, more than reducing the task completion time.
2. Design to maximize efficiency. Some users might take a little more time to complete tasks.

3. Include options for in-app help in settings.
4. Design an error message as such that you also point a way out. Jargons and error codes must be avoided; they might be insignificant to users.
5. Help users quickly get back after any interruption.
6. Hide information that is irrelevant at the time and does not require any action.

37.6.1.2 Design for Consistency

1. Design a reusable solution—a design pattern to keep consistency.
2. Design for UI reusability to share common screen layouts, media, icons, and support text, ensuring predictability.
3. Avoid creating new and unique experiences. Keep things familiar.
4. Avoid too frequent updates.

37.6.1.3 Design for Clarity Rather Than Simplicity

1. Avoid using flashy media.
2. Over simplifying UI may also increase cognitive load; too much interaction simplification by hiding gestures also strains memory.
3. Design interactions that are clear and understandable. Avoid using abstract or unfamiliar metaphors.
4. Define the role of a link and clarify what will happen if it is activated.
5. Avoid full justification of text (creating extra spaces) and pure decoration.
6. Design with general rules for color contrast and text size in mind.
7. Initiatives that recapture tasks are good but must be implemented carefully.

37.6.1.4 Design with Focus on User Goals

1. Design should help users perform tasks easily and efficiently than the way they did previously.
2. Design notifications and feedback that can be acknowledged.
3. Design with active keyboard states to avoid any traps.
4. Preserve user data on time-outs and untimely exits.

37.6.1.5 Design for Accessibility Needs

1. Design ensuring that UI is compatible with assistive technologies.
2. Limit the use of multiple and multisensory feedback in any form viz. audio, visual, or vibrating alerts.
3. Design to include easy ways to adjust color contrast, text size and language, but implementation should be done carefully so as to not overwhelm users.

37.6.1.6 Design to Reduce Cognitive Load but Increase Cognition Engagement

1. Design to prevent distractions and help users focus.
2. Design to proportionately distribute cognitive engagements in an extended UI.
3. Use appropriate straightforward content or messaging.
4. Keep in mind how users learn about your application.
5. Avoid any interference and or over-stimulation by keeping the number of feedbacks at an optimum.
6. Cautiously proceed with the use of gamification. It may help tackle a task quickly but ensure getting users to do so without going overboard.
7. Focus on designing simple processes and products.

37.7 Conclusion

This research completes with a set of UX design guidelines for mobile as a platform, the aim of which is to ensure that the application is more accessible and usable by all people. These guidelines have been developed by taking up a case study to redesign BlackBuck BOSS application, while the application has been redesigned using these guidelines. Therefore, both have been simultaneous activities; development of one has supported the development of other. Recognizing the need and adapting to change, encouraging the growth of the UX design community is the hope of this project. UX designers could leverage this research, supplement, and modify the UX design considerations above. Also, as the mobile platform is continuously evolving feedback from the wider UX community will be required to keep the current list of considerations updated.

Appendix A. Expert Interview

Questions asked and answered in an 'expert interview.'

1. What does BlackBuck do?
2. What other products in the market offer similar services?
3. Are any of your products paid?
4. Tell me about the current design team at BlackBuck.
5. What does a typical day in your user's life look like?
6. What are some of the other apps your user's use?
7. What are the main challenges while designing user interfaces?
8. How is BlackBuck trying to solve these challenges?
9. How has user research been done so far?
10. What are the sources of reference for design?

Appendix B. Usability Testing

This appendix contains a list of supporting questions used while conducting a usability test.

1. Tell me about the last time you tried to [problem/ task].
2. What do you like about how you currently [problem/ task]?
3. What is the biggest pain point related to [problem/ task]?
4. What is most appealing about this product?
5. What is the difficult part of using this product?
6. Is there anything surprising or unexpected about this product?
7. What could be done to improve this product?
8. Is there anything missing from this product that you expected?
9. Would you keep using this product after what you saw today?
10. What other products have you tried?

References

1. Disability. (n.d.). https://web.worldbank.org/archive/website01291/WEB/0__CO-43.HTM
2. Fact sheet on Persons with Disabilities. (n.d.). https://www.un.org/disabilities/documents/toolaction/pwdfs.pdf
3. Cognitive Impairment: A Call for Action, Now! (n.d.). https://www.cdc.gov/aging/pdf/cognitive_impairment/cogimp_poilicy_final.pdf
4. Convention on the Rights of Persons with Disabilities (CRPD). (n.d.). https://www.un.org/development/desa/disabilities/convention-on-the-rights-of-persons-with-disabilities.html
5. Mobile Accessibility at W3C. (n.d.). https://www.w3.org/WAI/standards-guidelines/mobile/
6. Mobile Accessibility: How WCAG 2.0 and Other W3C/WAI Guidelines Apply to Mobile. (n.d.). https://www.w3.org/TR/mobile-accessibility-mapping/
7. Park, J., Han, S.H., Lee, M., Jang, H.: A literature survey on UX design properties and principles of smart device design for the disabled: from the perspective of the value/quality of human life. Japan. J. Ergon. 422–425 (2015)
8. Serra, L.C., Carvalho, L.P., Ferreira, L.P., Silva, J.B., Vaz, Freire, A.P.: Accessibility evaluation of e-government mobile applications in Brazil. Proc. Comput. Sci. **67**, 348–357 (2015)
9. Dias de Oliveira, C., Fioravanti, M., Fortes, R., Barbosa, E.: Accessibility in mobile applications for elderly users: a systematic mapping. In: 8th Annual Frontiers in Education Conference (FIE 2018) (2018)
10. Retief, M., Letšosa, R.: Models of disability: a brief overview. HTS Teologiese Studies/Theol. Stud. **74**(1) (2018)
11. Accessibility guidelines, accessible websites, people with disabilities. (n.d.). https://www.un.org/en/webaccessibility/
12. Stone, M., Bentley, F., White, B., Shebanek, M.: Embedding user understanding in the corporate culture: UX research and accessibility at Yahoo. In: Proceedings of the 2016 CHI Conference Extended Abstracts on Human Factors in Computing Systems, pp. 823–832 (2016)
13. Sultan, N., Siddiq, K., Rashid, T., Farooque, M.: Evaluation of smartphone applications accessibility for blind users. Int. J. Comput. Appl. (0975–8887) **127**(3), 9–16 (2015)
14. Petrie, H., Bevan, N.: The evaluation of accessibility, usability and user experience. In: The Universal Access Handbook, pp. 1–30 (2009)

15. Leporini, B., Buzzi, M.C., Buzzi, M.: Interacting with mobile devices via voiceover: usability and accessibility issues. OZCHI'12, pp. 339–348 (2012)
16. Kane, S.K., Wobbrock, J.O., Ladner, R.E.: Usable gestures for blind people: understanding preference and performance. CHI **2011**, 413–422 (2011)
17. Sierra, J.S., Togores, J.S.: Designing mobile apps for visually impaired and blind users using touch screen based mobile devices: iPhone/iPad. ACHI 2012. In: The Fifth International Conference on Advances in Computer-Human Interactions, pp. 47–52 (2012)
18. Kapoor, P.: Guidelines for Mobile User Experience (UX) Design considering the Social Needs for Accessibility. NIFT Kannur (2019)
19. Pokinko, T.: Designing Mobile Applications for Older Adults with Cognitive Decline. OCAD U (2015)
20. Knapp, J. et al.: Sprint, 1st ed. Transworld Publishers Ltd. (2016)

Chapter 38
i-HAWA: An Interactive Device for Providing Cognitive Breaks in the Workspace

Anamika Bhatt, Saurav Vaishnav, Muskan Surana, and Naveen Kumar

Abstract New ways of living require new ways of working in the offices. There is a shift in work culture towards the new needs of the company administration and management. With time, it will be necessary to understand physiological, psychological, and emotional needs of the company workers. Thus, this knowledge can form a basis to develop the progressive office spaces of the future. The research focuses into employee 'burnout' and 'mental fatigue' and how a cognitive break can be provided to desk job employees such that it has a positive effect on their work performance. As humans have made immense environmental developments while making their immediate surroundings more and more unnatural, it is understood that employees who are forced to work in a man-made environment experience an increase in stress levels due to a lack of exposure to nature. The research was conducted through a personal interview format that covered topics related to employee aspirations, stress levels, relaxation/break time patterns, etc. The data collected was qualitative and quantitative in nature. After analysing the data, it was inferred that employees are most relaxed by natural abstractions. To meet this need, i-HAWA was created. i-HAWA is an interactive device to be installed in the common areas of a workspace with the intention to provide cognitive breaks to employees dealing with excessive mental activity. The aim of this paper is to introduce the concept of cognitive breaks in workspaces and present design of i-HAWA along with its design methodology, product features, and functionality.

A. Bhatt · S. Vaishnav · M. Surana · N. Kumar (✉)
Department of Industrial Design, Karnavati University, Gandhinagar, India
e-mail: naveenb8@gmail.com

A. Bhatt
e-mail: hianamikabhatt@gmail.com

S. Vaishnav
e-mail: saurav.vaishnav@gmail.com

M. Surana
e-mail: suranamuskan99@gmail.com

A. Chakrabarti et al. (eds.), *Design for Tomorrow—Volume 1*, Smart Innovation, Systems and Technologies 221, https://doi.org/10.1007/978-981-16-0041-8_38

38.1 Introduction

The future of work is portable computing devices and use of mobile applications, and the office spaces of the future will strive to move with this change [1]. However, what will remain unchanged is the people working in these spaces and their needs. Office work usually involves sedentary task flows. This is harmful both for the physical and cognitive system of the workers. Long hours working in the same posture lead to musculoskeletal discomfort and thereby performance of the worker reduced significantly [2]. Few attempts have been made for reducing workers' cognitive load in the workspace [3]. Therefore, workers should ideally take continuous breaks to break a monotonous workflow. However, employees usually turn to their smartphones during break time. Chatting, casual gaming, or browsing through social media are some of the tasks they engage in and these tasks require focused attention. This does not decrease their screen time as most work also happens on-screen. It is ideal that they indulge in activities which replenish their cognitive function in these breaks, to ensure sustainable performance at work. Hence, it is imperative for office spaces to enable their workers to take distraction free breaks.

With this premise, the research team investigated the following:

(a) What is an ideal break?
(b) What would future office spaces look like?
(c) How could the office space design ensure the productivity of its employees?
(d) What are the work pressure points and factors influencing the employee's performance at the workplaces?

Objective of this research paper is to identify the need of cognitive breaks in workspaces, identify the factors which influence workers performance, and design of an interactive product to provide cognitive breaks in workspaces. Contribution of this paper is to perform a case study of office environment and scenario in order to find the factors for cognitive breaks and proposed a design of an interactive product called i-HAWA.

38.2 Research Background

Most of the research was conducted in co-working spaces in Ahmedabad, Gujarat, India. To explain the ethnographic, demographic, and ergonomics knowledge of the co-workers and co-working spaces, a systematic literature review was done. The collated literature is related to co-workers' interactions in a co-working ecosystem, workers performance, ergonomics, and factors influencing the performance at the workplace. Below subsections will give a detailed description of the research background.

38.2.1 Workspace and Environment

Traditionally defined by a maze of cubicles and desks, the modern-day workspace is slowly embracing an open-plan design [4]. However, the elements which make up these indoor environments are still very artificial.

Humans have an innate tendency to connect to nature and they feel psychologically happy when connected [4]. Our built environments need to acknowledge this innate human desire to connect to nature. As green spaces recede in our cities, there is an opportunity to bring nature indoors. Much of the time spent by office goers is in unnatural environments which reduces their exposure to greenspaces. Biophilic elements have been proven to reduce stress, lower blood pressure, and steady the heart rate [5]. This can help maintain physical and mental well-being. This has tangible effects on an employee's performance. People experience greater attentiveness and concentration which results in greater productivity [6]. Because of this, employees can derive greater satisfaction from their job and can help companies retain talent in the long run.

38.2.2 Factors Affecting Performance at Work

Conditions which increase cognitive load such as interruptions, disruptions, and information overload are related to diminished performance at work [7]. Workers can attain a higher sense of productivity and a more sustained one if opportunities are created for them to relieve work stress. The means by which people relieve psychological stress such as smoking, and caffeine are not extremely healthy in the long run. Continuous physical activities can exert our muscles [8]. Similarly, continuous and tightly focused attention due to office workload can exert our mental muscle and create stress in our mind [8]. It is essential that we take a break to break this continuous strain on our mental muscles. However, it is even more imperative that we let our minds heal in this time and to replenish cognitive function.

Following sections will discuss the design methodology, proposed product design, and discuss the findings of this paper.

38.3 A Case Study for Designing i-HAWA: Interactive Device for Cognitive Break

This section will describe the design methodology of this paper, subject information, data collection and analysis methods and final research findings.

38.3.1 Design Methodology

Figure 38.1 shows the building blocks of the design methodology. Design methodology consists of four major steps which includes systematic literature review, data collection and data analysis, product design, and research finding.

Research data was collected through survey, interviews, and metaphor elicitation techniques. Figure 38.2 depicts the research process. Survey is a quantitative data technique to collect data in numeral format whereas interviews and metaphor elicitation are qualitative methods. Metaphor elicitation is a technique to gather subject insights about the domain by showing them some form of abstract. The team showed some office workspace pictures to collect the subject's qualitative data to understand their requirements and thoughts about the co-working spaces and environment. Design audits were also conducted across five co-working spaces to understand the amenities provided, nature of work the professionals there were engaged in, aspirations of managers and owners of the spaces, and the requirements of the people working there.

The researchers spent a day at a co-working space with ten subjects for close observation of their daily activities at the workplace. This allowed for organic conversations with other co-workers and have a first-hand impression of what it is like to operate from a co-working space. Independent professionals were interviewed to learn about their preferences of workspace and understand why they shifted to remote working.

The focus of data collection through interviews are:

- Current nature of work and their behaviours and attitudes towards it.
- Expectations from their current workplace.

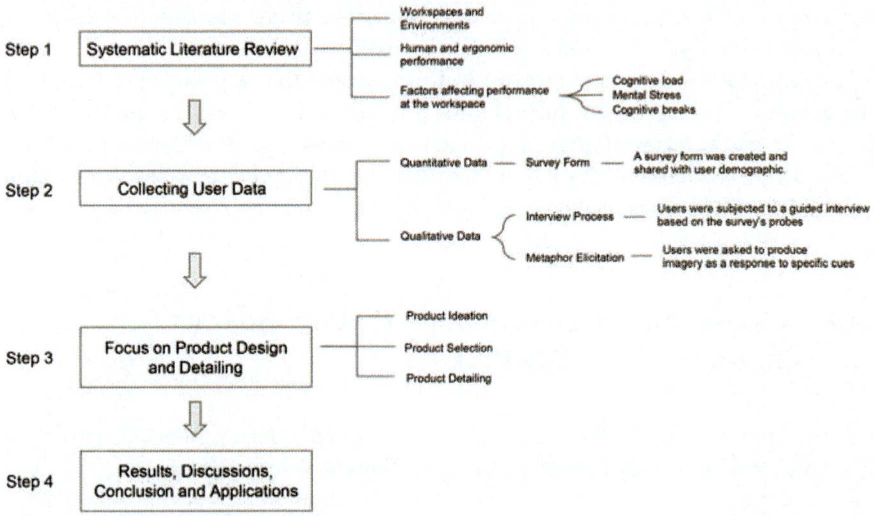

Fig. 38.1 Design methodology of i-HAWA

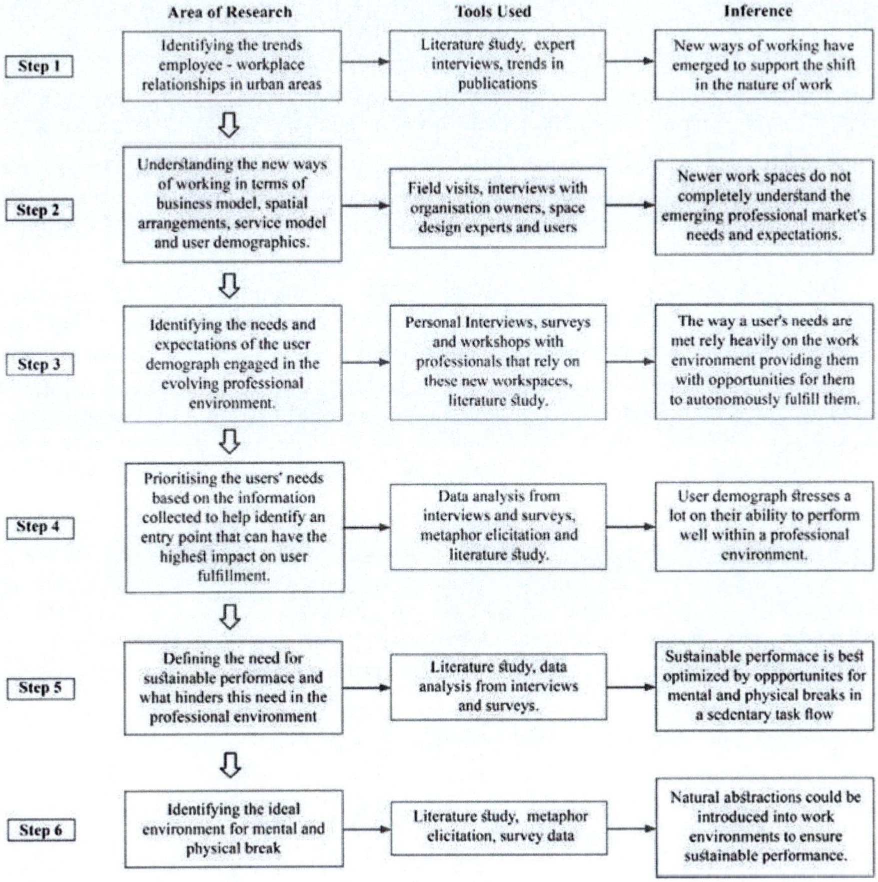

Fig. 38.2 i-HAWA research process

- Aspirations from professional life.
- Perception of an ideal workplace.

To get a holistic view on the subject experts such as interior designers, psychologists, co-working space owners and managers were interviewed to capture the relevant insights from their experience and perception about the future of remote working and co-working spaces.

38.3.2 Subjects, Data Collection, and Analysis Method

Total of '92 subjects' data was collected through qualitative and quantitative methods. For quantitative data, a survey was done with '30' subjects. For qualitative data, interview and metaphor elicitation were done with a total of '62' subjects. Through both qualitative and quantitative data analysis, we try to understand the user's demographic, pain points, frustration, motivations, and the requirements of cognitive breaks.

After the research process, the team combined the data points to arrive at insights which could reveal opportunities for design. This gave the team an understanding of problems which were consistent and needed to be addressed. Not only were pain points uncovered but the insights also pointed towards what theme could be introduced into the workspace infrastructure that would help provide for the users' aspirations.

38.3.3 Research Findings

All major research topics were spread to form a Web by association (affinity diagram). Each major node in the Web shown in Fig. 38.3 represents an influencer within a co-

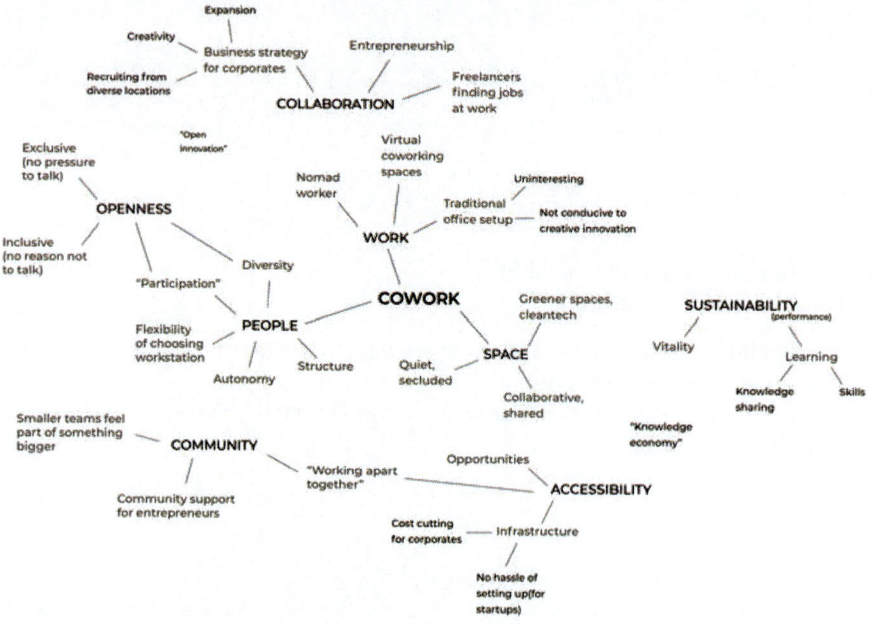

Fig. 38.3 Affinity diagram

working system which was the environment that the subjects were a part of. There are three influencers namely 'flexibility in timing', 'productivity and fulfilment', and 'socializing' which are main parameters to understand the core requirements of the employee and their influencing factors at the workplace. These three influencing parameters were calculated based on a survey conducted with 30 subjects (employees). The survey probes the subjects on the role of each influencers.

In terms of each influencers' priority, flexibility in timing (58%), 'productivity and fulfilment', (55%) and 'socializing' within the workspace (51%) emerged as the three most important factors that influence the user's perception of the workspace. Yet, 40% of the subjects are not able to find ways to relax within their workspace.

The subjects explained their workstyle, i.e. the frequency and duration of breaks taken by them on an average day. It was found that short breaks (under 15 min) were taken every 2–3 h by 50% of subjects while longer breaks (under 40 min) were taken by subjects only after reaching exhaustion or completing all assigned tasks.

To further understand the nature of the break, subjects were asked about what activity they would take part in during their breaks. 44% subjects mentioned taking a walk, out of which 10% responses specified taking a walk outside. 36% subjects mentioned consuming stimulants and snacks.

In an exercise on metaphor elicitation, subjects were asked to describe their ideal space to take a break. 67.8% of subjects preferred a setting that was exposed to nature. Figure 38.4 shows the two most popular images selected by the subjects.

35.7% of respondents chose (a) and 32.1% chose (b). From the findings, the subjects' goals (Flexibility in Routine and Productivity and Fulfilment) are mostly worked towards by the subjects alone without much intervention by the workspace itself. Moreover, workspaces do not encourage active release of stress and anxiety among workers, as seen in a gap between investment in infrastructure and workspace community management as their focus lies on stimulation and networking rather than relaxation.

Fig. 38.4 Image (**a**), Image (**b**)

Sustainable Performance. Sustainable performance in this context is being established as the description of the main objectives of the user demographic, i.e. flexibility in timing, productivity and fulfilment, socializing within the workspace, as determined by the results from interactions with research subjects.

58% of the world's population spends one-third of their adult life at work [9], and there is a need for the workspace environment, as an influencer within the workspace system, to facilitate its users to reach this level of performance. One way to achieve this is by enabling users to autonomously manage their time to reduce the build-up of stress and prepare the environment in a way that reduces the opportunities of mental fatigue to ensure that the goal of sustainable performance is being met.

We have identified the key points based on the qualitative data analysis to ensure sustainable performance:

- Providing sensory catalysts
- Voluntarily built personal boundaries
- Curb distractions
- Encourage active release
- Optimum personal comfort by ambience control.

Cognitive breaks. Just as physically taxing tasks lead to fatigue in the body's musculoskeletal make up, extended exposure to mental stimuli, as observed in the work environment of subjects, leads to mental fatigue which could deprive users of the space to carry out tasks that demand their motivation and autonomous decision-making capabilities.

Cognitive breaks, in this context, are being established as the period of mental rest or relaxation between events of heightened mental stimulus.

Evidence of Biophilia Within the Work Environment. A study on the effect of nature contact at work suggests that contact with nature, 'particularly indoor nature contact, is significantly associated with less job stress, fewer subjective health complaints and fewer days of sick leave.' [10]. Another study focused on how office workers compensate for a lack of windows revealed that windowless workers also had three times greater odds of having brought pictures of nature into their workspaces [11].

It is clear that humans continue to have a strong connection with natural elements in their work environment but a study aimed at understanding the link between indoor environment quality and occupant productivity [12] concluded that:

> 'Office design development should include biophilic design features from the concept design stage in the design process. A greater connection with nature would yield higher productivity results in an office environment. Biophilic design strategy can be used to infuse natural life into the office environment.'

Hence, it can be inferred that the indoor environment of the workspace is one of the most efficient points of contact from which an office worker can autonomously derive a cognitive break. The use of biophilic design can especially enhance this period of rest to ensure that the worker is able meet their goal of sustainable performance by relieving stress in a healthy manner.

38.4 Proposed Design of Interactive Device: i-HAWA

38.4.1 Design Ideation and Conceptualization

The ideation process for the resulting product i-HAWA was carried out following a set of ideal features and experience criteria derived from the study (Fig. 38.5):

- The product must have a multi-sensorial appeal.
- The product must be designed such that it can be appropriately integrated into the office environment, preferably at a public collision point (i.e. break room, lobby area, transit areas).
- The product must provide each user with a unique user experience and avoid having repetitive feedback as much as possible.
- The infusion of HCI in the task flow must create a seamless connection between nature and the office environment.

38.4.2 i-HAWA: Feature and Functionality

The concept of i-HAWA follows a three-step engagement using HCI at the core of its functionality. The intent of the product is to reduce fatigue and cognitive load

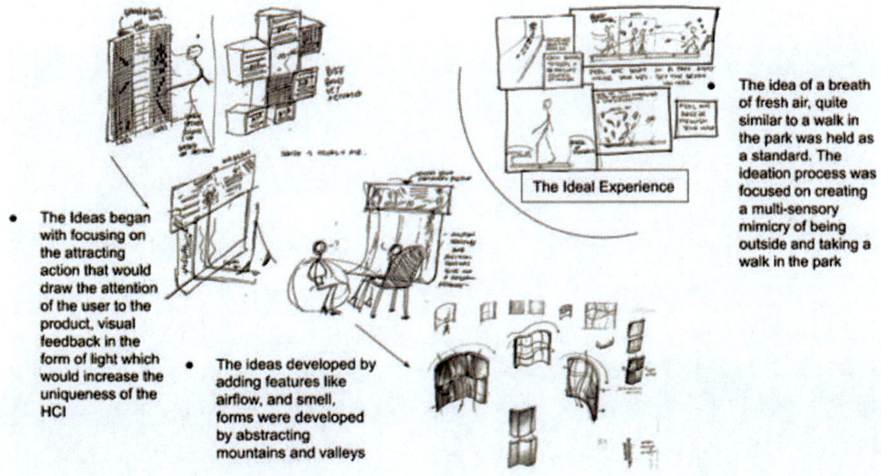

Fig. 38.5 Ideation sketches for i-HAWA

by redirecting the senses to natural abstractions that are created as a reaction to the users' presence (Fig. 38.6).

The Task flow of the device is as follows:

1. The product schedules a release of natural and neutral scents into the surroundings.
2. The product engages the user with stimulating lights which react to the worker's movements
3. The product will then track the user using proximity sensors which continue to engage them. The lighting system helps link people to the outdoor environment and keeps them on track with their natural 24-h cycle.
4. As the person is engaged in the experience, the fans in the panel activate to provide a cool humidifying. This is refreshing because of the dehydrated air-conditioned environment that is maintained in a typical corporate workspace setting.

i-HAWA will be placed on the pathways towards washrooms or cafeterias, allowing workers to move away from their seat to stretch or exercise within the workplace.

Figure 38.7 states the experience criteria that were derived from the study and their correlation with the product features. This explicitly defines the requirements for the interactive product and establishes a clear flow from the research findings to the product features.

Fig. 38.6 A software visualization of i-HAWA when activated

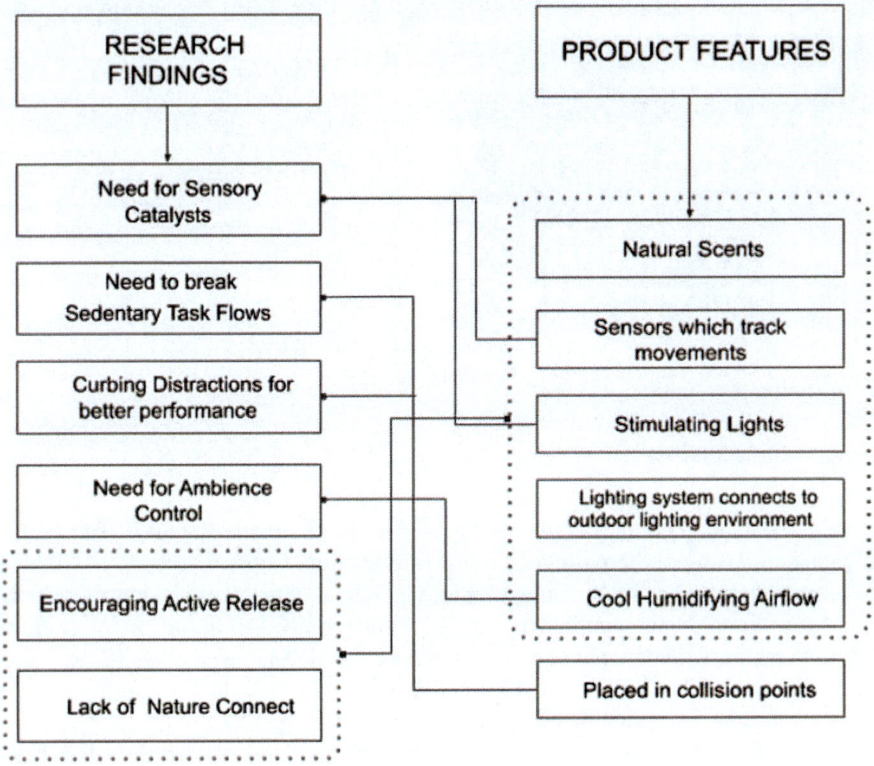

Fig. 38.7 Research finding–Product features correlation

38.4.3 *i-HAWA: Product Design and Detailing*

Technical components. To support the features and functionality offered by i-HAWA, aroma diffusers along with fans have been placed behind the vents on the front panel. Multiple infrared proximity sensors detect the human body and light up the front panel as a response along with mist diffusers and fans to recreate the experience of walking in nature. LED flexible strip lights help in creating the desired effect (Fig. 38.8).

Space, form and structure. i-HAWA has a sleek structure that can be easily wall mounted or placed on a stand to fit the narrow lanes/spaces. The front panel has an organic 3D pattern mimicking naturally formed steps and curves. Vents for air circulations have been provided on the back and bottom of the structure. Technical components are housed in the back tray and the front panel.

Material and texture. i-HAWA has a space grey, satin finish, fibre-reinforced bio-plastic body that compliments the required functionalities. Water and liquid aroma inspired from nature is needed to fuel the diffuser tanks.

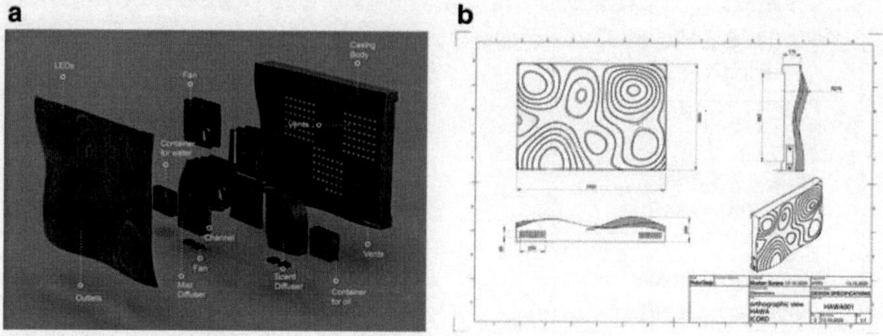

Fig. 38.8 **a** i-HAWA exploded view, **b** Detailed drawing

38.5 Discussion

True worker satisfaction can only be achieved when companies invest in the psychological well-being of their workers. For this to happen, future workspaces will have to become more innovative, productive, and inclusive of the worker's needs. To optimize workflows of the employees, cognitive breaks need to be provided. And the workspaces need to be designed in such a way which enable workers to take rejuvenating breaks. The synergy of HCI and biophilic design can be leveraged to aid workers in achieving such breaks.

The rapid increase in urbanization and lack of green spaces within urban landscapes have led to disengagement with nature. As people spend most of their day indoors, the built environment in which they exist is of paramount importance. Biophilic design and architecture have great potential to reconnect people to nature. In conjunction with technology and HCI, they can be used to create healthy living environments in otherwise high stress areas such as universities and courtrooms. Some common environmental stressors which strain human cognition are noise, crowding, and air quality. In areas with high concentration of these, such as airports and shopping malls, concepts such as i-HAWA can be used to create rejuvenating precincts. In enclosed spaces where growing and maintaining of plants are not viable, such as hospitals and laboratories, i-HAWA can be used to create natural abstractions.

In future, it will become imperative to use the synergy of HCI and biophilic design at the service of healthy built environments. i-HAWA is a case study of a small step towards this change.

References

1. De Stefano, V.:The rise of the "just-in-time workforce": on-demand work, crowdwork, and labor protection in the "gig economy". Comparative labor law journal: a publication of the

U.S. National Branch of the International Society for Labor Law and Social Security [and] the Wharton School, and the Law School of the University of Pennsylvania. **37**, 471–504 (2016).
2. Foley, B., Engelen, L., Gale, J., Bauman, A., Mackey, M.: Sedentary behavior and musculoskeletal discomfort are reduced when office workers trial an activity-based work environment. J. Occup. Environ. Med. **58**(9), 924–931 (2016)
3. Lindblom, J., Gündert, J.: Managing mediated interruptions in manufacturing: selected strategies used for coping with cognitive load. In: Advances in Neuroergonomics and Cognitive Engineering, pp. 389–403. Springer, Cham (2017)
4. McGregor, W.: The future of workspace management. Facilities (2000)
5. Klotz, A., Bolino, M.C.: Bringing the great outdoors into the workplace: the energizing effect of biophilic work design. Acad. Manage. Rev. (2020)
6. Ware, J., Grantham, C.: The future of work: Changing patterns of workforce management and their impact on the workplace. J. Facil. Manage. **2**(2), 142–159 (2003)
7. Kalakoski, V., Selinheimo, S., Valtonen, T., et al.: Effects of a cognitive ergonomics workplace intervention (CogErg) on cognitive strain and well-being: a cluster-randomized controlled trial. A study protocol. BMC Psychol **8**, 1 (2020)
8. Vischer, J.C.: The effects of the physical environment on job performance: towards a theoretical model of workspace stress. Stress Health: J. Int. Soc. Invest. Stress **23**(3), 175–184 (2007)
9. Rantanen, J., Lehtinen, S., Salmenius, R. (1994) Global strategy on occupational health for all: the way to health at work. In: Recommendation of the Second Meeting of the WHO Collaborating Centres in Occupational Health. Beijing, China, 11–14 October 1994
10. Bjørnstad, S., Patil, G., Raanaas, R.: Nature contact and organizational support during office working hours: benefits relating to stress reduction, subjective health complaints, and sick leave. Work (Reading, Mass.) **53**, (2015). https://doi.org/10.3233/WOR-152211
11. Bringslimark, T., Hartig, T., Patil, G.P.: Adaptation to windowlessness: do office workers compensate for a lack of visual access to the outdoors? Environ. Behav. **43**(4), 469–487 (2011)
12. Al Horr, Y., Arif, M., Kaushik, A., Mazroei, A., Katafygiotou, M., Elsarrag, E.: Occupant productivity and office indoor environment quality: A review of the literature. Build. Environ. **105**, 369–389 (2016)

Chapter 39
Speculating Cybernetic Cities in a Posthuman Era: Design of Internet of Bodies (IoB)

Srinjoy Ghosh and Mohd Saim Nasim Lari

Abstract This paper incorporates a quantum design approach to design possibility through the lens of Techno-humanism (interchangeably used with transhuman) and Internet of Bodies (IoB). Functionally, depending on how the present is characterized, both past and future possibilities are defined simultaneously is the basis of a quantum design approach. Please keep in mind that 'past', present', and 'future' are utilized here in a literal sense of lived temporality. Techno-humanism is seeking to utilize and upgrade the human mind to enable access to hitherto unknown experiences and unfamiliar states of consciousness. (Harari, Y.: Homo Deus—A Brief History of Tomorrow. Penguin Random House (2016)) At the same time, with humankind incorporating a posthuman phase and largely expected to extend to a transhuman one, supporters of trans-humanism argue that we have arrived in a post-Darwinian era in which we can shape our own evolution (Shanks in Speculative Design and the Posthuman Body, Museum Boijmans Van Beuningen, Rotterdam, 2013) and the inevitability of Internet of Bodies (IoB). A classic example of IoB application is the Matrix universe in The Matrix Trilogy films. The present harbours IoB through the Internet of things (IoT) portfolio, through its data centric lens of monitoring and optimization of networked infrastructural processes (as part of the human body) which seems as the cornerstone towards a new techno-social era. The aim of the paper is to add to the crucial emerging body of 'Internet of Bodies (IoB)' (Matwyshyn, A.: The Internet of Bodies, 61 Wm. & Mary L. Rev. 77 (2019). https://scholarship.law.wm.edu/wmlr/vol61/iss1/3) literature in HCI design by leveraging a speculative design outlook to the future of urbanism in a more integrated and connected era. Additionally, through an experimental study about the future of a connected IoB future, the paper shall expound ways of enabling, creating, and presenting the probable design architecture routes to help the conceptualizing and imagining of future possibilities in a connected IoB smart city. Alongside, we will illustrate the results of the experiment through the lens of probable outcomes as a result of co-opting and co-designing the homogeneous evolution of HCI, IoT, and IoB in future smart cities.

Note: All authors have made equal contribution to the paper.

S. Ghosh (✉) · M. S. N. Lari
CT RDA UXD-IN, Siemens Technologies and Services Pvt. Ltd, Bengaluru, India
e-mail: ghosh.srinjoy@siemens.com

A. Chakrabarti et al. (eds.), *Design for Tomorrow—Volume 1*, Smart Innovation, Systems and Technologies 221, https://doi.org/10.1007/978-981-16-0041-8_39

39.1 Future Cities: Concretized Data

> 'Design can give experts permission to let their imaginations flow freely, give material expression to the insights generated, ground these imaginings in everyday situations and provide platforms for further collaborative speculation.' [11].

Notwithstanding the myriad of counterarguments regarding the terms 'design' and the usage of 'expert' in Yelavich's aforementioned lines [11], the focus in this paper is the ability of speculative design to be 'collaborative' in nature that channelizes the potency to what Anderson [1] calls 'imagined communities'. We have tried to amalgamate the nuances of the socio-political core of Anderson's concept along-with the discursive practice of speculative design applied on the conceptual use case of 'Integration of Internet of Bodies (IoB) in a future Smart City Infrastructure. The paper is divided into four parts: Sects. 39.1 and 39.2 investigate the landscape of the techno-social aspects of speculative design and their movement towards integrating IoB. Section 39.3 explores the experiment, and Sect. 39.4 introduces the some of the present technologies that might prove invaluable in realizing a cybernetic city.

Mitrovic [8] in his explanation of the speculative design practice limits our understanding towards the impact of technology being the nucleus of societal development in future—'speculation on possible futures and the design of an alternative present. Speculation on the future generates scenarios of the future that critically question the concept of development, the implementation and use of new technologies and their wider social implications. The concept of an alternative present refers to the creation of parallel urban technological realities. These specific approaches offer a rich narrative potential for the questioning and criticism of technological development, but also of contemporary society as such. The issues dealt with can be exceptionally broad, from big socio-political topics to ordinary everyday activities.'

The establishing of the thought that cities of the future are purely technological in nature and that independencies of micro, meso, macro levels of social collectiveness be reduced to the abilities of the technology that the city would employ is also echoed by speculative architect Liam Young [12]—He opines that, 'there is a shift in the spatial experience in cities, from metropolis shaped by physical buildings and objects to digital networks, which are fundamentally in shaping the futures we are all racing toward. What the network has done to the planet is utterly radical. It defines new forms of communities, and even new forms of cities. We can imagine new cities organized around satellite sight lines, or inversion of property values as getting a good mobile phone signal becomes more important than having access to natural light. We navigate cities in a very different way from the ways we were used to. I look at my phone, in the maps interface I follow the pulsing blue dot of Google around the city. City form could become infinitely complex, intricate and labyrinthine because I always know where I am. Cities used to be organized based on human modes of perception and scaled according to the measurements of the body. Now cities are designed to be inhabited by the algorithms and technologies that have been tasked with reading and managing them.'

Auger [2] through Fig. 39.1 recalibrates the role of technology as the nucleus of the present and the force of the tomorrow—'the technology element on the left hand side represents research and development work, the higher the line the more emergent the technology and the longer and less predictable its route to everyday life. As we move to the right of the diagram and into the future we see that speculative designs exist as projections of the lineage, developed using techniques that focus on contemporary public understanding and desires, extrapolated through imagined developments of an emerging technology. Alternative presents step out of the lineage at some poignant time in the past to re-imagine our technological present. These designs can challenge and question existing cultural, political and manufacturing systems.'

Mitrovic, Young, and Auger each have specific arguments on the aspect of what role technology plays in the community of the future. By the process of 'a rolling stone gathers no moss', we might expect that Smart Cities which are so technologically sourced would easily slot in as a jigsaw piece to the whole 'networked communities' storyline we are dealing with here. CSIRO [3] have attempted to give us a glimpse of a future where data is the blood that fuels the design of city streets through what they call as the 'by-design' approach (this tangentially supplements with Young's anticipation on what can be loosely said that the requirements of technology and not humans will shape such a future city architecture) through their whitepaper 'Future Cities Thinking—Future Cities and Communities by design'—By way of contrast strategies that focus on technology first or solving point problems with point solutions, naturally result in fragmentation and walled gardens of solutions that do not interoperate together. This hampers further innovation and wastes investment dollars. In addition, our 'by design' approach eliminates the gamble and uncertainty behind Open Data approaches to innovation.

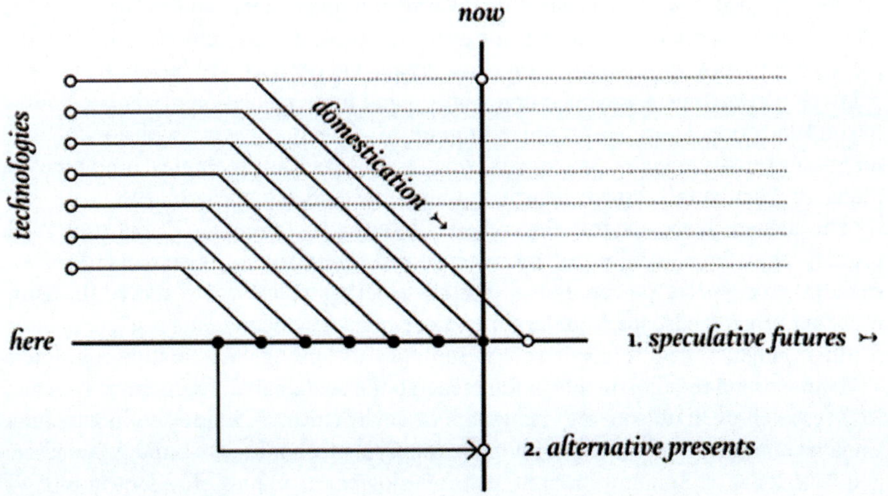

Fig. 39.1 Graphical depiction of the creation of speculative futures

With future cities that that are in need of *collaborative* design to create a future that the society would want and not leave it to chance, CSIRO is also mindful of the dangers that intercity rivalry might fosters a culture of isolation, mistrust, and secrecy. Their proposal to handle such a socio-political crisis is to make data sharing a national resource and make it one of the key building blocks of a future smart city. This begs the question whether communities of the future will be more of a concrete data construct and less of a de-concretized cultural construct.

39.2 Future Cities: Connected Bodies

Techno-humanism seeks to upgrade the human mind and gives us access to unknown experience and unfamiliar states of consciousness [5]. With humankind incorporating a posthuman phase and largely expected to extend to a transhuman one, supporters of trans-humanism argue that we have arrived in a post-Darwinian era in which we can shape our own evolution. The inevitable symbiosis of computer technology and the human body was the subject of The Matrix film trilogy (The Matrix, The Matrix Reloaded and The Matrix Revolutions from 1999 to 2003) but little by little it now appears to be becoming reality, as espoused by Shanks [9]:

Such techno-philosophical ideologies pave way towards the Internet of Bodies (IoB). Most of our world is outside of our human frame. And the outside infiltrates: we have been colonized by bacteria essential to our digestive well-being, and, of course, we must consume to survive. We are born into and inherit worlds of tangible things, intangible symbolic systems that generate meaning, infrastructures of convention and agreement like language, networks of association with others that make us who we are. And which we reproduce in our everyday living: society and culture, tangible and intangible, only exist through our everyday projects and activities, and yet are the very condition of possibility for those plans and designs. We have always been cyborgs, part organism part machinic control and implementation system, extended beyond ourselves, hybrid, in organized assembly with other species, with nonhumans, things, and environments. As humans, we have always been ecologies, oikologies in old Greek, where the oikos is hearth and home, extended household. [9].

The present harbours IoB through the Internet of things (IoT) portfolio. IoT through its data centric lens of monitoring and optimization of networked infrastructural processes is the cornerstone for this ushering techno-social era. At the same time, our urban landscape is undergoing drastic change due to humanity's impact on it. Environmental health has driven an influx of technological and anthropological adjustments and re-adjustments in the race to build sustainable smart cities. HCI and BCI researchers in unison with designers of smart cities are interested in a techno-social community that is technobiological (applying technology to biology) in nature and functionality. Some prominent technobiological examples prevalent in today's society include neurostimulators, enhanced DNA sequencing with quantum dot and chips, nanoparticles for cancer treatment, prosthetic and bionic devices.

Matwyshyn [7] in her canon defining work ' The Internet of Bodies' speculates that in the future 'as bits and bodies meld and as human flesh becomes permanently entwined with hardware, software, and algorithms, IoB will test our norms and values as a society', as 'a network of human bodies whose integrity and functionality rely at least in part on the Internet and related technologies, such as artificial intelligence'. Although her article primarily looks at IoB from the lens of the legal, she suggests social ramifications that extend towards 'what is means to be a 'human' in an age where bodies are the 'things' connected to the internet'.

Marr [6] contemplates on the 'connected-ness' and states that IoB is an extension of the IoT and basically connects the human body to a network through devices that are ingested, implanted, or connected to the body in some way. Once connected, data can be exchanged, and the body and device can be remotely monitored and controlled. There are three generations of Internet of Bodies that include:

Body external: These are wearable devices such as Apple Watches or Fitbits that can monitor our health.

Body internal: These include pacemakers, cochlear implants, and digital pills that go inside our bodies to monitor or control various aspects of our health.

Body embedded: The third generation of the Internet of Bodies is embedded technology where technology and the human body are melded together and have a real-time connection to a remote machine.

39.3 Future Cities: The Methodology, the Workshop, and the Findings

Speculative design instigates discussion by building prototypes in a future world—exploring probable, plausible, possible and preferable artifacts, services, experiences, opening up new possibilities, not only for technology, materials and manufacturing but also for narrative, meaning and the rethinking of everyday life. Asking - What if? How might we? And building, making to learn, trying things out [4, 9].

The experimental study was conducted in an online collaborative workshop. A total of 17 participants and two facilitators participated in it. The participants had varied backgrounds—experience designers, human scientists, and technologists. The objective was to ideate, discuss, debate, and create alternative ways of IoB integration in a smart city Infrastructure in a near future; to contemplate the measure of the reception of such a future in the immediate present through the lens of the socio-cultural and also to probe upon the ethical aspects of the same.

The conceptual brief given to our participants in the workshop was:

'A company in Wisconsin had a 'chipping party' in 2017 to implant microchips in some of its employees to make it easier for them to access the buildings and systems and to buy food in the company break room. Those employees joined a growing number of workers in other countries—Belgium, the UK, and Sweden, to name a few—who use microchips for workplace security, convenience, and commuting.

Microchips are just one example of the increasing variety of smart devices that are near to, attached to, or reside inside the human body. These devices are no longer relegated to science fiction and spy thrillers [10]. They are in use today—Welcome to the Age of Internet of Bodies and the future of urbanism'.

Speculative use case: Integration of Internet of Bodies in a smart city infrastructure.

There were two primary research aims addressed:

1. To ideate, discuss, debate, and create alternative ways of IoB integration in a smart city infrastructure in the recent future.
2. To contemplate the measure of the reception of such a future in the immediate present through the lens of the socio-cultural and the ethical.

39.3.1 Study Outcome—Speculative Futures

The study revealed interesting findings on how people perceive IoB being an integral foundational cornerstone for cybernetic cities. The findings direct us towards ways in which human 'meatware' is can be connected in a future cybernetic city by leveraging the technology of the present:

Connected posthuman as a healthy entity. Building upon the present behaviours of sensors and implants for health data collection, participants explored further into concepts which helped in humans overcoming illness. The ideas ranged from mood detection and mood enhancements to food shopping experience based on connected health data.

Connected posthuman as a social influencer. The participants discussed the ideas of technologies augmenting human social cues which helps them anticipate social cues during interactions with each other. The ideas further illustrated how this could be used to persuade someone or enhance social networking to make friends or strengthen relationships. Further to this were ideas of technologies which could influence human empathy in order to create 'equal empathy' between the citizens.

Connected posthuman as an effective communicator. In future cybernetic city, posthumans will have advanced brain computer interfaces which will help them interact with technology. The participants speculate that such a world riddles with 'brain-tech interfaces' will resolve issues around effective communications by removing the ambiguity and conflicts in semantics. A posthuman in a cybernetic city will be an effective communicator.

Connected posthuman as an urban superorganism. The participants also speculate that the cybernetic city with advanced 'brain-tech interfaces' will evolve as an urban superorganism with collective intelligentsia. Such an urban superorganism will have neuronal connections that will connect every brain to a centralized system resulting in a knowledge pool with open access to content and data.

Having set our future narratives, we used speculative design to debate potential ethical, social, and political implications. By exploring ideas before they become products or even technologies, designers can look into possible consequences of technological applications before they happen. [3] In a posthuman cybernetic city, interesting debate came up around ethical issues of personal data sharing and accessibility, draft of referendum on equal rights of posthumans, transhumans and humans, emerging of new social class and political power, evolution of political leadership as we know today, and the ‘rights’ and ‘access’ to technologies. For example, in a future cybernetic city, people might not go to hospitals to cure ailments but rather they go for implantations to ‘upgrade’ themselves leading to a higher social status. The wealthy might even go for technology upgrades to have ‘a cure for emotions’ which can treat negative emotions and bring more balance to emotional responses. Or, for example, such cybernetic cities could fracture social structures as some humans, termed as ‘purists’, could see posthumans as explicit or implicit threats to their own survival. Similarly, a political leadership with a decentralized government versus ‘connected minds’ government was debated upon showcasing polarity in political implications of such narratives.

The reflection of the narratives of the aforementioned proximate futures, places an importance on speculating the cybernetic cities and IoB as a design approach for smart cities. More than the design concepts and artifacts on IoB, the co-opting and co-designing of ideas and debates on their implications on the ethical, social, and political aspects of the smart cybernetic city, try to explore further and shed some light on the yet evolving field of speculative design as a research methodology applicable to the design of future cybernetic cities and integration of IoB in them.

39.4 Conclusion

According to Dunne & Raby [4], speculating more, at all levels of society, and exploring alternative scenarios, reality will become more malleable and, although the future cannot be predicted, we can help set in place today factors that will increase the probability of more desirable futures happening. Additionally, factors that may lead to undesirable futures can be spotted early on and addressed or at least limited [3]. Speculation always has a grain of authenticity wedged firmly in its pursuits towards the highways of constructed realities. Likewise, the speculation on smart cities looking to integrate the expanse of Internet of Bodies concept in a connected and networked infrastructure unearths surprisingly routes towards newer forms of psychosomatic experiences whilst keeping a firm grip on its possible consequences. For instance, current trends on utilization of the visual sense modality in terms of AR and VR tech stack or customized dash boards to further ‘install’ the human inside a ‘smart’ process are but a start for IoB enthusiasts to think, create, and realize the next form of society.

39.5 Glossary

Concretized data—Adapted from Deleuze and Guttiari's concept of concretization (1994) that deals with the ontological engineering of technological artefacts. For the purposes of this paper, we have extrapolated the same into the semantics of the present reality on the capabilities of data science and in its interaction with the tiny details that technologists and researchers need to tackle cybernetic futures.

Cybernetic Cities—A geographical urban system that is a nucleus of population clustering and co-habitation via a control loop governed by big-data utilization of information, algorithms, and agents.

De-concretized cultural context—Adapted from Deleuze and Guttiari's concept of concretization (1994) that deals with the ontological engineering of technological artefacts. For the purposes of this paper, we have taken the liberty of treating culture as a speculative constant during technobiological artefact building.

Design Architecture—For the purposes of this paper, this phrase has been utilized to depict cross-functional big-data involvements as architectural cornerstones into designing of a cybernetic city.

Posthuman—This paper adheres to Nick Bostrum's definition of the posthuman as 'maximum attainable capacities by any current human being without recourse to new technological means'.

Technobiological—Applying technological-based enhancements to the human body to transcend its innate physical and cognitive abilities. For example, Neuralink.

Techno-philosophical ideologies—or philosophy of technology (first used by Ernst Kapp) looks at the social effects that technology unearths.

Techno-humanism—A concept that posits that technology inclusion in social systems can increase the capacity and extend the threshold of abilities like intelligence, robustness, and cognition.

References

1. Anderson, B.: Imagined Communities. Verso, London (2006)
2. Auger, J. (2013). Speculative design: crafting the speculation. Digit. Creativity 24 (1)
3. CSIRO: Future Cities Thinking—Future Cities and Communities by Design: A Fresh Look at Data and Future Cities. Commonwealth of Australia (2018)
4. Dunne, A., Raby, F.: Speculative Everything: Design, Fiction, and Social Dreaming. MIT Press (2013)
5. Harari, Y.: Homo Deus—A Brief History of Tomorrow. Penguin Random House (2016)
6. Marr, B. What is the internet of bodies? and how is it changing the our world? (2019). https://www.forbes.com/sites/bernardmarr/2019/12/06/what-is-the-internet-of-bodies-and-how-is-it-changing-our-world/#238c248968b7

7. Matwyshyn, A.: The Internet of Bodies, 61 Wm. & Mary L. Rev. 77 (2019). https://scholarship.law.wm.edu/wmlr/vol61/iss1/3
8. Mitrovic, et al.: Introduction to Speculative Design Practice—Eutropia, a Case Study (booklet). Croatian Designers Association, Arts Academy (2015)
9. Shanks, M.: Speculative Design and the Posthuman Body. Museum Boijmans Van Beuningen, Rotterdam (2013)
10. Sitton, J.: What IG professionals should know about the internet of bodies. (2020). https://magazine.arma.org/2020/02/what-ig-professionals-should-know-about-the-internet-of-bodies/
11. Yelavich, S. (2015) Speculative everything: design, fiction, and social dreaming. J. Des. Hist. 28(2). https://doi.org/10.1093/jdh/epv001
12. Young, L.: Interview: Liam Young on Speculative Architecture and Engineering the Future (2015). https://nextnature.net/2015/03/interview-liam-young

Chapter 40
How Ethical Are Persuasive Design Practices? A Proposal for Assessment of Ethics in HCI Design

Sanju Ahuja and Jyoti Kumar

Abstract Ever since bounds on human rationality and cognitive biases in decision contexts have been reported, designers have exploited these weaknesses to yield conversion by creating persuasive HCI designs. Such design practices have been widely reported to be effective in influencing user decision making. However, the exploitation of a cognitive bias compromises the cognitive autonomy of an individual. This paper argues for the need of ethical assessment of persuasive design practices which undermine a user's cognitive autonomy. The paper proposes a model for persuasive information design in human–computer interaction (HCI PID model) and derives from it a framework to assess the ethics of persuasive design practices. In this framework, five design parameters and their twelve subcomponents have been proposed as measures of an HCI system's conduciveness to autonomous decision making without unduly influencing a user. The paper proposes a scoring methodology to assess design features of HCI systems on the proposed parameters. The proposed assessment framework was used by 20 participants to evaluate five mobile applications on features that are relevant to autonomous decision making. It was observed that the proposed framework has effectively helped the assessors to identify unethically persuasive design features.

40.1 Introduction

Persuasive HCI technology refers to interactive computing systems designed to change people's attitudes and behaviors [1]. With the growing knowledge of neuroscience and the reducing cost of neuroimaging technologies, commercial interest in persuasion design has risen [2]. In information design literature, some reported examples of persuasive practices include creating the illusion of scarcity of a product to increase its demand, creating price anchors, inconveniencing users to increase the perceived value of an item, foot-in-the-door technique, and rhyming jingles in advertisements to increase believability [3–6]. These are general persuasion principles that

S. Ahuja (✉) · J. Kumar
Indian Institute of Technology Delhi, New Delhi, India
e-mail: sanju.ahuja@design.iitd.ac.in

A. Chakrabarti et al. (eds.), *Design for Tomorrow—Volume 1*, Smart Innovation, Systems and Technologies 221, https://doi.org/10.1007/978-981-16-0041-8_40

can be used online or offline. In recent years, persuasive design practices have gained traction online with the capability to collect large amounts of user data, which can be harvested for data-driven personalization and targeting to persuade more effectively. Personalization algorithms can target and influence users through believable misinformation and selective filtering on social media platforms [7, 8]. They can also nudge users toward behavior that they themselves deem to be harmful, such as impulse purchases, unhealthy eating, and binge video streaming. Recently, persuasion tactics were claimed to have been used on a large scale to influence voter decision making in the 2016 presidential election in the USA through political advertisements on Facebook which targeted individual psychological vulnerabilities [9].

40.1.1 Cognitive Autonomy and Cognitive Biases

Cognitive autonomy is defined as the psychological freedom to be the person one wants to be, to pursue one's goals without unjustifiable hindrances or interference, to be self-governing [10]. It refers to an individual's freedom to make their own decisions of their own free will. Autonomy is considered to be a fundamental right in the Universal Declaration of Human Rights adopted by the United Nations [11].

Persuasive information design is effective in influencing user decision making because the human brain is not a rational information processor or decision maker. Simon [12] proposed that the human brain is boundedly rational and routinely takes mental shortcuts to decision making. These mental shortcuts were formalized by Kahneman and Tversky [13] as heuristics of judgment and decision making. Kahneman and Tversky's prospect theory [14] detailed and explained cognitive biases such as the endowment effect, status quo bias, and gain–loss asymmetry. Human cognitive biases are irrational by definition, in the sense that they violate the axioms of rationality set forth in economic theory [15]. These cognitive biases are normally useful, making decisions easier and faster than they could be if the decision process involved only rational evaluation. However, the same biases can be exploited by persuasive HCI technology to influence user decision making for the benefit of monetization, making humans vulnerable to an invasion of their cognitive autonomy [16].

The premise of this paper is that humans have a right to self-determination, based on the long-standing philosophical and political importance of autonomy [10, 11]. Therefore, there is a need to ensure that HCI systems are conducive to autonomous decision making by their users, without unjustifiable hindrances or interference. There is a need to understand the lines of divide between ethical and unethical persuasive design practices and to create methods of assessment that can be used by designers, policymakers, and users to assess how HCI systems can exert unethical influence on human decision making.

40.2 Current Research in Ethics of Persuasive Design

In UX Design, the term 'dark patterns' has been coined for unethically persuasive design practices that aim to trick users into behaving in a certain way. Gray et al. [17] decomposed dark pattern design practices into five categories: nagging, obstruction, sneaking, interface interference, and forced action, and argued for the need to integrate ethics into broad HCI design practice. In neuromarketing research, it has been argued that it is unethical to influence human decision making in a covert or hidden manner [2]. A covert influence undermines cognitive autonomy because it is not rationally discernible or attributable as the cause of the decision, sometimes not even in retrospect. Even when the covert influence is beneficial to the user, such as the nudges reported in behavioral economics, they still might undermine users' autonomy because of their paternalistic nature [16, 18]. Paternalistic design is not always unethical, especially when it serves the collective good and is deployed into socioeconomic systems through democratic procedures [19], such as traffic systems designed to nudge drivers to drive on the correct side of the road. However, Susser et al. [16] argue that if paternalistic covert persuasion intends to personally benefit the user by nudging them to eat better food, exercise or work harder, such persuasion harms the user by rendering them opaque to themselves. Even if the persuaded user becomes a better individual by any reasonable measure, he is not a product of his own reflection, imagination and powers of discrimination and analysis [20].

In behavioral economics, Thaler and Sunstein [21] have argued that paternalistic nudges are ethical if they do not attempt to make alternate decisions unreasonably difficult. According to this argument, even when alternate decisions are made slightly difficult with some moral rationale, the individual should not be denied the choice of making them. On the commercial technology front, the addressal of the issue of users' autonomy has been largely reactive, that too in response to the imposition of large government fines on various technology companies. This is because despite the ubiquitous presence of persuasive HCI technology, there is little policy regulation to ensure that persuasive influences in the digital world do not undermine users' autonomy.

40.2.1 Research Gap and Research Goal

The literature on ethics of persuasive design in HCI is quite broad, as outlined in the previous section. However, the philosophical understanding of autonomy is yet to be rigorously formulated within the HCI design context. Therefore, the ethical frameworks pertaining to persuasion design are harm focused [22, 23]. Although a few reports in literature do touch upon aspects of autonomy in emerging persuasive technologies [24], the authors could not find any reported framework for assessment of ethics which explicitly assesses persuasive design from the perspective of user autonomy. Authors were not able to find any HCI information design models either which help formulate the concept of cognitive autonomy within the HCI design context.

The research goal of this paper is to develop a model of persuasive information design within the HCI context, which will help formulate the concept of the user autonomy for persuasive HCI systems. Also, this paper aims to propose and test a framework for assessment of ethics in HCI design from a user autonomy perspective.

40.3 Proposed Framework for Assessment of Ethics in HCI Design

This section first develops a model for persuasive information design in Sect. 40.3.1, then develops a framework for assessment of ethics of HCI design in Sect. 40.3.2 and then finally empirically tests the usefulness of the proposed framework in Sect. 40.3.3.

40.3.1 Persuasive Information Design for HCI Systems—HCI PID Model

In Fig. 40.1, the authors propose a model of persuasive information design for HCI systems (HCI PID model). This model was created by the authors to build a step-by-step overview of the interactions and decisions involved in the process of human–computer interaction. The aim of the model is to identify various stages of interaction between the human and the computer and to subsequently identify the persuasive design elements in each stage. By outlining the interaction in stages, it became possible for the authors to identify the prominent threats to user autonomy in each stage separately. The model superimposes on each stage (in italics) the threats to cognitive autonomy due to persuasive design practices. These threats were identified from the surveyed literature and were mapped on the stage of interaction during which they can theoretically occur. The terminology used in the model is defined below:

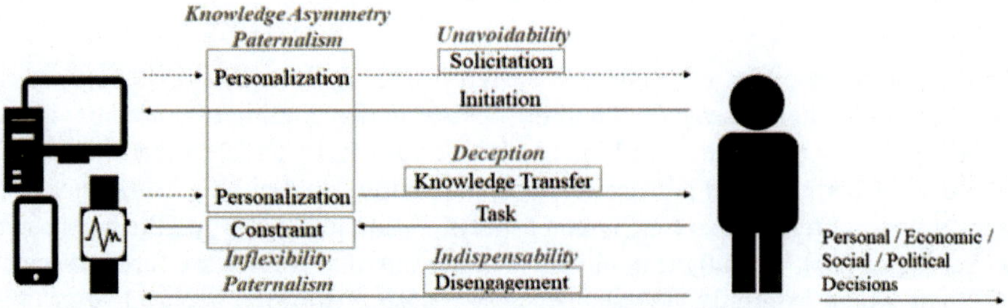

Fig. 40.1 Proposed HCI persuasive information design (HCI PID) model

1. *Solicitation*: Solicitation refers to an interaction initiated by the system for the user, thereby exposing the user to unsought information.
2. *Initiation*: Initiation refers to the user's action to initiate an interaction with the system, either in response to a solicitation or to fulfill an intention.
3. *Knowledge transfer*: Knowledge transfer refers to the presentation of information by a system to the user.
4. *Personalization*: Personalization refers to categorized or individualized interactions between a system and a user. Personalization typically relies on individual or demographic data to tailor user interactions to maximize a parameter of interest such as revenue, engagement, and clicks.
5. *Task*: A task refers to an active engagement with the system in which a user follows a series of steps on the system to achieve a goal. A task is inherently mediated by design constraints of a system.
6. *Constraint*: Constraint refers to a limitation on the ways in which a user can interact or engage with a system, communicated to the user through affordances, default settings, restricted access, etc.
7. *Disengagement*: Disengagement refers to a user leaving, disconnecting, or terminating an interaction with the system, on a temporary or permanent basis.

Threats to Cognitive Autonomy of Users in HCI PID Model. It has been argued before that a user's cognitive autonomy is threatened when a design is coercive, manipulative, or deceptive [16]. Coercion refers to influencing someone by constraining their options. Deception refers to influencing decision making by planting false beliefs. Manipulation is a form of trickery, often hidden, intended to influence someone through seduction, guilt, temptation, emotions, and desires. At each stage of human–computer interaction in the HCI PID model, a system can coerce, deceive, or manipulate user decision making. In Fig. 40.1, the authors have mapped (in italics) these autonomy undermining methods onto the HCI PID model:

1. *Unavoidability (solicitation)*: Unavoidability is defined here as the coercive design of a solicitation, limiting the ways in which a user can avoid engaging with it, both through outright and planned rejection.
2. *Deception (knowledge transfer)*: Deception refers to an HCI system dishonestly influencing users' beliefs, through inaccurate information, information misrepresentation, selective exposure to information or hiding information low in the visual hierarchy of the user such that it is not factored in the decision due to inattention.
3. *Knowledge asymmetry (personalization)*: Knowledge asymmetry refers to an HCI system having knowledge about the user that the user cannot conceive of or cannot rationally factor in the decision process, such as the information collected through digital surveillance. Knowledge asymmetry can be used to manipulate users by exploiting their vulnerabilities through personalized and targeted information.
4. *Inflexibility (constraint)*: Inflexibility refers to a coercive limitation on the ways in which a user can interact with a system due to design constraint. For example,

forcing the user to accept third-party cookies to browse a Web site, auto-enabling a subscription after a free trial period of a service, etc.

5. *Paternalism (personalization and constraint)*: The threat of paternalism emerges from a system being designed to influence user behavior in a manner which the system claims is beneficial for or in the best interest of the user, especially without their knowledge or consent to acquire the behavior desired by the system [18].
6. *Indispensability (disengagement)*: Indispensability refers to a system's coercive design features that make it difficult by design to quit, leave, or disengage with, on a temporary or a permanent basis.

40.3.2 HCI PID Ethics Assessment Framework

In this section, the HCI PID model was translated into the HCI PID ethics assessment framework, complete with parameters that are indicators of the extent to which an HCI system supports or allows autonomous decision making by its users. The rationale behind this assessment framework lies in the philosophical formulation of autonomy previously adopted by Friedrich et al. [25] in BCI research and Thaler and Sunstein [21] in behavioral economics. Friedrich et al. [25] used a three-component account of mental competence to assess the impact of brain computer interfaces on autonomy. The three components which are the foundation of autonomous decision making are as follows: (1) ability to use information and knowledge to produce reasons, (2) ability to ensure that intended actions are realized effectively (control), and (3) ability to enact intentions within concrete relationships and contexts.

In behavioral economics, Thaler and Sunstein [21] argued that persuasive nudges are ethical when they are beneficial to either the individual being nudged or the society which benefits from the nudge as long as the nudges do not make alternative choices difficult. Nudges advocate for soft or libertarian paternalism. This argument formulates autonomy as an ability to execute alternate decisions as opposed to those intended by the persuader. This argument resonates with the third component of the three-component account used by Friedrich et al. [25], which posits that the ability to execute or enact intentions within the design of the external world is an essential component of autonomous decision making.

The development of the HCI PID assessment framework in this paper follows the argument of Thaler and Sunstein [21] to generate parameters that are relevant to autonomous decision making. This paper argues that persuasive information design is unethical when it makes it difficult for users to exercise their autonomy. Therefore, this paper covers only one aspect of autonomy, which is the aspect of executing or enacting one's intentions within the HCI context (the third component in the three components from Friedrich et al. [25]). Building upon the HCI PID model of Fig. 40.1, we have derived a framework of five parameters and their twelve subcomponents that are integral to making HCI systems conducive to autonomous decision making (Fig. 40.2). Each of these parameters was created as a response to the threats

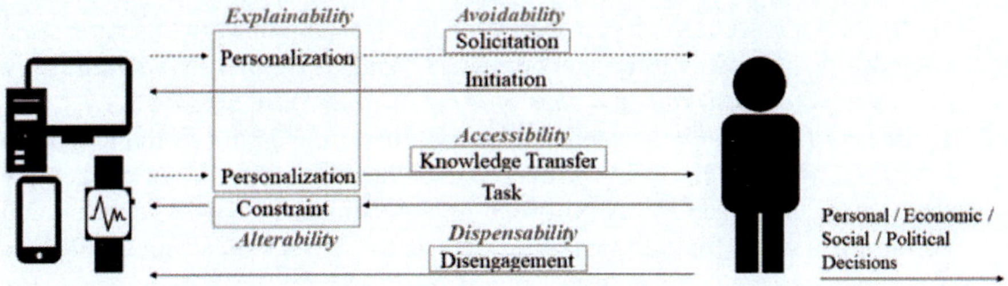

Fig. 40.2 Proposed HCI PID assessment framework for cognitive autonomy of users

to users' autonomy identified in Fig. 40.1. These parameters indicate the design requirements from HCI systems to allow users to execute or enact certain intentions if they so choose. A high rating on these following parameters by the intended users signifies how difficult HCI systems make it for users to exercise autonomous choice.

1. *Avoidability (solicitation)*: Avoidability is a measure of the ease with which a user can avoid interacting with a solicitation initiated by a system, either through advance planning or outright rejection. Avoidability is proposed as a parameter of cognitive autonomy because a user should not be coerced into engaging with a system. In this parameter, we propose to measure two subcomponents: *ease of planned rejection and ease of outright rejection or reasonable delay.*
2. *Accessibility (knowledge transfer)*: Accessibility is a measure of the ease with which relevant and accurate information can be accessed on an HCI system. Accessibility is proposed as a parameter of cognitive autonomy because autonomous decision making requires complete and accurate information. We propose two subcomponents of this parameter: *ease of accessing relevant information* and *ease of verifying information accuracy or system credibility.*
3. *Explainability (personalization)*: Explainability is a measure of ease with which a user can understand the dynamics of the infosphere, such as understanding the flow and usage of one's personal information, understanding the nature of the personalization of one's interactions with the system and understanding the differential nature of outcomes for oneself and other users. Explainability is proposed as a parameter of cognitive autonomy because information about the infosphere might be relevant to the user's decision process. In this parameter, we propose to measure three subcomponents: *ease of locating and understanding personal data flows, ease of understanding the role of personal data in personalization*, and *ease of understanding the difference between oneself and others as users.*
4. *Alterability (constraint)*: Alterability is a measure of the ease with which a user can circumvent a constraint to reach a desired state of a system, unless there is an explicit understanding and acceptance of the constraint as a part of the transaction. It is argued here that constraints, unless agreed upon, are a form of coercion, especially if there is no rational alternative than to accept them, even if hesitantly or unwillingly. Therefore, alterability outside the bounds of the

contractual transaction is a parameter of cognitive autonomy. In this parameter, we propose to measure three subcomponents: *knowledge of the constraint, ease of altering the constraint,* and *ease of accessing alternate systems.*

5. *Dispensability (disengagement)*: Dispensability is a measure of the ease with which the user can disengage with a system without being subject to an undue hindrance. Dispensability is proposed as a parameter of cognitive autonomy because it reflects the freedom to be off from electronic engagement [26]. In this parameter, we propose to measure two subcomponents: *ease of temporary disengagement* and *ease of permanent disengagement*.

40.3.3 Empirical Test Using HCI PID Assessment Framework

The proposed framework was used to assess design features from five popular mobile applications. Each feature selected for assessment qualified as one stage of human–computer interaction in the HCI PID model and was mapped to the corresponding parameter in the HCI PID assessment framework by the authors. Facebook was chosen for all five parameters of assessment to show that the framework can be used for the complete assessment of one single system. For each parameter, the authors also chose one other mobile application for assessment to show that ethics assessors need not assess each system on all five parameters and that each design feature can be assessed independently based on the stage of human–computer interaction it maps to. The following features were arbitrarily selected by the authors for assessment:

1. *Solicitation*: Mobile app notifications (Facebook and Gmail)
2. *Knowledge Transfer*: News on social media (Facebook and WhatsApp)
3. *Personalization:* Interest-based recommendations (Facebook and YouTube)
4. *Constraint:* Privacy default settings (Facebook and Google)
5. *Disengagement*: Quitting a mobile application (Facebook and WhatsApp).

Research Methodology. Each of the five design features selected above was assessed on the corresponding parameter of cognitive autonomy in the HCI PID assessment framework (see Table 40.1 for mapping). A 12-point questionnaire (Table 40.1) was formulated in which the questions corresponded to the 12 subcomponents of the five parameters of cognitive autonomy. This questionnaire was filled by 20 participants voluntarily (7 F, 13 M, mean age = 26.7 years). All questions were answered on a 5-point Likert-type scale for level of difficulty (very easy to very difficult) [27]. All the participants did the assessment by using their mobile app or mobile Web site.

Analysis. We argued in Sect. 40.3.2 that if cognitive autonomy is a fundamental right, it is unethical for HCI systems to make it difficult for its users to exercise autonomous choice. Therefore, during analysis of the data, it was decided that the parameter of interest is the fraction of users who scored each question ‘4’ or ‘5’ on the Likert scale. This means that they had found the design to be infringing on

Table 40.1 Ethics assessment questionnaire and results

Scoring guideline 1 (very easy); 2 (easy); 3 (neutral); 4 (difficult); 5 (very difficult)		X (No. of users who answered '4' or '5')	
Mobile app notifications (solicitation → avoidability)		*Facebook*	*Gmail*
1. Ease of planned rejection	1. How easy/difficult is it to customize notifications for this mobile application from app settings?	7	3
2. Ease of outright rejection or reasonable delay	2. How easy/difficult is it to avoid or delay attending to a notification from this app?	**9**	**9**
Fake news (knowledge transfer → accessibility)		*Facebook*	*WhatsApp*
3. Ease of accessing relevant information	3. How easy/difficult is it to find relevant news on this application?	**10**	**17**
4. Ease of verifying information accuracy or system credibility	4. How easy/difficult is it to verify the accuracy of news on this application?	**17**	**17**
Interest-based recommendations (personalization → explainability)		*Facebook*	*YouTube*
5. Ease of locating and understanding personal data flows	5. How easy/difficult is it to locate and understand who has access to what types of your data?	**14**	**14**
6. Ease of understanding the role of personal data in personalization	6. How easy/difficult is it to ascertain how your data are used to show you recommendations on this Web site?	**12**	7
7. Ease of understanding the difference between oneself and others as users	7. How easy/difficult is it to understand why your recommendations are different from other people's?	**9**	4
Privacy default settings (constraint → alterability)		*Facebook*	*Google*
8. Knowledge of the constraint	8. How easy/difficult is it to know and understand the privacy settings of this Web site?	**8**	**12**

(continued)

Table 40.1 (continued)

Scoring guideline 1 (very easy); 2 (easy); 3 (neutral); 4 (difficult); 5 (very difficult)		X (No. of users who answered '4' or '5')	
9. Ease of circumventing the constraint	9. How easy/difficult is it to change the privacy settings of this Web site?	6	**10**
10. Ease of accessing alternate systems	10. How easy/difficult is it to find alternate Web sites for this purpose in case you are unable to change the privacy settings as per your preferences?	**11**	**13**
Quitting a mobile application (disengagement → dispensability)		*Facebook*	*WhatsApp*
11. Ease of temporary disengagement	11. How easy/difficult is it to not check this application or logout of this application for a brief period of time?	3	**17**
12. Ease of permanent disengagement	12. How easy/difficult is it to delete your account on this application?	**13**	5

their ability to enact their intentions during application usage. It was decided that fraction of users who rated a '4' or '5' will be taken as a measure of how ethically acceptable or unacceptable the design was. In other words, if a 'large number' of users found it difficult to exercise autonomous choice (means rated '4' or '5'), it would be considered as 'unethical.' For this assessment, the 'large number' was decided to be at least 20% of the user population. Thus, a cutoff of 0.2 was selected. A cutoff of 0.2 signified that, a feature was ethically acceptable if no more than 20% of the user population found it 'difficult' or 'very difficult' to exercise autonomous choice. After the cutoff was selected, the following test procedure was applied:

Participants: N = 20.

Test statistic: X = Number of participants who scored each question '4' or '5'.

Null hypothesis (H0): The number of participants who found it 'difficult' or 'very difficult' to exercise autonomous choice is ≤ 0.2 N and the design feature was ethically acceptable.

Rejection criterion: The null hypothesis is rejected if the empirically observed number of participants who found it 'difficult' or 'very difficult' to exercise autonomous choice exceeded 0.2 N with >95% confidence. Using the formula for cumulative binomial probability distribution [28], the rejection criterion for the null hypothesis H0 was calculated for 20 participants. It was calculated that for $X \geq 8$, the null hypothesis can be rejected with 96.8% confidence.

Results. It was observed that the participants were able to understand and score the questions easily according to the scoring guidelines provided. Table 40.1 shows the observed values of X (the number of participants who scored each question '4' or '5'), and values of $X \geq 8$ are highlighted in bold. For an ethically acceptable cutoff of 0.2 N for X, this framework identified with >95% confidence the following persuasive design issues as unethical with respect to the users' cognitive autonomy: difficulty of ignoring or delaying a notification on Facebook and Gmail (Q.2), difficulty of finding relevant news on Facebook and WhatsApp (Q.3), difficulty of verifying the accuracy of news on Facebook and WhatsApp (Q.4), difficulty in finding and understanding personal data flows on Facebook and YouTube (Q.5), difficulty in understanding data-driven recommendations on Facebook (Q.6), difficulty in understanding data-driven user differences in Facebook (Q.7), difficulty in knowing and understanding the privacy settings on Facebook and Google (Q.8), difficulty in changing the privacy settings on Google (Q.9), unavailability of alternates to Facebook and Google in case of unacceptability of privacy policies (Q.10), difficulty in temporary engagement from WhatsApp (Q.11), and difficulty in permanently deleting one's account on Facebook (Q.12).

Limitations of this Analysis. The analysis presented here is a preliminary analysis to showcase the use of the framework. Within this assessment framework, the parameter of interest (number of users who answered 4 or 5) as well as the cutoff criteria are not fixed. They are dependent upon the moral rationale behind the persuasive design. The cutoff criteria may be lesser than 0.2 if in a certain context, it is moral for all users to be provided with the option of autonomous choice. The cutoff criteria may be higher than 0.2 if it is morally desirable for only the most skilled users to have the option of autonomous choice. For example, a desktop user may not have the motivation or the ability to make choices about desktop settings and may end up making suboptimal choices if provided with all the options. However, it needs to be considered whether the settings recommended by the manufacturer are truly optimal for users or whether they are optimal for business, making it difficult for users to access the options that meet their intended requirements. This conundrum can be incorporated in the assessment framework through an appropriate formulation of the questions. Because the authors did not argue the moral rationale behind each of the five features that were chosen for assessment, a common cutoff criterion was chosen to showcase the use of the framework. The richness of the 5-point Likert scale was reduced for the same reason, and ethics assessors in practice can use separate cutoff criteria for each score on the Likert scale based on the moral defense of the design feature.

Second, with certain framing of the questions, there is room for ambiguity in their interpretation by the participants. For example, the answers to Q.2 and Q.11 could reflect a psychological 'difficulty' due to self-control issues, instead of a difficulty in 'interface design,' even though the framework intends to capture issues of difficulty with persuasive interface design. Therefore, caution is needed while framing the questions to ensure that the participants interpret the questions as intended by the ethics assessor. On the other hand, it is also possible for the ethics assessor to intentionally leave the question ambiguous, or even to explicitly attempt to capture the psychological 'difficulty.' This may be valuable if the motive of assessment is to redesign an HCI system for beneficence instead of determining whether the interface design is ethical or not.

40.4 Discussion and Conclusions

The main contributions of this research are the HCI PID model and assessment framework. The HCI PID model outlines the stagewise interaction between the human and the computer and provides a systematic way to traverse the decision stages in human–computer interaction. The model was instrumental in the identification of the five parameters indicative of autonomous decision making by users in persuasive design, leading to the creation of the HCI PID assessment framework. To the best of our knowledge, this framework has provided one of the first methodologies to quantify the ethical issues with persuasive design pertaining to user autonomy. The framework was able to identify specific violations of persuasion ethics, and therefore, it was observed that this framework can help identify and prioritize redesign requirements to make persuasive design practices ethically acceptable.

The limitations of this framework are the following. By its design, the framework only touches upon one aspect of autonomy, which is the ability to enact or execute one's intentions. It does not speak of the ability to reason or the ability to effectively control one's actions, which are the other two components of autonomous choice. The framework is yet to be validated and checked for reliability. It needs to be used across different HCI systems on larger sample sizes to be established as a method of ethics assessment. In its current form, the framework has not been checked for dependencies between the five parameters and their subcomponents, theoretical or statistical. Another characteristic of the framework is that the identification of ethical issues is dependent on the demographic surveyed, which is to be expected. More skilled users of HCI systems will find it relatively easier to make different choices on the system, and hence will not find it 'difficult' to exercise autonomous choice even when the system is designed to be difficult. Therefore, the framework needs to be tested on the intended user of the HCI system to be able to accurately identify ethical issues.

In conclusion, this paper has argued for the need of ethical assessment of persuasive HCI design practices. Further, this paper has argued that beyond the philosophical conceptualizations of autonomy, there is a need to develop assessment frameworks which designers and policymakers can use to assess the ethics of persuasive HCI systems from the perspective of users' autonomy. Results of the empirical test using the proposed ethics assessment framework suggest that many design features of popular mobile applications may not be conducive to autonomous decision making by their users. In light of the findings, it is argued here that persuasive HCI design practices have become legitimized by virtue of their pervasiveness, omnipresence, and a myopic infatuation with technology. Whether such unethical persuasion is intentional or not, it is the moral responsibility of designers to identify these unethically persuasive practices and redesign them in a manner explicitly conducive to autonomous choice. If a designer's interests conflict with users' interests, the framework can help identify design practices that need regulatory or policy intervention. Despite its limitations, the proposed framework is one of the first to our knowledge to provide a methodology to explicitly assess persuasive design practices from the perspective of users' autonomy. We hope that the framework gets used across diverse usage contexts and evolves through future research contributions to become established in the much-needed ethical assessment space.

References

1. Fogg, B.J.: Persuasive technology: using computers to change what we think and do. Morgan Kaufmann Publishers, San Francisco (2003)
2. Murphy, E.R., Illes, J., Reiner, P.B.: Neuroethics of neuromarketing. J. Consum. Behav. **7**(4–5), 293–302 (2008)
3. McGlone, M.S., Tofighbakhsh, J.: Birds of a feather flock conjointly (?): rhyme as reason in aphorisms. Psychol. Sci. **11**(5), 424–428 (2000)
4. Guéguen, N.: Foot-in-the-door technique and computer-mediated communication. Comput. Hum. Behav. **18**(1), 11–15 (2002)
5. Dodonova, A.: An experimental test of anchoring effect. Appl. Econ. Lett. **16**(7), 677–678 (2009)
6. Aggarwal, P., Jun, S.Y., Huh, J.H.: Scarcity Messages. J. Advert. **40**(3), 19–30 (2011)
7. Shin, J., Jian, L., Driscoll, K., Bar, F.: The diffusion of misinformation on social media: temporal pattern, message, and source. Comput. Hum. Behav. **83**(1), 278–287 (2018)
8. Pariser, E.: The filter bubble: what the internet is hiding from you. Penguin Press, New York (2011)
9. Timanta Tarigan, J., Zamza, E.M.: Post Cambridge analytica fallout: observing Facebook users awareness regarding data security. Int. J. Eng. Technol. **7**(3.32), 123–126 (2018)
10. Roskies, A.: Neuroethics. In: The Stanford Encyclopedia of Philosophy (2016). https://plato.stanford.edu/archives/spr2016/entries/neuroethics/
11. The Universal Declaration of Human Rights. United Nations (1948)
12. Simon, H.A.: Theories of bounded rationality. In: McGuire C.B., Radner, R. (eds.) Decision and Organization. North-Holland Publishing Company, Amsterdam (1972)
13. Tversky, A., Kahneman, D.: Judgment under uncertainty: heuristics and biases. Science **185**(4157), 1124–1131 (1974)
14. Kahneman, D., Tversky, A.: Prospect theory: an analysis of decision under risk. Econometrica **47**(2), 263–291 (1979)
15. Von Neumann, J., Morgenstern, O.: Theory of games and economic behavior, 1st edn. Princeton University Press, Princeton and Oxford (1944)
16. Susser, D., Roessler, B., Nissenbaum, H.: Technology, autonomy and manipulation. Internet Policy Rev. **8**(2), 1–22 (2019)
17. Gray, C.M., Kou, Y., Battles, B., Hoggatt, J., Toombs, A.: The Dark (Patterns) Side of UX Design. CHI '18. Montreal (2018)

18. Dworkin, G.: Paternalism. The Stanford Encyclopedia of Philosophy (2020). https://plato.stanford.edu/archives/spr2020/entries/paternalism/
19. Ahuja, S., Kumar, J.: Surveillance based persuasion: the good, the bad and the ugly. CHIRA'20. Budapest (2020)
20. Scanlon, T.M.: The significance of choice. Tanner Lect. Human Values **7**, 149–216 (1986)
21. Thaler, R.H., Sunstein, C.R.: Nudge: improving decisions about health, wealth and happiness. Yale University Press, New Haven and London (2008)
22. Berdichevsky, D., Neuenschwander, E.: Toward an ethics of persuasive technology. Commun. ACM **42**(5), 51–58 (1999)
23. Bösch, C., Erb, B., Kargl, F., Kopp, H., Pfattheicher, S.: Tales from the dark side: privacy dark strategies and privacy dark patterns. Proc. Privacy Enhancing Technol. **2016**(4), 237–254 (2016)
24. Greenberg, S., Boring, S., Vermeulen, J., Dostal, J.: Dark patterns in proxemic interactions: a critical perspective. DIS' 14. New York (2014)
25. Friedrich, O., Racine, E., Steinert, S., Pömsl, J., Jox, R.J.: An analysis of the impact of brain-computer interfaces on autonomy. Neuroethics (2018)
26. Frischmann, B., Selinger, E.: Re-engineering humanity. Cambridge University Press, Cambridge (2018)
27. Vagias, W.M.: Likert-Type Scale Response Anchors. Clemson University (2006)
28. Blythe, C.R.: Approximate binomial confidence limits. J. Am. Stat. Assoc. **81**(395), 843–855 (1986)

Chapter 41
Assessment of Knowledge, Behavior, and Practices of Maternal and Child Health of Rural Pregnant Women

Rohit Kumar and Shatarupa Thakurta Roy

Abstract Globally, every day about 2,92,000 women die from pregnancy and childbirth-related issues, which are mostly preventable. India accounts for approx. 20% of this maternal death, which is around 44,000 death per day. More than thirty percent of the Indian population lives in rural areas with a poor economy and low literacy rate. The data from UNICEF India Maternal Health also indicates that women from this region with a poor economy and education background account for two and a half times higher mortality rate compared to the region with a better economy and literacy rate. Despite many government initiatives, the maternal and child health of rural India shows mixed results. Information communication through graphic design and illustrations, often supported through television, cell phone, community awareness campaign, print media, etc., shows potential in health awareness. However, research on the effectiveness of these community awareness methods across the rural region with low literacy and limited resource has not been sufficiently explored. This research is the initial phase of a design project. It will outline the outcome of eight-month of field-based studies in the rural region of Kanpur (in the state of Uttar Pradesh, one of the states performing worse with a high burden of maternal and child deaths). Data was collected from qualitative analysis of user (pregnant women) and its context. A total of 106 pregnant women were included for the study, where user need, information gap, different communication mode, behavior, sociocultural issues, are observed, studied, analyzed, and interpreted.

41.1 Introduction

Globally, every day about 2,92,000 women die from pregnancy and childbirth-related issues, which are mostly preventable. India accounts for approx. 20% of maternal

R. Kumar (✉) · S. T. Roy
Design Programme, Indian Institute of Technology Kanpur, Kanpur, India
e-mail: rohitkr@iitk.ac.in

S. T. Roy
e-mail: stroy@iitk.ac.in

A. Chakrabarti et al. (eds.), *Design for Tomorrow—Volume 1*, Smart Innovation, Systems and Technologies 221, https://doi.org/10.1007/978-981-16-0041-8_41

death [1]. In India, it is estimated that, annually, 44,000 women die due to preventable maternal and child-related issues. The data from UNICEF India Maternal Health also indicates that women from the low economy and education background account for two and a half times higher mortality rate.

India is the second most populous country in the world, with a total population of 1.37 billion people and a population density of 416 per square kilometer [2]. More than one-third of the population lives in rural areas. For these rural people, the Government of India launched the National Rural Health Mission (NRHM) in 2005 under the Ministry of Health and Family Welfare [3]. This mission aims to reduce maternal and child mortality through various measures and initiatives to control preventable diseases, maintain population and gender balance, and provide public health service for all. One of the essential features of the NRHM is Janani Suraksha Yojana (JSY), through which financial assistance is provided to poor pregnant women to ensure institutional safe delivery and various maternal and child care needed during pregnancy and after childbirth. Despite this government initiative, the maternal and child health of rural India shows mixed results. Though a decline can be seen in maternal mortality rate from the years 2007–2016, which is 212 death (per lakh live birth) to 130 death in the respective years [4], the decline is way behind the Sustainable Development Goal which is 70 death per 1,00,000 live birth. To date, only three states—Tamil Nadu, Kerala, and Maharashtra—have achieved the desired target of maternal mortality rate. The data from government health Web site of Ministry of Health and Family Welfare on 2018 Health Index conducted by NITI Aayog and Think-Tank has highlighted the uneven and wide gap between states on health performance, with some of the states such as Uttar Pradesh, Madhya Pradesh, West Bengal, and Bihar performing worse with a high burden of maternal and child health.

41.1.1 The Challenge

The health of the mother and child is of critical importance, as it represents the individual present health condition, communities around it, along with the nation, and next-generation health. As recent studies on child development have shown, early years during the growth is more crucial as an opportunity and challenge in enhancing the child's physical, emotional, and social growth. In these early years, there is a need for quality maternal and child health care, along with awareness and knowledge on maternal and child health to pregnant women. The awareness and knowledge are more crucial for the rural population who live in poverty, have low levels of literacy, and social isolation. A wide disparity can be seen in the healthcare outcome, reaching to these rural populations have had, at best, mixed results, in the surveyed region of rural Kanpur as well as elsewhere.

Many researches show that when women have greater awareness and knowledge of maternal and child care, there is a greater likelihood that they will have good pregnancy health and better outcome [5]. This is likely as being aware and having knowledge through education and other means equips women to take proper health decisions, including pregnancy and childbirth. A well-informed woman is more likely to make proper decisions on the obstetric emergency [6]. But many developing countries, including India, have women with low literacy rates and health awareness, which is more dominant in rural communities [7]. These pregnancy-related health risks could be reduced by providing knowledge and awareness of safe motherhood practices and promote better maternal and child health [8].

41.1.2 Objectives

- Identifying how aware are rural pregnant women of safe motherhood, how they manage and avail health checkups at the rural hospital, identifying problems faced by rural women during and after pregnancy.
- Identifying key information gaps related to a healthy pregnancy, government scheme, and services.
- Identifying social and cultural norms in a rural region with the context of pregnancy, how community health workers are involved in the facilitation of rural health services.
- Identifying the significance of community health workers in rural health.
- Identifying available sources of health information for awareness, how familiar are pregnant women of these sources of health information, what are the preferred mode.

41.2 Methodology

41.2.1 Study Design

India, being a country of many cultures, languages, beliefs, education levels, and economic status, demands an understanding of target users from different angles and their context to get a picture of the problem. To consider the sociocultural context while designing a particular product or service is critical when its users are from different cultural backgrounds. Identifying cultural preferences such as color, the spatial orientation of information, literacy level, and culturally suitable contents help in connecting with users and influences usability of design intervention [9]. Hence, qualitative analysis to understand the user and its context with the help of semistructured interviews and discussions was conducted. It helped in understanding problems associated with pregnant women and their behavior, along with an understanding of the existing system. This further helped in making contextual design decisions for intervention for maternal and child health awareness.

41.2.2 Study Setting/Area

A study was conducted over seven months from June to December 2019 on knowledge and awareness of safe motherhood among the rural women of Kanpur region, who had recently given birth. Kanpur is the largest city in the state of Uttar Pradesh, India. Uttar Pradesh, a state located on the west bank of Ganges River, is the twelfth most populous state in India.

Both fertility and mortality rates are higher in Uttar Pradesh compared to other states, with a total fertility rate of 4 birth per women (compared to 2.9 for all of India) and infant mortality rate of 87 death per 1000 lives (compared to 68 death per 1000 lives for all of India).

A contextual inquiry assisted by the nurse and midwives was conducted with the rural women (reproductive age between 18 and 49 years) of Kalyanpur location (state—Kanpur) attending the Shyam children and maternity center in the locality. This hospital, which was selected purposively for the study, acts as both maternal healthcare center and referral to higher health facilities in the Kanpur city. Antenatal care is provided by the trained nurses and midwives present in the clinic.

41.2.3 Questionnaire Overview

A semistructured questionnaire was prepared for interviewing with the selected sample group. The questionnaire consists of 52 questions covering a wide range of issues related to maternal and child health awareness. The survey questionnaire was prepared in English and then translated into the Hindi language. Later on, translation of answers into English was needed for some of the open-ended questions.

The questionnaire was made into five parts: The first part is related to the socioeconomic, personal, and family details of the respondents. The second part consists of knowledge, practice, and attitude toward safe motherhood, different complications, various government schemes, self-care, diet, etc. The third part consists of social and cultural norms in the rural region, the involvement of community health workers, and their assistance during pregnancy. The fourth part consists of sources of health information, familiarity, and acceptance of different communication sources. The fifth part consists of questions related to awareness and attitude after giving birth. Indirect measures are used to construct determinants for knowledge on common disease signs, its prevention, and nutrition requirements. We included questions on information about oral rehydration therapy (ORS), a remedy for diarrhea (common childhood ailment). Another four indirect measures on maternal health are used, which include knowledge on different contraceptive methods, AIDS, various nation health initiatives by government, and knowledge on the ovulatory cycle.

The doctors reviewed the questionnaire and the necessary changes were made to finalize the survey. This structured questionnaire also includes participant consent along with an information sheet about the study to make the participant aware and to participate in the study actively.

Midwives, who were acquainted with the culture and norms of the region, assisted the interviews at the health center. Each meeting lasted around 40 min to have sufficient engagement with the sample group. Notes were taken at the end of the interviews when necessary during the field visits.

41.3 Results

41.3.1 Participants Characteristics

A total of 106 participants was interviewed during their visits to rural hospital at Kalyanpur. The majority of PW were either housewives or laborers and spoke Hindi as their native language. The mean age of PW was 27.3 years, which ranged from 21 to 37 years. Out of 106, 12 PW had education level up to graduation, 72 PW had education level up to 10 grade, 16 PW had an education level below 10 grade, and 6 were illiterate. Only 27 PW participants owned cellphones and used it for calling purpose; no PW had ever used other features such as messaging, alarm, and calculator.

41.3.2 Experiment 1: Knowledge and Awareness of Safe Motherhood

Setting: Questions were related to awareness on different tests and checkups, complications, and danger signs during pregnancy, nutritional diet, exercises, correct sleeping posture, hygiene to be maintained, etc.

Observation: It was observed that very few PW have awareness on different tests and checkups such as ultrasound tests and blood test urine test. Even the women who are pregnant for second and third time lack such awareness. One PW who is pregnant for second time responded: "I do not know about the different tests, but go through whatever the doctor suggests." In other instances, it was observed that the doctor mentioned the disbursement of iron and calcium tablets without the actual disbursement. Similar cases were found related to different tests and checkups necessary during the pregnancy.

Apart from tests and checkups, it is also observed that PW lacks proper information on complications, signs, and symptoms during pregnancy, intake of iron-folic acid tablets, necessary nutritional diet, exercises, and sleeping postures. One consistent observation was a lack of awareness of how and when to take IFA tablets. The doctor recommends not taking the IFA tablet before and after taking tea, coffee, or caffeinated drinks since it reduces the absorption of iron; however, no PW was found aware of this.

Inference: Need for creating awareness on maternal health.

Under various schemes, the National Rural Health Mission provides free medical tests and services to the rural community, such as blood tests, urine tests, and ultrasound along with other benefits such as TT injections, iron tablets, calcium tablets, and monetary benefits for pregnant women. Most of the women enrolled under the NRHM were found unaware of these medical benefits and the importance of such medical tests and services during pregnancy.

41.3.3 Experiment 2: The Significance of Community Health Workers (Accredited Social Health Activist (ASHA))

Setting: Questions were related to community health workers (ASHA workers) which included how familiar pregnant women are with community health worker, what are the services and information they provide, does the health information provided by them are significant.

Observation 1: It is observed that ASHA workers assisted PW in registration under the NRHM, reminding PW of ANC visits at the hospital for tests and checkups. They also assisted in availing services under rural health missions along with encouraging PW for institutional delivery and post-pregnancy care necessary for newborns, such as timely vaccines, nutritional diet information, common disease signs, and symptoms. The majority of PW shared their pregnancy-related issues with them and trusted their assistance.

Excerpts from pregnant women and researcher conversation

Researcher—Have you met or heard about ASHA workers?

PW—Yes, I have; she visited our house five months ago and asked me questions on the regular menstruation cycle and asked me to attend hospital for the pregnancy test.

Researcher—after that, how she assisted?

PW—On my first visit to the hospital, she recorded my details and gave me a medical book. After that, during my second visit, she came a day before to remind me to visit the hospital.

Similar instances are observed, when asked about their (ASHA) availability during an emergency, one PW replied: "During the seventh month of my pregnancy, I got pain in lower abdomen. I called my ASHA, who had given her number on the first visit to our home. She came and assisted me to the hospital." Another woman stated, "ASHA helped me in regular visits to the hospital for necessary tests and checkups. Whenever ASHA called us, we visit the hospital and follow whatever she instructs."

Additionally, ASHA workers are selected from the community in the rural area with a minimum qualification up to 8 grade to ensure their acceptability among rural women. It is observed that PW and ASHA share a trustworthy relationship. The majority of women responded that they received pregnancy health information from ASHAs. From these observations, it can be mentioned that building trust with the help of community insider becomes prominent for considering any design intervention among semiliterate members of the rural community.

Inference 1: ASHA workers acting as a bridge between rural PW and rural health services.

Observation 2: Apart from ASHA assisting PW with different services provided by the rural health mission for safe pregnancy, many PW responded that they found the healthcare information provided by ASHA to be inadequate.

Excerpts from pregnant women and researcher conversation

- *Researcher*: What kind of information ASHA provided during the home visit?
- *PW*: She asked me to take an iron tablet regularly and not to do heavy work.
- *Researcher*: Any information on proper diet, exercises and sleeping postures, hygiene, etc.
- *PW*: She instructed me to take green vegetables in the diet, but gave no information on exercise, sleeping postures, and hygiene.

Inference 2: Lack of proper information and health awareness communication gap between ASHA and PW.

41.3.4 Experiment 3: Source of Information

Setting: Questions were related to different sources of information available for awareness among PW. It included questions on the use of cellphones, exposure to television, informative graphics at rural hospitals, medical booklet, etc. It also includes their communication with the elderly at home, on sharing information and experience, how often they have health talks with doctors, nurses, and ASHA workers.

Observation: Usage of cellphone has increased over the years, but still, its usage was found low among PW. When asked about how they use it, one PW responded, "I don't operate; whenever I need to call someone, my husband dials. Most of the time, he keeps the cellphone and takes it to the workplace." Most of the PW were found to use cellphones only to contact relatives and friends. No PW had ever explored and used other features such as saving contacts, messaging, setting the alarm, etc. Only 27 PW had featured phone and carried it along with them.

Medical booklet or maternal and child health card provided to PW helps in receiving timely health services and also act as an information resource. These health cards contain illustrated graphic information on dos and don'ts during pregnancy, checkboxes for regular receiving of iron and calcium tablets, child vaccination checkbox, baby weight graph, etc. The illustrations in these booklets are observed to be brief, unclear, and difficult to understand. Hence, medical books are used mostly for seeking health services and not as an information resource. The following are the observed problems with medical health book information.

- Medical book health information is brief, unclear, difficult to read and understand.
- Too much information printed on a single page creates information overload.
- Aesthetically unpleasing graphic images.
- Lack behavioral change qualities such as summarized statements, reader's response, frequently asked information, etc.
- The health card lacks some of the critical health information contents, such as proper diet, antenatal exercises, sleeping postures, danger signs and symptoms, breastfeeding techniques, baby food, etc.

Inference: Medical booklet acting as a source of information is not adequately designed, hence discourages its use as an information resource among PW.

41.3.5 Experiment 4: Knowledge and Awareness of Child care and Barriers to Effectively Communicating with the Doctor

Setting: Questions were related to knowledge and awareness on child care, such as the ability to read the vaccination card, exclusive breastfeeding, baby food after six months, child hygiene to be maintained, common disease signs and symptoms, time gap for another pregnancy, post-partum depression, contraceptive, and childcare discussions with doctors.

Observation: PW were found to have poor knowledge of exclusive breastfeeding, diet after six months, and common disease signs and symptoms. It was also observed that women rarely discuss chid health other than the current health problems they were facing. When asked about their difficulties in communicating effectively with doctors during hospital visits, one PW responded: "Since pediatrician visits two days a week and patient numbers were high at the hospital, we barely get time to discuss child health other than the problem we visit for." Another woman responded, "A doctor saw two–three patients at a time; hence we got divided attention from the doctor." Apart from these, similar instances were observed on knowledge of symptoms and care of post-partum depression.

Excerpts from one of the women who had given birth and researcher conversation

- *Researcher*: Did you suffer from any sort of mood swings, stress, or depression?
- *Women*: Yes, I had faced. It was after I gave birth. Many times I cried because I felt of worthless, and sometimes I thought of harming my baby and myself. It was challenging to concentrate on something else.

- *Researcher*: Did you discuss your feelings and sufferings with the doctor?
- *Women*: No, I never discussed these things with anyone. This is the first time someone is asking about it, and I feel glad.
- *Researcher*: Why did you not discuss it and seek counseling?
- *Women*: I thought that I was thinking a lot about it, and it would go away. Also, I didn't know whom to discuss it with and seek help.

Inference: Need for creating awareness on child care and other post-natal care information such as depression, contraceptive, the minimum gap to maintain for another pregnancy, etc.

41.4 Discussions

Many healthcare providers now understand that effective communication is one of the major factors in the promotion of public health behaviors. However, many do not consider that effective health communication is a fragile and complex human process that needs strategic design, listener feedback, and careful monitoring of the communication process [12–14].

Design intervention for maternal and child health needs to effectively communicate with a diverse spectrum of users, community health workers, rural health centers, and policymakers. The tools must be designed such that it covers right and adequate information for the intended audience. The tools need to be easy to use, require low technology acquittance, seek attention and engaging, and accessible by the rural health population [15, 16].

41.4.1 Existing Issues

Low literacy. The study indicates rural pregnant women of Kanpur region (India) had low literacy along with low usage of technology such as cellphones and the Internet. The majority of women participants had an education level up to grade 10.

A study conducted on participants having a 10-grade education level showed that they had difficulties in comprehending and accessing health information from Web site [17].

Although low literacy/health literacy can lead to poor health outcomes, the interlink between these two is not clearly understood. Some studies highlighted that literacy help in self-efficacy and good health outcome [18]. Other research highlighted that education attainment of patients and proper communication between patient and care provider correspond to health outcomes [19–21]. A study on the impact of low literacy on the choice of information sources found that women only used doctor's instructions or television advertisements over written health information as to their health information source [22]. Another study performed on patients with hypertension and rheumatic pain found that patent with low literacy had no interest in written materials for seeking information [23].

Information sources and barriers. In our study, the qualitative analysis showed rural women were significantly dependent on ASHAs (community health workers) for information and health care. However, women also responded that they (ASHAs) lack specific health information such as antenatal excises, danger signs and symptoms, post-partum depression, child care, etc. Also, few women (12 women, Fig. 41.1) responded to had had health talk with doctors, lack of time, and multiple patient presence during clinical visits were cited as a major reason. Apart from that, usage of cellphone and Internet was low in the rural region; financial constraints associated with owing a cellphone were one of the reasons cited.

A study on information source by Grason, Nicholsan et al. [24] categorized communication medium into three groups, intentional medium (newspaper/magazine, broachers/books, computer, cellphone, and Internet), less intentional medium (radio and television), and interpersonal medium (family, friends,

Form		Description	Users	Mode
Graphic BOOK		Information categorized graphically into chapters representing health information such as pregnancy exercises, diet, dos and don'ts, breastfeeding, child care, etc.	Doctors, Community health workers, Pregnant women	Graphics, Text, Verbal
Pregnancy CARDS		Different sets of cards with illustrated health information. With every visit of pregnant women to the health center, a collection of specific illustrated cards could be provided. These cards act as reminders of health information received.	Doctors, Pregnant women	Graphics, Text
Pregnancy milestones CALENDAR		A calendar with each month or group of months corresponds to different milestones of pregnancy, with each milestone consisting of a set of health information and checkups necessary during that month.	Pregnant women	Graphics, Text

Fig. 41.1 Description of design intervention strategy

neighbors, and doctors). They highlighted in their study that women using intentional medium (such as books, written materials, and Internet-based sources) are more aware of health issues and visit health centers for regular checkups.

41.4.2 Suggested Design Intervention Strategies

A categorical description of the design intervention strategy is tabulated in Fig. 41.2.

41.4.3 Design Guidelines for Print Materials

Careful design of printed graphic health information materials. Houts et al. [25] highlighted in their studies that health information with graphics and illustrations increased the attention span to patients toward materials, along with comprehension and recall of the information. However, such health materials need to be carefully designed such that it is readable and suitable for the intended population.

Content inclusion. While making an intervention, proper attention should be given to information such as rural government schemes available, nutritional diet, antenatal exercises, sleeping posture, and hygiene. Apart from these, proper childcare information such as exclusive breastfeeding, baby food after six months, common disease signs and symptoms, vaccination, child hygiene, post-partum depression, and contraception should be considered.

Readability of health material. Text included in the graphic book to complement illustrations must be easy to read and understand. Proper framing of text to enhance the readability of designed material may require testing. For example, a study [26] highlighting formative evaluation study on ways to enhance written text on injury prevention based on pretest with the patient, researchers found that comprehension increased from 42 to 65% among patient. Other studies [27] also highlighted that comprehension of childcare information on infant immunization among low literate women increased with the inclusion of easy to read text in health materials.

References

1. UNICEF India: Maternal health (2019). https://unicef.in/whatwedo/1/maternal-health. Accessed 24 Aug 2019
2. World Population Review. India Population (2019). https://worldpopulationreview.com/countries/india-population/. Accessed 24 Aug 2019

3. Boateng, K.: India—Janani Suraksha Yojana Safe Motherhood Program (2011). https://www.rbfhealth.org/resource/india-janani-suraksha-yojana-safe-motherhood-program. Accessed 31 Aug 2019
4. Maternal Mortality Ratio (MMR) (Per 1,00,000 Live Births) (2019). https://www.niti.gov.in/content/maternal-mortality-ratio-mmr-100000-live-births. Accessed 17 Sept 2019
5. Harrison, K.A.: The importance of the educated healthy woman in Africa. Lancet **349**(9052), 644–647 (1997)
6. Jammeh, A., Sundby, J., Vangen, S.: Barriers to emergency obstetric care services in perinatal deaths in rural gambia: a qualitative in-depth interview study. ISRN Obstetrics Gynecol. (2011)
7. Sharma, M., Sharma, S.: Knowledge, attitude and beliefs of pregnant women towards safe motherhood in a rural Indian setting. Social Sci. Directory **1**(1) (2012)
8. Mushi, D.L., Mpembeni, R.M., Jahn, A.: Knowledge about safe motherhood and HIV/AIDS among school pupils in a rural area in Tanzania. BMC Preg. Childbirth **7**(1), 5 (2007)
9. De Angeli, A., Athavankar, U., Joshi, A., Coventry, L., Johnson, G.I.: Introducing ATMs in India: a contextual inquiry. Interact. Comput. **16**(1), 29–44 (2004)
10. Braun, V., & Clarke, V. (2006). Using thematic analysis in psychology. Qualitative research in psychology. *Qualitative Research in Psychology*, *3*(2), 77–101.
11. Holloway, S.L., Valentine, G.: Spatiality and the new social studies of childhood. Sociology **34**(4), 763–783 (2000)
12. Kreps, G.L.: Strategic use of communication to market cancer prevention and control to vulnerable populations. Health Mark. Q. **25**(1–2), 204–216 (2008)
13. Kreps, G.L.: The pervasive role of information in health care: implications for communication policy. In: Anderson, J. (ed.) Communication Yearbook, vol. 11, pp. 238–276. Newbury Park
14. Kreps, G.L., Bonaguro, E.W.: 16 health communication as applied communication inquiry. In: Routledge Handbook of Applied Communication Research, p. 380 (2009)
15. Neuhauser, L., Kreps, G.L.: Online cancer communication: meeting the literacy, cultural and linguistic needs of diverse audiences. Patient Educ. Couns. **71**(3), 365–377 (2008)
16. Neuhauser, L., Kreps, G.L.: The advent of E-Health. How interactive media are transforming health communication. M&K Medien & Kommunikationswissenschaft **51**(3–4), 541–556 (2003)
17. Birru, M.S., Monaco, V.M., Charles, L., Drew, H., Njie, V., Bierria, T., Steinman, R.A.: Internet usage by low-literacy adults seeking health information: an observational analysis. J. Med. Internet Res. **6**(3), e25 (2004)
18. Schillinger, D., Barton, L.R., Karter, A.J., Wang, F., Adler, N.: Does literacy mediate the relationship between education and health outcomes? A study of a low-income population with diabetes. Publ. Health Rep. **121**(3), 245–254 (2006)
19. Bennett, I., Switzer, J., Aguirre, A., Evans, K., Barg, F.: 'Breaking it down': patient-clinician communication and prenatal care among African American women of low and higher literacy. Annals Family Med. **4**(4), 334–340 (2006)
20. Paasche-Orlow, M.K., Clarke, J.G., Hebert, M.R., Ray, M.K., Stein, M.D.: Educational attainment but not literacy is associated with HIV risk behavior among incarcerated women. J. Women's Health **14**(9), 852–859 (2005)
21. Schillinger, D., Bindman, A., Wang, F., Stewart, A., Piette, J.: Functional health literacy and the quality of physician–patient communication among diabetes patients. Patient Educ. Couns. **52**(3), 315–323 (2004)
22. Ishikawa, H., Yano, E.: Patient health literacy and participation in the healthcare process. Health Expect. **11**(2), 113–122 (2008)
23. Koo, M., Krass, I., Aslani, P.: Enhancing patient education about medicines: factors influencing reading and seeking of written medicine information. Health Expect. **9**(2), 174–187 (2006)
24. Nicholson, W., Gardner, B., Grason, H.A., Powe, N.R.: The association between women's health information use and health care visits. Women's Health Issues **15**(6), 240–248 (2005)
25. Houts, P.S., Doak, C.C., Doak, L.G., Loscalzo, M.J.: The role of pictures in improving health communication: a review of research on attention, comprehension, recall, and adherence. Patient Educ. Couns. **61**(2), 173–190 (2006)

26. Trifiletti, L.B., Shields, W.C., McDonald, E.M., Walker, A.R., Gielen, A.C.: Development of injury prevention materials for people with low literacy skills. Patient Educ. Couns. **64**(1–3), 119–127 (2006)
27. Wilson, F.L., Brown, D.L., Stephens-Ferris, M.: Can easy-to-read immunization information increase knowledge in urban low-income mothers? J. Pediatr. Nurs. **21**(1), 4–12 (2006)

Chapter 42
Enhancing User Experience of E-commerce Platforms—A Case Study of B2C Applications in the Indian Market

Megha Agrawal and Debayan Dhar

Abstract The Indian e-commerce industry is rapidly growing. One of the major contributing factors toward the success of these companies is the quality of user experience their platforms provide. The study reported in this paper primarily focuses on India's top B2C organizations and the role of definitive user experience elements in shaping the success of these organizations. A heuristic-based comparative study was conducted among the top e-commerce Web applications in order to identify unique features that contribute to a sumptuous user experience. The analytical study was backed up by user studies that report the expectation of the customers and their frustrations. The collective insights were then correlated to each platform's market ranking and share with an objective to identify and analyze features, design cues and elements that contribute to their standing in the Indian market. In a nutshell, the qualitative study explores connections between market ranking, usability heuristics, and user study insights in order to prescribe design features, cues, and elements that benchmark the user experience framework for B2C applications in an Indian context. The insights elaborate on the current trends, gaps, and opportunity areas for B2C applications.

42.1 Introduction

The e-commerce industry is growing massively each year. It is estimated that there were 1.92 billion global digital buyers in 2019 [1]. Further, the market size was valued at USD 3.35 trillion in 2019 [2]. This trend is predicted to only get better with the years—it is estimated that 95% of purchases will be made online by 2040 [3]. The Indian e-commerce market, owing to an increasing Internet and smartphone penetration, is also expected to grow colossally to USD 200 billion by 2026, from the

M. Agrawal · D. Dhar (✉)
Department of Design, Indian Institute of Technology, Guwahati, Assam, India
e-mail: debayan@iitg.ac.in

M. Agrawal
e-mail: megha170205015@iitg.ac.in

A. Chakrabarti et al. (eds.), *Design for Tomorrow—Volume 1*, Smart Innovation, Systems and Technologies 221, https://doi.org/10.1007/978-981-16-0041-8_42

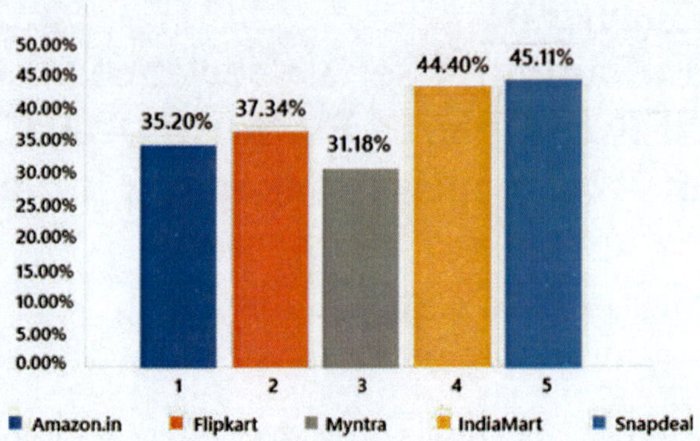

Fig. 42.1 Bounce rate of the five platforms [7–11]

USD 38.5 billion in 2018 [4, 5]. It is largely dominated by companies like Amazon, Flipkart, Myntra, IndiaMart, Firstcry, Nykaa, and Snapdeal [6]. Given the huge profit and scope of growth this industry promises, cutthroat competition among companies is inevitable. Thus, any B2C e-commerce company that wants to survive needs to be on the top of its game in all aspects.

The success of B2C e-commerce platforms can be generally judged on the basis of certain metrics that measure traffic and Web site usage, such as Web site country-ranking, bounce rate, page views, and daily time users spend on the platform, as these parameters directly affect revenues [12]. Figure 42.1 shows the bounce rates (from an unbiased source) of the five e-commerce platforms which rank the highest in India. Bounce rates help understand how many users navigate away from the platform after only viewing one page. Amazon India, which ranks the highest in the country, has a low bounce rate compared to Flipkart which, despite ranking second highest, has a higher bounce rate compared to Myntra [7–9].

Further, in a preliminary study we conducted, 52.6% respondents reported choosing Amazon India from the alternatives for primarily buying electronics and lifestyle products. 47.4% reported choosing Flipkart for finding cheaper alternative to products on Amazon. This study was conducted as informal unstructured interviews with 19 young adults of ages ranging from 20 to 23 years who have smartphones and steady access to Internet and primarily reside in the northeast region of India.

The market standings of the e-commerce companies, combined with the findings from the user study, can be attributed to many factors, such as the types of services provided by the company, nature of products they sell, their support systems and accessibility to their products, services available to users and the user experience provided by the platform. While it is not clear whether user experience solely contributes to the success of these companies, it is nevertheless an interesting perspective for HCI researchers to consider.

All major organizations are focused on user experience and are heavily investing in it. Good user experience accounts for how customers feel while interacting with a product and thus prevents customers from having bad experiences [13]. For a company, it implies a higher conversion rate and subsequently, a higher return on investment [13]. Furthermore, the general trend in digital economy is shifting from being goods-and-services-driven to experience-driven, and any company that fails to keep up faces a threat of being rendered irrelevant [14]. This has an implication of enhancing user-experience-fronts for e-commerce organizations.

The research questions tackled in this paper are—What are the differences between the top e-commerce companies in India? Further, does the user experience provided by these platforms play a role in the varied market rankings of the companies or is there perhaps no relation between the two?

In an attempt to answer these questions, this paper compares top five B2C e-commerce Web applications in India, identifies their key user experience features, and connects them with the respective organization's standing in the market. Furthermore, using Snapdeal as a case study, we propose and demonstrate a list of design features, cues, and elements that benchmark the user experience framework for B2C applications in an Indian context.

42.2 Literature Review

Heuristic evaluations are very beneficial in identifying major usability issues in an interface [15]. While numerous frameworks exist for assessing the usability of various types of interfaces, there is still a huge scope in the domain of e-commerce. Moreover, a combination of different evaluation methods should be used to evaluate e-commerce platforms in order to achieve multiple unique insights [15].

While frameworks developed specifically for e-commerce evaluations do consider other factors such as reliability, system quality, and maintainability [16], they fail to accommodate for the specific demands and expectations of a diverse group of users, such as that of the Indian subcontinent.

Further, there is a lack of frameworks that not only cater to user needs but also take into account the demands of the industry from a competitive perspective. Thus, in this paper, we hope to provide heuristics that take into consideration the unique demands and challenges that are faced by e-commerce Web application users in the Indian subcontinent and the application of which would aid the e-commerce company in achieving a better market standing.

42.3 Aim

To study the connection of market share and ranking, usability heuristics and user study insights in order to suggest and prescribe design features, cues, and elements that benchmark the user experience framework for B2C applications in an Indian context.

42.4 Objective

1. To study the market by conducting a heuristic-based comparative study on India's top B2C applications to identify unique features that contribute toward the application's ranking.
2. To correlate the insights gained from the market study with the applications' market ranking to identify the design features that contribute toward the market standing.
3. To understand user expectations and frustrations to support the insights gathered from the market study and prescribe design features, cues, and elements that benchmark the user experience framework for B2C applications in an Indian context.

42.5 Methodology

For the purposes of all the studies reported in this paper, mobile platforms of India's top five e-commerce companies were used. These included a diverse group of e-commerce companies, such as the ones that list large volumes of products across product categories (Amazon, Flipkart, IndiaMart and Snapdeal) and the ones that list products in only a single category (Myntra).

Firstly, in order to gain an understanding of what exists in the market and to gauge the industry benchmark, a heuristic-based market study was conducted on the five products. This involved a comparative study based on predetermined attributes and subsequent heuristic evaluations on the top two and the bottom ranking products out of the five. This helped significantly in identifying the presence of good and bad experience elements and their frequency in a particular product, thus helping answer questions about their potential correlation to the product's market ranking.

Secondly, in order to verify the findings from the market study and to deeply examine if user experience elements affected a user's choice of platform, a semi-structured user study was conducted with users of the mentioned platforms, aimed at finding out what made the users choose one product over the other. Finally, all the insights from the aforementioned steps were collated and a set of prescriptive heuristics specifically for e-commerce Web applications were created. These heuristics aim to guide in designing successful and user-centered e-commerce platforms.

42.5.1 Comparative Analysis

The five competing companies were analyzed on the basis of predetermined attributes. Studies report that initial perceptions about the aesthetics of a Web site can persuade users to explore it in detail [17]. Further, navigation plays a significant role in determining for how long the interaction is sustained, along with any persuasive strategies employed in order to prevent the user from ceasing interaction [17]. Thus, considering these factors, the attributes determined were: color palette, grid systems, icons, features (primary, regular, unique, and cosmetic), anticipated users (primary and secondary), unique selling proposition, persuasion cues and navigation.

Each platform was individually used by three experts who have a minimum of 4 years of experience in field of HCI. The platforms were used to carry out the tasks of buying a product by searching for it, buying a product by exploring on the platform, reviewing past orders, and looking up customer care information. The tasks were decided such that all the primary functionalities of the platforms were considered, without being too specific or restrictive for the experts. New user accounts were created for each evaluation session in order to eliminate recommendation changes with the progression of number of evaluations. The analyzed data was recorded in a matrix format. Further, recordings for each platform were compared and the following insights and inferences were drawn with respect to each attribute.

Color Palette. The five platforms typically use dark, cool colors with base as whites or light grays. Further, use of one or more contrasting colors that are analogous to each other is common for accents. Overall, platforms have a color scheme that is strictly identifiable to them (Fig. 42.4).

Grid System. Two-column grids are most commonly used in the platforms. Typically, single row horizontal grids are used to display categories, offers, and recommendations (Fig. 42.4).

Icons. The platforms used icons that are detailed yet easy to recognize at a glance. Large, hollow, and monotoned icons are common but using an additional accent color makes them more interesting (IndiaMart) (Fig. 42.4).

Features (Primary, Regular, Unique, and Cosmetic). A strong recommendation system to get the users back to what they considered buying, but did not, is most commonly observed. Users prefer seeing user-generated content such as customer reviews and images as these are given due importance in all the five platforms. Users should also be able to navigate to customer service easily. Further, straightforward and simple navigation systems with icons and images that are easier to use and make it easy for the user to quickly arrive at what they were looking for. Different kinds of offers, such as limited time offers, category-based offers, and context-based offers, are crucial in getting users to purchase products from the platform. Users prefer good quality product images. Convenient and multiple payment options strengthen user's trust on the platform. Information about the product and offers given upfront and simply makes the user feel they are being catered to justly.

Persuasion Cues. Persuasion cues are elements or features present in an interface that nudge the user toward performing a desired action—purchasing products in the context of this paper.

It was observed that platforms commonly had offers that demanded quick and impulsive actions by the user were widely used by the top platforms. Personalized messages and action recommendations that reduce the number of steps involved

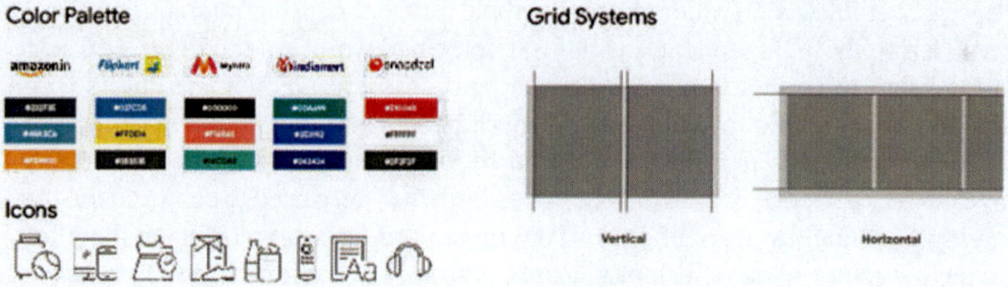

Fig. 42.2 Visualization of insights from comparative analysis

in buying a product were widely used as persuasion tools. Subscription plans that promise and good quality service seemed to add value to the platform and garner more user trust. Additionally, other features provided by the company such as streaming services, payment channels, and various subscription plans can make users trust the platform more.

Navigation. Analysis of the navigation systems of the five platforms revealed various cases such as confusing use of side arrow buttons, unintuitive grouping of categories, and lack of consistency—all of which can hamper a user's experience of the platform. Thus, well segmented and organized navigation combined with easy and quick to understand icons and images remove friction and frustration from the process of browsing. Appropriate use of call-to-action buttons facilitated primary tasks and made them easy and highly efficient. A consistent system of buttons and navigational markers followed all throughout the platform was important in not confusing the user. Further, a consistent categorization system across categories gave a perception of huge inventory. However, while listing the various categories and sub-categories, user can get confused with too many items in a single list (Fig. 42.2).

42.5.2 Heuristic Evaluation

After obtaining insights about current market trends, opportunities, and gaps from the comparative study, two heuristic evaluations were conducted to understand and examine any possible correlation between the usability of an application and its relative standing in the market. Therefore, the choice of platforms for the evaluations was Amazon India, Flipkart, and Snapdeal, that is, the two products that ranked highest and the product that ranked lowest, respectively. For the first evaluation, Jakob Nielsen's widely followed '10 Usability Heuristics for User Interface Design' was used [18]. The evaluation was conducted by same experts who conducted the previous study. The evaluation was carried out by listing out all the ways in which each specific heuristic was obeyed or violated in each platform, in a tabular format. Finally, upon weighing out the obedience against the violations with respect to frequency and severity, a generalized 'Yes' (for overall obedience) or 'No' (for overall violation) was

Table 42.1 Evaluation with Nielsen's usability heuristics

Heuristic	Amazon.in	Flipkart	Snapdeal
#1. Visibility of system status	Y	Y	Y
#2. Match between system and the real world	Y	Y	N
#3. User control and freedom	Y	Y	Y
#4. Consistency and standards	Y	N	N
#5. Error prevention	Y	Y	Y
#6. Recognition rather than recall	Y	Y	Y
#7. Flexibility and efficiency of use	Y	Y	Y
#8. Aesthetic and minimalist design	N	Y	N
#9. Help users recognize, diagnose, and recover from errors	Y	Y	Y
#10. Help and documentation	Y	Y	Y

assigned for each heuristic for each platform (Table 42.1). Secondly, a list of heuristics specifically defined for an e-commerce context was compiled from existing literature [15, 16] and used for conducting a second evaluation of the three platforms. The definitions provided by the authors of these heuristics were used while conducting the evaluation. The platforms were evaluated by the same experts by listing observations against each heuristic in a tabular format for the three platforms (Fig. 42.3).

As can be seen from Table 42.1, out of the ten heuristics, Amazon.in, which ranks the highest among the five, only fails to follow Heuristic #8—Aesthetic and minimalist design. Insufficient white space, some immensely cluttered pages, and

Heuristics	Amazon	Flipkart	Snapdeal
System Quality	The system is adaptable, easily available, fast and easy to use.	The system is adaptable, easily available, fast and easy to use.	The system is adaptable, easily available and fast. Some parts are confusing to use due to non-standard interactions.
Information Quality	Information is complete, easy to understand, personalised and relevant.	Information is complete, easy to understand, personalised and relevant.	Information is easy to understand, personalised and relevant. Some products and sections have incomplete information.
Assurance	Has good reputation, information appears clear and truthful	Has decent reputation, information appears clear and truthful	Excess offers for each product give an impression of false prices
Contact	Multiple contact channels clearly labelled and accessible	Multiple contact channels clearly labelled and accessible	Multiple contact channels clearly labelled and accessible
Interactivity	Facilitates good user interactivity via FAQs, reviews and alternate product suggestions	Facilitates good user interactivity via FAQs, reviews and alternate product suggestions	Facilitates user interactivity via reviews and alternate product suggestions
Flexibility	Very flexible; has multiple searching, ordering, payment, shipping, delivery and return options.	Very flexible; has multiple searching, ordering, payment, shipping, delivery and return options.	Very flexible; has multiple searching, ordering, payment, shipping, delivery and return options.
Findability	Good navigation, products are easily findable.	Good navigation, products are easily findable.	Navigation is decent, some products are difficult to find.
Communication Ability	Good personal communication channels and new offer/product communication.	Good personal communication channels and new offer/product communication.	Good personal communication channels only.
Security and Privacy	Addresses security and privacy adequately	Addresses security and privacy adequately	Addresses security and privacy adequately
Empathy	High individual attention via personal communication	Lacks individual attention features	High individual attention via personal communication
Error Compensation	Offers great error compensation	Offers great error compensation	Offers great error compensation
Net Benefits	Lots of net benefits are present	Lots of net benefits are present	Lacks time savings
Price Knowledge	Product, delivery and handling price are visible upfront	Product, delivery and handling price are visible upfront	Inclusion of delivery charges in product price is not explicitly stated.
Grammatical Accuracy	Maintained throughout	Maintained throughout	Maintained throughout
Alternate Language Support	Multiple Indian languages supported	Multiple Indian languages supported	Multiple Indian languages supported
Personalized Culturally Relevant Persuasive Strategies	Culturally relevant offers, personalized recommendations and content organization are present.	Culturally relevant offers and content organization are present.	Culturally relevant offers and content organization are present.
Responsiveness	Instant response via chatbot and personal assistance	Instant response via chatbot and personal assistance	Instant response via personal assistance
Responsiveness across Devices and Channels	Optimized for all platforms	Optimized for all platforms	Optimized for all platforms

Fig. 42.3 Evaluation with e-commerce-specific heuristics

unnecessary information are seen. Similarly, Flipkart, ranking second, also only violates Heuristic #4—Consistency and standards. The variance in action button radius, colors, and location of header icons without clear purposes can confuse users and lead them to errors. Snapdeal, which ranks the lowest, on the other hand, violates various heuristics such as Heuristic #2—Match between system and the real world, Heuristic #4—Consistency and standards and Heuristic #8—Aesthetic and minimalist design. Inconsistency in location and mapping of icons to associated labels, use of jargon, extensive use of attention-grabbing colors, etc., drastically hamper the user experience. Similarly, from Fig. 42.3, Snapdeal fails to uphold many heuristics as compared to Amazon and Flipkart. Snapdeal lacks in system quality, information quality, assurance, findability, communication ability, net benefits, and price knowledge, whereas Amazon upholds all heuristics and Flipkart only lack in empathy and personalized culturally relevant persuasive strategies.

Thus, there is possibly a direct connection between the number of heuristics followed by a platform and its relative market standing. The greater number of heuristics a platform follows, the better market standing it appears to have. A contributing factor could be that following usability heuristics ensures that a platform provides an experience that is at par with the market standards and provides the basic experiential features required toward a good user experience.

42.5.3 User Interviews

After the market analysis, with an objective to corroborate the insights drawn, actual users of the mentioned platforms were approached to take their desires, demands, and frustrations into account in the process of understanding the level of engineering of user experience elements into the products. The interview was aimed at understanding what makes users choose between different e-commerce platforms. A questionnaire was prepared to conduct a semi-structured interview of 10 such users. The questions were also aimed at identifying a correlation (if any) between the choice of platform and the category of product the user wanted to buy. The questions included specific inquiries about which platform they preferred for specific category of products—clothes, electronics, stationary, etc. Broader questions that investigated the reasons why users chose one platform over the other were also included.

Figure 42.5 shows a visual representation of the insights gathered from the user study in the form of a word cloud. Larger words denote the degree of importance of such factors to users. Smaller words indicate secondary desires of the users and considerations. This data revealed a few patterns. Users either preferred buying all categories of products from a single platform or bought specific categories of products from platforms that only specialized in that category. They also prefer a wide range of products, good quality product photographs, fast delivery speed, and quality products. In terms of navigation, an easy to understand organization structure of the platform and fastest route to desired product were the top priority.

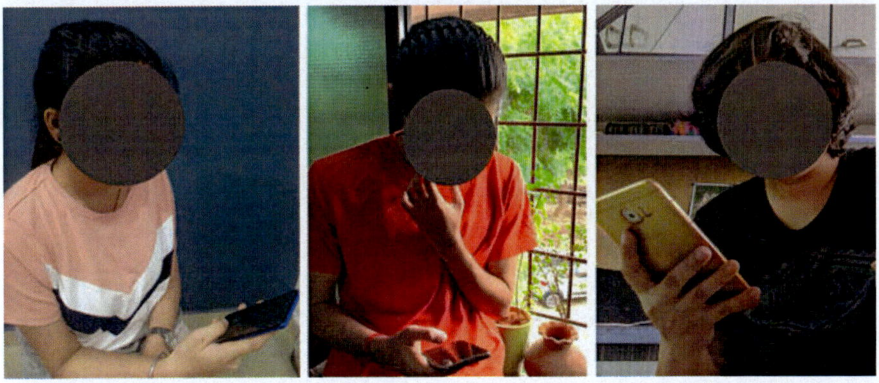

Fig. 42.4 Users of the platforms being interviewed

Fig. 42.5 User interview insights word cloud

42.6 Results

All insights from the comparative analysis, heuristic evaluations, and user interviews were card-sorted into groups on the basis of affinity. Each group of information was then summarized, considering all the data within. These summaries were then consolidated into a single list, signifying all insights from the three studies discussed (see Table 42.2). By comparing the five platforms against these insights, it was observed that a direct correlation was apparent between the degree of listed user needs a platform met (implying a superior user experience) and its market ranking. This implies, if a platform delivers a better user experience, i.e., it accounts for a greater number of user needs and desires, then it has a better chance of ranking higher.

Further, Table 42.3 prescribes some heuristics which are drawn from detailed insights from all three studies. These heuristics are prescribed for designing an e-commerce Web application, while taking into account the demands and behaviors of Indian e-commerce customers. If followed, it aims to ensure a user experience that is at par with the industry standards as it exploits the trends, opportunity areas, and gaps that the leading platforms in the industry provide.

Table 42.2 Insights

S. No.	Consolidated insights
1	Users want to buy exactly what they like; they do not want to compromise
2	Users want to be able to find a product they like in considerably less time
3	Users want to be sure about the authenticity, quality, appearance (compared to the product image), and fitting (in case of apparel) before buying a product. They also want to avoid the hassle of returns
4	Users want a fast and convenient delivery and return process
5	Many users want a one-stop solution for all their needs
6	Users want quick and easy ways of completing tasks
7	Users want the flexibility of trying multiple products before choosing to keep the one(s) they like
8	Users like to explore from a wide variety of products if they do not have anything specific in mind
9	Users often want other people's opinions on a product before making a purchase
10	Users want to buy new and unique things and want to be able to find them easily
11	Users want value for their money
12	Users want to be able to change their minds about an order without penalties

Table 42.3 Prescriptive heuristics for e-commerce web applications

Heuristic no	Heuristic
Heuristic #1	Users should be assured about the quality and authenticity of the products with ample information on the product page
Heuristic #2	A large variety of products within a category and across categories should be available to the customer
Heuristic #3	Categorization of products should be consistent and intuitive. The list of categories should not have too many items
Heuristic #4	A simple grid system with intermittent rows with horizontal scroll should be followed
Heuristic #5	The interface should allow the user's bad experience to be addressed immediately
Heuristic #6	Personalized action recommendations should be given to the user
Heuristic #7	Multiple kinds of offers, such as limited time, category-based, context-based, and personalized offers, are important to persuade users to make purchases
Heuristic #8	Multiple other services that help the user complete goals in app should be provided
Heuristic #9	Color palette should be strictly identifiable to the platform
Heuristic #10	Consistent, detailed yet easy-to-recognize icons should be used
Heuristic #11	Multiple regional languages should be supported

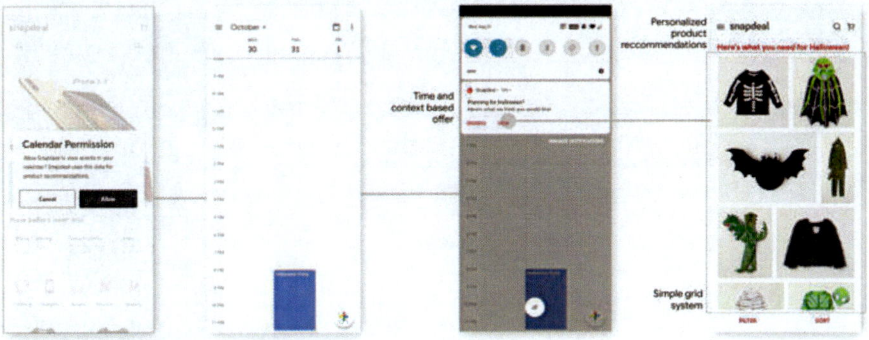

Fig. 42.6 Task flow: event-based product recommendations

42.6.1 Prescriptions

To demonstrate the usefulness of prescribed heuristics, they were applied to the platform that ranked the lowest among the five studied—Snapdeal. The resulting insights were taken into account and wireframes were made. Further, a small study was conducted using the wireframes, the results of which revealed that people preferred the new redesigned interface over the current one. This indicates that people genuinely desire new unique features which enhance the user experience of the current e-commerce Web applications.

The unique features proposed, as stated, to enhance the user experience of Snapdeal's Web application are as follows. The solution concepts and interface elements are both closely modeled on the user desires and pain-points.

1. Users do not want to spend a lot of their valuable time on e-commerce platforms and usually have to schedule time to shop online, making it a chore. Thus, Snapdeal can send push-notifications to suggest products as and when the need arises (see Fig. 42.6), such as for birthdays, festivals, and upcoming social commitments.

2. To address the problem of judging the authenticity of the product, as shown in Fig. 42.7, an additional step can be added in the shipment assembly line of the products. A photograph of the product, just before putting it in the packaging box, can be sent to the user. The user then has the option to cancel or proceed with the order within a limited frame of time.

3. Understanding if an apparel will fit appropriately or not is a major issue, despite the detailed sizing information provided on the platform. This is because every product has its own stylized fitting which is often difficult to judge only from the given measurements. Thus, Snapdeal can collaborate with local tailors that users visit to get detailed body measurements of the user (see Fig. 42.8). This measurement data can be matched with the product specification and fit-guides. Then, the closest match of the clothing can be sent to the user. Alternatively,

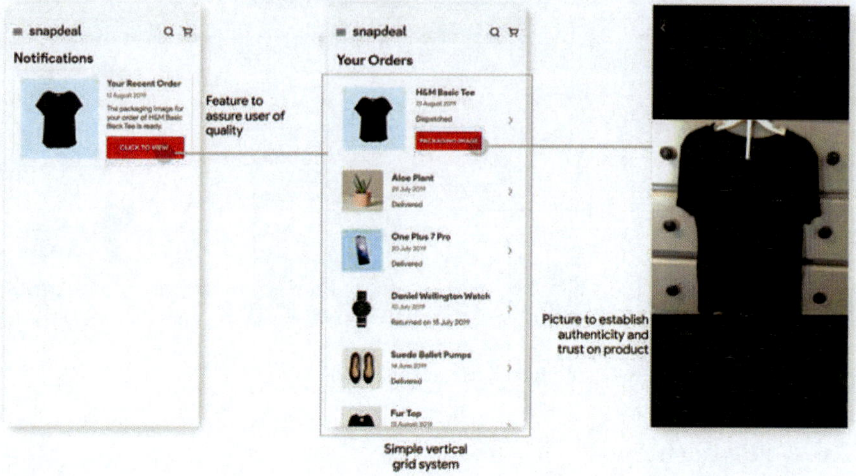

Fig. 42.7 Task flow: showing the user what is being sent

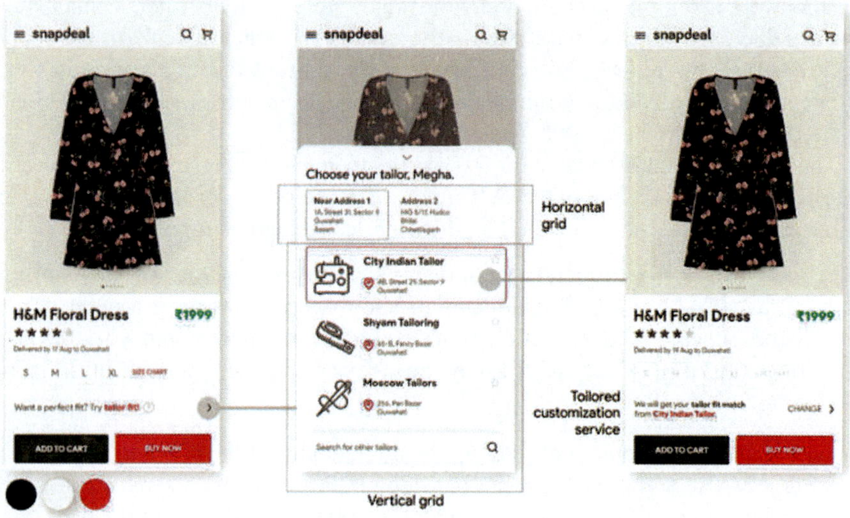

Fig. 42.8 Task flow: tailor-fit customization option

users can also choose to visit nearest tailor shops that Snapdeal has tied-up with, themselves.

4. Snapdeal can also provide an emotionally positive experience to its users by delivering products from local shops in the user's city that user prefers and trusts. This also facilitates fast deliveries, increasing user satisfaction (see Fig. 42.9).

Fig. 42.9 Task flow: viewing products from local vendor

42.7 Conclusion

Given the rapidly increasing growth of the B2C e-commerce industry, providing an experience that takes into account both the technology and the user is crucial for any company in the space to be successful against the huge number of competitors. This case study identifies the features that potentially contribute toward the successes of leading companies in the market and proposes their application on any e-commerce Web application to enhance its user experience to match the standards of the industry. Further, drawing from an understanding of user desires and frustrations, we propose unique and novel ideas which when applied will aid the company in surpassing its competition. We hope that this paper helps people in the human–computer interaction, usability engineering, user-interface designing, and consumer communities in providing B2C e-commerce solutions that are user-centric.

Acknowledgements We thank all the volunteers who helped us in this project by participating in the user interviews. We also appreciate the facilities provided at the Department of Design, Indian Institute of Technology, Guwahati which helped us in smoothly carrying out this project.

References

1. Law, T.J.: 19 Powerful ecommerce statistics that will guide your strategy in 2020 (2019). https://www.oberlo.in/blog/ecommerce-statistics-guide-your-strategy. Accessed 25 June 2020
2. B2C E-commerce market size: industry report, 2020–2027 (2020). https://www.grandviewresearch.com/industry-analysis/b2c-e-commerce-market. Accessed 25 June 2020
3. Osman, M.: Ecommerce statistics for 2020—chatbots, voice, omni-channel marketing (2020). https://kinsta.com/blog/ecommerce-statistics/. Accessed 25 June 2020
4. E-commerce industry in India. Accessed 25 June 2020

5. E-commerce market growing at a rate of about 17% in 2018–19 (2018). https://www.business-standard.com/article/news-cm/e-commerce-market-growing-at-a-rate-of-about-17-in-2018-19-118121700794_1.html. Accessed 25 June 2020
6. Top 10 e-commerce sites in India (2019). https://disfold.com/top-e-commerce-sites-india/. Accessed 25 June 2020
7. Traffic Analytics: amazon.in, https://www.semrush.com/analytics/traffic/overview/amazon.in. Accessed 12 Oct 2020
8. Traffic Analytics: flipkart.com, https://www.semrush.com/analytics/traffic/overview/flipkart.com. Accessed 12 Oct 2020
9. Traffic Analytics: myntra.com, https://www.semrush.com/analytics/traffic/overview/myntra.com. Accessed 12 Oct 2020
10. Traffic Analytics: indiamart.com, https://www.semrush.com/analytics/traffic/overview/indiamart.com. Accessed 12 Oct 2020
11. Traffic Analytics: snapdeal.com, https://www.semrush.com/analytics/traffic/overview/snapdeal.com. Accessed 12 Oct 2020
12. Ghandour, A., Deans, K., Benwell, G., Pillai, P.: Measuring ecommerce website success. In: 19th Australasian Conference on Information Systems (ACIS); University of Canterbury, Christchurch, paper 24 (2008)
13. Garrett, J.J.: The elements of user experience: user-centered design for the web and beyond. New Riders, Berkeley (2011)
14. Pine II, B.J., Gilmore, J.H.: The experience economy (1998). https://hbr.org/1998/07/welcome-to-the-experience-economy. Accessed 25 June 2020
15. Hasan, L., Morris, A., Probets, S.: A comparison of usability evaluation methods for evaluating e-commerce websites. Beh. Inf. Technol. **31**(7), 707–737 (2012)
16. Youngsin, H.: Develop an inclusive UX evaluation framework. B2C E-commerce heuristic metrics (EHM) (2019). https://hdl.handle.net/10589/149236. Accessed 12 Oct 2020
17. Sutcliffe, A.: Assessing the reliability of heuristic evaluation for web site attractiveness and usability. In: 35th Annual Conference on Hawaii International System Sciences, pp. 1838–1847. IEEE (2002)
18. Nielsen, J.: 10 Usability heuristics for user interface design. https://www.nngroup.com/articles/ten-usability-heuristics/. Accessed 25 June 2020

Chapter 43
Affecting Technology Consumption—Role of Designers in Ushering Behaviour Change

Arzoo Khare and Debayan Dhar

Abstract User-centred designers study user behaviours, to create products that fetch maximum adoption and longer loyalty. All this has led to an increased emphasis on user engagement. Often the ill-impact on the user gets overshadowed by business goals. The study presented in this paper focuses on digital well-being of the user and how designers can aid in developing tools to equip the users. A detailed literature review is conducted so as to highlight the inconsistencies of existing digital media that fail to address the issue of digital well-being. They highlight factors other than lacking self-control that lead to uncontrolled tech consumption. The literature review is then followed by a qualitative research study that investigates current mobile usage and its discontinuation pattern. Combining insights from the literature and the exploratory study with users, the study finally suggests unique approaches that designers and the mobile industry can adopt to design for disengagement. The findings of the study can contribute to understanding the middle ground that looks out for the user's well-being without compromising on businesses and platforms.

43.1 Introduction

The struggle with technology addiction is something that a lot of people deal with but in different ways. The root cause of this struggle is businesses that run on a user's attention [1]. They have turned the user into the product in the garb of providing free services [2]. Other factors that aid tech addiction are extrinsic motivations like rewards, popularity and external validation that bring the user little to no value

Selected Topics: Design of/for User Interfaces, User Experience and Human–Computer Interaction. Design for X (Safety, Manufacture & Assembly, Cost, Reliability, Robustness, **Social Interaction**, etc.)

A. Khare · D. Dhar (✉)
Department of Design, Indian Institute of Technology, Guwahati, India
e-mail: debayan@iitg.ac.in

A. Khare
e-mail: arzoo.khare@gmail.com

A. Chakrabarti et al. (eds.), *Design for Tomorrow—Volume 1*, Smart Innovation, Systems and Technologies 221, https://doi.org/10.1007/978-981-16-0041-8_43

[3]. They give the user short-term pleasure but end up affecting their psychological well-being in the long run. In addition, we do not factor in the responsibility of the platforms, the creators of the apps or the environment [3, 4]. As technology is designed keeping the users' innate behaviour in mind, going against it is like a fight with oneself. There is a need to equip the user and the designer with tools that can help them overcome and eliminate the addiction, respectively.

The objective of the investigation initially revolved around understanding how people experience tech overuse, users' interaction with technology, and their pattern of distraction. This will help us suggest changes that align with the user's mental model.

43.2 Related Works

Literature highlights that smartphones are becoming problematic due to excessive dependency on them. As a result, the fear of not being able to use one's phone or nomophobia leads to stress of varying degrees [5, 6]. Problematic smartphone usage is capable of hampering other pleasurable activities and social activities which may lead to depression [7]. Excessive dependency on smartphones stems from extrinsic motivations like rewards, popularity, likes, comments, etc. Extrinsic motivations cause people to engage in controlled, stressful and ego-centric activities [8]. They are also associated with lower self-esteem, lower quality of relationships with people around them, and increased consumption of artificial entertainment [9, 10]. Everything is designed to keep the user connected and dependent on technology. Aranda and Baig [5] observed four methods of disconnection: lifestyle choice, infrastructure constraints, unplanned outage and a short break. Out of these, the users were most comfortable with a short break as it provided mental relief and the option to undo their action. Hassenzahl, et al. [11] designed for a 'good' shared consumption of technology, hoping that meaningful experience will replace our need for extrinsic goals.

The problem arises because people are unable to stop consumption. This happens because each fulfilled extrinsic motivation gives the user a reward (dopamine hit) and makes them want to repeat their action to keep getting the reward. The moment the user stops the brain interprets the lack of a reward as a punishment, thus making the process of putting one's phone down so difficult. The study presented here is a user-centred design-based approach undertaken to address the issue of smartphone addiction. As highlighted in the literature review, it is high time that end-users are made conscious of their addictive behaviour which hampers their overall growth. The project undertaken is unique in a sense that instead of forcefully stopping a behaviour, the intention is to inform end-users about their current practice and then supporting users according to their personal characteristics for adopting a more meaningful behaviour towards smartphone use.

43.3 Aim and Objectives

The goal of the study is to understand how people get distracted and how they deal with it. The intention is to design strategies to bring about a positive behaviour change in users with smartphone addiction. Overall, through this project, we intend to motivate the user to be more conscious about their phone usage and how they spend their time.

43.4 Methodology

A qualitative research study was conducted including semistructured interviews to get familiar with the usage and discontinuation patterns of the participants. The sample included 15 participants: 8 female and 7 male (mean age: 23.46). All the participants were students pursuing either a bachelors or masters degree in fields including engineering, design, commerce, arts and business management. Their ages ranged from 18 to 29. They used a variety of devices including smartphones, laptops and tablets. Thereafter, affinity mapping was carried out to highlight the key problems areas and understand the underlying structure and method of usage. Finally, five potential measures were suggested that could aid the designers in disengaging the user.

43.4.1 Semistructured Interviews

The participants were students who were actively involved in some form of tech consumption. They used a combination of social media, gaming and video streaming platforms. In the interview, questions regarding the causes and effects of distraction were discussed. Findings suggest that the people found it very difficult to pull themselves out of their phones. Most of them had tried different methods to control their urge. We found that the closer the participants were to their work deadline the more successful they were in distancing themselves from their phones. They also found proximity to their phone to be the biggest challenge.

43.5 Analysis

43.5.1 User Behaviour Cycle

The participants described their cycle of addiction in the following way—It begins with a trigger, which can be an intentional or unintentional need to check their phone

[5]. Intentional need includes responding to an urgent call or text and unintentional need includes notification triggers. Sometimes there is a small window that allows their mind to escape and find easy refuge in their phones. Once triggered the user slowly gets into the zone of no thinking or a state of trance. Before they realize this, they end up spending a lot of their time. This is then countered by mental accounting–justifying the usage or mentally shifting schedules to accommodate time spent on phones, ultimately leading to the guilt of overusage. This vicious cycle affects their lives in different ways, and it affects their health (weaker eyesight, back pain), disappointment in oneself due to lack of self-control and poor impact on their work-life and family life as well.

43.5.2 It's not You (Dear Phone), it's Me!

When we asked our participants whether they believed their uncontrollable technology consumption was a result of their fault or their phone's fault. 14 out of 15 people believed they were to be blamed. They believed that they had the freedom to consume and therefore the power to resist technology. What fuels this kind of thinking is the numerous solutions and apps available. They assure the user that a combination of their service and the user's diligence can help the user get rid of their tech addiction. The concept that stays submerged in this wave of productivity apps is freedom violation. Freedom is not just consuming technology whenever you want, for as many hours you want, freedom is also being able to stop consumption whenever you want and for as long as you want. The problem with the way we consume tech is that there are no intelligently designed cues that nudge you to gradually stop.

43.5.3 Approaches to Phone

We studied the ways users approached their phones. There were primarily three ways with which they got stuck *in* their phones (Fig. 43.1).

The 'Notification Plunge'. This is the process where one notification trigger leads to prolonged phone usage. Notifications were designed to alert the user about an incoming SMS or email [13, 14]. But then the potential in this method to reach a billion people at once was realized [15]. Suddenly an array of most relevant to the least relevant pieces of information was getting delivered to the user's mobile phone in the form of notifications. So, once the user turned on his phone he was more likely to fall into the trap of checking the irrelevant notifications and indulging in them.

The 'In-phone Vortex'. This is the process where one intentional task (like making a necessary call, setting an alarm) leads to multiple unintentional tasks (like indulging in social media, games, etc.). This is the point where users get sucked into their phones sometimes even before or in the middle of completing their intentional task.

Fig. 43.1 Position of a phone during the three approaches to phone [12]

The 'Boredom Reflex'. This is the process where the user gets bored with the work at hand and is tempted to skip work and indulge in his phone. Conserving energy played a crucial role in human survival. This energy was put to better use like gathering food, avoiding predators and finding shelter. Thus, our brains are innately attracted to sedentary behaviours, as active behaviours use up more brain resources [16, 17]. Social media is the 'easy' thing to do. Often when users get stuck during a task that involves higher cognitive load (like working on an assignment), the brain looks for an easy escape. This part of our brain finds shelter in mobile phones (social media, games, video content, etc.). This happens because our brains are hardwired for laziness—for choosing the easier alternative [16].

43.5.4 Internal Needs and External Hooks

People rely heavily on their phones for productivity, information seeking, social interaction, etc. [18]. There is a vicious cycle that exists between internal needs and external hooks. Internal needs included emotions like loneliness, boredom, dependency, fear of missing out (FOMO), etc. They trigger interaction with one's phone. Once the user reached out for his phone, the external hooks like customized feeds, recommendations, visual distraction, notification sounds, etc., took over from there, that sucked the user deeper into his phone. The combination of internal need and external hooks led to:

Borrowed sense of control: Making to-do lists, being on top of all their messages, etc. This translated into an illusion of control that may or may not result in a productive outcome.

Digital emotional reserve: Participants accepted being digitally polite (responding to messages as soon as they entered the inbox, engaging in long but unwilling conversations, etc.). This ensured that in times of their need the other person would return the digital etiquette/favour.

Curiosity: Whenever there is an **information gap** (the difference between known and unknown), humans try to cover it as quickly as possible [19]. With the 'explore' pages where there is an abundance of information, people are unable to resist themselves.

A breather from reality: People can learn what is happening in the lives of their favourite celebrities, old classmates or relatives and even random people.

43.5.5 Users' Model of Defence

In an attempt to understand the user's mental model, we studied the ways which the participants used to resist or tackle their excessive device usage (Fig. 43.2).

Third-party help: One of the common methods applied by the participants was seeking help from an external source—this source was either a person or a device.

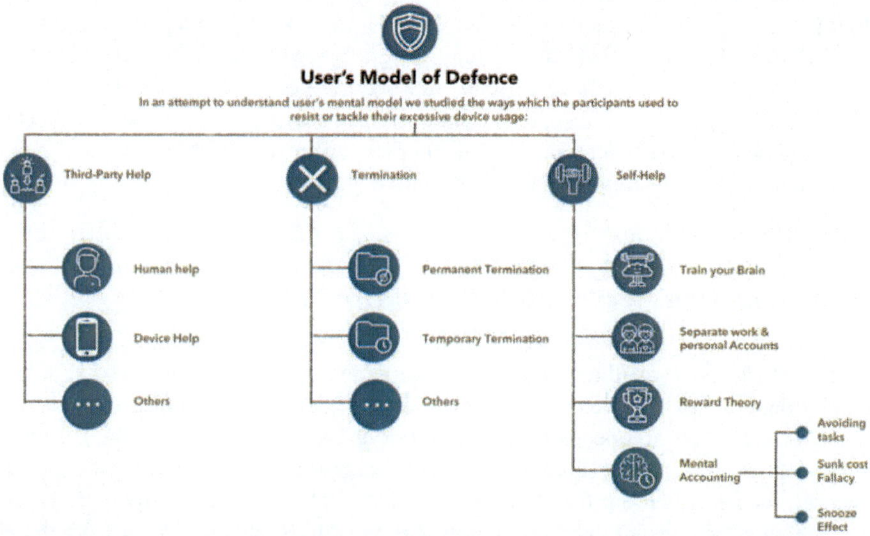

Fig. 43.2 User's model of defence [20]

Human help: For those who sought human help, they would turn to friends or family members in times of need (for example, when they had to meet deadlines). They believed they lacked self-discipline and needed another responsible person to enforce control over them. Someone who would not negotiate with them. Participants were very comfortable with this method as it got the job done.

Device help: Others who turned to their devices for help made use of features like the do not disturb (DND) mode, setting time limits on certain apps, setting alarms and reminders manually, and switching to audio modes over video modes (YouTube Music instead of YouTube).

Others: The participants reported another source of indirect third-party help, when a particular person or group of persons or featured trends or post did not meet their preference. In such a situation, they were more likely to stop using the platform.

Termination

Permanent termination: Upon realization of excessive usage, some participants were likely to uninstall the app. These were often apps for entertainment purposes like games and social media.

Temporary termination: This includes making conscious efforts to pull yourself away from your device. It ranged from shutting down the device, hiding it, putting it on flight or silent mode and disabling notifications—anything that would kill the trigger. One participant stopped recharging her phone as her exams were approaching and she did not want any distraction.

Others/external: When the disconnect is out of the user's control—for example, some students had their hostel rooms in areas where they got poor to no cell reception. As a result, they were forced to disconnect. Other situations included when they gave up using a certain social media platform because most of their friends were off it.

Self-help. Some participants relied on indigenous methods or *'jugaads'*.

Self-realization: Their primary method was to rely on self-realization and training their minds. They agreed this method required a lot of self-control and self-discipline and while it worked for the strong-willed it failed for the others.

Separate accounts: Creating separate work and personal accounts. One of the participants created separate YouTube accounts. This way YouTube curated separate playlists for him—this way he managed to be less distracted by his work account.

Reward theory: Often users would dedicate these distracting services as services that needed to be earned. So they would indulge in them only after they had finished the more important work.

Mental accounting: After having indulged in their phones for more than the stipulated time, the participants would start juggling deadlines and time in their minds.

One participant admitted to mentally striking off a couple of activities on her schedule to accommodate her urge to binge-watching a TV series.

Sunk cost fallacy: Continuing to invest in behaviour as a result of previously invested resources [21]. Having already exceeded their time limit, the participants tried to justify spent time by investing more time. For example, *'I've already missed my deadline, I can delay the submission by another 20 min'*.

Snooze effect: Snooze effect is when you delay your deadline by a short time interval. Often users lied to themselves—'just one more video', 'just 5 more minutes'. Some users would extend their deadlines in multiples of 5—'I will restart work at 7:30'.

43.6 Discussion

43.6.1 Action Over Awareness

Informing the user about his usage pattern is a great first step towards digital detox. Statistics help the users come face to face with their reality. But it is an incomplete step. It needs to be followed by some doable action. One participant complained, *'Just stats don't work. I know I use my phone for more than 5 h a day yet I continue to use it'*. It is important to consider alternative ways in which they can use the information they just received. Without an action plan, the stats become futile.

43.6.2 Finite Play

By finite play, we mean something that comes to an end. The concept of finite play can be seen in board and card games, like monopoly, ludo, uno etc. There is an endpoint in view—survive with the most money or to beat the other players. To add to these, the game also provides strategic checkpoints (consumer fatigue), in the form of penalties, that make the game interesting as well as bring it closer to the end.

In casinos, the carpet plays an important role as it draws the user into the gaming area. These carpets rarely have right angles in them, because right angles make the user stop and think, and it puts them in the position of a decisionmaker, as opposed to linear carpets that pull the consumer mindlessly into the gaming area, more often towards the slot machines [22, 23]. Compare this to 'mindless scrolling' that makes it very difficult for the user to come out of. As designers, it is very important for us to define end goals for the user, a point after which the user can simply stop and after which there is nothing more left to do.

43.6.3 Designing Consumer Fatigue

Despite having an enjoyable experience with an app, the users leave with resentment and regret. This happens due to their inability to disengage [5, 24]. Participants stressed on the fact that once involved in their phones it was difficult to find an exit route.

Consumer fatigue can be in the form of small deterrents that make the user come out of his state of flow. The cumulative sum of the elements of consumer fatigue leads to the larger goal of finite play. For example, a user wants to stream YouTube for 45 min. At the end of 45 min, technology can gradually create consumer fatigue. The buffering period increases which eventually breaks the user's flow. This helps the user to resume work. As designers, we need to design for easy disengagement too.

43.6.4 Transfer of Burden of Control

The burden of unintentional device usage weighs heavy on the user [5]. To top that, we also place the burden of not being able to quit on the user. We hold the user completely responsible for his addiction or excessive usage. When we look closely, our tiff with our phones begins at an innate level. At the base of it all is our need to fulfil our intrinsic needs—belonging, intimacy, mastery, and autonomy [3]. Thus, it becomes easy for anyone to fall in this trap of excessive usage.

The burden of not being able to stop makes the user feel that they lack self-control. But there are a lot of factors that contribute to our inability to stop other than poor self-control. These include our innate attraction to screens, our brains being hardwired for laziness, etc. Thus, we need to design a solution where the user does not have to monitor himself. Instead, a third party takes control or induces delayed gratification to unburden the user.

43.6.5 Social Prescribing

On studying different kinds of addiction, we found out that addiction is never just the result of the compound (in this case apps, technology and devices). It is often a concoction of the compound and the environment. In the words of JanTønnesvang, 'Different environments have different ways to meet your psychological needs to different degrees. Your ability to get your needs met is the amount of psychological oxygen in your environment' [3]. We got our dose of *'psychological oxygen'* even before gadgets existed and its source was the natural world—being part of the community, etc. As designers, we can create environments that bring value to the user, prescribing people to take part in social groups in positive, non-anonymous ways, within the device or outside.

43.7 Conclusion

When we started our research we wanted to find the perfect recipe to get people off their phones. As the research progressed, we realized it is not that simple. It is because there is not just one way to change behaviour. Tech addition differs from person to person and is dependent on and triggered by various factors. So we studied user behaviour, mapped their expectations and analysed their actions. Finally, we came up with five potential measures that can be used by designers to create responsible technology and change user behaviour in different scenarios with different kinds of users.

References

1. Vox: How ads follow you around the internet [Video]. Youtube (2020). https://youtu.be/HFyaW50GFOs
2. Oremus, W.: Are You Really the Product? The History of a Dangerous Idea (2018). https://slate.com/technology/2018/04/are-you-really-facebooks-product-the-history-of-a-dangerous-idea.html. Accessed 15 Oct 2019
3. Raskin, A., Harris, T., Hari, J.: Your Undivided Attention Podcast Episode 8: The Opposite of Addiction [Audio podcast] (2019). https://humanetech.com/podcast/
4. Raskin, A., Harris, T., Eisenstat, Y.: Your Undivided Attention Podcast Episode 3: With Great Power Comes… No Responsibility [Audio podcast] (2019). https://humanetech.com/podcast/
5. Aranda, J.H., Baig, S.: Toward "JOMO" the joy of missing out and the freedom of disconnecting. In: Proceedings of the 20th International Conference on Human-Computer Interaction with Mobile Devices and Services (2018)
6. Tams, S., Legoux, R., Léger, P.M.: Smartphone withdrawal creates stress: a moderated mediation model of nomophobia, social threat, and phone withdrawal context. Comput. Hum. Behav. **81**, 1–9 (2018)
7. Elhai, J.D., Levine, J.C., Dvorak, R.D., Hall, B.J.: Fear of missing out, need for touch, anxiety and depression are related to problematic smartphone use. Comput. Hum. Behav. **63**, 509–516 (2016)
8. Schmuck, P., Kasser, T., Ryan, R.M.: Intrinsic and extrinsic goals: their structure and relationship to well-being in German and US college students. Soc. Indic. Res. **50**(2), 225–241 (2000)
9. Kasser, T., Ryan, R.M.: Be careful what you wish for: optimal functioning and the relative attainment of intrinsic and extrinsic goals. In: Schmuck, P., Sheldon, K.M. (eds.) Life Goals and Well-Being: Towards a Positive Psychology of Human Striving, pp. 116–131. Hogrefe & Huber Publishers (2001)
10. YildizDurak, H.: Investigation of nomophobia and smartphone addiction predictors among adolescents in Turkey: demographic variables and academic performance. Social Sci. J. **56**(4), 492–517 (2019)
11. Hassenzahl, M., Eckoldt, K., Diefenbach, S., Laschke, M., Len, E., Kim, J.: Designing moments of meaning and pleasure. Experience design and happiness. Int. J. Des. **7**(3) (2013)

12. Shutterstock Images. https://www.shutterstock.com/search/pro-lock
13. Aranda, J., Ali-Hasan, N., Baig, S.: I'm just trying to survive: an ethnographic look at mobile notifications and attention management. In: Proceedings of the 18th International Conference on Human-Computer Interaction with Mobile Devices and Services Adjunct (MobileHCI '16), pp. 564–574. ACM, New York, NY, USA, (2016). https://doi.org/10.1145/2957265.2957274
14. Shah S. (2013, Oct).ThePreciseArtofMobilePushNotifications. Retrieved on Dec 21, 2019. https://techcrunch.com/2013/10/13/the-preciseart-of-mobile-push-notifications
15. Seidman, A.: Fixing mobile push notifications (2013). https://arielseidman.com/post/62564939335/fixing-mobile-push-notifications. Accessed 27 Dec 2019
16. Cheval, B., Tipura, E., Burra, N., Frossard, J., Chanal, J., Orsholits, D., Radel, R., Boisgontier, M.P.: Avoiding sedentary behaviors requires more cortical resources than avoiding physical activity: an EEG study. Neuropsychologia **119**, 68 (2018). https://doi.org/10.1016/j.neuropsychologia.2018.07.029
17. University of British Columbia: Hardwired for laziness? Tests show the human brain must work hard to avoid sloth. Science Daily (2018). www.sciencedaily.com/releases/2018/09/180918090849.htm. Accessed 14 May 2020
18. Van den Eijnden, R.J., Meerkerk, G.J., Vermulst, A.A., Spijkerman, R., Engels, R.C.: Online communication, compulsive Internet use, and psychosocial wellbeing among adolescents: a longitudinal study. Devel. Psychol. **44**, 655e665 (2008). https://doi.org/10.1037/0012-1649.44.3.655
19. To, A., Ali, S., Kaufman, G.F., Hammer, J.: Integrating Curiosity and Uncertainty in Game Design. In Digra/fdg (2016)
20. Icon made by Freepik, Pixel Perfect, Goodware, Vitaly Gorbachev, Those Icons, Noomtah, Becris from www.flaticon.com
21. Dobelli, R.: The art of thinking clearly. Hachette UK (2013)
22. Schüll, N.D.: Addiction by design: machine gambling in Las Vegas. Princeton University Press (2014)
23. Raskin, A., Harris, T., Schüll, N.D.: Your undivided attention podcast episode 1: what happened in Vegas [audio podcast] (2019). https://humanetech.com/podcast/
24. Portwood-Stacer, L.: Care Work and the Stakes of Social Media Refusal (2014). https://www.newcriticals.com/care-work-and-thestakes-of-social-media-refusal/print. Accessed 17 May 2020

Part III
Design Aesthetics, Semiotics, Semantics

Chapter 44
Visual Features of Ethnic Handloom Products for Retention of the Unique Traditional Signatures Along with Detection of Authenticity

Chirapriya Mondal and Sougata Karmakar

Abstract Indian handlooms are rich in their traditional styles and popular across the world. Each handloom technique has its unique ethnic features. However, the ethnic features are constantly undergoing change due to acculturation, innovation, and various other factors, which can significantly affect the authenticity of the products, and thereby consumers' purchase intention. The present research was carried out to identify the ethnic features and judging the authenticity of handloom products from handloom clusters of Assam through visual analysis. A Mixed methodology was adopted by incorporating a critical review of literature, in-depth ethnographic study, and photographic documentation. Images of handloom products were collected from ethnic clusters of Assam, namely Sualkuchi, Bijoynagar, Chaygaon, and Boko. Semiotic analysis of the collected 27 images (stimuli) was performed by expert designers (n = 20) to achieve the goal of the current research. The outcome of this study was a detailed matrix of smallest elements and their arrangement that form the syntactic analysis. Details in terms of influence or source of inspiration were studied in syntagmatic analysis. It was found that the motifs were traditionally inspired by naturally available flora and fauna, culturally significant elements, and mythological characters. Presented analysis technique can be referred by design practitioners, artisans, and researchers to identify ethnic features of handloom products based on traditional signatures during their design ideation. Identified visual design features would facilitate the retention of traditional inspiration during the process of product development and diversification.

44.1 Introduction

Indian textile industry, including handloom sector, contributes 7% value of the industry output, 2% of India's GDP and 15% of the country export earnings [1]. Handloom sector is the second largest employment provider to the rural India [2].

C. Mondal · S. Karmakar (✉)
Indian Institute of Technology Guwahati, Guwahati 781039, India
e-mail: karmakar.sougata@gmail.com

A. Chakrabarti et al. (eds.), *Design for Tomorrow—Volume 1*, Smart Innovation, Systems and Technologies 221, https://doi.org/10.1007/978-981-16-0041-8_44

It demonstrates the richness and diversity of Indian culture where each handloom technique has its unique ethnic features [2]. Ethnic features are the derivatives of the culture of a community. These ethnic features can be identified through sensorial experiences like the look, touch, smell, etc. Ethnic features are embedded in handloom products as design features like motif, pattern, color, texture, etc. Consumers of handloom product connect with these ethnic design features and develops an emotional affinity toward them. Wang et al. mentioned that the emotional aspects of a product create an effective resonance between the consumer's emotion and the product [3]. This effective resonance influence consumer's purchase behavior toward any product. Literature study reveals that there is a significant need to develop a design guideline for product innovation which will reflect such congruence. This guideline should incorporate scope for reflecting users' identities and sensitivity toward their cultural values represented through design features [4].

Design as a discipline has played the dual role of a mirror and an agent of change due to which its association with culture has seen many twists and turns [5]. Users are sociocultural beings, so products cannot be developed in cultural isolation. Design features must be extracted from target user's culture [6]. Thus, users should not be considered as just physical but as biological beings [7]. Literature study also reveals that designers are yet to explore the cultural phenomenon to the extent where they can imbibe the cultural phenomenon into products with an efficiency that has been achieved for cognitive and physical factors [7]. So, during product innovation, significant importance should be given to the user's perspective toward its sociocultural dimension [8].

Culture and ethnicity are interrelated and often used interchangeably. Culture is itself an intrinsic component of ethnicity. Bell [9] explains ethnicity as one of the most significant factors to define any self-ascribed ethnic group. Further research [10] shows that ethnic identity is a significant predictor of cultural values. Various studies explain the methods to identify and extract the design features [3]. Some other models for understanding consumer emotions are Kansei Engineering and SOR Model. Kansei is a Japanese term applied to express one's impression of the artifact, situation, and surroundings [11]. It is also identified through literature that sensory analysis is a significant method to understand human senses. Zoecklein [12] explained sensory analysis/evaluation as a scientific discipline that applies empirical methods to evaluate consumer goods. Human senses like smell, touch, vision, taste, and touch can be analyzed through experimental designs and statistical analysis. From the literature review, it can be summarized that cultural traits play a significant role in the creation of identity and purchase intention of a product. It also reflects the lack of researches that has been done in this domain. A number of researches have also highlighted the theoretical model where attempts have been made to develop products with cultural traits. It has also been found that ethnicity and cultural traits are correlated. However, there may be distinctions between culture and ethnicity based on how we define boundaries between groups and subgroups [9].

The rationale of the present research is the extraction of design features from traditional products (handloom textiles) to form a cultural memory on consumers. Based on the literature study, a model, as shown in Fig. 44.1, has been developed

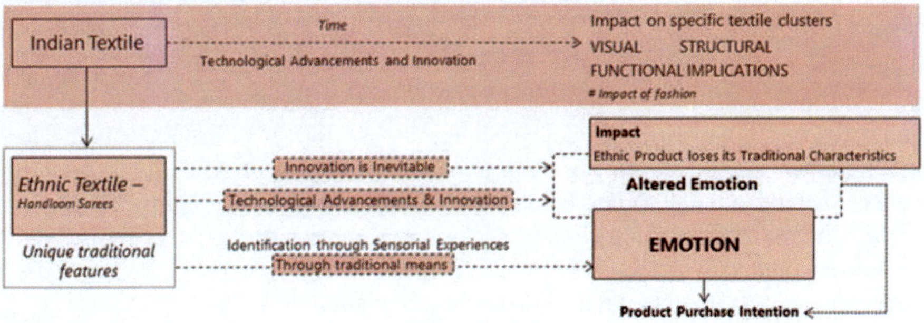

Fig. 44.1 Significance of extraction of design features from traditional products to form a cultural memory on consumers

to show how ethnic product features are altered by various influencing factors and thereby affect customer's emotion toward purchase decision.

Ethnic consumer's internal motivations and attributions are related to the consumption of cultural products [13]. Visual elements play an indispensable role in identifying ethnic products. Visual analysis can be extended in developing products to incorporate ethnic features and consequently influence consumer behavior positively for the people who are inclined toward ethnic products. The growth of global trade and the evolution of global consumer segments have increased awareness of and interest in the effects of ethnic product labels on consumer evaluations [14]. In line with common attitude models [15], a consumer can be favorable or have a positive attitude toward a foreign product but still may not buy it due to normative reasons such as a feeling that purchasing foreign products is immoral. Due to globalization, people have access to different kinds of products, wherein one can choose between ethnic origin and products from other regions. Researchers have shown that globalization provides a cultural identity [16] and maybe one of the major factors that influence the consumption behaviors [17].

Stakeholders of handloom textile products like artisans, weavers, consumers/customers, sellers, design practitioners, researchers, govt. experts (officials of Development Commissioner of Handloom/Handicrafts, Govt. of India and patent attorneys) dealing with innovation and promotion of handloom and others rely on these features during the identification of traditional or ethnic products. However, with globalization, these features are getting altered due to the influence of modernization, change in customer profiles, and the advancement of technologies as well as manufacturing processes. Consequently, it is difficult to judge the authenticity of ethnic handloom products both by Indian and foreign customers. It is imperative to study and identify ethnic handloom products' visual features for the retention of their unique traditional signatures and the detection of authenticity. Among the numerous Indian ethnic handlooms, the current paper deals with 'handloom products made from Assam Silk' as the case study to demonstrate how to identify the design features through semiotic analysis, to facilitate retention of traditional inspiration during the process of product development and diversification.

44.2 Methodology

A mixed methodology was adopted in the current research. Present research was conducted through a critical review of literature, in-depth ethnographic studies, and photographic documentation. Images of original samples of Assam silk products were collected from Sualkuchi, Bijoynagar, Chaygaon, and Boko. These are the four major Assam Silk producing clusters of Assam, a northeastern state of India.

Semiotic analysis of the collected 27 images (stimuli) was performed by expert designers ($n = 20$) to achieve the current research's goal.

44.2.1 Data Collection

Data collection in the form of a high-resolution photographic image of handloom products was collected from four ethnic clusters of Assam. All the clusters are situated in the lower Assam area with inhabitants belonging to same ethnic groups. The chosen clusters have been practicing the craft for many years and are recognized as mega or major clusters by Development Commissioner (DC) Handicrafts. Artisans (craft practitioners) are recognized by govt. and issued artisan cards in recognition. Artisans with an experience of more than 12 years were identified. In-person visitation was made to collect stimuli in the form of photographic images of the handloom products. During data collection, not more than three stimuli were collected from a single artisan. It was also carefully ensured through in-person interaction that the images collected were of their own creative ideation and not given by any commercial organization. This was done to ensure variation in the products and their style. In all, 73 stimuli were collected.

The collected stimuli were then edited (cropped) to fit in the form of cards for further analysis. A focus group of 20 participants for semiotic analysis was formed. Participants for the semiotic analysis were experienced designers with at least a degree in design from a recognized institution and an experience of at least 5 years in the domain of textiles and handloom product development. Stimuli were randomly selected from the complete pool of 73. It was ensured that one participant, at least, performs one analysis of the stimuli.

44.2.2 Data Analysis

Data analysis by an individual participant was performed in an a structured manner as represented in Fig. 44.2. In Stage 1, the image was considered as a complex structure comprising various elements. The target of this stage was to identify the elements which cannot be segregated or broken further. If the elements are further broken or segregated, they will lose their original structure and identity. The identification

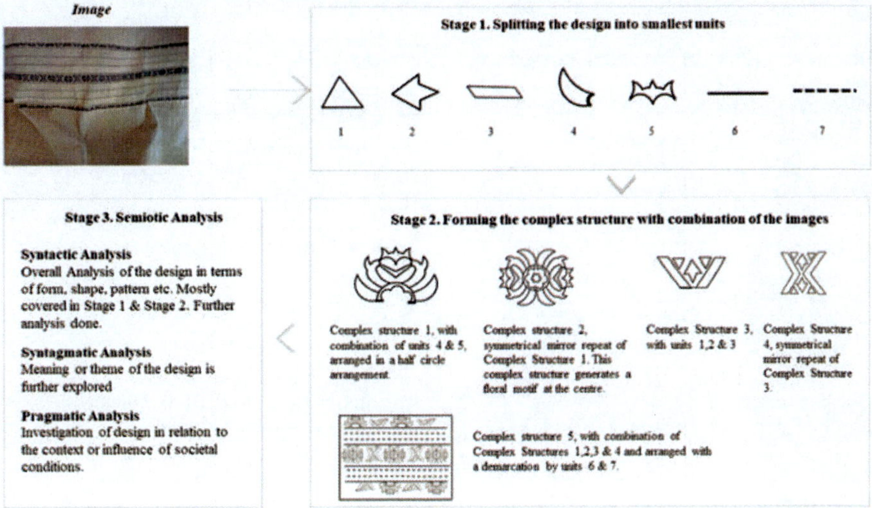

Fig. 44.2 Sample analysis through semiotic method: image of an ethnic handloom saree

is manually done by enlarging the complex structures and tracing the elements in vector graphics editor application of the expert user's suitability. As an example, the elements extracted are shown in Fig. 44.2. Stage 1. In Stage 2, the complex arrangement with the smaller units was analyzed. The aim of this stage was to identify the way elements were grouped or arranged together to form the complex pattern. The study of arrangement for multiple complex structures gave a pattern in arrangements of the smallest elements. This forms the syntactic analysis. Subsequently, in Stage 3, the theme or meaning of the complex structures was explored, forming the syntagmatic analysis. The influence of societal structures was explored to identify the possible influences of surrounding including flora, fauna, architecture, etc.

After completion of the analysis by the participants, the analyzed data was further arranged in tabular fashion (Table 44.1). Each analysis by an individual took at least 2.5 days of time. In all, 20 participants analyzed 27 stimuli collected from the handloom clusters of Assam.

44.3 Observation

The outcome of this study was a detailed matrix of syntactic, syntagmatic, and paradigmatic analysis. The visceral elements of the aforesaid handloom products are tabulated [18]. The smaller units, along with their individual characteristics, form the

Table 44.1 Analysis of images

Image	Structural arrangement	Syntactic	Syntagmatic
	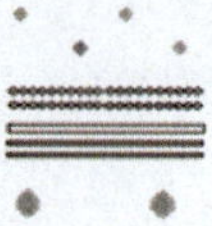	Repetitive shapes in the horizontal direction; geometric shapes; separation of shapes with lines	Inspiration from local flora; abstraction due to technical constraints
	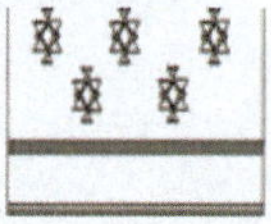	Triangle and square repeated in a structured manner	Muted color story; inspiration from summer
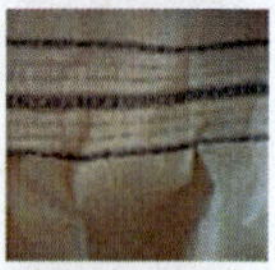		Mirrored pattern; separation of spaces through lines	Similarities with Jamdani motifs
	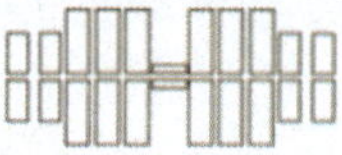	Geometric structure in horizontal and vertical repeats	Exploration under technical constraints
		Repetitive patterns; distinct space segregation	Inspiration from natural flora and fauna; peacock; cultural element like Jhapi
		Geometric shapes; rectangle and square; repetition	Exploration under technical constraints
		Repetitive pattern in a structured arrangement; floral and geometric	Local floral influence; cultural motifs
		The repetitive pattern in a structured arrangement	Inspiration from local flora and other cultural elements

(continued)

Table 44.1 (continued)

Image	Structural arrangement	Syntactic	Syntagmatic
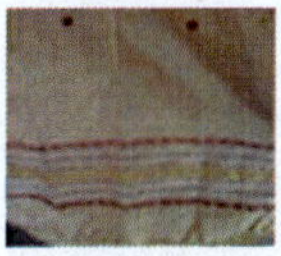		Geometric and organic shapes in a structured arrangement; the distinction of space with horizontal lines	Inspiration from basic geometric structures; floral motifs with an abstraction; muted color tones
	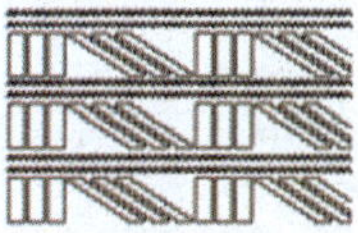	Repetitive horizontal patterns. Overall pattern symmetry; the distinction of space with horizontal lines	Inspiration from basic geometric structures; the play of colors inspired by local flora
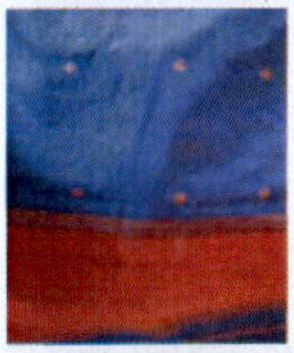	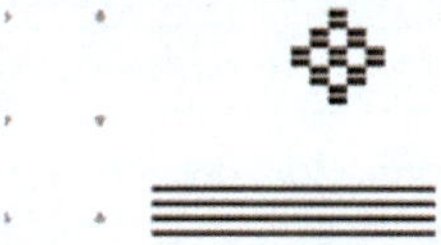	Geometric patterns; structured organization; vertical and horizontal repeats; horizontal lines segregate the space	Symmetry; muted tones; inspiration from existing range of products
		The complex structure repeats to create a larger pattern	Modern material usage; synthetic threads used instead of traditional gold threads
		Horizontal lines segregate the space Motifs are a geometric abstraction	Influence from other cultural objects; inspiration from across other objects
		Horizontal lines segregate the space; floral motifs with geometric abstractions	Influence of local flora and fauna; red and white color combination
		Structured horizontal and vertical repeats with geometric and abstracted floral motifs	Influence of local flora and fauna; dominant pattern; religious symbolism

(continued)

Table 44.1 (continued)

Image	Structural arrangement	Syntactic	Syntagmatic
		Structured horizontal and vertical repeats with geometric and abstracted floral motifs	The patterns are strongly influenced by local flora and fauna; color combination is on muted tones
	 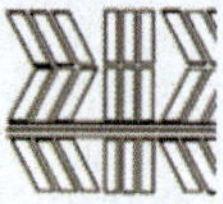	Structured horizontal and vertical repeats with geometric and abstracted floral	Inspiration from nature, local flora; abstracted waves; muted multicolors
	 	Structured horizontal and vertical repeats; geometric; organic and geometric abstractions	Motifs are inspired from nature; Kalki; the most dominant motifs used in traditional
		Structured horizontal and vertical repeats with geometric patterns	Inspired from traditional customs; e.g., Jhapi motifs
	 	Geometric abstraction; spaces segregated with horizontal lines	Inspired from nature; e.g.,. Kalki motifs; resemblance with a local flower
	 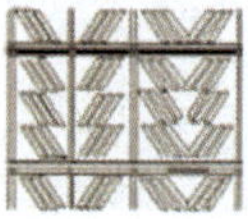	The spaces are segregated with horizontal lines Organic shapes; geometric abstractions	Inspired from nature; resemblance with bamboo; local floral motif
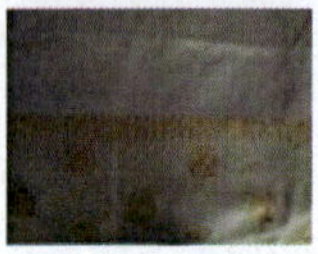	 	Spaces are segregated with horizontal lines Organic shapes are modified with geometric abstractions	Motifs are inspired by nature; e.g., Kalki motifs

(continued)

Table 44.1 (continued)

Image	Structural arrangement	Syntactic	Syntagmatic
		The spaces are segregated with horizontal lines Geometric shapes are repeated to form symmetrical patterns	Motifs are highly influenced by their traditional customs, strong resemblance with other customary and religious elements
		Space segregation through lines; motifs used are lion, peacock, Jhapi, and butterfly	Influence of traditional customs; resembles natural flora and fauna; auspicious white and red
		Spaces are segregated with horizontal lines Kingkhap, along with ornamental motifs, are used	Influenced by traditional customs. E.g., Use of Kingkhap motif; ornamental style
		Spaces are segregated with horizontal lines; geometric abstraction	Influenced by traditional customs; strong resemblance with customary and religious elements
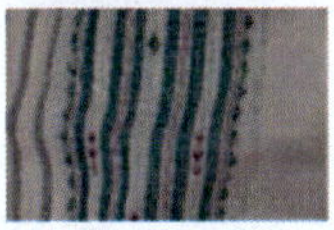	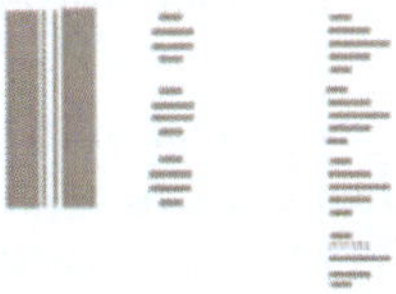	The spaces are segregated with horizontal lines; geometric abstraction	Influenced by traditional customs; strong resemblance can be found in other customary and religious elements; muted color tones

complex structures to give a combined effect were generated by their interrelationships [19]. Through the semantic approach, it was observed that the design inspirations were influenced by the surroundings and abstracted considering the technical constraints. In other words, the designs convey meaning and based on a theme [18].

It was found through the analysis that the motifs are inspired by naturally available flora and fauna, culturally significant elements, and mythological characters. The motifs were geometrically abstracted to suit their technical parameters and subsequent development. The space of the entire product was segregated into various sections which were usually demarcated by (a) change in color of the section (b) a border with thick line may with a different color or thread or (c) both, with a change in color and as well as a border. Motifs in any product are: (a) either geometric (b) organic with geometric abstraction (c) organic or (d) with a combination of (a), (b), and (c). Few cultural elements (Jhapi, Kingkhap, Pepa, etc.) were found to be consistently present across all the handloom products. Color application was limited to two to a maximum of 7–8 colors, with major usage of colors can be found in the range of 4–5 colors. However, there were instances where the number of colors exceeds 7–8 colors. The range of colors applied was wide and difficult to define in a spectrum. A major factor in the application of color was its derivation from nature. In majority of the instances, the color tones were found to be muted and subtle. Alignment of the motifs was found predominantly in horizontal or vertical, rarely in diagonal direction during pattern making. The symmetrical or mirrored pattern was significant in the designs. Exploration of similar motifs in scalable patterns was also found in various designs.

44.4 Discussion

Visual strength is a critical component of brand strategy [20]. Visual stimuli can help in building a strong affinity toward a brand [21]. However, there is very little literature available on the influence of design on the meaning of the product [20]. In the context of Indian handloom products, visual stimuli create their identity, and the literature regarding its design is minimal. As mentioned by [22], the cultural influences, i.e., the visual stimuli, will have an impact on the product and the development process.

A lot of emphasis has been given on following cultural designs and developing a design framework. Among them, [23] showed a detailed design development process of developing cultural products with an example of 'The Linnak (twin cup).' In the Indian context, few works like: [24] focused on the plight of handloom and emphasized its revival through automation which itself might have a negative impact on its traditional styles; [25] documented the dye yielding plants of Assam; an analysis of visual elements in ethnic products is also carried out [26] and issues pertaining to authenticity in handloom products is highlighted with Banarasi sarees as a case study [27]. Policies like assigning geographical indications are also not adequate and have a lot of inherent flaws [27]. However, literary sources could not reveal a practical guideline that is implemented for developing a cultural product by extracting ethnic, cultural features in the Indian context.

Designers follow various design guidelines like Promostyl, WGSN, f-trend, to name a few for international markets apart from individual manuals provided by clients. Major components for any design guidelines [28], 29 cover materials that include surface texture; color; designs that can be incorporated on the surface through various techniques like printing, weaving, etc., form of the product. NIFT, a premier design institute in India, has taken the initiative to develop a design guideline [30], especially in India's context. However, the initiative is still at a nascent stage.

The detailed analysis carried out in this research can be extracted in the form of design inspiration, especially for motifs, an arrangement of the motifs and colors used in the product. Design practitioners, artisans, and researchers can identify and refer these features during design ideation. This process, wherein the inspiration is restricted to ethnicity, ensures its visual authenticity. Products developed through the combinations of these design features will have the unique traditional signatures which can be identified visually. The adopted methodology for the ethnographic study and subsequent semiotic analysis can be replicated for other ethnic handloom products. The identified visual design features can be sources of inspiration for product diversification and help retain a cultural memory.

As the present study was focused on analyzing the visceral features of handloom designs of a specific location through the semiotic method, other sensory criteria like touch, smell, auditory, and taste sensations were out of the scope.

The outcome of this study can be replicated for other locations as well as considering other indigenous practices to retain their ethnic identity. It will also add significantly to the present practices of assigning geographical indications to particular products.

Acknowledgements Authors acknowledge the support of TATA Trusts, Sualkuchi Institute of Fashion Technology, all the craft persons and experts involved during the data collection.

References

1. Ministry of Textiles, Go. I: Annual report 2017–18. PIB, New Delhi (2018)
2. Limited, N. H. D. C.: Report on Market Research for Promotion of India Handloom Brand. O/o D C Handlooms. New Delhi (2016)
3. Wang, Y.H., Qin, S.F., Harrison, D.: Culture-inspired design principles, methods and tools in current products, pp. 1–12. Tokyo, SIT (2013)
4. Moalosi, R., Setlhatlhanyo, K.N.: Cultural memory, an asset for design-driven innovation within the creative industries sector: lessons for design education. Design Technol. Educ. Int. J. (2016)
5. Moalosi, R.: The Impact of socio-cultural factors upon human-centred design in Botswana. Thesis (Ph.D.). School of Design, Queensland University of Technology (2007)
6. Margolin, V.: The politics of artificial, 1st edn. University of Chicago Press, Chicago (2002)
7. Baxter, S.: Deep design and the Engineer conscience: a global primer for design education. In: Crossing Design Boundaries, pp. 366–373 (2005)
8. Moalosi, R., Popovic, V., Hickling-Hudson, A.: Culture-orientated product design. Int. J. Technol. Design Educ. 175–190 (2010)
9. Bell, E.: Wiley Online Library. https://onlinelibrary.wiley.com/doi/abs/10.1002/9781118663202.wberen578. Accessed 13 June 2020
10. Desmet, K., Ortuño-Ortín, I., Wacziarg, R.: National Bureau of Economic Research (2020). https://www.nber.org/papers/w20989. Accessed 12 June 2020
11. Lokman, A.M.: Design & emotion: the Kansei engineering methodology. Malayasian J. Comput. **1**(1), 1–11 (2010)
12. Zoecklein, B.W.: Production Wine Analysis, 1st ed. Springer, s.l. (1990)
13. Neal, G.S.: African American womens professional dress as expression of ethnicity. J. Family Consumer Sci. (1998)
14. Fischer, P.M., Zeugner-Roth, K.P.: Disentangling country-of-origin effects: the interplay of product ethnicity, national identity, and consumer ethnocentrism. Market. Lett. **28**(2), 189–204 (2017)
15. Ajzen, I.: Belief, Attitude, Intention, and Behavior: An Introduction to Theory and Research (1975). https://people.umass.edu/aizen/f&a1975.html. Accessed 28 June 2020
16. Wang, Y.: Globalization enhances cultural identity. Intercultural Communication Studies XVI, pp. 83–86 (2007)

17. Jung, K., Kau, A.K.: Culture's influence on consumer behaviors: differences among ethnic groups in a multiracial Asian country. Adv. Consum. Res. **31**, 366–372 (2004)
18. Shahid, M.: Visual rhetoric of the title design in Bollywood movie posters. Thesis [Ph.D.]. Dept. of Design, IIT Guwahati, Guwahati (2019)
19. Dey, S., Bokil, P.: Sound symbolism in India comic books. In: Chakrabarti, A. (eds.) ICoRD'15—Research into Design Across Boundaries Volume 1. Smart Innovation, Systems and Technologies, vol. 34. Springer, New Delhi (2015)
20. Henderson, W.P., Cote, A.J., Leong, M.S., Schmitt, B.: Building strong brands in Asia: selecting the visual components of image to maximize brand strength. Int. J. Res. Market. 297–313 (2003)
21. Hutton, J.: The influence of brand and corporate identity on consumer behavior: a conceptual framework. J. Brand Manage. 428–439 (1997)
22. Gautam, V., Blessing, L.: Cultural influences on design process. ICED 07, 28–31 Aug 2007, Paris (2007)
23. Lin, R.T.: Transforming Taiwan aboriginal cultural features into modern product design: a case study of a cross-cultural product design model. Int. J. Design 45–53 (2007)
24. Naik, D.S., Vastrad, V.J.: Protection and revival of traditional hand embroidery, Kasuti by automation. Indian J. Tradition. Knowl. 197–203 (2007)
25. Kar, A., Borthakur, S.K.: Dye yielding plants of Assam for dyeing handloom textile products. Indian J. Tradition. Knowl. 166–171 (2008)
26. Mondal C., Karmakar S.: A study exploring the facets of visual elements in ethnic products: case study of sarees from West Bengal. In: Rebelo, F., Soares, M. (eds.) Advances in Ergonomics in Design. AHFE 2017. Advances in Intelligent Systems and Computing, vol. 588. Springer, Cham (2018)
27. Basole, A.: Authenticity, Innovation and the Geographical Indication in an Artisanal Industry: the case of the Banarasi Sari. J. World Intell. Prop. 127–149 (2015)
28. WGSN (2013), https://www.wgsn.com/en/micro/Art_Dept/2013/ffaw1415/ff_industrial_evolution_html/WGSN__Fashion_forecast_AW1415_Rendering_Reality.pdf. Accessed 27 June 2020
29. Promostyl (2020), https://promostyl.com/service-trendbook/?lang=en&cn-reloaded=1. Accessed 27 June 2020
30. The Hindu: The Hindu (2019). https://www.thehindu.com/news/cities/chennai/nift-to-study-and-forecast-indian-fashion-trends/article26274765.ece. Accessed 27 June 2020

Chapter 45
Visual Alankars: Toward a Decolonized Visual Design Framework

Raina Singh and Saurabh Tewari

Abstract The research aims to experimentally evolve visual design fundamentals by borrowing literary principles from Hindi Vyakaran (Hindi grammar system). Alankars, which are literary devices in Hindi Vyakaran, are used to enrich literary compositions through the playfulness of structure and enhanced meaning. This study methodologically explores Alankars in the context of visual design. The pragmatics of the research employs Madhubani folk-art tradition from the region of Mithila in India. Compositions using visual elements and styles of traditional Madhubani folk-art attempt to understand the structures and meanings of Shabd Alankar and Artha Alankar in the research. The study highlights the potential of indigenous knowledge systems to offer a methodological lens for visual design research.

45.1 Introduction

Graphic design education in postcolonial geographies, including India, is heavily inclined toward Bauhausian and Swiss schools of thought which are philosophically rooted in a western worldview. In contemporary discourse of design and its pedagogy, there have been only a handful of attempts, like by Asian Designers [1] and Asian Design paradigm [2], to explore alternative frameworks and understand the visual design culture through a non-western lens. This study underlines that indigenous knowledge can be instrumental in bridging a gap that exists in learning design and its principles.

R. Singh (✉) · S. Tewari
School of Planning and Architecture, Bhopal, India
e-mail: rainasingh2605@gmail.com

S. Tewari
e-mail: saurabh@spabhopal.ac.in

A. Chakrabarti et al. (eds.), *Design for Tomorrow—Volume 1*, Smart Innovation, Systems and Technologies 221, https://doi.org/10.1007/978-981-16-0041-8_45

45.2 Aim and Objectives

The paper aims to highlight a link between Alankars, a grammatical aspect of Hindi-Sanskrit language system, and structured approaches in graphic design to evolve a new visual design principle system. The primary objective of this exploration is how Alankars have a potential to develop as a visual design principle that can be studied, explored and spread across for the learning and interpretations in plural contexts of visual communication (VC). There is a vast literature on visual/design/computation grammar systems. However, the scope of the study here is to experiment the possible transliterations. Further, it is an attempt that underlines a decolonized method of reading VC and highlights the importance of intellectual wealth rooted in the Indian knowledge systems and cultures.

45.3 Conceptual Framework: Syntactics, Semantics, and Pragmatics

Since the focus of this study is to develop a visual design principle system by including the visual grammar of compositions, it works at three levels: **syntactics, semantics, and pragmatics**. Syntactics and semantics work at the visual element and composition level, and pragmatics defines the visual language and style. In this study, Alankars form the syntax and semantics of visual design, with Madhubani art, a folk-art tradition from the northern Bihar in India providing the pragmatics. The details of the elements of the conceptual framework are as follows.

45.3.1 Alankars

In Hindi Vyakaran [3] (Hindi grammar system) and literary tradition, Alankars, or grammatical ornaments, are a crucial literary device. As ornamental figures of speech, Alankars enrich literary compositions through playfulness in sentence structure or accentuated meaning. There are two forms of Alankars, Shabd (word) and Arth (meaning), with a variety of types and subtypes. The types of Shabd Alankar are *Anupras (alliteration), Yamak (homonym), and Slesh (pun).* The types of Artha Alankar are *Upma (simile), Rupak (metaphor), Utpreksha (imagining the other), Atishyokti (hyperbole), and Manvikaran (personification).* The presence of Alankar [4] has historically been there in the many compositions of Hindi literature in various dialects including Awadhi, Braj Bhasha, Khadi Boli, Bundeli, and Bagheli. Many medieval and modern poets have expressed their creations with Alankars as embedded ornaments in their structure and meaning, both (Fig. 45.1).

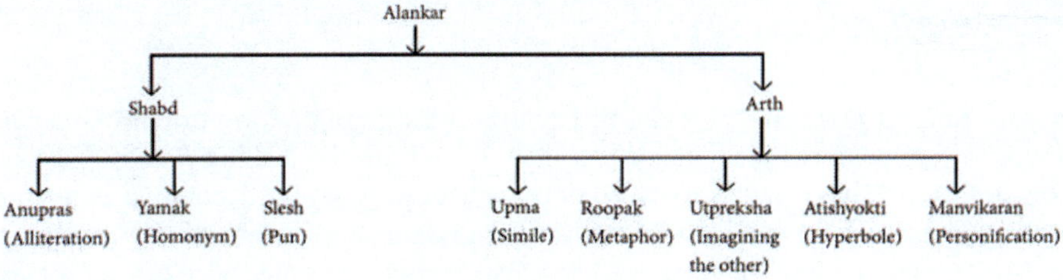

Fig. 45.1 Alankar and its types. *Source* Author

Fig. 45.2 Madhubani tree of life. The form, proportions, and details are organic as they used to be in early times with a heavy border. *Source* Author

45.3.2 Syntactic and Semantics

Traditionally in a visual design framework, syntactic relates to the physical elements, order and structure of the visual design [5]. On the other hand, semantics refers to the object in composition (signifier) and what they imply (signified). The two types of meta-Alankar, Shabd (word) and Artha (meaning), conceptually relate to syntactic and semantics, in their structure and meaning.

45.3.3 Madhubani Art

Madhubani art is a folk-art tradition from the Mithila region in northern Bihar in India.[6] From ancient times, majorly women have been the leading practitioners of this art form. These paintings were done on house walls as decorative elements during the time of weddings and festivities. Traditionally, there are mainly five types of Madhubani paintings that are practised. Every artisan has their way of visualizing it in different mediums as Bharni, Kachni, Tantrik, Godna, and Kohbar [6], which are the names given to these types of Madhubani style [7]. These styles are distinct in their appearance with recognizable features [8] (Fig. 45.2).

45.3.4 Pragmatics

Pragmatics relates to the interpretation of syntactic and semantics in a particular form and context. There are various methods to employ pragmatics, such as basic design from the Bauhaus tradition has employed basic shapes like a square, a circle, and a triangle as pragmatics in visual design exploration. Another example: visual design schools of thought such as the Basel School experimented with typography and its elements as pragmatics combined with syntactic and semantics to come up with a visual design language.

45.4 Methodological Framework: Explorations Through Transliteration

The methodological approach involves transliterating the syntactic and semantics of literary Alankars into the visual form of Madhubani art, i.e., composing Madhubani artwork illustrating both the forms of Alankars, Shabd, and Arth; the visual elements and styles of Madhubani forms as the pragmatics, through which artwork drawing explorations are made. The authors transliterate the composition structures and meaning in Shabd Alankar and Artha Alankar examples using subjective hermeneutics and skills. Shabd Alankar types: *Anupras (alliteration) and Yamak (homonym)* and an Artha Alankar type: *Manvikaran (personification)* transliterate into visual compositions along with their vernacular examples. However, within the scope of the paper, only three forms are illustrated.

45.4.1 Motif Library

The Alankars and Madhubani paintings have a lot in common as both the practices are decorative. As ornaments, they enhance the beauty of the composition. Importantly,

Fig. 45.3 Motif library. *Source* Author

Madhubani paintings are composed of several elements on contextual themes. They have their meaning at composition level as well as its components have a role at the individual level (Fig. 45.3).

To pursue the experiment of creating a visual language for Alankars, a library of motifs was created. The idea was to have a sufficient range of elements to develop further compositions to explain Alankars. The artworks were hand-drawn in an 8-in. square area with black color and then digitized for documentation. The motif's outlines are thick, the details inside are intricate, and no colors are used. The motifs created include natural objects like the sun and moon; fishes animals (deer, cows, elephants); birds (peacocks, parrots); trees; flowers (lotus); human figures and faces; geometric patterns (like lines, dots, and basic shapes with organic edges. All compositions are decorated with intricate borders as they play an essential role in this art form.

45.4.2 Transliteration Method

A famous phrase in Hindi, "रीझि रीझि रहसि रहसि हँसि हँसि उठै (ReeJhiReeJhiRaHaSiRaHaSiHanSiHanSiUThai)" is reconstructed here. It is to illustrate the method of transliteration from Form to Motif to Theme. Table 45.1 explains it further.

Table 45.1 Steps involved in transliteration. Source: Author

Step	Action	हिंदी *Transliteration*	हिंदी *Translation* Meaning Purpose
1	The basic shape, organic or geometric, which is the building block, is drawn with an outline	र R	व्यंजन *Vyanjan* Consonants Shape
2	The basic shape is added with patterns and/or colors. These intricate details not just enhance the appearance but also give it a legible form	री Ree	स्वर *Swar* Vowel Patterns/Textures
3	Similarly, another basic shape embedded with steps 1 and 2 is introduced	झि Jhi	
4	These decorated forms combine and make a motif	रीझि ReeJhi	शब्द *Shabd* Word Elements
5	These motifs are then combined and placed in such a way with visual balance creating harmony among all. This harmonized composition follows a particular theme	रीझि रीझि रहसि रहसि हँसि हँसि उठै ReeJhi ReeJhi RaHaSi RaHaSi HanSi HanSi UThai	वाक्य Vakya Sentence/Phrase Composition

The stepwise construction is followed by the reconstruction of visual theme through form and motif, as shown in Fig. 45.4. In Madhubani art, there are different themes which have specific motifs from a particular color palette. The placement of motifs plays a critical role in conveying a story. The explorations here are hand-drawn and then further optically corrected on a photograph editing software. The construction of the visuals here is done with the purpose to make readers understand and interpret the visuals appropriately.

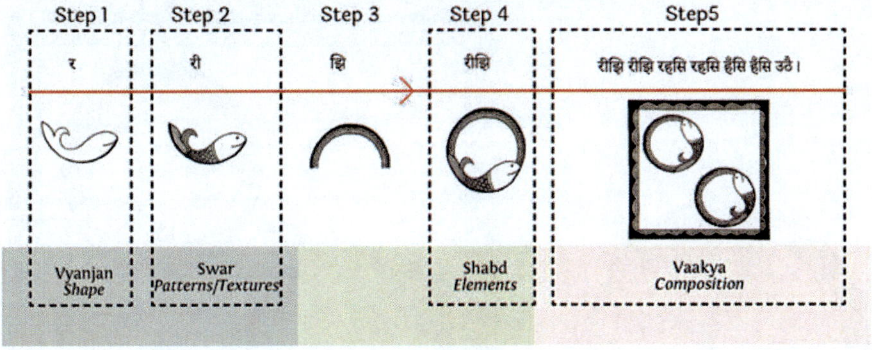

Fig. 45.4 Three stages of construction. *Source* Author

45.5 Explorations

A phrase each from three Alankars, Anupras, Yamak, and Manvikaran, is illustrated through the above-mentioned method to explain the transliteration of a literary to visual composition.

45.5.1 Anupras Alankar

In the Shabd Alankar, the accentuation happens due to the playful placement of letters and words in a phrase. One of the types of Shabd Alankar, Anupras is taken for illustration. Anupras Alankar is a form of alliteration which is primary identity, involving the repetition of the letters to enhance the poetic value.

"चारू चन्द्र की चंचल किरने, खेल रही थी जल थल में ।"

Transliteration: ChaRu ChanDra Kee ChanChaL KiRaNe, KheL RaHi Thi JaL ThaL Mein

One can see the repetition of 'Cha' in the phrase above. Its repeated presence in different words makes the phrase poetic. Similarly, the visual transliteration and interpretation of Anupras would involve the repetition of a shape (geometric or organic). One of the subtypes of AnuprasAlankar, 'Chhek-AnuprasAlankar', is taken for transliteration here.

"रीझि रीझि रहसि रहसि हँसि हँसि उठै"

Transliteration: ReeJhiReeJhiRaHaSiRaHaSiHanSiHanSiUThai

Here, the accentuation is created using the repetition 'ReeJhi' and 'RaHaSi' as words. The repetition as per Chhek-AnuprasAlankar in the visual composition can be illustrated with a repeat of the motif. As shown in Fig. 45.5, the meaning of the motif, its visual details and colors, is precisely the same when it repeats.

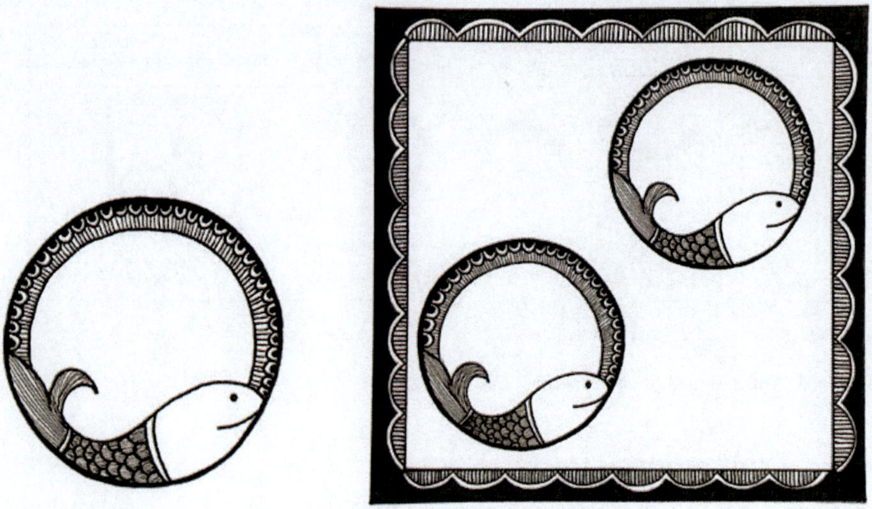

Fig. 45.5 Motif and the composition in Chhek-Anupras. *Source* Author

In this composition (Fig. 45.4), the motif contains two different forms, a geometric circle and an organic fish, that repeat only once. This whole composition is considered similar to a sentence, and the elements are different words coming together in harmony, making a balanced visual composition.

45.5.2 *Yamak Alankar*

One of the other types of Shabd Alankar, Yamak, exhibits the beauty of structure (syntax) and meaning (semantics) in its composition. The structural accentuation lies in two same words, often a homonym, employed at different parts in the phrase. The poetic beauty lies in the wordplay within the sentence, which has its deliberate meaning.

"तीन बेर खाती थीं ते तीन बेर खाती हैं।".

Transliteration: TeeNBeRKhaTiThiTeTeeN Ber KhaTi Hain.

The three words, 'TeeN', 'BeR', and 'KhaTi', are repeated in the latter part of the phrase, with the term, 'BeR', acting as the homonym. The first 'BeR' conveys the number of times, and the second expresses the fruit. As the visual interpretation for Yamak Alankar is a form repeated with different meaning with each time of its occurrence in its visual transliteration, one shape can further make multiple forms, when illustrated with different other shapes and forms. Its placement changes its meaning in all the contexts. Like here the same configuration is combined with a bird to make a bird motif, and then it repeats in water to show itself as a fish (Fig. 45.6).

Fig. 45.6 Motif and the composition in Yamak Alankar. *Source* Author

45.5.3 ManvikaranAlankar

On the other hand, in Arth Alankar, the accentuation happens due to the playful meanings of the words in a phrase. Now to explain the Arth Alankar (semantic) below is a subtype example of Manvikaran Alankar, where a sentence expresses personification.

"फूल हँसे कलियाँ मुसकाईं"।

Transliteration: PhooL HanSe KaLiYaan MusKaaYi

In the above phrase, 'PhooL' (flowers) are 'laughing', and 'KaLiYan' (buds) are 'smiling', both human expressions. To explain it through visuals here is another example of Manvikaran Alankar. In this example, a human face and its features are being used on a peacock body to personify it. So we can say that two forms (A peacock and a human face) are combined to form a motif here and placed inside a border forming a composition (Fig. 45.7).

45.6 Discussion

It was challenging to represent a text and its ornamental purpose through a visual throughout the study. Notably, in the Artha Alankar, where the meaning and its interpretation matters, the challenge was to represent the slight differences in visuals unlike the linear structures of/in the text. However, after the resolution of basic formation for visual principles, the artworks communicated their intended meaning. For example, there are minute differences in the semantics and syntactic aspects

Fig. 45.7 Motif and the composition in Manvikaran Alankar. *Source* Author

in the subtypes of Artha Alankar: Upma, Roopak and Utpreksha. There can be a confusion for readers in visuals due to the extreme similarity in physical appearance or characteristics of the two elements. In Upma Alankar, a leaf motif compares with a fish due to physical shape. In Roopak Alankar, the leaf claims to be a fish because of the extreme visual similarity with a fish. On the other hand, in Utpreksha Alankar, it is expected on the viewers' imagination to imagine a leaf-like fish.

Reflecting the process: Once the basic formation for visuals was concluded, the artworks were comfortable to express their sole meaning. The creative freedom for the artists is the unique part of this art form which allows the artisans to express their thoughts in any style of it, so we chose this art form because it allowed exploring Alankars with more freedom.

45.7 Epilogue

This study attempted to construct a visual grammar/visual composition framework through consciously working on syntax, semantics, and pragmatics of visual design. The first two components, syntax and semantics, were explored as Alankars while latter, pragmatics, was employed to assist comprehension of composition through contextual understanding, here Madhubani art. The pragmatics further allows exploring the subjectivity through the details in the artwork, reminiscent of the artistic freedom in the Madhubani, and other Indian art traditions. The research highlights the potential of indigenous knowledge to offer a methodological lens for visual design. This indigenous knowledge exploration can be instrumental in bridging the gap that exists in learning design and its principles in our cultural context and environment. Further, it has the potential to develop decolonised frameworks of visual design and its principles.

References

1. Kohei, S.: Books, Letterforms and Design in Asia: Sugiura Kohei in Conversation with Asian Designers. ADARG and Marg Foundation (2014)
2. Trivedi, K.: The Way of Asian Design. Nanyang Technological University (2010)
3. Sathyarthe, K., Gupt, R.P., Prakash, D.: Manak Hindi Vyakaran Evam Rachana Class 9 and 10. New Saraswati House (2018)
4. Winternitz, M.: History of Indian Literature. Motilal Banarasidas (1985)
5. Morris, C.: Foundations of the Theory of Signs. University of Chicago Press (1938)
6. Thakur, U.: Madhubani painting. Abhiinav Publications (2003)
7. Tejal. 5 Distinctive Styles of Madhubani Painting
8. Gambhir, S.: Madhubani paintings of Bihar. June **3** (2011)

Chapter 46
Typeface Design for Cultural Identity—An Exploration with Assamese Typeface Design and Its Future Scope

Abhijit Padun and Amarendra Kumar Das

Abstract The identity of culture also comes from its language. In India, which is known for its diversity in culture, each language that exists establishes a cultural identity. In this paper, an exploration conducted on reviving the identity of the Assamese script which represents Assamese language and culture of North-East India has been described. The exploration outlines initially the need, the resources, the applicability, and the design considerations. Then, further, it proposes a design alternative of the Assamese script in the form of a Typeface which may address the need to bring back the original essence of the script that was present in its historical form. The very idea of this exercise is to establish an identity for Assamese culture and language in the digital world with the help of its script which could be represented by a new Typeface that addresses the originality and tradition with historical essence.

46.1 Introduction

A language narrates the essence of a culture that represents its identity [1]. In a country like India which is known for its diversity in culture, each language that exists establishes a cultural identity, be it Hindi, Tamil, Bengali, Punjabi, Assamese, and so on. A language can be divided into two types: colloquial and written. Colloquial language encourages the identity of a culture within the community or a region more strongly than outside of it [2]. But written language represented by a script spreads the identity of that culture outside of the community or region by sharing information. A script can store and pass on the information beyond community, region, or generation. Hence, script is very important for a language to survive which in turn helps the culture to grow beyond its boundary. There are many languages in India which do not

A. Padun (✉)
Department of Multimedia Communication and Design, Central Institute of Technology, Kokrajhar, BTAD, Assam, India
e-mail: a.padun@cit.ac.in

A. K. Das
Department of Design, Indian Institute of Technology Guwahati, Guwahati, Assam, India
e-mail: dasak@iitg.ernet.in

A. Chakrabarti et al. (eds.), *Design for Tomorrow—Volume 1*, Smart Innovation, Systems and Technologies 221, https://doi.org/10.1007/978-981-16-0041-8_46

have their script. Due to which these languages are striving to establish their cultural identity and existence. In this era of technological advancement, many languages lost their identity and become extinct due to lack of written information. Many languages adopt other scripts to survive. Few examples in the Indian context are Khasi language of Meghalaya adopts Roman script, the Bodo language of Assam adopts Devanagari script, and many more. Therefore, script plays a vital role in establishing an identity of a culture and its language.

46.2 Background

With the advancement of technology, the ingenuity of a script has been re-structured to move out of its traditional boundary. In today's digital world, almost all the scripts around the world have been digitized. Indic scripts are also not an exception. Though digitization opens up a huge scope to explore the creative angles of a script, but on the other hand, the traditional essence present in the script in its antique form gradually disappeared. Hence, there may be a need to bring back the traditional essence of a script to preserve the original identity as well as uniqueness which in turn may help retain the identity of that culture in today's digital world. In this regard, an exploration on Assamese script and its scope has been conducted and presented in this paper.

46.3 Literature Review

Assam is a state located in the North-East part of India. Assamese is the official language of Assam, out of 23 official languages recognized by the Republic of India. Assamese script is the writing system of Assamese language [3]. The script has a rich history which evolved from fifth century to recent times. The evolution went through many phases from rock inscription to modern day digital publication. Both Assamese and Bengali script share same script characters with a few typographical differences [4]. Origination of both the scripts is not exactly known but they share similar historical background. During the progression phase, writing styles of both the scripts developed their own identity based on language, culture, and region [5]. Hence, ancient writing styles were different in both the languages even though they share similar script characters.

To revive the original identity of Assamese script, a revisit to the historical resources may provide evidences. Further, this may be exercised by designing a new typeface based on historical writing styles with unique identity that addresses the originality and tradition.

46.3.1 Historical Essence and Identity of Assamese Script

Assamese script was extensively used for writing manuscripts, royal chronicles, declarations, rituals, religious books etc. in the medieval time period fromfifteenth century to early nineteenth century. In that period, the script was further categorized under three different writing styles known as *Bamuniya* style, *Kaitheli* style, and *Garhgayan* style [5].

Bamuniya style of writing: This style was created by the traditional Sanskrit scholars from Brahmin caste (higher priest class who perform *Pujas*, social rituals and religious events) to write Sanskrit texts, rituals, making *Kundalis* (horoscope calculations), etc. Though *Bamuniya* style was followed by the Sanskrit scholars and Brahmins, yet they rather prefer *Kaitheli* because of the popularity of the style. This writing style was in trend among the scribes who were associated with the Vaishnava Satras of Assam [6] (Fig. 46.1).

Kaitheli style of writing: This style was initiated by the *Kayasthas* who used to be known as writer caste. It was also known as *Lakhri* style. This style used to be found in the manuscripts of various document writings, royal charters, and most of the treaties. The style was also followed by professional scribes employed in royal service. The *Kaitheli* style was very popular among the literature of lower Assam region at times. In comparison with the other two, this style was more popular due to its ornamental style and artistic patterns [6] (Fig. 46.2).

Garhgayan style of writing: This style was initiated and primarily followed by the people appointed by Ahom dynasty to write chronicles, various official documents, and public declarations. The name *Garhgayan* came from the capital of Ahom dynasty Garhgaon which was also the center of Ahom culture and located

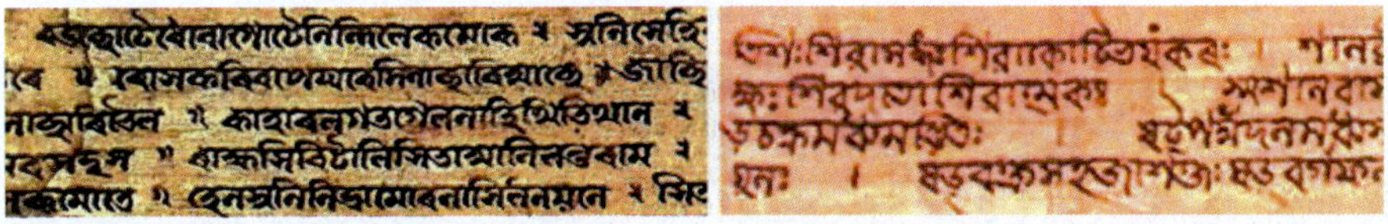

Fig. 46.1 Original image of *Bamuniya* style of writing from different time periods

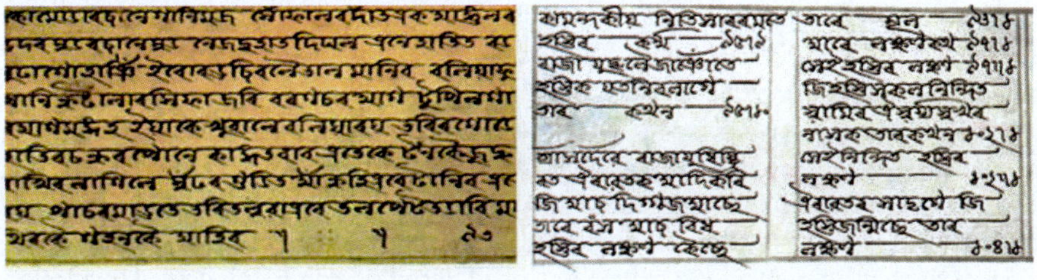

Fig. 46.2 Original image of *Kaitheli* style of writing from different time periods

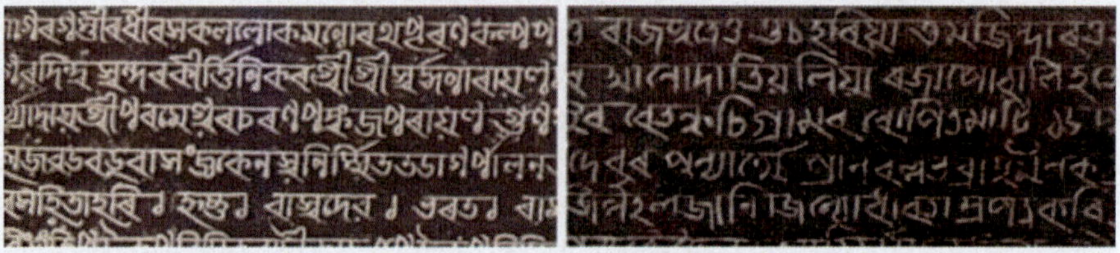

Fig. 46.3 Original image of *Garhgayan* style of writing from different time periods

in the current eastern part of Assam. The writing style later spread all over Assam due to royal metropolitan influence [6] (Fig. 46.3).

During that time period, there was a connection among the script, the script writing style, and the language. The people who initiated the script and its writing style were very much influential in their region and among their people. But due to the royal influence, the *Garhgayan* writing style was adopted on all kind of royal charters, documents, and other declarations all over the region of Assam which was ruled by Ahom dynasty. Hence, *Garhgayan* style portrays an identity of Ahom and Assamese culture during that time period [6].

46.3.2 Assamese Script in Digital Era

While progressing into modern era, the script was modified for machine printing to meet the commercial demand. Though Assamese and Bengali scripts have similarity in their characteristics, but due to accessibility and prominence, Bengali script was considered for machine reproduction during British regime by adding additional Assamese characters. In the early stage of letterpress printing, the crafting of metal typesets were made in Bengal region with machine friendly and modified script styles which addresses the commercial need of both Bengali and Assamese languages [7]. This happened due to the lack of resources in Assam during that time which used to continue till recent time.

Entry of digital typefaces: With the progression of technology, new typefaces or fonts entered into the business of printing by the use of computer. Most of the typefaces designed in its early times were based on old letterpress typesets and Linotype Bengali typeface. Those typefaces were used for printing purposes as well as computer screen display. Some of the examples of typefaces are Ramdhenu, Bijoy, Sankar, etc. After the arrival of digital technology and use of Web and mobile media, the world of typeface design to represent a script completely evolved out of its closed shell with the entry of hi-end software technology. Nowadays, there are lots of new typefaces available online to write or type Assamese script on digital media or devices. But none of the typefaces used to represent Assamese script with historical identity as those typefaces have been designed based on commercial need to address both Assamese and Bengali scripts.

46.4 Methodology

To revive the cultural identity and emotional connection through a new typeface to represent Assamese script, a methodology has been conceptualized to achieve the objective (Fig. 46.4).

Description of each phases of the methodology has been discussed as follows:

46.4.1 Assamese Culture, Language, and Script Writing Styles

The initial three steps of the methodology that describe Assamese culture, language, and script writing styles have been discussed already in the literature review section (refer Sect. 46.3).

46.4.2 Choosing a Script Style and Finding Unique Identity

Although not people's choice in the medieval time, but *Garhgayan* writing style was widely accepted for its basic characteristics of simplicity, legibility, and symmetry [6]. Also it contained royal metropolitan essence of Ahom era that represented medieval Assamese culture. This justifies to choose the script style to represent Assamese language and its culture in today's digital era.

46.4.3 Finding Unique Identity

Apart from its primary characteristics, few more unique features have been observed in this writing style (refer Fig. 46.5). Description of these features has been listed in the following table (refer Table 46.1).

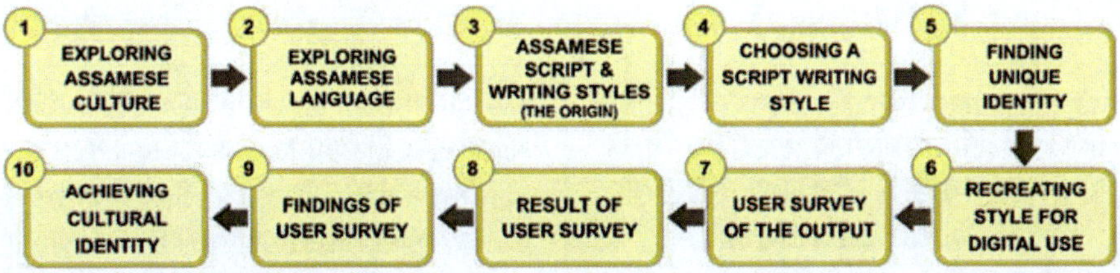

Fig. 46.4 Proposed methodology

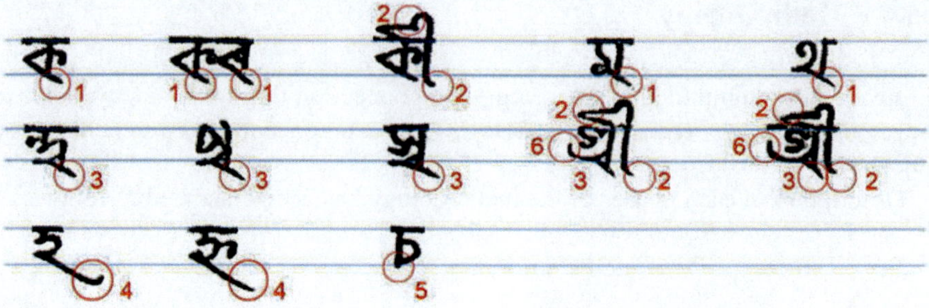

Fig. 46.5 Observation of unique features in medieval *Garhgayan* writing style

Table 46.1 Observation of unique features

Marking	Unique features observed
1	Letters have a sharp and long stretch at the end where two strokes intersect
2	The vowel diacritics have a decorative contour at the top and forms a tendril at the bottom
3	In the formation of conjunct characters where bottom letters used to be "Ba", "Ka" etc., ends with sharp and long stretch as in marking 1
4	Letters "Ha", "E" and "Ee" used to have an elongated ending at the bottom
5	Letters "Cha", "Ccha", "Ta" and "Dha" developed a sharp angle at the bottom
6	Letters "Ai", "Oi" and conjunct of "Ra" have a curvy flow at the end

46.4.4 Recreating Style for Digital Use

After a hand drawn practice session, the final sketches were scanned to recreate a new type style for digital version. The new type style has been designed with all the unique features as observed. Following figure describes the comparison of historical writing style and the digitally recreated style (Fig. 46.6).

A complete set of new Assamese script characters has been designed with monolinear form to render appropriately for digital display by implementing all the unique features. The new Assamese type style has been named as *Garhgayan Assamese* typeface (refer to Fig. 46.7). It includes vowels, consonants, consonant extras, numbers

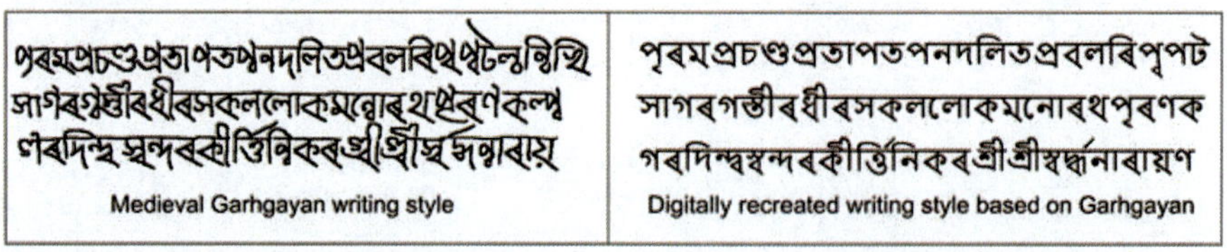

Fig. 46.6 Comparison of medieval *Garhgayan* writing style with digitally recreated style

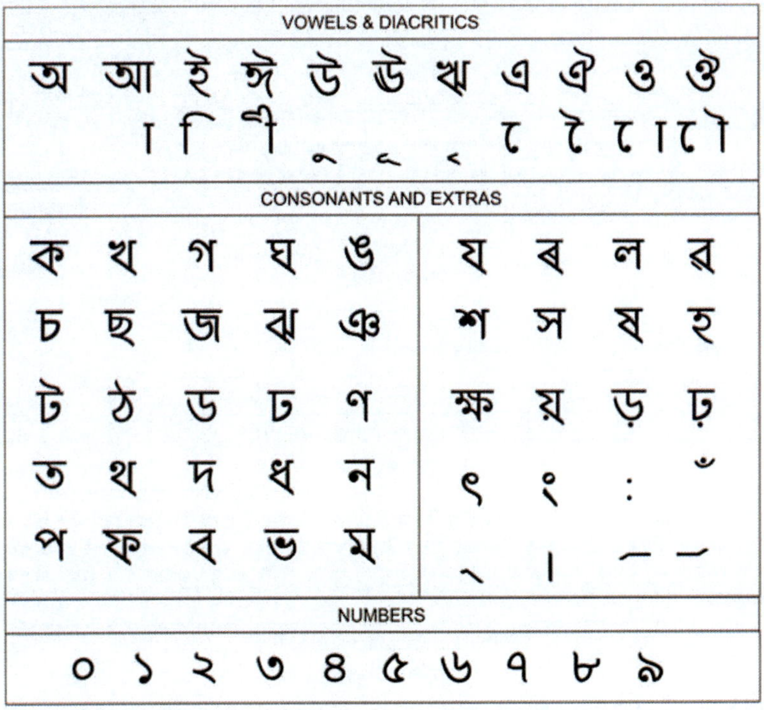

Fig. 46.7 Complete character set of new Assamese typeface based on Garhgayan style

and specially introduced start and end *matra*. The start and end *matra* used to be found in the framing of sentences or paragraphs as an indent style in ancient manuscripts.

46.4.4.1 Designing for Web and Computer Display

Setting up anatomical structure is an important task while designing a new typeface. It is a technical layout of the characteristics of each letters, symbols, and characters of the typeface. It describes the placement, sizing, letter formation, conjunct formation, and sentence formation, etc., of a typeface (refer to Fig. 46.8).

Based on the anatomical structure, vector layout of all the characters has been created using Adobe Illustrator. The vector layout has been then exported to Font Lab Studio for designing and producing a system compatible font by assigning unicode numbers. The new typeface has been then imported to Microsoft Volt for checking technical corrections such as upper and lower modifier positioning, conjunct positioning, etc. A True Type font (.ttf) has been produced as a final result after all the technical amendments and testing.

The new typeface has been then tested on various software such as MS Word, Adobe Illustrator, and Adobe In-Design with the help of in-script virtual keyboard

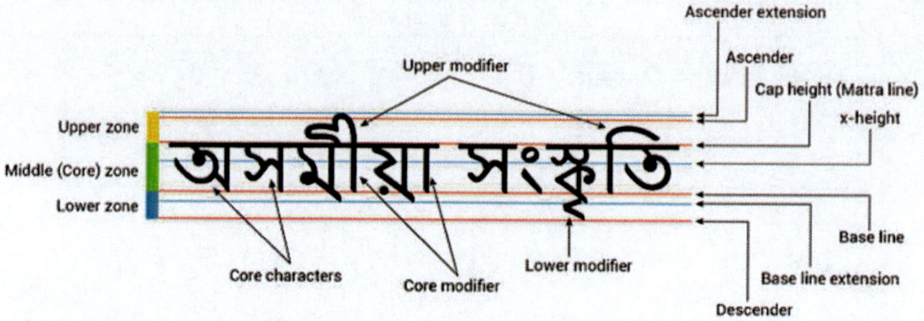

Fig. 46.8 Description of anatomical structure

Font Size: 10, Alignment: Justified
॥ “ঈশ্বৰহীনতা” ॥
“অসহজ সময়ৰ ধাৰাবাহিকতা চলি আছে। আগেয়ে কাহানিও নোপোৱা এক অভিজ্ঞতাৰ মাজেৰে দেশৰ মানুহে দিন কটাবলগীয়া হৈছে। কেৱল দেশেই নহয়, পৃথিৱীৰ ১৩০ কোটি মানুহৰ ঘৰৰ ভিতৰতে কেইবা সপ্তাহ সোমাই থাকিবলগীয়া এই বাধ্যবাধকতা মানৱ সভ্যতাত প্ৰথম। জীৱন ৰক্ষাৰ বাবে 'ভক্ততাৰ' এই সম্পৰীক্ষাত পক্ষ আৰু প্ৰতিপক্ষ এচাম মানুহ বনাম আন এচাম মানুহ নহয়। সমগ্ৰ মানৱ জাতি এফালে আৰু দানৱ অৰ্থাৎ ভাইৰাছবিধ আন এফালে”।
Font Size: 12, Alignment: Centred
॥ “ঈশ্বৰহীনতা” ॥
“অসহজ সময়ৰ ধাৰাবাহিকতা চলি আছে। আগেয়ে কাহানিও নোপোৱা এক অভিজ্ঞতাৰ মাজেৰে দেশৰ মানুহে দিন কটাবলগীয়া হৈছে। কেৱল দেশেই নহয়, পৃথিৱীৰ ১৩০ কোটি মানুহৰ ঘৰৰ ভিতৰতে কেইবা সপ্তাহ সোমাই থাকিবলগীয়া এই বাধ্যবাধকতা মানৱ সভ্যতাত প্ৰথম। জীৱন ৰক্ষাৰ বাবে 'ভক্ততাৰ' এই সম্পৰীক্ষাত পক্ষ আৰু প্ৰতিপক্ষ এচাম মানুহ বনাম আন এচাম মানুহ নহয়। সমগ্ৰ মানৱ জাতি এফালে আৰু দানৱ অৰ্থাৎ ভাইৰাছবিধ আন এফালে”।

Fig. 46.9 Word processing and layout check in MS Words

for Assamese tying. A paragraph was composed in MS word using the new typeface to check its readability, legibility, and visual appeal (refer Fig. 46.9).

46.4.4.2 Designing for Mobile and Digital Devices

A typing app has been designed with an Assamese in-script keyboard to display the new *Garhgayan Assamese* typeface for mobile phones and digital devices (refer Fig. 46.10) to address the user demand on digital platform.

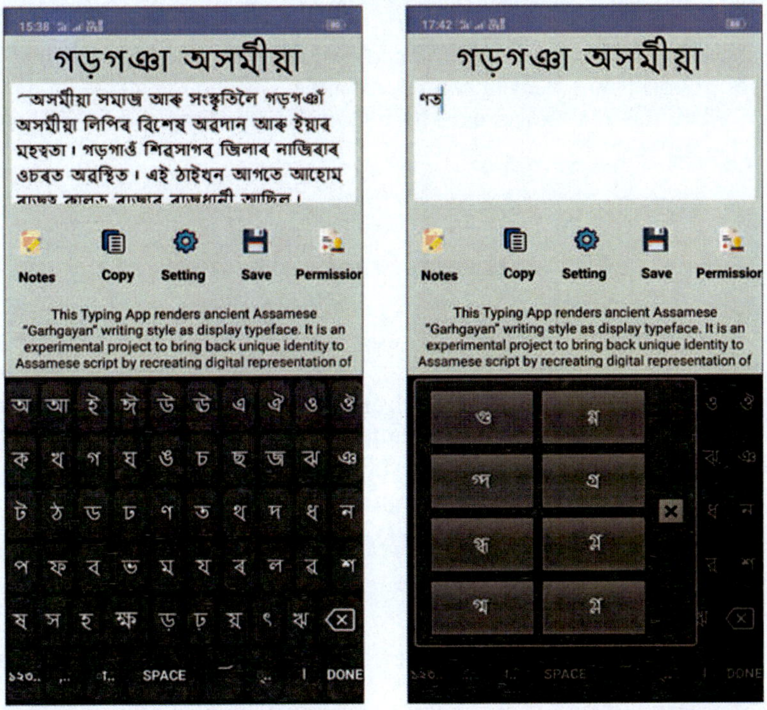

Fig. 46.10 Screenshots of *Garhgayan Assamese* typing app

46.4.5 User Survey of the Output—Garhgayan Assamese *Typeface*

A total of 53 users have been invited to take part in the user survey from across Assam as well as India with varied professional background with a keen interest for Assamese language and script. Out of 53, responses of ten users have been shortlisted based on their approach toward the survey and completion of the task. Key factors considered for the survey have been as follows:

- Factor-1: Carrying a legacy of the historical style
- Factor-2: Novelty in design
- Factor-3: Readability and legibility
- Factor-4: Appeal factor (emotional attachment)
- Factor-5: Ease in writing/typing (Unicode standardization)
- Factor-6: Effectiveness of new special characters (start and end matra)
- Factor-7: Effectiveness in document writing and publishing
- Factor-8: Keyboard customization for typing on digital devices
- Factor-9: Effectiveness of typeface on mobile screen or digital devices
- Factor-10: Recall value (ease of recognition among other typefaces).

All the users have been asked to give a rating point on a scale of 10. The rating points above 5 point has been considered as positive rating and below 5 points have been considered as improvement needed. After the collection of all the ratings from different users, the total points have been then combined together to evaluate the result in percentage.

46.4.6 Result of User Survey—Graphical Analysis

Following bar graph represents the user survey in percentage performance (Fig. 46.11).

On the *X*-axis, it shows total points given by the users on various factors and on *Y*-axis, it shows percentage performance of various factors after compilation. In the legends, color coding for percentage performance has been showed with description based on amount of percentage acquired. The bar graph shows the positive achievement of the experiment with all the factors scoring above 50%.

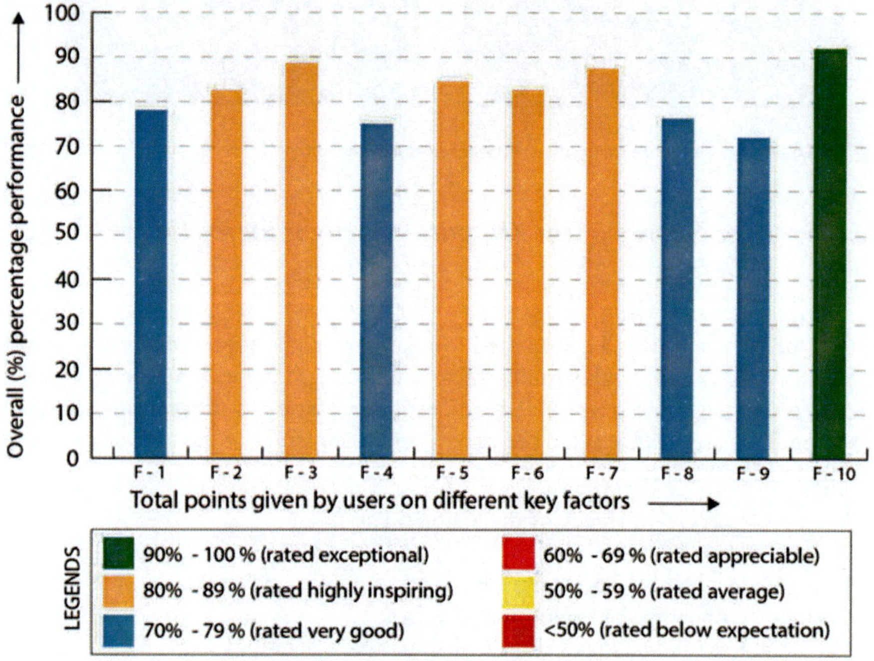

Fig. 46.11 Graphical representation of user survey on the new typeface

46.4.7 Findings from User Survey—Cultural Identity

The user survey result shows that factors F-1 to F-4 which addresses the primary objectives of identity and emotional connection resulted in a positive direction. Factors F-5 to F-9 which represent technical aspects of the typeface show promising outcome as per user feedback. The last factor F-10 receives an inspiring response from the users in terms of recognition among others. Hence, the findings can be concluded with a positive note that cultural identity could be established though typeface design that can connect the script and the language with historical essence.

46.5 Conclusion

This exploration establishes the fact that, to create an identity for a language and its culture, script plays a pivotal role. A script could be represented by many typefaces that expresses the emotions written in that script. Therefore, a typeface can play an important role to portray the emotions of a language and its culture through their script. Hence, it is very essential for a typeface to be unique and should have a historical essence as well as emotional connection among the users. The results of this experiment with *Garhgayan Assamese* typeface rendered those achievements. This experiment also opens up the scope of typeface designing for Assamese script to a wider platform irrespective any media. Future implication of this exploration may offer more promising and prolific results if conducted with an assenting approach.

References

1. Language and Culture: (2019). Available from: https://www.britannica.com/topic/language/Language-and-culture
2. Colloquialism: (2020) Available from: https://en.wikipedia.org/wiki/Colloquialism
3. Assamese language: (2019). Available from: https://en.wikipedia.org/wiki/Assamese_language
4. Assamese alphabet: (2019). Available from: https://en.wikipedia.org/wiki/Assamese_alphabet
5. Neog, Dr. M.: Prachya-Sasanavali, Publisher: Publication Board, Assam, 1 edn (1974)
6. Bhattacharyya, Dr. S.C.: Asamiya Lipitattva Adhyayana, 1st edn. Chandra Prakash Publishing, Panbazar, Guwahati, Assam (2013)
7. Neog, Dr. M.: The Resuscitation of the Assamese language by the American Baptist missionaries. Compiled by: Pranavsvarup Neog. Publisher: Omsons publications, New Delhi (2008)
8. Padun, A., Das, A.K.: Study of Assamese letterform and its structure: an exploratory approach. Conference paper. Conference: ICoRD'19—Research into Design for a Connected World, At IISc Bangalore, vol. 1 (2019). doi: https://doi.org/10.1007/978-981-13-5974-3_13
9. Goswami, Dr. U.: ASAMIYA LIPI (The Assamese Script), 9th edn. Assam Publication Board, Guwahati, Assam (2014)
10. Developing OpenType Fonts for Bengali Script. Available from https://docs.microsoft.com/en-us/typography/script-development/bengali. Accessed on 02 April 2020

11. Ross, F.G.E.: Digital Typeface Design and Font Development for Twenty-first Century Bangla Language Processing. doi: https://doi.org/10.4018/978-1-4666-3970-6.ch001 (2013)
12. Chandra, S., Bokil, P., Kumar, D.U.: Anatomy of Bengali letterform: a semiotic study. Conference paper. Conference: ICoRD'15—Research into Design, At IISc Bangalore, vol. 1 (2015). doi: https://doi.org/10.1007/978-81-322-2232-3_22
13. Bengali-Assamese script. (2016). Available from: https://en.wikipedia.org/wiki/Bengali%E2%80%93Assamese_script
14. Indic Scripts and Languages. (2018) Available from https://www.unicode.org/faq/indic.html
15. Abhijit Padun and Amarendra Kumar Das. 2018. Designing a new Assamese Typeface for digital media: Challenges and Possibilities. J. Assam Sc. Soc. Vol. 59; No. 1 & 2. December 2018; Pp 39–48. ISSN 0587–1921.

Chapter 47
Impact of Indian Culture-Oriented Color, Material, and Finish (CMF) Design on Contemporary Products

Supriya Ugale and Nachiket Thakur

Abstract Aesthetic appeal and cultural meaning of product are two key aspects that are becoming very popular in consumers and which are highly influencing consumer's buying decisions in today's competitive market. Culture-oriented Color, Material, and Finish (CMF) Design workaround aesthetic values and cultural connections are based on emotional values. Cultural CMF elements are considered to be a unique character to embed into a product for the enhancement of product identity in the global market and for the fulfillment of the individual consumer's experiences. The intent of this paper is to study the impact of Indian culture-oriented CMF design on contemporary product's aesthetic and semantic attributes. The research work has been carried out in two stages. Firstly, a visual audit was conducted to understand the impact of Indian culture-oriented CMF design on the visual identity of the product by re-visiting past and current trends. Secondly, an effort has been made to find out differences in the visual product perception of the designers and users by using semantic differential (SD) tests. A one-way ANOVA test was carried out to analyze the results achieved from the SD test. The result shows that Indian culture-oriented CMF design creates a significant impact on contemporary product aesthetic and semantic values. If contemporary product aesthetics is designed considering Indian users' perception and their needs, then products will have enhanced semantic values, and products will become more meaningful for users. Indian traditional CMF elements will help to create product identity in the global market.

47.1 Introduction

Consumers are getting highly attracted to products that are aesthetically appealing and culturally meaningful [1, 2]. To create a successful product and to provide a better user experience, it is essential for designers to understand the aesthetic perceptions of users and cultural backgrounds along with the context where the product will be placed. Culture-oriented Color, Material, and Finish Design approach gives the

S. Ugale (✉) · N. Thakur
MIT Institute of Design, MIT ADT University, Pune, India
e-mail: Supriya.ugale1@gmail.com

A. Chakrabarti et al. (eds.), *Design for Tomorrow—Volume 1*, Smart Innovation, Systems and Technologies 221, https://doi.org/10.1007/978-981-16-0041-8_47

opportunity to study the socio-cultural trends and come up with desirable products for the target users [3]. Culture-oriented CMF Design works around aesthetic values and cultural connections, which are based on emotional values.

Color design, material design, surface, and finish design are the key areas of CMF design. Surface and finish design focus are on designing of texture, patterns, graphics for the product surface and which creates a significant impact on the emotional and cultural values of the product [3]. Cultural CMF elements are considered to be a unique character to embed into a product for the enhancement of product identity in the global market and for the fulfillment of the individual consumer's experiences. The unique and most diverse traditional Indian cuisines and kitchen products reflect the culture of India and which are a valuable source of inspiration toward looking at and understanding contemporary design [4, 5].

According to the literature review, the study to transform cultural features to design elements has been conducted to generate a culture-oriented design. It is practice-based research limited to graphic design [6]. In the field of design, CMF design is still relatively unknown. The companies gradually started giving attention to CMF design as it fulfills consumers' aspirations, desires, and brings profitable business [7, 8]. Consumers are buying products for various reasons like cultural meanings, aesthetic appeal, and emotional values along with practical functionality. It needs to be discovered which category of culture-inspired products that efficiently stimulate users' emotions and creates a significant identity in the global market [1]. Culture-oriented product design will become a design trend worldwide [1, 9]. There are rare studies for culture-oriented CMF design. So, for more insight's interviews have been conducted with CMF experts who are working in the same area. It facilitated a clearer idea of the early days of CMF design to its advancement. According to experts, Indian culture is becoming more visible and spreading worldwide, and this is the right time to work on cultural CMF design with respect to the Indian context. Consumers are now looking for more personalized and customized products where the color, material, and finishes play a vital role.

Aesthetic preferences change with time and which are based on cultural background. Products need to support the aesthetic values according to different influences of socio-cultural trends [3, 10]. In the first part of the research paper, a visual audit was conducted to understand the impact of Indian culture-oriented CMF design on the visual identity of the product by re-visiting past and current trends. In the second part, semantic differential (SD) tests were performed to find out differences in the visual product perception of the designers and users. Culture-oriented CMF design is an emerging professional discipline and increasingly in demand across all design disciplines.

47.2 Method

This research work is based on a qualitative approach. As there are limited studies on culture-oriented CMF design, the information for the current work is gathered from written sources, online surveys of potential users, and interviews conducted with successful Indian CMF designers.

To investigate the proposed work, a water pot (Mutka) is selected as a sample product, which is one of the most essential and frequently used kitchen products in Indian homes. For products like water pots, users are also the potential buyers. In this study, the user and the consumer both are Indian homemakers. This research work has been divided into two stages. In stage I, a visual audit was conducted to understand the impact of culture-oriented CMF design on the visual identity of a product by re-visiting past and current trends. Old, middle, and new from each era, 15 water pots were selected for visual audit. A visual identity map has been developed, and trend analysis is carried out for all 45 water pots [11]. Out of which, ten water pots were selected from different eras for validation purposes. The validation was done with a sample size of 50 designers consisting of young designers, design students, design faculties, and design professionals because they are aware and well versed in design language. Responses received from designers were recorded and compared with study results. In stage II, the work is carried out to find the differences in visual product perception of designers and users by using the semantic differential (SD) test. From different eras, most preferred five water pots, images were selected for the SD test. SD scale was developed for each product. In the SD scale, a 7-point scale is used to evaluate the product and its preference perception score. In this evaluation scale, the 3-point score indicates that the user has a very strong positive impression, whereas the -3-point score indicates that the user has a strong negative impression of a given product. The SD scale shows a bipolar pair of an adjective, with the positive adjective on the right side, while its negative adjective on the left side, such as Ugly-Beautiful, Delicate-Sturdy, Useless-Useful. From 20 adjective pairs, ten adjective pairs were selected after doing pilot tests with five designers and five users. Before the SD test, the purpose of the SD test and SD scale with a bipolar pair of adjectives was explained to participants [12–14]. The sample size used for the SD test included 25 designers and 25 users. The responses received from the SD test were recorded and analyzed using a one-way ANOVA test. The ANOVA technique is used here because it allows to perform simultaneous test and compare between more than two samples [15]. Water pots Model 3, 4, and 5 having similar form but completely different CMF treatment were selected for a One-way ANOVA test to examine there is a significant effect of three different CMF designs upon how designers responded to all five bipolar adjective pairs. SD test response scores received from 25 designers are used to conduct the One-way ANOVA test for five bipolar adjective pairs, which are significantly correlated to look and feel.

47.3 Results

As discussed in methodology, a qualitative test has been carried out, and the following results have been recorded.

47.3.1 Visual Audit—Water Pot

Water pots are widely used across India. Water pots are mostly kept on the kitchen platform, and that is why it is an essential part of the kitchen interior. Visual audit for water pots conducted in two stages is as described below based on color, material, and finish designs.

47.3.2 Stage I—Visual Identity Mapping

Visual identity maps have been developed based on CMF study (Fig. 47.1). Five attributes were selected against three different eras, which are very commonly observed across all time periods. A visual identity map, which is developed based on CMF study, has been compared with a visual identity map based on a designer survey (Fig. 47.2).

The result shows that visual identity maps based on CMF study and visual identity maps based on designer surveys are 90% identical. Based on the result, it has been observed that traditional water pots are making a comeback in contemporary Indian

Fig. 47.1 Visual identity map based on CMF study

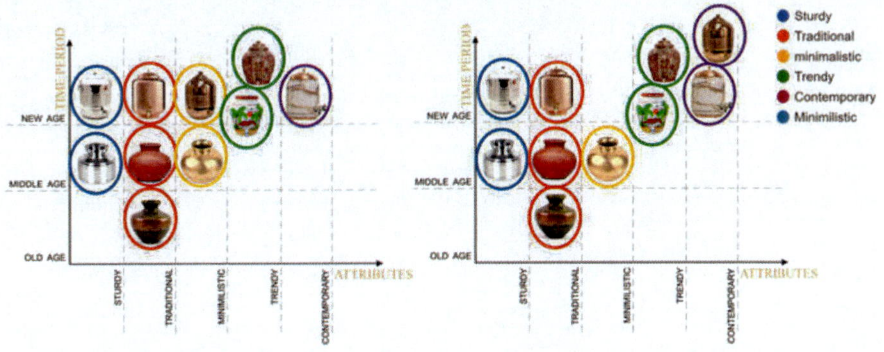

Fig. 47.2 Visual identity map based on CMF study versus Visual identity map based on designer survey

kitchens. The customers are now looking for water pots made in traditional material. Copper is observed as the latest trend in water pots.

47.3.3 Stage II- Perception Mapping

From different eras, the most preferred five water pots images were selected for the semantic differential (SD) test (Fig. 47.3).

The semantic differential (SD) scale was developed for each product having a 7-point scale, where a 3-point score indicates that the user has a strong positive impression, whereas a -3-point score indicates a strong negative impression for a given product. The SD scale shows a bipolar pair of an adjective, with the positive adjective on the right side, while its negative adjective on the left side, such as Ugly-Beautiful, Delicate-Sturdy, Useless-Useful [12, 13] (Table 47.1).

The results of the semantic differential (SD) test are recorded and analyzed (Figs. 47.4, 47.5, 47.6, 47.7 and 47.8). The scores for the ten adjective pairs rated by designers and users showing significant differences in adjective pair Rational-Emotional. The result shows that there is a perception of differences between

Fig. 47.3 Five sample water pots for stage II research work

Table 47.1 Semantic differential scale

Adjectives	Rating scale							Adjectives
Modern	−3	−2	−1	0	1	2	3	Traditional
Useless	−3	−2	−1	0	1	2	3	Useful
Ugly	−3	−2	−1	0	1	2	3	Beautiful
Meaningless	−3	−2	−1	0	1	2	3	Meaningful
Delicate	−3	−2	−1	0	1	2	3	Sturdy
Heavy	−3	−2	−1	0	1	2	3	Handy
Angular	−3	−2	−1	0	1	2	3	Rounded
Usual	−3	−2	−1	0	1	2	3	Unusual
Unoriginal	−3	−2	−1	0	1	2	3	Creative
Rational	−3	−2	−1	0	1	2	3	Emotional

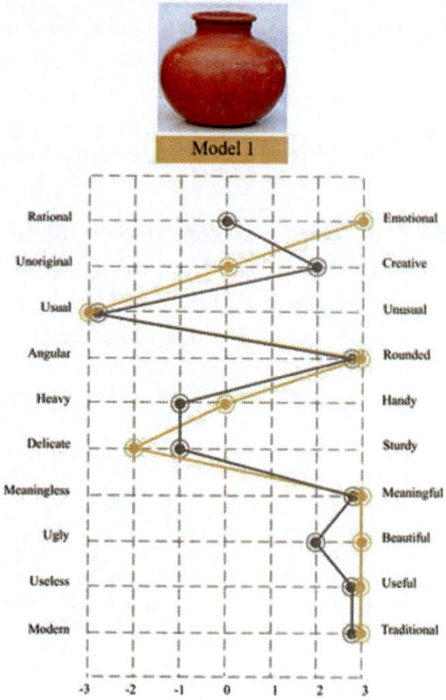

Fig. 47.4 Water pot model 1 semantic differential survey responses

designers and users when it comes to an understanding of the meaning and emotional aspect of a product.

Water pots Model 3, 4, and 5 having similar form but completely different CMF treatment were selected for a One-way ANOVA test to examine there is a significant effect of three different CMF designs upon how designers responded to all five bipolar adjective pairs. SD test response scores received form 25 designers are used

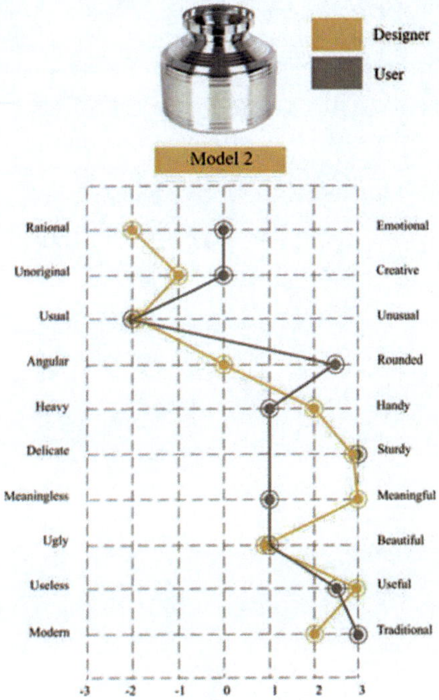

Fig. 47.5 Water pot model 2 semantic differential survey responses

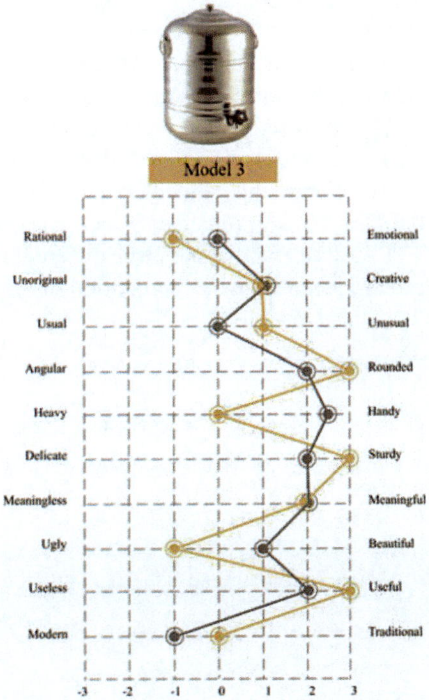

Fig. 47.6 Water pot model 3 semantic differential survey responses

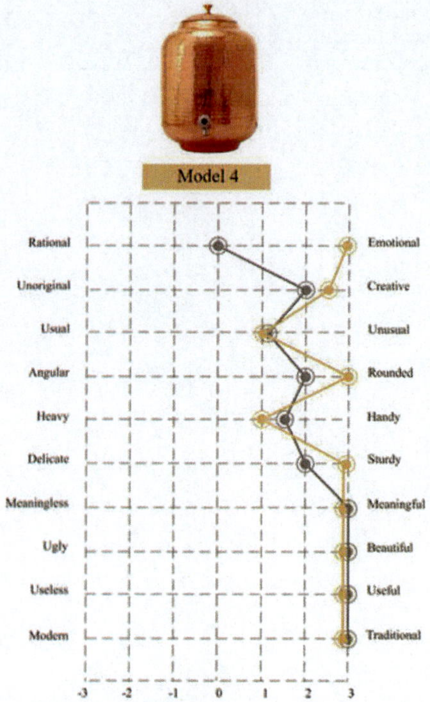

Fig. 47.7 Water pot model 4 semantic differential survey responses

to conduct the One-way ANOVA test for five bipolar adjective pairs, which are significantly correlated to look and feel (Tables 47.2 and 47.3).

Ho = No significance effect of three different CMF designs of water pots on the designers' perception for all five bipolar adjective pairs.

The compound value of $F = 6.059$ is higher than the critical value of $F = 3.89$ at 0.05 level of significance. Hence, the null hypothesis (Ho) is rejected.

Results show that the three different CMF designs of water pots have a significant effect upon designers' perception of all five bipolar adjective pairs.

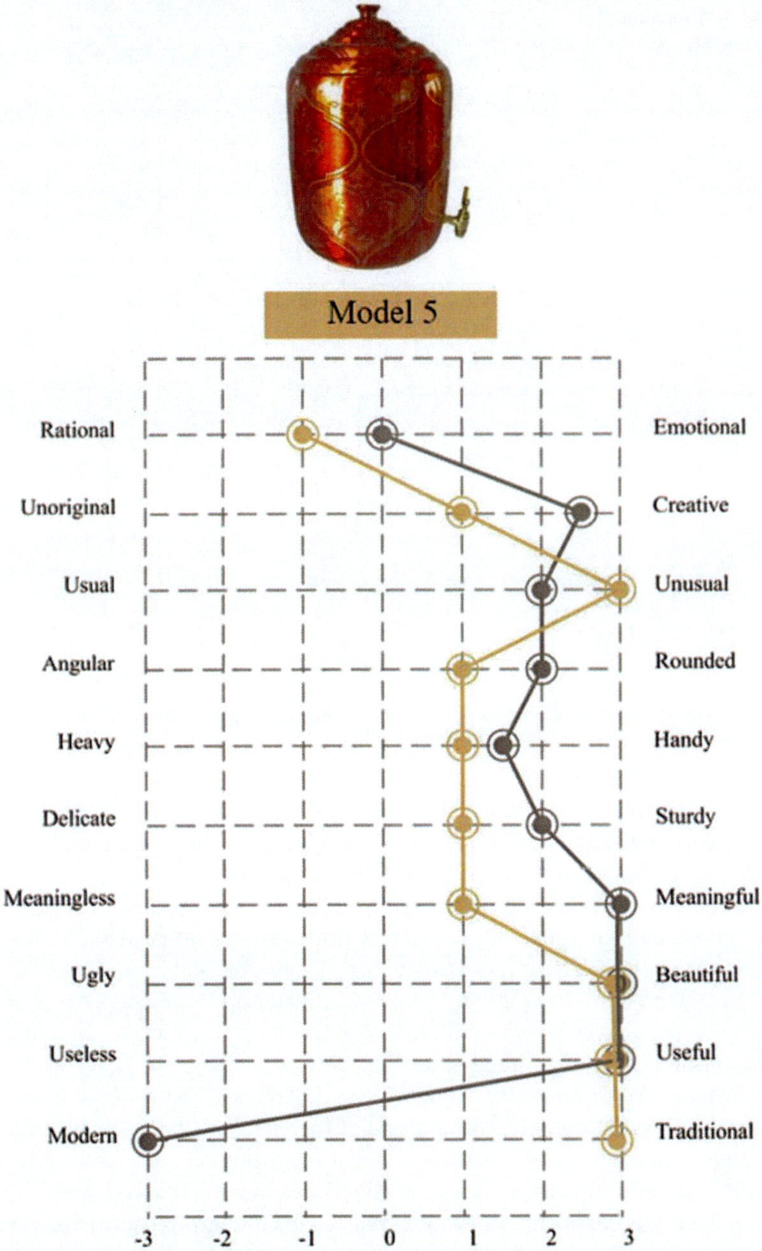

Fig. 47.8 Water pot model 5 semantic differential survey responses

Table 47.2 Bipolar adjective pair selected for water pot model 3, 4, and 5

Bipolar adjective pair	Water pot model 3	Water pot model 4	Water pot model 5
Modern–traditional	0	3	3
Ugly–beautiful	−1	3	3
Meaningless–meaningful	2	3	1
Unoriginal–creative	1	2.5	1
Rational–emotional	−1	3	−1

Table 47.3 Analysis of variance table for one-way ANOVA

Source of variation	Sum of squares (SS)	Degrees of freedom (d.f.)	Mean square (MS)	*F*-ratio	5% of *F*-limit (from *F*-Table)
Between sample	18.3	$(k - 1) = (3 - 1) = 2$	9.15	6.059	$F\ (2,12) = 3.89$
Within sample	18.2	$(N - k) = (15 - 3) = 12$	1.51		
Total	36.5	14			

47.4 Discussion

In research work, it has been observed that during old age, consumers used to prefer products that are sturdy and long-lasting. Clay, brass, and copper were the most preferred materials with intricate designs carved on it. During the Middle Age, consumers were used to preferring products that were very simple, sturdy, and minimalistic. Now, in the new era, a trend has been changed where users are looking for products that are appealing and culturally meaningful.

In perception mapping, it is observed that Model 1, water pot with clay material, brown color, and rough texture, gives an earthy and traditional look. In Models 2 and 3, water pots with stainless steel material, metallic color, and smooth finish give a sturdy and domestic look. In Model 4, a water pot with copper material and hammered texture gives a very traditional look. In Model 5, a water pot with copper material, floral print gives a trendy look.

The results of the One-way ANOVA test conducted for Models 3, 4, and 5 show that the three different CMF designs of water pots have a significant impact on designers' perception. Models 4 and 5 are from the new era; both designs are following Indian cultural approaches, but CMF treatment given to them is different. Based on the CMF treatment, Model 4 is creating the highest impact at the emotional level, whereas Model 5 is looking very superficial, fancy, and thus users are not able to connect emotionally. So, just applying a traditional motif's print on a product will not solve the purpose. Much more in-depth thought process and method required to implement Indian cultural CMF design elements onto contemporary products in order to create a strong emotional connection between product and user.

47.5 Conclusion

Aesthetic perceptions vary according to cultural background. It is essential for designers to design the products by understanding the cultural background and context in which the product is going to be positioned. The culture-oriented CMF design approach gives the opportunity to study the socio-cultural trends and come up with a solution which more relevant and desirable for users. Appropriate selection of color, material, and finish helps to satisfy the emotional requirement of users and will also reduce the risk of market failure.

In a result, it has been observed that traditional materials like copper and terracotta are in trend and making a comeback in contemporary kitchen products. The SD test result shows that perception varies between designers and users when it comes to an understanding of the meaning and emotional aspect of a product. ANOVA test results show that $F = 6.059$ is higher than the critical value of $F = 3.89$ at a 0.05 level of significance, which means CMF designs of water pots have a significant effect upon designers' perception for all five bipolar adjective pairs. There were a few limitations to this research study. First, it was a pilot study; the accuracy of the results here can be increased by increasing the sample size, and by conducting more tests. Secondly, in the semantic differential (SD) test, product images were shared with participants. The accuracy of the results here can be increased by showing them actual products.

Future research work may explore a deeper understanding of which product categories are more suitable for the application of cultural-oriented CMF and a detailed framework for how to apply cultural CMF elements on contemporary products. Cultural-oriented CMF design is an upcoming trend in product design, and thus, this study will be beneficial for both academia and industry.

Acknowledgements The author wishes to thank you CMF Experts form 3F Design Studio, Pune., Sarvasva Design Pvt. Ltd., Pune., and AKFD Studio, Jaipur., for sharing their knowledge, experiences and views through interviews, which gave thoughtful insights to go forward in this research area. Author would like to gratefully acknowledged the support received from MIT Institute of Design, Pune.

References

1. Wang, Y.H., Qin S.F., Harrison, D.: Culture-inspired design principles, methods and tools in current products. Corpus ID 51797494, Taiwan Aboriginal Research Website, Brunel University, UK (2013)
2. Mumcu, Y., Kimzan, H.S.: The effect of visual product aesthetics on consumers' price sensitivity. In: 4th World Conference on Business, Economics and Management, WCBEM, Elsevier Publisher (2015). https://creativecommons.org/licenses/by-nc-nd/4.0/
3. Becerra, L.: CMF Design the Fundamental Principles of Color, Material and Finish Design, Frame Publishers, Amsterdam (2016)
4. Tulsian, N.: What Brings 'Indian-ness' in the Indian Design? Issue 39, Creative Gaga (2017)
5. Kapur, H., Mittar, S.: Design intervention and craft revival. Int. J. Sci. Res. Publ. **4**(10) (2014)

6. Ser, S.: On culture-oriented product design: a study to transform cultural features to design elements. Int. J. Creat. Future Heritage (IJCFH) **6**(2), 163–197 (2018)
7. Isoaho, H.: Color, Material and Finish Design. KYAMK University of Applied Sciences (2016)
8. Kayina, H., Goel, S.: Significance of product values for an enriched retail experience. Int. J. Appl. Home Sci. **4**(3 & 4), 397–411 (2017)
9. Lin, R.T.: Transforming Taiwan aboriginal cultural features into modern product design: a case study of a cross-cultural product design model. Int. J. Des. **1**(2), 45–53 (2007)
10. Qu, Y., Mao, X., Li, D.: Research on the role of design aesthetics in modern design. In: MATEC Web of Conference 176, 02012 IFID (2018). https://creativecommons.org/licenses/by/4.0/
11. Karjalainen, T.M.: It looks like a Toyota: educational approaches to designing for visual brand recognition. Int. J. Des. **1**(1), 67–81 (2007)
12. Aros, M., Narváez, G., Aros, N.: The Semantic Differential for the discipline of Design: a tool for the Product Evaluation. In: 13th International Congress on Project Engineering, Badajoz, pp. 422–433 (2010). ISBN 978-84-614-0185-7
13. Hsu, S.H., Chuang, M.C., Chang, C.C.: A semantic differential study of designers' and users' product form perception. Int. J. Ind. Ergon. **25**(2000), 375–391 (2000)
14. Pham, B.: Design for aesthetics: interactions of deign variables and aesthetic properties. In: Proceeding of SPIE IS&T/SPIE 11th Annual Symposium- Electronic Imaging 99, vol. 3644, pp. 364–371, San Jose, USA (1999)
15. Kothari, C.R.: Research Methodology: Methods and Techniques, 2nd Revised Edition. New Age International (P) Ltd., Publishers (2004)

Chapter 48
A Visual Analysis of Motifs and Patterns of Ghanasyam House Sivasagar, Assam, India

Saurav Khuttiya Deori and Utpal Barua

Abstract The Ghanasyam House of Sivasagar district, Assam, India, is one of the detailed specimens in the ornamented brick architecture monument of the Ahom Dynasty. The monument is built during the reign of Ahom king Rudra Singha (CE 1696–1714), and it is protected under the Archaeological Survey of India, Guwahati circle, Assam (ASI). Multiple terracotta plaques are being used for the ornamentation that includes motifs, patterns, symbols and sculptures. This paper aims to present a visual analysis of the motifs and patterns of the monument. The collected data of ornamentation are sorted considering a categorization theory. This paper investigates the visual composition analysis and symmetrical analysis of the ornamentation. The results provide detailed insight into the visual attributes of the motifs and patterns. The paper identifies rudimentary principles in the composition of the ornamentations within a specific frame of reference. The quantitative data analysis identifies the dominating motifs and pattern types and multiple symmetry operations.

48.1 Introduction

Assam has a rich history of 600 years of the Ahom Dynasty. They ruled Assam from thirteenth century to nineteenth century. According to the Ahom chronicles, Ahom came from Mong-Mao, a Tai state in South-Western Yunnan, in 1215 A.D., then migrated to northern Burma crossing Patkai hills. Under the Ahom kingdom, the art and architecture flourished remarkably. Initially, the constructions were of woods and bamboo, but the buildings did not sustain for long durations; hence stones and bricks were used for construction purposes [1]. The Ahoms built many brick architecture buildings and temples in Assam, and most of the ornamentations include motifs,

S. K. Deori (✉) · U. Barua
Indian Institute of Technology Guwahati, Assam, India
e-mail: s.khuttiyadeori@iitg.ac.in

U. Barua
e-mail: utpalb@iitg.ac.in

A. Chakrabarti et al. (eds.), *Design for Tomorrow—Volume 1*, Smart Innovation, Systems and Technologies 221, https://doi.org/10.1007/978-981-16-0041-8_48

sculptures and patterns. To date, studies related to monument sculptures and iconographies are undertaken and are mostly descriptive in nature. The visual study on the ornamentation of the Ahom architecture is still yet to be explored and documented. Considering ornamentations in brick architecture, one of the most significant examples is the Ghanasyam House of Sivasagar. The monument is protected under the Archaeological Survey of India (ASI). It is built using brick and terracotta plaques, rich with motifs, sculptures and patterns. The present study focuses on the ornamental architecture of the monument and aims to examine the visual elements of motifs and patterns. The analysis will help give an insight into the structural symmetry of the ornamentation and help to understand the underlying design principles and craftsmanship of the architecture.

48.2 Method

To understand the context of Ahom and their art and architecture broadly, historical approach was undertaken. Multiple site visits were conducted to Ghanasyam House to study architecture type and the ornamentation in it. Although the monument's ornamentation is abundant with various motifs, patterns and sculptures of Hindu gods and deities, only motifs and patterns analysis were taken into consideration for visual analysis as Ahom's art and iconography of sculptures are extensively studied by many authors and scholars. Descriptive approach was chosen for analysis of the historical and cultural context of the ornamentation. To reduce the chances of data misinterpretation, some variations of animal and symbolic motif descriptions are accessed thought secondary data.[1] The collected photograph data are extracted in linear form using Illustrator software. Since the data were large, categorization theory[2] [4] is adopted to systematically categorize the motifs and patterns based on the ornamentation area, as shown in Fig. 48.1. Based on the categorization, coding on the types of motifs and patterns is done to differentiate the types. After the sorting, we extracted the motifs and patterns for its variation, symmetrical analysis and visual attributes. Quantitative data show the most common type of motifs and patterns with predominating symmetry types by analysing symmetry operations like rotation, reflection, translation and glide reflection.

[1]Das [2].

[2]Adopted by [3] Ph.D thesis, *Identification of Design Elements in Naamghar of Assam* for the charting the taxonomical structure and categorization of elements of Assamese *Naamghar. Naamghar* is social-cultural institution associated with Vaishnava faith of Assam.

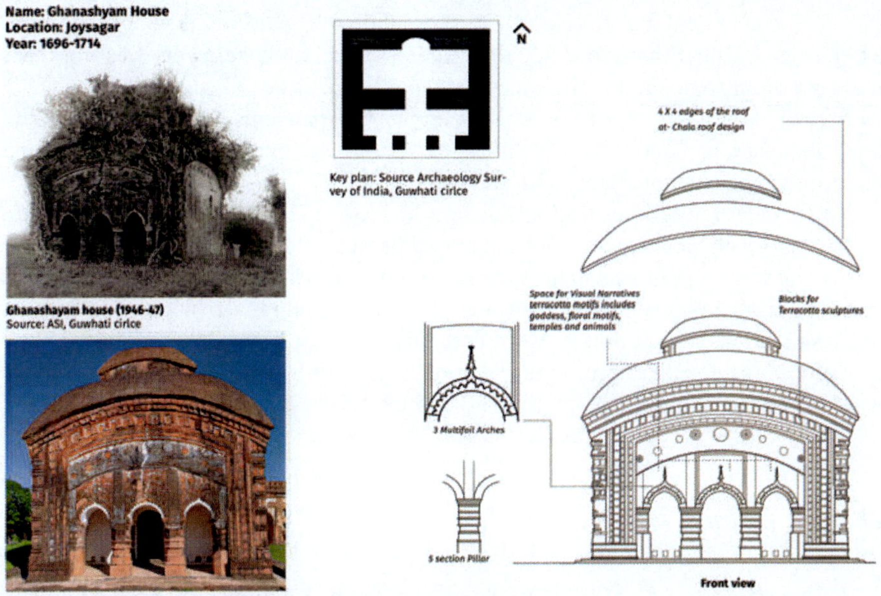

Fig. 48.1 Ghanasyam House with the key plan and structural study

48.2.1 Data Gathering

For primary data, photographic and video data of Ghanasyam House have been collected. We conducted semistructured interviews with Senior Conservation Assistant, Sibsagar Circle, Archaeology Survey of India (ASI), a local Museum Owner of Sibsagar Area. The secondary data were old photographs of the monument, which were collected from ASI (Guwahati circle) for comparison purposes. The other sources were the library and the Internet.

48.3 Results and Discussions

48.3.1 About Ahoms and Hinduism

King Chao Lung Sui-Ka-Pha was the founder of the Ahom Dynasty; he set up his first capital in Charaideu, which is 28 km away from Sivasagar city. According to history, in the later years, they shifted the capital to different places like Saragua, Gargaon, Rangpur and Jorhat, which are places near Sivasagar. As for which most of the monuments built by the Ahom kings fall within the Sivasagar area. The entry of the Ahoms was a turning point in the art history Assam. The Ahoms are Shans belonging to the great Tai Family [1]. For the initial two hundred years, the Ahoms practiced their traditional religious practices, and with socio-political and linguistic exigencies actuated them to gradually lean towards Hinduism. This was not mass scale nor because of Hindu religion as such, but in the wake of a plural society which was slowly emerging out of the Ahom conquest, expansions and absorption. With the acceptance of Hinduism, the Ahom rulers started following Hindu rituals and building temples to appreciate and practice it [5].

Under the reign of king Rudra Singha (CE 1696–1714), most of the temples are built, also locally known as 'Dol'. 'He was anxious to build palaces and city of bricks, but there was no one in his kingdom who knows how to do this. Therefore, he imported from Koch Bihar, an artisan named Ghanasyam under whose supervision numerous bricks and buildings were erected at Rangpur, close to Sibsagar, and at Charaideo' [6]. Furthermore, in the memoir of his contribution, Ghanasyam House was built in the Joysagar area, which falls in the Sibsagar district [7].

48.3.2 Ghanshyam House

Ghanasyam House or *Dol* (Temple) Fig. 48.1 is on the southern end of the west bank of the Joysagar tank in the Sivasagar district of Assam. The monument measures 7 m 23 cm × 6 m 35 cm. [2]. The architecture is constructed using bricks and terracotta plaques with sculptures, motifs and patterns. The architecture of the house or Dol has resemblance with the contemporary temple architectures of West Bengal and Bangladesh [2]. The style of architecture is known as the *Chouchala* structure. There are different types of *chalas*. The *do-chala* type has only two hanging roof tips on each side of a roof divided in the middle by a ridge line; *char-chala* type, four roof corners and have a dome-like shape; the double storey has *at-chala* type and has eight roof corners. Considering the roof corners and the double storey, the Ghanasyam House falls under *at-chala* architecture Fig. 48.1 (front view, roof). According to the Archaeological Survey of India [7], the monument reflects the influence of Islamic architecture because of the three arch type of the door entrance Fig. 48.1 (front view) and in inside the north wall contains a *Mihrab.* The edges of arches are ornamented with decorative terracotta tiles with multiple zigzag patterns. The monument is rich with detailed terracotta sculptures, various floral motifs, patterns, mythical creatures and visual narratives. 'The mosaic over the central arch depicts a lively scene of a number of monkeys that appear frolicking, but a closer study reveals that is a kaleidoscopic scene from the *Ramayana*[3] in which the monkeys are actually engaged in constructing the stone bridge to Lanka. In the upper left-hand side, two monkeys are carrying a stone block while others are giving directions. In the upper right-hand side *Sita* is seated in the *Asokavana*[4] while another tile shows a crude representation

[3]*Ramayana* is an ancient Sanskrit epic which follows Prince *Rama's* quest to rescue his beloved wife *Sita* from the clutches of *Ravana* with the help of an army of monkeys.

[4]The name of tree under which *Sita* Devi spent one year while she is in Lanka. That garden is called '*Ashokavanam*'.

of *Ravana* in meditation.' [2]. The monument radiates cultural and religious fusion with the craftsmanship of the artisan.

48.3.3 Categorization of Motifs and Patterns

As mentioned earlier in the paper, the Ghanasyam House is one of the most ornamented monuments of Ahom architecture. The data gathered were too large to compile together. For the ease of sorting the data, categorization theory [4] is used. His study of 'Categorization..Natural Language and Design' states that the organized categorization can lead to concepts, and with that visual information can be controlled. The source of the theory was from Roach [8], in *Cognition and Categorization*, clearly demonstrates that instead of categorizing objects into large numbers of finite discrimination and tightly organized concepts, it is better to let the mind automatically select the cognitively optimal economical option of neglecting the infinite differences among objects to behaviourally and cognitively usable proportions. The categorization process serves non-identical objects as equivalent when differences are irrelevant to the human response.

Hence, considering the theory under the superordinate level the exterior and the interior sections of the monument are considered, the data of the monument are sorted based on the entrance, wall, pillar and roof. At the basic level, the motifs and patterns types are included as these are the core subject of the research. Finally, the types and variation of each motif and pattern types are sorted. Using this categorization chart (Fig. 48.2), the data are easily traceable with the source, and it reduces the chances

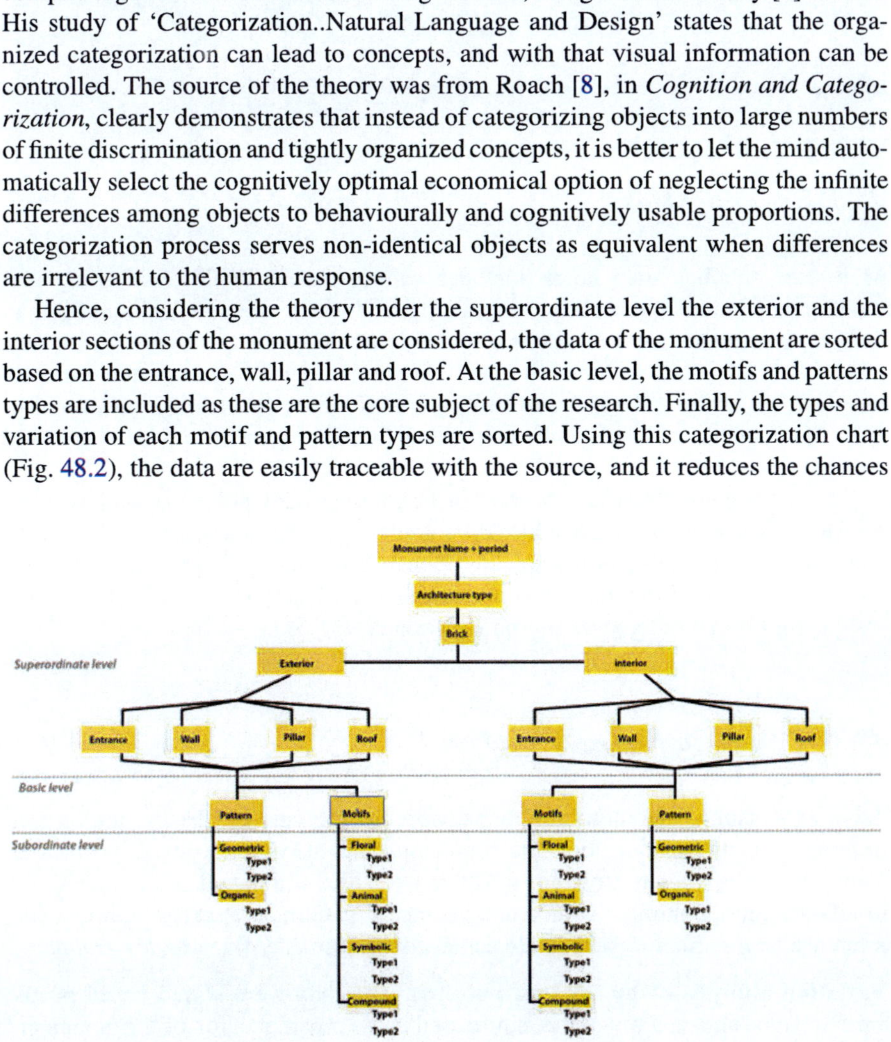

Fig. 48.2 Systematic categorization chart for the motif types and pattern types

Table 48.1 Coding on motif and pattern type based on the superordinate level

S. No.	Sections	Motif type	Code
1	Entrance (En) Pillar (Pi) Wall (Wa) Roof (Ro)	1. Floral	*M-En-F-01*
		2. Compound	*M-En-C-01*
		3. Symbolic	*M-En-S-01*
		4. Animal	*M-En-A-01*
S. No.	Sections	Pattern type	Code
1	Entrance (En) Pillar (Pi) Wall (Wa) Roof (Ro)	1. Organic	*P-En-O-01*
		2. Geometric	*P-En-G-01*

of data redundancy.

48.3.4 Coding Data

Motifs are subdivided to floral (includes various shapes of flowers), compound motif (motif united with multiple elements like flowers, creepers, animals, human or branching plants within a frame), symbolic (includes mythical animals or religious symbol) and animal (includes birds, fish, human, wilds animals and domestic animals). Patterns are divided into two groups—organic pattern (includes floral or abstract organic forms) and geometric patterns (includes geometric lines, curves and circular forms).

The motifs and patterns are observed in the entrance (En), pillar(Pi), wall(Wa) and roof(Ro) of the monument. To reduce the complexity of the data sorting and analysis, Table 48.1 is being coded with designated alphabets and numbers. For example, floral motifs in the wall of the monument are coded as *M-Wa-F-01, M* for motif, *Wa* for wall, *F* for floral motif and *01* for the variation type.

48.3.5 Data Sorting and Analysis

Table 48.2 displays the motif variations, with quantity and the description of each different type, information about the floral motifs and the variations with its quantity. Considering the similar structure of Table 48.2, the compound motifs, symbolic motifs and animal motifs, organic and geometric patterns are sorted with its *code*, reference *image*, *quantity* and its elemental compositions types with *descriptions.*

Variation analysis of motifs and patterns. The data were sorted based on the variation type and quantity of the motifs used in the ornamentation of the monument along with its structural descriptions. The quantified results of Table 48.2 indicate that there are seven different types of floral motifs with variation in shapes and sizes.

Table 48.2 Variations of floral motifs with descriptions

Code	Image	Quantity	Description
M-En-F-01		4	Terracotta Lotus motif in the Kaleidoscope Scene of Ramayana above the Multi-foil Arch entrance. The lotus has a circular base with 3 step petal expansion
M-Wa-F-02		3	Lotus motif on the front wall. The lotus has a circular base with 3 step petal expansion. The petals are bit sharp and the base is simple
M-Wa-F-03		3	Broken Lotus motif on the front wall. The lotus has a circular base with 3 step petal expansion. These are the largest lotus of all
M-Wa-F-04		5	Lotus motif on the front wall. The lotus has a circular base with 3 step petal expansion. 2 on the back wall of the monument
M-Ro-F-05		160	Lotus motif near the roof line. The lotus has a Square base with 3 step petal expansion with 4 leaves on each edge
M-Ro-F-06		5	Lotus motif near the top roof. The lotus has a Square base with 3 step petal expansion
M-Wa-F-07		1	4 petal flower with line details on it. The plaque is found to be on the back wall of the monument

It was observed that out of 248 motifs, 73% are floral motifs, 4% are compound motifs, 17% are animal motifs and 7% are symbolic. The variation analysis of the floral motifs indicates that the lotus motif is the predominant floral motifs of all. According to [9], motifs can be obviously representational (e.g. a rose) or abstract and non-representational (e.g. a circle is drawn around a pentagon), may possess an inherent symbolism. However, in the popular imagination, they may often be deemed to fulfil a purely decorative or ornamental function. Interestingly when the variation of the motifs is examined, apart from the symbolic motifs, the ornamentation includes animal motifs like parrot, elephant and lotus, which falls under some of the prominent Hindu symbolic motifs, and these motifs have been on use in early Indian manuscripts and textiles. This shows that the integration of symbolic elements and sculptures was an essential consideration for the ornamentation of the monuments. The monument includes pattern plaques in both the exterior and interior sections. For easy navigation of data, based on the interior and exterior sections, the data are analysed. It is observed that in the exterior surface of the monument, there are more than 709 motifs plaques, out of which 63% are organic patterns and 37% are geometric. In the interior, out of 89 pattern plaques, 69% are organic patterns and 31% are geometrical. The results of both the data indicate that the organic patterns are more commonly used pattern in the monument. In the analysis, it is observed that consideration of seamless design is visible in the patterns. The designs of the terracotta plaques are moulded in such a way that when placed next to each other, a visual continuity of the motif is observed. The pattern plaques display conscious design consideration for artistic need. This reflects the design understanding of the artisan in terms of architectural ornamentation.

Shape and form. The visual compositions of the motifs and patterns are analysed based on shape, form, texture and colour. Table 48.3 studies the elemental composition of the varied compound motifs. A similar structure has been adopted for floral motifs and varied patterns. The table identifies the different shapes which accumulate in the formation of the motifs and describes the compositions considering the types. The analysis extracts the rudimentary elements like the petals shapes, core, base, extension type in branching, abstracts shapes and repetition of shapes. Analysing the visual attributes, it is observed that in most floral motifs, the petals of the flowers are rounded in shape compared to pointed petals. The repetition of the petals is arranged in radial manner, and in some floral motifs, the arrangement of the petals do not follow a strict repetition pattern but vary individually or progressively to conform to overall shape and structure, which according to Wong in his book Principles of Form and Design describes it as Affinity, leading to unity in the composition [10].

The compound motif (Table 48.3) is an amalgamation of multiple organic shapes and forms. Some of the motifs follow branching, with flowers and leaves. The petals and leaves are pointed and rounded in shapes. Few leaves also follow metamorphosis and deformation in shapes and some display proliferations within the shapes. Interestingly all the compound motifs are composed within are bordered frame.

In animal motifs, the analysis considers the visual anatomy of the animals. Most of the animal motifs are represented in forms of human, animals and birds. Head, eyes, body, legs and hands are distinctly visible in human forms; apart from that, man-made and natural forms are placed within the frame. The animals form consists of head, legs, limbs, body, tail, tusk, trunk and ears based on animal species. The bird

Table 48.3 Element extraction of compound motifs

Motif type	Elements				Description
Compound	Core/ centre	Petals	Sepals	Border / Base	
		Rounded Floral Cut	Pointed		The floral motif has a rectangle border; The central floral motif has two layers of petals with rounded tip. There are two flowers on top and bottom of the motifs which are half cut from the center. The central flower has sepals extended towards each corners of the rectangle border. The sepal tips are pointed in shape. Proliferation is seen on the petals.
	Branching	Flowers	Leaves		
		Pointed	Rounded & Pointed		The compound motif follows symmetrical branching with flowers and leaves. It has two branch extensions with flowers and leaves on each sides. The branching ends with a larger flower with two leaves extended on both side from the top of it. Two of the flowers includes buds with them. The floral motif has a rectangle border
		Rounded	Rounded & Pointed		The compound motif follows symmetrical branching with flowers and leaves, It has two branch extensions. The branching starts from the base with flowers and leaves on both the sides. The larger flower at the top extends the other branch with pointed tip leaves. This motif has a rectangle border

representation includes head, beak, legs and feathers. The animals and birds both the compositions are associated with natural forms like trees, plants and reptiles (Table 48.5 displays the linear extraction of animal motifs). It is observed that most of the composition of the animal motifs display movement in the living beings' forms and posture.

The ornamentation patterns exhibit the repetition of the unit forms (Table 48.5 displays the linear extraction of pattern types). The organic pattern consists mostly of flowers and leaves. The petals are rounded and pointed in shapes. One pattern follows a spiral curve connecting the next spiral, which is just the inverse of it. The design ends with a half-cut spiral, which, if added, another block displays a seamless design. Angular line groves are seen all over the spiral curve, including the base's edges in multidirection. The centre of the spiral has a floral motif with rounded petals.

Colour and texture. The colour of the Ghanasyam House (Fig. 48.1) radiates a tint of red colour due to terracotta plaques, bricks and mortar compositions. The terracotta process includes baking of clay as for which it creates a brown and reddish tone in the brick plaques and according to ASI, the composition of the early mortar includes mixing of lime (limestone and snail shell), pulses, resin, hemp (*canarium resiniferum*), molasses, fish, etc. Again the fusion of molasses and limestone leads to the creation of red tint. In the ornamentation of the monument, the decorative texture is used for contrast in the composition. On the entrance arches of the front and Ramayana's visual narratives, zigzag texture is observed on the edges of each composition. The composition of the texture decorates the surface and remains subordinate to the shape. Though the zigzag lines of the texture have irregularity, the composition of the texture plaques maintains a degree of uniformity with overall ornamentation. This highlights the artisan's design consideration with the elemental composition of each form and shape of the ornamentation leading to a harmonious design.

Composition analysis of ornamentation in monument. Considering motifs and patterns as the elements of the ornamentation, the placement of motifs and patterns displays repetition, gradation, contrast, anomaly, similarity and concentration. Table 48.4 shows the details of each component type within the frame of reference. In compositions with repetition, two-way continuance type is mostly seen in the overall ornamentation style. The nine lotus motif in the front wall (Fig. 48.1: Ghanasyam House) displays gradation in shape and size, also leading to a concentration in composition towards the central lotus motif. The similarity in the composition is commonly observed considering the repetitive of use of motifs and patterns as ornamentation. Concentration in the composition is observed with the symbolic motifs (*Temple with Siva linga inside*) in all the three entrance arches from the front. The placements of the pattern plaques highly reflect the contrast in compositions; considering the thick and thin plaques, the placements create contrast within the ornamentation. Anomaly is observed in one of the pattern plaques where the plaques change direction from side of the front wall towards the centre, forming an angle. The pattern plaque in the corner changes its shape and size retaining its elements to display seamless design, with continuity with the pattern flow. The monument's overall composition radiates

Table 48.4 Composition types in ornamentation of the monument

Composition with	Spatial relationship
Repetition *Two way continuance*: Repetition of floral motifs, compound motifs are seen in the ornamentation of the monument. The forms follows two way continuance in repetition. The arrangement of the patterns are extended horizontally and vertically. This arrangement are extended from one corner all the way to other corner of the monument	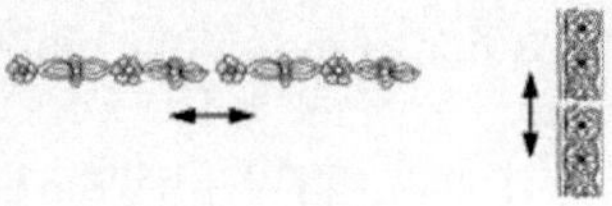
Gradation *Shape and size*: Gradation of the lotus motif is observed, as the shape and the form gradually varies which arc placed over the visual narratives of ***Ramayana***. Gradation in sizes of the lotus motifs is seen, from smaller to bigger towards the centre and again lo smaller in the end	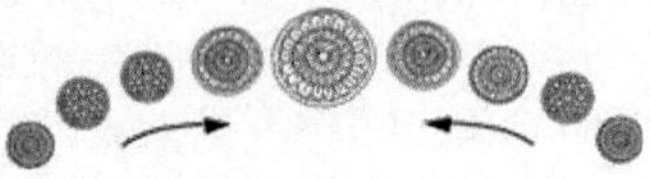
Similarity *Repetition*: The context of similarity is considered on the base of repetition considering the specific natural forms and animals. The peacock with snake in its beak and the spread wing peacock falls in the similarity of species yet the form and orientation are different, same with other similar animal motifs with same species	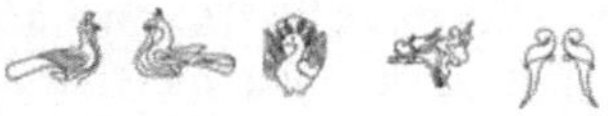
Concentration *Linear Concentration*: The formation of the band in the symbolic motif is seen, which together is inclined towards the tip of the arch type entrance of the monument	
Contrast *Appearance and Placement*: Considering the portion of the left and right section of the front wall as the frame of reference, the pattern composition exhibits the contrast in appearance and the placement of the patterns. The placement of thin pattern plaques between the thick pattern plaques helps to differentiate the variation of the ornamentation	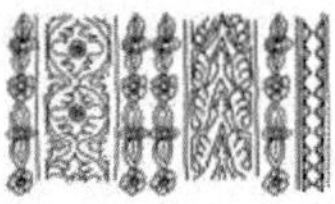
Anomaly *Shape*: Anomaly is observed in one of the pattern plaques where the plaques changes direction from side of the front wall towards the centre forming an angle. The pattern plaque in the corner changes its shape and size retaining its elements to display seamless design, with continuity with the pattern flow	

Table 48.5 Symmetry analysis of motifs and patterns

Motif Types	Rotational symmetry	Rotational symmetry: 4 fold Symmetry	Reflection Symmetry: Two direction reflection symmetry	Reflection Symmetry: One direction reflection symmetry
Floral				
Compound				
Animal	Asymmetric type with 360 degree rotation 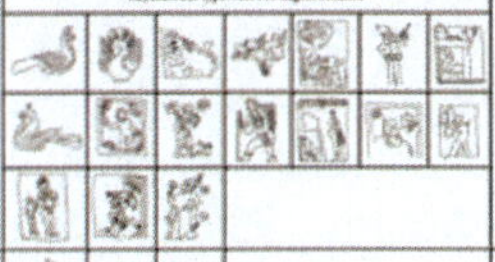			
Symbolic				

Pattern Types	Transitional symmetry: Two direction reflection symmetry	Transitional symmetry: One direction Reflection symmetry	Transitional symmetry: Asymmetric type with 360 degree rotation
Organic			
Geometric	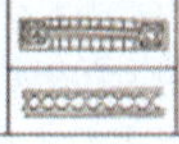		

balance, harmony and unity, irrespective of varied motifs and patterns with distinctive shapes and form.

Symmetry analysis. [9] defines that the symmetry implies balance of physical form, and, in everyday usage, can be applied to an image, figure, or object with two equal parts, each of the same size, shape and content, one a reflection (like a mirror) of the other. Symmetry is applicable to three-dimensional objects, the concept of symmetry is more easily understood in the context of two-dimensional design. Considering it, the linear extraction of the motifs and patterns of the Ghanasyam House is done. Using the four symmetry operations rotation, reflection, translation and glide reflection, the motifs and patterns of the monuments are analysed.

Symmetry in architecture. The architecture of Ghanasyam House itself retains reflection or bilateral symmetry. If an imaginary vertical line is drawn from the centre of the middle multifoil arch, most of the elements of the ornaments present in the front wall support the bilateral symmetry (Fig. 48.1) and considering the vertical axis it can be termed as one-directional reflection symmetry [9]. Similarity in motifs placement can be observed on both sides of the front wall. Animal motifs like a peacock with a snake in it's beak are placed on both the top corners of the front wall facing towards each other; similarly, the symmetry in the arches and the visual narratives over the smaller multifoil arches on each side supports bilateral symmetry. The floral motifs like the lotus motifs present in the front wall are also placed in such a way that it supports the reflectional symmetry. Considering the overall ornamentation and the architecture of the Ghanasyam House, it is observed that the structure follows the reflection symmetry operation, including placement of motifs and patterns. Rani [11] applied the core symmetry operations to study the Mural Art Forms of Santhal Pargana, Jharkhand, India. The collected data were categorized and analysed based on the selected districts. As the symmetry analysis of [11] seems relevant with the research analysis of this paper, the analysis method has been adopted. Considering the variation type, the data here are divided and analysed based on the categorization theory. With the help of theory, the data analysed are easy to trace back to the superordinate level, resulting in systematic data analysis based on the typology of the monument.

Symmetry analysis in motifs. Table 48.5 displays linear extraction of various types of motifs in the monument. From the symmetry analysis, it is observed that the single floral motifs follow rotational symmetry, and two of the types follow fourfold rotation symmetry. The compound motif has one-directional and two-directional reflection symmetry. Apart from it, most of the animal motifs are asymmetric in nature with 360° of rotation, there are only two types with one-directional reflection symmetry. The symbolic motifs are asymmetric in nature with 360° rotation symmetry. With the analysis of the data, it is observed that out of 284 motifs, 73% follow rotation symmetry, only 4% follow bilateral symmetry and 23% are asymmetric in nature with 360° of rotation symmetry. The analysis shows that rotational symmetry is the predominant symmetry, including some motifs variations with fourfold symmetry.

Symmetry analysis in patterns. Table 48.5 of the pattern type analyses the pattern plaques of the monument. Each terracotta pattern plaques are placed to provide transitional symmetry, i.e., one after another. In both organic and geometric pattern plaques, it is observed that the motifs plaques are grouped with more than one element. Both organic and geometric plaques possess one-directional and two-directional reflection symmetry, also some patterns are asymmetric in nature with 360° of rotation. With the data, it is observed that out of 798 pattern plaques, 50% of the plaques possess one-directional reflection symmetry, 32% possess two-directional reflection symmetry and 18% are asymmetric with 360° of rotation symmetry. The symmetry analysis indicates that the predominant symmetry in the pattern plaques is one-directional reflection symmetry.

48.4 Conclusion

After the study and analysis of Ghanashyam House, it is fair to conclude that the structure is rich with simple and complex motifs and patterns. The motifs like peacock, elephant, parrot and lotus are also a part of symbolic representation in the ornamentation. The sections like the multifoil arched entrances, Mihrab inside the monument, the visual narratives of Ramayan terracotta plaques, motifs of tribal man carrying fruits, Hindu symbolic motifs such as Shiva linga, definitely reflects the process of fusions and integrations of the culture and philosophical system of that era. The details of the ornamentation highly showcase the craftsmanship of the architects. The variations analysis indicates that the ornamentation of the monument is more inclined towards organic floral designs and patterns than geometric designs and patterns. The visual elements of the motifs and patterns inherit design principles like symmetry, repetition, radiation, affinity and unity. The ornamentation's composition analysis highlights the underlying principles like repetition, gradation, similarity, contrast, concentration and anomaly. The use of seamless design in the pattern plaques discloses the expertise and conscious decision in design consideration and development of the ornamentation by the artisans. The monument structure, including the ornamentation, follows symmetrical properties. The study highlights the design considerations with elemental compositions to maintain visual balance, retaining the sense of unity in the overall ornamentation. The symmetry analysis gives a new perspective to cultural integration in terms of visual art and design. With the analysis of the motifs, it is clear that the rotational symmetry is the predominant symmetry, and this because of the repetitive use of floral motifs as a part of the ornamentation. Hence, the comprehensive study on variation analysis, composition analysis and symmetry analysis of the ornamentation of this paper will contribute in understanding role of design and its principles in the conservation of historic buildings and traditional art.

Acknowledgements The author would like to thank Archaeological Survey of India(ASI), Guwahati Circle, Assam; the Directorate of Historical and Antiquarian Studies, Guwahati, Assam for the support in collecting secondary data.

References

1. Das Gupta, R.: Art of Mediaeval Assam. 1st edn, pp. 10–40. Cosmo, New Delhi (1982)
2. Das, P.: Terracotta Sculptures of Assam. 1st edn, pp. 74–84. Research India Press, New Delhi (2019)
3. Monga, C.: Identification of Design Elements in Naamghar of Assam (Ph.D). Indian Institute of Technology Guwahati, Assam (2019)
4. Athavankar, U.: Categorization… natural language and design. Des. Issues **5**(2), 106 (1989). doi:https://doi.org/10.2307/1511517
5. Barpujari, H.: The Comprehensive History of Assam, 3rd edn, pp. 247–275. Publication Board Assam, Guwahati (2007)
6. Gait, E.: A History of Assam, pp. 176–177. Lawyer's Book Stall/Saraighat Printers, Panbazar (1997)
7. Gupta, S., Panda, S.: Centrally Protected Monuments of NorthEast India. 1st edn, p. 33. Superintending Archaeologist, ASI, India, Guwahati Circle, Guwahati (2011)
8. Roach, E., Lloyd, B.: Cognition and Categorization, pp. 7–10. Lawrence Erlbaum Associates, Publishers, New Jersey (1978)
9. Hann, M.: Symbol, Pattern and Symmetry. 1st edn, pp. 23–34. Bloomsbury, London (2013)
10. Wong, W.: Principles of Form and Design. 1st edn, pp. 188–231. Wiley, New York (1993)
11. Rani, P.: Rural women artists: a visual analysis of the mural art forms of Santhal Pargana, Jharkhand, India. J. Int. Women's Stud. **18**(1), 80 (2016). Retrieved from https://vc.bridgew.edu/jiws/vol18/iss1/6

Chapter 49
Brand Style DNA in Consumer Products: Decoding Strategies from a Design Perspective

Shoubhik Dutta Roy and Wricha Mishra

Abstract Brands use style DNA strategies to differentiate from competition and enhance brand perception. It is, therefore, critical for design professionals to have a clear understanding of these strategies, so as to be able to deliver the benefits successfully. However, literature available on the subject is found to be varying in terminology and largely limited to certain product categories and iconic brands. Brand style DNA concerns the complete portfolio of a brand, but very few researchers have gone to the extent of analyzing a brand's product range and comparing it with that of competitors. Moreover, a bias toward the marketing and psychological perspectives necessitates research focused on the design perspective. This paper consolidates available knowledge into an integrated structure, something that does not exist at present. It comprises of four facets—parameters of brand style DNA strategies, factors to be considered, evolution over time, and deconstruction techniques. Further, this study deep-dives into one of the most prominent areas of decoding brand style DNA strategies—the analysis of explicit design cues. Two existing techniques, namely Design Format Analysis and 4DD Analysis, are critically examined. A systematic and definitive conceptual framework is proposed that addresses the weaknesses of existing techniques and makes significant improvements through introduction of image boards with multiple views, standardized design cues, and category-level analysis.

49.1 Introduction

'Brand Style DNA' can be defined as a set of building blocks of a brand's style, which is consistently used across its product offerings to strengthen its brand identity. In this highly competitive era, consumer product brands are emphasizing on defining a

S. Dutta Roy (✉) · W. Mishra
MIT Institute of Design, Pune 412201, India
e-mail: shoubhikduttaroy@mitid.edu.in

W. Mishra
e-mail: wrichamishra@mitid.edu.in

A. Chakrabarti et al. (eds.), *Design for Tomorrow—Volume 1*, Smart Innovation, Systems and Technologies 221, https://doi.org/10.1007/978-981-16-0041-8_49

proper strategy for the style DNA of their products [1]. The term 'consumer products' is commonly understood to encompass all man-made objects we use in our daily lives.

Brands develop a consistent and distinct style for their products for two reasons: to facilitate recognition and to transfer an existing product's impression on to a new product from the same brand. The style is manifested in certain visual references that are deliberately and consistently embedded in the brand's portfolio. Some famous examples are the 'waisted' bottle profile of Coca Cola, and the 'kidney-grille' seen on every BMW car [2].

Literature available on brand style DNA reveals parameters, considerations and techniques from different viewpoints and concerning different contexts. However, an overview carried out from the design perspective reveals some critical gaps. There is clearly a bias toward the psychological and marketing perspectives. Psychology-heavy papers delve into various aspects of consumer perception related to brands [3, 4]. Marketing heavy literature talks about issues of brand positioning and innovation [1, 5]. The drawback of this perspective bias is that there is lack of clarity on the practical implication of this knowledge for designers. This is particularly worrisome because of the importance of design in implementing brand style DNA. There are two aspects to this. First, design is that attribute of a brand's DNA that bonds together other attributes such as ease of use, technology, and dependability [6]. The second aspect relates to the role of designers. Although product managers and brand managers make strategic decisions regarding style DNA, the responsibility for its implementation, i.e., creating a distinct and consistent brand image, rests with product designers [7, 8]. It is for this reason that several authors recommend close collaboration between managers and designers [9, 10].

Moreover, literature available on brand style DNA is scattered and limited to specific product segments or specific brands. Approximately half of the available studies pertain to the automotive domain, while consumer products of everyday use do not get adequate consideration. Literature also shows a bias toward analyzing iconic or premium brands, while mass market brands get insufficient attention. Another gap encountered is that a large chunk of literature on brand style DNA discusses individual products, without looking at the product range offered by the brand. Whereas, deconstruction or reconstruction of a brand's style DNA requires proper understanding of not just the brand's portfolio but also that of its competitors in a particular category. On the whole, a glaring question that emerges is: which style DNA strategy works for what kind of brand, and in what kind of product segment? Existing literature does not provide a clear answer. Therefore, it can be assumed that there is paucity of studies toward a structured framework to decode brand style DNA.

In view of the above, the present study is carried out with two objectives:

1 To consolidate all facets of brand style DNA into an integrated structure
2 To propose an articulated conceptual framework to decode the style DNA of key brands in any consumer product segment.

49.2 Methodology

Brand style DNA strategies are complex and the purpose of this study was to systematically decode its various nuances. This was approached in two parts. In the first part of the study, an overview of relevant research papers was carried out, directed toward finding answers to the following questions:

1. What are the parameters of brand style DNA strategies?
2. What are the factors affecting brand style DNA strategies?
3. Does the style DNA of a brand change over time? If yes, why and how?
4. What are the techniques for analyzing and implementing brand style DNA?

Answers to these questions revealed different viewpoints and approaches. These were compared and consolidated into an integrated structure, presented in Sect. 49.3 of this paper. Building on the resultant structured body of knowledge, the second part of the study was focused on developing a conceptual framework to assess the brand style DNA of different brands in any consumer product category. The process is demonstrated taking a particular product category as an example. Finally, the benefits and applications of the framework, its limitations, and scope for future research are discussed in detail.

49.3 Facets of Brand Style DNA

The overview of literature on brand style DNA yielded valuable insights, although these were found to be scattered, and varying in terminology and contexts. The knowledge is summarized below under four sections (Fig. 49.1) according to the four research questions (Fig. 49.2).

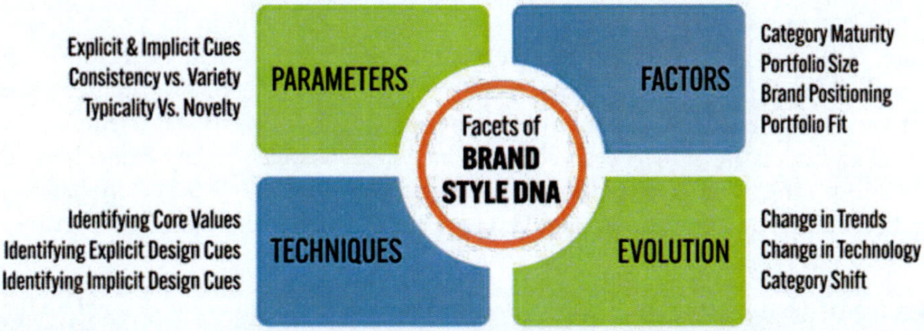

Fig. 49.1 Facets of brand style DNA

Fig. 49.2 Examples of explicit and implicit cues. *Source* https://www.bmw.com.my/ and Torres [15]

49.3.1 Parameters of Brand Style DNA Strategies

Style DNA strategies are too complex and case-specific to be classified into types. However, a review of strategies discussed by multiple authors across the world reveals three pairs of parameters, namely explicit and implicit cues, consistency versus variety, and typicality versus novelty. The style DNA strategy of any brand can be explained in terms of these parameters [1, 11–14].

Explicit and Implicit Cues. Brand identity is manifested in certain design cues that the brand uses consistently across its product range. Karjalainen (2007) classifies them as explicit and implicit cues [12]. Explicit or 'artificial' design cues are easy to identify and make products attractive. For example, BMW uses the explicit cue of kidney-shaped grille across its range to create recognition [12]. Implicit or 'value-based' cues, on the other hand, are subtle and express the core values of the brand. An interesting example is Alessi, the Italian company famous for its unique kitchen utensils and housewares. Alessi's brand personality lies in its fun value combined with functionality, and its high visual and emotional appeal [15].

Consistency versus Variety. Design cues of a brand, whether explicit or implicit, are used across its product range. This creates consistency within the brand's portfolio. However, individual products of a brand need to be distinguished from each other, calling for some level of variety. Brand style DNA calls for the right balance between consistency and variety. Too much consistency hampers categorization of a brand's products and affects brand excitement, while too much variety reduces brand reliability [16]. Simoni et al. [11] call this phenomenon 'stylistic heterogeneity' and describe it as product portfolio language homogeneity versus product portfolio language heterogeneity.

Typicality versus Novelty. The previous two pairs of parameters are internal, in the sense that they look at products within a brand's portfolio. Typicality and novelty are external parameters, because they refer to similarity or difference with respect to competitors in a particular product category. Too typical designs may fail to stand out from the competition and communicate the brand message, while too radical designs may face low acceptability. Therefore, typicality and novelty must coexist in a tension to facilitate brand style DNA [14]. According to Keller et al. [17], point of parity (POP) refers to design elements typical of a product category, while point of difference (POD) refers to unique elements used by a particular brand to create identity. Ranawat et al. [18] use the terms product category descriptors (PCDs) and brand identity carriers (BICs), respectively, to refer to the same.

49.3.2 Factors Affecting Brand Style DNA

Literature has brought out a plethora of factors that need to be considered while defining or following style DNA for a brand. The first three factors described below are broadly based on Person et al. [5] classification. The fourth one, portfolio fit, draws from Andersson and Warell [7] study on brand extensions.

Category Maturity. In product categories that are in an early phase, brands tend to keep designs similar to competitors to avoid the risk of recognition of the purpose of the product. At this stage, they compete on technology and functionality. These aspects get more or less standardized by the time the category reaches maturity. Then brands bank more on design to stand out in the market [5].

Portfolio size. Brands with a smaller portfolio prefer style consistency within their products, for easy recognition by customers. Brands with larger portfolios mostly serve a wider customer base, so they need to address different preferences and contexts. Therefore, these brands are likely to feature greater variety [5].

Brand positioning. Brands can be classified according to their approach to innovation (incremental versus radical) and in their orientation to the marketplace (market-driven versus driving markets). If brands do not respect their positioning during new product development, it may lead to customer confusion regarding brand meaning, and loss of competitive advantage [9].

Portfolio fit. Whenever a brand introduces a new product, it has to give due consideration to where that product would fit into its existing portfolio, so as to maintain the brand's style DNA. Line extension is the most basic type of brand extension, wherein a new product is added to an existing product range in the same product category. Category extension refers to extension of a brand into a category it was non-existent in. Vertical extension happens when a brand introduces a professional, luxury, or low-cost version of one of its existing products [7].

49.3.3 Evolution of Brand Style DNA Over Time

As we see from the above sections on parameters and factors, strategizing brand style DNA is a complex process. Literature throws up another dimension to the entire exercise, that of time. Style DNA of a brand rarely stays the same over the years, it evolves over time. This can be attributed to three reasons. Firstly, technologies change drastically in many product categories, necessitating design changes. Secondly, trends change over time. Keeping the product image updated with the times is necessary to prevent brand aging [19]. Thirdly, brands exit less profitable categories and step into more lucrative ones.

Brand style DNA can be changed in two ways, redesign and revitalization [7]. In redesign, the brand modifies its DNA incrementally based on trends and competition. Revitalization is more drastic, wherein product messages need significant reinterpretation of meanings. They take time to diffuse and achieve success.

49.3.4 Techniques for Deconstructing Brand Style DNA

This is the primary focus area of this study and very critical because it forms the basis for future strategies of a brand. Karjalainen [12] proposed a style branding process involving three stages of deconstruction: identifying core values, identifying explicit design cues, and identifying implicit design cues.

Identifying core values. This stage involves a study of the brand's vision and mission statements, positioning, target customers, and heritage. For brand s that do not have well-defined core values, the Brand Translation Framework introduced by Mulder-Nijkamp and Eggink [20] seems to be appropriate. The framework starts with compiling physical characteristics of the brand's product designs in the form of pictures, then derives associated keywords and finally decodes core values of the brand.

Identifying explicit design cues. McCormack et al. [21] shape grammar approach accurately captures the geometry behind brand-typical shapes, but fails to capture elements of color and texture. Moreover, the use of parametric rules makes the process too tedious and mathematical for analyzing product style. Somewhat similar is the case with Ranscombe et al.'s 'degree of similarity' method. It is useful only for certain product categories that have subtle differences between products [22]. The most promising method seems to be Design Format Analysis (DFA) developed by Warell [23]. This technique is discussed in detail in Sect. 49.3.5.

Identifying implicit design cues. Implicit cues are more difficult to identify, because they belong to the language domain and are highly subjective in nature. Therefore, this stage is ideally based on consumer research. A widely used method for this is Semantic Differential, wherein bipolar parameters are rated on a Likert scale to decode the personality characteristics of individual brands [11, 14].

Literature primarily deals with identifying explicit design cues, as this is considered to be the most important and most elaborate step in decoding brand style DNA. Therefore, having overviewed the three key stages above, we delve deeper into analysis of explicit cues.

49.3.5 Analyzing Explicit Design Cues

The most widely accepted technique for analyzing explicit design cues is Design Format Analysis. This subsection critically analyzes existing versions of DFA and proposes an improved framework to address its limitations.

Existing versions of Design Format Analysis. Karjalainen's [12] DFA plots the occurrence of characteristic design features of a brand against selected products of the brand. This seems to work for iconic brands like Apple that Karjalainen exemplifies. However, in case of most mass market brands, the characteristic design features are not articulated by the brand or by researchers. Karjalainen does not explain how to use the technique for such cases. Some more shortcomings surface as we examine it in detail. In Karjalainen's examples, product visuals used are not in the same view, and several of them are shown in flat-front view that does not communicate the form fully. Moreover, Karjalainen's explorations are limited to individual brands, he does not analyze a product category.

Ranawat et al. [18] presented an improved version of DFA called 4DD analysis, which addressed some of the shortcomings mentioned above. They expanded the original method to analyze different brands in a particular product category. In their example of power tools, shown in Fig. 49.1, they do a summation of scores in the right-most column and identify product category descriptors (PCDs). They also identify brand identity carriers (BICs) as all the cues that are not PCDs. This is not convincing, because a design cue that occurs in only one of the four products of the brand cannot be an identity carrier for the brand. Another concern with this technique is that some design cues change from brand to brand. This makes it hard to compare the tables side by side. Figure 49.3 shows a comparison of the two techniques described above.

The proposed framework. The conceptual framework seeks to address concerns with existing techniques and provide an elegant and definitive approach for decoding brand style DNA in any given product category. First of all, top brands in the product category are identified and key products selected for each. Image boards are prepared for all the selected products, comprising of up to three key views of the products. Images of all selected products have comparable views, and the same are scaled to comparable dimensions.

The image boards are examined in detail for variations in aesthetic elements, i.e., form, color, texture, and space, and explicit designs cues are identified for the product category. Next, a brand-level DFA table is prepared for each of the brands. Summation of scores is done on the last column. If a brand is found to have several high scoring cues, it can be said to have high consistency. Next, a category-level DFA table is made, wherein total scores of all brands are further added to a grand total. Here, high overall scoring cues are identified as product category descriptors. The remaining cues, that have high scores for individual brands but not at the category level, are identified as brand identity carriers. The overall technique is summarized in the conceptual framework shown in Fig. 49.4.

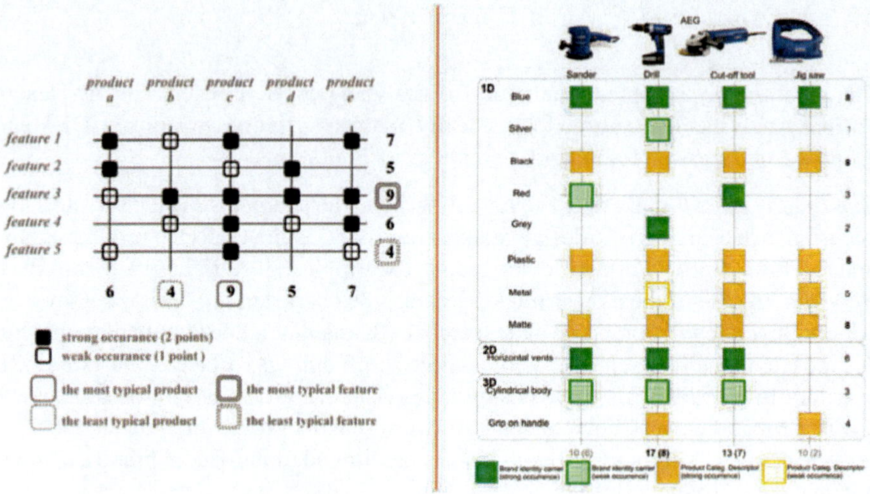

Fig. 49.3 Comparison of design format analysis (left) and 4DD analysis (right). *Source* Karjalainen [12] and Ranawat et al. [18]

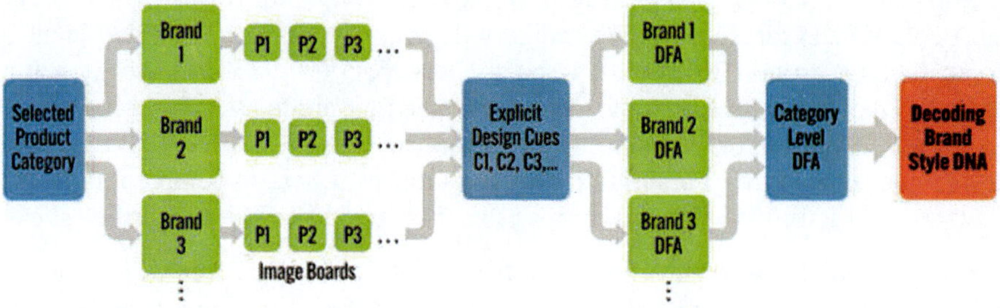

Fig. 49.4 Proposed framework for decoding brand style DNA

Example. Let us understand the conceptual framework with the help of an actual example. The category chosen was semi-premium watches for men. The price range for semi-premium was defined as ranging from 5000 to 20,000 rupees in the Indian market. Top five products of top five brands were shortlisted, based on ratings and popularity (as on March 19, 2020) on amazon.in, the top ecommerce retailer in India [24].

The brands selected were Fossil, Casio G-Shock, Casio Edifice, Titan, and Invicta. Image boards of all 25 products were prepared, with three key views of each product (Fig. 49.5). This step is completely missing from existing techniques, and this step is vital for meticulous identification of explicit design cues. A total of 21 cues were identified, based on band material, case material, case thickness, bezel design, dial shape, dial complexity, color scheme, display type, and hour markings (Fig. 49.6). The cues are standardized across brands, unlike in Ranawat et al.'s 4DD analysis.

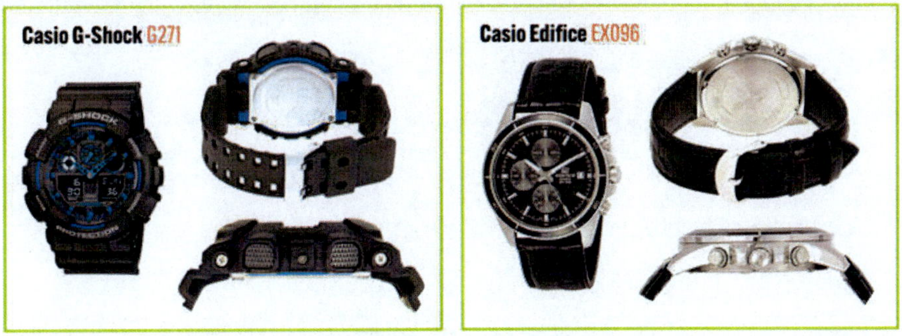

Fig. 49.5 Image boards of 2 of the 25 watches

Fig. 49.6 Twenty-one explicit design cues identified for the product category of watches

Unlike prior techniques, here the DFA table was constructed at two levels to bring transparency to the process (Fig. 49.7). It was first constructed for each brand. Results showed that Casio G-shock had very high consistency, because 11 out of 21 design cues were found to be consistent, while only 3 were inconsistent. Casio Edifice had high consistency, and the other three brands were found to have low consistency. Next, the category-level DFA table was created. PCDs were identified as metal case, markings on bezel, circular dial, and analog display. Casio G-Shock was found to have the most BICs, while Fossil and Invicta were found to have the least.

BRAND LEVEL DFA TABLE FOR CASIO G-SHOCK

	Casio G-Shock G310	Casio G-Shock G271	Casio G-Shock G346	Casio G-Shock G441	Casio G-Shock G660	Σ
Metal band						0
Leather band						0
Plastic band	●	●	●	●	●	10
Metal case						0
Plastic case	●	●	●	●	●	10
Thick case (16-18mm)	●	●	●	●	●	10
Slim case (9-11mm)						0
Exposed screws/rivets	●	●	●	●	●	10
						0
Segmented bezel	●	●	●	●	●	10
Markings on bezel	●	●	●	●	●	10
Circular dial	●	●	●		●	8
Complex dial	●	●	●	●	●	10
Light coloured dial						0
Dark coloured dial	●	●	●	●	●	10
One-colour product	●	●	●	●	●	10
Analog display		○	○		○	3
Digital display	●	○	○	●	○	7
Roman hour markings						0
Number hour markings					○	1
Rectangular hour markings		●	●		○	5
Circular hour markings						0

CATEGORY LEVEL DFA TABLE

○ Product Category Descriptors

○ Brand Identity Carriers

	Fossil	G-Shock	Edifice	Titan	Invicta	Σ
Metal band	2	0	(8)	6	(8)	24
Leather band	(8)	0	2	4	0	14
Plastic band	0	(10)	0	0	2	12
Metal case	(10)	0	(10)	(10)	(10)	40
Plastic case	0	(10)	0	0	0	10
Thick case (16-18mm)	0	(10)	0	0	4	14
Slim case (9-11mm)	0	0	2	(10)	0	12
Exposed screws/rivets	1	(10)	0	0	0	11
Segmented bezel	1	(10)	2	0	0	13
Markings on bezel	1	(10)	(10)	4	(10)	35
Circular dial	(10)	(8)	(10)	(10)	(10)	48
Complex dial	5	(10)	(8)	5	3	31
Light coloured dial	4		0	(8)	6	18
Dark coloured dial	6	(10)	(10)	2	4	32
One-colour product	0	(10)	2	1	5	18
Analog display	(10)	3	(10)	(10)	(10)	43
Digital display	0	(7)	0	0	0	7
Roman hour markings	6	0	0	2	0	8
Number hour markings	0	1	(0)	3	0	4
Rectangular hour markings	4	5	10	5	4	28
Circular hour markings	0	0	0	0	6	6

Fig. 49.7 Brand-level DFA table for Casio G-Shock and category-level DFA

49.4 Discussion

This study has two important achievements. Firstly, it presents an integrated structure for understanding the facets of brand style DNA. This was possible through a careful assembly of scattered knowledge, equating different terminologies for similar phenomena, and comparing different techniques used by different authors. Secondly, it proposes a conceptual framework that systematically decodes brand style DNA of key brands in any product category. This was achieved through identification of explicit design cues, in a transparent and diligent manner, as demonstrated in Sect. 49.3.5.

The proposed structure and framework are created to benefit product design professionals, researchers, and even marketing professionals concerned with the aesthetic aspect of a product brand. Since the responsibility for implementation of brand style DNA rests with product designers [7, 8], this structure and framework would give them a formal justification for product style. Beverland et al. [9] warn against misalignment between brand positioning (led by marketing) and new product development (led by designers). The proposed method is neither heavy on psychology nor on statistical tools. Therefore, it can serve as a common discussion platform for design and marketing professionals. Debates between the two sides regarding design strategy can be formalized with this method, urging both sides to take a focused look at each product offering, right from an early concept phase to the final launch phase.

Brand style DNA continues over generations of products, therefore it is critical to articulate explicit design cues [21]. The proposed framework does exactly that, thereby ensuring that a brand's style DNA remains consistent irrespective of which designer works on it at which point of time. Further, the framework can be used by brands to do a style status check with respect to competition and ease the decision on whether to move toward greater similarity or greater differentiation. This decision is a primary aspect of strategic product styling, as confirmed by Person et al. [5]. By clearly identifying characteristic explicit design cues for a brand, the proposed framework would help it to keep a check on whether its style DNA matches with its core values. After all, consumers would recognize a brand and its associated values only if they have been translated correctly into explicit design cues [13]. Moreover, the propositions of this paper would be of immense benefit to new brands and young designers, helping them unearth style DNA strategies of successful brands for benchmarking and inspiration.

The proposed framework addresses the shortcomings of existing techniques and suggests significant improvements. Brand-level and category-level application of DFA are presented as separate tables for transparency in the process and clear calculation of PCDs and BICs. Category-level analysis was not present in Karjalainen [12] examples and ambiguous in Ranawat et al. [18] version. Another significant improvement is that Ranawat et al.'s 4DD analysis uses varying design cues for different brands, while the proposed technique standardizes them into a common set. This facilitates direct comparison of the selected brands and also consolidates all design cues in use within the category, into one list.

The existing techniques analyze style attributes based on one image per product. For several products, one view may not do justice to the design and important design cues may get ignored. Therefore, this framework suggests the use of up to three images of the selected products. Further, the proposed framework recommends creation of image boards with multiple views, so that the identification of design cues can be approached with rigor and appreciation of design details. This is evident from the fact that 21 design cues were identified for a medium-complexity product like wrist watches, while examples found in literature, even for cars, do not exceed 14 design cues.

The scope of this paper was to the extent of developing a framework for analyzing explicit design cues. While it facilitates vital insights about style DNA of brands in a particular category, decoding individual brands' strategies would remain incomplete until we decipher implicit design cues, through consumer surveys on different product categories. Analyzing explicit and implicit cues together can help us assess brand personality. There is immense scope for further study, and we hope this paper helps design, research, and marketing professionals to make further contributions to the knowledge base, in the domain of brand style DNA.

References

1. Dell'Era, C., Verganti, R.: Strategies of innovation and imitation of product languages. J. Prod. Innov. Manag. **24**, 580–599 (2007)
2. Karjalainen, T.M., Snelders, D.: Designing visual recognition for the brand. J. Prod. Innov. Manag. **27**, 6–22 (2009)
3. Warell, A.: Identity references in product design: an approach for inter-relating visual product experience and brand value representation. Procedia Manuf. **3**, 2118–2125 (2015)
4. Kreuzbauer, R., Malter, A.J.: Product design perception and brand categorization. In: Fitzsimons, G., Morwitz, V. (eds.) Advances in Consumer Research, vol. 34, pp. 240–246. Association for Consumer Research, Duluth (2007)
5. Person, O., Schoormans, J., Snelders, D., Karjalainen, T.M.: Should new products look similar or different? The influence of the market environment on strategic product styling. Des. Stud. **29**(1), 30–48 (2008)
6. Takamura, J.H.: Molecular branding: a study on the nature of the relationship between product design and brand loyalty. In: Conference Proceedings of DeSForm, pp. 4–55. Northumbria University, Newcastle (2007).
7. Andersson, T., Warell, A.: Aesthetic flexibility in the management of visual product branding. Procedia Manuf. **3**, 2191–2198 (2015)
8. Andersson, T., Warell, A., Holmlid, S., Ölvander, J.: Product gist: An approach to identifying form characteristics of the current product sign. In: Proceedings of the 10th European Academy of Design Conference—Crafting the Future. Göteborg University, Gothenburg (2013)
9. Beverland, M., Napoli, J., Farrelly, F.: Can all brands innovate in the same way? A typology of brand position and innovation effort. J. Prod. Innov. Manag. **27**(1), 33–48 (2010)
10. Kreuzbauer, R., Malter, A.J.: Embodied cognition and new product design: changing product form to influence brand categorization. J. Prod. Innov. Manag. **22**, 165–176 (2005)
11. Simoni, M., Cautela, C., Zurlo, F.: Product design strategies in technological shifts: an explorative study of Italian design-driven companies. Technovation **34**(11), 702–715 (2014)
12. Karjalainen, T.M.: It looks like a Toyota: educational approaches to designing for visual brand recognition. International Journal of Design **1**(1), 67–81 (2007)
13. Mulder-Nijkamp, M., Eggink, W.: Unravelling the secret of successful brand extension: a case study to explore consumer response. In: Proceedings of the 19th DMI: Academic Design Management Conference, pp. 479–504. The Design Management Institute, London (2014).
14. Lee, S., Self, J., Andrietc, E.: Most advanced yet acceptable: a case of referential form-driven meaning innovation. In: Lloyd, P., Bohemia, E. (eds.) Proceedings of DRS2016, pp. 157–178. The Design Research Society, Brighton (2016).
15. Torres, P.: Creating Brand Recognition Through Product Design. In: International Conference on Research into Design, pp. 205–214. Bangalore (2015).
16. Fischer, M.S., Holder, D., Maier, T.: Evaluating similarities in visual product appearance for brand affiliation. In: Fukuda, S. (ed.) Advances in Affective and Pleasurable Design, pp. 3–12. Springer, Cham (2019)
17. Keller, K.L., Apéria, T., Georgson, M.: Strategic brand management: a European perspective. Pearson, Harlow (2012)
18. Ranawat, A., Tuteja, S., Hölttä-Otto, K.: Contribution of visual design elements to the perceived product family look. J. Des. Res. **10**(3), 189–205 (2012)
19. Hangyue, C., Shijian, L.: Product visual image design based on product family DNA. In: 9th International Conference on Computer-Aided Industrial Design and Conceptual Design, pp. 783–787. Kunming (2008)
20. Mulder-Nijkamp, M., Eggink, W.: Brand value by design: the use of three levels of recognition in design. In: Proceedings 5th IASDR 2013 Tokyo, pp. 5639–5650. Tokyo (2013)
21. McCormack, J.P., Cagan, J., Vogel, C.M.: Capturing the essence of a product's brand through a shape grammar representation. In: Folkeson, A., Gralen, K., Norell, M., Sellgren, U. (eds.) Proceedings of ICED 03, pp. 175–176. Stockholm (2003)

22. Ranscombe, C., Hicks, B., Mullineux, G., Singh, B.: Visually decomposing vehicle images: exploring the influence of different aesthetic features on consumer perception of brand. Des. Stud. **33**(4), 319–341 (2012)
23. Warell, A.: Design syntactics: a functional approach to visual product form theory, models, and methods. Chalmers University of Technology, Gothenburg (2001)
24. Leading online stores in India in 2017, ranked by net e-commerce sales. Survey by Statistica. https://www.statista.com/statistics/297355/leading-e-retailers-in-india-ranked-by-annual-web-e-commerce-sales/

Chapter 50
Product Semantics: The Emotional Design Language

Lakshay Gaur and Subhajit Chandra

Abstract This study explores the co-existent nature of emotions and cognition in humans to build and propose a framework that helps correlate 'likeability' and 'sellability' of a product by introduction of a relatively new term—unique selling factor (USF). The framework runs on context-based logical correlations among its constituents. The aim of this framework is to qualitatively express the emotional characteristics of a product. As emotions work alongside with cognition, the design attributes of the product under the lenses are first analysed as per the three levels of our brain's processing. Each design feature corresponds to the processing level based on the consumers' probable preferences to choose that feature in the first place. After this cognitive breakdown, we further diverge the semantic analysis at emotional levels. Each design feature when stated with the consumers' probable preference and the cognition level involved can now help develop context of the scenario. This context that triangulates the connect between the design feature, consumers' preference and processing level plays significant role as the backbone of *emotionality* in the analysis overall. To apply the understanding built, a logical study is done considering a black V-neck T-shirt as the product under the lenses. For this product, we define the likeability, sellability and the unique selling factor. For analysis, we create a feature analysis table that subdivides the product features first, into its design characteristics. Second, against these characteristics are explored the probable reasons the consumer might have had for opting for those characteristics. Third, each reason for the preference for its respective design characteristic is assigned to its corresponding levels of brain processing. Fourth, for the context developed so far, we can assign emotions involved. The co-existence of emotion and cognition paves way for this product-semantic design language. Thus, the framework proposed works evidently on emotion and cognition and helps provide a novel perspective—that of the most

L. Gaur (✉) · S. Chandra
Shiv Nadar University, Greater Noida, UP, India
e-mail: lg985@snu.edu.in

S. Chandra
e-mail: subhajit.chandra@snu.edu.in

A. Chakrabarti et al. (eds.), *Design for Tomorrow—Volume 1*, Smart Innovation, Systems and Technologies 221, https://doi.org/10.1007/978-981-16-0041-8_50

significant stakeholder of all—the customer and the people we design for. The framework follows an ecosystemic approach that provides it with appropriate literature and a holistic approach.

50.1 Introduction

It is for the co-existence of human emotions and cognition, researchers are exploring the paradigm of 'Emotion in Design' [1–3]. We can define emotions as our response to stimuli in order to survive, and it is these responses that require processing power of brain to build constructs to help take decisions accordingly in respective situations [4], 5. Emotions can also be defined as a 'state of mind' an individual is in throughout the whole scenario as well as during each stage of such scenario. Studies in the domain of 'emotions' identify that emotions are always context-driven [4]. Emotions and cognition hence work in tandem arrangement to interpret the context of the situation one is in and respond accordingly [5]. Psycho-evolutionary theory of emotions divides emotions as primary and secondary emotions. There are eight basic emotions that exist as four pairs of two opposite emotions—joy and sadness, trust and disgust, fear and anger, surprise and anticipation. It is possible to experience the same emotion(s) through complex network of other emotions but in relatively similar context [4]. These 'feelings' or 'emotions' stimulate our brain to process at three levels—visceral, behavioural and reflective. At visceral level of processing, we experience what can be called as 'precursors to emotions' therefore these set the base of the situation an individual is experiencing [5]. Visceral levels lead to immediate perception and result in the feeling of either attraction or repulsion. As compared to the 'immediate perception' at visceral level, at behavioural level actions are associated with an 'expectation'. Achievement of expectation gives the feeling of control to the individual. This expectation is important in any task because when the outcome is up to the expectation, it releases tension of the outcome and creates a sense of *relief*. Emotions at behavioural level are that of hope and joy, frustration and anger. As compared to the abovementioned two levels, reflective level of processing is the most impactful as this is the stage where reasons are assigned to build mental constructs in mind. A positive reflection over a product or experience drives the individual to recommend the same further. The superior emotions of guilt and pride, blame and praise and anger and pleasure are felt as reflection over 'something' [5].

Insightful studies have been conducted in the domain that may differ in goals but share common grounds—emotion. Most of them conclude that designing for emotions is not about aiming to induce emotions or cater to human emotions straight ahead, but about building a product that is balanced in all of its key components along with its emotional aspects [1, 5]. This research adopts design research method—II (DRM—II) as for which following phases were adopted in iterative fashion (when and where needed) instead of a linear approach—(i) review-based *research clarification*: to create a tentative structure for the overall research in; (ii) *comprehensive descriptive study*: that is aimed to develop in-depth knowledge through literature review and detect key features to build prescriptive framework; and (iii) *initial prescriptive study*: that proposes a framework which is context-driven and is an update on 'emotionality' to the previous model of analysis [6].

The research work adopts a small-scaled study with concept representation and analysis method applied on a T-shirt. An example of radio as a product is also undertaken to draw analogies with real-life examples. This approach for designing for emotions is multi-faceted, diverse in content and balanced overall.

50.2 The Need of Human-Centred Framework

'It is said the single best predictor of an individual's behavior is the persons' intention to perform that behavior' [7]. It is for this reason, marketing managers survey applicants to help measure consumer's 'purchase intentions' (PI) through their probability of making the respective purchase in time to come, termed as 'purchasing behaviour' (PB). The results of the survey on purchase intentions are correlated with the purchase behaviour of the applicant population. It becomes crucial to note whether the applicants behave in accordance with their 'intent' (stated during the survey) or in contrast. Though the process of surveying and correlating the results appear to be unidirectional, it reveals many facets to itself when viewed from a broader perspective. These revelations guide researchers into deep insights of their research. Though the correlation between purchase intention and purchase behaviour is found to be 0.53, it is in 95% confidence in a range of 0.15–0.92 [7, 8]. Further research works show that the relationship between the two terms depends on several factors—how intentions are measured, type of product, mere act of measuring purchase intentions, certainty, specificity, etc. These factors are to be held responsible for having an affect on consumer's preferences. Such cases are:

(i) Survey type
(ii) Question-behaviour effect
(iii) Customization.

Studies show that type of survey done with applicants can change their responses accordingly. There are multiple ways of surveying the applicants. These may differ—in how they ask the applicant the same question in different ways; in the way of questioning and in the variety of responses the applicant can respond in. Of these, *purchase probability scale* predicts most reliable results. It happens so because when applicants are asked to respond in binary (yes/no) [7], their *psychological process* weighs whether their probability of purchasing is high enough to respond 'yes' as their intention or not. Hence, it is better to *directly* ask and measure their probability of purchasing [7, 9]. This change in approach for surveying is a step towards 'humanizing' the research being conducted because it shifts the perspective of the research towards the consumers' end. Hence, how a survey measures intention can significantly make an effect on consumers' behaviour.

Similarly, the mere act of measuring ones' intentions can change their behaviour. This is the impact of the '*question-behaviour*' effect. It makes people tend to predict that they will be more likely to perform *socially desirable behaviours* rather than socially embarrassing ones when their intentions are measured as compared to those who have not been surveyed in likely fashion. These people not only predict positively but also tend to perform the behaviour in real. This rigour of 'socially desirable' act is what triggers a response in positive state and makes the individual experience the feeling of 'satisfaction' before even performing the act because the act itself brings you closer to your desired-self, here, in a social context [10]. Desire drives us to stretch beyond what we would do to fulfil our needs or wants. The achievement of 'something desired' gives immense pleasure and satisfaction, whereas 'need' or 'want' of something is more fundamental to us as compared to upper hierarchical relativity of 'desire'. To have a mobile phone to connect through calling is a need as compared to added features of high resolution camera, good speakers, touch screen technology, and customized mobile phone cover. The same features can be basic requirement to some people but for others they may be *desired*. This journey towards the desired-self is what is termed as '*customization*', which caters to individuality and uniqueness of each individual [3]. The emotions that we feel on achieving this desired-self are felt in higher intensities (than other emotions) which make 'customization' more significant to the consumer. Customization not only leads to feeling of emotions of more intensity but also stimulates cognition at all levels of brain processing—visceral, behavioural and reflective—in our case. When individuals make a prediction, they follow through on it and hence behave differently as to how they would have if not have made a prediction or had been surveyed in first place.

These transitions from 'being asked' to 'perform' are where our personal choice shape our user experience. First, the mere act of being asked corresponds to *visceral level* of brain processing which makes one anticipate the feeling of *satisfaction.* Second, the follow-up on the task to be performed, i.e., a desirable one, corresponds to *behavioural level* of processing which makes one dig deeper into the task in whole and realize its merits even further. Third, the completion of the task makes one's mind process on *reflective level* and feel the achievement of the desire.

Hence, the underlying map of the whole system tends to run and situate itself around the consumer's psychology in order to fit marketing decisions in alignment to this psychology. The most effective type of survey, the question-behaviour effect, the desire-driven cognition as well as the tandem working of emotions and cognition itself, all are facets of the common aim-driven research—measure of purchase intention and purchase behaviour. Hence, we need to balance the approach overall from both the industry's as well as consumer's perspective. The terms 'purchase intentions' and 'purchase behaviour' stand on analytical and empirical grounds. Their correlation relates them in binary grammar—whether the purchase is made or is not made, whereas to a consumer, the psychological process from 'intent' to 'behaviour' is relatively more complex. From a research point of view, the understanding of consumers' psychological process is a must but from the other end, consumers' sub-conscience weighs and analyses the product under the lenses. Here, product semantics in emotional language becomes important. The attributes of 'a design' are not just to be decoded but must be related in emotional context too. Hence, the product also carries a significant value in the resonant perspective as does the consumers' behaviour of purchase. It is the semantic analysis of a product that helps it fit into the constructs built over time in our minds, hence guiding our behaviour in turn. Product semantics aims to simplify complexities in order to improve user–product interactions as well as enhancement of self-exploration through products [11]. The understanding developed so far further extends the envelope of 'design semantics'. It brings about the importance of literal terminologies, the balance of multiple perspectives and the framework proposed that acts as a tool. This tool helps in semantic breakdown and analysis through what new it brings to the design society. As discussed in related works, we can certainly believe that no manufactured products stay contextually neutral. These have a statement of their own to make [1] reference.

50.3 Ecosystemic Approach for the Resonant Framework

Finding new aspects of utility of emotional expression in real life, it is required for designers to strong, articulate and thorough with their work (reference) [1]. In order to build up a framework that caters to human emotions and cognition, first we need to define and achieve subgoals to create an output in the end—that is, to propose a framework as a tool for designers to incorporate emotional characteristic in designing. Hence, it is required to adopt appropriate literature to define key aspects of the system and set rules, relations and assumptions for the framework to process. (i) We redefine the available literature; (ii) we introduce new terms that are crucial to the framework being built; (iii) we impart relations in between them; and (iv) we analyse to gain output.

For the purpose of the research, we first update what has previously been defined as 'purchase intention' and 'purchase behaviour' to 'likeability' and 'sellability' of the product. Here, likeability can be defined as viewer's preference of choosing a certain product out of other options available. This preference comes due to the design elements of the product the consumer prefers 'personally' than in an applicant population. Such preferences can also be seen when our perception regarding what we see comes from our past experiences. Designers must design with this knowledge (of customer through 'their' experiences) and draw metaphors by mirroring or abstracting the *referential semantics.* Alteration in proportions and roundness contributes to the 'age' of the product. Similar to the phenomena, it has also been observed that baby features induce a feeling of warmth and protectiveness [1]. Similar approaches have been seen in the case of radios wherein form of the radio, trend, culture and emotional styles of products are significantly crucial to sales as well as to customer experience. It is awarded to the design process and decisions in marketing, outreach, etc., that has led radios to exist as a successful product—or more like as an electronic equivalent of chair—over decades [2]. Through the consideration of such 'personal preferences' the view point of the approach first becomes more humane and secondly, caters to 'customization' indirectly where the consumer is not directly choosing over other options but the analytical system takes the 'preference' into consideration. Also, the shift from the term 'intention' to 'liking' is that of context, which is now more consumer-centric. Similarly, 'sellability' of the product advocates whether the product through its design characteristics is resonant enough to resonate with the consumers' psychological process and therefore differs from 'purchasing behaviour'. If it is so, that is if the design characteristics in whole resonate (as a product) with what the consumer consciously or subconsciously may be looking for, then there will more chances of stronger liking of the product. Similar phenomenon is evident in human brain, as a resultant of the same we are able to allocate deeper meanings to our own body ratios and proportions. For example, if head is bigger it looks cute and child-like [1]. Stronger intent or liking during the survey (the survey should be defined in accordance to the framework proposed) may even be followed by positive consumer behaviour—its purchase. Since it is human emotions the whole framework is built upon, it becomes essential to set the context of the same. This context is the correlation of 'likeability' and 'sellability' of the product which is analogous to the correlation of purchase intention and purchasing behaviour as discussed earlier. To define this correlation, the term 'unique selling factor' is introduced. The context that the correlation brings about is the vision through the lens of common resonant perspective between the industry and the consumer.

The three terms—'unique', 'selling' and 'factor'—in the entity of USF stand separate grounds as well as a common one. The term 'unique' represents the individuals' personal preference from their own perspective, 'selling' attributes to researchers' point of view as 'purchase intentions' and 'purchasing behaviour' of the applicant and 'factor' conjoins the two perspective into a common shared spectacle of the consumer and the researcher—by redefining purchase intentions and purchase behaviour as 'likeability' and 'sellability'. For this common spectacle and the context in overall, stands the entity of 'unique selling factor' that expresses the strength of the probable transition of the 'ability' of liking to be turned into a 'factor' of purchasing, subjected to and based on—indirect customization through semantic understanding of the product's design attributes. The contextual nature of the USF can also be shown graphically in Fig. 50.1.

The 'unique selling factor' *correlates* likeability and sellability through the contextual and qualitative relationship in between. It facilitates the correlation through analyses of design aspects of the product on the basis of structuring of emotions and processing levels of the brain but also catering to individuals'

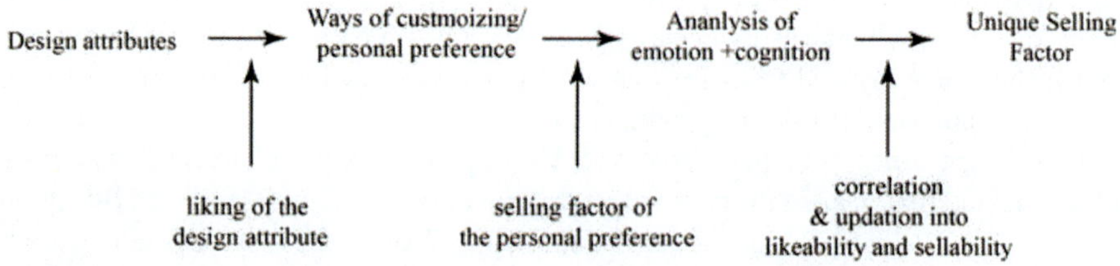

Fig. 50.1 Unique selling factor (self-created image)

probable preferences to stimulate this analysis in the first place. To demonstrate the significance of the USF, an appropriately scaled study has been undertaken in order to communicate and visualize the application(s) of the hypothesis. The realization of the unique selling factor follows a process to express/suggest a contextual-based understanding model of the product, rather than an equation that calculates the result. The preferences consumers make can be conscious or subconscious, volitional or non-volitional and are hence quantified accordingly in the framework. It is this blend of qualitative, quantitative and contextual pillars of the framework that is analogous to transdisciplinary nature of design and the methodology followed that makes the research 'scientific' [6]. Hence, the framework follows *ecosystemic approach* by taking advantage of the very limitation of the surveying system—that the product is not available in real but in imagination of the applicant's mind—and builds the method around it—to tackle the situation through psychological process in the consumers' mind. The approach henceforth followed explores the problem statement and maps the solution around it. To explain the framework further, an example is considered where the product to be purchased is a T-shirt with following characteristics:

Colour:	Black	Neck cut:	V-Neck
Size:	Medium	Brand:	ABC
Material:	Cotton	Price:	XYZ

50.4 Unique Selling Factor

The 'idea' of the feature analysis table in Table 50.1 is to use it as a tool to understand as well as convey the hypothesis of the emotional design language. It aims to explore and explain the product through relatively qualitative and quantitative semantics. The hierarchy in between (i) want, desire and need and; (ii) visceral, behavioural and reflective levels of brain processing act as indicators in quantification methodology. Hence, the feature analysis table helps represent products and quantify their emotionality in relative manner. The output the table provides is the view to an 'understanding' from the resonant perspective of the researcher/designer and the consumer. It is this understanding that we can term as the *unique sellability factor*. For the same, it is out of the scope of this research to develop a scale to measure emotions. The approach followed does not try to mimic human emotions but rather attempts to define its own definition of emotions. It is done by considering constituents of the framework in generic sense. For the sake of the most significant stakeholder—the consumer—designers must respect this privacy and design around it. Instead, as designers we can create products or services that provide an emotional interaction. Henceforth, in the approach to feed data to the table, we explore the probable reasons for consumers' decisions and impart holistic view through semantic analysis.

Table 50.1 Feature analysis table

Design element	Reason for choice/preference	Analysis parameter										
		Cognitive			Emotional							
		V	B	R	J	T	F	S	Sa	D	A	N
Colour	Person. prefr	•	•	•	R			V	B			V
	Trend	•			V							
	Regnl. bckgr		•									
	Availability	•	•									V
Size	Comfort	•		•		VR						
	Usage type	•	•			VB		VB				
Neck	Person. prefr	•		•	R			V				V
	Trend	•			V							

Cognitive:
V visceral level processing, *B* behavioural level pros, *R* reflective level pros
Emotions:
J joy, *S* surprise, *A* anger, *T* trust, *Sa* sadness, *N* anticipation, *F* fear, *D* disgust

Decisions can be made due to different reasons for different consumers in different situations—personal preference, trend, regional background, type of usage, brand awareness, comfort (which can again be a personal preference), availability, etc. All these reasons vary in the way how our mind works on them and under which construct(s) are these analysed. Some chosen design elements may be personal preference hence may have higher impact while filtration like—colour, if the same colour is in trend then its impact may change. Product form, culture that the product is engulfed in or the one it stimulates, emotional styles and narrative of the product–user relationship are similar elements. Similarly, if price and material are elements the consumer may not have direct control over, they may be of less impact whereas price may have higher impact than material. If someone is more concerned for their comfort, material may have higher impact accordingly. This volatility in the system requires complex structuring of emotion–cognition mapping and setting hierarchical relationships and is hence subjected to a different research in itself. For the sake of relative ease, the reasons for choices made can be viewed under general spectrum though it is crucial to express the USF in qualitative and quantitative terms for which the scale of the USF is set in relational terms. These terms are the design elements, their choices and corresponding processing level involved along with the probable emotions in play. Table 50.1 shows the working of the proposed framework followed with description of the same.

The work done does not put boundaries to product semantics, instead it provokes a call for deeper research and analysis. Regarding the same, further research being pursued aims to legitimize the framework for which carefully designed surveys are being conducted and anew insights are coming to light. The higher degree of the framework can also be turned into a computational model. Through a generic spectrum, we can demystify products of various natures—tangible, intangible, services, corporate entities, etc. Through this approach, we can find where a product stands strong and where it must be reinforced. As from the example chosen for the analysis table, we can draw insights:

(i) Colour, when chosen as personal preference at reflective level may be a joyous experience but at visceral level, it may lead to anticipation of the outcome to the consumer. On behavioural level, the anticipation may further give joy if met with satisfaction or sadness if not. If the same is chosen due to one's regional background, behavioural level maybe at work. Even though personal preference relates to all three levels of processing in different scenarios, it can holistically and emotionally be more valuable than other probable reasons.
(ii) Neck pattern, at reflective level maybe chosen due to self-recommended design attribute hence may relatively be more significant to the customer as 'satisfaction', though at visceral level, one may anticipate how the product will fit if it is a new experience with the chosen neck pattern.

The brief discussion is a demonstration of the 'how to' read the table as well to relate design attributes to emotion and cognition, followed with their *relative* quantification.

50.5 Discussion

The feature analysis table *expresses* design elements of a product on the basis of consumers' reasons to choose them. The psychological process in action for these choices is further broken down into cognitive and emotional parameters. The analysis first adopts cognitive breakdown as Stage—I and then subdiverge it into emotional structuring as Stage—II. It is crucial for initial stage to be cognition based in order to systemize the research and then contextually enrichen it with emotional structuring. As discussed earlier, the three levels of brain processing are associated with their respective emotions. Though these can and surely are complexly interconnected, for the sake of simplicity in correlating them, we analyse in direct or rather linear fashion.

For each design characteristic, there can be several reasons to prefer that characteristic. The preferences are further divided into their respective processing levels as—(i) colour: being visual may correspond to visceral level but being in trend, can also be associated with reflective level. If it is chosen due to regional background, the analysis complicates and becomes even more divergent, it can also be processed as visceral levels if an individual while shopping online comes across a new colour and likes it instantly; (ii) size: can be a matter of comfort as well as the type of usage one intends to put the product in. If someone knows their size of preference, reflective level processing is likely to be involved also, the same can be analysed at visceral level if the product visually looks *comforting* and at behavioural levels if someone has *experienced* the similar product or design characteristic before. Similarly type of usage such as party-wear, night clothes, official purpose, etc., will be processed accordingly; (iii) neck type: can be a personal preference due to ones' own inclination to buy the certain type of neck T-shirt hence associated with reflective level and if trend brings something new to the plate then maybe visceral.

Connecting these directly with emotions is a matter of context-based structure but for developing a generic understanding, a simpler correlation can be considered. Colour, when processed at visceral levels, may be due to element of surprise which may lead to sudden attraction towards the product, one may even anticipate how the particular product will suit them when worn. It can also be joyous to find a certain colour when ones' personal preference drives to choose it due to reflection over the same. Similarly trend when processed at visceral level, can lead to a joyous moment to have found the product in trend. Regional background can certainly guide the consumers' choices also. Availability on the other hand cuts down options available but one can surely anticipate the fit of the T-shirt. Comfort can be a very important requirement for many consumers therefore when processed at reflective levels, can lead to purchase due to trust based on previous experiences with the cloth or the brand as well as can be visceral upon the first look. Usage of the product can similarly be processed at both visceral and reflective levels leading to the same emotion of trust that may be due to attraction in the first look or due to prior experiences of the consumer. The neck cut is a peculiar aspect one may be more conscious while opting for hence may be joyous to have bought or found. It can also be due to anticipation one draws for the same neck cut.

The study demonstrated the emotion–cognition research work through a framework as its own application. Through the analysis of Table 50.1, we fetch the unique selling factor of the product. It is for this new direction, the 'emotion in design' paradigm can be viewed and followed on the emotional design language. Several products can be designed and analyse in similar fashion, e.g. a computer mouse. Its design attributes can be analysed for the overall form, colour, usage and ergonomics. For probable preferences, we these attributes can be explored as—(i) overall form: current trends, personal preference, given as a gift; (ii) colour: can be one's favourite, in trend, eye catching print; (iii) usage: normal daily usage, gaming, designer/artist work, corporate client. In similar way, the context of the correlation can be mapped onto and subdivided into emotions and cognition (as brain processing levels). While in the designing phase the framework and the same feature analysis table can also be used as a tool to analyse one's own designed product. It facilitates for the designer to realize the emotionality of the product and take appropriate decisions in the work ahead.

The correlations between emotions and cognition as discussed above are only meant to provide an understanding of how to correlate the two. For more realistic view, this correlation calls for further research and development of building mechanism that can weave these in complex yet structured lattices. Every such correlation is built up on a context of its own which can lead to same or similar emotions in various constructs. These emotions can further vary in their degree or intensity but are sure to exist even in the very simplest of tasks, goals or subgoals. These lattices are not one-to-one connections but hold virtue when viewed as a single entity comprising of its components—in our case, emotion and cognition. Hence, the framework so prescribed expresses the overall essence of the product through the semantic analysis of its design characteristics.

50.6 Conclusion

Designers struggle with the 'how to' incorporate the emotionality into a product as they see 'emotions' as a stand-alone entity to be catered to, whereas emotions are a state of mind which tells us that emotions and cognition work in tandem. If this tandem arrangement can be explored, considered and fitted into the process of designing well enough, designers can achieve what we term as 'Emotion in Design'. This paper hence takes the advantage to develop an understanding via semantic analysis of products' key features. The grammar of semantic analysis dictates the correlation that acts both ways on 'emotion' and 'cognition' and plays no bias. To understand a product's emotional value, it is evident to realize its cognitive parameters. On these parameters, design attributes of a product can be analysed for which semantic breakdown is necessary. It can provide appropriate grammar to describe product longevity beyond simply physical attributes, expanding the horizons of designing towards emotional value and human relationships with the product. On analysis of all the features proposed in the framework above, we can qualitatively and quantitatively define *the understanding* as 'unique selling factor'.

More is the richness of this entity, more can be the chances of a product being liked or even bought. In the approach other way round considering the framework, if a designer is to design a product with certain emotional features in mind, the roadmap of USF can be followed in the opposite direction. To do so, the feature analysis table can be used as a score board that can help design a product instead of analysing it. Its qualitative aspects can quantitatively be considered while designing. Though this process is another application of the framework so developed and is hence the future scope of this research, yet it displays ambidextrous nature of context-driven systems. Since the framework 'correlates' emotions and cognition, it gives scope to pick either of these to start the process with. When we define emotions as 'our response to stimuli in order to survive', we can certainly say that each scenario we are in is a whole sum of many of its components. For each such situation there is an input our brain processes, and an output it decides to act on. These inputs can be through elements of nature, perceived through our senses, their analysis is context-driven and the outcome is survival (which can again be contextually significant, not restricted to literal definitions). Each such iteration from input to output is a part of an emotion or is an emotion in itself. This further diverges the scope of this research into more generic model of 'Emotion in Design'.

References

1. Demirbilek, O., Sener, B.: A design language for products: designing for happiness. In: Proceedings of the International Conference on Affective Human Factors Design (2001)
2. Yagou, A.: Is everyday technology serious or fun? Reflections Emotional Styles Prod. Des. **17**, 40–56 (2011)
3. Morwitz, V.: Consumers' purchase intentions and their behaviour. **7**(3) (2012)
4. Chakrabarti, A., Blessing, L.T.: DRM, a Design Research Methodology (2009)
5. Plutchik, R.: A General Psychoevolutionary Theory of Evaluation (1980)
6. Goldsworthy, R.: Designing instruction for emotional intelligence. **40**(5), 43–48 (2000)
7. Chapman, J.: Design for (emotional) durability. **25**(4), 29–35 (2009)
8. Gaver, W.: Designing for emotion (among other things). Philos. Trans. Roy. Soc. B 3597–3604 (2009)
9. Athavankar, U.: From product semantics to generative methods (2009)
10. Norman, D.: The design of everyday things. Basic Books (2013)
11. Juster, F.T.: Consumer buying intentions and purchase probability: an experiment in survey design. J. Am. Stat. Assoc. **61**(315), 658–696 (1966)

Chapter 51
Design of a Railway Station: Creative Expression of Cultural Heritage Identity

Somya Mishra and Debkumar Chakrabarti

Abstract By the mid-nineteenth century, the railway had established itself as an essential part of life in the industrial era, and the railway station took on an increasingly characteristic appearance. Architects have amended the industrial aspect of the train sheds through a more traditional façade design. The station buildings are fighting for a place among the league of theatres, museums, and city halls. The role of embellishments is mostly neglected in governance policies for railway stations, even though the architecture and spatial planning are considered. The visual design is monographic and ignores the context of visitors. Establishing the diverse iconography that must develop for decoration of railway stations is a part of an extensive study to showcase cultural heritage with creativity. City marketing has grown into an established field of research and an academic subdiscipline. This article explores visual design elements of the Kamakhya Railway Station that acts as a gateway to the Indian city Guwahati of Assam for people with the intent of visiting the Kamakhya temple. The study proposes a suggestion for the railway station's visual design with a focus on tourism. Finally, it points to areas for further research and exploration.

51.1 Introduction

Indian Railways owns and manages one of the largest railway networks of the world, with over 64,000 route kilometers and 7,000 stations. The Ministry of Railways under the Government of India oversees the operation of this 64,000 km railways, and there are 16 zonal railways headed by their general managers. The Indian Railways carries more than 17.5 million passengers every day, and some of the major railway stations handle 100–200 million passengers per annum [1]. Most of the railway stations have been built over 100 years ago and have a limited and aging infrastructure that handles

S. Mishra (✉) · D. Chakrabarti
Indian Institute of Technology Guwahati, Guwahati, India
e-mail: smishra@iitg.ac.in

D. Chakrabarti
e-mail: dc@iitg.ac.in

A. Chakrabarti et al. (eds.), *Design for Tomorrow—Volume 1*, Smart Innovation, Systems and Technologies 221, https://doi.org/10.1007/978-981-16-0041-8_51

an ever-increasing number of passengers. The railway stations are also located in the middle of the cities and offer enormous potential for redevelopment and commercial expansion. Throughout its vast network, every railway station carries an identity of its kind relevant to their locations.

51.1.1 Indian Railway Stations Development Corporation Limited

Indian Railway Stations Development Corporation Limited (IRSDC) is a special purpose vehicle and a joint venture company of IRCON International Limited (IRCON, A Govt. of India Undertaking, under Ministry of Railways) and Rail Land Development Authority (RLDA) (a statutory authority under the Ministry of Railways). IRSDC has been mandated to develop/redevelop the existing or new railway station(s) in India. The main objective of the company is to make Indian railway stations of international standards and quality by upgrading the level of passenger amenities [1]. It should also maintain location-specific identity.

51.1.2 The Design Brief by the Government

The team of IRSDC has come up with design and planning principles for railway stations, proposed for renovation/redevelopment. These principles are taken from a design competition invitation and include, but not limited to, the following [2]:

Identity and context. The competition seeks designs that will assist in improving the identity of respective stations in the broadest possible way. The design must be a positive place to use, stay, or pass through, be a significant landmark for the area, and act as a positive gateway element for the respective cities. The design must make a positive contribution to its location and engender a sense of civic pride. The cultural context—setting, landscape, built history, memory, indigenous cultural conditions, understanding of the past, and continuity into the future—should all inform the design outcome.

Site design. The concept must propose a well-designed use of the site and respond appropriately to factors including orientation, natural light, maintenance, and servicing. Visibility, lighting, signage, and wayfinding will all be important factors.

The public realm and urban fabric. The design must contribute to its place by careful integration into the existing urban fabric. Designs should create a distinctly urban character and sense of place which complement the identity of respective cities and other citywide strategies, without cliché.

The organizational diagram. A brief organizational diagram must underlie the design concept, with intuitive patterns of movement that invite pedestrian activity within and through the site. Identification of connection networks on and beyond the site is encouraged.

Station area mobility aspects. The railway station is a transit hub, design of operational areas, and comprehensive planning for mobility, access, and circulation is of utmost priority.

Form, massing, appearance. The created space must be attractive with appropriate relationships about scale, form, mass, materials, finish, and function of all elements. The revision site must mediate between the station and the rest of the city.

Functional planning. The planning must be done intelligently together with a demonstrated understanding of amenity and convenience.

Materials. The material used must be intelligent, keeping in mind its usage and economic factor. It should be able to withstand harsh conditions associated with such a public infrastructure project.

User Perspective. The design must recognize and respond to the needs of all users. The proposal must include universal design for people of all ages and abilities, and traveling by any mode.

Community Priorities. In general, all stakeholders and the local community aspire that the station precinct is safe, vibrant, sustainable, accessible, and connected, comfortable, convenient, with an inspiring design that is reflective of respective cities.

The current study addresses on the issues mentioned above in a cultural context linked with railway station as an example.

51.2 Cultural Context

This study focuses on renovating the Kamakhya Railway Station of Guwahati, Assam, and the probable elements of identity of the location to anchor with. The contextual inquiry is made for the Kamakhya temple, where the railway station finds its name. And the city context in terms of the Smart City Proposal for Guwahati is also studied.

51.2.1 The Kamakhya Temple

Kamakhya Devi temple is one of the most famous Hindu temples in India. It has an exceptional status as the seat of Goddess worship. Kamakhya is considered significant also as the center of culture and the capital of the *puranic* period *Pragjyotishpur* or

Kamarupa Desh. Despite a very long and rich active tradition of visitation, worship, and celebration of festivals here, the fundamental nature and character of this sacred complex have hardly changed over the centuries [3].

In India, the system of the Goddess (*Devi*) worship and pilgrimage to their seats (*Shaktipithas*) is one among many ancient living traditions. The absolute total number of the *Shaktipithas* in India, which runs into thousands, attests to the importance of Goddess worship contemporarily. However, the levels of importance vary as per the textually prescribed and perceived greatness and superiority. Due to its glorious past and highly revered status, Kamakhya is most venerable and is called *Adipitha.* There is enough historical evidence to support the fact that Shiva worship was widespread in Assam. The whole mythological development about the beginning of Goddess Kamakhya's worship revolves around a mythological character called *Naraka* [4].

Like other ancient Indian tales, the image and presentation of Goddess Kamakhya are full of symbolic meanings. The essential attributes of Her imagery symbolism are described under different aspects. Since Kamakhya represents the genitals of the Goddess, referred to as the *yoni-pitha* or *yoni-mandala,* it is partly covered with cloth and garlands of flowers. The Goddess Kamakhya, represented simply by her *yoni,* expresses the creative aspect of the power and energy that is an integral element of *Shakti.* Additionally, it could also be explained as a symbol of menstruating, representing the female's fertility. Also, female fertility is regarded as equally sacred by the cultivators concerned with the Earth's fertility.

The Sacred Complex. The sacred complex has different meanings for different people, according to their understanding of the temple boundaries. For some, the physical boundary wall of Kamkhya Temple is the complex, while for others, it is the temple cluster. As per the present study, the sacred complex of Kamakhya extends up to the foothills of Kamagiri/Nilachala. The maps present a synoptic landscape view of Kamakhya and the environment. The central part of the complex is around 5 km away from Guwahati City Centre. One may interpret the pilgrims' cognitive map as the popular image of any sacred place, in this case, Kamakhya, coming to a devotee's imagination. Here, we may add that these maps play a substantial role in perceiving the sacred space, and the mythological narratives play a specific central role in this process. The temple complex houses a variety of 25 Gods and Goddesses (Table 51.1).

Table 51.1 Divinities housed in sacred complex

S. No	Divinities	Numbers	Percentage
1	Goddesses (one for each)	13	52
2	Gods	11	44
	Shiva	05	20
	Ganesha	04	16
	Vishnu	02	08
3	Others (Vetala, Demigods)	01	04
Total		25	100

The Temple Architecture. The temple complex has witnessed several major renovations and now exhibits a mix of medieval and contemporary styles of construction. The temple is divided apparently into four interconnected parts: the *Vimana*, the *Chalanta*, the *Pancharatna*, and the *Natamandira*. Of these, the first three parts' ground plan is original and of the same time, whereas the *Natamandir* was added later. The first three chambers are known as *Garbhagriha*, *Chalanta*, and *Pancharatna* and were reconstructed in 1565 during the reign of Naranarayan. Still, according to archeological studies, the reconstruction of this may have done over the ground plan of an older temple. However, some decorative designs in the wall below the main dome date to the eleventh–twelfth century.

Festivals. In India, most festivals have religious underlining. In the Kamakhya temple complex, the festivals–celebrations are observed at two levels: in the Kamakhya temple and among the common people. The main festivals and their features are discussed below [5]:

Ambuvachi. It is held for three days, starting from the end of *Mrigashira* star up to the end of *Adra* star in the early *Ashadha* (June–July). On the first day, the *Pithasthana* is ceremoniously covered by a red cloth, and the temple is closed, and on later days, purification rites are performed. Red color dominates the visual palette at this time. A big fair is organized during this period, which is supported by the government and voluntary agencies. A vast number of ascetics, monks, traders, and commoners assemble in the fair.

Durga Puja. This *puja* lasts for five days, 6th–10th day of the light half of *Ashvina* (September–October) only. A clay idol of Goddess Durga, in *Mahishasura-Mardini* form, and other companion divinities (*Ganesha, Lakshmi, Sarasvati,* and *Karttikeya*) are also placed in the temple and worshipped specifically on *Mahasaptami, Mahashtami, and Mahanavami*. Finally, on *Dashami* (the 10th day), *Jaya-Vijaya* and *Aparajita puja* are done. The clay idols are taken out in a procession to the Brahmaputra river for immersion in the afternoon. This procession is marked with the sound of *dhol-nagada,* drums, and trumpets, a celebration of power, in the form of *Shakti* and its victory of evil.

Pilgrims and their Motives. Kamakhya, being a Goddess place of the pan-Indian level, naturally attracts people from different corners of India. The presence of people from different parts of the country in connection with business and trade, services (government and private both), etc., in the northeast region facilitates the diversified presence of pilgrims. However, the highest percentage of pilgrims is from Assam, especially Lower Assam. And other pilgrims are from West Bengal, Bihar, Jharkhand, Uttar Pradesh, Orissa, etc., in decreasing order [6].

51.2.2 Guwahati City

Guwahati is the largest city in the Indian state of Assam and the largest metropolis in northeastern India. A major riverine port city and one of the fastest-growing cities in India, Guwahati, is situated on the south bank of the Brahmaputra. The ancient cities of *Pragjyotishpura* and *Durjaya* (North Guwahati) were the capitals of the old state of Kamarupa. Guwahati lies between the Brahmaputra and the foothills of the Shillong plateau. The Guwahati Municipal Corporation (GMC), the city's local government, administers an area of 328 square kilometers. At the same time, the Guwahati Metropolitan Development Authority (GMDA) is the planning and development body of the greater Guwahati Metropolitan Area. The Guwahati region hosts diverse wildlife, including rare animals such as Asian elephants, pythons, tigers, rhinoceros, gaurs, primate species, and endangered birds. The key features highlighted in the Smart City Development Proposal for Guwahati include the quality of life indicators, transportation conditions, water availability, etc. [7].

51.2.3 SWOT Analysis

Strength. The following are the strengths of development in and around the city.

Location and topography. Guwahati, because of its strategic location, is the gateway to the northeastern region. It is a regional hub, and all other states in the region depend on Guwahati for connectivity within the region as well as with the rest of the country. It has a unique wetland ecosystem that should be preserved and restored. Wildlife Sanctuaries and Ramsar Convention listed sites like Deepar Beel are exclusive to the region and can be main tourist attractions that can boost the tourism economy of not only the city but the region as well.

Heritage and Culture. Often referred to as "The City of Temples," Guwahati has a rich historical, cultural, and religious heritage that has the potential to be showcased for tourism. Festivals, temples, archaeological sites, local food, music, dance, arts and crafts, etc., can play a role in the city's physical as well as economic development.

Weakness. The weakness experienced by the city due to its history and geography is discussed below.

Flash Floods and Landslides. Guwahati city has started experiencing flash floods on a yearly basis, which in many cases can be attributed to heavy rains and general degradation of the city's natural ecosystem. Changes in the pattern of land use of the city and unplanned urbanization are one of the main reasons for this reoccurrence of the phenomenon.

Opportunities. With the abovementioned strengths and weaknesses in mind, the opportunities identified for development in the city are immense.

To develop into a Regional Hub—Education, Commercial and IT. Apart from locals, the city attracts most of the students persuading higher education who are from the northeastern states and other states. Also, because of its strategic location and its connectivity with the other northeastern states, it has an innate potential to serve as a commercial and IT capital of the east.

To Become an Ecotourism Destination using the Riverfront. As per tourism plan for the northeastern region, Assam alone received around four lakhs domestic tourists, which is highest in the region and 15, 557 international tourists in 2010. This reflects that Assam's potential to be developed as an ecotourism destination.

Threats. The major threats faced by any type of development in the city, with context to the abovementioned opportunities, are mentioned below.

Urban Expansion. Unauthorized urban expansion near the hills has led to frequent landslides, which, when coupled with soil erosion, results in reducing the capacity of these stormwater drains further as the silt flows directly into these drains and inundates the city.

51.2.4 Kamakhya Railway Station

This railway station is a junction with almost 45 trains per day stopping here. Out of which almost 10 are local trains. Including the weekly trains, train traffic can be up to 60 trains a day. The railway station beautification has been done recently. To reach the station from the city using public transport, one has to drop down at Assam trunk road at Kamakhya Railway Station stop. A 200-m walk inside leads to the station (Fig. 51.1). The station has an eatery outside the station (Fig. 51.2) and several smaller shops and food stalls on the platform. There are five platforms and are accessible from both sides of the city.

Fig. 51.1 Access road to station

Fig. 51.2 Railway station

51.3 The Survey

51.3.1 Site Study Methodology

The users of the Kamakhya Railway Station can be recognized under five categories, namely the pilgrims, the regular local travelers, daily commuters, the shopkeepers on the station, and the railway staff. An open-ended questionnaire was designed to develop an understanding of the requirements of each user group. A sample size of 30 people was chosen, and purposive sampling was used. This method was selected due to the availability of samples and language issues. The questionnaire had two sections, one focusing on the railway station design and the other on the cultural perception of the Kamakhya region. In the first section, the respondents were asked questions about the railway station's identity and its link to the city and other railway stations. The second section inquired about people's perception of the temple and ways it can be replicated in the railway station design.

51.3.2 Survey Results

The survey results are classified according to the five human factors that play an important part in experiencing a place. The factors are physical, cognitive, social, cultural, and emotional [8].

Physical. People feel comfortable and do not want any changes in the spatial planning of the railway station. **Cognition**. Sometimes, the language of signage was a barrier for people with different cultural backgrounds. So, signage for indicating the station is needed, which everybody can understand. Also, the ambiance of the station did not reflect the environment of the Kamakhya temple. **Social**. Due to the social norms, the visuals from the temple cannot be used here. **Cultural**. Cultural significance is high, hence more culturally significant decorations need to be used. **Emotional**. The emotions of the place are attached more to Assam state, particularly to Guwahati city. Thus, the current decorations cannot be scraped off totally. These paintings are significant for people.

51.4 Analysis and Results

51.4.1 Analysis

The user behavior and response analysis have shown that people see the Goddess as an abstract symbol. Although they have a robust relationship with the Goddess, the link is established more strongly with the temple architecture. The offering-activity-culture map is drawn for the railway station and the temple to understand the relationship between them (Fig. 51.3). It is evident that the railway station is related to the activities and those activities generate the culture for the place. For example, the activity of waiting generates the culture of eating, reading, roaming around, sleeping, etc. These activities thus affect the architecture of a place by creating a culture that needs infrastructural support. The offering-activity-culture map of the temple shows that there are lesser number of activities but more cultural impacts. It is seen that celebrating festivals in a particular way, generates the activity of yearly fairs, which is a result of the festival culture of the place. The abstract idea of the Goddess runs in the culture of the place, and the architecture reflects that abstraction.

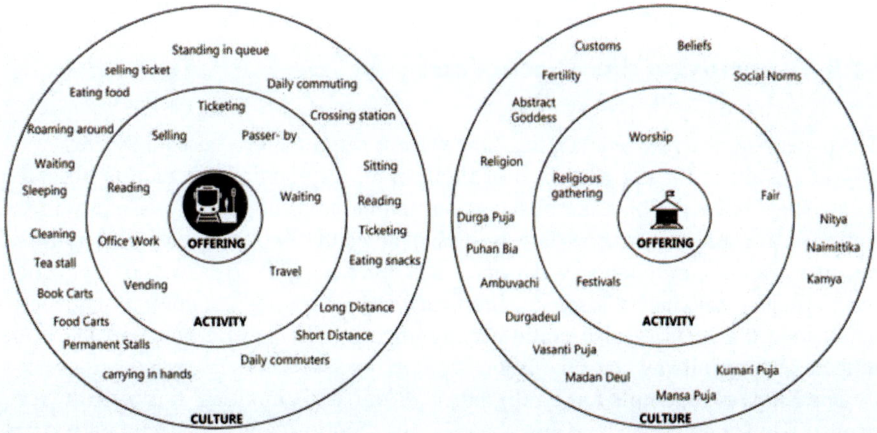

Fig. 51.3 Offering-activity-culture map for Kamkhya Railway Station (left) and The Kamkhya Temple (right)

There is little effect of the Goddess' figurine on the activities. The temple complex majorly caters to the festivals and their celebration. Thus, it can be interpreted that the temple's culture can be combined with the activities of the railway station to generate a more cohesive design that represents the city.

51.4.2 Results

Physical. Although the design of the station includes a lot of red colors and natural settings, it is out of the line of sight. The offering-activity-culture chart shows that only waiting is an overlap in the activities for the two places. So, if the wall paintings are brought down at human eye level and added to the waiting areas too, then the visual connectedness can be achieved.

Cognition. The suggestion from people was to create a visual as well as auditory scene by the usage of more red color in façade design. For acoustic simulation, soft music from the temple, and recitation of *shlokas* and prayers can be added. Even without an understanding of the language, it will give a distinct identity to the station.

Cultural. Cultural significance is high. Hence, more culturally significant decorations need to be used. These decorations include but are not limited to the terracotta murals, native handicraft of the area. These murals may depict stories related to the temple.

Emotional. It is found from the survey that people are more linked to the natural setting of the temple and its festivals. The station may have an exhibition space featuring photographs and paintings of the place.

51.5 Conclusion and Discussion

The paper begins with the analysis of the various requirements laid out by the Ministery of Railways for the new face of modern railway stations. It can be observed that theoretically, the brief emphasizes on maintaining the identity of a city in the station visual and experience design. It also states the importance of perception of people toward their city and its identity. But, somewhere in the process of realization of the project, this factor loses its significance. In this study, an attempt is made to understand the identity of a culturally significant heritage city and how it can be reflected in the railway station design.

The Kamakhya temple has a religious, cultural, and emotional link with the residents of Guwahati city. The Kamakhya Railway Station fails to reflect their perception of the temple into the design. They see the railway station as a transit point and a gateway to their city. But this gateway, according to them, is not a correct representation of the temple's image.

The designers and planners put effort into maintaining a connection between the station and the heritage and culture of the city. The attempts are made to reflect the identity in the interior as well as exterior design. Yet the failure of connecting people's perceptions can be accounted for keeping them out of the planning process. From this paper, we would like to suggest that the place identity of a city is created by the people of that city. If it needs to be appropriately reflected, people must be at the center of the design and be involved in the planning process.

References

1. Government of India Homepage. https://innovate.mygov.in/. Last Accessed 25 Nov 2019
2. Government of India Homepage. https://innovate.mygov.in/irsdc/. Last Accessed 15 June 2020
3. Foulston, L., Abbott, S.: Hindu Goddesses: Beliefs and Practices. Sussex Academic Press, Brighton UK (2009)
4. Singh, R.P.B., Singh, R.S.: Sacred Places of Goddesses in India: Spatiality and Symbolism. Singh, R.P.B. Sacred Geography of Goddesses in South Asia. (Planet Earth & Cultural Understanding Series, Pub. 5. Newcastle upon Tyne, Cambridge Scholars Publishing, U.K, pp. 45–78 (2010)
5. Singh, R.S.: The Kamakhya Devi Temple: Symbolism, Sacredscapes and Festivities. In: Singh, R.P.B. (ed.) Holy Places and Pilgrimages: Essays on India. Planet Earth and Cultural Understanding Series, Pub. 8. Shubhi Publications, New Delhi, pp 81–104 (2011)
6. Assam Life Homepage. https://www.assamlife.com/2019/10/kamakhya-temple.html. Last Accessed 15 June 2020
7. Government of Assam, Guwahati Development Department, Guwahati Smart City Limited Homepage. https://gscl.assam.gov.in/. Last Accessed 24 Nov 2019
8. Kumar, V.: 101 Design Methods: A Structured Approach for Driving Innovation in Your Organization. Wiley (2012)

Chapter 52
Language and Colour

Shekhar Bhattacharjee

Abstract Point of discussion started during colour course At NID with B.Des students. In one of the discussions, it was found that in India different languages have different colour names and different language has slightly different name for same colour. This was discovered when a group discussion was happening on an assignment, collection of colour vocabulary in mother tongue. During the time it was came to my mind that along with the difference in name of a colour, is there any difference in the colour itself also. For example, colour orange is called Narangi, Keshri, etc. In Hindi and in Bengali it is called Kamala, Kusumi and so on. But is Kusumi as colour, different from Narangi or orange? Other way also it can be said that is the word orange produces different visual perception of the colour while pronounced than the word Narangi or Kusumi. Also, as dealing with visual art it was always made me feel that word has connection with visual perception. Prof V. S Ramachandran established the fact that the sound human vocally produces as word has connection with visual form or shape, it could be other way around also, sound of any pronounced word can bring visuals in mind [1]. Prof. Dan Everett argued that language consists of index, icon and symbols with grammar. Now index, icon and symbols are also connected with visual. So, visualization is unavoidable when language comes [2] Also, it was felt that different language creates different visual or multisensory experiences and sensations for the same phenomenon. For example, Bhay (Bengali) creates different visualization than Fear (English). It is also assumed that the language which is been used during zero to five years of human development can give the deeper appeal because, during the age range most of the cognitive development happens, which has lifelong impact [3]. Couple of experiments were conducted to investigate two assumptions, one was 'name of the same colour in different languages create different visualization of the colour' and the second one was 'name of the same colour in mother tongue creates different visualization than foreign language. Experiments were conducted during the colour course with B. Des and M.Des Students. The paper is about relation in between language and visualization, particularly language and colour.

S. Bhattacharjee (✉)
Faculty, National Institute of Design, Gandhinagar, India
e-mail: sekharbhattacharjee76@gmail.com; shekhar_b@nid.edu

A. Chakrabarti et al. (eds.), *Design for Tomorrow—Volume 1*, Smart Innovation, Systems and Technologies 221, https://doi.org/10.1007/978-981-16-0041-8_52

52.1 Introduction

As discussed in abstract, discussion was taking place in NID Foundation studio, that India has so many languages, and each language has slightly different vocabulary for colour. To understand the differences, initially an assignment was designed. At NID students are from different part of the country, majorly four parts, north, south, east and west are covered. So, students were asked to write down colour name along with colour swatches. They were asked to cover primary, secondary and black and white. NID B. Des Foundation deals with more than hundred students. Colour swatches with names were pasted on A3 size paper. It was found that there are similarities and differences in naming of colours in different reginal languages. For example, Bengali, Hindi, Gujarati have same name for the red colour, which is Lal but In Odia it is called Nalee, in Ladakhi is called Marpo or in Konkani it is called Tambddem and in chakma it is called Ranga. During the time it was observed that for same colour, along with different names the colours painted were also slightly different. This difference brings into notice that may be because of difference in name causing different visualization of the colour.

52.2 Investigations

To investigate this possibility, two explorations were conducted. One is with M. des batch, with approximate 88 participants. Another one was with B. des students, approximate 45 students. First experiment was to investigate "name of the same colour in different languages create different visualization of the colour". And second one was, "name of the same colour in mother tongue creates different visualization than foreign language". Both the samples were investigated and figured out patterns. Also, to see relationship between language and visualization, particularly colour, work of linguists and neuroscientists were explored. To see meaning and association of colour terms, a number of dictionaries and etymology dictionaries were seen.

52.2.1 Investigation-1

Name of the same colour in different languages creates different visualization of the colour.

Number of participants—88, Swatches taken of 72 participants. Age groups—in between 20 and 30; profession—design students (M.Des); male/female radio—50/50; language speaking—Hindi, Bengali, Odia, Punjabi, Gujarati, Tamil, Malayalam. States—West Bengal, Odisha, Uttar Pradesh, Gujarat, Madhya Pradesh, Chhattisgarh, Tamil Nadu, Kerala, Karnataka. Culture—Hindi, Bengali, Odia, Punjabi,

Gujarati, Tamil, Malayalam, Kannada. Materials—Poster colour, cartridge paper; Swatch size—1 cm × 2 cm.

A random colour orange was taken for the investigation. Participants were asked to pronouns the colour name in their mother tongue and visualize the colour and paint it on paper and present it in a form of rectangular swatch of given size. Approximate eighty.

Eighty swatches were collected and categorized according to languages. It was found that in one language there were synonyms for the colour. Each synonym was associated with different things. The categories are Hindi, Bengali, Odia, Guajarati, Punjabi, Tamil and Malayalam. Hindi had total 9 synonyms and 2 major synonyms for the colour, Narangi and Kesari, under Narangi 28 Responses came as colour swatches and under Kesari 11 Responses came. Then third synonym was Bhagwaa, Bhagwaa had 5 responses. Forth synonym was Suryast with 1 response, fifth synonym was Pakka Pila with 1 response, sixth synonym was Champai with 1 response, seventh synonym was kamla with 1 response, eighth synonym was Satrangi with 1 response, nineth synonym was Santri with 1 response. In Bengali, there were 2 synonyms for the colour, Komla with 2 and Kusumi with 1 response. For Odia 3 responses came for 3 synonyms, kamala, Hanuman rang and Narangi. In Gujarati 2 responses came for Kesari. In Punjabi for synonyms Narangi and Kesari 1,1 response came. In Tamil the colour is called Semmanjal, for that 4 responses came. For Malayalam, Kirmeeram 8 responses came.

If we look the entire picture of the responses in terms of number of colour swatches, under Hindi language 50 colour swatches were there. And 9 different synonyms were presented. So Hindi as language was taken for investigation. In Hindi, Naragi and Kesari have major number of swatches. Narangi has 28 swatches and Kesari has 11 swatches, so to be precise Narangi and Kesari were taken for further investigation. Meaning of the word Narangi is colour of orange. It has association with orange fruit. Little bit history of the word. Orange as fruit came from China to India, in Sanskrit the tree is called Narang, the word Narangi for the fruit came from that in Hindi. The fruit went to Persia and Arabia from India, in Parsia Narang became Nareng and in in Arabia it became Naranj, from there the fruit travelled to Spain and France and became Auranja in France and Naranja in Spain, then it came to Britain and became orange [4, 5]. The colour (Narangi) associated with the fruit orange, now seeing the 28 swatches it was found that 9 swathes are little bit odd to connect with orange, other than that 19 colour have resemblance with the fruit orange, spatially Indian orange which is a colour in between colour red and yellow. The word Kesari has 11 swatches, Kesari came from Kesar, which means Saffron [6]. so Kesari is associated with saffron. Among 11 swatches 6 swatches has close resemblance with saffron, which is dark orange, near to red. So we see that the colours of the swatches associated with the word Narangi were very different from colour of the swatches associated with Kesari. One is orange another is red orange. Why this difference? (Figs. 52.1 and 52.2)

Word is consisting of meaning and meaning is multisensory experiences means different associations with visual, tactile, auditory (and more) of object, phenomenon, dynamic experiences. For example, orange means Narangi in Hindi means a fruit

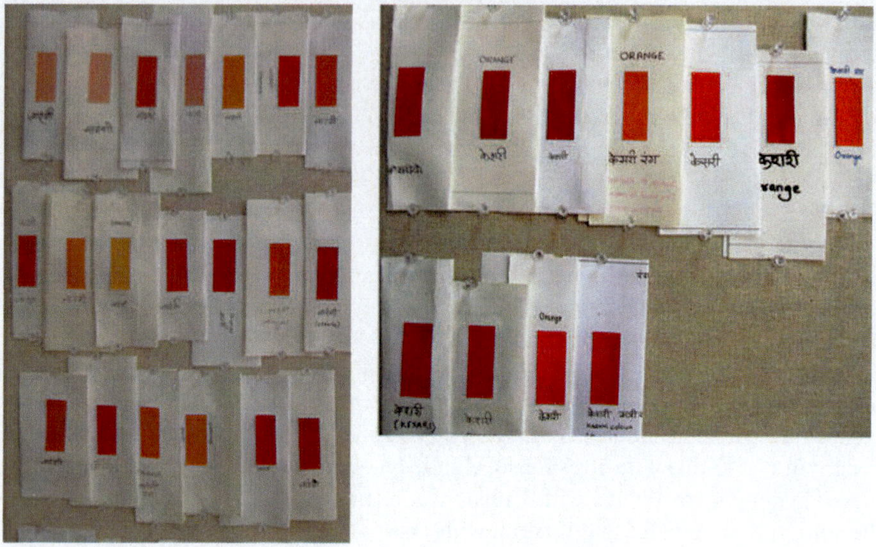

Fig. 52.1 Swatches at left are Narangi and right are Kesari

Fig. 52.2 Close view of Narangi and Kesari swatch

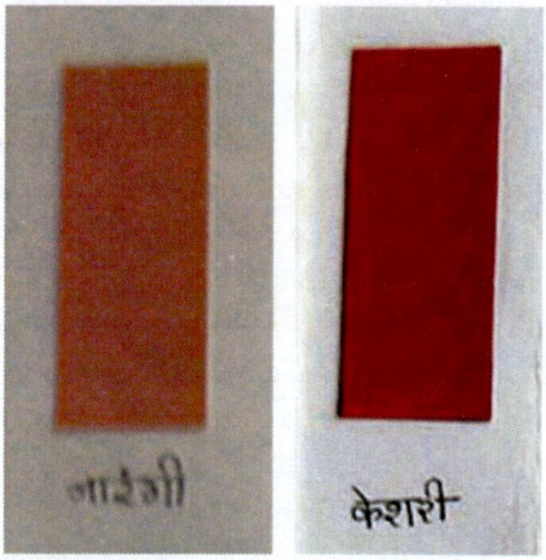

with multisensory experience of colour, taste, smell, form and texture. And the word Narangi is symbolizing these experiences. Now pronouncing the word means recalling all the experiences. This is the connection between pronunciation and visualization.

Around the area of Language and cognition since long thinking and work were happening and the discussion is ancient since pleto's time, thinker like Gorgias who argued that "physical world cannot be experienced except through language" later on in nineteenth century, thinker like Wilhelm von Humboldt, Franz Boas, Edward Sapir were discussing around language as the expression of the spirit of a nation, later students of Sapir Benjamin Lee Whorf observed consequences that linguistic differences put impact, human cognition and behavior. This hypothesis is known as Sapir–Whorf hypothesis. Later on 1920s German linguist Leo Weisgerber, developed relativist theory on the topic and later Roger Brown and Eric Lenneberg who conducted experiments designed to find out whether colour perception varies between speakers of languages that classified colours differently and by doing that reformulate the Whorf hypothesis into testable hypothesis [7].

But till now there was no framework of principles. In this paper, I referred some contemporary work on the topic through couple of lectures and papers.

Prof V. S Ramachandran established the fact that the sound human vocally produces as word has connection with visual form or shape, so, sound of any pronounced word can bring visuals in mind [1]. Prof. Dan Everett argued that language consists of index, icon and symbols with grammar. Now index, icon and symbols are also connected with visual. So, visualization is unavoidable when language comes [2].

Couple of other experiment shows there is link in between language area and visual area in brain. Recent findings by a research team at the State Key Laboratory of Brain and Cognitive Sciences of The University of Hong Kong (HKU) suggest that the language people speak influence their perception of the world.

"By using neural imaging, we have succeeded in showing that brain regions mediating language processes participate in neural networks activated by perceptual decision", explained Dr Luke, one of the researchers from the team. This may be re establishing the Whorf hypothesis, the evidence found through neuro-imaging techniques [8, 9].

Quoting part of the report-

> In their experiments, seventeen subjects were asked during neuro-imaging sessions to decide whether two squares were of the same colour. Some of the squares were filled with easy-to-name colours (such as 'red' or 'blue'); others with hard-to-name colours. The result shows that the perception of both kinds of colours involved the same cortical regions which have long been known to be associated with colour vision. However, in comparison with the hard-to-name colours, perception of the easy-to-name colours evoked significantly stronger activation in two additional brain areas that have been found independently to be responsible for word searching suggesting that with colours that have names in a language, there is a close link between language processing and colour perception. [8, 9].

Another paper is stating that there is bidirectional relation in between Language and perception, and they influence each other.

> linguistic cues guide interpretation of visual scenes, while perceptual information shapes interpretation of linguistic input. [10]

The bidirectional relation can be understand like, initially visual clue helps to learn language and language later can help to communicate visual experiences. A visual object with certain colour helps to learn the name of the colour. I early language development stage, visual clues put stronger impact in learning language.

> Visual information has been shown to activate (prime) language related information early in development. [10]

The visual reference stays in mind as memory to associate with the colour name. That is why pronouncing the colour name brings visualization of the colour associated with objects. Now for a colour name in different languages or in different synonyms of a language, association with colour objects varies based on geographical environment. That is why visualization and representation of the colour vary language to language and in between synonyms of a language.

So different word in different languages or different synonyms of a language of same meaning gives different visualization. Now taking the same example of colour orange, orange means Narangi in Hindi, Narangi is a fruit with multisensory experience of colour, taste, smell, form, texture. Kesari the word is also representing same meaning, colour orange. But Kesari means saffron in Hindi, because of association with saffron, all the multisensory experience changes because colour, texture, smell taste of saffron are different from fruit orange.

52.2.2 Investigation-2

Name of the same colour in mother tongue creates different visualization than foreign language.

Number of participants—45, age groups—in between 20 and 25, profession—design students (B.Des), male/female radio—50/50, language speaking—Hindi, Bengali, Odia, Tamil, Malayalam, Kannada, Assamese Marathi, Kashmiri, Telegu. Culture—Hindi, Bengali, Odia, Tamil, Malayalam, Kannada, Assamese Marathi, Kashmiri, Telegu. Materials—poster colour and cartridge paper, swatch size—1 in. × 1 in..

In this exploration, participants were asked to take any random colour of their choice. At First, they had to pronounce the colour in mother tongue and whatever they are visualizing as colour, they had to paint on a swatch of said size. Then they had to pronounce the colour name in English, and whatever they are visualizing as colour, they had to paint on a swatch of said size. English was not mother tongue for majority of the participants. Next task was to put this colour swatches together on A5 paper and mention what association they visualized when pronouncing names of same colour in two different languages. 45 samples submitted. In these samples, there were mainly four groups, red (including pink and orange), green, blue (including purple or violet) and yellow. And languages as mother tongue taken were Hindi, Bengali, Odia, Tamil, Malayalam, Kannada, Assamese Marathi, Kashmiri, Telegu. For investigation, samples were taken from four colour groups and tried to cover major languages submitted. Colour blue got a greater number of swatches, total 12 that is why blue is taken for discussion.

Blue—At first glance each sample represented the colour in English which is different from the colour represented in Indian language. Two samples were taken for example, in first sample in Hindi, blue is represented as Asmani Neela and in English it is represented as blue. Asmani Neela is lighter than English blue. In Hindi, it is associated with summer sky and in English it is associated with winter night sky

Fig. 52.3 At left Asmani Neela and at right blue

Fig. 52.4 At left Neel and at right blue

with star. In second sample in Hindi, blue is represented as Neel and in English it is represented as blue. Neel in Hindi is associated with powder blue, which is used in clothing industry to make clothes whiter, this comes from indigo plant. This is presented darker than English blue representation (Figs. 52.3 and 52.4).

To understand this phenomenon elaborately, colour blue was taken, and as language English and Hindi are taken to investigate. A number of dictionaries were explored to find out the synonyms and the associations of the said colour in English and later compared with Hindi. In English, blue is associated with sky, different material like cobalt, indigo, associated with organization like Navy, Royal. etc., and in Hindi Neela/Neel (blue) is associated with Sky, Gems, Indigo, a member of God Ram's army, a Hill, poison and so on. In association there are similarity and dissimilarity. And dissimilarity causes the difference in visualization, also in similarities because of different geographic location atmosphere colour is different and colour of object and material, etc., also different. As we discussed in previous investigation that language developed in particular geographical area interacted with its environment and these environments varies geographically. Taking the example of first blue and Asmani Neela both are associated with sky. But the skies are different in the geographical areas from where these names emerge. That is why colour visualization is different.

Another pattern was seen after investigating 12 blue colour responses. Under English language, the degree of differences in between colour responses is less than the colour responses under mother tongue. In mother tongue, we see extreme dark and extreme light, that extremity is not visible in English colour swatches (Fig. 52.5).

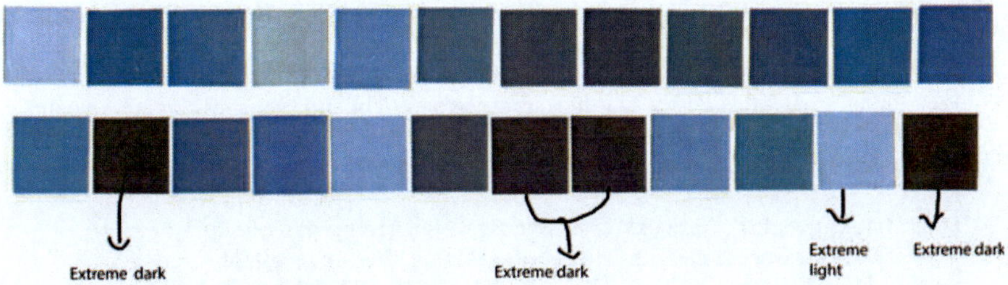

Fig. 52.5 In upper row colour for English and in lower row colour for mother tongue

This might occurred because in childhood when colour is introduced in English language, the references are mostly a colour swatch with a particular fix colour in books, even in the illustration, shown objects are also has similar colour, but in mother tongue which is introduced far before the book, gives a variety of association of objects from life with diverse appearance of the colour. Also, different language with different geographical location adds further more variety in appearance of the same colour. That puts effect in visualizing a colour in mother tongue and in English (Foreign language).

52.3 Conclusion

In this short investigation with smaller sample size it will be difficult to draw any conclusion. It needs more deeper research with neuroscience. But some patterns of understanding are emerging.

Patterns—Word and language are deeply connected with colour, and words (colour names) of different languages can change the visualization and representation of a colour. Language relates to visualization and different language can visualize differently a visual representation. Mother tongue has associations with wide objects and phenomenon than foreign language because we learn foreign language books and diagrams totally disassociated from the environment the language evolved where we learn mother tongue verbally before reading and writing with the direct experience of the object from the environment where the language evolved. Visualize in mother tongue can bring uniqueness and novelty in visualization. So, in education and creative education, mother tongue could have a vital role to play.

Acknowledgements I would like to Acknowledge, Mr. Immanuel Suresh, (former faculty NID), Ms Chaula Patel (Former Faculty NID) Ms Swasti Singh Ghai (Faculty NID) Students of B,Des and M.Des Batch 2019

References

1. Language of Synesthesia. https://www.youtube.com/watch?v=VjFCaGK7gX8 29/1 /2020 11.04 am
2. How language Began. https://www.youtube.com/watch?v=qFxg5vkaPgk 30/1 /2020 5.20 pm
3. First thing first. https://www.firstthingsfirst.org/early-childhood-matters/braindevelopment/ 31/1/2020 4.56 pm
4. Simple English Wikipedia. https://simple.wikipedia.org/wiki/Orange_(fruit) 13/12/2019 11.47 am
5. https://books.google.co.in/books?id=9ZWDBgAAQBAJ&pg=PA260&lpg=PA260&dq=association+with+the+word+geoluhread&source=bl&ots=WEwTVmWstc&sig=ACfU3U1RDadMPXJK5y0MpUG59NFwDeTvSw&hl=en&sa=X&ved=2ahUKEwiUsLWGwrLmAhXryDgGHUHWBmYQ 6AEwCnoECAkQAQ#v=onepage&q=association%20with%20the%20wor d%20geoluhread&f=false 18/12/2019 3.52 pm

6. Shabdkosh. https://www.shabdkosh.com/search-dictionary?lc=hi&sl=en&tl=hi&e=kesar 3.02 pm, 13/6/2010
7. Wikipedia. https://en.wikipedia.org/wiki/Linguistic_relativity 4.44 pm. 16/10/2020
8. Science Daily. https://www.sciencedaily.com/releases/2008/04/080407201846.htm 2.43pm 15/10/2020
9. University of Hong Kong.: Language And Color Perception Linked In Human Brain. ScienceDaily (2008, April 10). Retrieved November 9, 2020 from www.sciencedaily.com/releases/2008/04/080407201846.htm
10. Vulchanova, M., Vulchanov, V., Fritz, I. et al.: Language and perception: introduction to the special issue speakers and listeners in the visual world. J. Cult. Cogn. Sci. **3**, 103–112 (2019). https://doi.org/10.1007/s41809-019-00047-z
11. Tan, L.H., Chan, A.H., Kay, P., Khong, P.L., Yip, L.K., Luke, K.K.: Language affects patterns of brain activation associated with perceptual decision. PNAS **105**(10), 4004–4009 (2008). https://doi.org/10.1073/pnas.0800055105

Chapter 53
A Review: Representation of Women in Commercial Advertising and Impact in Indian Society

Swarnadeep Nath and Bhaskar Saha

Abstract The emergence of print media and television advertisements played a significant role to bring the national consciousness among Indian people. Focus of this research is to consider how advertising has been represented by the time of sociocultural evolution in the form of visual communication in modern identity. Visual arts and design occupy a unique position as a medium of expression in the field of mass media communication. This paper is considered as the discourse analysis on representation of women in the contemporary space of Indian television commercials and its impact on society. Women are often associated with stereotypical gender roles, sexually objectified as sexual rewards to the user of the product. The study deconstructs the discourse built by the brand through propagating a prejudice of 'women's beauty' as only one kind of beauty, i.e. fair skin and skimpy clothes—which becomes imperative to attain anything in life.

53.1 Introduction

> There is nothing more rare, nor more beautiful, than a woman being unapologetically herself; comfortable in her perfect imperfection. To me, that is the true essence of beauty [1].
>
> —Maraboli.

The establishment of printing press, print media and television advertising played an important role to bring national awareness in the midst of Indian people. Visual arts or design techniques occupied a unique position as a medium of expression in the field of mass media communication. The establishment of British Empire in the eighteenth century and the following westernization of India covered the way for essential change of artistic experience, and a new style of exploration has emerged in the field

S. Nath (✉) · B. Saha
Department of Multimedia and Communication Design, CIT Kokrajhar, Assam, India
e-mail: s.nath@cit.ac.in

B. Saha
e-mail: b.saha@cit.ac.in

A. Chakrabarti et al. (eds.), *Design for Tomorrow—Volume 1*, Smart Innovation, Systems and Technologies 221, https://doi.org/10.1007/978-981-16-0041-8_53

print media and other aspects of art practice [2]. Advertising is a powerful and influential form of visual communication. Advertisements have become an essential tool for producers to increase their sales, to improve their brand image, to attract the attention of the people towards their products, services and ideas to remain in the minds of the masses. Kotler [3] and his team have defined that advertising is to raise the sale of the concerned product, service or idea by convincing the perspective consumers, by capturing their attention within a short span of time. Advertising is omnipresent where nobody can escape from its influence and some consider advertising as a mere economic activity with the single purpose to sell [4].

Television commercials have developed essentially throughout the last many years. In their attempt to sell their products, they have created various images and similarly which after some time have accepted solid socio-social dimensions. As advertising has both positive and negative impact on the society in its various forms, its role in the society is a debated topic. Though it has a number of benefits, it is still a mere instrument in hands of the marketers. Its benefits can be measured on the basis of its use. The influence and impact of television are manifolds. But its influence is likely to be even more significant in the country like India. The contents and image of the advertising are changing by the time with the public needs and desires. Indeed, these images have figured out how to shape a structure of 'stereotype' regarding various constructs including sex, class, religion, sexuality, nationality among others [5]. In the present day, Indian women have voices revealing to her how she should look, from television and Bollywood to fashion magazines to her family. As such, it is important to go directly to sources like television advertisements, magazines and other mass media platforms to figure out how Indian women are considered as beauty. The media not only plays a significant role in setting these norms, but also subsequently distressing these women, as they see promotions which encourage them to look in a specific way to fit a new beauty standard.

53.1.1 Women in Indian Commercial Advertising

Throughout history, women were portrayed as obedient, fearful and shy and in all such attributes of the so-called feminine. In art and literature, women were subject to the 'male gaze'. They were seen only in and around the secluded places like kitchen or private gatherings as opposed to the male dominance in the drawing rooms and the public places and the advertising field is not an exception [6]. Advertisements often depict the woman as an object of attention to be admired by men for his sake, particularly in advertisements for health and beauty products.

Earlier print ad of 'Cinthol soap' is an example of objectification of women and gender inequality in advertisements. Following is the older print advertisement of 'Cinthol soap' for gents and ladies.

In Fig. 53.1, one can see the sexual orientation gap in the model itself. Here the male model is energetic, happy and is in full dress. However, the female model is almost bare and exposed in a kind of seductive posture. Most of the toilet soap

Fig. 53.1 Old print advertisements of '*Cinthol*' [6]

advertisements follow suit. Both men and women use soap. At that point why have the advertisement produced entirely different ideas with respect to a similar product? Some ads stereotyped women as always trying to impress men. Exposing female physique was also one of the common aspects in earlier Indian advertisements.

Gender roles and objectification of women in mass media are one of the most common practiced strategies that have been used by advertisers from the very beginning in television commercials and other print media platforms. Similarly, pertaining women in manners is also considered useful for selling products. Some advertisers feature women as capable, smart and confident to please others in order to attract male or to create a good impression on their employer. To investigate this phenomenon, Radzi [7] and his team in their study explored gender representation in advertising. They investigated how advertisers use language to represent women, and how influential this language is to attract the consumers. Result of this study shows that symbolic language is being used which draws attention but also represents the stereotypical role of women. The gender difference was also shown and the concept of male dominance (patriarchy) was also prominent. Further they suggested that advertising designing and the concept of the particular product which is advertising need much improvement and awareness should be spread among consumers. Advertising represents an ideal image of beautification and so-called perfection in the context of Indian social viewpoint. Andrews [8] and his team in their study named 'Women objectification by consumer culture' state that invention of 'beauty ideology' attracts a massive number of female consumers but who are working in media have become more venerable victims of this beauty ideology and thus, they get more appealed by such advertisements. In her studies, she found that consumer culture has a perverted concept of beauty and they feel 'inadequate'. This sense of inadequacy leads them towards unhealthy beauty treatment and thus, this beautification presents the sexual objects for men. The argument of this paper is to explain how portrayal of 'female body' offers an open door for its visual interpretation in justification of spectatorship and gaze.

53.2 Research Gap

The purpose of this research is to investigate the gender stereotyping of women in the contemporary space of Indian television commercials which is the core of this research. Mostly advertisements represented woman in the role of an attractive object or in a sexual content. Some have portrayed them as a homemaker or a role of being dominated by men. The opinion by the respondents brings out a clear picture of the need for the change in content of advertisements. The issue of spectatorship and gaze has been a major concern of first-generation feminist scholarship, and it needs to be addressed in the Indian context. There is a lack of research about this particular analysis; there is hardly any study which has conceptualized above analysis based on certain dimensions and propositions.

53.3 Aim and Objective

The aim of this research is to explore how advertising plays a significant role in constructing images of women. This paper also deconstructs the stereotypical representation and the discourse interpreted by the brand through propagating a myth and prejudice of women beauty as only one kind of beauty, i.e. fair skin, skinny figure which becomes crucial to attain anything in life and demonstrate how representation of 'female body' offers an opportunity for its visual interpretation in account of spectatorship and gaze.

53.4 Methodology

The study was exploratory in nature and is based on methodical review of the literature to find out the types and trend of the portrayal of women in Indian television advertisements and the perception of the audience regarding the different portrayal of women in commercial advertisements. This research comprises multiple sources of data including reviews, interviews, open-ended questions, field studies, documents and journals to gain insight into people's thoughts and beliefs around certain behaviour and systems. The discussion was carried out through virtual platform and physical meeting as well. The prominent reviews were considered from several existing research articles that aligned with the subject matter and concept as well as taking the present market scenario into consideration. The television commercials that portray women have been discussed and raised issues have been taken into consideration while proposing the entire concept. Along with the digital commercials, the available mass media prints that provide facts stating the issue of discussion have been thoroughly reviewed.

53.4.1 Representation of Women in Indian Contemporary Advertising: Analysis and Interpretation

Television advertising is a form of visual communication which pursues to attract attention, inspire desires and encourage people to respond to the communication messages by formulating a favourable impact on viewer's perceptions [9].

Women are often associated with stereotypical gender roles, sexually objectified, subservient to the males and used to add glamour to the advertisements. The role of women has been changing over the years in various fields around the world like sports, academics, politics, etc. In recent years, advertising has witnessed a significant transformation in attracting customer to buy or use their products or services. The need for advertisements has risen to great levels in print and visual media. One among the transformation is the way women are depicted in advertising, with changing times, reflecting the transformation taking place in society but has the representation of women in advertising modified over a period of time in commercial advertising? Or does it confirm to a number of the conventional thinking about women and their position in society? That is the question wish to deal with in this discussion.

Over the years, several studies have shown that portrayal of women in television advertisements has not been satisfactory. Women are often associated with stereotypical gender roles, sexually objectified, obedient to the males and used to enhance attraction to the advertisements. In India, similarly the trend of representing women as object of attention and conquests has been observed. The concept in such ads has been the portrayal of women as sexual rewards to the user of the product.

53.4.1.1 Analysis and Interpretation of Contemporary Advertising in India

Visual media plays a significant role in promoting various products. Often we are presented to various commercials through different media like newspapers, magazines, TV, web and different outdoor media. Yet, there has been a lot of criticism against promotions and advertisements like: body spray, male undergarments to automobiles. These worldwide representations impulse individuals to consider sex as commodity and these may contribute to violence against women [9]. In the contemporary urban Indian setting, the media-driven consumer culture mirrors western beliefs of the so-called ideal body—slim, fair and tall. Young girls have become the objective of different skin-lightening creams, encouraged to become fair to either appeal a husband or to get a job. It appears to be that light complexion and slim and tall figure is the main source and all other qualifications and qualities are irrelevant [10]. The redefined, idealized thought of Indian feminine beauty represents a thin, fair skin and tall women who can relate with western idea of beauty and mix effectively among these beliefs. The entry of models and heroines of Indian origin in western fashion and entertainment industries is also partly responsible for this huge change [11]. In

Fig. 53.2 'Wild Stone' perfume body spray (Fine Fragrance deo) Retrieved from: https://youtu.be/_1QMX0aLPkE

the advertisements of perfume and deodorant it is often pictured that women represent a thin, fair skin and tall and they are always sexually driven. Being intoxicated in the odour of a perfume/deodorant body spray, they always undergo an inclination to indulge themselves in a sexually intimate relationship within man.

Take an example of 'Wild Stone' deodorant advertisements (Fig. 53.2). In this ad of 'Wild Stone' perfume body spray, the hero usually gets the girl because he has had the foresight to spray the deodorant on his body. The woman gets attracted by the smell of the deodorant and is hooked on to the man. As the formula for such advertisements goes, women in this advertising are seen getting attracted to a male model who has applied the 'Wild Stone' deodorant. It has been observed the most of the deodorant advertising the essence of the particular product and its specification is simply driven in sexual and sensual contents and represents women as one aspect that is objects of sexual desire.

The depiction of women in advertisements often represents unrealistic and sometimes irrelevant to the primary concepts. For example, in the 'Jack and Jones' advertisement, shown in Fig. 3a, a popular Bollywood actor did an advertising campaign based on an international menswear brand, which is 'Jack and Jones'. This particular advertising regretted for shows that actor posing slinging women on his shoulder, suggesting women are nothing more than an object of entertainment and an apparent punchline by the side that says 'Don't Hold Back, Take Your Work Home'. Advertising agency 'Marching Ants' made this particular hoarding, which has been placed up to twelve cities across the country to promote brand's formal shirts. The particular male figure was properly dressed in formal attire carrying a woman on his shoulder

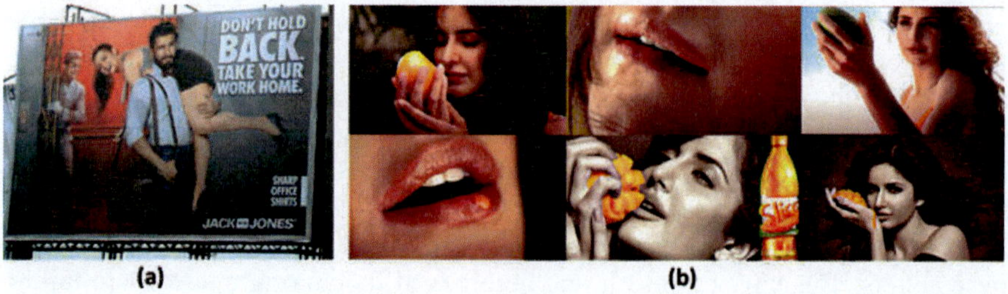

Fig. 53.3 **a** Jack and Jones ad, for menswear 'Don't Hold Back, Take your work home' [12]. **b** Mango slice advertisement Campaign [13]

and about to enter an elevator as the evocative statement 'Don't Hold Back, Take Your Work Home'. In other hand woman exposed in half-dressed is shown smiling while the elevator boy, smirking, is shown opening the door. Social media users enraged for such kind of representation of women by the deliberate sexism in the advertising. The advert was slammed by many for being sexist and for showing insensitive towards psychological sexual harassment at the public place [12]. The Slice commercial advertisement shown in Fig. 3b uses the sexual implication associated with kind of sweet food items such as chocolates and applies it to mangoes. This sexuality is also evident through the semi-bare model in the advertisement. This particular protagonist associated in the advertising has featured in many TVCs of the brand and it has been unforgettable, indeed. The actor caught attention of everyone for her sensuousness and romancing a mango in it. Women are often associated with stereotypical gender roles, sexually objectified, subservient to the males and used to add glamour to the advertisements. In India, the trend of showing women as sexual objects and conquests has been observed. The concept in such ads has been the portrayal of women as sexual objects, as sexual rewards to the user of the product.

Concept of 'fair skin' is often said to transcend all other aspect of beauty. Tumato describes it as 'Colonial Hangover' [14]. Indian girls are taught from a young age that Fair & Lovely go hand in hand, which will lead to a successful marriage and career, while, on the other hand, darker Indian girls are berated for their darkness and compared to their lighter-skinned family. Marketers have exploited this phenomenon to their advantage in promoting fairness creams, lotions and soaps, spreading the wrong message that fairness is associated with increased marriage eligibility, career achievement and other positive outcomes. Thus, the Indian market is flooded with these fairness creams. Overall, the ads promised not only the lighter skin, but also clear and radiant skin that glows, prevention of skin dryness, sunscreen protection and long hair. The ads assert that users will achieve fairness in four to six weeks [15].

Yashomatimaiyya se bole Nandlala, Radhakyungori, main kyunkala.

Singer: Lata Mangeshkar, Manna Dey 1987.

A devotional song where Lord Krishna demanding his foster mother to find why his lady love is fair while he is dark. Much before India was connected into fair-skinned people, the nation has become curious about the lighter shade of skin. India, among the oldest civilizations of the world, did not discriminate on the basis of colour of the skin. Indians, as we realize of them these days, are an accumulation of various races and distinct cultures with many equivalents that increasingly formed a nation-state known as present-day India [16]. Indians have varied degrees of colour complexion just as well as facial uniqueness dependent on the geological territory to which they belong. For instance, Indians from the northernmost area are lighter skin tone in other hand from the north-eastern background are generally known as having a yellow complexion and facial appearance increasingly likened to our Southeast Asian counterparts. Southern Indians, or those from the Dravidian's family, generally have a darker skin tone. These examples show that physical condition has enormously formed the physical features of Indians. Thus, there was truly an acknowledgement of assorted variety in physical traits, and beauty was not accorded on the basis of

Fig. 53.4 **a**, **b**, **c**, **d**, Fair & Lovely commercial advertisement, retrieved from www.youtube.com

skin colour [17]. It is essential to apprehend how and when skin tone or complexion became a vital aspect for acceptance or popularity in Indian society; therefore, a brief empathetic of Indian society is vital.

53.4.2 A Viewpoint Towards Beauty Product and Its Impact: Analysis 'Fair & Lovely' Television Commercials

Fair & Lovely, the 'largest selling skin-whitening cream in the world' marketed by Unilever in Asia, Africa and India in particular, illustrates the issue [18]. Fair & Lovely manufactured by Hindustan Unilever Ltd. (HUL) which is the Indian subsidiary of Unilever, launched in 1975. Dermatologists dispute the efficacy of Fair & Lovely since it does not contain chemical ingredients that can create a permanent change; nor does it claim to be a pharmaceutical product. Currently it is the largest brand of fairness products with a market share of 53% [19]. Advertisements for the Fair & Lovely fairness products are underpinned by the hidden guarantee of dramatic skin-whitening results. The packaging design of the particular product sold in India, cover design shows that a dusky women succeeding stages of skin-lightening, changes into a wonderfully fair-skinned person. The metamorphosis of the woman's complexion is accompanied with the simultaneous transformation in her expression from sober to joyous as shown in Fig. 4a. The visible contrast between the 'before and after' appearance of the model is also enhanced by her resultant heightened confidence. This change is both external and emotional–psychological emphasizing the message that fair skin is essential for femininity and thus fuelling the continued cycle of female consumerism.

53.4.3 Fair and Lovely' Television Advertisements and Interpretation

Figure 4b builds a story of a dance group where the choreographer chooses one artist over the other because of her fair skin. The selected dancer advises her colleague that despite talent, she would not be able to shine out as does not have the physical charm. The physical appeal is then brought in by handing over the 'Fair & Lovely' tube to the protagonist. The last shot of the advertisement shows the male choreographer being charmed by the dancer who he had earlier discarded. Figure 4c tells the success story of a young middle class girl who has flair for cricket commentary. She practices her skill at home with a fake mike in her hand which is replaced by a 'Fair & Lovely' tube by her mother. The next part of the advertising is where the girls' immediate selection as a television cricket commentator where her glowing skin tone becomes the centre of attention for co-commentators and the television spectators. Figure 4d shows 'dark complexion' daughter of a priest being ridiculed by the staff of an 'air hostess' training academy for not having the physical beauty to even be in that space. The following scene suggests the priest being hurt by way of the lamenting and using all his knowledge in 'Ayurveda' to develop something on the way to change his daughter's skin colour. The advertisement situates 'Fair & Lovely' cream to be that concoction. As soon as daughter makes use of it and attains fair skin the air hostess academy not only trains her but she additionally becomes an absolute successful air hostess. The narrative of each of the discussed ads identifies following common narrative agents like (i) her abilities and potential, (ii) her rejection, (iii) dark skin as the reason for rejection and (iv) an advisor/friend who brings 'Fair & Lovely' into the protagonist's life—The immediate success after the use of the product enhances skin tone and gives fair skin.

In this study advertisement in account of Indian context contains clipping of scene that portrays representation of female body prominent object of attention on account of spectatorship and gaze and that also portrays colour discrimination. A thorough discussion and interaction sessions (Table 53.1) have been carried out about the stated agenda with professional designers, artists and faculty belonging to various backgrounds since they have better understanding of the subject matter.

In the context of above-mentioned discussion session, the feedbacks and outcomes are listed below.

The participants were selected from various cities and states of India which includes men and women from different age groups, both married and unmarried, along with expertise from various professional fields. The participants opinion have been taken to evaluate whether Indian advertisements still need to redefine the way it portrayed, stereotype perception of women's beauty and its continuation up to the contemporary advertisements. Gender-based discrimination was also further discussed. The participants consist of 60% male and 40% female. Most of the participants agreed that the portrayal of women in various advertisements seems to be bisexually objectified and has its impact on cultural values. Majority of the feedbacks indicates for the need of ethics in the content of such advertisements.

53.5 Result and Discussion

The question of spectatorship and gaze has been one of the major concerns in social history of first-generation feminist scholarship, and it must concentrate on the Indian context as well. Through extensive periods of history, the mode of human intelligence

Table 53.1 Feedback and discussion from various eminent personalities

Name	Designation/Organization	Feedback of the discussion
Mr. Y. Suresh	Asst. Professor Department of Multimedia Communication & Design, CIT Kokrajhar, Assam, India	The topic can be explored in depth for improvement of social outlook on objectification of women representation in television commercial
Dr. R. K. Mazinder	Assistant professor, Department of Visual Arts, Assam University Silchar, India	The desire for fair skin is not new in India and existent since centuries. Dark-skinned god and goddess in the scriptures are often shown as fair in present time. This can be also an imperative study in account of God and Goddesses who are in dark skin in mythology and with the current representation of so-called prejudice women beauty and colour discrimination in Indian advertising
Mr. Sandipan Bhattacharjee	UI/UX Designer, India (Freelancer)	Essential aspects raised and explained. Could be used for proper portrayal of content in advertisements
Mr. SudiptoNath	Animation Designer and 3D Character artist in Anibrain VFX India	The importance of the particular product and its specification is more important rather than eroticizing the particular advertisement which may not require or can be portrayed in different approach. Such ads are like—body spray, soap and cold drinks
Mr. Sudip Das	Asst. Professor, Art History and Aesthetics, Maharaja Sayajirao University, Baroda, India	This particular topic is found to be very interesting and serious; this is an issue in our society. Such advertising contents need to be redefined
Ms. C. Brahma	Registrar CIT Kokrajhar, Assam, India	While discussing, the topic seems to be very interesting and never thought of an idea while watching any advertisement
Mr. K. K. Sinha	Senior Art Faculty Ednet Art and Design School. India	The topic is appreciated and a noble approach. Several advertisements for example: body spray the contents of the product could be represented in a different way rather highlighting and diverting the particular advertisement in sexual contents

(continued)

Table 53.1 (continued)

Name	Designation/Organization	Feedback of the discussion
Ms. S. N. Shasmal	Former 3D artist, VFX and look development at Rhythm and Hues Studio, India	Psychological impact has been imprinted in such a way which is difficult to liberate from the current situation of our society. All these notions are socially constructed and this is fact. To understand beauty and its real significance is one's responsibility to understand all the social circumstances in proper ways
Ms. M. Ghosh	Post-Doctoral Research fellow. Department of Bengali, Assam University, Silchar, India	Though the role of women in our society is changing, but still colour discrimination and objectification exists in movies and TV commercials. Stop objectifying women's biological structure for amusement
Dr. P. Kalita	Associate Professor, Department of Chemistry CIT Kokrajhar, Assam, India	Dark skin as the cause of professional rejection is represented in many advertising of cosmetic products. These particular advertising and representation could be improved as we are living in a nation with multicultural and different ethnicity, class and cast. Skin colour should not be a concern for any profession
Ms. S. Bhaumik	VFX Artist in Television commercials and movies, Technicolor India	'Dark skin also matters' she also expressed her own experience as her complexion is dark. Darker-skinned women especially facing discrimination at work, at school, even in love
Ms. Kanika Sinha	Yoga Teacher, B. Ed collage	The mass is encouraged to look fair and lighter skin tones and in many cases choosing surgeries to achieve this. Rather than wanting a fairer complexion one should opt for healthier skin and body. Ads should promote healthier skin not fair skin
Mr. AsharAlam	Pre-production Artist at DNEG India	The current trend involves the usage of advertisements falling in the alleged category, while the base idea could be investigated
Ms. K. Das	Creative Director, Crestra Communications Private Limited	The initial idea was impressive and satisfactory. Particular research can be studied further
Mr. K. Vanaparthi	Character Modelling artist in live film entertainment, DNEG India	To some extent it is a never ending process until and unless we change ourselves individually

(continued)

Table 53.1 (continued)

Name	Designation/Organization	Feedback of the discussion
Ms. A. Choudhury	Rural Socialist, AFPRO	To increase the seal or promote any advertising one should not use any men or women that are often objectified, body shaming or any discrimination

insight changes with humanity's entire mode of existence. The manner in which human notion is organized, the medium wherein it is accomplished, is determined not only by nature but by historical circumstances as well.

The advertisement put up a discourse on women's objectification because someone is constantly evaluating how she seems. The narrative agents and the values they have held have not changed entirely at the ideological level. As a result, the idea of 'male gaze' emerges as a powerful subtextual content right here. Dark skin as the cause of professional rejection is also repeated in many narratives. While on the one hand it constructs a very racist discourse and disregards the dark woman further more in the broader framework of women as the marginalized gender. Fair & Lovely's television advertising campaigns have caught to propagating hegemonic ideologies in their narratives over the past many years. The discourse constructed by their narratives not only objectifies the woman but also looks down upon her by situating fair skin much above capability and talent that she possesses. The above discussion of the advertisement campaign has reclaimed the narrative of a woman's success—personal or professional—as being dependent on physical parameters like skin colour. No wonder the discourse has been so extensively domesticated that the identical organization launched another product 'Fair and Handsome' for men a while back. While such a development indicates a more egalitarian or far more unequal discourse that needs serious scholarly engagement.

53.6 Conclusion

There are certain representations of women in Indian television commercial that often depict women as an unrealistic and sometimes irrelevant to the main concepts. In several advertisements, the objectification of women is typically the focal point of consideration rather focusing on the specific products and its significance. She is normally wearing something very revealing and the screenplay usually takes the view of the male eye. This speculative framework can impact in general and effects of advertising images in particular which can influence audiences as well as encourage normative ideas on sociocultural constructs including body image and gender roles. The representation of women in Indian commercial advertisements often shows in an unrealistic and unattainable standards of feminine so-called ideal beauty. The issue of spectatorship, gaze and representation of female body and the colour discrimination needs to be addressed in Indian context. Though it has not enclosed all the advertisements, future research could be covering more number of advertisements in different perception. This topic can be explored in depth for improvement of social and cultural outlook on objectification of women representation in television commercial. Further, the interpretation of men could also be studied and a comparative analysis also would add more insight to the study. The ethical aspect of Indian advertisement is extremely significant for restoration of our Indian culture, norms, ethics and heritage.

References

1. Maraboli, S.: Unapologetically you: reflections on life and the human experience. A Better Today (2013)
2. Kumar, S.: Postcolonial identity in a globalizing India: case studies in visual, musical and oral culture
3. Kotler, P.: Marketing management. Pearson education (2009)
4. Rekha, V.S., Maran, K.: Advertisement pressure and its impact on body dissatisfaction and body image perception of women in India. Glob. Media J. Indian Edn. **3**(1) (2012)
5. Klein, K.M.: Why don't I look like her? The impact of social media on female body image (2013)
6. Saranya P.S., Vijayalakshmi, P.P.: Women in Indian Advertisements then and now: a feminist reading. Int. J. Eng. Lang. Lit. Trans. Stud. **4**(2) (2017)
7. Radzi, N.S., Musa, M.: Beauty ideals, myths and sexisms: a feminist stylistic analysis of female representations in cosmetic names. GEMA Online® J. Lang. Stud. **17**(1) (2017)
8. Andrews, M.R., Talbot, M.M. (eds.): All the world and her husband: women in the 20th century consumer culture. Bloomsbury Publishing (2000)
9. Patowary, H.: Portrayal of women in Indian mass media: an investigation (2014)
10. Munshi, S. (ed.): Images of the 'modern woman' in Asia: global media, local meanings. Psychology Press (2001)
11. Zimik, C.: Women and body image: a sociological study of women in India. Int. J. Eng. Lang. Lit. Humanit. **4**(6), 66–76 (2016)
12. PTI, Jack and Jones ad row: Would never do anything to disrespect women, Nov, 2016. https://economictimes.indiatimes.com/magazines/panache/jack-jones-ad-row-would-never-do-anything-to-disrespect-women-says-ranveer-singh/articleshow/55631219.cms. Accessed 8 Sep 2020
13. Mango Slice Ad Campaign- TVC with Katrina Kafe. https://youtu.be/JfB2mo5juOk
14. Tumato, M.: Real Beauty... If You're White: Unilever Tell Western Women They're Beautiful as Is. In India, It'sa Different Story. Tyee Online News (2007)
15. Karan, K.: Obsessions with fair skin: color discourses in Indian advertising. Adv. Soc. Rev. **9**(2) (2008)
16. Parpola, A.: The roots of Hinduism: the early Aryans and the Indus civilization. Oxford University Press, USA (2015)
17. Mishra, N.: India and colorism: the finer nuances. Wash. U. Global Stud. L. Rev. **14**, 725 (2015)
18. Karnani, A.: Doing well by doing good—case study: 'Fair & Lovely 'whitening cream. Strateg. Manage. J. **28**(13):1351–7 (2007)
19. Khan, S., Khan, B.M.: Brand power of fair & lovely (a case study on fair & lovely fairness cream). Asia Pacific Mark. Rev. **1**(2):104–114 (2012)

Chapter 54
An Approach to Identify Indigenous Color Palette: A Case Study of Majuli

Sunanda Sahu

Abstract The main aim of the paper is to create a methodology for extracting the color palette to create a brand identity of a location through the study of artifacts/spatial architecture. In this paper, a case study of Majuli, a river island of Brahmaputra river, is considered. Majuli has 22 Satras; the religious institutions which came into existence after the Neo Vaishnavite movement lead by Sankaradeva in the sixteenth century. These Satras are center for Majuli's performing arts and handicrafts. The thriving tourism industry of Majuli is dependent on the identity and restoration of these Satras. They also play a very crucial role in Majuli's history and culture. The present study was done through extensive field visits and observations, ethnographic study, closed user group interviews, photographic documentation, and analysis of collected visuals through an image processing application image color summarizer. An approximate number of 150 colors were extracted from these visuals using image processing image color summarizer. These colors were clustered using their hues. Clustering of the colors showed major clusters of brick red, earth yellow, amber, camouflage green, tapa gray, stone gray, walnut brown, and bahama blue. Segregation of clustered colors can act as a color guideline for any practitioner. While designing, these colors can be taken into consideration, and the designers can create different color palettes that suit their context and product applications.

54.1 Introduction

In the state of Assam, Majuli, a river island, and a newly declared district is known to be the seat of Vaishnavite culture with its unique contribution to the evolution of a subculture in the region. Majuli is a river island, on the river Brahmaputra, with historical and cultural significance. Geomorphologically, Majuli forms a part of the flood plains of the river Brahmaputra. Majuli had an area of 1,256 km^2 in 1991, but due to extreme bank erosion, it has been reduced to 924.6 km^2 [1].

S. Sahu (✉)
Indian Institute of Technology, Guwahati, India
e-mail: sunanda.sahu1607@gmail.com

A. Chakrabarti et al. (eds.), *Design for Tomorrow—Volume 1*, Smart Innovation, Systems and Technologies 221, https://doi.org/10.1007/978-981-16-0041-8_54

Majuli came into prominence in the sixteenth century with the arrival of Sankardeva and the propagation of Neo—Vaishnavism. Majuli became the asylum of various art forms like Bhaona (theatrical performance) folk culture, music and dance forms, handicraft, pottery, ethnicity, sacred Satra architecture, mask making, wood carvings, and paintings (illustrated manuscripts), etc. All these developments are typically based on the religious and Vaishnavite ideology, to serve and uplift the collective society, just apart from contemptible and palate entertainment [2]. Satras play a significant role not only in the religious sphere but also in sociopolitical and economic aspects of life [3].

54.2 Color Meaning and Semiotics

Color is significant in generating an identity. Every color has a symbolic assigned to it. However, other non-chromatic factors influence the way colors is interpreted in different social environments. The formulation of a chromatic identity includes identity, differentiation, tradition, and innovation. Color is amongst the key features of globalized brands. So, it is essential to study it as a component in the process of meaning making.

Color has a symbolic meaning. It is considered significantly transparent when studied as an aspect of visual semiotics. Among color, written texts, and iconographies, color is considered more transparent [4].

54.3 Purpose

Color is one of the most effective design tools to create an atmosphere in design since its direct relations with sensations. Designers and architects use color codes as a tool for corresponding a theme or a message to the user [5]. In addition to this, the characteristic of a space is experienced by the user as an atmosphere that surrounds and encloses him.

Color influences the cognitive processing of the viewer as it carries distinct psychological meanings. It also acts as a moderator of perceived similarity. Hence, use of colors is pervasive in consumer contexts and carries varied applications in branding [6]. This paper aims at identifying and documenting chromatic characteristics associated with various aspects of Majuli. For this particular research, three essential factors have been focused on comprising of Majuli handicrafts, performing arts, and Satra architecture.

Satras are chief institution of Assam to practice Vaishnavism in a *guru-shisya paramapara* (teachers and disciples living and learning together). Various art forms like *Bhaona* (theatrical performance) as folk culture, music and dance forms, handicraft, pottery, ethnicity, sacred *Satriya* architecture, mask making, woodcarvings, and paintings (illustrated manuscripts), etc. have been originated and flourished in these Satras under the guidance of Srimanta Sankaradeva.

54.4 Materials and Methods

54.4.1 Literature Review

The first part of the literature review sought to identify research papers and published journals to understand the history and culture of the river island. Search terms covering different aspects of Majuli (Majuli, Satras, Assam crafts, River Island, Majuli Mukha, Salmora Pottery, etc.) were included in database searches of Scopus and Web of Science. A total number of 30 papers and published journals were reviewed.

Majuli, a sublime island of river Brahamaputra, has a unique rich culture consisting of numerous Satras that have their socio-religious significance. These institutions introduced a much simpler version of religion to the disciples preaching them selfless devotion and salvation. These traditional practices have evolved with time and are struggling to preserve the cultural essence, arts, and crafts. Hence, there is a need to identify artifacts that still carry the essence of Majuli and convert them to have more relevance in modern times.

54.4.2 Primary Research

The second phase of the study encompasses field research to conduct contextual inquiry and visual ethnography. The contextual inquiry was made with 35 stakeholders, who can be categorized under personas of teaching professionals, research scholars, craft practitioners, religious practitioners, Majuli administrators (Majuli administrative staff) (Deputy Commissioner Majuli), guides, tourists, hospitality staff locals of Majuli, and NGO workers. It was focused on understanding the origin, inspirations, processes, occasions, and applications of the handicrafts. The insights regarding the evolution of methods, materials, and representation of these craft articles were also achieved during the study.

Several Satras (Dakinpat Satra, Kamalabari Satra, Garamur, and Shamaguri Satra) and Salmora village were the main focus for visual ethnography to get the essence of Majuli's history and craft culture. Apart from this, silk production and weaving in villages of Majuli were also covered in the field study. A number of 915 visuals were gathered through visual ethnography.

Table 54.1 Grouping of visuals collected in Majuli

Group name	Number of visuals
Majuli landscape	78
Mask making	213
Salmora pottery	107
Satra architecture	158
Bhaona	64
Majuli festivals	113
Other crafts (boat making, weaving, sericulture)	182

54.4.3 Data Extraction

The data extraction was done in several steps. The first step included the segregation of visuals for color extraction. In the second step, an open-source platform, Image Color Summarizer, was used to extract prominent colors from the visuals.

Segregation of Visuals. A total of 35 visuals were chosen for image processing. These visuals were a good set of representative images of various different factors associated with Majuli. These factors included Satra architecture and attire, Majuli handicrafts (mask making, Salmora pottery, weaving, and boat making), performing arts, festivals, and landscape.

Image Processing. Image color summarizer is an open-source platform that was used to process visuals of Majuli.

The color summarizer produces descriptive color statistics for an image. Reported are the average, median, minimum, and maximum of each component of RGB, HSV, LCH, and Lab. Average hues are calculated using the mean of circular quantities. The color clustering function tells us the representative colors of the image and shows us how the pixels in the image partition into groups [7].

Below is an example of the result produced using this platform. While processing the image output format, statistics, the number of color clusters, delimiter, and precision are chosen as HTML, color clusters, 5, space, medium (100 px), respectively.

Colors of the image were grouped into five clusters (k-means: is a method of vector quantization, originally from signal processing that aims to partition n observations into k clusters in which each observation belongs to the cluster with the nearest mean [8]). For each of these clusters, the average color of the colors is shown. The name is the closest named color, and its distance is shown using ΔE (a number that is representative of the distance between two colors [9]) [7] (see Fig. 54.2).

Hue Segregation and Grouping. More than 150 colors were extracted through image processing. RGB values of these colors were used to categorize them. The categorization was a broad level grouping based on 12 hues of the tertiary color wheel.

Fig. 54.1 Visuals selected from Majuli

Fig. 54.2 A mukh mukha of Hanumana from Samaguri Satra, Majuli

In the first step, color variants were grouped together using the color picker of a graphic editor. Further, the variants were arranged using color value and saturation.

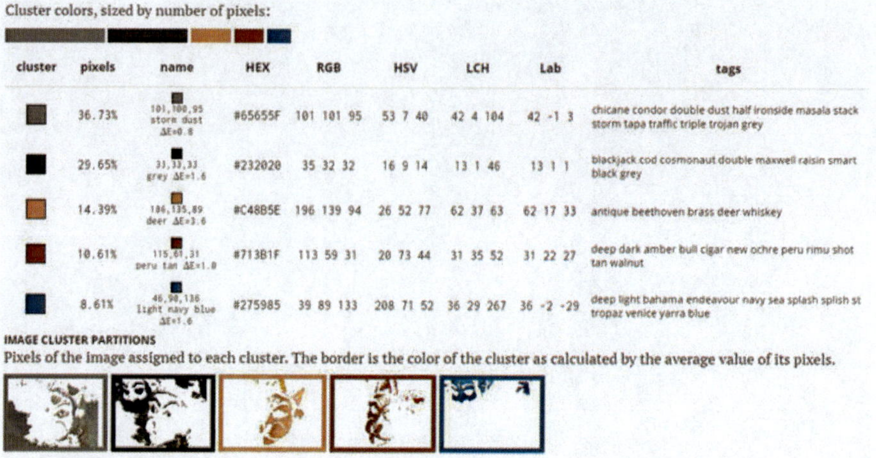

Fig. 54.3 A mukh mukha of Hanumana from Samaguri Satra, Majuli

54.5 Findings

Figure 54.3 is a visualization that represents a set of colors which are extracted from Majuli's indigenous color palette, through image processing. The majority of colors are variants of brick red, earth yellow, amber, camouflage green, Tapa gray, stone gray, walnut brown, and Bahama blue. The only monochromatic craft is Salmora Pottery, which comes in brick red brown color.

54.6 Practical Implications

In today's modern context, when professionals struggle to create regional essence in their works, i.e., branding, promotion, textile, architecture, etc., this study can be a base for understanding and implementing the spirit of colors in those contexts.

Various color schemes (Complementary, Monochromatic, Analogous, Triadic, Tetradic) can be generated referring to this study that will contain the indigenous essence. These schemes can then be used in different soft and hard products associated with a particular region.

In the context of Majuli, this study can partially contribute to different visual aspects of Majuli cultural tourism, i.e., textile tourism, craft tourism, etc. focusing on deriving a color palette most suitable to the modern context and while bearing the regional and cultural resemblance. One such application has been covered here as an example. In this example, color tool by material design [10] has been used to create a scheme for Majuli (see Fig. 54.6).

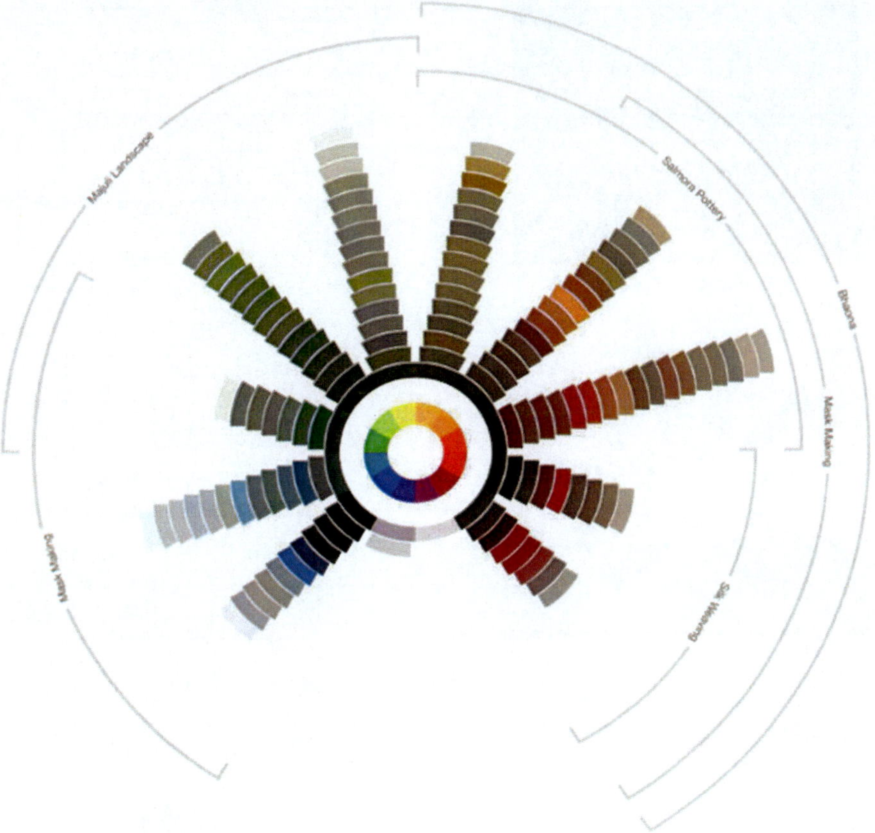

Fig. 54.4 Visualization of colors extracted through image processing

54.7 Justification

Before the modern period, craftsmen used to use natural dies for the beautification of the masks (Mukha). They kept colors in the bamboo tubes of different sizes; small brushes were made from pigeon's feathers or bamboo sticks by affixing goat's hair as bristles. Colors and their application were completely ingenious and amalgamated with indigenous folk flavor [11]. Our field study revealed that the craftsperson of Majuli has started using other mediums such as acrylic and poster paints.

In modern times as we are moving toward digital solutions that are more technologically advanced, the majority of creative personals are adapting to new techniques. Therefore, this study will contribute to mending the bridge between older and newer techniques of the creative industry.

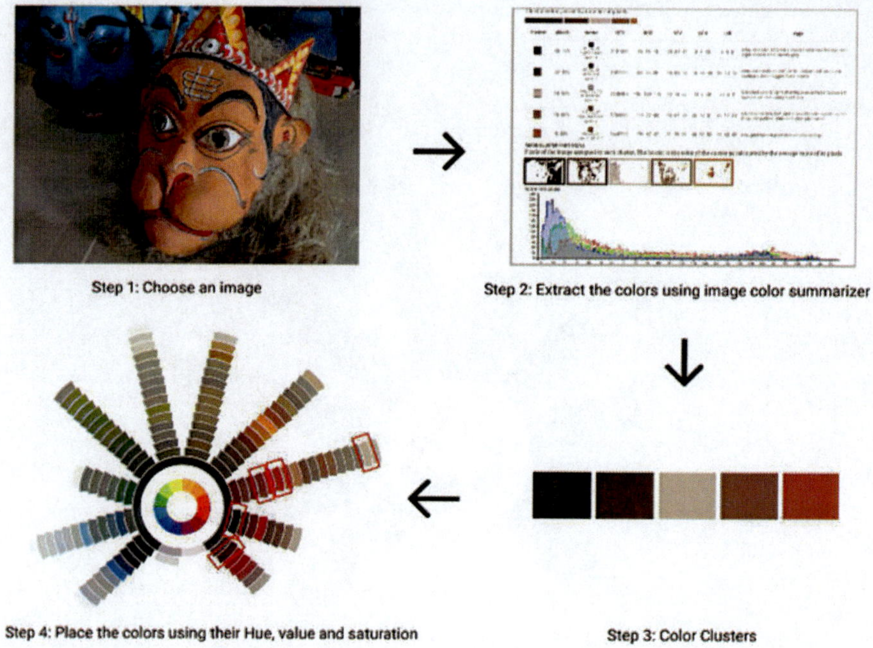

Fig. 54.5 Step-by-step process of extracting the colors and placing it in visualization

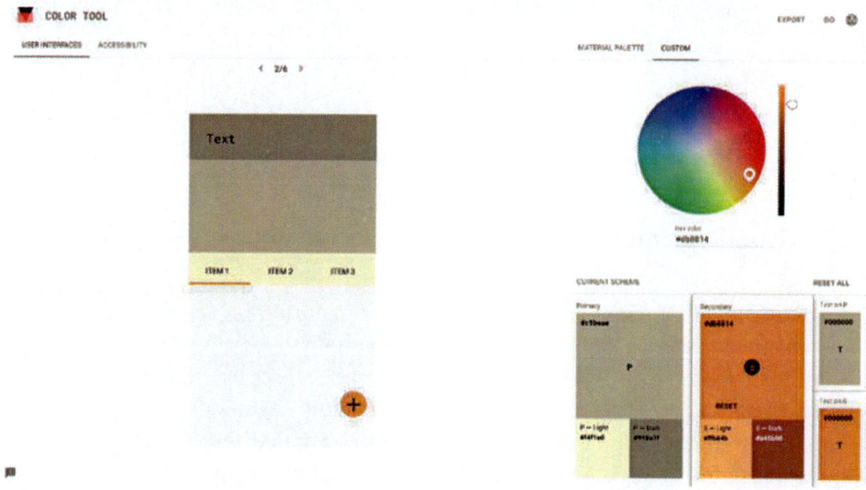

Fig. 54.6 A view of color theme generation using color tool by material design

54.8 Originality/Value

The existing literature on Majuli does not cover visual description in detail, especially related to its chromatic aspects. Hence, this study having the potential to contribute to recreating the regional essence of Majuli is a value addition for future researchers. They can take this study and apply the same methodology to craft the chromatic regional essence of other place that are rich in their tradition and craft culture.

54.9 Research Limitations

The research has been conducted with a small sample size of people and articles. Different festive seasons of Majuli have only been covered through the literature, not by field study. Hence, the research needs to cover broader aspects of Majuli in order to achieve exhaustive results.

References

1. Pranalee, S., Asha, S.: Changing notions of "Ideal" Monkhood: a case study from a Satra of Majuli. Space Cult. India **4**(2), 29–38 (2016) (2)29. doi:https://doi.org/10.20896/saci.v4i2.199
2. Nath, D.: The Majuli Island—Society, Economy and Culture. Anshah Publishing, House, Delhi, 2009
3. Bora, S.: Indian J. Gender Stud. **25**(3), 331–350, © 2018 CWDS. SAGE Publications (2018). sagepub.in/home.nav. doi: https://doi.org/10.1177/0971521518785665, https://journals.sagepub.com/home/ijg
4. Caivano, J.L., Lopez, M.A.: Chromatic identity in global and local markets: analysis of colours in branding. J. Int. Colour Assoc **1**(3), 1–14 (2007)
5. ITU A|Z: Cultural color codes in interior, Demet ARSLAN DİNÇAY **17**(2), 63–72 (2020). doi: https://doi.org/10.5505/itujfa.2020.72621
6. Kim, Y.-J.: The influence of color on brand extension: mediation role of processing style. Jpn. Psychol. Res. **62**(4), 268–278 (2020). https://doi.org/10.1111/jpr.12269
7. Image Color Summarizer 0.76 © 2006–2020 Martin Krzywinski, mkweb.bcgsc.ca
8. Kriegel, H., Schubert, E., Zimek, A.: The (black) art of runtime evaluation: are we comparing algorithms or implementations? Knowl. Inf. Syst. **52**, 341–378 (2017). https://doi.org/10.1007/s10115-016-1004-2
9. Sharma, G.: The CIEDE2000 Color-Difference Formula: Implementation Notes, Supplementary Test Data, and Mathematical Observations (2004). Retrieved from https://www.ece.rochester.edu/~gsharma/ciede2000/ciede2000noteCRNA.pdf
10. Color Tool. https://material.io/resources/color/#!/?view.left=0&view.right=0
11. Vaidhya, R.K.: Art approaches of Majuli: an analytical study of illustrated manuscripts, mask-making and Wood-carvings. Department of Visual Arts, Assam University, Silchar (2011)

Chapter 55
Competitive Branding of Sustainable Domestic Products: A Product Semantics Approach

Rik Bhattacharjee and Saurabh Tewari

Abstract A study of packaging for three different types of sustainable domestic products was conducted to understand the design strategies by existing brands and the consumers' perception about the same. The objective of the study was to know if all the additional values of a product of such kind are effectively communicated through their packaging. This study utilises product semantics as a tool to analyse the packaging design at two broad parameters of the conceptual framework; packaging semantics and branding semantics; encompassing a range of factors like form, shape, material and texture of the packaging and the use of typeface, colour and graphics on it. Multiple user surveys and reviews conclude that there is an evident gap in the effective communication of the values of sustainable products with the existing design strategies. New strategies were then devised to design packaging for the three different types of products to overcome the shortcomings in the existing strategies.

55.1 Introduction

The word 'sustainability' might trigger multiple expressions and definitions. In the present market scenario, numerous brands claim to make sustainable products and provide sustainable services. Most of these products and services, however, are branded to convey their environmentally sustainable attributes, very similar to 'green branding' [1] which targets to brand environmentally sustainable/green products by bringing out their functional and emotional values. Sustainability, however, is more than just the environmental value of a product/service. Environment accounts for just one out of the four pillars of sustainability which also includes human, social and economic factors as well [2].

R. Bhattacharjee (✉) · S. Tewari
School of Planning and Architecture, Bhopal, India
e-mail: bhattacharik75@gmail.com

S. Tewari
e-mail: thinksaurabh@gmail.com

A. Chakrabarti et al. (eds.), *Design for Tomorrow—Volume 1*, Smart Innovation, Systems and Technologies 221, https://doi.org/10.1007/978-981-16-0041-8_55

Our country has a lot of products that are more than just the tangible attributes associated with them and hold an altogether different semiotic value apart from their functional value in terms of the different actors and network involved in their production, marketing and consumption. *Khadi* clothes are perhaps the best example. *Khadi* in its cycle of production, distribution and consumption do almost zero damage to the environment. Apart from not consuming fossil fuel in its production, it generates employment for a considerable number of individuals working to produce these fabrics. It also holds a different semiotic meaning from the pre-independence time when it became a symbol of the struggle for freedom in India. Products like *Khadi* hence stand out from the other conventional products available in the market and have a distinct value.

A user associates' product with brands and the way they communicate the various values of the products. The usual way for brands is to utilise strategies to bring out the functional benefits of the products which communicate the essential functional attributes of the products and how it serves the purpose better than the other brands out there. Another way is to communicate the emotional benefits of the product through various branding strategies and packaging [1]. A user feels a different kind of connection and satisfaction with a product when the brand communicates a feeling of well-being and brings out the other values associated with the product that the user associates to and feels even others need to know. Users also respond to benefits involving nature and products that show a connection with natural elements, which again triggers a feeling of happiness and well-being in the users. Both, namely the functional and emotional values of a product, need to be effectively communicated to the users for them to have a positive attitude towards the brand [1].

Branding strategies differentiate one product from another in the market and communicate the values associated with a product to the users. Every brand would advertise and portray their product as the best in solving the intended purpose. A lot of them would also go a few steps ahead to prove how the other products fail in solving the problem in consideration. Competitive branding strategies [1, 2] define user groups and target to convey the different functional and emotional attributes of the products over similar available products in order to build a positive brand attitude in users.

55.2 Research Methodology

The study includes two user surveys and visual analysis of existing products from various brands to understand how users perceive the different products and the efficiency of the existing packaging and visual semantics in communicating the sustainable values associated with the products. Products from three different segments have been chosen for the study to validate the research and make the interventions/guidelines more general rather than specifically directed towards a particular product segment, where terms like 'Indian products' are debatable, and the subject products here are being used in India since before attempts were made for their commercialisation. 'Digestive Amla Candy', 'Khadi Clothes' and 'Herbal Tooth Powder' have been selected as the three products for the study which belong from three different product categories, namely 'digestive confectionary', 'clothing' and 'dental care'.

The first study is about knowing how users perceive the different products in a segment under different parameters in two different Cartesian spaces, namely product category space and product expression space [3], a method used by Athavankar [4] in his semantic studies. The second study is in the form of a survey to bring out the image of the 'ideal packaging'. This is to cater to users' imaginations when they are informed about the three different product types and are given a few different attributes of the packaging like shape, colour, texture and opacity to think further. The third tool is used to analyse the visual semantics of the packaging based on attributes like fonts, colours, graphics and texture.

55.2.1 Product Categorisation and Expression

Product category space devised by Athavankar [3] is built on Wittgenstein's [5] ideas of the way humans put in objects that are not identical into the same space under some common word label. In case of the indigenous sustainable products, utilising this process would mean a sorting process, where we intend to understand the mental construct and understanding of the users of the core member and the overall expression of the product category. Particular word labels like 'Health-candidness' were provided to users to set the category space, and a few conventional and well-known branded products were placed alongside few not-so-popular products.

Another aspect utilised in this study is the product expression space. Built on Eleanor Rosch's [6] ideas, besides the category itself, numerous expressions can be associated with a product category. 'Youthfulness' was selected as an expression here in case of the three products chosen to understand how users identify the different non-identical products under the expression. This was a pilot study done to understand the various drawbacks of this method and find the most significant aspects of packing to be used to create a framework for further study with a bigger sample size (Fig. 55.1).

55.2.2 Ideal Packaging Survey

Complete images as clues as used in the previous tool could hinder with the exploration of the new visual potential of the respondents. This self-devised tool hence simply uses linguistic labels to ask about various attributes of an ideal package with no visual clues. This lets the respondents explore their visual potential and define their unique core member. Of course, every response would be a result of a lot of prior experiences and associations of respondents with different brands. Still, a simple description of the product, it is various functional and emotional benefits,

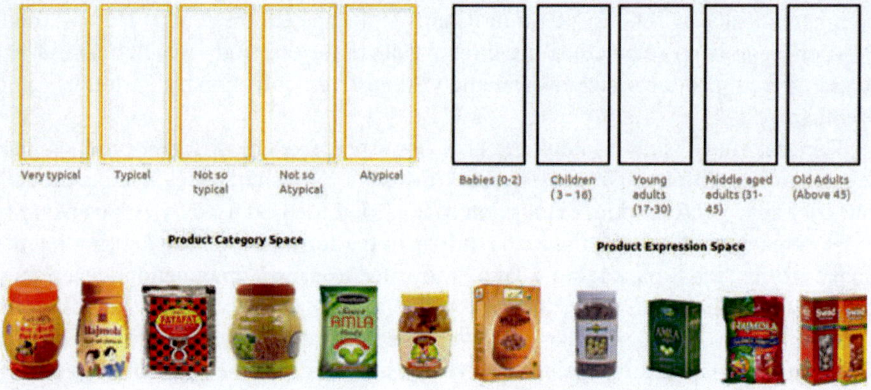

Fig. 55.1 Product category space, product expression space and the different candy packages for sorting. *Source* Author

would set a respondent to select an ideal packaging for each of the three different product categories.

This tool utilised Google Forms to hand out surveys with straightforward questions regarding the following four aspects of packaging of the products: a. the shape of the packaging, b. the colour of the packaging, c. the texture of the packaging and d. the opacity of the packaging material, which was indicated to be the most significant aspects noticed by users in the pilot study.

55.2.3 Visual Analysis of Existing Products

Visual attributes like font(s), colours used in packaging, graphics used and the texture of the packaging were studied after selecting two popular brands from each segment to understand the visual semantics associated with these products and how these brands convey the different values and benefits of the products. This helped understand the gap in the expression of the non-functional values associated with products through their packaging (Table 55.1).

55.2.4 Sampling

The product categorisation and expression survey were done with six individuals of varied age groups ranging from 23 to 40 years of age in an offline medium of card sorting. These were individuals who have been using the subject products for a long time and are aware of the different functional values of the products. A brief idea of the other underlying values was given to them before the process of card sorting.

Table 55.1 Visual analysis of existing products

Packaging image	Product name and brand	Typeface used	Image/graphics	Colours	Texture
	Patanjali dant kanti	Sans serif–neo grotesque	Ingredients of manjan	Yellow, white, earthy	Smooth
	Vyas Dant Prabhakar manjan	Sans serif–grotesque	No image	White, yellow, earthy, red	Smooth
	Vyas amla candy	Sans serif–humanist	Ingredient (amla)	Green, orange, yellow	Smooth
	Patanjali amla candy	Sans serif–neo grotesque	Ingredient and the final product	Green,orange, yellow	Smooth
	Fab India bag	calligraphic and serif	No image	Brown	Rough
	The industry of all nations packaging	Sans serif– neo grotesque	Photograph of worker	Brown, blue	Rough

The ideal packaging survey was done using the Internet with Google Forms which recorded a total of 32 responses which had a mix of individuals from varied age groups and geographical locations to avoid age and location-based trends in results.

55.2.5 Materials

The images for the packaging were collected from Amazon India website [7]. Renowned brands dealing in the respective products were selected for the category and expression space mapping. Brands with iconic products, for instance, ‘Dabur Lal Dant Manjan’ and ‘Patanjali Amla Candy’ were also included in the survey and the user review. The product categorisation and expression survey were a pilot study and were done on a tablet PC with a pen stylus. The ideal packaging survey and the user survey were done using Google Forms which were circulated using WhatsApp messenger and email.

55.3 Responses and Results

The results of the first card sorting survey showed that users associate the shape, colour and texture of the product with the packaging. Candy packages which were radially symmetrical and had a form somewhat similar to amla (Indian gooseberry), and colours found in nature (green and yellow) were selected as most typical options to be packaged for health/digestive candy bottles. More colourful packages were chosen to be ideal for kids, and easy opening packages were sorted for adults and the elderly. For herbal tooth powder (dant-manjan) on the other hand, again apart from the radially symmetrical and close to cylindrical forms with red and white coloured packages were selected to be most typical. Still, the colours here either reflect the colour of the product itself or of the teeth (on which they are meant to be used).

The outcomes of the visual analysis conducted showed a trend in the existing packages, and the way shape and visual semantics have been used in conveying the values which are very similar to how the users perceive packaging to be. A great emphasis on functional values and their benefits is given, where the shape, colour, graphics, texture and typeface used only to try to bring out the functional values of the product. Bottles of herbal tooth powders, for instance, use earthy colours with white, which either represents the colour of the content or the teeth (on which it is meant to be used). Even the graphics would present either the ingredients used (herbs, barks, leaves) or the picture of ideal families using the product. However, there are significant values which are often ignored in the packaging and visual semantics of products. In case of herbal tooth powders, they have been around in India for a long time in the form of a branded product right from when like Dabur [8] and Karkhana Zinda Tilismath[9] started producing them. There is a specific aspect of sustainability involved with it which is clear from the fact that they thrive among the numerous kinds of toothpaste available in the market today by different brands, and almost every major dental care brand still produces herbal tooth powders as well. Again, there is a certain 'Indianness' to how they are used without a conventional toothbrush and are either applied/massaged directly to the teeth using fingers or using a Datwan (A twig of neem or other medicinal trees used to brush teeth) along with the fact that these products are very cost-effective.

Similarly, packages of amla candy fail to show how it is traditionally started as a homemade recipe and is suitable for all age groups equally. Even Khadi cloth packaging needs to convey the idea of Swadeshi, the concept of collective identity and the fact that they are not just clothes meant for the elderly and are available in fewer colour choices but have options for all age groups and moods and perhaps is the most sustainable product in the list here.

55.4 Design and User Surveying

The result of the surveys indicated that functional and physical attributes is mostly translated to the packaging design of a product. At the same time, a lot of the emotional and contextual aspects are ignored. The design process involved clubbing together all the physical, functional, emotional and contextual use-based attributes to ideate packaging for the three different types of products, namely 'Amla candy', 'Khadi clothes' and 'Dant manjan'. Decisions for specific parameters of design like shape and colour were taken based on the insights from the surveys. Hence, mostly conventional shapes and colours close to the product/ingredients were used. However, an entirely different set of considerations were to be made to convey the underlying values. A list of such emotional values was first listed. Amla candy is being prepared in a lot of Indian households for a long time in a straightforward manner by boiling dried amla pieces in sugar syrup. Similarly, Khadi clothes have always had this meaning, where it became a symbol for 'self-sustainability' and something that unified the whole of India and are available in varied colour range. Herbal dant manjan meanwhile is a cost-effective dental care formula that is used in this unique style with fingers and has been made into commercial products with iconic packaging by many brands.

A few common considerations for all the designs, however, have been the choice of material and the concern for an after use. With products that are sustainable in their life cycle, even their packaging should reflect that attribute and must literally 'sustain' the test of time. Durable materials have been used in all designs that make it appropriate for after use as well. Even the typeface used for the product name falls under the 'vintage' category in case of amla candy and herbal dant manjan with a specific hint of fun to make it more appealing. A neutral typeface was used for conveying the necessary information. Figure 55.2 shows the final designs, and Fig. 55.3 shows the labels and graphics for the three products.

An online survey was done using Google Forms to validate the effectiveness of the designed options. The designed options were placed along with five other products of the same category from different leading brands. All the products were de-branded by hiding the brand name with a black strip to avoid biased responses based on prior association to a brand. A five-point marking scale was provided to the respondents to mark each of the packaging based on how appropriate they think it might be for the product mentioned with 1 being least appropriate and 5 being most appropriate. The respondents were also asked to mark factors that they considered while marking the options. A total of 75 responses were received with respondents from varied age groups and professions. Figure 55.4 shows the results in the form of average marks obtained by each option.

Fig. 55.2 (Left to right) packaging for *amla* candy, packaging for carrying bag for Khadi clothes and herbal dant manjan. *Source* Author

Fig. 55.3 Label for *amla* candy packaging showing spatulas indicating it to be a homemade recipe, a label for herbal dant manjan with fingers along with the ingredients showing the way manjans are commonly used and graphic for a khadi carry bag with multi-coloured hands pulling yarn towards a central charkha showing how khadi united the diverse Indian population. *Source* Author

The designed options were marked the highest by the 75 respondents of the survey. The shape of the packaging, the material of packaging and opacity of packaging were the most looked at factors by the respondents for marking the amla candy packaging. For the Khadi cloth packaging, the shape of the packaging, material of packaging and texture of packaging were the most looked at factors. For the herbal tooth powder, respondents paid attention to the shape, colour and graphics of the packaging.

55.5 Discussion

The outcomes clearly show that utilisation of emotional values in packaging and branding design can have a significant impact on the competitive branding of sustainable domestic products. This would also help bring about a positive brand attitude in consumers about domestic products and help build value for the brand with an after use of the packaging as well. Brands dealing with products of a similar kind can utilise the findings of the study to design their packaging. Moreover, it can also be concluded that packaging offering possibilities of prolonged after use were selected to be most suitable for sustainable domestic products. The concept of sustainability hence lasts beyond the life of the product itself and continues living as the packaging is reused

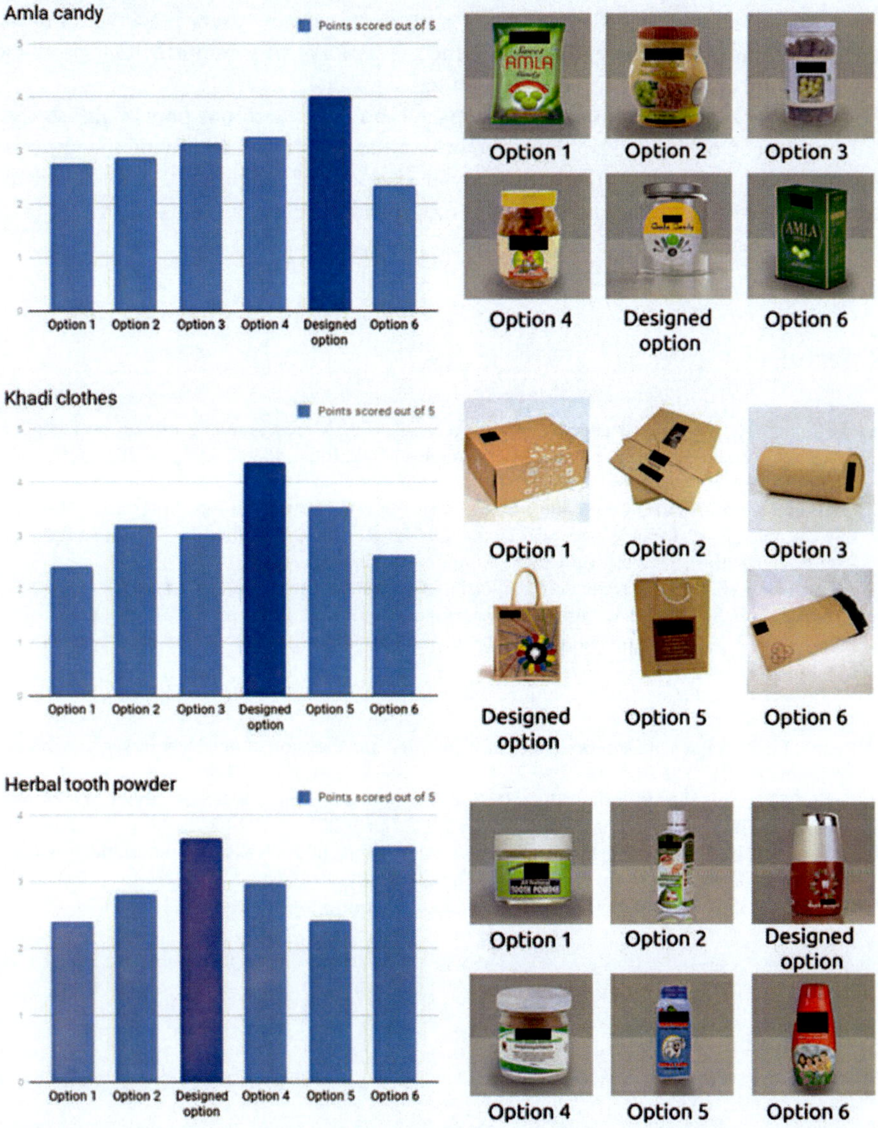

Fig. 55.4 Charts showing points scored by all options out of 5 by the packaging of amla candy, Khadi clothes and herbal tooth powder. The designed options have been shown in dark blue

by the consumers. Sustainable products are supposed to benefit people, environment, society and economy and are different in many ways from the conventional products and hence need different strategies for their packaging and branding design as well. Even functional designs can fail if the overall meaning is not communicated in an efficient and undiluted manner. The product semantics approach utilises study of the various physical and visual attributes of a product and how users perceive them to form an opinion about a product and hence is perhaps very important for product designers working towards making meaningful products.

Utility-based mechanisms like dispensers and caps were not part of the study, and an impact of change in these factors to the user's perception of the overall product needs to be studied. Factors like the texture of the package are subject to tactile feedback while buying products from a shop. Hence, a survey with physical prototypes should be done.

References

1. Hartmann, P., Apaolaza Ibáñez, V., Forcada Sainz, F.J.: Green branding effects on attitude: functional versus emotional positioning strategies. Mark. Intell. Plann. **23**(1), 9–29 (2005). doi: https://doi.org/10.1108/02634500510577447
2. Grubor, A., Milovanov, O.: Brand strategies in the era of sustainability. Interdisc. Description Complex Syst. **15**(1), 78–88 (2017). doi: https://doi.org/10.7906/in-decs.15.1.6; Author, F., Author, S., Author, T.: Book title. 2nd ed. Publisher, Location, (1999)
3. Athavankar, U.: From Product Semantics to Generative Methods. Accessed 12 Jun 2020. (Online). LNCS Homepage, https://www.springer.com/lncs. Last Accessed 21 Nov 2016
4. Athavankar, U.A.: Categorisation... natural language and design. Des. Issues **5**(2), 100 (1989). doi: https://doi.org/10.2307/1511517
5. Wittgenstein, L.: Games and definitions. In: A Modem Book of Aesthetics, 5th edn. New York (1979)
6. Rosch, E.: Principles of categorization. In: Cognition and Categorisation. New Jersey, Lawrence Erlbaum (1978)
7. Online Shopping site in India: Shop Online for Mobiles, Books, Watches, Shoes and More-Amazon.in. https://www.amazon.in/. Accessed 12 Jun 2020
8. Dabur Lal Dant Manjan—Ayurvedic Tooth Powder |Dabur. https://www.dabur.com/in/en-us/products/dabur-lal-dant-manjan. Accessed 12 Jun 2020
9. ZINDA TILISMATH. https://zindatilismath.co.in/aboutus.html. Accessed 12 Jun 2020

Chapter 56
PowerPost: A Framework for Designing Visual Political Communication

Anik Ghosh and Saurabh Tewari

Abstract Visual media is an essential medium of political communication which is used by various stakeholders in democracies. However, most people engaging in political discourse are not professionally trained in design and are unable to communicate their messages effectively through visual media. The process of designing visual political communication (VPC) can be made more effective by incorporating domain-specific design principles in the design process. These design principles are identified by an examination of existing research on perception and cognition of VPC and analysis of its useful examples. An application, PowerPost, is proposed based on the basis of this analysis.

56.1 Introduction

Visual media is an essential medium of political communication in modern democracies. The first uses of visual media for political communication were in the nineteenth century when political parties started employing multimodal posters (containing both textual and pictorial information) for election campaigns [1]. Since then, political communication through visual media has evolved and now includes a variety of forms such as billboards, television commercials, and election posters, along with growing usage of the social media [2].

In the contemporary context, the usage of visual modes of political communication is not limited to political parties and governments but also being used by broader stakeholders in democracies. Members of civil society are increasingly using visual means of political communication for various causes, including expression of dissent against their governments and those in positions of power. For such kinds of uses and

A. Ghosh (✉) · S. Tewari
School of Planning and Architecture, Bhopal, India
e-mail: anik.think@gmail.com

S. Tewari
e-mail: saurabh@spabhopal.ac.in

A. Chakrabarti et al. (eds.), *Design for Tomorrow—Volume 1*, Smart Innovation, Systems and Technologies 221, https://doi.org/10.1007/978-981-16-0041-8_56

users, there lies an opportunity for designers to develop a framework for designing effective visual communication.

Through analysing historical and contemporary visual political communication (VPC), this project aims to develop a framework for designing effective new political communication in the visual medium. At its foundation, this framework consists of an understanding on three levels. First, an understanding of the process of cognitive reception of VPC among recipients. Second, the syntactic choices made regarding framing, usage of colour, iconography, typography, etc. Third, the semantics and pragmatics of political communication, which ties the syntactic elements mentioned above to the meaning conveyed through them. It further looks into how context, the communicator of the message, the intended recipients, and the socio-political background adds to the meaning. These parameters in the study objectify otherwise subjective political communications. The research deals with a vast landscape of visual culture to empower the values of democracy. Through a historical and contemporary understanding of related visual culture, it enriches the literature resolution. The purpose is to develop and design a product for prospective users and enable them to design their unique political communication like posters and digital campaigns.

56.2 Making Design Accessible

In nearly, all political movements led by citizens, whether it be an environmental movement, a human rights movement, or protest movements against a government; the use of visual modes of political communication is a common theme. However, most people who engage in political discourse are not professional designers. As a consequence, they are often unable to communicate their messages effectively. Many of the placards and posters used in these movements fail at several levels, such as: being unable to attract the attention of recipients; failure in identifying target audiences; being unable to create interest or positive associations; being hard to memorise, and so on (Fig. 56.1).

Domain-specific design principles that enable the design of more effective visual communication can be developed and identified by analysing existing examples of visual communication and examining prior research on perception and cognition on visual communication [3]. Designers use their tacit knowledge and understanding of design to create artefacts that effectively convey ideas and information. The next section of this paper, using examples, discusses how designers use design principles to address socio-political context, appeal to target audiences, and leave a lasting cognitive impression on the audience. The design principles identified here can be used to frame guidelines for designing VPC which politically active members of society can utilise. This research aims to empower citizen participation in political discourse by proposing a tool that guides citizens in the process of creating VPC.

Fig. 56.1 People holding placards at a protest in Bhopal against a controversial Citizenship Amendment Act in January 2020. *Source* Author

56.3 Framework for Analysing Visual Political Communication

This project begins by looking at existing research on the cognition and perception of VPC and creates a theoretical framework for the analysis of historical and contemporary artefacts. The process of analysis is as follows:

1. **Understanding the process of cognitive reception** among recipients
2. **Outlining the context** (social and political) surrounding an artefact which includes identifying the source of the communication, whom the intended recipients are, understanding the desired effects that the communication intends to achieve and the socio-political background of the time.
3. **Identifying syntactic elements** used in visual communication such as colour, iconography, textual content, typography, and composition.
4. **Understanding semantics and pragmatics** associated with these syntactic elements, i.e. the meaning associated with them.

This understanding is further used to develop a framework for designing more effective VPC, which can then be used to create a tool for enhancing VPC.

56.3.1 Cognitive Reception

Geise [4] proposes a theoretical model for understanding the cognitive reception of VPC, which can be explained in the following five distinct steps: (Fig. 56.2).

Fig. 56.2 Stages of cognitive reception of visual political communication

1. **Pre-attentive Stage**: This stage occurs when an artefact exists in a recipient's peripheral vision, the recipient not yet consciously aware of its presence. The first impression is generated in this stage and may lead to an emotional response to the artefact. A positive emotional reaction is likely to result in a higher possibility of the artefact capturing the conscious visual attention of the recipient.
2. **Allocation of Attention**: This is the stage in which an artefact captures the conscious visual attention of a recipient. This may last for as little as 1–2 s. Information intake from the artefact takes place in this stage.
3. **Reflection**: After the conscious intake of information, recipients go through a stage of *post-attentive evaluation* in which they reflect upon the information that they have been exposed to. The recipient chooses whether or not to accept the message communicated through the artefact. Messaging that generates a positive emotional reaction has greater acceptance among recipients.
4. **Memorisation**: This stage is one in which the recipient reconstructs the message received from an artefact in their memory. Visuals are more easily processed and retained in memory than textual information [5].
5. **Cognitive Effects**: This is the stage in which the artefact leaves an effect on the recipient, which influences their opinions and evaluations of political information that they receive from other sources.

This model underlines the importance of attracting the attention of a recipient towards an artefact, without which none of the other stages would materialise. It also shows the importance of forming a positive emotional response among recipients to be effective in successfully delivering a message and having a sustained effect. These are the points that need to be taken into account while designing VPC.

56.3.2 Context

The socio-political context of the time and place plays a significant role in perception of VPC by its recipients. The literature suggests three basic and distinct units in the process of visual communication: (i) the communicator, (ii) the message, and (iii) the recipient [6]. These three units and the relationship between them (which includes the socio-political background in which the artefact exists) need to be understood to employ a pragmatic analysis of the syntactic elements used within the artefact (Figs. 56.3 and 56.4).

Two election posters, one used by Eugene McCarthy's campaign during the 1968 campaign for the Democratic Party's presidential nomination in the US, and another used by the Bhartiya Janata Party in India's 2014 general elections, are used here to illustrate a framework for analysis. These posters have been chosen to illustrate how artefacts of VPC can utilise the principles described in this paper to effectively communicate a message and reinforce positive associations among their target audience.

In 1968, Eugene McCarthy's campaign was being run with the Vietnam War at the forefront of political discourse in the US and the rise massive anti-war protests and

Fig. 56.3 Poster used by Eugene McCarthy's Presidential campaign in 1968 [7]

Fig. 56.4 A poster used by BJP in 2014 Indian elections [8]

demonstrations associated with it across the country. There was growing discontent among voters regarding Lyndon B. Johnson's (the erstwhile President) handling of the American war effort. McCarthy's campaign positioned him to be the primary representative of Democratic Party voters who were anti-war and were seeking a peaceful resolution of the war. An unexpectedly strong showing by McCarthy in the first leg of the primary elections encouraged other candidates like Robert Kennedy to join the race and forced President Johnson to withdraw his candidature [9].

Similar to McCarthy's campaign, the Bharatiya Janata Party's (BJP) campaign for the 2014 general elections in India were made in the backdrop of growing discontent against the existing government. In the lead up to the elections, there were a series of protests and demonstrations against the government on issues ranging from growing inflation, women's rights, and allegations of corruption against ministers. The BJP's campaign relied heavily on projecting Narendra Modi, the erstwhile Chief Minister of Gujrat, as their Prime Ministerial candidate [10]. These posters leveraged the dissatisfaction among the citizens with the government. By using Modi's image alongside these complaints, they implicitly suggested to voters that Modi would be the person to solve their problems. These posters broke away from the stereotype of the smiling politician [11] and instead utilised a 'serious' image of Modi gazing at a distance.

56.3.3 Syntactics, Semantics, and Pragmatics

The study of signs in communication can be divided into three parts: syntactics, semantics, and pragmatics [12]. Syntactics refers to the structure and order of a language, i.e. the relationship of signs to one another. Semantics refers to the relationship between signs (signifier) and their meaning (signified). Pragmatics is the relation between signs (signifier) and their human interpreters (in this context, the receivers of VPC). For this framework, syntactic elements in VPC have been divided into the following categories:

- **Iconography**: Images, symbols, or icons used. These are often the primary carriers of information in artefacts of VPC.
- **Text**: The textual content in an artefact. This is often in the form of large headlines or catchy slogans.
- **Typography**: The style of typeface used, the size of type, and its positions around other elements on an artefact can be used to convey information to recipients.
- **Colour**: Prevalent colour associations can be used to convey emotions and information to recipients.
- **Composition**: The positioning and framing of elements are also an essential part of the grammar of visual communication.

Using this framework, the illustration with the example of the 1968 Eugene McCarthy's campaign poster, and the 2014 BJP campaign poster is continued: (Table 56.1 and 56.2)

Table 56.1 Analysis of the McCarthy **poster**

Syntactic element	Object (signifier)	Semantics and pragmatics (signified)
Iconography	Hand-drawn dove, coloured with paper-cut type strips	(i) the dove represents peace, symbolic attribution of McCarthy as the candidate for peace, (ii)symbolic suggestion by the hand-drawn and paper-cut styles that were associated with posters used in anti-war demonstrations
Text	McCarthy (top), PEACE (bottom)	Symbolic attribution by drawing a direct association with McCarthy's candidature as president and the concept of peace (concerning the Vietnam war)
Typography	Bold, capitalised text in a paper-cut style	Repeating the association between McCarthy's campaign and the language of posters used in the anti-war demonstrations
Colour	Red and blue, along with shades of grey on the dove	Red and blue are the colours of the American flag; they are used here for the positive emotional associations those colours have among potential voters. The grey is used on the bird to further emphasise it as a dove, a symbol of peace.
Composition	The dove is in the centre, enveloped by words: McCarthy and Peace	Emphasises the centrality of getting a peaceful resolution as the central plank of the McCarthy campaign

56.4 Incorporating the Framework in the Design Process

The examples given in the previous section show how designers use syntactic elements and the associations that those elements have to convey their messages effectively. In both examples, the designer has used the semantic principles of symbolic attribution and symbolic suggestion [13] to construct relationships between syntactic elements. Another common theme in effective VPC is the repetition and reinforcement of a concept through multiple syntactic elements. We see this being used by both the McCarthy and BJP posters. The association of McCarthy with peace was suggested not only through the literal uses of the word 'Peace' alongside his name but also through the use of the dove, a symbol of peace, and the use of typography and imagery associated with the peace movement. The use of Modi's image with a serious appearance, gazing into the distance, reinforces the idea of the campaign slogan by providing a symbolic suggestion to recipients that Modi is the answer to the everyday grievances of the people.

Table 56.2 Analysis of the BJP poster

Syntactic element	Object (signifier)	Semantics and pragmatics (signified)
Iconography	Image of Modi gazing into the distance, a lotus symbol	Modi is established as the primary face of the party and its Prime Ministerial candidate for the election campaign, leveraging Modi's growing popularity for the campaign
Text	'Had enough of high petrol–diesel prices' on top, followed by the campaign slogan 'Abki Baar Modi Sarkar' (This time, Modi's government)	By the juxtaposition of Modi's image with a complaint against inflation, the poster implies that Modi will solve the problem—symbolic suggestion
Typography	Formal Devanagari letters	The formality of the typeface suits the message that the poster is communicating
Colour	Most of the poster is in saffron, with a hint of green	These represent the colours of the party; their usage seeks to increase brand recognition
Composition	Modi's image takes priority in the visual hierarchy, followed by the party symbol.	The visual hierarchy of the poster reinforces the idea of Modi being the face of the campaign; Modi's posture and his gaze towards the distance are intended to make the symbolic suggestion of Modi as a visionary leader with the answer to the complaint made in the poster

56.4.1 *PowerPost: An App for Designing Visual Political Communication*

Several digital applications exist which can be used to create visual media, ranging from sophisticated tools such as Adobe Photoshop and Illustrator to online applications and even simpler tools like MS paint. While several such tools exist, there are no tools that guide users in the design process. There exist some online options which are targeted at amateurs, such as Canva and Adobe Spark. Still, these apps rely primarily on the usage of pre-made templates to help the user move forward in the design process. Apps that guide users through the process of making design decisions are not available to user. This presents an opportunity to create a product that can address this gap.

PowerPost is envisioned as an application that will provide prompts to a user during the design process (derived from the framework discussed in the previous sections) so that users are more consciously aware of the design choices which they make when designing VPC. This is achieved by adding steps in the design process which provide users with prompts and questions concerning the points discussed earlier.

Interventions in the Design Process

The proposed interventions in the design process are explained as: (Fig. 56.5)

1. **Setting Context**: The examples of VPC analysed in Sect. 3 highlight the importance of being aware of and addressing the social and political context of the artefact. The failure to address the context is a common issue in many of the posters or placards used by citizens in demonstrations and protests (Fig. 56.1). Most design apps currently take users directly to a blank design canvas and provide them with tools to draw on and edit things on a canvas. As a result, a user may fail to consciously keep in mind that the need to identify and address the context in which they are designing for. With PowerPost, users will be first given a questionnaire regarding the context in which they are designing an artefact (Fig. 56.6). The intention is make them more *consciously* aware of the decisions that they need to make.

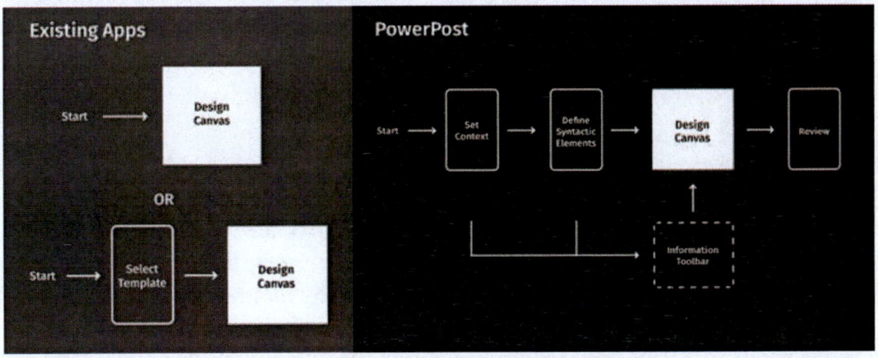

Fig. 56.5 Comparison of user flow in PowerPost to existing design apps

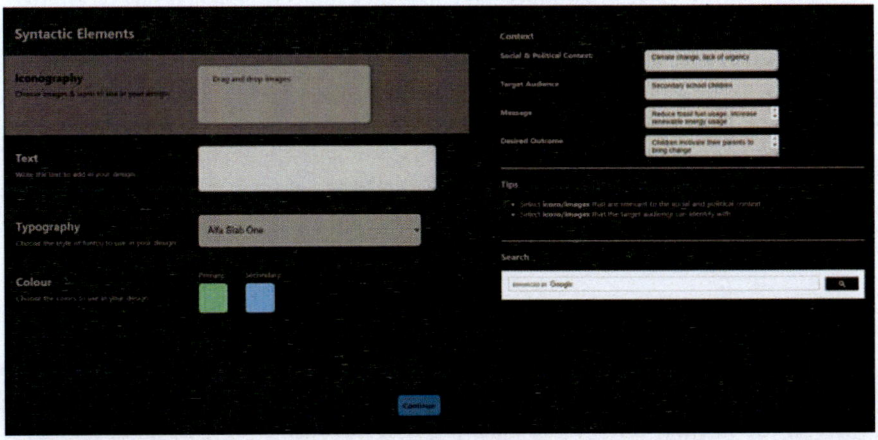

Fig. 56.6 Screenshot of step 2, defining syntactic elements

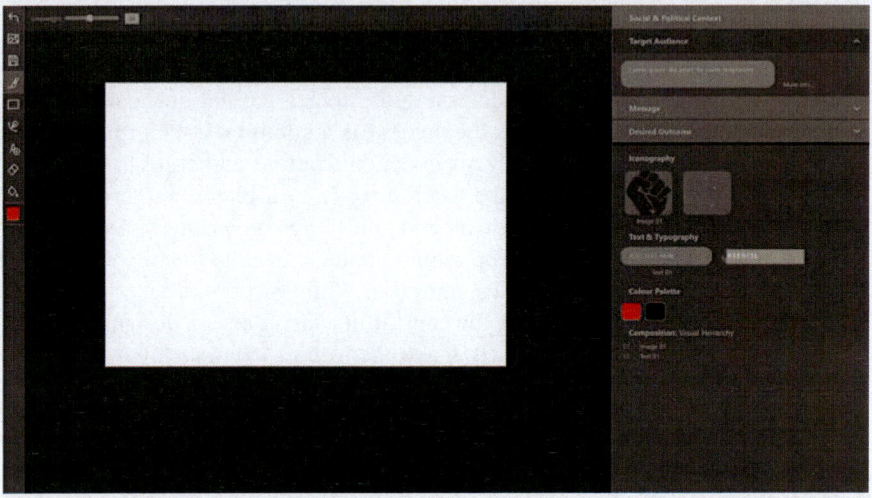

Fig. 56.7 Screenshot of step 2, editing on the canvas

2. **Defining Syntactic Elements**: In this step, taking information from the answers given by the users in the previous step, users are directed to pre-select different syntactic elements that they think will be most effective to address their context and purpose (Fig. 56.7). Users will be asked to select the images or symbols that they think will be appropriate, the typography that will be appropriate, and so on. In these steps, users will be given prompts which guide them on the decisions that they need to make to attract the attention of potential recipients and gain a positive response in their cognition.
3. **Editing on Canvas**: On making these selections, users are taken to the canvas on which they can draw on or edit like other design applications. Users will have access to a toolbar showing them information they had previously entered and things they need to keep in mind.
4. **Reviewing**: On completion of the drawing/editing process on the canvas, users will be taken to a review process. Here, they will be given checklists concerning the different syntactic elements used, and whether they believe that they have been able to address the context appropriately or not. Based on this review, users can choose to either end the design process at this stage if they are satisfied with their result or go back to the previous step to alter their design.

56.5 Epilogue

PowerPost looks at a new way of enabling users to design VPC by making it more accessible to the ordinary citizen. This is done by providing users with tools that will empower them to make more conscious choices in the design process. The process used by the app is derived from a theoretical framework based on a review of existing literature associated with the field. There exists the possibility to take this research further by validating the suggested framework through qualitative user feedback. It is also possible to strengthen further the design process suggested by this paper by enhancing the scope of the theoretical framework for the analysis of visual political communication.

References

1. Holtz-Bacha, C., Johansson, B.: Posters: from announcements to campaign instruments. In: Election Posters Around the Globe: Political Campaigning in the Public Space, pp. 1–12. Springer International Publishing (2017). https://doi.org/10.1007/978-3-319-32498-2_1
2. Ekman, M., Widholm, A.: Political communication in an age of visual connectivity: exploring instagram practices among swedish politicians. Intellect **15**, 15–32 (2017)
3. Agrawala, M., Li, W., Berthouzoz, F.: Designprinciples for visual communication. Commun. ACM **54**, 60–69 (2011)
4. Geise, S.: Theoretical perspectives on visual political communication through election posters. In: Election Posters Around the Globe: Political Campaigning in the Public Space, pp. 13–31. Springer International Publishing (2017). https://doi.org/10.1007/978-3-319-32498-2_2
5. Paivio, A.: Mind and its evolution: A dual coding theoretical approach. In: Mind and Its Evolution: A Dual Coding Theoretical Approach (2014). https://doi.org/10.4324/9781315785233
6. Waller, D.S.: A proposed model of political communication: expanding the communication model to incorporate multiple sender s/receivers. J. Mark. Commun. **1**, 209–219 (1995)
7. Shahn, B. In: McCarthy Peace Poster. (1968). Available at: https://www.flickr.com/photos/iisg/8160531960. (Accessed: 11th June 2020)
8. Bharatiya Janata Party.: In: BJP Election Poster. (2014). Available at: https://hindi.newscentral24x7.com/maharashtra-petrol-price-rs-89-90-parbhani/. (Accessed: 11th June 2020)
9. Zeitz, J.: The Myth of Eugene McCarthy. In: The New York Times (2018). Available at: https://www.nytimes.com/2018/03/08/opinion/eugene-mccarthy-lyndon-johnson-vietnam.html. (Accessed: 25th May 2020)
10. Willnat, L., Verghese, R., Mammadov, R.: Symbols, slogans, and charisma: political posters in India's 2014 national election. In: Election Posters Around the Globe: Political Campaigning in the Public Space, pp. 187–209. Springer International Publishing, (2017). https://doi.org/10.1007/978-3-319-32498-2_10
11. Novelli, E.: Themes, styles, and tendencies of the European political poster. analysis of the 2009 and 2014 European election campaigns. In: Election Posters Around the Globe: Political Campaigning in the Public Space, pp. 91–114. Springer International Publishing, (2017). https://doi.org/10.1007/978-3-319-32498-2_6
12. Morris, C.: In: Foundations of the theory of signs. University of Chicago Press, (1938)
13. Kress, G.R., Van Leeuwen, T.: In: Reading images : the grammar of visual design. Routledge, (2006)

Chapter 57
Principles of Complementarity in Designing Balanced Urban Space: Cases of Sociocultural Spaces of Kolkata

Tanima Bhattacharya and Joy Sen

Abstract Unplanned skewed urbanization trend of developing countries has made the cities and their constituent urban spaces vulnerable that are struggling to retain their identity. In the case of the cities like Kolkata, with a colonial past, striving hard to maintain the physical, environmental, and cultural fronts and the acceptable yardstick of urban experiences. Hence, the study underscores the re-envisioning of urban regeneration practices, suggesting a paradigm shift from the standpoint of'either-or' to a liveable'principle of complementarity' rather than exploring the two constituent parameters, i.e., urban design and urban engineering in isolation. To explore the principles of complementarity, the paper explores two areas of Kolkata, i.e., area around Rabindra Sadan and Saltlake City Center1. The study has arrived at an integrated toolbox based on required technological (tangible) and artistic (intangible) Elements of Design, to devise stepwise analyses of bringing the complementarity between the two design elements, i.e., one, action plans and possibilities of urban regeneration to induce happiness and well-being two, the experience of space.

57.1 Introduction

Recent urbanization trends have skewed the balance distribution within a system of cities in a nation. According to the report of the United Nations (2015) [1], 54% of the world's population has continued to reside in urban areas and it will soon be 66% by the year 2050 [2]. As a result, rapidly expanding urban spaces with steep population growth start to make city-systems imbalanced. In this context, compared to the cities in developed countries, cities and their constituent urban spaces of

T. Bhattacharya (✉)
Ranbir and Chitra Gupta School of Infrastructure Design and Management, Indian Institute of Technology Kharagpur, Kharagpur, West Bengal, India
e-mail: Tanimabhattacharya@iitkgp.ac.in

J. Sen
Department of Architecture and Regional Planning, Indian Institute of Technology Kharagpur, Kharagpur, West Bengal, India
e-mail: Joysen@arp.iitkgp.ac.in

A. Chakrabarti et al. (eds.), *Design for Tomorrow—Volume 1*, Smart Innovation, Systems and Technologies 221, https://doi.org/10.1007/978-981-16-0041-8_57

developing countries are struggling more to retain their identity [3, 4], resulting an unmanageable complexity depleting the possibility of liveability in urban mosaics [5].

Considering this context, cities like Kolkata, with its colonial past, though known as the cultural capital of the country, striving hard to retain its identity. Even the predominant cultural and social spaces of Kolkata sometimes failed to balance the physical, environmental, cultural fronts and lack an acceptable yardstick of urban experience [6]. Hence, a consequent sense of disjointedness is physically evident and normatively detrimental [7]. To mitigate such issues, the paper explores the possibilities of reinstalling the balance between the need and aspirations of the stakeholders and the capacity of the space. The paper aims at earmarking a process of balanced regeneration based on the application of principles of complementarity of urban design and urban engineering [8]. To best explore the principles of complementarity, the paper investigates two areas of Kolkata, i.e., area around Rabindra Sadan and Saltlake City Center1, which are the sociocultural lifeline of the city. The study has underscores the re-envisioning of urban regeneration practices, suggesting a paradigm shift from the standpoint of 'either-or' to a liveable 'principle of complementarity.'

57.2 Urban Spaces of Kolkata: Historical, Sociocultural Context

Kolkata, the administrative capital of the Eastern Indian state of West Bengal, is the fifteenth largest city in the world and was India's first metropolitan entity [9]. It has grown from a colonial port city to a giant megalopolis [10, 11] with an escalating horizontal and vertical urban expansions [9, 15]. Located on the east bank of the Hooghly River, it is the principal commercial, cultural, and educational center of East India [9, 12, 13] with a population of 4.5, making it the third most populous metropolitan area in India [16].

Kolkata is popularly known as the 'cultural capital of India' and the 'City of Joy' [6, 11, 14], and the cradle of culture and intellectualism [12]. Kolkata is known as the 'City of Furious, Creative Energy' [11, 14], and the two selected urban spaces are the prime locations celebrating Kolkata's contemporary literary and artistic thought [8]. Figure 57.1 shows the geographical location of the two spaces. The former one, i.e., Rabindra Sadan, falls under the jurisdiction of Kolkata Municipal Corporation (Ward 63), and the latter one is the part of the first eastward extension of the city, Saltlake (DC Block). Both areas are well-connected to the city through transportation and cater to the social and cultural needs of the city, devoid of class-economic distinction.

The following subsections will explore the social, cultural, traditional features of both the places in detail.

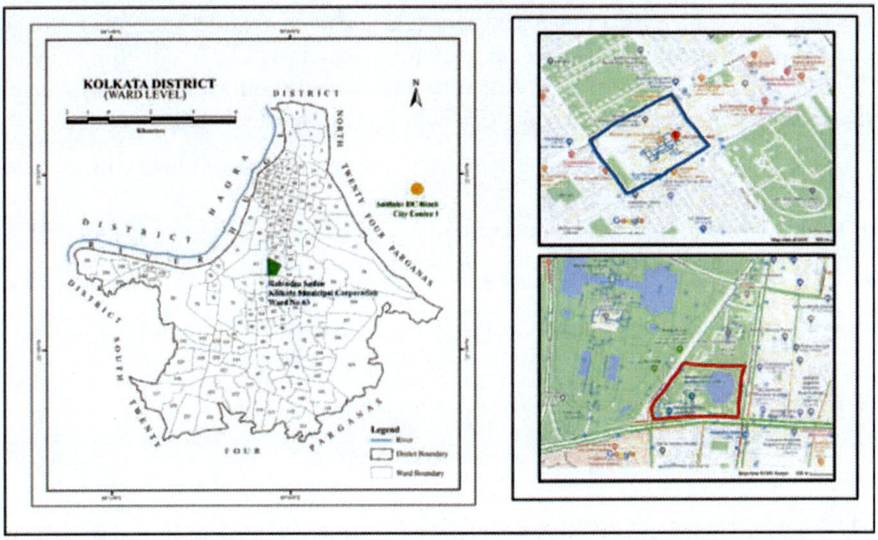

Fig. 57.1 Ward wise map of Kolkata Municipal Corporation (Left), Map of Rabindra Sadan (Top right), and Map of City Center Saltlake (Bottom right). (*Source* Kolkata Municipal Corporation and Google maps)

57.2.1 Rabindra Sadan: The Prime Cultural Node

One of the most celebrated space, receiving the highest footfalls, including artists, musicians, painters, and art enthusiasts, is Rabindra Sadan. Bubbling with passion and enthusiasm, Rabindra Sadan is one of the most prominent cultural locations in Kolkata [11], named after the memory of the Noble laureate poet Rabindranath Tagore. Rabindra Sadan epitomizes the fascinating culture, fine art, and rich traditions of Bengal [13]. It is a premier venue for theater and cultural activities of the city, attracting footfalls from around the world (Fig. 57.2).

Rabindra Sadan is located on Belvedere Road, Bhawanipur, Kolkata. Rabindra Sadan stage, Nandan Pashcimbanga Bangla Akademi (West Bengal Bengali

Fig. 57.2 From left Nandan, Rabindra Sadan, Academy of fine arts. (*Source* Author)

Academy), Gaganendra Pradarshanshala (Gaganendra exhibition center) abovementioned four are the prime centers of cultural activities [17]. In addition to these, Sisir Mancha, Academy of Fine Arts, and The Calcutta Information Center make Rabindra Sadan a treasure trove for rich cultural heritage and exuberance [18].

- Rabindra Sadan is a premier location for staging Bengali theaters and the traditional location for Kolkata Film Festivals.
- Nandan Film Center is a complex that showcases films, disseminate knowledge, raises awareness, and promotes global cinema with specific emphasis on Bengali cinema [11, 12].
- Academy of Fine Arts is one of the finest art galleries of India, established in 1933. It attracts art lovers from different corners of the state and beyond.

57.2.2 City Center 1: An Interactive Space

Saltlake City Center is a comparatively young place established in the year 2004 [19], offering multifaceted entertainment and cultural opportunities [20, 21]. The place offers an exceptional interplay between two distinct traditions of shopping—the traditional open markets of Indian streets and the controlled box of the American style mall that has evaded Indian metropolis in the near past [11, 20, 21] (Fig. 57.3).

Charles Correa's design for the Saltlake City Center represents a sociable connection between the mall as a shopping space and of the culture, the city stands for [19, 20]. According to Correa, '*The City Center in Kolkata is a very special place…a microcosm of the whole metropolis, catering to multiple land-uses and diverse income profiles…a kaleidoscope of contrast, colour and energy. We have in the City Center a wide range of different-sized residences, entertainment centers, offices and shops varying from the smallest "dukaans" to the most glamorous air-conditioned boutiques and large department stores* [19].' The design juxtaposed complex systems of spaces, from narrow lanes to large terraces, from westernized colonnades to a very traditional concept of *kund* placed in an open space, from concrete to strategic location of the trees and topiary [19]. Hence, it can be called a microcosm of the cities tradition and contemporary cutting-edge design innovation [19].

Fig. 57.3 Juxtaposition of open and closed spaces of CC1 promoting human interaction, and adding dynamicity to the place. (*Source* Ambuja Neotia Group)

Design features of City Center1 exemplify the integration of normative and physical dimensions [22–24], providing spaces for cultural, social congregation, and festivals. The features that make CC1 unique are mentioned below.

- The **context is very Indian**, and the experience is very contemporary and lively [20]
- A fair amount of **greenery** is included throughout the complex to provide warmth and intimacy [19, 22]
- The **informal format**, intensely personalized seller–customer interaction and pluralistic form of a bazaar, is juxtaposed with mammoth retailers of branded products [13, 20]
- The **proximity** of shopping, food, entertainment, and cultural programs cater to a variety of population [25, 26]
- **Inclusiveness** of the spaces caters to all age, ethnic, social–cultural, and economic groups [25, 26]
- **The relaxed and hybrid ambiance** created through an interplay between natural and conditioned design [19, 20]
- That **'boundless' feeling and sense of receptiveness,** generated through an open format of the design [19, 27]

57.3 Methodology

Both of the cultural nodes of Kolkata, i.e., Rabindra Sadan and City Center (CC1) have been surveyed considering the two sets of elements of design, i.e., Art elements of Design and Technological elements of Design. The design elements were identified by the authors through an extensive literature survey and validated through inducted case studies on Chandigarh and Delhi Haat, which is a part of the larger ambit of the study. The identification and validation process are left out of the scope of the present paper, which deals with the cases of sociocultural spaces only.

The design elements are tapped to map their impact on the space and the people. A total number of 120 people (60 in each place) have participated in deep engagement surveys. The scores have been assigned on a 9-point Likert scale to assess the present imageability. Assigned scores have been tabulated through weighted sum analysis. The weightage of the individual elements is assigned even before by experts (A team of 14 people encompassing Architects, Urban Planners, Designers, Visual Artists, Performance artists, Civil Engineers, Administrative) opinion survey. Economic capability and freedom choices have a variable impact on the space and its people. Therefore, the scope of economic possibilities is consciously left out of the study.

The weightage of both elements of design is calculated individually and groupwise. Individual obtained scores help to identify the hierarchy of the design elements; hence, the sequence to implement them for the regeneration of urban spaces in its best possible way. Likewise, the cumulative assessment of weighted scores of Rabindra

Sadan and City Center substantiate that the two sets of design elements perform better in complementarity than of individual implementation.

57.4 Physical and Normative Dimensions of Urban Spaces

Exploring the two cultural nodes of Kolkata gave a brief idea about accomplished spaces. There are effectual design parameters or elements of design common to both the nodes that secure the imageability and have made these places acceptable to the users [6, 11]. Drawing on the international research that identifies the key domains impinging the imageability, acceptability of urban cultural spaces, the indicators contextualized at the space level of Kolkata have been cross-validated and used to assess the link between the city's current urban forms its problems and the way to address it [15, 28]. Hence, the aim is to identify the design elements of sociocultural space, modified and applied on the local scale in Kolkata for balanced rejuvenation.

Along with deep engagement surveys of the users and in-depth literature study helps to identify the design parameters of a thriving place [7]. Table 57.1 shows the identified design elements, broadly divided into two sections, namely **a.** technological elements of design and **b.** art elements of design. The prior one denotes the primary and foundational elements that deal with the basic needs a human being to

Table 57.1 Artistic and technological elements of design conceived through deep literature and stakeholders' survey

Art elements of design		Technological elements of design	
1	Presence of green space	1	Transportation facility
2	Presence of open spaces	2	Drinking water
3	Availability of recreational space	3	Sanitation facility
4	Colour scape	4	Waste disposal
5	Mural	5	Healthcare facility
6	Graffiti	6	Education facility
7	Painting	7	Electricity
8	Sculptures	8	Information center
9	Art installations	9	Proper lighting
10	Topiary	10	Pollution check
11	Art performances	11	IoT facility
12	Performing art		
13	Presence of water body		
14	Heritage structure		
15	Scope of economic regeneration from the heritage		
16	Availability of community space		

survive [29–31], like drinking water, sanitation, waste disposal, healthcare facility, education, and transportation facilities to satisfy the physiological needs [32, 33]. However, to make a place livable, acceptable, and sustainable, the sociocultural, normative-intangible, artistic [24, 27] elements need to be integrated. As suggested by Maslow in his 'Hierarchy of Needs' [32], fructification of the physiological needs are followed by the need for safety security, sense of belongingness and ownership, the idea of self-esteem, and actualization [29–31]. All these steps consider intangible elements to be integrated with the physiological one. Thus, this paper aims to achieve the complementarity of technology-driven (physiological) infrastructure and the intangible normative design elements to make an urban space accommodative, receptive, acceptable, and happy to its users.

57.5 Analysis, Result, Discussion: Process of Balanced Regeneration

This section would demonstrate the need and impact of elements of design. Both the sets of elements have been surveyed to design balanced urban space. The impact of technological and art elements of design on the imageability of the two case study areas, i.e., Rabindra Sadan and Saltlake City Center have been tapped and discussed below.

57.5.1 Technological Elements of Design

To understand the need and impact of the technological elements of design, engagement surveys have been done amongst 60 stakeholders in each area. Table 57.2 shows the survey results need, impact (obtained score), and weighted scores in detail. As can be seen from Table 57.2, the weightage of the individual elements is assigned by experts based on importance. The highest weightage is given to the availability of drinking water (1.7), followed by the transportation facility (1.5) and availability of electricity (1.4).

The obtained scores (refer Fig. 57.4) of Rabindra Sadan shows the availability of electricity got the highest score of 8.31, followed by healthcare facility with a score of 7.78 and availability of transportation with a score of 7.63. As Rabindra Sadan is one of the prime nodal points for transportation and housed government-aided hospitals, the total obtained score is quite high as 68.08.

On the other hand, in the case of Saltlake City Center, the highest weightage was also assigned to the availability of electricity (6.8) followed by the availability of drinking water (6.65) succeeded by close scores of the educational facility (6.36) available and the sanitation facility (6.31). Though the area of Saltlake City Center

Table 57.2 Calculation of the weighted sum from the obtained scores to rank the elements of design

Technological elements of design	Weight assigned	Rabindra Sadan (obtained score)	Rabindra Sadan (calculated weight)	Saltlake (CC1) (obtained score)	Saltlake (CC1) (calculated weight)	Weighted sum	Rank
Transportation facility	**1.5**	**7.63**	11.445	5.78	8.67	**20.12**	3
Drinking water	**1.7**	6.12	10.404	**6.65**	11.305	**21.71**	1
Sanitation facility	0.8	7.39	5.912	**6.31**	5.048	10.96	5
Waste disposal	0.8	6.26	*5.008*	5.97	*4.776*	9.78	7
Healthcare facility	1	**7.78**	7.78	5.82	5.82	13.60	4
Education facility	0.8	6.75	5.4	**6.36**	5.088	10.49	6
Electricity	**1.4**	**8.31**	11.634	**6.8**	9.52	**21.15**	2
Information center	0.3	3.64	1.092	1.53	0.459	1.55	10
Proper lighting	0.5	7.04	*3.52*	5.92	*2.96*	6.48	9
Pollution check	0.7	5.68	*3.976*	3.92	*2.744*	6.72	8
IoT facility	0.5	1.48	0.74	1	0.5	1.24	11
TOTAL	**10**	**68.08**	**66.911**	**56.06**	**56.89**	**123.80**	

The 1st, 2nd and 3rd highest scores obtained by the elements of design is in bold, along with the total scores

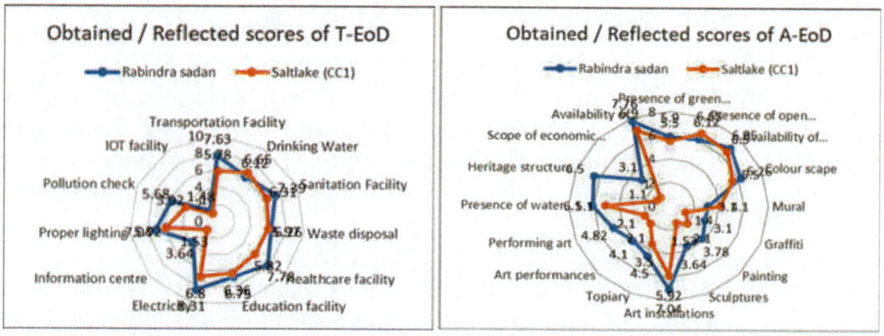

Fig. 57.4 Visual representation of obtained scores of technology elements of design (left) and art elements of design (right)

is nascent in comparison to the Rabindra Sadan area, still the overall score obtained by the technological elements of design are as low as 56.06.

Surprisingly the obtained score of Rabindra Sadan reduces while calculating considering the weightage of the elements. From the score of 68.08, It goes down to 66.91. On the contrary, in the weighted sum, the scores have a bit increased from 56.06 to 56.9 in the case of Saltlake City Center.

The tendency of the weighted score exemplifies the actual scores are lower than the visible or reflected scores. Increased reflected scores represent high imageability, expectation, and acceptability of the space among the stakeholders. However, in the case of City Center, the actual and the reflected scores are almost equal or even better.

The total calculated weighted sum assigned the priority to the design elements. The highest priority is given to the availability of drinking water (21.71), followed by the availability of electricity with the weightage of 21.15 and the availability of adequate transportation facilities with a score of 20.12. Obtaining the lowest scores, the need for pollution check (3.976/ 2.7444), proper lighting (3.52/ 2.96), and waste disposal management (5.008/ 4.776) demand immediate consideration (while keeping IoT and the role of information center as the lowest priority).

57.5.2 *Art Elements of Design*

Likewise, an analysis of the artistic elements of design has been done in both places based on the assigned weights by the experts (Refer Fig. 57.5). The highest weightage has given to the presence of green space (2), followed by the presence of water body (1.4), economic regeneration potentiality (0.8), and availability of community space (0.8).

As demonstrated in Table 57.3, in the case of Rabindra Sarovar, the highest obtained score is 7.76 attributed to the community space design. Specifically, art and craft installations got the second highest prominence with a score of 7.04, followed by the recreational ability offered to its users (6.95) by the space itself.

Surprisingly, in the City Center, which is conceptualized to be a shopping center, has obtained the highest score for the availability of space for interaction (6.9); the presence of open space and art installations successively obtained the scores of 6.65

Fig. 57.5 Strategic placements of art elements to enhance imageability as fountains, colonnade, installations, trees and topiary, sculptures at Rabindra sadan (1,2,3 from left) and CC1(4,5)

Table 57.3 Weighted sum calculation from the obtained scores to rank the art elements of design

Art elements of design	Weights assigned	Rabindra Sadan (obtained score)	Rabindra Sadan (calculated weight)	Saltlake (CC1) (obtained score)	Saltlake (CC1) (calculated weight)	Weighted sum	Rank
Presence of green space	**2**	5.9	11.80	**5.5**	11.00	**22.80**	**1**
Presence of open spaces	0.7	6.12	4.28	**6.65**	4.66	8.94	**4**
Available recreational space	0.5	**6.95**	3.48	6.5	3.25	6.73	**6**
Colour scape	0.7	6.26	4.38	5.5	3.85	8.23	**5**
Mural	0.3	3.1	*0.93*	4.1	1.23	2.16	**12**
Graffiti	0.3	3.1	*0.93*	1.4	*0.42*	1.35	**14**
Painting	0.4	3.78	1.51	2.1	*0.84*	2.35	**11**
Sculptures	0.5	3.64	1.82	1.53	*0.77*	2.59	**10**
Art installations	0.2	**7.04**	1.41	**5.92**	1.18	2.59	**10**
Topiary	0.5	4.5	2.25	3.5	1.75	4.00	**7**
Art performances	0.2	4.1	0.82	2.1	0.42	1.24	**15**
Performing art	0.2	4.82	0.96	2.1	0.42	1.38	**13**
Presence of water body	**1.4**	6.1	8.54	5.1	7.14	**15.68**	**2**
Heritage structure	0.5	6.5	3.25	1.1	0.55	3.80	**8**
Economic regen from the heritage	**0.8**	3.1	*2.48*	1	**0.80**	3.28	**9**
Community space	**0.8**	**7.76**	6.21	*6.9*	5.52	**11.73**	**3**
TOTAL	**10**	**82.77**	**55.053**	**61.00**	**43.794**	**98.85**	

The 1st, 2nd and 3rd highest scores obtained by the elements of design is in bold, along with the total scores

and 5.92, which is very rare. Attributing the artistic-normative dimension, an aimed to be shopping complex transformed not only to cater to the shopping spree but to add a socioculturally unique dimension to it. This triumph of aesthetics over economics only been possible by integrating the elements of design-driven by art and aided by technology.

Weighted scores have been calculated from the obtained score using the weighted sum method. The weighted scores are quite lower than the reflected score, which is

55.053 compared to 82.77. The tendency of declining calculated weight represents that the social, cultural reputation of the space is affecting the actual scores.

Considering the calculated cumulative weightage mentioned in Table 57.3, the presence of green space is the highest in the rank order (22.80), followed by the presence of water body (15.68) and availability of command space (11.73). In the case of art elements, including paintings (1.51/ 0.84), murals (0.93/ 1.23), sculptures (1.82/0.77), and specifically, economic regeneration possibility (2.48/0.80) as part of space design demands immediate consideration.

57.5.3 *The Paradigm Shift from the Standpoint of 'either-or' to a Liveable 'principle of complementarity'*

Most of the time, the two genres of elements, i.e., technology-aided elements of design and art-based elements of design, faced an either-or problem. However, as revealed from the study of sociocultural spaces of Kolkata, i.e., Rabindra Sadan and CC1, that successful, thriving, and dynamic space reflects the complementarity between the two. The same idea of complementarity is reflected through the obtained scores, calculated weights, and the cumulative scores mentioned in Table 57.4.

- The idea extracted from the engagement surveys is that the individual set of elements cannot fulfill the aspirations of the stakeholders.
- Technological elements of design can be regarded as the foundational, physiological elements that conform to the minimum prerequisite of a space. Successively, the art elements of design accredit a space with aesthetics, dynamicity, and an acceptable benchmark of imageability [32].
- Even the obtained and weighted scores referred to Table 57.4 elucidate that cumulative scores (8.90 and 11.36) are more effective than the individual weighted scores (7.91 and 9.52) in augmenting the imageability and acceptability of a space. Hence, both sets of design elements should be contemplated in sync while designing a balanced urban space.

Table 57.4 Calculation of the cumulative scores of technological (T-EoD) and art (A-EoD) elements of design

Scores assigned	CC1 (T-EoD)	CC1 (A-EoD)	CC1 (TA-EoD)	Rabindra Sadan (T-EoD)	Rabindra Sadan (A-EoD)	Rabindra Sadan (TA-EoD)
Obtained score	5.09	3.81	**8.90**	6.19	5.17	**11.36**
weighted score	5.17	2.74	**7.91**	6.08	3.44	**9.52**

The 1st, 2nd and 3rd highest scores obtained by the elements of design is in bold, along with the total scores

57.6 Conclusions

Establishing the idea of complementarity, shifting from 'either-or' paradigm to the 'paradigm of complementarity,' has been the primary concern of the paper. By investigating two prime, popular cultural nodes of Kolkata, Rabindra Sadan, and City Center, the paper attempts to establish the principle of complementarity between technical and artistic elements of design for sustainable urban regeneration. The ideology tried to establish a new framework for urban regeneration practices, processing a paradigm shift away from a conventional focus on technology-aided infrastructure to reaching an equilibrium of urban engineering and urban design.

- Both the areas exemplify strong tangible (technical) and intangible (art and normative dimension) heritage, having strong physical as well as cultural connotations.
- The study has identified a set of design parameters of required technological (tangible) and artistic (intangible) elements of design, to devise stepwise analyses of bringing the complementarity between the two design elements, i.e., one, action plans and possibilities of urban regeneration to induce happiness and well-being; two, the experience of space.
- Intensive engagement surveys, analysis, and result bolstered the attempted to approach the principle of complementarity rather than exploring the constituent parameters of the two, i.e., urban design (normative dimensions) and urban engineering (physical infrastructure) in isolation.
- Finally, the result obtained from the sociocultural spaces testifies that technology-aided urban infrastructure is the primary constituents to make a space habitable, but, artistic normative dimension adds liveliness to it, making the place liveable and happy.

Hence, the significant design elements recognized from the present study can be implemented as a prototype to other sociocultural urban nodes to ensure the principle of complementarity and device balanced urban regeneration.

Acknowledgements I want to acknowledge my supervisor Prof Joy Sen for his constant support and guidance, and IIT Kharagpur for providing the opportunity to continue my research.

References

1. UN-HABITAT.: Inclusive urban development policies the challenge : in a context of growing urbanisation (2019)
2. U. N. H. S. Programme: In: The State of the World's Cities 2004/2005: Globalization and Urban Culture. Earthscan (2004)
3. Johnson-Roehr, S.N.: City Planning in India. In: Selin, H. (ed.) Encyclopaedia of the History of Science, Technology, and Medicine in Non-Western Cultures, pp. 1–12. Springer, Netherlands, Dordrecht (2008)
4. Ahluwalia, I.J., Kanbur, R., Mohanty, P.K.: In: Urbanisation in India: Challenges, Opportunities and the Way Forward. SAGE Publications (2014)

5. NITI Aayog.: Sustainable development Ggoals–India : mapping of central sector schemes and ministries of Government of India. no. August, (2018). [Online] Available https://niti.gov.in/writereaddata/files/SDGMapping-Document-NITI_0.pdf
6. Shaw, A.: Inner-city and outer-city neighbourhoods in Kolkata: their changing dynamics post liberalization. Environ. Urban. ASIA **6**(2), 139–153 (2015). https://doi.org/10.1177/0975425315589157
7. Bhattacharya, T., Dasgupta, S., Sen, J.: An attempt to assess the need and potential of aesthetic regeneration to improve walkability and ergonomic experience of urban space, vol. 966. (2020)
8. Bhattacharya T., Sen, J.: In: An exploration of design: bsed on Pan-Asian semiotics and semantics, vol. 66 (2017)
9. Parilla, J., Leal, T.J., Berube, A. Ran, T.: In: Global Metro Monitor 2014: An Uncertain Recovery, pp. 44. The Brookings Institution, (2015). [Online] Available: http://www.brookings.edu/~/media/Research/Files/Reports/2015/01/22globalmetromonitor/bmpp_GMM_final.pdf
10. Haque, I., Das, D.N., Patel, P.P.: Spatial segregation in indian cities: does the city size matter? Environ. Urban. ASIA **9**(1), 52–68 (2018). https://doi.org/10.1177/0975425317749657
11. Sinha, S. (ed.): Cultural Profile of Kolkata. The Indian Anthropological Society, Delhi (1972)
12. N.O.V.J.A.N. Mar India: calcutta, the capital of culture (2020)
13. Wikipedia, F., District, C.B., Setu, R., Line, C.T., Setu, V.: Kolkata
14. Bose, N.K.: Calcutta: a premature metropolis. Sci. Amer. **213**(3), 90–102 (1965). https://doi.org/10.1038/scientificamerican0965-90
15. Bardhan, R., Kurisu, K. H., Hanaki, K.: Linking urban form and quality of life in Kolkata, India. In: ISOCARP Congress, vol. 47, no. section 2) pp. 1–12. (2011). [Online] Available http://www.isocarp.net/data/case_studies/1923.pdf
16. Directorate of Census Operations West Bengal.: In: District Census Handbook. West Bengal, October, pp. 296 (2011). [Online] Available http://www.censusindia.gov.in/2011census/dchb/1911_PART_B_DCHB_
17. Haque, I., Mehta, S., Kumar, A.: Towards sustainable and inclusive cities: the case of Kolkata. Observer Res. Fond. (ORF) 83(83), (2019). [Online] Available https://www.orfonline.org/series/urbanisation-and-its-discontents/
18. Part I The kolkata gazette, extraordinary, September 9: 1:1–92, (2009)
19. Ambujaneotia group: City Center. Retrieved from https://www.ambujaneotia.com/business_vertical/city-centre-salt-lake/
20. Center, T.C., Lake, S., Correa, C., Das, P.R., Center, C.: Correa changed iconic City center design at eleventh hour (2015)
21. Bengal, W., Up, S.: Busiest metro hub in making TT App (2017)
22. Petermans, A., Pohlmeyer, A.: Design for subjective well-being in interior architecture. In: Proceedings of the 6th Symposium of Architectural Research 2014: Designing and Planning the Built Environment for Human Well-Being, no. October, pp. 206–218. (2014). https://doi.org/10.13140/2.1.1584.2241
23. Macklin, B.L.: Creating healing spaces with facility design (2014)
24. Marcus, C.C., Marcus, F., Francis, C.: In: People Places: Design Guidelines for Urban Open Space. Van Nostrand Reinhold (1998)
25. Dobrivoje, T.: A review on salt lake city, Kolkata, India: master planning and realization. Spatium **434**(17–18), 98–105 (2008)
26. Correa, C., Das, P.R., Center, C.: Correa changed iconic city center design at eleventh hour (2015)
27. Sepe, M.: Liveable and healthy city design. WIT Trans. Ecol. Environ. **217**, 177–189 (2018). https://doi.org/10.2495/sdp180171
28. Bhattacharya, T., Banerjee, A., Sen, J.: Urban rejuvenation through techno art intervention: a case of kolkata. **5**(1), 472–477 (2017). https://doi.org/10.5176/2301-394x_ace17.63
29. Burton, N.M.D.: Our hierarchy of needs psychology today. pp. 1–3. (2012). [Online]. Available http://www.psychologytoday.com/blog/hide-and-seek/201205/our-hierarchy-needs
30. McLeod, S.A.: CT2 Paper 1 Maslow's hierarchy of needs. Simply Psychol. **1**(1), 1–8 (2007). https://doi.org/10.1016/b978-0-88415-752-6.50250-2

31. McGuire, K.J.: In: Maslow's Hierarchy of Needs: An Introduction. GRIN (2012)
32. Maslow, A.H.: A theory of human motivation. Psychol. Rev. **50**(4), 370–396 (1943). https://doi.org/10.1037/h0054346
33. Sadri, G., Bowen, C.R.: Meeting employee requirements: maslow's hierarchy of needs is still a reliable guide to motivating staff. **43**(10), 44+ (2011)

Chapter 58
Dots and Lines: Indian Folk and Tribal Art Inspired Activities for Kids

Rinki Sambhani and Saurabh Tewari

Abstract Moving beyond the models of passive learning through secondary sources, the design aims at employing an interpreted form of indigenous design tradition to engage children in various creative art and craft activities. In this case, the Gond paintings of Madhya Pradesh and its visual forms are explored to create an activity kit. The methodology in designing the learning tool involved a series of steps to conceptualise, develop and synthesise a designed product. The dots and lines activity book and the Gond art learning kit inspire kids to pursue creative art and craft activities at an early age.

58.1 Introduction: Art and Craft in Education in India

Children relate to and discover the shapes and colours of the world around them through painting. The human impulse to paint is related to the need to communicate, express and make sense of the world around. At the school level, art education constitutes an elemental area of curricular activity for the development of the wholesome personality of the students [1]. Today, many schools do promote art, dance, theatre and crafts at the level of primary education. However, it mostly relates to ideas of contemporary culture or the west. The indigenous models and frameworks are hardly included as experiential learning. It is gloomy in the higher classes, where the competitive academic and professional subjects take the front stage. Moreover, art is limited as an extracurricular activity. Traditional and indigenous art forms are rarely discussed. Through techniques of teaching the art of children, educator Devi Prasad underlines an understanding of the holistic growth of the individual [2]. The paper here develops on the idea and advocates a look into adopting active learning kits

R. Sambhani (✉) · S. Tewari
School of Planning and Architecture, Bhopal, India
e-mail: rinkysambhani16@gmail.com

S. Tewari
e-mail: saurabh@spabhopal.ac.in

A. Chakrabarti et al. (eds.), *Design for Tomorrow—Volume 1*, Smart Innovation, Systems and Technologies 221, https://doi.org/10.1007/978-981-16-0041-8_58

based on experiential and progressive learning for kids, employing the indigenous art forms.

58.2 Art and Craft Learning in the Schools

In the contemporary art and craft curriculum and its pedagogy in schools, the focus lies on description (through descriptive books and reading materials) instead of the process and experience of learning [3]. Aligning with school frameworks has been the agenda of famous publishers who work on the textbook domain. However, there have been several efforts by various organisations including Katha, Pratham, Eklavya and Ektara which look beyond the existing curriculums in education. They argue for holistic, more vibrant and rooted content from the regional knowledge systems of India as they have contributed to the awareness of the vibrant art and craft traditions of India through their publications. Books like bioscope [4], with minimal text and lively illustrations by the Madhubani artist, Shyam, present an opportunity for children to look at folk art. Another example is Bulli and the Tiger [5], a bright and colourfully illustrated book in the Gond style, which predominantly uses dots and dashes to create art. The art and craft book on Gond art [6] provide information and engagement as hand skills activity. However, many of this publication are not dialogic and progressive. The information is provided for learners' understanding. However, there is no mechanism to address the engagement of the learners' skills and techniques. Further, most of the artworks are a stylised form of Gond art paintings does not provide the essence and elemental understanding of the tribal artwork. Though the efforts above provide an entry to the world of Indian art and craft to the young learners, they also lack the experiential aspect of the art and craft, which makes learning a passive experience.

58.3 Objectives

Reflecting upon curriculum structures of various education boards and a pilot study at local urban schools in Bhopal, it was realised that there is a need to bring in the values of art and craft in primary school education. To move beyond the models of passive learning through secondary sources, the approach aims at employing a form of indigenous design tradition to engage children in various creative activities. It revisits the curriculum and art activities for children to facilitate how they can reimagine design, including the primary forms, shapes, colours and textures around them. This paper includes a feedback approach employed during the development of the design. It looks into how kids discover the values of indigenous craft traditions through creative engagement. The design process focuses on the process of adapting the visual language of Gond painting into new age learning to reach a broader target audience.

58.4 Focus

58.4.1 Folk and Tribal Art Form in India

In India, community painting reflects the identity of a region or a particular culture and follows common characteristics. Every region, city and village boast of its local traditions and skills; the centuries-old knowledge is embedded deeply in its culture. Folk and tribal art form a significant component of this repository. Some of the renowned traditions from India are Madhubani paintings of Bihar, Phad paintings of Rajasthan, Patachitra paintings of Odisha, Kalamkari paintings of Andhra Pradesh, Saura paintings of Odisha, Warli paintings from Maharashtra, Santhal paintings of Bihar, Bhiland Gond paintings of Madhya Pradesh and more. Most of them derive their substance from nature, depicting birds, animals and plant life, as one of the purest forms of expression. In the indigenous traditions, the paintings on floor, wall, ceiling, paper, pots and various places connect to myths, unique festivals and special occasions. The philosophy of society, rural life and realism is the subject matter of these paintings. It includes simple lines and curves, but the profoundness and timelessness of the motifs are overwhelming and showcase the country's rich heritage [7].

58.4.2 Design Focus: Gond Painting Concerning Central India

Practised by the Gond centred around the central region of India, Madhya Pradesh, Gond paintings form a parallel world created by the dots and lines. Here, the lines convey a sense of movement to the still images [8]. Dots and dashes impart a greater sense of movement to increase the amount of detail. One of its distinguishing features 'signature pattern' is used as 'infill' in the motifs. These infill arrangements are distinguishing recognisable symbols used by the Gond artist, and every Gond artist has established their signature style.

Gonds decorate their walls with beautiful paintings, in which they capture their everyday life, celebrations, rituals and the local flora and fauna. According to the belief system, their paintings reflect human kind's close connection with nature and their surroundings. 'Viewing a good painting brings good luck', this inherent belief forms the basis of Gond art. Themes of Gond paintings build on the local festivals on which a new painting is created. Gond paintings also preserve a remarkable resemblance with aboriginal art from Australia as both styles use dots and tiny lines to embellish the artwork.

Interestingly, the Gond artist is not much bothered about the surface, texture and quality of the material, medium and style as they aim to draw and decorate the surface with readily available materials. However, in recent times, poster colours, acrylic and oil paints are also being used. Handmade paper and canvas are the new media of storytelling for the artisans. For a modern eye, they also relate to the innocence of a child while selecting primary colours to create bright, multi-hued paintings regardless of tonal balance and colour harmonies [9].

58.4.3 Design Derivation

The consensus among the educators in the surveyed school was that they considered art and craft as a step-by-step activity. Unlike Gond art that represents story, tradition and skills, these prevalent practices are devoid of meaning as they exist merely as educational art. The information is meant to occupy students and complete the curriculum. Ideally, practical art education should include methods that satisfy children's need to develop their capabilities and to fulfil their desires. Interest should be more significant in the materials that support children's growth, develop their creative thinking and provide them with new methods to deal with various art and craft forms. There is a visible gap not just in the structures (theory) but also the pedagogy (practice).

58.5 Design Concept

With the gap identified, the authors attempted to address the design objectives by transforming initial ideas in the representation of the visual forms of Gond painting tradition of Madhya Pradesh. The most captivating feature of Gond art is its simplicity, in terms of the tools required to make the painting, and as the painting itself. A canvas or chart paper along with original paint is all that is required to start making a Gond painting. However, in the simplicity lies incredible intricacy and care. One of the most noticeable and easily recognisable features of Gond painting is the line and dot technique used to create entire paintings. These lines and dots impart a sense of dynamism and movement on the canvas, immediately capturing the attention of its viewers. Gond paintings reflect the innocence of creation in children's art, whether it is the selection of colours, primary colours to the motif rendered from day to day life.

Similar creative drawings were noticed during our preliminary study (Fig. 58.1) with students of primary classes. It was found that Gond art can be easily explored with this age group as Gond art has characteristics of using contrasting colours, storytelling, two dimensional, overall patterns and fluid contours. In Fig. 58.2, a student's thinking skills can be seen; the student was asked to draw a village scene, so he had included all the things which he has already seen or experienced.

Moreover, a story can also be made out of his drawing, which shows similar characteristics in Gond art. Use of bright colours and two-dimensional figures shows that they have no rules of perspective and composition, and they compose their colour palette. Gond paintings are a portal into a surreal world, a world where there are no

Fig. 58.1 Artwork by the students of class 1 and 2

Fig. 58.2 Workshop and product testing

restrictions and imagination run free. There are no mundane colours or void space; there is no blockage, and the imagery is fluid and rhythmic. Based on the literature review, interpretations were made into three categories—visual language, motifs and medium. Concepts were to be based on the visual language, which included design exploration with a variety of repeated patterns and textures. Motifs were to be inspired by nature such as trees, birds and imaginative creatures addressing the user group. In the medium category, different mediums were to be explored with children since a child is not given opportunities to try different materials during the early stages, and he might find it challenging to try new materials at a later stage.

58.6 Design Methodology

The design process involved a series of field interviews in private schools in realising the apathetic gap towards art and craft in pedagogy. The authors employed the Gond painting and its interpretation to develop a visual language, motifs and medium in the problem-context. The authors analysed art and craft activities from various publications books to know how Indian folk and tribal art forms are taught and practised with students of different levels of classes from first to the eighth standard. It was also to evaluate the recent attempts of folk and tribal art on early childhood learning; it was noticed that information available on folk and tribal art in students' art education books was more, and less focus was given on the development of practical skills in students.

Gond patterns are among the most elaborate and intricate patterns of the tribal art form [10]. This art is more of creative thinking and drawing skills. Therefore, a need was felt to study and explore folk and tribal painting with students. For the selection of this area, the data collected was from primary as well as secondary sources. The literature study also revealed that many artists from the Gond Pradhan community are living and practising the Gond art primarily from their homes in Bhopal since the last 20–30 years. Then, a couple of artists who practice in the city were interviewed to gather knowledge about this art form. Based on the interpretation, design thinking methodology was used to explore and generate concepts based on Gond art.

58.6.1 Ideation

From the literature and preliminary study, design ideation was involved infusing the experimental print media with traditional Gond art. The ideation included creating various artworks enquiring how the process can be made accessible, easy and fun for young students considering the broader purposes of creativity, education and self-learning.

58.6.2 Developing Tools and Techniques

The following tools, techniques and materials were designed and explored to capture the essence of Gond paintings (Table 58.1).

58.6.3 Constructive Approach

A constructive approach was employed in the making of drawing as in the following Table (58.2).

58.6.4 Reflection

The Gond art is based on essential aspects of storytelling, and often, the work is based on folktales or their tribal customs. One can always find a seed of reality visible or

Table 58.1 Tools and techniques

Image	Tools and technique
	1. Stencil Often the children find it challenging to draw forms such as animals and trees. So, a stencil technique is used to assist the users to draw the shapes and forms. With a paper cut technique, the design of the stencil tool was explored
 	2. Stamp A set of frugal tools and stamping technique allows the patterns of Gond art to be interpreted through them. The tools employ the modification of rubber-based erasers with a basic pattern block engraved at one end. The stamping activity through the rubber stamp is playful and familiar for kids as it involves the modification of everyday items of primary class students. The image on the left shows a Gond-styled stamped elephant illustration by a participant using the stencil and rubber blocks

Table 58.2 Various levels in the approach

<table>
<tr><th>Image</th><th>Level</th></tr>
<tr><td>
</td><td>1. Drawing in several steps
Beyond the stencil and stamp technique, a step-by-step constructive drawing approach was explored for the fluid forms of animals. The progressive construction of the image allows the process to remain comprehensive for children to draw. It also adds to their drawing confidence</td></tr>
<tr><td>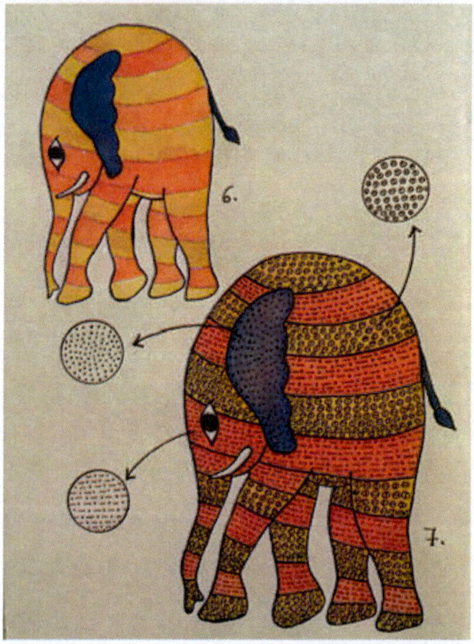
</td><td>2. Introducing complexity through patterns
The patterns are introduced gradually. The three compositions of patterns were employed to fill in the image on the left by a participant. This allows the user to distinguish even smaller patterns</td></tr>
</table>

(continued)

Table 58.2 (continued)

<table>
<tr><th>Image</th><th>Level</th></tr>
<tr><td></td><td>3. Finding utility
Building on further, this level focused on creating useful art and craft pieces. Products like small greeting cards, fridge magnets and framed paintings to increase user's interest and attention. The utility aspect adds further motivation to complete the drawing in detail as it may become a work to be gifted or framed</td></tr>
<tr><td>

</td><td>4. Adding knowledge
Finally, the introduction of knowledge-based activities like crossword and story-writing with various image compositions adds meaning to the activity kit. It looks into the creative aspect of the learners beyond the drawing. It further addresses the concerns of mainstream learning need by constructively addressing the issues regarding the 'theoretical understanding' of the subject. Various quiz-based exercises can also be added here</td></tr>
</table>

abstracted in the paintings. There are several paintings by the various artists on the same stories which are expressed in different forms and artists' signature pattern filling style. For example, with the story of Udta Hathi or the Flying Elephant, which has been interpreted by various artists, the authors let children explore the style with creative storytelling, where they have to see the picture and write their own story. The response was excellent with the children coming up with their own versions. The opposite was done with an exercise in which they had to draw with their cognition and imagination to complete the missing parts of the artwork.

58.6.5 Testing

The authors tested the pilot concepts, discussed in Sect. 6.3′s levels 1 and 2, with the ten students of standard third to fifth. Their approach gauged not just the typical methodologies but also offered the clues to develop the basic tools. Post-conceptualisation, they developed prototypes and further refined it through another round of user testing as the authors organised an open workshop with a prominent Gond artist from the city. The thirty students from different schools from the city agreed to be part of this experimental workshop.

58.6.6 Feedback and Improvement

The result of the workshop showed that the children were able to make Gond artworks. It was realised that for the participants, constructive forms were easy to draw. However, some of the younger users had difficulty in drawing dense patterns like the multi-branch trees. Few participants also created their patterns; by using the back tip of the brush provided, they created perfect dots and different patterns which made their art subjective. Their excitement reflected in their artwork as they began to paint concepts like a peacock fridge magnet, wall paintings and greeting cards as these utility concepts could be beneficial as teaching tools for learning indigenous traditional art (Fig. 58.3).

58.6.7 Final Design

Building on the insights and feedback received; another round of additions was to address the pragmatics of consumption by the users and production by the probable printing house. The designed product "Gond Art Activity Kit' was created, which has stimulating activities for kids. It comprises stories and life forms which not only provoke the creative interest of a learner but also informs them towards the cultural aspects of India's vernacular art forms. The kit contains an activity book, dots and

Fig. 58.3 Participants with the outcome

lines, for all the above-mentioned explorations by a user. It has all the essential equipment such as colour set, drawing set and glue and utilitarian accessories like a painting frame, fridge-magnets, bookmarks, greeting cards, which the learners can keep as a memorabilia or gift to the dear ones. A self-completion certificate was also kept in the box which will be filled by the user on exhaustion of all the art and craftwork (Fig. 58.4).

Fig. 58.4 Final gond art activity kit: box and activity book

58.7 Discussion

The project promotes the creative capacity of children through enhancing the engagement in art and craft activity books of the primary classes by providing quality participatory activities and promoting Indian folk and tribal art. The results, after the workshop and product testing, suggest exposure to the visual language of a particular folk and tribal with new media techniques and interventions can enhance user's art and craft learning towards a subject. The authors came across many interesting visual responses that depict the potentials of tribal and folk art in early childhood education. Observing their artworks and simplicity of the folk and tribal paintings, mainly flat surface-based paintings with bright colours make it easy for them to draw and explore.

The process proved that fun and utility product outcome is an essential element in designing for children's learning experience. The most important aspect of tribal and folk art is its ability to tell a story, in its tempting forms and figures of visual expression. The authors sense the responsibility to communicate the folk art as not just through its prominent dimension but through its visuality of patterns. They also feel the need to convey its rich, embedded values. Some of the typical explanations are given in the kit. However, the kit has to be complemented, if possible with the tribal artists to offer the multiple but authentic descriptions regarding the art form. The authors realised the same during the testing phase through one of the prominent Gond artist's engagement. However, this whole effort is just one of the early steps to realise the larger indigenous and cultural landscape of folk art.

The efforts through this exploratory approach also offer educators, illustrators, publishers and all parents interested in promoting early childhood education in this field. The authors also got ample opportunities to discuss, review and learn about the liking and moods of curious art learners.

The Gond art activity kit along with the dots and lines activity book has the potential to be included in school art and craft curriculum as it provides not just the experiential learning of the form with the necessary details but builds the theoretical understanding of the subject. The useful by-products and mix media techniques used add further meaning to it. The activity kit as a model has the potential to develop other forms of indigenous arts which can further contribute to decolonising the curriculum, content and learning methods.

References

1. CBSE: Art Education (CODE No. 502) (2020). [Online] Available https://cbseacademic.nic.in/web_material/CurriculumMain20/Coscholastic/ART_EDUCATION.pdf. Accessed 15 June 2020
2. Prasad, D.: In: Art-The Basis of Education. New Delhi, National Book Trust (1998)
3. NCERT.: Craft Traditions of India: Past, Present and Future, New Delhi, NCERT (2011)
4. Nainy, M., Devi, S.: Bioscope. Katha, New Delhi (2020)
5. Reys, S., Shyam, N.: Bulli and the Tiger. Pratham, New Delhi (2005)

6. Fatima, N.: Start-Up Art and Craft. Acevision, Lucknow (2017)
7. Malik, G., Sethi, S.: Gond paintings: a historical evolution. Int. J. Appl. Home Sci. **5**(4), 832–836 (2018)
8. Bora, S.A.S.: Recent trends in gond painting. Int. J. Appl. Home Sci. **3**(7), 279–283 (2016)
9. Bharadwaj, K.: Colors in gond tribal art: an interpretation and critical evaluation of colors in gond paintings of Madhya Pradesh. Int. J. Res. Granthaalayah **12** (2014)
10. Hans, J.: Five Beautiful Tribal Painting Forms You Should Know About. Youth Ki Awaaz 26 Nov (2019). [Online] Available https://www.youthkiawaaz.com/2019/11/five-beautiful-tribal-painting-forms-you-should-know-about. [Accessed 15 June 2020]

Chapter 59
The Creative Prints of Somnath Hore: A Review

Namrita Sharma and Utpal Barua

Abstract The use of printing as a technique was in propagating religious images in multiples in different parts of world, including India. With the technological advancements, the techniques of printing gained momentum and became an important way to disseminate news and illustrations by printing newspapers, periodicals, magazines, and books which made it a commercially viable method to reach out to masses. A turning point in the history of Indian printmaking came as introduction of graphic art in the curriculum at Kala Bhavan, Santiniketan which was not just to train students in the technique of printing but to express themselves through this medium. One of the prominent printmaking artist from Bengal who played an important role in setting up Graphic Art Department at College of Art (New Delhi) and later contributed to Santiniketan as a teacher is Somnath Hore. His continuous experimentation with the form, textures, and its relevance to the theme of wounds defined Indian printmaking in terms of giving it an identity. This paper reflects upon various perspectives (aesthetical, technical, thematic, etc.) through which the artist's prints have been looked at and understands the combination of intuition and planning as printmaking requires technical knowledge.

59.1 Introduction

Indian printmaking has been shaped by the contribution of certain artists who explored new possibilities of image making through different techniques other than painting or sculpture [1, 2]. Somnath Hore is one amongst those artists and educator whose curiosity led him to experiment with different mediums of printmaking itself, where he dissected the possibilities of attaining new pictorial grammar and defining his own symbols which could be understood by others. He worked in linocut,

N. Sharma (✉) · U. Barua
Indian Institute of Technology Guwahati, Guwahati, Assam 781039, India
e-mail: namri176105004@iitg.ac.in

U. Barua
e-mail: utpalb@iitg.ac.in

A. Chakrabarti et al. (eds.), *Design for Tomorrow—Volume 1*, Smart Innovation, Systems and Technologies 221, https://doi.org/10.1007/978-981-16-0041-8_59

woodcut, etching, lithography, and paper pulp [3–5]. Not only he developed his own visual language but setup Graphic Art Department at College of Art, New Delhi as well as at Kala Bhavan, Santiniketan which contributed in the growth of printmaking scenario during 1960′s and 70′s in India [4]. In his young days, he was an active member of Communist Party of India, and during 1943, Hore reported the horror of Bengal famine through visuals under the tutelage of Chittaprasad Bhattacharya, another prominent printmaker of 1940′s. Due to his urge to learn and a skilled hand for drawing, the communist party helped him in securing a seat in the Visual Art School in Kolkata, where he was trained by Artist Zainul Abedin and Safiuddin Ahmed. Safiuddin Ahmed was instrumental in initiating Hore into the world of printmaking through woodcuts and wood engravings. His thematic concern ranged from politically motivated Bengal famine to peasant movement called Tebhaga—A revolt in North Bengal by the sharecroppers to claim the $2/3^{rd}$ of the produce in place of 1/2nd share in the times of colonial India and then slowly progressed to a single metaphor called wounds which he depicted through different mediums [5, 6]. We find through various literature and interview conducted with his students—Dr. Nirmalendu Das (Former Professor—Printmaking, Kala Bhavan, Santiniketan and eminent Artist) and Kavita Nayar (Eminent Printmaker and founder member of Indian Printmakers Guild and Multiple Encounters Group) that the artist (Somnath Hore) chose to explore printmaking in its different forms and techniques due to its interesting graphic qualities and mastered them as well as improvised the processes to suit his imageries as well as climate of the place which indicates his approach of including the nature friendly components in his practice. By studying various aspects of pictorial qualities in his work, one may learn permutations and combinations of crystallizing the thoughts into concrete visuals and open up doors for understanding and imbibing emotions into design. This paper shall review various themes, techniques, and aesthetical approaches practiced in his creative prints.

59.1.1 Background

Art critic Pranab Ranjan Ray who had written extensively on Somnath Hore discusses in detail about the socio-political situation of 1940′s, the response of people through cultural means, personal experience, and visual expression of the above mentioned artist in his article—The Political in Art. He has written from the perspective of how the visual and performance artists emerged out of this situation which is termed as "objectification of a politically motivated response to societal events in the arts." Ray wrote this article by reviewing the illustrated diary of Somnath Hore, intended to give attention to politicization of non-performing visual arts in Bengal. Thereby, understanding the historical and political backdrops is necessary to relate to the design of his visuals. The artists of that period were called "socially responsive" due to the content of their works. Tebhaga episode (December 1946) illustrated/documented by Hore holds an important place as it gives an understanding into the development of pictorial elements used by the artist in his early years of artistic journey because he had transcribed the episodes from this diary in his woodcuts in early 1950′s [6, 7].

59.2 Literature Review

59.2.1 Creative Journey—Various Perspectives: Theme

Although **Wounds** became the central concept of his expression, but the artist had to traverse through various subjects to arrive at this focal point of his experience and its visual attainment. These subjects dealt with his observations in daily life around the artist and connecting those events with empathy [6]. It is evident from the perspectives of various writers that Somnath Hore represented either the direct encounter of the consequences of the social or political upheaval, the economic conditions or the assimilation of his experiences in later stages of his life through interesting assemble of pictorial elements in a two-dimensional or three-dimensional space (in case of his sculptures). Japanese Bombings on Chittogram (native place of Somnath Hore) during Second World War, Bengal Famine (1942–43), The great Calcutta killing (1946–47), American aggression over Vietnam, Naxalite Movement, Bangladesh war, etc. moved the human in him, and perceiving the social function of art, he communicated the pain and suffering into a recognizable entity of wounds which exists visually to be seen by us [5–7].

We find this clue about the rationale behind calling the works as wounds but not as pain in his book chapter—Somnath Hore: His Life and Art written by well-known critic—Mrinal Ghosh, where the author has quoted certain lines by Somnath Hore which were published in magazine Desh (November 1985), where the printmaker says:

"No I don't call them suffering or pain. My term is 'Wounds' Pain and suffering is rather abstract; they exist on a mental plane. They are the feelings of pain… To me, the tracks etched by bicycle wheels on a dirt track, is the idea of a wound. The pain factor does not exist as such…. A tree which has been axed, this also suggests a wound. This leaves scar, hence it is real. Or the girl hits by a bomb, the innards of her stomach tumbling out… this is a wound. Wounds exist, they are not abstractions."

59.2.2 Aesthetics

With passage of time and continuous experimentations, Hore comprehended that attainment of skills and a theme is not the "only" ingredients to produce a meaningful work of art. In an article of a published catalog of his exhibition called "WOUNDS" by CIMA Gallery in the year 1995, Hore has explicitly mentioned that the theme should not prevail because if this happens, sentimentalism tends to saturate the work which builds up an obstacle to thorough understanding of the pictorial values. This connotes that pictorial values holds outmost importance in the making of a visual language for him. According to him, the actual events diminish when the time progresses but the aesthetics overcomes phases, placing the genius of the artist over the content [8]. Lev Manovich acknowledges the kind of revolution which the human race has witnessed with the progress of digital technology, where software now can manipulate/refine any image in the form of pixels into a desired aesthetic perspective particularly in the visual sphere. To relate to the arrangement of elements in a rhythmic manner, attaining balance in composition, creating patterns, forms, and its variations, planned use of color or texture, use of grids, or other methods in a space is an innate part of cultural activities of the civilizations. Manovich emphasizes regarding the study of the concepts of aesthetics and beauty being under-utilized references in media studies and related industry and that it becomes vital to decipher the interrelation of form and content which makes art, modern mass media as well as user generated content a distinctive kind of human communication and experience [9].

Art and design went on hand in hand for Hore. During his direct association with CPI, Somnath Lahiri, the editor of Janayuddha—weekly journal approached Hore for creating cartoons, which the artist made and they got published in the journal [5, 8].

A chaos can be observed in use of angular lines which results in formation of the figures, where the form (positive space) resides in a widespread area or negative space along with one or two supporting elements. This use of minimal and compact forms and exclusion of too many details in the composition is said to have arrived from the designing of cartoons which the artist did for Janayuddha around 1945. His earlier works have realistic figures but progressing further, those figures starts to distort and reduced to more condensed visual [5]. There is a rigor in the lines used to create those structures and compactness which defines the interaction of the elements with each other in a rhythmic manner. The artist seems to have used a limited color palette in his prints there by putting the focus upon the form, generally using tones of red, blue, brown/sepia, and black. The technical reason might be of availability of limited variety of colors at the time when these works were executed but definitely it is not to say that the artist got stuck himself to a specific color combination. It is rather to state the circumstances in which the artist has executed his works [5, 7, 10, 11].

59.2.3 Techniques

During 1940′s not many artists were practicing graphic art as a medium for personal expression but Hore did not limit himself to one style or technique of graphic's language [5]. Major shifts in his techniques and mediums are shown in the flowchart as follows, and sample images are attached to illustrate those shifts visually too (Figs. 59.1, 59.2, 59.3, 59.4, 59.5, 59.6 and 59.7).

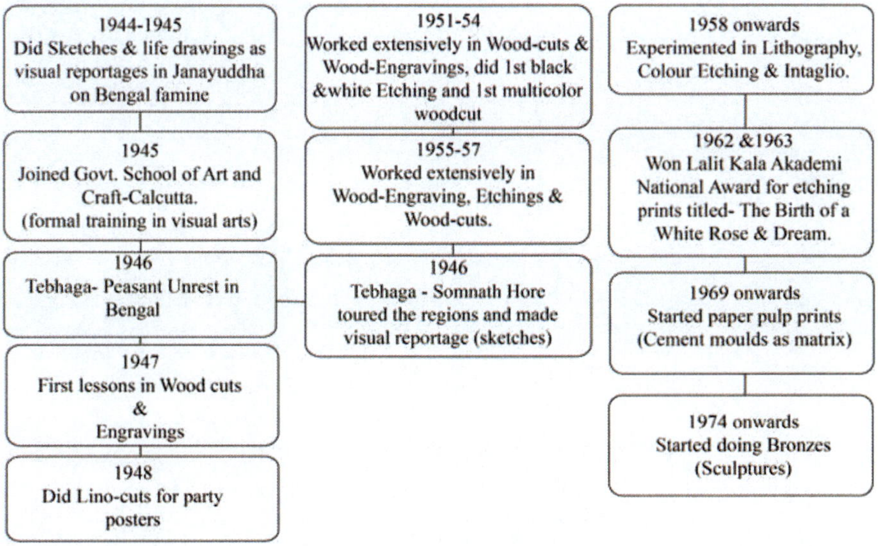

Fig. 59.1 Flowchart of major events and shifts in techniques in career of printmaker Somnath Hore

Fig. 59.2 Untitled Technique: Woodcut Year: Early 1950′s

Fig. 59.3 Birth of a white rose Technique: Etching Year: 1960

Fig. 59.4 Untitled Technique: Etching Year: 1967

Fig. 59.5 Untitled Technique: Lithography Year: 1969. *Image source* Seagull Foundation for the Arts

Fig. 59.6 Wounds technique: paper pulp year: 1977. *Image source* Seagull Foundation for the Arts

Fig. 59.7 Untitled technique: etching year: 1983. *Image source* Seagull Foundation for the Arts

59.3 Analysis of the Prints

By looking at works from various periods, one could understand that Hore used the elements according to the **mood** and **demand** of the composition. The technical explorations went on in an overlapping manner. His ways of rendering the forms seem to changed with method and materials as every material has its own innate character which Sommnath Hore utilized to his advantage. For instance, while doing woodcut, he created forms with single **bold line** scooped off the woodblock in one go, whereas with etching, the forms emerged with slow biting of the metal plate in the acid with subsequent dips in the acid bath tray. In the woodcuts executed during 1950s, the lines were thick and bold one reason being the use of tools and the nature of matrix/surface, i.e., wood, yet well balanced, clean in terms of distribution of blacks and whites. These lines were well scooped/carved out of the wood block to give **naturalistic** touch of light (whites) and shades (blacks) to all the entities in given space. This requires preparation of drawings and understanding of possible visual outcome of the carvings being executed because once the carving is done, it cannot be reversed. But later in his career, Hore's forms became suggestive in order to express suffering irrespective of a particular event. As observed in lithographs, the forms are executed as silhouettes but the brush strokes provide the figures with required volume through its tonal variation of the applied hue. He also created several other lithographs with conte crayon giving the figures a dry, chalky character. The figurative structures have been reduced to skeletal with a thin layer of flesh over it not to exagerrate but to portray the sufferings in its true sense. In many compositions, the facial expressions are gloomy, lost in thought. The figures shown were sometimes **static** and sometimes showed rigorous **movements** freezed in that moment. He utilsed combination of hard single and soft multiple lines to represent the figures and also create **scratchy textures** giving an impression of **bruises** in and around the form. He has used **distortions** on the forms, not keeping them too realistic just like what picasso deviced to dissect his images by using various angle onto a two-dimensional surface. The spatial distribution in his prints changed from filling up the area with various motifs to reduced, compact, detail-less yet curvaceous, and striking forms being placed on a flat two-dimensional surface of the matrix. The forms were a combination of **angular** as wel as **strong fluid** lines. By strong, the researcher means to focus on the tonal value of the lines used in the art works.The contours of the forms are dark, entails hardness and are executed in single stroke displaying the command over the medium and years of practice. The artist brings out the contrast in positve and negative space through the use of dense lines (slanting as well as cross hatching lines) in the background, rendering the figures in soft and lighter tones of aquatint and etching. The figures overlap each other to create a unity in the disturbing space making them look haunting as if a tragedy has struck them. The colors in his compositions varies in hue and tones. The artist has majorly used tones of black, burnt umber, burnt sienna in permutations and combinations of bright red, green, orange, and blue, visually setting an environment of dense, pensive, and seriousness in the work. The textures seems to become an important visual device in his works initially by supporting the composition's forms and conveying a meaning/evoking certain emotions as observed. The textures are somwhere little soft and majorly crude and become dominant at later stages particulary in his paper pulp series, where he used objects directly to pierce into the soft paper fibers in order to yield a subtle visual impression called The "**Wounds.**" The visuals had been designed so as to suggest and not directly pin point a particular issue which makes them relevant in today's scenario too. Artist Hore has used the pictorial elements carefully to depict the vulnerability of life yet being resilient to hardships and strive forward. It is vital to go deeper into the rationale of choosing these particular ways of expressions and plan courses in order to increase the visual literacy and sensitize people about visual art and design.

59.4 Research Gap

The literature written has focused immensely on the political aspects, where the artist was progressing in understanding of the visual grammar at the art college in Calcutta as well as social and philosophical understanding of his work in a cohesive manner. Interestingly, there is a mention about Hore's use of pictorial elements—lines, shapes, color, space, texture, and evolution of his forms along with his experimentation of various conventional printmaking methods in horizontal progression, giving insights into certain characteristics of his works and psyche in different periods of his creative journey. Since Somnath Hore himself wrote in his article—Wounds, where he outlined the importance of developing pictorial values over theme/content to produce a meaningful work, it becomes important to go deeper into the design of his visual characteristics to appreciate how he innovated and balanced the interplay of pictorial elements—variations in use of lines, colors, texture in his compositions which further may enrich the visual design industry.

59.5 Aim

The aim of his paper is to review various themes, techniques, and aesthetical approaches depicted in his creative prints in order to develop deeper into the pictorial qualities by understanding the usage of the visual elements—form, colors, textures, and space in the design of his prints.

59.6 Methodology

The literature reviewed for this paper, consisted of data sources, i.e., articles and papers, thesis available on online platforms—Shodh Ganga, Research Gate, Google Scholar, e-newspaper articles, The Critical Collective, Seagull Art Foundation (it has a huge collection of artist's works, also has published two of Somnath Hore's books My Concept of Art and Tebhaga Diary), articles published in the form of books, by various artists and art critics of national repute, Unstructured Personal interviews of students (02) and colleagues (02) of Artist Somnath Hore, who are themselves noted printmakers and teachers with an experience of more than three–four decades.

59.7 Discussion

Somnath Hore has worked extensively in variety of conventional mediums of printmaking and found new directions in terms of developing his own ways of playing around the image and its characteristics. His prints contain organic forms. These are executed in prominent—bold lines. Compositions have either central or side alignments and are monochromatic (single color) or dichromatic (two colors) in nature. There is a concept of permanency and reality as compared to abstract feelings, e.g., he had used wounds instead of pain to denote his expressions. Texture does break the monotony of the space and create an interest in the compositions. A lot has been discussed by critics about his philosophy of "Wounds" but there are various facets of his body of works which can be studied further to develop ways of designing the visual plane and improvising the process and planning. This is significant in modern times, where digitization significantly contributes in the image making process in an indirect manner. However, in the Indian printmakers' context, it is noticed that even in present generation, artists prefer conventional strategies, to which they explore digitization to satisfy their urge for building a physical surface. It is observed that in many of Hore's prints' surface, i.e., the textures created, evoke emotions and, hence, appear to build image more prominently through his use of strong linear quality, brush strokes, and color palette. This knowledge provides us with the possibilities of meaning making and refining the aesthetics of contemporary times.

59.8 Conclusions and Recommendations

Somnath Hore was a visual designer and artist who built up his forms with essential elements, eliminating the extra or the non-essential, keeping the space interacting with the forms and colors or textures in harmony. The artist chose to explore printmaking in many techniques due to its interesting graphic qualities and mastered them as well as improvised the processes to suit his imageries (as done in case of paper pulp prints, where he had used cement blocks as surface for getting the impression onto paper) according to the local climatic condition. Through this study, it was found that the overall impact of the art work was a combination of not only intuition and constant planning but also the understanding of the material's own characteristics by the artist/printmaker, wherein he utilized knowledge of handling the visual elements at every stage to achieve the necessary impact. The artist had attempted to break the monotony of the given space by balancing the image with minimal motifs. So, in order to have deeper insights into the rationale of visual designs, it becomes essential to study various facets of pictorial values. Through this study, one may learn permutations and combinations of crystallizing the thoughts into concrete visuals and open up doors for understanding and imbibing emotions into design. His works can further be studied to understand the storage, preservation of prints in the form of digitization of his works as well as study of the materials to make them less prone to damage due to sunlight, dust which still remains an area to be taken care of in order to protect the cultural heritage/lineage.

Today's art world cannot remain isolated from digitization technology. While it is essential to maintain/retain the essence of physicality of the created object, it is also important/advantageous to take the aid of digitization "to further the impact" and for long-term preservation. Somnath Hore is distinctly famous in past as well as present periods for exploring the path of "his own," breaking away from the conventional norms prevalent in those times. A deeper insight into his ways of exploration shall serve as a guiding beacon for the modern generation artists to express their own creative ideas in their own innovative ways, for which digital technology can be gainfully utilized. This study is an attempt in the same direction, to understand the various possibilities created and explorations by Somnath Hore in some of his works.

References

1. Mukhopadhyay, A., Das, N.: Graphic Art in India since 1850, 5th edn. Lalit Kala Akademi, New Delhi (1985)
2. Das, N.: A brief history of printmaking at santiniketan: Artetc. News and Views (August 2011). https://www.artnewsnviews.com/view-article.php?article=a-brief-history-of-printmaking-atsantiniketan&iid=23&articleid=593. Last Accessed on 12 June 2020
3. https://www.criticalcollective.in/ArtistInner2.aspx?Aid=225&Eid=617. Last Accessed 04 June 2020
4. https://criticalcollective.in/ArtistInner2.aspx?Aid=225&Eid=200. Last Accessed 04 June 2020
5. Ray, P.: Hunger and the painter: somnath Hore and the wounds. Cressida Transactions **1**(2), 203 (1981)
6. Ray, P.: The political in art: review of Somnath Hore's Tebhaga: an artist's diary and sketchbook. J. Arts Ideas **22** (1992)
7. Ghosh, M.: Somnath Hore: His Life and Art. Samakalin Bhaskarya, 4th edn. pp 101–122. Pratikshan Publications, Kolkata (1995)
8. Hore, S.: Wounds exhibition catalogue. CIMA Gallery, Kolkata, W.B (1995)
9. Manovich, L.: Aesthetics. In: Ouellette, L., Gray J. (eds.) Keywords for Media Studies, pp. 9–11. NYU Press. New York. (2017). Last Accessed on 09 June 2020 from www.jstor.org/stable/j.ctt1gk08zz.5
10. Mago, P.N.: Contemporary art in India: a perspective. National Book Trust, New Delhi, India (2001)
11. Mallik, S.K.: An educative experience. Indian Contemporary Art J. **3**(1&2), 58–61 (2011). https://www.indiaartfestival.com/assets/Indian-Contemporary-Art-Journal-pdf/art%20journal%20may2011.pdf. Last Accessed on 13 June 2020
12. Ganguly, N. (ed.).: Readings–Somnath Hore. 1st edn. Lalit Kala Akademi, New Delhi (2010)
13. https://www.seagullindia.com/exhibitions/portfolio/hunger/#&gid=1&pid=134. Last Accessed 12 June 2020
14. https://www.seagullindia.com/exhibitions/portfolio/hunger/#&gid=1&pid=137. Last Accessed 12 June 2020
15. https://www.seagullindia.com/exhibitions/portfolio/somnath-2011/#&gid=1&pid=7. Last Accessed 12 June 2020
16. https://www.seagullindia.com/exhibitions/portfolio/somnath-2011/#&gid=1&pid=10. Last Accessed 12 June 2020
17. https://www.seagullindia.com/exhibitions/portfolio/somnath-2011/#&gid=1&pid=8. Last Accessed 12 June 2020
18. https://www.seagullindia.com/exhibitions/portfolio/somnath-2011/#&gid=1&pid=23. Last Accessed 12 June 2020
19. https://www.seagullindia.com/exhibitions/portfolio/somnath-2011/#&gid=1&pid=62. Last Accessed 12 June 2020
20. https://www.seagullindia.com/exhibitions/portfolio/somnath-2011/#&gid=1&pid=71. Last Accessed 12 June 2020
21. https://www.seagullindia.com/exhibitions/portfolio/somnath-2011/#&gid=1&pid=78. Last Accessed 12 June 2020
22. https://www.seagullindia.com/exhibitions/portfolio/somnath-2011/#&gid=1&pid=92. Last Accessed 12 June 2020
23. https://prinseps.com/research/somnath-hore-prints-india-art-fair-2020/. Last Accessed 12 June 2020

Chapter 60
Androgynous Fashion from the Concept to Consumers: An Empirical Study

Indranil Saha, Akanksha, and Diotima Basu

Abstract The concept of androgynous or gender-neutral fashion is known for its distinctive attribute that blends both conventional masculine and feminine design characteristics. In the history of fashion, the notion of androgynous fashion has been evolving since the 1920s, although it was irregular at times. In the post-modern Western cultures, androgynous aesthetic in fashion is increasingly accepted, encouraging the multiplicity of gender expressions. With significant influencers of the generation identifying themselves as gender-neutral and speaking out on the topic, the concept of being gender fluid is catching a lot of attention recently in the international fashion industry. Androgynous fashion is an emergent trend, which reflects in fashion ramps with models showcasing silhouettes and design elements that breakdown gender stereotypes. With this in mind, the current research aims to study androgynous fashion from both conceptual and user-centric perspectives in the Indian context. Data were collected through primary and secondary sources. Relevant secondary data were gathered from various books, research papers and fashion publications to set the conceptual context of the research. Additionally, to gather primary information about the Indian LGBTQ consumers' perception of androgynous fashion, a questionnaire was circulated amongst young Indian fashion consumers using convenience and snowball sampling methods. The results and analysis of the study reveal the aspirations behind the gender-neutral design genre. This study also brings out the emotional needs of the Indian LGBTQ community members, who are the primary consumers of androgynous aesthetic.

I. Saha (✉)
Indian Institute of Technology, Hyderabad, India
e-mail: md18resch01001@iith.ac.in

Akanksha
Amity University, Kolkata, India
e-mail: akanksha.diksha@gmail.com

D. Basu
Biswa Bangla Marketing Corporation Limited, Kolkata, India
e-mail: basudiotima@gmail.com

A. Chakrabarti et al. (eds.), *Design for Tomorrow—Volume 1*, Smart Innovation, Systems and Technologies 221, https://doi.org/10.1007/978-981-16-0041-8_60

60.1 Introduction

The construction of social identity is significantly influenced and shaped by the selection of style, colour and form of the clothes that one puts on according to the age, social status, identity and role [1]. With the concept of gender expressions gradually being accepted to be non-binary, the current fashion industry is experiencing a charming presence of androgyny [2]. Androgynous fashion, aka gender-neutral fashion, is an emergent trend, which lets its wearer feature design elements that break gender stereotypes. Androgynous fashion is steadily conquering the mainstream runway. Hence, the topic of androgyny in the context of fashion becomes relevant to study [3]. However, as the orthodox communities still find it inapt due to the lack of understanding of the actual meaning of androgyny, the Lesbian–Gay–Bisexual–Transgender–Queer (LGBTQ) community members are the initial key adopters of this trend [2, 4]. There are around 45.4 million documented LGBTQ people in India [5], which seems to be much less than the actual number due to the social stigma of openly being an LGBTQ person in India [4]. Existing studies on the LGBTQ community members have partially examined the experiences of fashion and clothing from generic [6] and clothing fit [7] perspectives. Although researchers have highlighted the gap in addressing the functional, aesthetic, and expressive fashion needs of the LGBTQ community members [2], there is a lacuna of academic empirical research work in the context of the Indian market. Due to the increased acceptance of androgynous design elements in fashion, LGBTQ community members, especially the youngsters, are predicted to be the key target consumers of such trend. Therefore, the current study aimed to explore the concept of androgynous fashion on the constructs and workings of fashion creators neighbouring this subject. Furthermore, the objective of this study was also to examine Indian LGBTQ consumers' primary emotional needs concerning androgynous fashion.

60.2 Framework for the Research

The theory of consumers' design-specific needs is the basis of the conceptual framework used in this study. Mere protection of the body as the only purpose behind the consumption of clothes was first refuted by Veblen (1953, originally published in 1899) [8]. It was further established that people do not buy things only for what they can do, but for what they want to express about themselves [9]. Lamb and Kallal (1992) conducted a study to come up with a model that would focus on clothing needs from consumers' perspective [10], which lead to the development of the FEA consumer needs model. The model broadly classifies consumer needs into functional or utility-based needs (protection, style, cut, fit, insulation, breathability, ease of movement, etc.); need for individual expression or communication (values, personality, culture, lifestyle, etc.); aesthetic needs or desire to look attractive (art elements, principles, design, etc.). The FEA model is a consumer-centric model which is encircled by various cultural, social, and biological complexities and considerations. Previous studies have established that more inclusive and versatile social and cultural factors have a more significant influence on clothing choices than other prevailing factors [11]. Additionally, researchers argued that consumers' interests and opinions are largely reflected by their native's collective values and attitudes [12]. The quantitative section of the current study deals with the androgynous fashion category, which lacks exploration from the consumers' needs perspective, specifically in the Indian context. Hence, we followed a reconstructed model based on Lamb and Kallal's (1992) FEA framework [13], which considers the cultural context of each category of need to explicate emotional design criteria that fulfil the target consumers' need.

60.3 Review of Literature

60.3.1 Concept of Androgyny and Fashion

The term "Androgyny" is originated via Latin (*androgynous*) from Biblical Greek words *anēr* and *gunē*, which mean man and woman, respectively [14]. It signifies the combination of male and female. The concept of androgyny is explored from the physical, psychological and biological perspectives [15]. However, American philosopher and gender theorist Judith Butler indicated that the cultural norms construct gender rather than science [16]. Studies suggest that androgyny can also be integrated with psychological and biological perspectives [17]. The concept of androgyny started gaining popularity in the area of fashion around the 1970s, especially in the USA [18]. Economic and cultural factors were the two main reasons which influenced people to adopt something different, like androgynous styles [19]. Androgynous fashion is about nondiscrimination and nonconformity related to birth-assigned sex [20]. In the era of gender neutrality and equality, Indian society is also playing its role in enhancing this at the grass-root level. Tridha school in Mumbai took an exemplary step by introducing genderless uniforms (a short cotton kurta paired with trousers of their individual choice) for the students to educate students and parents on treating everyone equally irrespective of their gender expressions [21].

60.3.2 Designers as Fashion Creators and Innovators

Traces from the early twentieth century agree to the point that androgynous fashion was not considered to abolish gender distinction, but it was only to reconnoitre the boundaries of gender discrimination [22]. Western influence in the global fashion contributed to women trousers by Coco Chanel in the early and mid-twentieth century as a strong bending in feminine dressing. Around the same time, two famous fashion labels Yves Saint Laurent and Giorgio Armani launched their androgynous-themed fashion collection to the public. In the 1960s, designer Yves Saint Laurent presented the first suit as couture wear for women's evening wear [22]. Many fashion editorials over time have showcased fashion silhouettes that blur the gender lines, mentioned about androgynous fashion, setting a new pace for a forward-thinking society [23]. As trendsetters, ambassadors from the entertainment and fashion industries have also played an influential role in advancing a challenging perspective on human sexuality for modern times [24]. Rohit Bal introduced the concept of gender neutrality in fashion back in 2003 Lakme Fashion Week to the forefront by showing men on the ramp wearing *sindoor*, as a strong statement [25]. Anjali Lama, a transgender model, walked down the ramp for Gaurav Gupta's sculpted gown replete with rainbow lights at the Lakme Fashion Week 2017 [26]. Sohaya's collection for LFW 2018 blended the rules of gender norms, as the models walked the runway in a pair of grey, wrap-around pants, detailed with mock metallic buttons at front, and a detailed oversized black jacket with highlighted sleeve [27]. Various Indian fashion labels, such as Bloni (Akshat Bansal), The Pot Plant, Anam, Bobo Calcutta (Ayushman Mitra), showcased their designs in the Gender Bender cluster at the Lakme Fashion Week 2018. It was a true representation of androgynous fashion with loose silhouettes and minimal prints [28].Designer Akshat Bansal says, "*Androgynous Fashion is at its good growth. Fashion labels adding non-binary fashion line and all of it together is making India proud and a free country which hails gender equality*" [29].

60.3.3 LGBTQ Community and Their Interaction with Fashion

The continuous effort of designers, artists, and fashion activists has built a gradual acceptance towards androgynous sensibilities in India. With the Indian fashion industry becoming more prominent globally, it is also getting its much-awaited recognition in all related sectors, and the LGBTQ community is also being supported and accepted largely [4]. Late Indian designer Wendell Rodricks was openly gay and always showed support for the LGBTQ community through his shows. For Lakme Fashion Week 2017, Rodricks introduced India's first plus-size transgender model, Mona Veronica Campbell [30]. However, existing studies suggest that defining androgynous fashion is difficult as androgynous fashion is still accepted by a small section of the mainstream consumers [17]. On the other hand, after the abolishment of Section 377 in the Indian Penal Code, Indians are expressing their gender identity and fashion preferences more vividly than before, and the number of followers and practitioners off the ramp to follow androgynous fashion are gracefully increasing [31].

60.3.4 Historical and Traditional References of Androgynous Fashion in India

In Indian society, fashion has been used to emphasize power, to challenge the system, and to initiate social change [32]. Indian LGBTQ community traces its existence from the Indian mythology, which has been filled with multiple instances of mysterious characters, for example, *Shikandi*, a character in the Hindu epic, the *Mahabharata*, written by Veda Vyasa) [33]. Another such example is *Ardhanareshwar* (an androgynous form of the Hindu deities), as the masculinity in the Goddesses, such as Kali, and the femininity in the Gods, such as Krishna [34]. Elements of traditional Indian clothing, such as drapes and silhouettes, have typically overlooked the specificity of gender. Unstitched *Saree, Lungi, Kediyu* top from Rajasthan, shirts paired with *Ghaghras* from Haryana, and *Kedia* from Gujarat worn by men have always been instances of blurring lines between the male and female forms of clothing and dressing styles [35]. Rich Indian textile is full of fabrics, embroideries, embellishments and cuts that have never thought on gender discrimination, and studies have suggested that these aesthetics should be revisited for androgynous fashion inspirations [32].

60.4 Methodology

60.4.1 Approach

The approach of the current study was empirical and descriptive in nature. The study involved both secondary (qualitative) and primary (quantitative) methods of data collection and analysis. Firstly, relevant secondary data were reviewed extensively from sources like published research papers, reputed fashion publications, books and articles. The objective of this qualitative method was to theorize comprehensive information regarding the concept and the design essentials of androgynous fashion using a semi-systematic review approach [36]. Secondly, primary data was collected through an online survey to acquire firsthand information about consumer needs related to androgynous fashion. The objective of this quantitative method was to capture the factors that exemplify the target consumers' emotional needs based on functional, expressive and aesthetic needs [13] for androgynous fashion.

60.4.2 Sample and Data Collection

The secondary study began with collecting eleven published academic research papers ([2, 10, 13, 15, 17, 19, 37–40]), nine articles from fashion publications ([16, 20–22, 26, 29–31, 41]), three books ([32, 34, 42]) and forty nine photographs which described the androgynous trend, its evolution, perception, and relevance in the historic and contemporary society. To carry out the review, the abstracts of the research papers were read before going through the full-text articles in full to confirm that they meet the research objective. The primary research used convenience and snowball sampling methods and involved participants from the LGBTQ community of India, who are recognized to be the major consumer segment by existing literature [4]. Data were collected through an online questionnaire that was shared on social media platforms. Out of 131 recorded responses, 120 were found to be complete and valid. 70% and 16% of the respondents were male and female respectively. 11% of the respondents identified themselves as non-binary. 56% of the respondents were from the age group 19–25, whereas, only 5% of the respondents were from the age group 36–40.

60.4.3 Research Instrument

To collect the secondary data, relevant terms, such as *androgynous fashion; androgynous aesthetics; gender-neutral designers in India; androgynous fashion adoption and conceptual framework; fashion and LGBTQ community; clothing and gender identity; gender-neutral fashion events; androgynous movements; androgynous fashion vs. unisex; the evolution of androgynous fashion India; perception of androgynous fashion; androgynous fashion trends acceptance; design elements of androgynous fashion*, were used to search for textual and imagery references from online platforms. For the primary data collection, the questionnaire began with a short description of androgynous fashion and the research objective. The participants were asked to indicate if they want to take part in the survey, after which they were directed to the questions regarding their needs and choices related to androgynous fashion. Materials for measuring the data included twenty-five sets of adjectives validated by previous studies [13, 43], on a six-point bipolar semantic scale accompanied by the adverbial labels, i.e. slightly (−1/1), quite (−2/2) and extremely (-—3/3) [44].

60.4.4 Data Analysis

The secondary data were analysed concerning the current research objectives, which included the conceptual design-specific elements (such as silhouette, material, and styling) as well as a general direction of the history and evolution of androgynous fashion in the Indian context. The primary data were analysed through descriptive statistics of the demographics, and the reliability was confirmed through Cronbach's alpha (0.735). Finally, principal components analysis (PCA) was conducted for its established potential in narrowing down a large set of measurement variables into the most relevant set [45]. A Keyser Meyer Olkin (KMO) score of 0.79 indicated a passable degree of sampling adequacy. After visually examining the eigenvalue plot or scree plot, four principal factors could be extracted. While interpreting factors, 0.40 was fixed as a minimum cut-off for loadings, as recommended by most classical models. In order to maintain orthogonal factors, varimax factor rotation was followed [46].

60.5 Results

Findings of the qualitative research component suggested that androgynous fashion is not only a trend at the moment in Indian fashion, but it is taking the shape of a social movement. Blending fabrics, floral shapes, minimal traditional embroideries and prints, Western influence in foraying and loose silhouettes with experimental cuts have been the highlight of many leading Indian designers in their latest collections at major fashion events. Time structured silhouettes have taken a break as flowy fabrics like linen, muslin, crepe and georgette with ruffles, frills, and deep necklines took over. New silhouettes and mix-matching like skirt style pants, red nail paint, *sindoor* on the head, lipstick on male models, kitschy jewelery as flamboyant ensembles with beard and moustache were seen on the ramp.

Four components explained 61% of the variance in the original data. Out of the 25 components, ten distinctly classified product attributes; 5 broadly reflected buyers' style statement; 5 highlighted physical and mental predisposition; and the last 5 reflected primary motivation. *Component* 1: *Product attributes:* Very high loading for "natural and light attributes" opposed to "technical and heavy" show an inclination towards comfort in design. Medium-sized loadings on "delicate, fluid and multicultural" attributes vs. "rough, stiff and traditional" attributes indicate creative expression through comfortable fashion. *Component* 2:*Discussion on buyers' chosen style statement:* The component 2 identifies that the LGBTQ community prefers a dynamic and global approach towards fashion and may not shy away from clothes that accentuate corporeal features and give them a certain sex appeal. This reveals the intent of staying well-informed (au courant) with international trends and a zeal to stand out amongst the members of this community. ***Component 3: Physical and Mental Predisposition:*** The loadings in the third component show a strict resolve and a degree of complexity or edginess in behaviour—a relatively high degree of negative loading on soft vs. hard, point towards strength in mental make-up. The most important observation is a medium-size loading towards the scale measuring feminine vs. mannish predispositions, implying non-adherence to a rigid gender make-up. *Component 4: Primary Motivation:* Component 4 shows two equally high loadings on motivation to be perceived as new-age and modern as opposed to being perceived as old and classical in appearance. A high negative loading on warm vs. cool shows an inclination towards being perceived as cool, and a high positive loading on brave vs. shy indicates an intent to took bold through choices in clothing..

60.6 Discussion and Conclusion

The perspective of fashion creators on androgynous fashion was explored by reviewing books, review papers, online publications and visuals. There was a time when genderless fashion was in the closet, and it has now come to see the light of the day, with opinion leaders trying to pull it up from the sidewalk (street culture) to the catwalk (haute couture) [47].

The goal of the primary study was to assess the emotional needs concerning androgynous fashion amongst the consumers, i.e. members of the LGBTQ community in India. Out of the 25 bipolar adjectives selected to measure the attitude of the respondents, four factors emerged, leading the way to following key observations. *Theme 1: the need for comfort in androgynous fashion:* According to research directed towards ascertaining the usefulness of unisex fashion, comfort in clothing was established as an important need [48]. Comfort complete with flexibility, personality, and professionalism are key demands present even in bold and brave expressions of fashion [49]. High factor loadings on natural and light product attributes in this research further strengthen this underpinning need for comfort amongst the respondents. *Theme 2: Making a style statement through dynamic and flattering clothes:* We see an inclination towards dressing dynamically with careful attention to internationally prevalent trends and a desire to highlight one's bodily features amongst the respondents. This opens a challenge for fashion creators to design clothes that are comfortable to wear yet in tune with changing trends and desire to look good. *Theme 3: complex and strong mental make-up:* As opposed to the common stereotype of gay men exhibiting and idealizing effeminate features, many gay men may find such views to have little consequence to them and rather value traditional masculinity [50]. Our research shows, most respondents displaying strength over weakness and relating to complex mental make-up. Sexual and gender identities are multifaceted and historically situated [51]. The already available literature on the mental health of the LGBTQ community concurs with the findings of our research. *Theme 4: Primary motivation for buying is to be perceived as cool and modern:* The concept of androgynous fashion had begun to be explored from the 1990s, however, and modern androgyny is about blurring gender lines [22]. The finds of our research concur with the need to be perceived as cool, modern, and brave in one's fashion choices. Therefore, all communication around it should be targeted towards making the user appear to be modern and cool.

The current study fulfils the research objectives by theorizing the conceptual constructs from the fashion creators' perspective, and analysis of the survey lead to relevant findings on the emotional needs of the LGBTQ community in India concerning androgynous fashion. The findings of the study may lay the basic groundwork for both design practitioners and researchers directed on the distinct design aesthetics, and consumers need perspective or other such topics revolving around the central theme of androgynous fashion. However, to meet the first objective of the study, exploring secondary data had its limitation to develop, choose, and adapt primary data collection protocols. Hence, future studies can focus on interviewing androgynous fashion designers to acquire firsthand and precise conceptual interpretation of the topic. In addition to that, although the sample size of the primary study met the criteria for a PCA, a larger sample size might have positively influenced sampling adequacy. There is always a possibility of comparing the findings of this study by choosing a sample from different demographics using the same set of variables.

References

1. Tseelon, E.: Communicating via clothes (1989)
2. Ranathunga, G.M., Uralagamage, S.R.: An investigative study of the androgynous fashion concept and its impact on the Sri Lankan fashion market (2019)
3. Chapek, G.: Genderless Fashion (2017)
4. Chauhan, V., Reddy-Best, K.L., Sagar, M., et al.: Apparel consumption and embodied experiences of gay men and transgender Women in India: variety and ambivalence, fit issues, LGBT-fashion brands, and affordability. J. Homosex 1–27 (2019)
5. Kealy-Bateman, W.: The possible role of the psychiatrist: the lesbian gay bisexual, and transgender population in India. Indian J Psychiatry **60**, 489 (2018)
6. Reilly, A., Catalpa, J., McGuire, J.: Clothing fit issues for trans people. Fash Stud **1**, 1–22 (2019)
7. Reddy-Best, K.L., Pedersen, E.L.: Queer women's experiences purchasing clothing and looking for clothing styles. Cloth. Text. Res. J. **33**, 265–279 (2015)
8. Veblen, T.: The Theory of the Leisure. Cl (New York New Am Libr 1953) (1899)
9. Veblen, T.: 77ie Theory of the Leisure Class. Allen, London (1959)
10. Lamb, J.M., Kallal, M.J.: A Conceptual Framework for Apparel. 42–47 (1992)
11. Sudha, M., Sheena, K.: Impact of influencers in consumer decision process: the fashion industry. SCMS J. Indian Manag. **14**, 14–30 (2017)
12. De Mooij, M.: In: Consumer Behavior and Culture: Consequences for Global Marketing and Advertising. SAGE Publications Limited (2019)
13. An, H., Lee, I.: A conceptual framework for Asian women's emotional needs in fashion design. Int. J. Fash. Des. Technol. Educ. **8**, 206–213 (2015)
14. Dictionary.com: AndrogynelDefinition of Androgyne at Dictionary.com. In: dictionary.com (2020). https://www.dictionary.com/browse/androgyne. Accessed 5 Jun 2020
15. Watchravesringkan, K., Hodges, N.N., Kim, Y.-H.: Exploring consumers' adoption of highly technological fashion products. J. Fash. Mark. Manag. **14**, 263–281 (2010). https://doi.org/10.1108/13612021011046101
16. Duignan, B.: In: Butler, J. (ed.) Biography, Gender Trouble, and Facts, Britannica (2019). https://www.britannica.com/biography/Judith-Butler. Accessed 9 Jun 2020
17. Crepax, R.: In: The Aesthetics of Mainstream Androgyny: A Feminist Analysis of a Fashion Trend (2016)
18. McKeage, K., Crosby, E., McKeage, J.: The roots of genderqueer identity through consumption: androgynous and unisex fashion in the 20th Century. Genderqueer Identity 206–208 (2015)
19. Venter, M., Chinomona, R., Chuchu, T.: Factors influencing fashion adoption among the youth in Johannesburg, South Africa. J. Econ. Behav. Stud. **8**, 92–108 (2016)
20. Sarkar, A.R.: In: 2018, Gender-Fluid Fashion Dominated Indian Runways and How!-The New Indian Express (2018). https://www.newindianexpress.com/lifestyle/fashion/2018/dec/23/in-2018-gender-fluid-fashion-dominated-the-indian-runaways-and-how-1915572.html. Accessed 9 Jun 2020
21. Jamooji, S.: In: Gender-Neutral Uniforms Are The Norm At This Mumbai School-Homegrown (2017). https://homegrown.co.in/article/801006/gender-neutral-uniforms-are-the-norm-at-this-mumbai-school. Accessed 9 Jun 2020

22. Chrisman-Campbell, K.: In: A Brief History of Unisex Fashion-The Atlantic (2015). https://www.theatlantic.com/entertainment/archive/2015/04/when-unisex-was-the-new-black/390168/. Accessed 9 Jun 2020
23. Radin, S.: 6 Indie fashion magazines highlighting queer storytelling in an authentic way-fashionista. In: fashionista.com (2018). https://fashionista.com/2018/12/queer-fashion-magazines-2018. Accessed 9 Jun 2020
24. Atkinson, M.: Battleground: Sports [2 volumes]: Sports. ABC-CLIO (2008)
25. Talati-Parikh, S.: Dark side of the moon. In: livemint (2018). https://www.livemint.com/Leisure/5Gn1fwt8y5m9PVuPg5sFGL/Dark-side-of-the-moon.html. Accessed 9 Jun 2020
26. Mankani, S.: Transgender model Anjali Lama is breaking gender barriers in the modelling industry|VOGUE INDIA. In: Vogue India (2018). https://www.vogue.in/magazine-story/anjali-lama-transgender-model-lgbtq-rights-vogue-india-interview/. Accessed 9 Jun 2020
27. D'Souza, O.: The rise of genderless fashion in India. In: DNA India (2018). https://www.dnaindia.com/just-before-monday/report-the-rise-of-genderless-fashion-in-india-2657979. Accessed 9 Jun 2020
28. Ladha, S., Shaikh, S.: Highlights from day 3 At Lakmé fashion week winter/festive 2018, Verve Magazine (2018). https://www.vervemagazine.in/fashion-and-beauty/highlights-from-day-3-at-lakme-fashion-week-winter-festive-2018. Accessed 9 Jun 2020
29. Tripathi, N.: Lakmé Fashion Week Winter/Festive 2018: Gender-neutral fashion concept takes centre stage, Forbes India. In: Forbes India, pp. 6 (2018). https://www.forbesindia.com/aperture/slideshow/lakme-fashion-week-winter/51121/6. Accessed 9 Jun 2020
30. Negi, S.: Wendell rodricks only one who supported LGBTQ community, says india's first plus size queer model-News18. In: News18 (2020) . https://www.news18.com/news/lifestyle/wendell-rodricks-only-one-who-supported-lgbtq-community-says-indias-first-plus-size-queer-model-2499031.html. Accessed 9 Jun 2020
31. Katju, A.: One year after freedom: how are LGBTs faring today, since Section 377 was struck down by the Supreme Court? In: The Times of India (2019). https://timesofindia.indiatimes.com/blogs/toi-edit-page/one-year-after-freedom-how-are-lgbts-faring-today-since-section-377-was-struck-down-by-the-supreme-court/. Accessed 9 Jun 2020
32. Tarlo, E.: In: Clothing Matters: Dress and Identity in India. University of Chicago Press (1996)
33. Lee, Y.: Shikhandi: and other tales they don't tell you, by Devdutt Pattanaik, India: Penguin and Zubaan 2014, pp. 179 (2016). ISBN 9789383074846. Motifs An Int J English Stud 2:156–158
34. Devy, G.N.: In: Indian Literary Criticism: Theory and Interpretation. Orient Blackswan (2002)
35. Deb Roy, L.: LGBTQ: queer fashion challenges rigid stereotypes. In: Outlook India (2018). https://www.outlookindia.com/website/story/queer-fashion-challenges-rigid-stereotypes/316132. Accessed 9 Jun 2020
36. Snyder, H.: Literature review as a research methodology: an overview and guidelines. J. Bus. Res. **104**, 333–339 (2019). https://doi.org/10.1016/j.jbusres.2019.07.039
37. Arvanitidou, Z., Gasouka, M.: Construction of gender through fashion and dressing. Mediterr. J. Soc. Sci. **4**, 111 (2013)
38. Van den Heuvel, T., van Trijp, H., van Woerkum, C., et al.: Linking product offering to consumer needs; inclusion of credence attributes and the influences of product features. Food. Qual. Prefer. **18**, 296–304 (2007)
39. Arvanitidou, Z., Gasouka, M.: In: Fashion, Gender and Social Identity. First Fash Colloq, London, pp. 1–19 (2011)
40. Marcangeli, S.: Undressing the power of fashion: the semiotic evolution of gender identity by coco chanel and Alexander Mcqueen (2015)
41. Dhar, S., Narayanan, K.A.: Gender neutrality starts with school uniforms, India News-Times of India (2019). https://timesofindia.indiatimes.com/india/gender-neutrality-starts-with-school-uniforms/articleshow/71326053.cms?utm_source=contentofinterest&utm_medium=text&utm_campaign=cppst. Accessed 9 Jun 2020
42. Sandler, P.C.: In: The language of Bion: A Dictionary of Concepts. Routledge (2018)
43. An, H., Lee, I.: Extraction of fashion sensibility vocabulary for globalization. J. Korea Soc. Basic Des. Art. **14**, 133–141 (2013)

44. Hartley, J., Trueman, M., Rodgers, A.: The effects of verbal and numerical quantifiers on questionnaire responses. Appl. Ergon. **15**, 149–155 (1984)
45. Balasundaram, N.: Factor analysis: nature, mechanism and uses in social and management science research. J. Cost Manag. Accountant Bangladesh **37**, 15–25 (2009)
46. Tabachnick, B.G., Fidell, L.S.: Using Multivariate Statistics. Northridge, Cal Harper Collins (1996)
47. Vänskä, A.: From gay to queer—or, wasn't fashion always already a very queer thing? Fash. Theory **18**, 447–463 (2014)
48. Vasan, S., Pujar, S.K., Gopalakrishnan, S.: Perception on unisex clothing among apparel designers: a study. Int. J. Text Fash. Technol. (IJTFT) **7**(2), 17-26 (2017)
49. Lupano, M., Vaccari, A.: Fashion at the Time of Fascism. Damiani, Bol Italy (2009)
50. Riggle, E.D.B., Whitman, J.S., Olson, A., et al.: The positive aspects of being a lesbian or gay man. Prof. Psychol. Res. Pract. **39**, 210 (2008)
51. Diamond, L.M.: What does sexual orientation orient? a biobehavioral model distinguishing romantic love and sexual desire. Psychol. Rev. **110**, 173 (2003)

Chapter 61
Spectacles of the Temporal City: An Analysis of the Effects of Festivals on Identity in an Urban Context

Anusmita Das and Amarendra Kumar Das

Abstract Cities around the world are becoming complex systems wherein every city possess unique identity manifested through its people and activities that occur in its urban spaces. The sense of urban space is an ever-changing concept and to understand the fabric of urban spaces, knowledge of the concept of space and its interrelations is required. Reading a city in a time where change is the norm, with urban spaces gaining new meanings and functions with changes in cultural practices conveys the hybrid urgencies of metropolitan India. An analysis is thus, particularly relevant in India, as in the post-industrial scenario, cities in India have become some of the largest urban conglomerates with a dualist identity of static and kinetic pervading their urban landscape. The static city is the conventional physical entity of a city whereas, the kinetic city is perceived as a city in motion, temporary in nature and the emerging symbolic image of urban India. Understanding these multitudes of identities in the backdrop of traditional events and festivals, as temporal transformations of urban spaces will thus highlight the dominant visual culture of Indian cities. This Temporal City enables a better understanding of the blurred lines of contemporary urbanism and the changing roles of social space in the urban fabric. This study focused on the ancient festival of Ambubachi in the historic Kamakhya Temple in Guwahati, India, through visual ethnography, participant observations, open-ended interviews, photography and document analysis. The central concern of this paper is to develop a critical stance on understanding the identities of cities through the study of urban spectacles such as festivals in a bid to illustrate the fluid and dynamic aspect of one's identity that is being constantly redefined and reconstructed with time, space, activities and context. Through the analysis of this ancient festival of Ambubachi in the light of 'identity' and 'place attachment' of the festivalgoers, the emerging notions of identity will be explored. The fieldwork sought to analyse two major aspects: The first was the generic, spatial characteristics of festival spaces and the second was the temporal, fleeting events that occur in their physical settings following

A. Das (✉) · A. K. Das
Department of Design, Indian Institute of Technology Guwahati, Guwahati, Assam, India
e-mail: anusmitadas.design@gmail.com

A. K. Das
e-mail: dasak.iitg@gmail.com

A. Chakrabarti et al. (eds.), *Design for Tomorrow—Volume 1*, Smart Innovation, Systems and Technologies 221, https://doi.org/10.1007/978-981-16-0041-8_61

the parameters of event-space relationship by Frenchman in a bid to understand how such eventful urban spaces manifest in the negotiation of one's identity.

61.1 Introduction

The city is a representation of not just the physical built fabric, but an assemblage of various intangible characteristics that contribute to its identity, transformation, history, etc. Various socio-spatial factors are at interplay that manifests themselves contributing to the spatial identity of the city. In the present age, as cities expand and continue to grow, there has been a mass influx of people that not only stretches the physical boundaries of cities, but the psychological periphery as well. With the confluence of ideas, philosophies, and culture, cities increasingly encounter greater challenges for the negotiation of their identity which is a dynamic process of adaptation and reconstruction. The adaptation is observed notably in the intangible cultural heritage. Urban spaces are gaining new meanings and functions with changes in cultural practices. Thus, a city can no longer be seen as uni-directional, but as a constantly changing, evolving organic entity where fluidity and temporality are observed as some of the key aspects. In a time where change is the norm, cities are increasingly facing issues of identity. As a city transforms over time, intangible culture has become one of the principal factors for examining how a city metamorphoses with time and context. In a bid to bolster their image and an upward surge of cultural resurgence, cities are resorting more and more towards the tangible and intangible cultural heritage of a city to rebuild and redefine their identity. As culture is a representation of the people, their beliefs and values with a stronghold in the past and present, festivals as representations of the intangible cultural heritage can be a powerful tool in analysing their role in the creation of multifarious identity of the urban fabric. The research delves into understanding the sense of relatedness that people develop with a place through interaction in the socio-spatial realm as a way to adapt themselves towards place engagement which reconstitutes their identity over time. Though there is no dearth of literature on identity in various disciplines, from the perspective of urbanism, however, due to the lack of perceived importance of space as an integral aspect of identity construction and the unaccounted link to architecture, most of the available non-spatial research on identity remains incoherent. In order to have a deeper and complete understanding of our spatial environments, designers and architects need to accost these missing threads between identity and urban fabric. This entails for a more interdisciplinary approach that would take into account both the non-spatial and spatial perspective, to throw light on the intricacies of identity, thus providing a platform for designers and architects alike to have a better understanding which could aid and enhance their interaction with their environment.

61.2 Identity: Key Concepts and Complexities

According to Snow [1], three levels of identity may be perceived: personal identity, social identity and collective identity. To situate oneself in the social space, these three levels of identity interplay according to the context. Personal identities relate to the self when situated in relation to others, self-attributed and regarded as personally distinctive. They are self-attributes and designations imparted to others in the social space, 'those aspects of the self that differentiate the self from all others' [2]. The extension of this sense of identity to social self or social identity becomes a fundamental process after realisation of one's personal identity [2, 3]. Social identity can be inferred at two levels 'those that derive from interpersonal relations and interdependence with specific others, and those that derive from membership in larger organisations' [2]. They are, respectively, interpersonal and collective identity capturing a shared essence of belonging together in a group. The level of intimacy varies as in interpersonal identity the level and frequency of actual contacts is high as compared to collective identity which may not require personal relationships.

As cities in India move towards rapid urbanisation, certain urban complexities become indispensable components of its urban environment. One of the major challenges faced by Indian cities is the increasing need to reinvent their image and identity. Indian cities have grown organically over the decades and most of them have a traditional core. However, with globalisation and subsequent urbanisation, certain salient features of these cities have undergone adaptations more in sync with the global aspirations. It is surprising to note that despite the homogenising forces of globalisation around the globe, Indian cities have been able to retain their individual flair owing largely to the cultural resurgence observed across many cities, especially the tier two cities [4]. There are multifarious factors that influence identity and make it a complex domain of study. Identity is a basic human need that forms the seed of interaction between self, others and society and forms a central theory in social psychology. There is, however, another entanglement that is frequently overlooked, the relationship between identity and place. This overlook is more apparent when it concerns the notion of identity within festival research. This paper specifically sought to explore this particular aspect of identity by analysing how festivals influence one's sense of identity individually and socially, the sense of belongingness to a collective identity and a place. In this regard, the various interdisciplinary concepts of identity will be presented to emphasise these aspects of socio-spatiality embedded in identity. Discussions around identity, place and experiences will aim to comprehend this interrelationship and bring to light how it influences notions of identity. Drawing insights from these multiple aspects, this paper will provide various perspectives offered by the interdisciplinary study and their contribution to the understanding of place and identity relationship through the analysis of the festival of Ambubachi in the city of Guwahati Assam.

61.3 Emerging Notions of Identity and Place

One of the key concepts investigated in this research is the concept of space or the notion of space and its significance in terms of identity. Physical environment has hardly been the focus of studying identity and largely considered only in the backdrop wherein activities occur. Many seminal authors have emphasised the need to consider space as one of the driving forces behind manifestation of identity. Philosophers Edward Casey [5] and Jeff Malpas [6] consider space as the primal and fundamental concept in the study of any domain and our very existence. Christian Norberg Schulz [7] further adds to it by reiterating how identity is embedded in our subconscious and a part of our existential meaning production. He gives reference of how people usually mention their place of origin when talking about themselves instead of mentioning their occupation or their beliefs, thus placing human identity as a function of places and things. He concludes by saying that human identity presupposes the identity of place.

In recent discussions within environmental psychology, Dixon and Durrheim [8] hypotheses that places are dynamic arenas that are 'both socially constituted and constitutive of the social' and addresses the fundamental question of 'who we are' as being intimately related to 'where we are'. This adds a new dimension towards the study of identity concepts wherein place plays an ever-increasing role in our notions of who we are. Korpela [9] defines place identity as the cognition of our physical environment in which an individual consciously or subconsciously regulates one's sense of self and does not limit a person's place identity only to familiar environments such as one's home and its surroundings but expands to every physical environment that has a role in one's self-regulation. Thus, it becomes the core around which social, cultural and biological definitions and cognitions of the place become a part of the individual's place identity.

Dixon and Durrheim [8] views place identity as 'a psychological structure that arises out of individuals' attempts to regulate their environments' thus situating place identity as a part of personal identities with dynamic needs towards place attachment. This interrelationship may further extend to affiliation with the environment and other people through place–people interactions. In conclusion, place attachment maybe composed of place identity, dependence, familiarity, belongingness and rootedness [10]. As long as people carry impressions of their languages, culture and histories, their personal identities will remain rooted in culture and to place [11]. According to Lefebvre, the very roots of social relationship are embedded in spatiality itself [12]. Conversely, Richard Steadman [13] states that place itself is laden with meaning and aids in the construction of social meanings derived from experience with the environment forming a strong 'social construction'.

61.4 Festivals, Identity and Place Attachment

In this research, identity is observed in the context of how temporal events such as festivals, which are rooted in time and place, can enhance, add meaning and manifest in the formation of identity through the myriad of activities and interactions that occur in urban spaces during festivals. This study is part of a research that builds on the hypothesis that identity is shaped by socio-cultural processes such as festivals as much as the socio-spatiality of the environment. Harcup [14] argues that festivals support the redefinition, rediscovery and expansion of urban social life and imparts meanings to a place through specific appropriations of urban spaces. These temporal events provide opportunities for social engagement through experiences in festival spaces and can be key aspects to enrich identity and build social bonds [15]. The study of festivals has been explored as a tool in this paper to understand how they function as informal arenas wherein social identities metamorphoses in the context of the changing social and spatial organisation of the city [16]. In the last few decades, festivals have become a popular instrument to create a sense of distinctiveness and unique identity of cities around the globe. Festivals have become the medium through which community life is renewed and hold the potential as a tool for cultural reconstruction and transferring of knowledge to younger generations over time [17]. Eventually, these festivals will contribute to citizens' well-being by providing spaces for social recreation [18] and allowing them to participate in experiences that are distinct from the everyday experiences. By compressing people and activities within these limited spaces, festivals in urban areas can lead to heightened sensory and emotional experiences.

61.5 Methodology

This study focused on one of the ancient religious festivals in Guwahati, India: Ambubachi Mela which has its roots in the origin of the city. Through participant observations, open-ended interviews, photography, and document analysis, the fieldwork of the research was carried out. The fieldwork sought to analyse two major aspects: The first was the generic, spatial characteristics of festival spaces and the temporal, fleeting events that occur in their physical settings. The second was discreet participant observation of the social behaviour of festivalgoers within the visual narratives of these festival spaces. Together these aspects would be crucial in understanding the role that religious festivals can play in the formation of identity through informal human interactions that is inscribed in rituals, memory and place attachment.

A broad spectrum of people across different roles was sampled, and a total of 43 in-depth interviews were conducted with the devotees, along with interviews with the Head of Bordeuri of Kamakhya Temple, Assam Government employees in various accommodation camps, Assam Tourism and PWD Contractors. The first set of interviews with the various stakeholders were conducted to get an awareness about the existing scenario of the festival and the planning that went behind the entire gamut of activities that prevailed during festival days. Some of the interviews were also conducted at the site during the duration of the festival to get an insight on the current situation and management aspects of the festival by the authorities. The second phase of interviews was focussed on analysing the various parameters of festival spaces and their impact on the festivalgoers in a bid to understand the event-space-participant relationship and their role in place attachment and identity formation.

61.6 An Ancient Festival in an Ancient City in the Age of Urban Complexities

Ambubachi Mela is the most auspicious and sacred festival of Kamakhya Temple. This ancient temple, situated atop the Nilachal hills in Guwahati, is one of the oldest Adi peetha (earliest shrine) in India. Among all the Shakti peethas in the world, Kamakhya is considered the foremost from which a network of other peethas and shrines emerged. There are many myths surrounding the origin of Kamakhya Temple as mentioned in the Kalika Purana. According to the Department of Geological Sciences, Gauhati University, as per evidences found, the age of Kamakhya Temple dates back to as early as the seventh to eighth century [19]. Ambubachi has been celebrated in the city of Guwahati since time immemorial. Guwahati, or ancient Pragjyotishpura (the Light of the East), is the capital city of Assam in the north-eastern part of India. It is an ancient city that finds mention in the epics and Puranas. As mentioned in the Kalika Purana (tenth century A.D.), Brahma created the stars in Pragjyotishpura, and hence, the city was a reference to the city of Indra [20]. An explanation put forward by Gait, emphasises the importance of Pragjyotishpura as the land where the Tantrik form of Hinduism originated, thus breaking down the etymology as 'Prag' meaning former or eastern and 'Jyotisha', meaning a star, astrology or shining. Kamakhya constituted the core of the ancient city. However, due to natural decay the temple lost its importance until it was rediscovered in the early part of the sixteenth century. The temple was rebuilt by King Biswa Singha (1549–1550), the founder of the Koch kingdom in the western part of Assam (Table 61.1).

During the field work of Ambubachi Mela in the years 2017 and 2018, it has been observed that the travelling pilgrims can be divided into three categories: Ascetics/Sadhus, Performers and New Devotees. The first category of devotees has been visiting Ambubachi for more than a decade now and is the regulars who have been involved with this festival before the surge of its popularity had begun. The second category of pilgrims is mostly Baul singers who usually move in groups performing in the festival. They have been a part of Ambubachi for a long time. The third category of devotees is relatively new devotees who have been visiting this festival for less

Table 61.1 Case study festival characteristics

Ambubachi mela: context and size	Festival characteristics
Ambubachi Mela is a weeklong yearly festival held around the middle of June in the Kamakhya Temple Complex celebrating the harmony between the feminine reproductive energy and fertility power of nature This mela witnesses a footfall in the range of millions of devotees every year since 2012 and has been heralded the 'Mahakumbh of the East' as it is the largest festival in the entire North-East of India	• The concept of Ambubachi has in its origin, formative influences and elements of agricultural, social and religious ideas that have contributed to the emergence of the phenomenon. It is thus symbolic, supported by religious sanction • Daily worship is suspended during this period. All agricultural work like digging, ploughing, sowing and transplanting of crops are forbidden • It is a common belief that during this period, a natural stream of water flowing through the Yoni mandala located in the Garbagriha turns red and hence is a symbolic representation of the Goddess's menstruation • Red is a very important colour for the devotees of Goddess Kamakhya and represents Shaktism. This colour also symbolises creativity • Widows, Brahmacharis and Brahmins avoid cooked food during these days • On the fourth day, Ambubachi being over, household items, utensils and clothes are washed, cleaned and purified by sprinkling sacred waters, worship of Goddess Kamakhya begins after cleansing and other rituals are performed • Entry to the Shrine is considered to be auspicious after this

than three years. They mostly consisted of day labourers working in fields from West Bengal. Many foreigners also visit the temple during to partake in its activities. As per interview with K. Sharma, the head of Bordeuri Samaj of Kamakhya, it came to light that with the involvement of various organisations and availability of free food, accommodation, free transportation, devotion is not necessarily the primary focus which attracts the visitors. Despite this, Ascetics and Sages who visit this mela still observe fast during this period, which was the original way of displaying one's devotion to Kamakhya Devi. This has lent a character of a fair to this festival. As was observed during the fieldwork, majority of the travelling pilgrims were from West Bengal. This goes back to history as in another study of the shakti peethas, Ramos [21] discusses how the region of Assam became such an integral part of the Bengali community during the nineteenth century amidst the British colonial rule. Assam was separated from Bengal by the British in 1874 [22], despite which, a strong kinship remained between these two regions owing to their cultural, linguistic and religious links with Assam and embodied in religious places such as Kamakhya. Apart from these inherent threads binding Assam and Bengal, there were many authors, Haliram

Phukkan [23] being chief among them, who actively pursued the spread of awareness about the region and encouraged people to undertake a pilgrimage to the holy site of Kamakhya. With the changes in present times and growing number of devotees, the District Administration of Kamrup Metro, along with the Tourism Department, Government of Assam, has been involved with the organisation and implementation of this Mela every year since 2014. The surge in popularity of the mela does not spell excitement for all as many believe the spiritual core of Ambubachi has somewhere been lost in the cacophony of creating a 'spectacle' mostly targeted at tourists.

61.7 Analysis

Festivals are inseparable from their spatial settings. Neither can they be separated from the participants that occupy those spaces. There is an intuitive relationship between festivals, their spatial settings and the participants. Even though these spaces appear to be liberated from the hegemonies of the formal static structure of a city, for a brief period in time and space, they articulate the behaviour and experiences of the participants by promoting some social arrangements and identities while inhibiting others. Nonetheless, such spaces encourage informal interactions where multifarious identities can be expressed and negotiated to an extent [16]. Most of the literature on urban festivals is through the consumer gaze and lacks any specific analytical focus on the spatial settings of these festivals [15]. There are, however, limitations when it comes to studying Indian religious festivals and their physical settings. It is said that these religious festivals are the gateway to know India. Still majority of the studies on festivals in India are from the standpoint of denoting them as spectacles. As cities in India continue to urbanise, the rituals associated with the festivals continue to evolve while still having a stronghold in the past.

The study of the religious festival of Ambubachi in Guwahati has its own particular narratives, space and spatial movement (rituals), creating an urban imagery that is simultaneously formal (process and steps of the ritual) and informal (place-based associations). This study would, however, focus only on the experience of these spaces by the participants and not on the constraints of planning and management. This study has its limitations as they cannot be the yardstick to explore the full scope of human behaviour during urban festivals. It can nevertheless provide a basic framework for analysing socialisation in spatial settings of festivals. Further study and discourse on these aspects would be conducive for refining this area of enquiry. Fieldwork was conducted across the hotspots covering the nine accommodation zones of Ambubachi Mela across the city and the Kamakhya Temple in Nilachal hill. Observations were undertaken before and during the main festival events, so that local spaces could also be studied in their everyday use. Observations were focused around the following key concerns as discussed below.

61.7.1 Festive Settings (Spatial Focus)

The study of the religious festival of Ambubachi Mela highlights how urban spaces can be transformed for varied uses. In Ambubachi Mela, the central core is the historical site of the Kamakhya temple. The other urban spaces that are associated with this festival create an ephemeral city with elastic boundaries that spreads over the entire Nilachal hill of Guwahati. Even though the site for this festival is spatially segregated, and opportunities for social interaction are confined within the zones, the culmination occurs at the site of the mela where all the devotees congregate. The analysis of Ambubachi mela was conducted following the parameters of event-space relationship by Frenchman (2004) [24]. These parameters are: Territoriality, Intimacy, Granularity, Triangulation, Movement, Scale and Sensuality.

1. Territory: Event-places demarcate a special phenomenon that is set off from everyday life. The event space of Ambubachi has the ability to create a sense of separation within the physical fabric of the urban space. Having defined edges, Ambubachi can be separated in space or time from the common terrain. The road leading up to the Kamakhya Temple has an entrance gate which clearly demarcates and designates it from the urban fabric of the city. It creates a territory which allows people to separate themselves emotionally from everyday life and give themselves over to the festival.
2. Intimacy: Good event-places create a social and emotional bond between the event and the individuals involved. By providing a sense of enclosure and limit to the event as people and activity are compressed together during the mela, these urban spaces can facilitate intimacy. Intimacy is further enhanced by sharing the same physical space allowing people to see and being seen. Finally, intimacy can be facilitated by focus, where attention and views are directed by landmarks, lighting or circulation towards others participating in the event.
3. Granularity: Good event-places are not static as they accommodate multiple nodes of activity and groups allowing experience at many levels. The site of Ambubachi mela facilitates granularity by providing a series of spaces or a series of experiences both within the space and throughout the event: from personal interaction with the event, to social interaction with the larger urban setting and the crowd. The road leading to Kamakhya temple is interspersed with various pockets of activity zones like food stalls, people dressed up in costumes of Gods and Goddesses and hawkers selling various trinkets.
4. Triangulation: Interaction or a sense of community, between two strangers in a public space is encouraged by a third activity or object they can both share. Triangulation can become more difficult to achieve, if the events grow in size as the focus can be overwhelmed by the crowds. During Ambubachi Mela, the culminating point is the temple which is the spiritual focus of the devotees and visitors.
5. Movement: Good event-places are dynamic, facilitating and even promote movement and participation. During the Ambubachi Mela, dynamic activities occur with the physical setting of the festival. The road leading to the temple

was open only to pedestrians; hence, they were able to engage themselves thus making the walk uphill less laborious.

6. Scale: Good event-places balance the size of the event and the place. When events get too large for their place, the above parameters of intimacy, granularity and triangulation can be lost. While it is advised not to change the physical scale and setting when attendance grows as it can dramatically change an event and destroy its very essence. Festivals in India are grand spectacles with thousands of participants. Hence, to conceive such festivals in an urban setting of a city is a daunting task. During Ambubachi Mela, more than 15 lakh devotees had participated. This proved to be a hindrance as a majority of the devotees could not enter the temple premises due to this large gathering of masses.
7. Sensuality: Good event-places are fundamentally sensual as they can artfully engage all of the senses: the feel of movement through space, warmth, or cold; the sound of music; the smell of burning wood and flowers; the taste of prasad and food; the sight of costumes and colour, blended by the imagination into an unusual experience. During Ambubachi Mela, the visual setting of the festival transports one to a surreal experience. As emotions are heightened, these urban spaces gain special importance as sites for collective memory and attachment.

61.7.2 Festive Actions (Behavioural Focus)

These socio-spatial spaces where festivals are celebrated can be compared to public spaces both in terms of spatial settings and the meaning attached to them. The five dimensions/ aspects of public space as developed by Mehta [25] as a toolkit for Public Space Index were considered during the research which included: inclusiveness, pleasurability, meaningful activities, safety and comfort as an integral part of meaning generation in festive spaces. Based on observations and interviews with the respondents, these five parameters were explored to understand the interrelationship between festival space and participants and meaning attached to them.

Structured and unstructured observations showed that Ambubachi Mela provided ample opportunities for inclusiveness as people belonging to variable ages, different genders, diverse classes, different geographical locations, diverse regionalities, diverse physical abilities were participants in the festival. Participants/ Devotees also responded regarding the diverse activities that occurred in the mela such as the fairs, traditional Deodhani Nach performed by the male group dancers, presence of various 'Baul' singers, which added a broad spectrum of variety and flair to the festival. The devotees intermingled in all these activities thus becoming an integral aspect of the festival. Ambubachi thus facilitated interactions that aided in the development of an extended self or a social identity or a social self.

The festival, at the same time also became an arena for cultural revitalisation through the meaningful activities occurring in its domain. People connected to the narratives, symbolism and myths of the Ambubachi, thus carrying on the history of the festival percolating down to generations. Their personal stories were also were interwoven along with the stories of the festivals allowing them an opportunity to be a part of the narrative. Thus, their personal identities intertwined with the culture and origin of the festival creating a collective sense of identity or collective identity.

The festival provided a range of facilities for the devotees which contributed to their feeling of comfort especially for the devotees who have travelled away from their own place of origin. As the festival is spread over a week, such facilities were imperative to create a sense of belongingness with the festival. Though some devotees faced certain maintenance issues, such as scarcity of water and hygiene, the spirit of the festival did not dampen despite such inconveniences. When enquired if they would reconsider about visiting this festival again the next year in the light of the problems faced by them, they unanimously responded in the affirmative. This highlights an interesting observation that the heightened sense of celebration during festivals overshadows the lack of comfort and amenities as people demarcate such experiences from their everyday lives. This temporal dimension of festivals is an integral aspect that aids in the creation of mutual kinship thus transcending to their sense of identity.

The sense of safety relies on multiple factors such as the environment, people, legibility of the elements of space, amenities which all combine together to create this dimension of social space. As was observed in the festival, there were volunteers and officials in various nodes of the festival arena to guide the devotees in the event of getting lost. Emergency services were also stationed at strategic points across the venue. The participants mostly travelled in groups and in such situations as getting separated from the group, they usually travelled back to the last known spot. These facilities along with the familiarity of being part of a group, thus strengthened their perceived sense of safety. Many of the devotees have been frequent visitors to this festival and hence were acquainted with the festival site and the activities. They in return served the role of a guide for their new fellow festivalgoers. Familiarity fostered the feeling of safety and a shared sense of being part of a community.

Ambubachi is celebrated atop the Nilachal hill, surrounded by lush green hills and a view of the Brahmaputra. During the festival, devotees mentioned that they are drawn by the natural beauty of the site apart from the spiritual essence of the festival. The presence of such sensory stimulations thus derived increases the pleasurability of the festival, thus becoming a factor for repeated visits. Apart from the devotees, there are many participants who throng the festival in their costumes representing Hindu Gods and perform variety of activities that keep everyone enthraled. This increases the attractiveness of the festival and manifests in the sense of shared experiences.

The site of Ambubachi constitutes an arena that strengthens continuity by creating opportunities to draw on: 'shared histories, shared cultural practices and ideals, as well as creating settings for social interactions. They are arenas where local knowledge is produced and reproduced, where the history, cultural inheritance and social structures that distinguish one place from another, are revised, rejected or recreated' [26].

61.8 Conclusion

The study of the ancient festival of Ambubachi Mela in the urban fabric of Guwahati illustrates that festivals may contribute to the formation of individual and collective identities, create a sense of attachment and belongingness to a place, thus strengthening bonds through shared meanings. These identities are interchangeable and are observed in relation to how one situates oneself and others in the social space. As described by Baumeister and Leary [3] and Brewer and Gardner [4], events such as this festival contribute to social identities and place bonding [10] to encompass place identity, dependence, familiarity, belonging and rootedness. The myriad of complex interaction between historical, cultural, social and psychological factors within a limited time and space, this festival creates greater impacts at identity than any other entity of urban environments. Through the narratives of the festival, the participants share a sense of kinship which is further endowed in their own cultural identities [27].

Festivals influence the regeneration of places and identities as McCabe reiterates: 'legend or recreation of events, but in a determined effort to continue the event as a means to provide a symbolic system of continuity to the past' [28]. As seminal authors have already established that the negotiation of one's identity is a process of human motivation [3, 4, 10], individuals will tend to search for avenues such as festivals, to reinterpret and reinvent their identities through the experience of tradition, cultures, history, nature and shared values. Through meaningful activities, celebrations, narratives, these festivals attract the participants thus creating a unique bond that surpasses boundaries of time and place. Thus, these places also develop identities in the process that are parallel to the personal and social identities of the participants. During festivals, urban spaces create a temporal experience which transcends time and space as the senses are heightened due to the visual and spatial settings. These events thus invoke place memories that percolate down generations which impart the distinct identities to urban places.

References

1. Snow, D.: Collective identity and expressive forms. In: Smelser, N.J., Baltes, P.B. (eds.) International Encyclopedia of the Social and Behavioral Sciences, pp. 2212–2219. Elsevier Science, London (2001)
2. Brewer, M.B., Gardner, W.: Who is "we"? levels of collective identities and self representations. J. Pers. Soc. Psychol. **71**(1), 83–93 (1996)
3. Baumeister, R.F., Leary, M.R.: The need to belong: Desire for interpersonal attachment as a fundamental human motivation. Psychol. Bull. **117**(3), 497–529 (1995)
4. Singh, B.: Cultural resurgence, place making and urbanism: towards a conceptual framework, Indian Anthropologist. Special issue on Culture Place Ecol. **48**(2), 7–20 (2018)
5. Casey, E.S.: In: The Fate of Place: A Philosophical History. New Ed (University of California Press), vol. 9 (1998)
6. Malpas, J.: Putting space in place: philosophical topography and relational geography. Environ. Plann. D: Soc. Space **227** (2012). https://doi.org/10.1068/d20810

7. Norberg-Schulz, C.: Genius Loci : Towards a Phenomenology of Architecture, p. 50. Rizzoli, New York (1980)
8. Dixon, J., Durrheim, K.: Displacing place-identity: a discursive approach to locating self and other. Br. J. Soc. Psychol. **39**, 27–44 (2000)
9. Korpela, K.M.: Place-identity as a product of environmental self-regulation. J. Environ. Psychol. **9**, 241–256 (1989)
10. Hammit, W.E., Backlund, E.A., Bixler, R.D.: Place bonding for recreational places: conceptual and empirical development. Leisure Stud. **25**(1), 17–24 (2006)
11. Hall, S.: A place in the world?: places, cultures and globalization. In: Massey, D., Jess, P.M. (eds.) Oxford University Press in association with the Open University, Oxford, vol. 178, pp. 175–213 (1995)
12. Lefebvre, H.: In: The Production of Space. vol. 129, Malden, Mass, Blackwell (1991)
13. Stedman, R.S.: is it really just a social construction?: the contribution of the physical environment to sense of place. Soc. Natural Res. **16**(8), 671–685 (2003). https://doi.org/10.1080/08941920309189672
14. Harcup, T.: Re-imaging a post-industrial city: the Leeds St Valentine's Fair as a civic spectacle. City **4**(2), 215–231 (2000)
15. Stevens, Q., Shin, H.: Urban festivals and local social space. Plan. Pract. Res. **29**, 1–20 (2012)
16. Willems-Braun, B.: Situating cultural politics: fringe festivals and the production of spaces of intersubjectivity. Environ. Plann.: Soc. Space **12**, 75–104 (1994)
17. Owusu-Frempong, Y.: Afrocentricity, the Adae festival of the Akan African American festivals and intergenerational communication. J. Black Stud. **35**(6), 730–750 (2005)
18. Nelson, G., Prilletensky, I.: Doing Psychology Critically. Making a difference in diverse settings, EEUU Palgrave Mcmillan, New York (2002)
19. Deka, P.J.: Nilachal Kamakhya. Her history & Tantra, Guwahati (2004)
20. Barua, B.K.: A Cultural History of Assam, p. 12. Lawyer's Book stall, Guwahati (1969)
21. Ramos, I.: In: Pilgrimage and Politics in Colonial Bengal, pp. 60–1. Routledge (2017)
22. Sharma, J.: Empire's Garden: Assam and the Making of India, pp. 3–4. Duke University Press, Durham, NC (2011)
23. Phukkan, H.: Assam Buranji. Guwahati: Moksada Pustakalaya **1962**, 95 (1994)
24. Frenchman, D.: Event-places in North America: city meaning and making [Research and Debate]. Places **16**(3) (2004)
25. Mehta, V.: A toolkit for performance measure of public space. In: 43rd ISOCARP Congress (2007)
26. Quinn, B.: Changing festival places: Insights from Galway. Social Cultural Geograp. **6**(2), 237–252 (2005)
27. Bird, E.S.: It makes senses to us: Cultural identity in local legends of place. J. Contemp. Ethnograp. **31**, 519–547 (2002)
28. McCabe, S.: The making of community identity through historic festival practice: the case of Ashbourne royal Shrovetide Football. In: Picard, D., Robinson, M. (eds.) Festivals, Tourism and **Social** Change, pp. 99–118. Channel View Publications, Bristol, UK (2006)

Chapter 62
Implementation of Digital Techniques Process Through the Storyboard for Better Understanding in Visual Narratives

Yendrembam Suresh Singh and Bhaskar Saha

Abstract Storyboard is one of the design processes which include creativity, concept, and skill. Storyboard design is required in the form of drawings or illustrations for the content of animation production in the first stage of the production pipeline. It acts like a blueprint for the movies or series or animated clips and depicts the final design. In the production process, some creative directors, animators have the creative skill and can visualize a concept but lack the artistic skill of drawings to execute the storyboard. In this regard, the other team member or artist could not understand the concept of the director. This papers focus on those people who lacks the sketching skill and provides a technique that can be implemented to execute the concepts. For this, one survey was conducted on the design-based students studying in Central Institute of Technology Kokrajhar and experts. This paper shows that the techniques can also meet the demand of skills through the design thoughts, and the process of the methodology has given the new design technique to give a better production output for better understanding between the team members working on the project.

62.1 Introduction

The form of storytelling through images is one of the oldest form of artwork as is visible in early cave drawings, and more developed sequential art can be traced back to the Egyptian hieroglyph and the Bayeux Tapestries and the later are claimed as the first storyboard by some [1]. According to Hart [1], Storyboards is the pre-visualization or conceptual imaginary form of the written words or phrases and further defined in his research article that storyboarding process starts with the rough concept or idea sketches in the early stages of the development of the pre-production

Y. S. Singh (✉) · B. Saha
Department of Multimedia and Communication Design, CIT Kokrajhar, Kokrajhar, Assam, India
e-mail: Ys.singh@cit.ac.in

B. Saha
e-mail: b.saha@cit.ac.in

A. Chakrabarti et al. (eds.), *Design for Tomorrow—Volume 1*, Smart Innovation, Systems and Technologies 221, https://doi.org/10.1007/978-981-16-0041-8_62

pipeline. It shows that the story narratives are shown through a sequences of drawings or illustration instead of verbal or written form. This is the early form of the modern storyboard and is a very vital part of the pre-production part in animation, film, and series production process. Without this content, it is difficult to explain the concept to the animation artist who is working at the animation projects/film production, etc., in studio or industries. The production team consist of many expert professionals who will work together for a common goal, so an amazing storyboard for the film or series is a must for enhanced and better communication and transparency of concept. Hart [1] gives the credits to Winsor MacCay who established animated cartoon as an art form and paves the way for the other animators/studios and also mention about famous celebrated Director Alfred Hitchcock who also used storyboard for many of his successful films.

Mou [2] and his team specified that storyboarding is not a mere transformation of the story into series of visual images but it is an imaginative creative process to represent words into image sequences with the help of words, sound, music, and sketching effects. Georges Melies is known for using pre-visualization or imaging detailed sketched illustration for his films, and it is considered to be the first step of the dignified formalized storyboard [3]. Hitchock rarely looked through the camera viewfinder on the set, since it was merely a photographic equivalent of a storyboard that had been finalized earlier [4]. At the Disney Studio, Ub Iwerks is believed to have produced the first primitive storyboard for the short film's Plane Crazy, 1928 [5]. Pallant [3] and his group delivered in their research article that earlier form of storyboard in the Disney studio are sketches mounted on corkboards that are discussed and rearranged before the actual production. The turning point of the process comes when the storyboarding of Snow White and Seven Dwarfs, 1937 was created in the Disney Studio. It redefines the production process not only for animation film but also has an impact on live film as well [3]. William Cameron Menzies designed storyboard for the live film, Gone with the Wind, 1939, and is considered to be influenced by Disney [3]. Consequently, storyboarding continues to be an important part of the pre-production section of film making and is not limited to the medium of animation only.

It is already known that storyboard artist must not only has to be creative but also possess the essential artistic skills and thoughts capability to draw out those images through appropriate proper planning and thought process. According to Simon [6], storyboarding not only solves production problem which saves time and money but also can be used for testing an idea and also approached that different artist prefers their individual and specific choice of materials and standard structure in the development process. Different styles of storyboards require strong drawing abilities, methodical empathetic, and thoughtful of the cinema content [5]. The rough planning of visual thumbnail sketches of a good storyboard encompasses of trial and error methods which is a very creative part but challenging and time consuming. Glebas [7] describes the use of a beat board which are a series of single drawings that represents each scene of the movie before the actual storyboarding begins. Mou [2] and his team approached that, a good storyboard can be accomplished only by the experience and talented persons who learns with years of effort behind the art

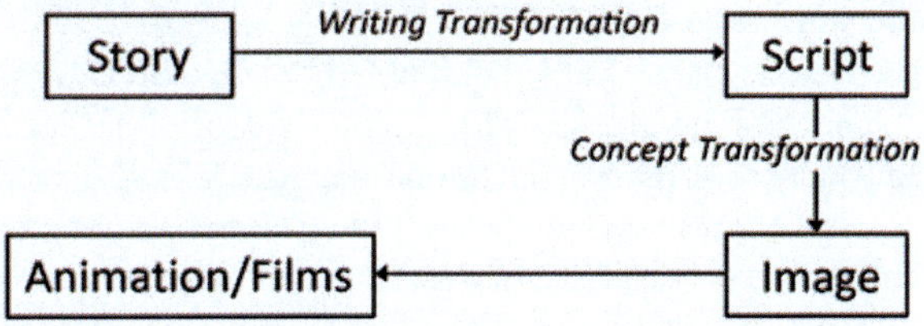

Fig. 62.1 Animation is transformation from words to imaginary

and researcher also defined storyboard as a visualized sequence of illustration made for the animated or live-action film. She also defined that the process of animation film making as a transformation from the word to the image/illustration as shown in Fig. 62.1 [2]. There are other basics fundamental that also needs to be considering at the time of the transformation. With the advance of technology, storyboarding has become digital while the traditional way of manual drawing also remains.

The storyboard is the visual model of the script and consists of a number of panels that show the visual action of a sequence in a logical narrative [8]. Miss Chatz [9] approached that storyboard is drawn on the panel (Fig. 62.2A), which represent the rectangular screen aspect ratio, in a sequence of drawings or illustration based on the story or script. The most common aspect ratios are 4:3 and 16:9. According to Dermott [10] and his team, a shot is a continuous sequence of frames filmed by one camera. Each frame represents an individual shot which can be a type of camera shot, angle, movement, action, or special effect, to effectively tell a story. Goldman [11] and his team specified that storyboards typically depict several "key" moments in the time span of a shot. The final storyboard is done in details starting with rough planning as shown in Fig. 62.2b and [9]. The focus is on the storytelling, general composition, actions, and camera motions rather than technical details [12].Chen [13] and his team defined that storyboard creation is nevertheless a challenging task even for professional artists, which requires cinematic view, relevant details, and visually consistent with coherent styles and characters across all images.

In the animation film production process, each member of the team is working on their specific role and so clarity of the concept is requires before the actual production. Storyboarding not only simplifies the understanding of textual stories with visual aids, but also makes following steps in story production go more smoothly via planning key images in advance [13]. Thus, the storyboard will serve as the backbone of the project where each individual will rely upon. Already referred to in advance, there

Fig. 62.2 (A, B, C), Existing design planning of storyboard [9]

are innovative and creative directors who have the vision and the talent but lacks the artistic skill. Hart [1] explained how David O. Selnick will deliberate and makes the necessitate changes along with the story boarder for the final implementation of production of the movie "Gone with the Wind".

Hart [1] referred to approximately Ben Wooten concept of operating collectively for a more powerful, effective, representative, and better design output. It is observed that in the production or studio there are digital animators, especially the 3D animators who cannot draw but have the techniques and the knowledge of tools of the trade. They confronted the hassle while explaining their ideas to other members of the team. Mao [14] and his team described about an intelligent storyboard, which enables each one who can draw, to "sketch-out" 3D virtual humans, and their animations, as well as intercommunication. But it will work only for those who can draw. Mou [15] described about the process of transformations from script to storyboard which requires years of effort, experience, and even talent to accomplish the work. "Good physical gags can't be written, only drawn. Gags are made funny by the expressions of the characters and the actions they go through..." [16]. Musburger [17] discussed the choices in production regardless of the technology we work within and must be comfortable. Haesen [18] approached about the tool that supports the communication in multidisciplinary teams by providing features to digitize a storyboard and to make connections between the storyboard and other artefacts.

62.2 Research Gap

Many studies have explored the various aspects of assistance for developing a good storyboard but an elegant solution for the complicated situation that arises due to lack of drawing skill in creating a storyboard remains. In this study, the questions about creation of a good storyboard and implementation of digital techniques to develop a better visual narrative without inadequate drawing knowledge was investigated. These were based on the feedback and discussion with professionals, animators and design students. Some question arises like how to find out an alternative solution for creating a good storyboard without having appropriate drawing skill.

62.3 Aim and Objective

The aim of the research is to develop an effective digitalized form of storyboard to support the professional designers and students with poor drawing skill. And the objective for the research can be describe into the following two points like to make better empathetic of the user or designer and new intangible method of digital tools and technology for creating better storyboard.

62.4 Design Methods and Procedures

See Fig. 62.3.

62.4.1 Sampling

There were 100 numbers ($n = 100$) participants who joined in this survey experiment which includes the industry experts like 2D artist, 3D artist, visual designers, and Directors from the entertainment industry field along with design based students and academicians. The evaluation indicates that design process technique could be implemented and resolve the drawing problem.

62.4.2 Methodology

The goal of the study was to verify whether digital based storyboarding using available digital content could be the solution for developing a better visual narrative. The questionnaire was created based on the basic need of the research and was carried out through survey to explore and find out the difficulties of drawing problems and the consistency of digital implementation. The validation of the design technique process is needed, so a survey was used to define the drawing skill problems and the implementation of digital techniques in visual narrative for solving the issue. The survey mode through questionnaire was used for the investigation as the participants consist of mixed professionals working in different areas or studios along with the design students from Central Institute of Kokrajhar as being an institute where students of

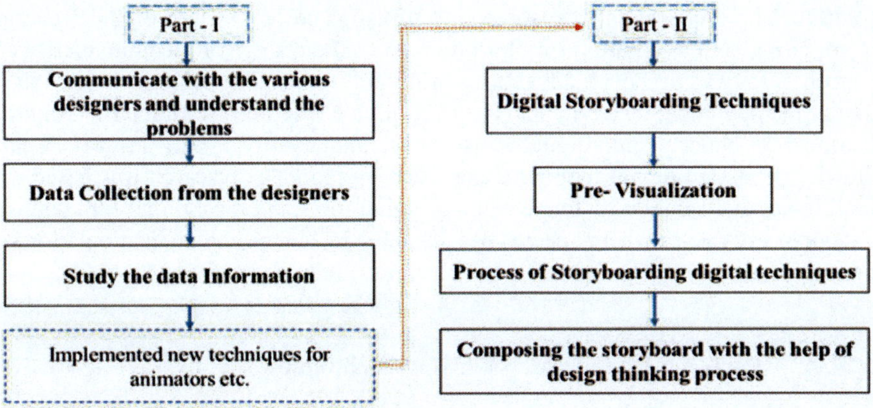

Fig. 62.3 Flowchart of design process method

diverse backgrounds studies. A total of 15 questions was used for the study in Part I. The first set of questionnaire was carried out to verify whether the lack of drawing skill is a genuine problem. The second set was to find out the possibilities solution related to the research topic. Based on the study, the design propose method was discussed and develop based on the point of view and unbiased input of the participants by communicating and conducting an experimental collaborative workshop for creating the storyboard through design thinking process with the implementation of the digital tools and technique which is shown in Part II.

62.4.2.1 Part—I: Data Collection from Various Designers

A survey was carried out with design students, educators and art teacher, and various industry experts which includes 3D and 2D artist, visual designers, UX designer, animation supervisor, production head, lighting supervisor, motion graphic artist, creative director, etc., to identify and validate if the drawing problem existed and acceptance of a new digital system to assist the process in production.

It was verified from the response of the first set of questionnaire from Questions 1 to 6 that were based to validate the need of drawing skill in storyboarding, significance of design thinking and visualization of concept, advantages of storyboard in production, ineffective storytelling due to lack of skill, knowledge of film language, and preference of paper work regarding authenticity. 3D artist, directors, and students were likely to be the most affected, and so there was a need for an alternative solution.

The phase discussed participant's acceptance and views on the digital forms as an original creation. However, artworks based on the traditional paper work seem to have more value than the digital ones because of its originality and its acceptance as a pure form of creation by many. The information gathered was verified and was taken into consideration for the follow up question from Questions 7 to 15 which were topic related questionnaire to understand the implementation of digital techniques to make things easier and more convenient as well as forming an elegant solution. These questionnaires were set to investigated on whether digital creation is not considered as a pure form of art, impact of the change of medium, clarity of concept through storyboard, advantages of digital over traditional process, trending digital manipulation, creativity and use of available resources from internet, digital platform for better output, digital tools and techniques to aid designers with weak drawing skill, and digital tools as a user friendly tool. The research was based on finding out that solution by implementing digital tools and technology for creating a modern storyboard where one can use available resources and its manipulation as per the requirement.

62.4.2.2 Part—II: Digital Storyboarding Techniques

In relation to these, a new system was proposed based on the feedback and views of the participants. Digital medium is gaining popularity due to its convenience, and the same techniques can be implemented in creating a new form of storyboard. The new digital technology system was discussed with professional animators and design experts to come up with an efficient solution. The system that was proposed is based on the unbiased feedback is called the digital storyboarding technique (DST) which is innovative and creative. The main problem for the participants is the creation of storyboard due to lack of conventional drawing skills based on the survey. Digital technique (DST) aims to bridge the gap that was uncovered by this survey. In this section, the digital techniques that can facilitate the visual narrative of storyboarding are described. The DST process was designed and developed by communicating designs professionals, and the user's feedback was taken for further improvement of the technique.

62.4.2.3 Pre-visualization

Storyboarding art is created on the blank panels and intends to convey visual information to the team members based on the story or script. Here, a new system involving digital manipulation in the creation was introduced. Before the final storyboard images, the artist usually sketches out the different layouts in the traditional process but as designers who does not have the skill of drawing can use the manipulation of available resources from the Internet. Let us take an example to create a storyboard for an action sequence with a simple scenario where the character jumps and hang on the branch of a tree. So, for this scenario, the character poses are required and the backgrounds can be taken from the internet. For the case, live photo shoots were taken and traced them in light box. These photo shoots were based on the needs of the story and will act as the initial thumbnail sketches that are usually done in the early phase of the creation of a traditional storyboard process.

62.4.2.4 Process of Storyboarding Through the Digital Techniques

The images are kept above the light box or either imported in digital software like Adobe Photoshop, Illustrator, Flash, Animate, Toon Boom for tracing. The traced images Fig. 62.4 are then assemble using digital platform for the final visual narrative.

62.4.2.5 Composing the Storyboard

For the final visual narratives, the traced images are arranged in a logical sequence within the blank panels for final presentation as shown in Fig. 62.5.

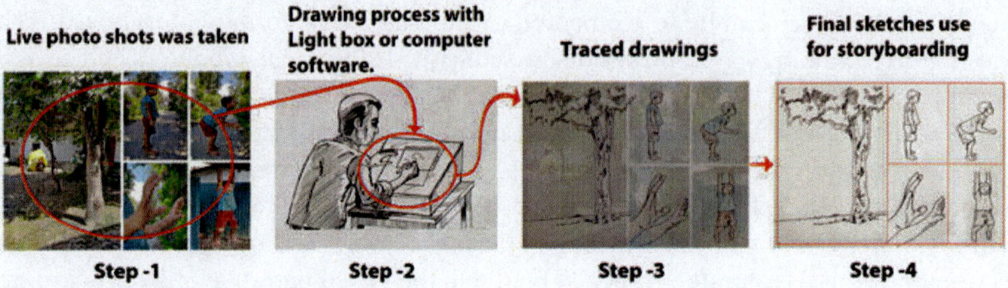

Fig. 62.4 Applying digital techniques process

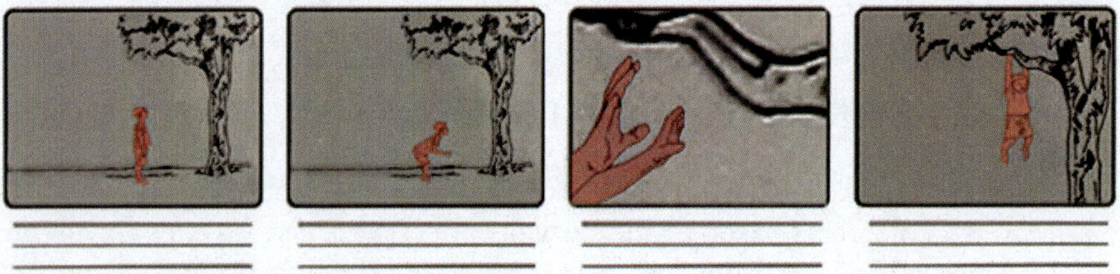

Fig. 62.5 Final storyboard which is made by using traced images and digital software

62.4.3 Result and Discussion

The following is the result analysis of the 100 participants ($n = 100$) that was taken based on the survey through questionnaire shown in Fig. 62.6. It was confirmed from the feedback data from the first set of questionnaire based on the basic information

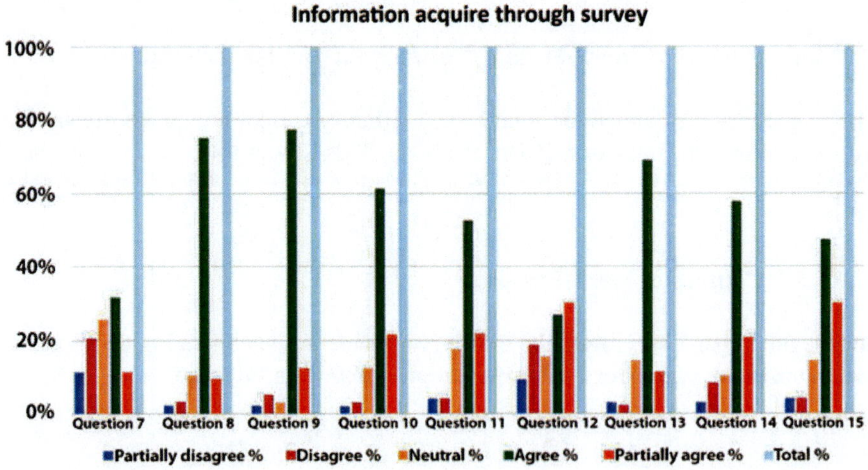

Fig. 62.6 Graphical representation of responses

and the participants agreed that drawing skill problems were faced by design students, 3D artist and creative directors. In this study, it was also found that apart from drawing skill, design thinking, and knowledge cannot be overlooked. And designers were also inquisitive about a brand new digital techniques system that will help and aid in the creation of storyboard.

Based on the inputs and the unbiased feedback from the respondents, the follow-up second set of questionnaire was digital technique and tools implementation that can enhance the storyboard in the later part. The study realized how designers regarded the creativity and advent of the storyboard with the use of digital techniques can be carried out in an effort to make things easier and more convenient for the storyboard artist.

The participant's feedback was taken into account, and the second set of questionnaire was survey through online platform, and here, it can be seen that majority of the respondent favor on the implementation of digital technique to make the storyboarding creation easier and convenient.

The seventh question was to validate whether designers preferred paper work for originality instead of digital to which 32% agreed, 11% partially agreed, 26% neutral, and 20% disagreed while 11% partially disagreed. The eighth question examined the participants that the storyboard will not be affected by the change of mediums and 75% agreed, 09% partially agreed, 10% neutral, and 03% disagreed while 02% partially disagreed. In the ninth question, clarity of the concept for the team through storyboarding was opined and 77% agreed, 12% partially agreed, 03% being neutral, and 05% disagreed while 02% partially disagreed. The tenth question investigated whether the digital has advantages above the traditional process to which 61% agreed, 21% partially agreed, 12% were neutral, and 03% disagreed while 02% partially disagreed. Based on the concept of digital content, manipulation practices which is becoming popular was discussed in the eleventh question, and 53% agreed, 22% partially agreed, 18% neutral, and 04% disagreed while 04% partially disagreed. In the twelfth question, whether the use of available resources restricted creativity was discussed and 77% agreed, 12% partially agreed, 03% being neutral, and 05% disagreed while 02% partially disagreed. The thirteenth question was to analyzed whether digital manipulation techniques helps in creating better output and here, 69% agreed, 11% partially agreed, 14% neutral, and 02% disagreed while 03% partially disagreed. In the fourteenth question, the participants were evaluated on the application of digital technique as a helping tools for designer with inadequate drawing skill and 58% agreed, 21% partially agreed, 10% neutral, and 08% disagreed while 03% partially disagreed. The fifteenth question was to substantiated whether the new digital tools are more user friendly than the traditional process to which 47% agreed, 30% partially agreed, 14% neutral, and 04% disagreed while 04% partially disagreed.

The findings further clarify the advantages of the digital techniques being superior and appropriate leading to its popularity in the field. The study emphasized the application of digital tools and techniques in modern storyboard.

Form the survey, it is clear that the new process can benefit both in the academic and industry/studios which will result in good productivity. Though there seem to be flaws in the process when we are talking about creativity, and so for some artist, the new techniques will be regarded as less ingenious and not very creative.

After validation of the problems and discussion with design experts communicated through Skype, email, and mobile phone, it was evident that requires the possibilities to be carried forward. Development of a new DST process and techniques of creation using the digital technology were the result from the overall response and feedback taken into account so as to preserve the creative thinking and design approaches in the new system as well.

The study shows prospective where the new process can be implemented in design studios and helps designers who in spite of acquiring strong vision lacks the drawing skill. The process can also help novice designer and students studying in the design institute who have the same problem as mentioned above. The proposed design will add a new dimension to the design field and can serve the need to quicken the production process in the entertainment studios and industries.

62.5 Conclusion

The DST process was introduced, and it will allow designers with insufficient drawing ability in communicating the visual narrative which can be better shown by the use of the digital techniques or strategies. The posing and the base model of the storyboard can be trace and compose in the digital platform. Therefore, it can be concluded that DST procedure or process will aid design students and professional.

Further study can be carried out in the near future as tools, and technology will get updated with time and more sophisticated advance tools may be introduced to enhance the productivity and find a better elegant solution. This digital technique is not meant to be a substitute for the traditional process but is meant to provide a solution for those designers who lacks the professional drawing skills.

References

1. Hart, J: In: The Art of Storyboard: Storyboarding for Film TV and Animation, Focal Press (1999)
2. Mou, T.Y., Jeng, T.S., Chen, C.H.: From storyboard to story: animation content development. Educ. Res. Rev. **8**(13), 1032 (2013)
3. Pallant, C., Price, S.: In: Storyboarding: A Critical History, Palgrave Macmillan (2015)
4. Katz, S.D., Katz, S.: In: Film Directing Shot by Shot: Visualizing from Concept to Screen. Gulf Professional Publishing (1991)
5. Paez, S., Jew, A.: In: Professional Storyboarding: Rules of thumb. Taylor & Francis (2013)
6. Simon, M.: In: Storyboards: Motion in Art, Elsevier Inc (2007)
7. Glebas, F.: In: Directing the Story: Professional Storytelling and Storyboarding Techniques for Live action and Animation, Elsevier Inc (2009)
8. Tumminello, W.: In: Exploring Storyboarding, Cengage Learning (2005)

9. Chatz, M.: How to Storyboard: A Basic Guide for Aspiring Artists, May (2018). https://design.tutsplus.com/articles/how-to-storyboard-basic-guides-for-aspiring-artists--cms-30962. Accessed Date on 30 March 2020
10. McDermott, S., Li, J., Bares, W.: Storyboard frame editing for cinematic composition (2018)
11. Goldman, D.B., Curless, B., Salesin, D., Seitz, S.M.: Schematic storyboarding for video visualization and editing. Acm Trans. Graph. (Tog). **25**(3), 862–871 (2006)
12. Shin, M., Kim, B.S., Park, J.: AR storyboard: an augmented reality based interactive storyboard authoring tool. In: Fourth IEEE and ACM International Symposium on Mixed and Augmented Reality (ISMAR'05) 2005 Oct 5, pp. 198–199. IEEE (2005)
13. Chen, S., Liu, B., Fu, J., Song, R., Jin, Q., Lin, P., Qi, X., Wang, C., Zhou, J.: Neural storyboard artist: visualizing stories with coherent image sequences. In: Proceedings of the 27th ACM International Conference on Multimedia 2019 Oct 15 (pp. 2236–2244) (2019)
14. Mao, C., Qin, S.F., Wright, D.K.: Sketching-out virtual humans: from 2D storyboarding to immediate 3D character animation. In: Proceedings of the 2006 ACM SIGCHI International Conference on Advances in Computer Entertainment Technology 2006 Jun 14, pp. 76-es. (2006)
15. Mou, T.Y.: creative story design method in animation production pipeline. In: DS79: Proceedings of the Third International Conference on Design Creativity, Indian Institute of Science, Bangalore (2015)
16. Simon, M.: In: Producing Independent 2D Character Animation: Making and Selling a Short Film. Taylor & Francis, Focal Press, (2003)
17. Musburger, R.B.: In: Animation Production: Documentation and Organization. CRC Press (2017)
18. Haesen, M., Meskens, J., Luyten, K., Coninx, K.: Draw me a storyboard: incorporating principles and techniques of comics. In: Proceedings of the 24th BCS Interaction Specialist Group Conference 2010 Sep 6, pp. 133–142. British Computer Society (2010)

Chapter 63
Influence of Sync Sound Recording Technologies on Aesthetics of Voice and Sound Design in the Hindi Cinema

Hitesh Chaurasia

Abstract Voice conveys information as well as emotion in the film. The film sound production starts with voice recording, and it informs the subsequent sound design. This paper examines the influence of sync sound recording technologies on the aesthetics of voice and sound design in the Hindi cinema since 2001. In the landscape of Indian cinema with the practice of dubbing, a unique aesthetics of voice evolved, privileging the human voice over all other sounds, imbuing it with 'God-like' acoustic properties. However, at the turn of the millennium, one observes a shift in the pattern of sound design with the re-emergence of sync sound recording technologies in the Hindi cinema. The present paper analyzes ten case studies of Hindi films through critical listening and comparative analysis of dubbed and sync sound films between 2001 to present. Observations from the analysis reveal that sync sound recording has caused a profound shift in sound design principles within mainstream Hindi cinema conforming more closely to global sound design aesthetics of realism.

63.1 Introduction

Sound design is a term used for recording, editing, processing, and re-recording sound design elements such as voice, ambience, sound effect, music, and silence to create the desired cinematic effect. In the film soundtrack, voice is of foremost significance as it conveys information and emotion through speech and non-speech expressions. The film sound production starts with voice recording on location; hence, it also informs the subsequent sound design. Michel Chion in his seminal book *Voice in Cinema* argues 'In actual movies, for real spectators, there are not all the sounds including the human voice. There are voices, and then everything else' [1]. This statement sets the hierarchy of voice and in turn, of voice recording technologies in film sound design.

H. Chaurasia (✉)
Associate Senior Faculty, Film and Video Communication, National Institute of Design (NID), Ahmedabad, Gujarat, India
e-mail: hitesh_c@nid.edu

A. Chakrabarti et al. (eds.), *Design for Tomorrow—Volume 1*, Smart Innovation, Systems and Technologies 221, https://doi.org/10.1007/978-981-16-0041-8_63

63.1.1 The Voice Recording Technologies

The talkie era started in the 1920s with synchronized voice recording along with the moving picture. It is also known as direct, live sound or sync[1] sound recording. In this technology, the voice is recorded with a microphone placed outside the frame of the camera. Hence, the voice changes along with visuals and reflects the authentic acoustical properties of the location. The ambience and sound effects also get captured along with the human voice. The same voice is used in the final soundtrack of the film.

In dubbing,[2] the voice recorded on location is replaced and re-constructed in sound post-production. Actors re-record their voice looking at the picture while listening to the original voice in a process known as post-synchronization. Since the voice is completely replaced, the acoustical properties and ambience recorded along with it are lost. The voice and sound effects recorded in the sound studio need to be processed and mixed to match the perspective as per the cinematic space.

Since the advent of talkies, sync sound recording remains the global norm of voice recording while dubbing became the norm in India from the 1950s onwards.

63.1.2 Sound Design Theories and Aesthetics of Voice

The debate about the role of sound and its technology also started with talkies. There was concern about the quality of sound recording and re-production. H.B Marvin in the journal of Society of Motion Picture Engineers (1928) defined a good sound system as,

> The chief requirements of a good system combining sound and pictures are: first.
>
> Synchronism between sound and picture: and second. Distortion-less recording and reproducing: That is a high degree of faithfulness of sound both in loudness and quality [2].

Marvin further explained the *high degree of faithfulness* as the re-production of perspective, tonality, spatially, and the sonic environment of the shooting space along with synchronized audio–video. His observations are in the context of film sound technologies but the idea of sound being *real* also reflect in the early film theories of Kracauer [3], Balázs [4], and Bordwell and Thompson [5]. According to James Lastra, the idea of *representational correctness, realism*, and *audience as an invisible auditor* shaped the early *Hollywood representational model.* He argues.

> The Hollywood representational model keeps the fidelity and representational.
>
> correctness of the dialogue as supreme, but the sound elements constantly change proportionate to the image in order to maintain a correct listening experience for the hypothetical audience [6].

---

[1] Abbreviation of on-set synchronized sound recording as popular in India.

[2] The popular Indian term for automated dialogue replacement (ADR) or Looping.

This idea of ‘realistic’ and ‘natural’ sound has defined the aesthetics of sound in western cinema; hence, sync sound remains the norm of voice recording while dubbing is used in exceptional situations.

63.1.3 Voice in Indian Cinema

Indian cinema also used sync sound in the early talkie era. With the arrival of post-synchronization technology, dubbing became the norm in India cinema by the 1950s. As, dubbing allowed Indian filmmakers to control the aesthetics of voice performance and manipulate its hierarchy with sound effects, ambience, and music to create the desired cinematic effect in the final mix. According to the sound designer, Andrew Belletty, who has worked in both Indian and western cinema, the close microphone dubbing technique was preferred in Indian cinema over sync sound recording ‘to create a sense of intimacy between the actor and the audience members’ [7]. Additionally, to match the on-screen avatar of ‘God-like,’ larger than life characters, excessive reverb, echo, and spatialization techniques were used, for example, the synthesized voice of Amitabh Bachchan in the film *Shehanshah* (1998).[3] This was further re-enforced with the use of loud sound effects such as thunder, rain, bangs, laughter, and music with minimal use of ambient sounds. The aesthetics of the sound design that evolved with such practices was *expressionist and melodramatic* as observed by Budhhaditya Chattopadhyay [8].

Ashish Rajadhakshya in his paper, *An Aesthetic for film sound?* (2007), highlights the lack of any theory on the aesthetic of film sound in India [9]. In the absence of any Indian sound design theory, scholars base their studies on western sound design theories. However, Belletty in *Cultural Distortion* argues that the Indian film sound needs to be examined in its larger cultural contexts. He claims.

> The Indian audience places little importance on the Hollywood representational model and its sound design theory of realistic representation and is more concerned with having a range of Rasas invoked by the soundtrack, than conforming to Hollywood notions of realism [7].

While Belletty’s argument holds in the context of aesthetics of dubbed films it needs to be re-assessed with the re-emergence of sync sound in films *Lagaan* (Ashutosh Gowarikar) and *Dil Chahta hai* (Farhan Akhtar) since 2001. The commercial success and Oscar nomination[4] of *Lagaan* encouraged Indian filmmakers to use the globally accepted norm of sync sound recording to further expand their market globally. This triggered a shift in sound design with mainstream Hindi cinema conforming more closely to the global aesthetics of realism. However, many Indian filmmakers continued to use dubbing while adapting to other modern film sound technologies such as digital audio post-production and surround sound mixing. This provides a unique opportunity to study Hindi films which differ primarily in terms

---

[3]Shehanshah is a 1998 Hindi Feature film directed by Tinu Anand.

[4]In the final nomination list of the Best Foreign language film category in 2002 academy awards.

of dubbing or sync sound. There is no systematic study available focusing on the influence of sync sound recording on the aesthetics of voice and sound design in Hindi cinema since the re-emergence of sync sound in 2001. The findings of the paper shall bring in a much-needed focus on the relationship between sync sound recording technologies and sound design in the field of film studies and sound studies in the Indian context.

63.2 Methodology

The methodology used in this paper is case studies and comparative analysis through critical listening of the film's soundtrack. Ten Hindi films from the last two decades were accessed via digital streaming platforms and from the author's collection. Hence, the analysis is based on a stereo sound mix.

63.2.1 Selection of Films

The selection criteria applied to a large number of Hindi films to narrow down on the films used in this study include the voice recording technology, audio post-production, and the format of the final soundtrack. This information was obtained from the films' credits and re-confirmed with the films' sound designers and film-makers. Ten Hindi films were selected for the analysis comprising five dubbed and five sync sound films of varying genres. Films spread across time are selected to represent the diversity of Hindi cinema over the last two decades and observe the shift in sound aesthetics from 2001 to 2019. In limiting the analysis to popular Hindi films, the objective is to keep parameters such as economics, film sound technology, and target audience similar across the selected case studies. All the film selected used digital audio post-production and mixed in surround sound, with only difference in terms of dubbing or sync sound recording. See Table 63.1 for the list of selected films with brief observations.

63.2.2 The Frameworks of Analysis

The voice in cinema and sound design is a vast subject of study beyond the scope of a single paper. Therefore, the analysis in this paper is limited to change in aesthetics of voice and sound design with the re-emergence of sync sound recording in the Hindi cinema since 2001. The framework of analysis is based on the sound design theories discussed in Sect. 1.2, especially Marvin's idea of *faithfulness*. Hence, the analysis is carried out for synchronization, pitch, tonality perspective, and spatiality of sound

Table 63.1 The list of selected films with brief observations

Set	Film	Director	Sound designer	Genre	Year	Voice recording	Observations
A	Gadar:Ek Prem Katha	Anil Sharma	Narendra Singh[a]	Period drama	2001	Dub	Period drama films released in 2001. Underlines the contrast of Indian aesthetics evolved over the years with dubbing against global aesthetics of sync sound
	Lagaan	Ashutosh Gowarikar	Nakul Kamte	Period drama	2001	Sync Sound	
B	Lage Raho Munna Bhai	Rajkumar Hirani	Bishwadeep Chatterjee	Comedy	2006	Dub	Comedy films by the same director and sound designer. Marks difference in sound design with dubbing and sync sound and shift of filmmakers toward sync sound
	3 idiots	Rajkumar Hirani	Bishwadeep Chatterjee	Comedy	2009	Sync Sound	
C	Rahasya	Manish Gupta	Jayant Vajpayee	Mystery thriller	2015	Dub	Thriller films are based on the same incident released in the same year. The difference in aesthetics of sound design narrows down in dubbed and sync sound films
	Talvar	Meghna Gulzar	Sajith Koyeri	Mystery thriller	2015	Sync sound	

(continued)

Table 63.1 (continued)

Set	Film	Director	Sound designer	Genre	Year	Voice recording	Observations
D	Dil Chahta Hai	Farhan Akhtar	Nakul Kamte	Comedy-drama romance	2001	Sync Sound	New voice aesthetics and use of site-specific ambience
	Saas Bahu Aur Sensex	Shona Urvashi	Andrew Belletty	Drama	2008	Dub	A new approach to dub voices and sound design in dubbed films
	The Lunchbox	Ritesh Batra	Michael Kaczmarek[b]	Romance	2013	Sync Sound Dub	A new taste of realism in voice and ambience, even with a partial dub
	Manikarnika: The Queen of Jhansi	Kangna Ranaut	K. Moran, N. R. Samal	Period drama	2019	Dub	A complete departure from the aesthetics of voice and sound design of dubbed films

[a]Credited as Sound Recordist
[b]Credited as Head of Sound and Re-recordist mixer

design elements such as voice, ambience, and sound effects. The present paper analyses only dialogue and vocal expressions, as voice over, narration or commentary is recorded in a sound studio even in sync sound films. Likewise, music is also excluded from the analysis. In this way, the analysis is a very focused study and by no means comments upon the film sound design as a whole. The comparative analysis of a dubbed and sync sound film is carried out in three sets of films. Set A includes films released in the same year and of the same genre, Set B includes films of the same genre, director, and sound designer, and Set C includes films of the same genre, story, and year of release. This is followed by individual case studies of four films in an additional set, D.

63.3 Analysis with Case Studies

63.3.1 Comparative Analysis

63.3.2 Gadar- Ek Prem Katha *and* Lagaan*: Period Drama Films Released in 2001*

Gadar represents the aesthetics that evolved with the practice of dubbing since the 1950s while *Lagaan* is one of the first mainstream films to use sync sound in 2001.

Voice

Gadar: The synchronization is missing at more than one place, for instance, dialogues of *Ashraf Ali* (Amrish Puri) when his house is attacked (08:00:00)[5] and the dialogues of people in the crowd at the Lahore Airport scene (01:35:38). Similarly, non-speech elements such as crowd reactions, breath, and grunts sound are disembodied and do not match the action. The voice is constantly in close-up ignoring the perspective of the shot, for instance, the voice of *Tara Singh* (Sunny Deol) and *Sakina* (Amisha Patel) is in close perspective in a long shot at (00:34:00) and (01:17:00). The room reverb is audible on more than one occasion in the open space, for instance, *Ashraf Ali's* voice has a heavy big room reverb while talking in an open makeshift tent. (02:54:00).

Lagaan: Being an authoritative character speaking a foreign language of Hindi, the British actor's voice is aptly calm, composed, and low pitched; on the other hand, the oppressed Indian villagers' spontaneous voice which is generally high pitched is portrayed as is. There is a change in voice as per the magnification of shots and acoustics of space. In one scene, *Lukha* (Yashpal) is hiding to save himself from an angry crowd inside a small village temple. His breath and crying reflect the acoustics of the small temple and the voice of *Bhuvan* (Aamir Khan) also changes as soon as he enters the temple (02:36:00). In terms of spatiality, reverb is audible only in interior spaces, justifying the vastness of King's palace (00:58:00), the headquarters of British Officers (01:22:00), and the small size of the village temple (02:36:00). The reaction of the crowd and player's breath and grunt sounds are embodied which creates the excitement of a live cricket match.

Ambience and Sound effects

Gadar: Background music dominates the film's soundtrack. Ambience and even the prominent sound effects like the truck and train sound are diffused and muted. The synthesized wind used in multiple scenes at different locations sounds similar. In the fight sequences, the punches, the body falls, and other foleys are loud and sound processed. In one of the scenes, *Tara Singh* extirpates the foley as the sound does not

---

[5]The timecode of the scene in HH:MM: SS, the same format is used in this paper.

reflect the true sound of the material of the solid metallic pump hitting the human body (02:15:51).

Lagaan: The winds used at the cricket ground, village temple, and the hillock have distinct tones and texture. Similarly, foley in the cricket match such as the bat hitting the ball and fielders falling on the ground reflect the materiality of objects and surfaces. The clapping of British officers has some reflection as they are sitting under the shade while the clapping of villagers sitting in the open has none. Such detailing creates an authentic and immersive soundscape for the audience. In one of the scenes just before the Janmashtami celebration at the temple, the chanting of the crowd, the flame in the plate, the flag fluttering, and temple bell sound situated in the space reflect the true nature of the sound (01:30:28).

63.3.2.1 *Lage Raho Munna Bhai* and *3 Idiots*: Same Genre, Director and Sound Designer

Lage Raho Munna Bhai (2006) represents the aesthetics of dub films with better synchronization, the tonality of voice. *3 idiots (2003)* underlines the difference in sound design with sync sound and the shift of filmmakers toward sync sound recording.

Voice

Lage Raho Munna Bhai: The film sound is synchronized except for minor out of sync bits. There are several scenes in which the voice is broadcast on radio and telephone, but the change in voice tonality displayed between the original voice and the broadcasted voice is minimal. The voices sound situated in the space, but remains in the foreground, for instance, in a scene *Munna* (Sanjay Dutt) hangs a man upside down from the balcony of a high-rise building in a long shot, yet the man's voice is in close perspective without any breath or struggle (00:54:00).

3 idiots: The voice changes as per the perspective, interior or exterior spaces, for instance, when *Pia* (Kareena Kapoor) is riding the bike and talking to *Rancho* (Aamir Khan), her voice is projected to overcome the surrounding noise of traffic (01:07:00). In the crucial childbirth scene, *Pia* is instructing *Rancho* on a video call and her voice is processed to match the texture of the laptop (02:19:00). The speech of *Principal Viru* (Boman Irani) reflects the acoustics of the open space surrounded by walls. (00:17:45).

Ambience and Sound effects

Lage Raho Munna Bhai: The sound design is background music-driven and the sound effects and ambience are diffused, for instance, in one scene where *Munna* and *Circuit* (Arshad Warsi) are riding a bike on a Mumbai road, the frame is full of vehicles, but the bike and traffic sounds are muted, and the projection in voice is also missing. (00:08:00). The foley such as punches, sound synthesized and does not match the materiality of the surface/object.

3 Idiots: Music dominates the soundtrack, however, ambience, and sound effects have been used effectively as well, for instance, in the childbirth sequence, the rain, water, and thunderstorm create a sense of emergency and anxiety. The foley and sound effects reflect the materiality of objects, giving authenticity to the procedure (02:19:00).

63.3.2.2 *Rahasya* and *Talvar*: Thrillers based on the Same Incident, Released in 2015

Both films are based on the Noida double murder case and voice play an important role as both films have long interrogation sequences. The difference in aesthetics of voice and sound design narrows down in dubbed and sync sound films by 2015.

Voice

Rahasya: The film sound is fully synchronized except for minor out of sync bits. Interestingly, Actor KK Menon's vocal performance changes as per scene/space in terms of both perspective and tonality while other's voice remains in close-up and their tonality does not follow the action or space, for instance, in the police station scene, *Inspector Malvade* (Nimai Bali) is eating chickpeas, but his voice does not reflect the same (00:17:00). The dialogues are situated in the space except for a few scenes where the reverb is used for dramatic effect, for instance, the courtroom scene (00:20:16).

Talvar: The voice is situated in the space and change as per the perspective. In one of the scenes, *Inspector Dhaniram* (Gajraj Rao) moves from a small study to the main room (00:24:00), and the voice changes within the same shot. Also, a clear change in voice tonality can be observed when he is chewing betel nut (00:08:45) and when he is not (00:17:18). In the Narco test sequence, the voice has been processed to reflect the room, laptop, and mobile phone texture. The voice sounds situated spatially, for instance, the reverb in the courtroom scenes (00:15:10). The reporters and media person's dialogues, reactions are random and overlapping creating a television news feel.

Ambience and Sound effects

Rahasya: In the exterior scene, the use of site-specifics ambience and sound effects create a realistic soundscape. The ambience changes as per space within interiors locations as well. At places, pre-recorded library sounds are noticeable for instance the same vendor call is used at two different times of the day in the slum (00:15:50) and the sound of a local train. The foley such as slap, punches, and footsteps are in close perspective in interiors yet reflect the materiality of objects/surface and sound situated in the space.

Talvar: In the film, a sequence is narrated multiple times from a different point of view, and each time the ambience changes accordingly. In the food street (00:03:30) and the last rites scene at the river (00:18:17), sounds are constructed with the actual location recording. Each room has a specific room tone, for instance, traffic rumble in the police board room (00:56:00) and electric humming in the interrogation room (01:21:00).

63.3.3 Individual Case study

63.3.3.1 *Dil Chahta Hai (2001)*: New Aesthetics of Voice and Sound Design

This film is one of the earlier mainstream films to use sync sound in 2001. It is set in present-day Mumbai and known for new aesthetics of sound design in Indian films.

Voice: The film has three main lead characters, and each one of them has a different voice texture. The voice sounds situated in the space and does not stand out. For instance, in the hospital scene, the voice of *Sammer* (Saif Ali Khan) and *Siddharth* (Akshay Khanna) is very low as they are seen through the glass (00:10:26). In another scene, *Sameer* and *Priya* (Suchitra Pillai) are jogging toward the camera. The voice with breath initially sounds distant gradually come to the foreground (00:24:58).

Ambience and Sound effects: Film has a site-specific sonic environment, for instance, traffic in Mumbai is full of horns and auto-rickshaw sounds while in Sydney the sound of cars passing silently is mixed with people chattering. In one of the scenes, *Sameer* (Saif Ali Khan) is trying to propose to *Pooja* (Sonali Kulkarni) in a garden in Mumbai in which there are constant crow calls in the background with traffic noise. In the Sydney underground train station scene between *Akash* (Aamir Khan) and *Shalini* (Preity Zinta) the voice, footsteps, and breath with the rumble, reflect the acoustical space.

63.3.3.2 *Saas Bahu Aur Sensex* (2008): New Approach to Dub and Sound Design

The author was part of the sound design process of this film as the sound editor. Sound designer, Andrew Belletty, used a shotgun microphone placed at a distance to avoid conventional aesthetics of close microphone dubbing. Actors were then asked to move and project their voices as per the perspective of the shot in the sound studio. This informed the entire sound design process including foley recording and mixing.

Voice: The film's sound is fully synchronized. Each character in the film has a different accent; their tonality and pitch also change. Voice follows the perspective of the camera, for instance, in the office scene where stockbroker *Feroz Sethna* (Farooq

Shaikh) is trying to stop *Binita* (Kiron Kher), his voice changes as he moves from his cabin to the hall through the corridor. The dubbed voices have been processed in the mix to be situated as per space, for instance, in the call center training room sequence, the voice of *Ritesh* (Ankur Khanna) has long reverb as he is addressing trainees (00:27:19).

Ambience and Sound effects: The sound design is site-specific and detailed. The office of *Feroz* is in the middle of a busy market in Mumbai. The footsteps and doors have a creaking sound as it is an old office with wooden flooring. There is a clear shift of ambience from the street to the staircase of the building to his office on the first floor (00:25:30). Even within his office, each room has a different soundscape. His cabin with an open balcony has a lot more traffic than the hall (00:30:10).

63.3.3.3 *The Lunchbox* (2014): New Taste of Realism

The film is shot with sync sound in Mumbai while certain parts being dubbed as per the requirement of the script. Sound design and mix of the film were done in Germany by Michael Kaczmarek and represent the global aesthetics of film sound design.

Voice: There is an exchange of letters between *Saajan* (Irfan Khan) and *Ila* (Nimrat Kaur) through the lunch box and the letter reading is dubbed. Also, Ila's neighbor *Auntie* (Bharati Achrekar) is never seen in the film and only her voice is heard; it was also dubbed. Despite dubbing for these parts, the voice sounds situated in the space, for instance, the voice of *Shaikh* (Nawazuddin Siddiqui), *Saajan,* and *Ila* changes as per space. (00:22:50) Also, the voice in the empty canteen has a lot more reverb (01:15:30).

Ambience and Sound effects: The soundscape of Mumbai, local train, office, house, and road is site-specific. The ambience changes from office to the canteen (00:11:00) and the tonality of foley in a crowded canteen and the empty one is different (01:15:30).

63.3.3.4 Manikarnika: The Queen of Jhansi (2019): Dubbed with Realistic Aesthetics

A biopic based on the life of the Queen of Jhansi Laxmibai. This is also a period film and represents the shift in aesthetics in the last two decades from *Gadar* in 2001.

Voice: The film is fully synchronized. The voice reflects the tonality, perspective, spatiality of the location. In one of the sequences, the British officer *Captain Gordan* (Edward Sonnenblik) visits the palace, and the dialogues with *Maharaja* (Jishu Sengupta) and *Manikarnika* (Kangna Ranaut) sound situated and reflect the space (00:28:00). In another sequence, *Manikarnika* and *Maharaja* are playfully fighting and with the change in space, the tonality and spatiality of voice also change (00:56:00).

Ambience and Sound effects: Site-specific use of ambience is prominent. In one of the sequences, *Maharaja* and *Manikarnika* are on the terrace of the palace and there are sounds of a distinct temple bell and prayer (00:43:12). The sound effects and the foley create the drama and evoke a mood in the final war sequence (02:03:03). Sound effects also reflect the space and materiality of the objects. The horse footsteps change as per the surface and perspective of the scene.

63.4 Discussion

The 2001 period film *Gadar* represents the aesthetics of voice and sound design that evolved with the practice of dubbing since the 1950s. In *Gadar,* the synchronization is missing in more than one place. Voice and sound effects do not reflect the space in terms of tonality, perspective, spatiality, and the materiality of objects. The voice is always foregrounded, ignoring the magnification of the shot; hence, the ambience is diffused to an inaudible level. The sound effects are synthesized and processed. In contrast, *Lagaan,* another period film released in the same year represents the global aesthetics of realistic sound design with sync sound recording. The voice and sound effects are situated in space. The foley, cheering of the crowd, the cry and shout of players creates an immersive experience similar to a live broadcast in cricket match sequence. Another sync sound film, *Dil Chahta hai,* set in urban India released the same year. The sound of crows and traffic which was earlier considered as noise is used effectively to create a site-specific soundscape. The voices sound situated in the space and do not stand out in the soundtrack which is a major shift in sound design from the traditional foregrounding of the voice of the 'Hero.' The success of *Dil Chahta hai* and *Lagaan* made the concept of realism popular in mainstream Hindi cinema. The synchronization improved even in a dubbed film like *Lage Raho Munna Bhai* and the subsequent films are near perfect. Gradually, filmmakers shifted to sync sound recording to capture the 'natural' performance of actors and to further expand their market globally, as in the case of Rajkumar Hirani with *3 idiots* (2009). Sound designers such as Andrew Belletty, experienced in the sound design of Western cinema changed the approach of sound design in a dubbed Hindi film, *Saas Bahu Aur Sensex* (2008). Likewise, the sound design and mix of *The Lunchbox* (2014) by Michael Kaczmarek done in Germany create a new taste of realism for the Indian audience with dialogue, letter reading, and voice of an invisible character that blends amidst the cacophony of Mumbai. The difference in aesthetics of sound design narrows down as reflected in films *Rahasya* and *Talvar* by 2015. Finally, with *Manikarnika* (2019) the analysis comes to full circle. This film is also a period drama like *Gadar* (2001) but reflects a complete shift in the aesthetics of sound design toward realism even in a dubbed Hindi film. The synchronization is perfect. The voice reflects the tonality and spatiality as per the perspective of the space. Occasionally, the voice is processed but is done as per the acoustics of the location. The sound effects also follow the same pattern and the detailing of each, and every sound is near perfection. The use of environmental sounds such as the sound of the wind, birds, and night cricket is distinct for each scene.

The sound design of a film gets informed by multiple factors such as genre, story, filmmaker, sound designer, technology, and larger cultural–social context. The sync sound recording is just one of the elements of film sound technologies yet the influence of sync sound recording is evident on aesthetics of voice and sound design over the last two decade as other film sound technologies; digital audio post-production and surround sound mixing, has remained the same in this time period.

63.5 Conclusion

Based on the findings of the analysis in this paper, it is observed that there is a significant change in the aesthetics of voice and sound design with the re-emergence of sync sound recording in the Hindi cinema since 2001. This new aesthetics situates the voice of the hero as 'human' and within a plausible cinematic space. The voice reflects the acoustical properties as well as the soundscape of the location. The voice projection, tonality, and pitch follow the change in perspective in visuals. Although the background music continues to dominate the soundtrack, the use of ambience and sound effects has increased to create a realistic soundscape for emotional impact and drama. The site-specific ambience and sound effects create an immersive experience for the audience through surround sound projection technologies. Many filmmakers continue to use dubbing, yet the synchronization has improved and a gradual departure from the 'God'-like voice aesthetics to 'natural' and 'realistic' global trends is evident.

The current paper is a preliminary analysis of the complex relationship between aesthetics and technology in film sound design which needs to be explored further. While the present paper analyzed mainstream Hindi films over the last two decades, this study can be extended to films in other regional Indian languages, independent films, web-series, advertisements, and other mediums. Furthermore, an in-depth analysis is required of individual sound design elements both diegetic and non-diegetic sounds such as speech, non-speech, voice over, narration, ambience, music, and silence in sync sound recording. The findings of this study can inform the aesthetics of voice in the human–computer interface and voice-based interface technologies to make them sound more 'natural' or 'human' and can further be extrapolated to 360-degree videos, virtual reality (VR), and mixed reality to create new modalities of immersive storytelling.

Acknowledgements I express my sincere gratitude to Dr. Andrew Belletty, Dr. Subir Dey, Dr. Nanki Nath, and Mr. Bishal Nath for their detailed feedback, support, and encouragement.

References

1. Chion, M.: The Voice in Cinema. Columbia University Press, USA (1999)
2. Marvin, H.B.: A system of motion pictures with sound. Trans. Soc. Motion Picture Eng. **12**(33), 86–102 (1928)
3. Kracauer, S.: Theory of Film. Princeton University Press, Princeton (1960)
4. Balázs, B.: Theory of the Film. Dover Publications, New York (1970)
5. Bordwell, D., Thompson, K.: Film Art, 8th edn. McGraw Hill, New York (2008)
6. Lastra, J.: Sound Technology and the American Cinema: Perception, Representation, Modernity. Columbia University Press, New York (2000)
7. Belletty, A.: Cultural Distortion: Rasa Aesthetics in Indian Cinematic Sound Design. Sydney, Australian Film Television and Radio School (2011). Accessed at https://andrewbelletty.com/writings/
8. Chattopadhyay, B.: Audible absence: searching for the site in sound production, Leiden University (2016). Accessed at https://openaccess.leidenuniv.nl/handle/1887/47914
9. Rajadhyaksha, A.: An aesthetic for film sound in India? J. Moving Images. Kolkata (2007)

Chapter 64
Protest as Performance: The Staging of CAA Protests

Neelakantan Keshavan, Saroj Boipai, and Devi Meghana Karanam

Abstract This paper employs a metaphoric reading of a political event of a protest from the perspective of design and space. It does this by treating the event as a performance, thus applying the notion of staging to the real life event. Protest stagings involve expressive spatiality and visualization. Visual images from the event are examined closely to reveal various vectors at work which upset the conventional opposition between performance and life; design and emergence. The method employs aspects of staging like actors, costumes, props, devices, scripts and dialogues to highlight the dramatic appropriation of everyday space for the sake of the event. The event creates spaces that do not otherwise exist outside the performance of democratic protest and pitches emergence against design and tactics against strategy. The widespread and intense protests that took place in India over the Citizenship Amendment Act in 2019–2020 period serve as the event under study.

64.1 Introduction

This paper examines the aesthetics of a political event by treating the event as a novel occurrence that makes visible things that would otherwise remain hidden. The focus is on how the event becomes visible in occupying and creating space. The method employed to conduct this examination is to read the political event, metaphorically, as a performance allowing the equation of space to a stage and the performance to the act of stagecraft. Spatial design, rooted in architecture and urban design, essentially is the deployment of space in relation to behaviours that occur within and outside it. From the perspective of design, what drives is paper is the desire to conduct a

N. Keshavan (✉) · S. Boipai · D. M. Karanam
Department of Design, IIT Hyderabad, Hyderabad, India
e-mail: neel@des.iith.ac.in

S. Boipai
e-mail: saroj.boipai@gmail.com

D. M. Karanam
e-mail: karanam.dm@gmail.com

A. Chakrabarti et al. (eds.), *Design for Tomorrow—Volume 1*, Smart Innovation, Systems and Technologies 221, https://doi.org/10.1007/978-981-16-0041-8_64

metaphoric reading of the 'usage' of a piece of architecture or spatial design. This paper will first set up the definitions of both metaphorical thinking and the political event (Sec 64.2). After describing the specific event under study (in Sec 64.3), the examination conducts some incisions into the body of the event to analyse public space. Section 64.4 shows how the event performance reinterprets space and extends staging in new ways. The conclusion uses these aspects of staging to revisit and discuss the meanings of design, emergence, performance and public space. The aim of the paper is to demonstrate a method of reading the shape of an event and to show how design as underlying structure and design as visible affordance collaborate.

64.2 Protest as Performance

Considered central to design in terms of the power of association, metaphors function as poetic jumps to gain insight into an object. This creative grasp of things is indirect and seems to stand in sharp contrast to the scientific approach, though this distinction has been severely contested by many [4]. Metaphors, with lightning speed make one thing appear as other things, allowing an insightful glimpse into the character of the thing which would otherwise be lost in a laborious analytic breakdown. According to Harman, objects have realities that cannot be understood by analytic breakdown or by seeing those objects simply as just smaller parts of a larger whole. According to him both approaches of undermining and overmining escape the face-to-face encountering of the object [12]. It is in this spirit of an indirect quick grasp of extra-routine behaviour that this paper revisits the metaphor of performance of a political event. Life as play or theatre has been used as a model to understand life [11]. The whole discipline of performance studies has been built on this porous boundary between performance and life as it unfolds [30]. In the case of protests, because of their primary agenda of visibility, performance is more than just a metaphor.

Protest is an interdisciplinary field of study from the perspective of motivations, impact and reaction that it can generate in the governing entities [33, 16, 17]. This paper examines instances or moments of an event from the not clearly separable and intertwined standpoints of design, urban design, aesthetics, visual culture, semiotics, and more importantly, performance studies. Performance studies as a discipline allow one to see the protest as an unfolding performance. Semiotics permits one to pay attention to signs and their rooting in referents and associational signs thus affording a reading of the signs as symptoms of deeper or larger events. Aesthetics homes in on the phenomenon of appearance—the modalities by which something makes its appearance. Urban design presents the framework of physical collective space of the city as a stage on which the performance takes place. Visual culture provides a reading of the event in terms of an activity of memory and miming. The perspective of design affords a pitting of the role of pre-thought intention as opposed to unpredictable emergence. Protests are visible spectacles in urban spaces [22]. They reclaim and create a space for various publics by agitating routine flows and literally hacking spaces entrapped in the everyday. According to Badiou, the event is that which creates radical shifts and unpredictable irruptions on the surface of the everyday [2]. It is in the sense of a visible expressive event that this paper examines instances from the event to grasp the shifts of meaning occurring in the event temporality. This paper focuses on the non-violent protests where expressions are critical but employ methods in which combat is not directed in grossly physical terms.

64.3 CAA Protests

The protests in India for and against the Citizenship Amendment Act (CAA) were unprecedented in India's recent history in terms of their scale, synchronicity and multiplicity. The active discourse and acute disagreement over the CAA were evidenced in the large scale and strong presence of university students evoking associations of protests from the times of the imposed emergency in the late 1970s. The recent CAA is based on an amendment which is only the most recent in a series of amendments ever since the institution of the Citizenship Act in 1955 [5, 6]. Since India does not have an immigration policy for refugees and scope for subsequent citizenship and has been working with long, ad hoc extensions to expiring visas, this amendment was pitched as an attempt to count its citizens. This amendment was acutely protested on the basis that it employed religion to discriminate against granting citizenship thus going against the Constitution. Those advocating the CAA claimed that it was meant to support the religious minorities fleeing religious persecution from the three neighbouring Islamic countries, while denying citizenship to Muslim refugees; the dodgy assumption being that those who have fled the neighbouring countries would legally have documents to prove the reason and fleeing and that they arrived in India on or before 31st December 2014 [9]. Some protests vociferously decried the act's discriminatory nature when it came to communities like the Buddhists who had fled Tibet and ethnic Tamils from Sri Lanka who have been living as refugees in India for years and would yet not be eligible to citizenship post-amendment. The National Register of Citizens (NRC) was an announcement to create a national document containing all the citizens in India. Such a register had already been implemented in the state of Assam, but that has been in the making for a few years due to migrant influx disrupting the habitat of the indigenous populace. The prospect of extending this to other states of the country was also aggressively protested by various parts of the country.

64.4 Staging

To reiterate the enquiry based on the premise of the protest as an event that is staged, the leading questions are: How does it literally 'take a place'? How does it occupy a visual-spatial field? How does the visual story appear and unfold in the image-documents in which it is registered? Staging involves the actual stage of the event, the bodily presence of the protestor, the objects that they display and how the protest is framed. Glimpses of the visual narrative of the protest are provided via sketches using the photographs of the event circulating on various media platforms. These are meant as purely illustrative impressions of the event and are not to be treated as documents.

64.4.1 Street-War

While a report in the Washington Post referred to 2019 as 'the year of the street protest', the Guardian called the 2010 s the decade of the street protest [37]. Streets form part of the infrastructures of urban flow, but, especially in urban India, streets are active public spaces due to the lack of enough designated public spaces and during the CAA protests they bristled with the snarl and bite of visual rhetoric (Fig. 64.1).

In a Public Interest Litigation filed against the Shaheen Bagh protestors for disrupting traffic, the Supreme Court asked the protestors to carry on protest activity in designated spaces. This raised questions related to land use planning which argued that when sufficient public spaces do not exist, how can protests occur in 'appropriate' spaces? Protest performances share semantics with Performance Art, Site-Specific Performance, Installation Art, Flash Mobs, and the more traditional Street Theatre. These performances are all political in terms of the fact that they set their performance beyond the pre-designated spaces meant for performance. There is a powerful element of appropriation, and at the same time, it demolishes the distance between the performers and audience opening it up to participation. Protest performances have characteristics of all these forms, but are not as tightly planned. They only have the semblance of a performance and in that sense are more 'folk' performances requiring no prior training except that of expecting the actors to assemble at a particular time and place and form a 'mob' or a collective body with placards. Accompanying this commonness of the folk figure of the protestor is also that of folk space made up of a plethora of images and instances of private spaces from which protests are beamed digitally. The public stage in this sense is a sort of patchwork quilt made of myriad

Fig. 64.1 Image on the extreme left illustrates the disruption of the everyday flow by protestors treating the whole street as a space to stage their protest (like in Shaheen Bagh), while the image on its right shows the usual flow of the street continue as the protestors take the median as a stage to enact their disagreement. The image on the extreme right shows the live streaming that marks the staging of protests

textures of various spaces. From the perspective of public space as a discourse, it is during the discourse or in the duration of the performance of a space being public that space is complexioned out of non-everyday meanings, sound and activity. On the contrary, from an urban design perspective public space is any non-commercial space of the city accessible to any dweller of the city, especially for leisure. Bhatia articulates public urban space as 'empty space' open to multiple gatherings and uses, with scope for varied meanings [3]. Public space in its' non-defined, not yet differentiated form, acts like an open stage on which different performances can occur. When a cluster of new meanings inhabit a space in those moments, in the life of that space, in that time-interval of duration, the space is active and reactive with heated voices, emotions, affect and meanings. According to Sassen, designed and defined public spaces demand the playing out of existing scripts without the scope for subversions. This is what performances which question power structures might need. According to her, the streets have a raw, yet unprocessed freedom as opposed to the ritual-bound European piazza-boulevard complex. While Bhatia laments the lack of designated public space, Sassen frames the street as an anti-establishment and a more real public space [29]. She coins the term the 'Global Street' to capture the similarities that street protests possess—including the non-western ones across the world, protesting for varied agendas. The semantic similarities between protests across the globe also lend to it a set of repeatable characteristics, a behavioural pattern which acts, in more serious parlance like a script and in a lighter, popular manner, a meme that gains in variation, strength and distribution via the repetition that it encourages across cultures and geographies. This is especially evident in the proclamations in the posters and the assemblage of material expression.

64.4.2 Street Gallery

A group of protests against the CAA released 10,000 black balloons from the streets of Delhi's Khureji Khas—similar to the protests in Central Park, New York, in 1969 against the Vietnam War [13]. The balloons made the urban proscenium vertical expanding the stage, creating a cloud billowing up, and a sign of aggregation made visible to many against the background of the sky. In another instance, in Shaheen Bagh, the protestors fenced-in the protest site, placed in rows, pairs of different kinds of footwear to stand-in for the many who could not be there in person due to the lockdown imposed due to the viral outbreak [19]. The precedent for this type of protest was seen in a climate change protest in France, where people sent or left their footwear with their names to protest in physical space [34]. These carve out a gallery-kind space, a stage for display, from the street. The balloons and the shoes are virtual stand-ins for the protestors. Similarly, a 35-foot map of India, fabricated out of sheet metal, by Pawan Shukla, uses the map as a canvas to display the text which states that we the people of India do not accept the CAA and NCR [1]. According to Shukla, the map of India has been 'held hostage' by a few in power and his is an attempt to reclaim the map as a space that equally belongs to the others (Fig. 64.2).

Fig. 64.2 Illustration on the left shows the India Map installation in Shaheen Bagh, the middle image is that of Rangoli protests and the image on the right shows footwear protests

Probir Gupta's installation, titled *A Poem of Instruments*, displays a large comb, typewriter, microphone and mixed grinder; the comb being a part of his memory of his own mother [24]. Probir dedicates this piece, displayed in the India Art Fair in Delhi simultaneous to the protests in Shaheen Bagh against the CAA, to the mothers of Shaheen Bagh. In an interesting contrast the organizers of the India Art Show disallowed halfway artists from 'performing' an art piece which was allegedly in protest of the CAA—a striking contrast with Probir's piece which is symbolically refined and in some ways so personal that is appears distant enough from the protest to be read as part of a heated protest [25–27]. A report in the Times described how protestors were detained for drawing anti-CAA Kolams [32]. Kolams, akin to Rangoli in the north of India, refer to the cultural South India wide practice of diagrammatic patterns drawn at the break of dawn at the threshold where the house or private residence meets the public or neighbourhood street. This is an everyday practice with religious, spiritual connotations of prosperity and cosmic goodwill for the residents. Usually drawn by women, it is a common sign of domestic practice which gain festive colour and size on special occasions. The protestors employed this cultural visual practice of floor drawing to weave in slogans against the CAA first in a public space in the Besant Nagar bus stand in Chennai. When the police disallowed them on grounds of inconvenience caused to others in the public space and failure to obtain permission for the activity, the protesters migrated to neighbourhood streets to draw up the Kolam based on permissions from the residents. The police then detained them for the inconvenience caused to the private realm due to encroachment. The protestors not only took the Kolam to the bus stand but also brought it back to the threshold of the house, as a diagram of protest. This incident points to how the everyday site specificity of the Kolam confined otherwise to neighbourhoods, was dislocated to a public space to become a politically visible sign of disagreement. Though the Kolam in its traditional avatar is mostly a visual image and only exhibits letter or word symbols with religious connotation, at times, in the protest Kolam, the protest was essentially contained in the letters or words adorning the visual image making it into a floor-poster. A minister decried the protest act as *alangolam* (ugly) as opposed to Kolam (beauty) [14]. The ritualized diagram-drawing can be thought in terms of graffiti when the white lines of the Kolam incorporate political protest and seek a canvas which is public. The event also points to the discomfort created between the private and public worlds, between the ordinariness of the everyday and the theatricality of disagreement, between a cultural practice and art, between tradition and subversion, in the act of hacking space via protest. This contrast raises questions as to whether galleried art can really protest or whether galleries will find enough semantic complexity within street protest art for gallery display. Assemblages render and amplify the protest via properties that have embodied the energy of the act of protest. The gallery space is a traditional space for interacting with art. So, when the streets carve out a space of display, they borrow the semantics of the gallery indicating that protest art is where the performative dimension of the protest is heightened.

64.4.3 Street-Speak

Amongst the usual rhetoric of the predictable heated, angry and direct assaults that protest slogans engage in were present a strange kind of placard text that projected cool, distant, essentially apolitical traits, demonstrating non-affiliation to any group, ideology or party and yet appearing in the particular protest confessing to the dire need of the tumultuous times. The three images in Fig. 64.3 evidence this kind of self-conscious apolitical trait amongst the protest articulations.

Though it might appear as a sign of middle class irony and sarcasm, Mishra aligns the Indian student protests with the 'anarchist' movement, comparing it to the 1968 student revolts in France [23]. His hypothesis is that as states become consistently repressive across the world, student and street protests countering these would increasingly become the norm. The act of coming together in a coordinated effort, in solidarity against authority should in itself be seen as a movement not necessarily aligned with the more conventional representative politics. Here Mishra echoes Sassen when he underlines that it is inappropriate to think of anarchist politics as just meaningless chaos pointing out that many of the protesting students are not easily identifiable in terms of political camps (like left-wing or right-wing). This seems to ring true as a universal phenomenon visible in the street protest images

Fig. 64.3 Placards which illustrate a certain distanced kind of articulation and approach in sharp contrast to the heat and anger of the predominant protest

from various parts of the globe. This casual trait is highlighted especially in a placard themed on the global furnishing and product giant Ikea: 'I've seen smarter cabinets in Ikea'. Circulating on Twitter and Facebook, amongst a swirl of protest images from Bandra, Mumbai against the violence that took place in Jawaharlal Nehru University in New Delhi that placard stands out due to its direct reference to Design via the mention of the Swedish Design Giant IKEA which is only a couple of years old in India. Though its appearance might be novel in an Indian protest site, it is an already well-circulated global meme from American and European street demonstrations. In the sense of being a clever and cheekily urban slogan with none of the heated revolt and anger that tends to dominate most of the posters it seemed appropriate to be found on the streets in the hip commercial urban milieu of Mumbai. IKEA in India reeks of middle/upper-middle class privilege unlike the European or American contexts in which it has an almost class-agnostic, common place complexion. In the poster, IKEA Design's everyday, functional and assembled smartness gets pitted against the clumsiness of the 'cabinet' made of political representatives pointing to the everyday, matter of fact performative elegance of IKEA design as opposed to the ill-performed cabinet of democratically elected political members. This global trait, though only a minor strain in the ocean of angry placards, stands out due to the noticeably strong contrast it generates against the dominant affect of the protest. This fringe-quality of a hesitant protestor, in terms of protest performance, points to a by-stander-like casualness; of a participation motivated only by the dire state of things. The laid-back attitude of an urban surfer or casual browser of places is an easy reminder of the behaviour of Benjamin's flaneur figure as he floats around in the city [7].

64.4.4 *Street-Guise*

One image which gained attention in social media and circulated widely was that of a girl costumed in a long black robe and hijab (headdress), holding a placard which read 'Mr. Modi, I am Indulekha, can you identify me by my dress'? [28]. This placard splits her image into a name and appearance which do not correspond; the visual rhetoric playing off the stereotype and its identity botching up the proposed state categorization of national belongingness in terms of religious identities (Fig. 64.4).

Fig. 64.4 Costumes to embody the protest. On the extreme right is the image from the stadium

An illustration, an example of play at multiple inseparable identities including gender, showed the Indian goddess Kali wrapped in a *hijab* (headdress) with the caption that 'Women will destroy the Hindu Rashtra' sporting a headnote clarifying that by Women, they meant 'anyone who identifies as a woman' (Fig. 64.4). One controversial image attempted to morph a red Nazi swastika symbol into an Om by breaking one of the four limbs of the Swastika to create an Om in Devanagari script [36]. This identity play was enacted differently in Shaheen Bagh when a burqa clad woman amongst the protesters was caught asking questions and filming the protestors. The police ascertained the identity of the detained woman to reveal that she was actually a YouTuber who had costumed herself in a burqa disguise as a protestor in order to gauge and report the situation of the protest movement from the inside [25–27]. In a humorous instance of semiotic misreading, stadium authorities in Wankhede stadium in Mumbai allegedly restricted cricket match entry, in accordance with instructions from the city police, to spectators wearing black t-shirts assuming they were protestors [25–27]. The protesters created a disjuncture between the symbol and its performance by slipping in wearing white t-shirts to form a protest-line in the stadium stands. Even political campaigns are subject to ridicule and scrutiny over the Internet due to visual gaffes—the UKIP party in the UK, for instance, were sharply criticized for the superficial gesture of employing an actor to play a destitute worker in one of their campaign posters of how the EU policy has hit the British workers hard [31].

This play of identity and appearance shows up in social media warnings advising protestors to costume themselves in masks and make-ups to avoid being tracked by facial recognition technology in possession of the police. Activists from the Internet Freedom Association complained to the Delhi Police Commissioner and the Internet Freedom Foundation (IFF) against the use of facial recognition technology and video surveillance to identify protestors rallying against the legislation [20]. These instances highlight the use of playful subterfuge in the appearance and expression of the protest upsetting the partition between the performative and the non-performative space by mixing reality and shifting identity. This play occurs through still and moving images in the field of vision of the camera.

64.4.5 Street-Vision

In a video song by Sumit Roy (though this is not really a CAA protest song, he has been actively responding to the CAA/NRC), he puts the home-space to use as a stage especially by referencing it in the video through animated white lines which bring forth parts of his domestic habitat to the narrative frame. This is prototypical of similar non-descript environments travelling through the web and appearing on its various platforms, joining up in curatorial hash-tag to become part of the stage on which performance of protests are framed—the privacy of the home as stage for public space through the camera's capture. The CAA protests were also mediated through the multitude of frames clicked using the cell phone, from within the thickness of

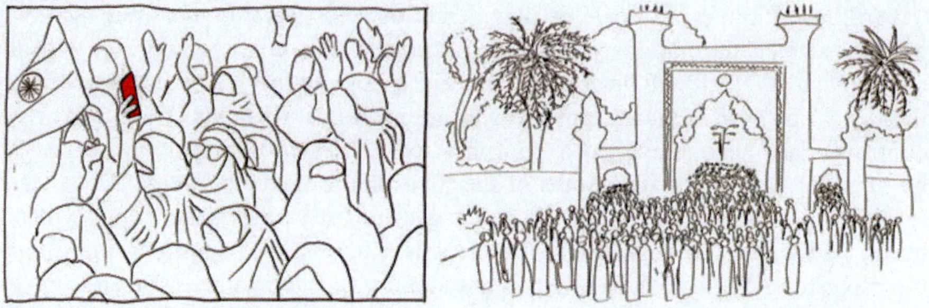

Fig. 64.5 Image on the left shows the cell phone perspective and the one on the right shows the perspective from the drone

mass protests. In addition, an interesting contrast to the close-up, a new kind of protest image also surfaced—that of the anonymous sweeping drone shot. Protestors complained that the Directorate General of civil aviation's guidelines were being flouted by the Delhi police's use of drones to monitor public space and behaviour. The cell phone camera's immediacy and real-time visibility along with newspapers and other modes affect policy at an everyday level influencing decision-makers in real time. The omnipresence of the eye of the camera and what it is framing is a dominant factor in the theatre of visibility. Anthony McCosker points at the camera-consciousness of protests in the drone-era [21]. Mccosker contrasts the mobility of the cell phone camera (in the thick of protests) to the mobility of the drone camera (sweeping top views). Drone activity and imaging he says are part of the contested 'vertical public space' (Fig. 64.5).

Camera Ocupa argues for taking images as seriously as words especially when it comes to youth cultures [10]. The cultural gap between the photographer as ethnographer and photographer as protestor suggests a retroactive mining of this gap between the two for insights. According to Gaby David, notions of intimacy, privacy and relationships between private and public, individual and collective, memory and experience co-evolve with emerging technologies. The camera, especially, he says blurs these boundaries and creates a feeling of life as fiction. Life streamed he called it—the instantaneous upload of events in life which smudges the line between the actual and the performed [8]. The camera is dislocated vision because it brings many eyes from various spaces and times, asynchronously, to focus on a particular incident occurring in a specific space-time. Both the drone camera and the cell phone camera create new molar and molecular levels of staging and performing the protest. The aim of the protest is to make a defiant expression visible. The currency of a protest in public space is the real space it traps which can be used to negotiate with the powers [18]. The fact that it is being enacted in real space needs to be made visible to the multitude and to the state.

The street is a site for warfare, rhetorical speech, expressive props and dramatic vision in its being part of the staging of protest. The design of the street is hacked to script performances. The performance model offers scope to bring into sharp contrast the two senses of performance in design.

64.5 Conclusion

The use of the urban space for a socio-political event affords a reading or an interpretation of space as a stage. From the perspective of performance, public space is used as an expressive entity, light in terms of design and script, but strong in terms of the power and scale of expression. The city-user as a protestor taps into the affordances offered by the features of existing structure—like roads, kerbs, footpaths, lawns, medians, houses, stadiums, airspace and traffic. They create a lightly designed choreography and communication to weave the site and script together. Features which would otherwise remain submerged in the ordinary become props for the spectacle. In this sense, the usage points to both conflict and collaboration between the performative and the non-performative aspect of the design of urban space. It brings these two senses of use beside each other. One is that of the use of the city as an 'infra' structural, background system maintaining smooth flow and the other is that of its use for visible expression. In this sense, design acts as both a background 'stage' and as radical 'staging' where the background becomes an active part of the theatre of making visible, where it goes beyond stage to become the performance itself.

This paper has been an exercise, a demonstration, of the possibility of using metaphor in more than poetic terms, to conduct an analytic reading of the non-ordinary use of a design. This demonstration has been motivated by the desire to systematically apply metaphor to spatial design behaviour, without reducing it to just a spatial mapping between theatre and urban space. While the purpose of the paper has been methodological and processual, it concludes with a set of subjective impressions that need further pursuit by pointing to innovative user-centric behaviour, in terms of using the street and its spatiality, its aural and oral theatricality, and its perceptual and optical visuality. Though metaphors in design thinking and practice tend to be treated with tool-like reverence, why is it uncommon to come across the methodological application of metaphor to produce new readings of design? Does not metaphor allow the possibility of 'seeing' affordances that would otherwise not be visible?

Acknowledgements We would like to thank Arun M, International Centre for Free and Open Source Software, Kerala, for providing critical insights into digital public space. We would also like to thank Prof. Nina Sabnani and Prof. Uday Athavankar, both from IIT Bombay, for providing critical inputs in shaping the paper.

References

1. Abbas, Z.: 35 feet 'Giant India map' comes up at Shaheen Bagh (2020, January 17). Retrieved from www.millenniumpost.in: http://www.millenniumpost.in/delhi/35-feet-giant-india-map-comes-up-at-shaheen-bagh-396178
2. Badiou, A.: In: Being and Event. Continuum International Publishing Group (2006)
3. Bhatia, G.: Cities for public enjoyment: where are the spaces in Indian cities that can take us away from daily routine? (2019). Retrieved from www.timesofindia.indiatimes.com: https://timesofindia.indiatimes.com/blogs/toi-edit-page/cities-for-public-enjoyment-where-are-the-spaces-in-indian-cities-that-can-take-us-away-from-daily-routine/
4. Brown, K.: In: The measured Word: On Poetry and Science. University of Georgia Press (2001)
5. Citizenship Amendment Act. (2019). Retrieved from The Hindu. https://www.thehindu.com/topic/citizenship-amendment-act/
6. Citizenship Amendment Act 2019: All you need to know.: (2019, December 15). Retrieved from Livemint: https://www.livemint.com/news/india/citizenship-amendment-act-2019-all-you-need-to-know-11576401546515.html
7. Coates, J.: Key figure of mobility: the flaneur. Soc. Anthropol. **25**(1), 28–41 (2017)
8. David, G.: Camera phone images, videos and live streaming: a contemporary visual trend. Visual Stud. (2010)
9. ET.: Citizenship (Amendment) Act 2019: What is it and why is it seen as a problem (2019). Retrieved from www.economictimes.indiatimes.com:https://economictimes.indiatimes.com/news/et-explains/citizenship-amendment-bill-what-does-it-do-and-why-is-it-seen-as-a-problem/articleshow/72436995
10. Fogarty-Valenzuela, B.: Camera ocupa. Current Anthropol. (2020)
11. Goffman, E.: In: The Presentation of Self in Everyday Life. Harmondsworth London (1978)
12. Harman, G.: In: Object-Oriented Ontology: A New Theory of Everything. Penguin UK (2018)
13. Hashmi, R.: (2020, January 22). Retrieved from www.siasat.com: https://www.siasat.com/caa-nrc-protest-10000-black-balloons-released-khureji-1800685/
14. Hindu, S.C.: Women, lawyers held for anti-CAA kolam protest. The Hindu (2019)
15. Hindu, S.R. (2020, February, 6). Retrieved from www.thehindu.com: https://www.thehindu.com/news/cities/Delhi/shaheen-bagh-protesters-hand-over-burqa-clad-youtuber-to-police/article30746550.ece
16. Johnson, V., Freeman, J.: In: Waves of Protest: Social Movements Since the Sixties. Rowman & Littlefield Publishers, Inc. (1999)
17. Johnston, H.: In: Culture, Social Movements, and Protest. Routledge (2009)
18. Juris, J.: Reflections on#occupy everywhere: social media, public space, and emerging logics of aggregation. American Ethnol. 259–279 (2012)
19. Khan, F.: Shaheen Bagh marks janata curfew by holding 'symbolic protest', petrol bomb hurled at it (2020, March 22). Retrieved from www.print.in: https://theprint.in/india/shaheen-bagh-marks-janata-curfew-by-holding-symbolic-protest-petrol-bomb-hurled-at-it/385612/
20. Mandavia, M. (2019, December 31). Retrieved from www.economictimes.indiatimes.com: https://economictimes.indiatimes.com/news/politics-and-nation/activists-rally-against-illegal-surveillance-of-caa-protests/articleshow/73039535.cms?from=mdr
21. McCosker, A.: Drone vision, zones of protest, and the new camera consciousness. Media Fields J. (2015)
22. McGarry, A.: In: The aesthetics of global protest: visual culture and communication (protest and social movements). Amsterdam University Press (2019)
23. Mishra, P.: The global wildfire of street protests has finally reached India (2019, December 19). Retrieved from The Print: https://theprint.in/india/the-global-wildfire-of-street-protests-has-finally-reached-india/337515/
24. Ornella, D.: Art activism at Delhi's India Art Fair (2020, February 22). Retrieved from www.newindianexpress.com: https://www.newindianexpress.com/cities/delhi/2020/feb/02/art-activism-at-delhis-india-art-fair-2097734.html

25. PTI.: (2020, January 14). Retrieved from www.timesofindia.indiatimes.com: https://timesofindia.indiatimes.com/sports/cricket/australia-in-india/1st-odi-anti-caa-protest-at-wankhede-stadium/articleshow/73253623.cms
26. PTI.: (2020, February 5). Retrieved from www.business-standard.com: https://www.business-standard.com/article/pti-stories/burqa-clad-woman-identified-as-youtuber-gunja-kapoor-detained-at-shaheen-bagh-120020501075_1.html
27. PTI.: Delhi Police interrupts India Art Fair after complaint of anti-CAA paintings (2020). Retrieved from www.firstpost.com: https://www.firstpost.com/india/delhi-police-interrupts-india-art-fair-after-complaint-of-anti-caa-paintings-artwork-was-about-indias-women-clarify-participating-artists-7993451.html
28. Rakesh, K.M.: (2019). Retrieved from www.telegraphindia.com: https://www.telegraphindia.com/india/girl-who-unmasked-the-modi-dress-detector/cid/1729569
29. Sassen, S.: The global street: making the political. Globalizations (2011)
30. Schechner, R.: In: Performance studies: An introduction. Routledge (2017)
31. Shariatmadari, D.: (2015). Retrieved from www.thegaurdian.com: https://www.theguardian.com/commentisfree/2015/jan/05/political-posters-that-backfired-why-you-cant-fool-the-internet
32. Sivakumar, B.: (2019). Retrieved from www.timesofindia.indiatimes.com: https://timesofindia.indiatimes.com/city/chennai/kolam-protest-against-caa-and-ncr-spreads-in-tamil-nadu/articleshow/73024170.cms
33. Suri, J.: In: Power and Protest: Global Revolution and the Rise of Detente. Harvard University Press (2009)
34. Times, N.: With Marches Banned, Shoes Carry a Message (2015). Retrieved from www.nytimes.com: https://www.nytimes.com/interactive/projects/cp/climate/2015
35. What is CAA, Citizenship Amendment Act? What is it and why is it seen as a problem? Everything you need to know (2019). Retrieved from The Economic Times: https://economictimes.indiatimes.com/news/et-explains/citizenship-amendment-bill
36. Yaajnaseni.: (2019). Retrieved from Swarajyamag.com. https://swarajyamag.com/news-brief/as-anti-caa-protesters-desecrate-om-as-a-nazi-symbol-heres-how-its-actually-islamism-that-idealises-hitler
37. Yi, J.S., Caryl, C.: The year of street protest (2019). Retrieved from Washington Post. https://www.washingtonpost.com/graphics/2019/opinions/global-opinions/2019-was-the-year-of-the-street/

Chapter 65
Reconstruction of Vanishing Indigenous Cultural Threads of Naamghar in Assam

Charu Monga and Amarendra Kumar Das

Abstract Rapid modernization taking place in the twenty-first century especially in developing countries like India having multiple religious faiths has seen major changes in social, economic, and environmental aspects. With these rapid changes, a major concern and challenge is to retain identity of sociocultural institutions found in India. Design details of architecture and assets used in these institutions with religious/social significance are hence, an important feature, which reveals the indigenous identity of these institutions. This paper focuses on understanding and documentation of the identity of *'Naamghar,'* a social–cultural institution associated with *Vaishnavism* found in Assam so that the essence of it as identity of *Vaishnavism* can be institutionalized to facilitate translating it for Naamghar to be established in future. It focuses on three major aspects of indigenous identity of *Naamghar*: Visual design details of the exterior and interior of a *Naamghar*, Relation of design details with *Vaishnava* faith, and spatial and temporal evolution of *Naamghar*. This paper unfolds the evidence-based research and results toward structured design approach through form study with typicality analysis in *Naamghar* of urbanizing Assam and its fragmented areas in different parts. This paper discusses and shows systematic way and how a unique identity has been transformed over the period of time in main gate and doors of *Naamghars*.

65.1 Introduction

The *Vaishnavite* Saint *Srimanta Sankardeva* between fifteenth and sixteenth century AD introduced *Naamghar* in Assam. *Srimanta Sankardeva* was a great social reformer. He initiated new ideas, and it gave birth to a new religious practices nurturing simple believable faith by the people. It is an institution associated with

C. Monga (✉)
Indian Institute of Technology Delhi, Delhi, India
e-mail: charum.design@iitd.ac.in

A. K. Das
Indian Institute of Technology Guwahati, Guwahati, India
e-mail: dasak@iitg.ac.in

A. Chakrabarti et al. (eds.), *Design for Tomorrow—Volume 1*, Smart Innovation, Systems and Technologies 221, https://doi.org/10.1007/978-981-16-0041-8_65

the basic building blocks of Assamese society. It is not only a place of worship but also a central meeting point where all the people discuss the issues related to their society. *Naamghar* function in the society as the central meeting points of congregations, as well as a place for theatrical and dramatic performances like *Bhawona, Raas lila,* and many other cultural activities. The *Naamghar* is also called as *Kirtanghar*. These are the central core structure of *Satra* of Assam (monasteries of the *Eka Sarana* religion). Also one can find these wide spread across Assam in almost every village. One can find more than a single *Naamghar* in some of the villages, which signifies many congregational communities (Bhattacharjee 2006). The design of the *Naamghar*, which is the core of Satra institution, has been also influenced by various external tribal cultures, which had travelled from different parts such as Bhutan, Thailand, Mongolia, and other parts of East Asia [5]. The movement of people [2] from different neighboring parts around Assam influenced many changes in design of Satra and Naamghar.

65.2 Background Study

65.2.1 Structure and Position in Naamghar

The *Naamghar* is a place of worship, a prayer hall, a community hall made by *Srimanta Sankardeva*, the great poet saint of Assam. The *Naamghar* is usually a rectangular building with a gable, giblet or a hip roof raised on pillars the length of which is aligned in the East–West direction Figs. 65.1, 65.2, and 65.3.

The traditional thatch roof has been replaced in the late twentieth century by corrugated tin sheets, and the timber or bamboo pillars by concrete ones. This hall usually has an outer wall, with the main entrance at the west end. There is a verandah around the outer wall with windows and minor entrances on the north and south sides. Two parallel rows of pillars usually run along the length, with the North East most pillar, called the *Lai Khuta*, which name has its roots in Tai language (which is the

Fig. 65.1 Overview of *Naamghar* in India. (Researcher)

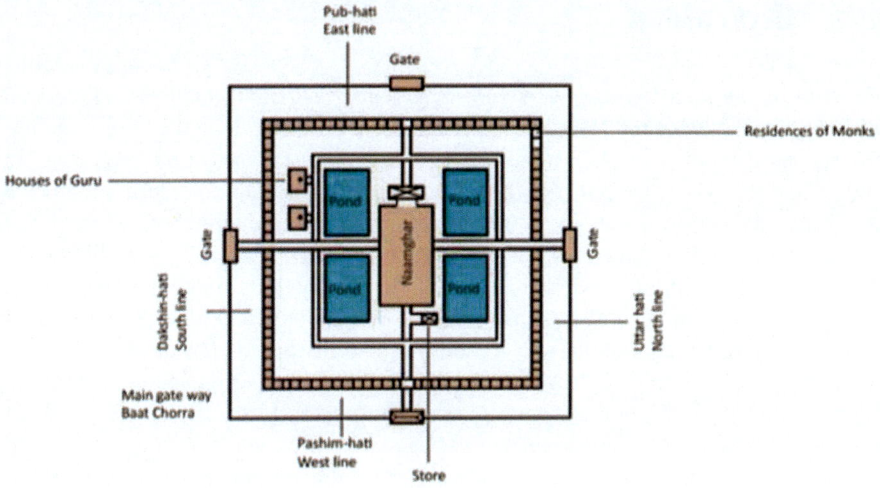

Fig. 65.2 Plan view of *Satra*

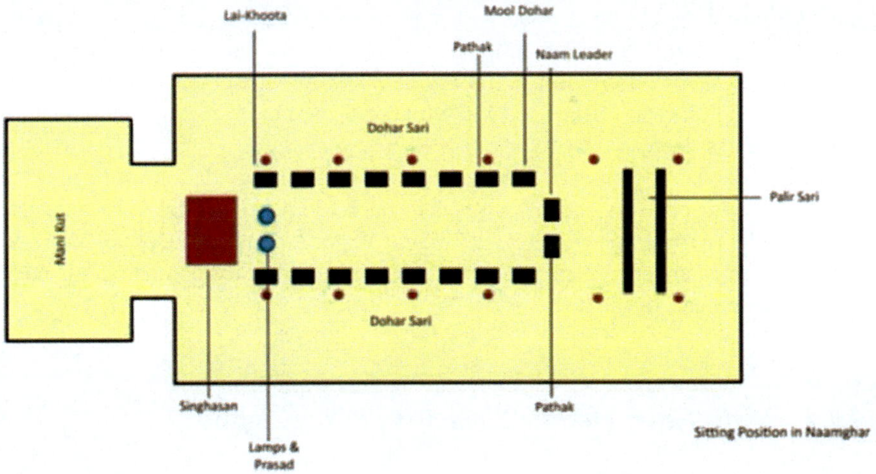

Fig. 65.3 Top view of *Naamghar* .(Researcher)

main language of Ahom people) of Ahom kingdom [4], carrying special significance. The floor is traditionally mud, replaced in recent times by concrete.

The east end of the hall does not in general have doors and windows, though very often it leads to an independent room called *Manikut* (the jewel hut), also called *bhajghar* in western Assam, with its own roof. It houses idols representing the worship of God, or a *Guru-Asana* (the Guru's seat). It is fully walled, with either no windows or small ones and also function as a repository of important articles. This room is a later addition to the basic *Naamghar* structure.

65.3 Methodology

The aim of the study is to understand and identify the main elements associated with gates of *Naamghar*. Overall approach is qualitative depicting the connotative meaning at each level. Primary and secondary researches have been done to understand the whole system. Further, methodology for data collection includes a comprehensive multiple set of visual survey and questionnaire (both at field and in studio level). For data analysis, softwares such as photoshop and illustrator were adopted. Literature review, analysis, field survey, data generated using field survey/questionnaire as well as final analysis of data have been done. Form analysis of main gate in context of social, cultural, and historical significance has been conducted with the help of documenting the gates of 26 Naamghars and extracting the form out of it. *Naamghars* were selected purposefully so as to cover all three regions such as rural, semi-urban, and urban region. In addition, they were also selected based on their chronological history of establishment. The intention was to cover *Naamghars* that shows range of establishment from 600 years old to 3 years old. This helped to provide a broader overview of the variations in design elements that were taking place in both spatial and temporal aspects. Once the data has been collected, sampling has been done with the help of card sorting and segregation process. Images were compared and visually analysed to identify and compare the forms and aesthetics of key design (on the gate and door) elements of *Naamghar*. Categorization theory has been used and with the help of card sorting, and typicality analysis data has been analysed.

Mapping of visual elements and categorization of various data have been done and also specific code has been assigned to each element (image). The data has been divided into rural, semi-urban, and urban category based on the population and location of the established Naamghar. Further, typicality analysis method has been used for the analysis of the data.

65.3.1 Main Gate: Design Elements of Naamghar

The *Naamghar* is a large open hall for the purpose of offering mass prayer and originally constructed with bamboo, reeds, and thatch. It has gabled roofs, the western facade being apsidal. A *Naamghar* of the ancient type (the modern *Naamghar* is simply an open rectangular hall) consists of a nave and side aisles with rows of wooden pillars separating the nave from the aisles. *Satra* has four gates on different sides but the main entry point Fig. 65.4 is the West line (*Pashim* side). It has many elements, which makes its identifiable in first glimpse.

Fig. 65.4 Layout of a *Naamghar* in a *Satra*, Assam. (Researcher)

65.3.2 Case Study

26 *Naamghars* Were Chosen from Rural, Semi-Urban, and Urban Areas of Assam.

(i) **Categorization theory**

One of the methodologies which was considered relevant was used by Athavankar [1] in the study of 'Categorization…Natural Language and Design.' He mentions that structured categorization can lead to concepts and with that visual information can be controlled. It has been adapted from the work of William Labov, 'The boundaries of words and their meaning' a linguist, illustrates this effectively. Labov's study demonstrates the relationship between the variations in product form and in conceptual meaning. Linguistic subjects were randomly presented, and on the basis of users choices, data was categorized. In the principles of categorization, Rosch [11] clearly demonstrates that instead of categorizing objects into large numbers of finite discrimination and tightly organized concepts, and it is better to let the mind automatically select the cognitively optimal economical option of neglecting the infinite differences among objects to behaviorally and cognitively usable proportions.

For the present study, charting the taxonomical structure and categorization of elements has been identified and adopted to further segregate the collected data of Assamese *Naamghar*. It has been divided into 'exterior' (main gate, outer gate) and 'interior' level in the 'super-ordinate level' of *Naamghar* and further classified into 'basic level' like motif, pillar, and ornamentation. Further, it is been classified into 'subordinate level' (shapes, signs, carvings, jali work, etc.).

(ii) **Significance of Design elements: Card Sorting**

The technique of card sorting method found suitable to cull-out the Typical from the A-Typical rating in order to bring out the significance of design elements. This technique requires the users to arrange the pictorial cards of *Naamghar* elements in present context (for instance, main gate and main door here) in a sequence of essentiality.

Focus group interviews were conducted, and the pictorial cards were given numeral coding to elicit and analyze the responses regarding degree of being 'Typical' as distinguished from 'A-Typical'. The sample size taken was 26 *Naamghars,* and the study had been conducted with 263 numbers of respondents from various sections. Two examples of pictorial data arrangement from Typical toward A-Typical are given below Figs.65.5 and 65.6.

(I) Typical to A-Typical of **Main Gate (Pratham Dwar)**

For the *Naamghars* main gate (also called *Dalam*), the typical ones are no. 10, 13, and 11 as numbered in the picture (this structure came from Ahom kingdom, and it is more similar to *Xorai*) has been chosen in a highest category. The respondents has given many reasons, and for choosing these particular gates, as it has many things like *Udanto singha* (particular kind of flying lion, which has been the symbol of Ahom kingdom and represents a mythical creature that is used as a metaphor for showcasing the might of the Supreme Hindu deity Lord *Vishnu*) on top of the gate. Also respondents mentioned that gate is an extension of *Thapana* in a symbolic manner, same *bhakat* with *taal* and *khool* (musical instruments) in their hand on the gate representing welcoming of the visitors. Respondents said that on the gate something in relevance to supreme (*Vishnu*) has to be there in form of *Thapana* or *Vishnu* avatar itself. *Xorai* with *Bhagwat* and *Gamosa* has been used to give respect as a symbol of Assamese culture. Also, it is auspicious and has been used in the main gate. Gate no. 21 has Tibetan influence over it, which is not so typical to the respondents. Second preference has been given to gate no 21 and 2 in which some of the elements are present like *singha* (lion) and *Jay Vijay* but *singha* is not the exactly *udantu singha* (flying lion). Also, image no. 14 has influence of Mohammedan architecture.

For the Main Gate (Pratham Dwar) preference has been given to the following by the respondents. It is been said that the following things have to be present in the Main Gate of Naamghar: (a) **Udantu Singha/ Naama Singha** (flying lion) denotes bhakti also called Naamghar rokheyas, i.e., the protectors of the Naamghar, (b) **Bhakat with Bhor taal and Khols** (drums with two sides): Khols are typically prepared with baked clay in Naamghar and used as a musical instrument, (c) **Xorai with Bhagwat and Gamosa** representing auspicious and respect to the supreme, (d) **Shape of dome structure,** (e) There is a **crown on the top of the gate** with semi-hemispherical shape.

The A-Typical ones (no. 1, 4, and 7) were identified for the following reasons: The respondents mentioned that the main gate should not look like a house gate; also, there is no symbolic elements present which represents the Naamghar. The lion, which has been used, is also not *Udantu singha,* and different *singha* has been used. Hence, the respondents choose them as A-Typical.

Main Gate			
Typical			
	SU9(MG10)	SU17(MG13)	SU9((MG11)
Not so Typical			
	R6(MG21)	U20(MG2)	
Fuzzy Boundary			
	SU10(MG12)	R13(MG18)	R11(MG22)
Not so A-Typical			
	R15(MG20)	R8(MG9)	R2(MG14)
A-Typical			
	R18(MG1)	SU26(MG4)	U25(MG7)

Fig. 65.5 Selected pictorial images of main gate of naamghar by the respondents from typical toward A-Typical arrangement. (Researcher)

(II) From Typical to A-typical Main Door of Naamghar (Saptam Dwar)

The same procedure was followed to elicit responses regarding main door of the *Naamghar*, also called *Saptam dwar*. The typical cards numbers were 9, 10, and 11. These have been chosen in the highest category. The respondents have given many reasons, and for choosing these particular doors as these have many depictions of *Dashavatara*, which represents *Vishnu* and its ten primary avatars (incarnations), the Hindu God of preservation. With these representations, visitors and *bhakats* instantly

Main Door			
Typical			
	R15(MG9)	SU9(MG10)	R2(MG11)
Not so Typical			
	U21(MG3)	SU1(MG13)	
Fuzzy Boundary			
	U25(MG7)		
Not so A-Typical			
	SU19(MG8)		
A-Typical			
	U24(MG5)	SU26(MG4)	

Fig. 65.6 Selected pictorial images of main door of naamghar by the respondents from typical toward A-typical arrangement. (Researcher)

connect with the power of God and its presence on earth. However, most draw from the following set of figures, omitting at least one of those listed in parentheses: *Matsya, Kurma, Varaha, Narsimha, Vamana, Parshurama, Rama, Krishna (Balrama) or (Budha), and Kalki.* All of them are *Vishnu* avatars. The stories of all *Vishnu's* incarnations inspire the *bhakats* and visitors. Respondents also mentioned that *Jay Vijay* present on right, and left side of the door is important. Also on top of the main door presence of *Vaikunth Dham (Anant Sajja),* where the God or *Vishnu* lives has to be there as it represents the supreme of all 'Vishnu' (the preserver God) which means he protects the earth from being destroyed and keeps it going, and he has come to earth with its ten forms.

Vishnu is most famous form of *Rama* and *Krishna. Vishnu's* wife Lakhmi was the Hindu goddess of luck and fortune. *Vishnu* is usually shown with light blue skin and four arms. He holds a lotus, mace (*gada*), conch (*shankha*), and *Sudarshan chakra* in each of four hands. The presence of peacock on the top of the door is a symbol of Sri *Krishna.* The whole main door should give the aura of *Vishnu's* stories/ narratives. Respondents also said that traditional influence is required on the main door.

Second preference has been given to door numbers 3 and 13, which again have *Vishnu* and *Dashavatar* but not in clarity as well as the carvings is minimal. Later the respondents have chosen A-typical doors as 5 and 4, which does not have any of the symbols or *Dashavatar* in any form, and it mainly looks like a house door.

For the Main door (*Saptam Dwar*), the respondent's preferences are the following: (a) ***Dashavatar*on main door:** It denotes the stories of Vishnu avatars with its incarnations, (b) **Vaikuntha (Unnat sajja/ Vaikuntha dham):** In Hinduism, Vishnu is the preserver and supreme of all. So its position is on the top of the main door, where the main *dham* should be represented, (c) **Jay Vijay:** They are the two gatekeepers of the abode of Vishnu known as *Vaikuntha* (meaning place of eternal bliss) [12, 9], (d) **Arch on the main door:** Polylobed arch of the pediment, (e) **Motifs:** It depicts the elements of nature like flowers and plants.

65.4 Mapping of Field Visual Content and Form Study

(i) **Main Gate**

The researcher has categorized the data, and with the emergence of the pattern, the Naamghars has been divided into three parts: (1) rural, (2) semi-urban, and (3) urban. It is been also done on the basis of other parameters like population and area covered. The form of the key elements in Naamghar was identified. Following subsections contain the details of such forms along with description. Illustrated form of each element has been identified clearly and detailed out (Figs. 65.7 and 65.8).

(ii) **Main Door**

By analyzing 26 case studies showed that urban Naamghars are significantly changing as compared to rural Naamghars in terms of socio-temporal differences. (Urban Naamghars: Zoo road Naamghar, Nizarapar Naamghar, Akashi Path Naamghar,

Fig. 65.7 Mapping of field data of main gate and door. (Researcher)

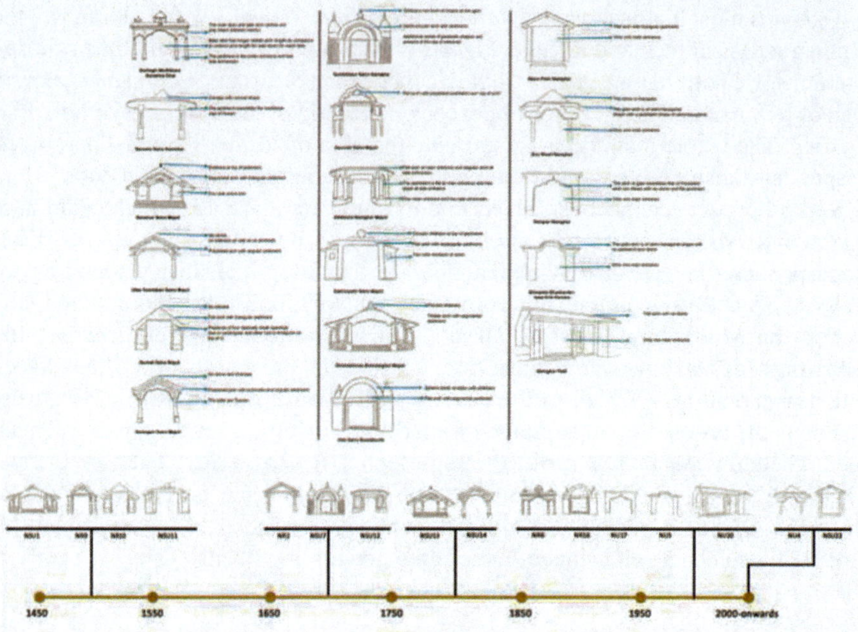

Fig. 65.8 Timeline showcasing the main gate changes and variations over the period of time. (Researcher)

Srimanta Shankardeva Kala Kriti Kendra, Rural Naamghars: Barpeta Satra, Auniati Satra, Bangshi Gopal, Hajo Pakhmela Naamghar, etc.) But they have strong correlation of similarity in its physical structure (Figs. 65.9 and 65.10).

From the form study, one can see that their is an extent of modifications happened in *Naamghars* over the period of time and with population as a variable:

(a) For instance, outer gate of Naamghar in (a) *Nizarapar* (Urban: 55 years old), (b) *Naamghar Path* (Urban: 20 years old), and (c) *Akasipath Naamghar* (Urban: 30 years old). It shows that the variation of total area of *Naamghar* and also decision-makers to corresponding *Naamghar* with year of establishment. It can be observed that area of gate appears to decrease but at a much smaller rate than that of total area of *Naamghar*. Reasons for the same were found that there is lack of sufficient space and sufficient funding which lead to these decisions.

(b) The main door for recently constructed *Naamghars* in urban area is usually plain and simple as compared to those of older *Naamghars*, which are adorned with characters from *Vaishnava* sect (*Bhakats*, *Xorai*, *Narasimha*), etc., with jali work on grilled gates.

(c) There is tendency for more emphasize on *Xorai*, *Bhakat,* and *Dasavatar* designs on main door or entrance in urban *Naamghar*. This might be also to distinguish themselves from temples in the city.

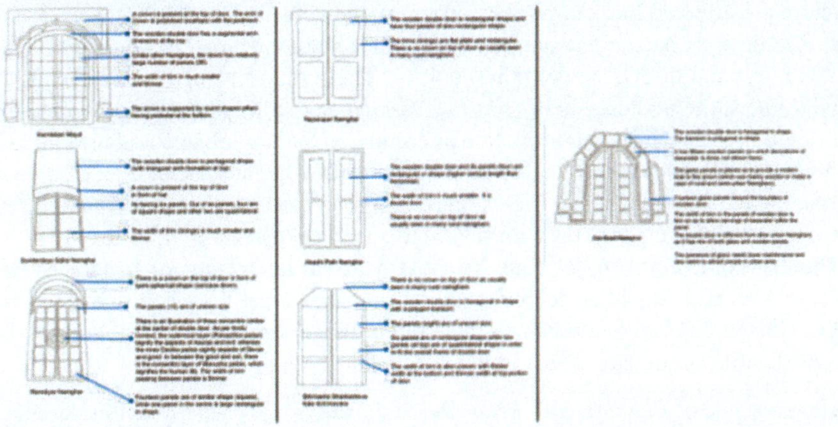

Fig. 65.9 Form analysis of main door of Naamghar at rural, semi-urban and urban areas. (Researcher)

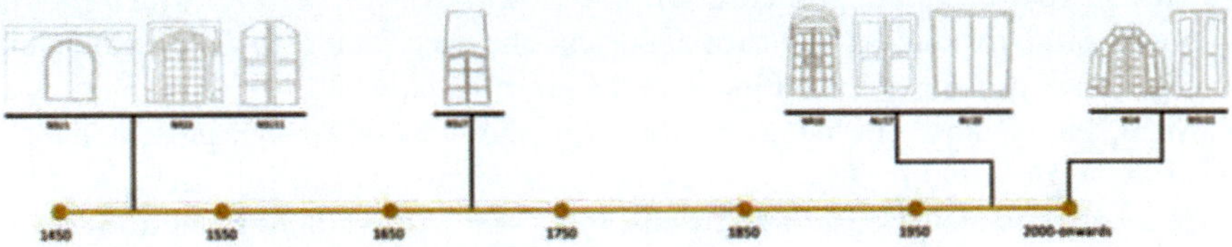

Fig. 65.10 Timeline showcasing the main door changes and variations over the period of time. (Researcher)

(d) Placement of *Garuda* and *Hanuman* statues is not that prominent in urban or recent *Naamghars.*

(e) Old age *Naamghars* appear to have more design elements in terms of floral, animal, and forms of God than their recent once.

(f) Outer gate elements are not considered that essential as that of *Kirtan* hall and *Manikut* in urban *Naamghars.*

65.5 Findings

The broad structure of the Main gate and door in *Naamghar* has survived over time and social changes. However, the social changes seems to have effected both the structural changes of *Naamghar* as also its functions in some aspects. The study shows that gate and doors of urban *Naamghars* have tried to modify the outlook in order to attract larger number of community members. The changes in the construction techniques have had their impact on the manner in which various elements of structure are affected like entrance, number of elements and assets on the door and gate. The designs have also changed partly due to compulsions of construction as also due to social faith of wider communities. The rural urban divide is reflected in highlighting some of these changes as the *Naamghar* in the rural setting retained the traditional structure and social function as compared to their urban counterparts. The funding of the *Naamghars* has affected the manner in which its social organization is adapting to the changing time. Inclusion of religious motifs, relief work, sculptures, carvings has also their imprint with regard to social changes and inclusion.

The shape of *Udantu Singha* found in most of the earlier *Naamghar* largely in rural and semi-urban settings has been changed in a manner that it looked like a regular *Singha.* In some urban *Naamghar,* it has totally disappeared. It may be because of non-availability of artisans or resources.

- The presence of *Bhakats* with *Bhor taal* and *Khols* is quite dominant in the older *Naamghars* mostly situated in rural setting but with the passage of time and particularly in urban settings, the image of *Bhakat* with *Khols* has substantially changed. Part of the reason is that the shape and the significance of main gate have not been sustained particularly in *Naamghars* located in urban settings.
- There used to be great significance attached to the presence of *Xorai* with *Bhagwat* and *Gamosa* together as indicator of the place being a *Naamghar*. The study found this design of main gate in almost all the older *Naamghars.*
- Significantly, the top of the main gate having the shape of *Xorai* is present in most of the *Naamghars* located in rural and semi-urban settings. But in some of the *Naamghars* situated in urban settings, the *Xorai* is missing, instead of that it is replaced by *jhapi* in some of the *Naamghars* located in urban settings.
- The top part of the main gate in some of the rural *Naamghar* has the shape of poly lobed (trilobed mainly), whereas in semi-urban, it has semicircular shape with archivolts with *shikhara* having *chakra* at the gate. In comparison with that the gate of *Naamghars* in urban settings is mostly having either the flattened arch with triangular pediment or arch is semi-hexagonal shape with a pediment. In most of the *Naamghars* situated in rural settings, there is a prominent presence of *Vakhuntha (Unnat sajja/Vakhuntha dhamm)* and *Naamghars* in urban settings this is typically missing. One of the reasons could be the influence of other religious groups and communities and non-availability of traditional artisans.
- Traditionally, the presence of gate keepers on the main door called *Jay Vijay* used to be invariably present in older *Naamghars* located mostly in rural settings but with the passage of time and we find in some semi-urban and most of the *Naamghars* situated in urban settings *Jay Vijay* are typically missing. This is another indication of dilution of the portrayal of religious designs. This may also be attributed to either non-availability of conventional artisans or lack of concern for maintaining the sanctity of *Naamghar*.

65.6 Conclusion and Recommendations

Since the basic philosophy of establishment of Naamghar was 'Vaishnava sect,' it can be concluded that the design elements of Naamghar may be similar to each other. The design elements such as Dasavatar, Xorai, Bhagvad are all related to Vaisnava sect, which is also documentation of stories related to each of these elements. Design elements are changing over time and location to a reasonable degree. However, some basic elements have sustained over time. Due to rapid and heterogeneous (space and time) urbanization, there is development of contrast in social and cultural aspects (including lifestyle) of people. Since Naamghar is considered the basis of binding people in earlier times, it is essential how the design elements are evolving with space as well as time. From the study, one can conclude that the changing social and cultural milieu had substantial effect on the design elements of Naamghars. However, some essential forms and functions are maintained.

The sociocultural forces had substantial effect in changing the design elements of Naamghar over time. The spatial constraints had substantial effect on design of Naamghar as evident from the study. However, in a number of cases the changes had been to adapt to the changing social times. The study concludes that there are adequate changes both in the structural and the design elements with regard to structural elements as also changes due to social–cultural dimensions over time. One can say that in various aspects under study, viz. organization, design and social relations, there is essential continuity underneath the apparent changes. Also, the social changes have had their impact, and at a number of cases, they have been amalgamated in the development of Naamghar as a structure as also its functioning.

References

1. Athavankar, U.: Categorization...natural language and design. Design **5**(2), 106 (1989) Spring
2. Bhagawati, A.C.: Ethnic identities in North-East India. In: N K Bose Memorial Lectures. Vihangama, IGNCA Newsletter, March–April 2002, vol. 2 (2002)
3. Bhattacharji, S.: In: Legends of Devi. vol. 16, Orient Blackswan (1998)
4. Dutta, S.: The Mataks and their kingdom castes and tribes of Assam (2011)
5. Guha, A.: Pre-Ahom roots and the medieval state in Assam: a reply. Soc. Sci. **12**(6), 70–77 (1984)
6. Mahanta, P.J.: In: The Sankaradeva Movement, Its Cultural Horizons. 1st edn. Guwahati, Purbanchal Prakashan (2007)
7. Monga, C., Das, A.K.: Semiotic analysis: a study to identify the design elements of Namghar in Assam. In: Research into Design for a Connected World, pp. 343–354. Springer, Singapore (2019)
8. Monga, C., Das, A.K.: A comparative visual study of design elements of social cultural institutions, 'Namghar' with monuments in South East Asia. In: International Society for the Study of Vernacular Settlements, vol. 5 (2018)
9. Maehle, G.: In: Ashtanga Yoga the Intermediate Series: Mythology, Anatomy, and Practice. pp. 34. New World Library (2012)

10. Neog, M.: Shankardeva and some aspects of Assam Vaisnavism. A creative vision: Essays on Sankardeva and Neo-Vaisnava Movement in Assam. 1st edn, pp. 43–59. Guwahati, Srimanta Sankar Kriti Bikash samite (2004)
11. Rosch, E., Lloyd, B.B. (eds.).: In: Cognition and Categorization (1978)
12. Sukumari, B.: Beder yuge kshuda o khadya (1998)

Chapter 66
A Thematic Analysis of Graffiti in Urban Spaces by Using Indic Scripts for Experiments and Exploration for Academic Educational Project

Experimenting with Indic Scripts as 'form' Culture to Speculative Educational in an Age Commercial Application

Siddhesh Sushil Shirsekar

Abstract "There is no expression without excitement, without turmoil"-Dewey. J The present study positions a research to analyze and attempt to experiment with Indic scripts with academic typography students. A thematic analysis of the graffiti is used to investigate a script on a superficial surface. As a single lifetime is barely enough to explore one script fully, especially since each one is so deeply rooted in a particular historical and cultural context. Many children are growing up in diverse bilingual or multilingual contexts and learn to speak and read in more than one language. India, home to vivid and beautiful scripts in urban spaces, is one of the most ideal environments to study and practice type design. As designers we are aware of our surroundings, things make a lot of sense to us as we always try to find meaning out of everything that we come across. Typography is one such thing that we are surrounded with, in today's world. The academic project deliberated here demonstrates how the scripts chosen from various sources. Sometimes they are taken from printed material, such as existing graffiti styles, newspapers, advertisements, and labels. Indic script can be treated as art form toward unconventional approach and without conscious reasoning. The output of the research shall be a system design being an academician to foster a wholesome learning environment for Indic scripts. The conference could itself be an opportunity to discuss possibilities and encourage design students in this endeavor.

S. S. Shirsekar (✉)
Srishti Manipal Institute of Art, Design and Technology, Bangalore, India
e-mail: shirsekarsiddhesh@ymail.com

A. Chakrabarti et al. (eds.), *Design for Tomorrow—Volume 1*, Smart Innovation, Systems and Technologies 221, https://doi.org/10.1007/978-981-16-0041-8_66

66.1 Introduction

When typographic communication created for a certain audience is consumed or encountered in a different cultural landscape, what are its effects? The acknowledgement and cognizance of this 'consumption' and 'encounter' lead to questioning of purpose and interrogating the merit of what we produce as visual communicators and type designers [1]. John Berger makes clear that images of social difference work not simply by the way they show but also by the kind of seeing that they invite. He uses the expression 'ways of seeing' to refer to the fact that 'we never look just at one thing; we are always looking at the relation between things and ourselves' [2]. The ideas put forth in this paper are generic and can be used as measures to assess any paraphernalia that communicates visually (Fig. 66.1).

To consider graffiti as popular culture, and as a teachable text, with some questions we asked each other as we worked with these images. In what ways is graffiti popular culture? In what ways is it a teachable text? Graffiti is a popular form in the most foundational sense: It is of the people, by the people, for the people. It is DIY in the sense that it is a personally crafted artifact, but it is also a communicative act meant to be seen or consumed by both friends and strangers as they tunnel through cityscapes, swing by playgrounds, or loiter on corners.

Are graffiti an urban popular form? Is graffiti art or politics? Is it free expression or vandalism? Is it an archive or a crime—or an archive of crimes? By evoking such questions, graffiti expresses tensions and contradictions that are inherent in cultural production more broadly. It is punk, street, youth culture. It can be fascist or feminist. It can be dense with meaning, or gestural, autobiographical, a tag, a quick flick of a sharpie [4].

Fig. 66.1 Frame from Inquilab: a documentary [3]

66.2 Analysis of Graffiti

The schematic view attempts to represent the structure of graffiti writing culture. The horizontal axis is the 'juridical'; the vertical axis is the 'aesthetic'. The broken line represents the structural shift, or 'development', that I think occurred during the 1990 s. To be sure, 'commercial' and 'legal graffiti' did exist prior to 1990, but it was much less common than it is today. The shaded area to the left of the 'aesthetic' dimension represents where, or how, graffiti writing culture generally existed for much of its earlier history. The shaded areas to the right of the 'aesthetic' dimension indicate the 'legal' and 'commercial' graffiti were marginal modes of production during the 1970s and 1980s [3]. Below are few graffiti studies from Western and Indian cultures. With these self-critical and purpose-driven reflections, our audiences surely have a lot to gain (Fig. 66.2).

Note ***Kitsch*** *(/kɪtʃ/ KITCH; loanword from German), also called* ***tackiness****, is art or other objects that, generally speaking, appeal to popular rather than 'high art' tastes. Such objects are sometimes appreciated in a knowingly ironic or humorous way.*

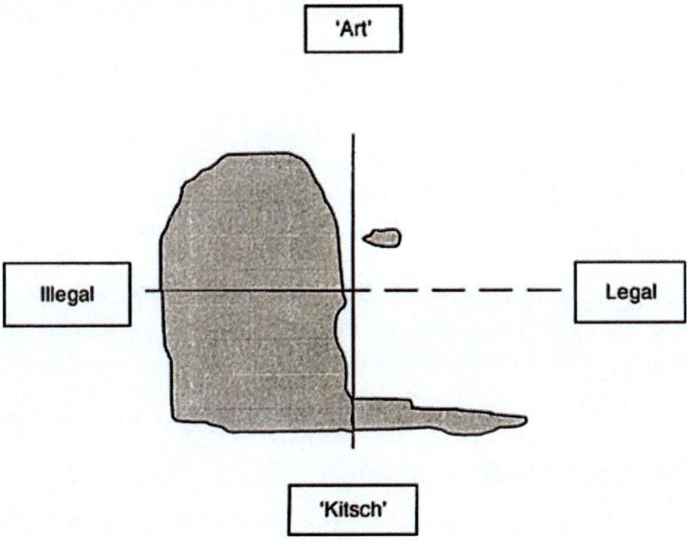

Fig. 66.2 Structure of graffiti writing culture [5]

Fig. 66.3 Graffiti on the wall at Oaxaca, in Southern Mexico [4]

66.2.1 Graffiti Case I

A stenciled poster declares 'TODX PRESX ES PRESX POLITICX.' ('ALL PRISONERS ARE POLITICAL PRISONERS.') Why is the 'OS' replaced with an 'X'? Why do x's appear so often in graffiti around here? (Figure 66.3).

The graffiti above are the true reflection from slogans scrawled and spray-painted on the walls and sidewalks, and more elaborate stenciled graffiti and artwork painted and plastered. This give rise to a new era of activist art tradition to convey a message to the community, which makes this media more powerful.

66.2.2 Graffiti Case II

A hand painted wall with simple sans serif font to write 'WE DO HAVE TIME TO STAND AND STARE' Why is it a pun intended sentence on the roadside? (Figure 66.4)

The graffiti above are the true reflection from slogans scrawled and spray-painted on the walls and sidewalks, and more elaborate stenciled graffiti and artwork painted and plastered. This give rise to a new era of activist art tradition to convey a message to the community, which makes this media more powerful [5].

Fig. 66.4 Graffiti writing on the wall at Fort area, Mumbai (From author's collection photograph Shirsekar.S)

66.3 Germ of This Research Paper

As designers we are aware of our surroundings, things make a lot of sense to us as we always try to find meaning out of everything that we come across. Typography is one such thing that we are surrounded with, in today's world. This paper is an attempt to discuss appreciating the concept of forms in typography to the students of design to develop their typographic sensitivity.

66.4 Objective

To study existing graffiti writing and incorporate size, color, texture, and lettering aspect in an Indic script and learning the art of replicating it or getting close to the overall impact of it.

66.4.1 Learnings from Letterforms

Letterforms are the structural elements of all typographic communication. The alphabet epitomizes articulated frequencies. The contemporary forms of letters are derived from a typographic evolution. It developed from handwriting and that is why the chief section that constructs each letterform is a linear stroke.

66.4.2 Word as an Image

We all possess a subtle vocabulary of style that compels us to read a composition involving letterforms on many levels. Subjective representation is conceptual interpretation. It is used on a theme or an idea that creates an experience for the viewer. It allows a greater scope for complexity and simplicity as well. It plays on a multiple layers of interpretation for the implied meaning of the design. It has an impact on the emotions of the viewer. In some cases, the perfect expression makes striking typography. The priority is to find the correct visual language which will communicate in a precise way. Once this concept is clear language is not a barrier, but it gives a greater scope for exploring ideas through typography [6].

Below are few examples of typography as an image where the concept of crowd in Mumbai local trains is expressed in an unconventional way. The project '*Gardi*'-Rush is an inseparable part of daily life. This project highlights the Devanagari letterforms in unconventional layouts with different poems for various situations. The ink opacities play a major role reflecting the depth of field in individual art form (Fig. 66.5).

66.5 Statement

Are graffiti something to be archived? Or to elevate and translate graffiti as 'street art,' something to be taught in the academy, produced and consumed for new purposes.

66.6 Approach

Is it necessary to capture graffiti, to record it, in order to teach it? Does the significance of the gesture change when we remove it from its context? When we take graffiti seriously as teachers, we use a range of overlapping perspectives through which we and our students can engage with popular culture. Graffiti is discourse, art, and visual culture.

66.6.1 Introducing Urban Spaces

The environment plays a vital role in grooming as a human and aids to observation sensitivity on different peripheries. In this exercise, we asked our students to capture various graffiti in vicinity as well as well-known areas. Absorbing the city from another point as the normative forms of understanding a city are differentiated zones

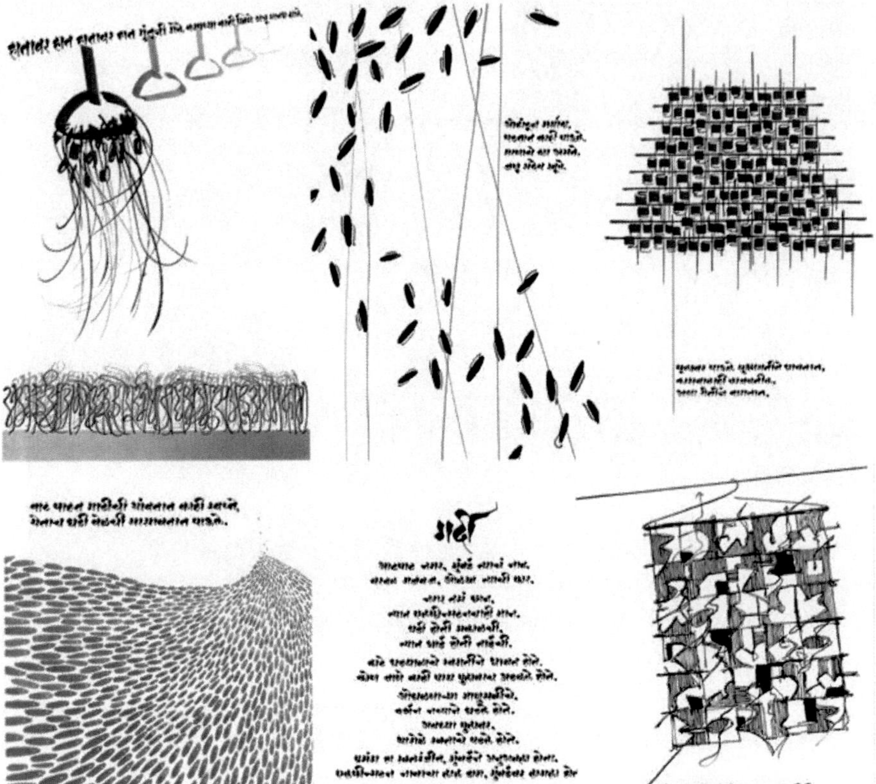

Fig. 66.5 Typography project of final year design student-Panchal P

Fig. 66.6 Documentation, brainstorming, and collage done by typography design students

of activity and flows of people and materials between the zones a collage was derived (Fig. 66.6).

66.6.2 Extracting Visuals and Interpretations

This is the crucial stage of this experiment where we try to redefine the meaning of compassion and caliber in relation to graffiti using different Indic scripts. The multi-scriptural choice and their respective graffiti from where these styles have been accumulated show affinities to create visual layouts in terms of emerging from-content relationships processed mainly in the visual simulations. In order to understand the core of visual cultural display and meaning-making processes involved as part of this exercise. Further, it modestly initiates a style of analytical overview over few known and other unknown influencing factors that bring forth the indigenous in the type expressions in the vivid strokes forming different scripts implementing in different graffiti. Below are the outcomes of this experiment done by typography design students (Fig. 66.7).

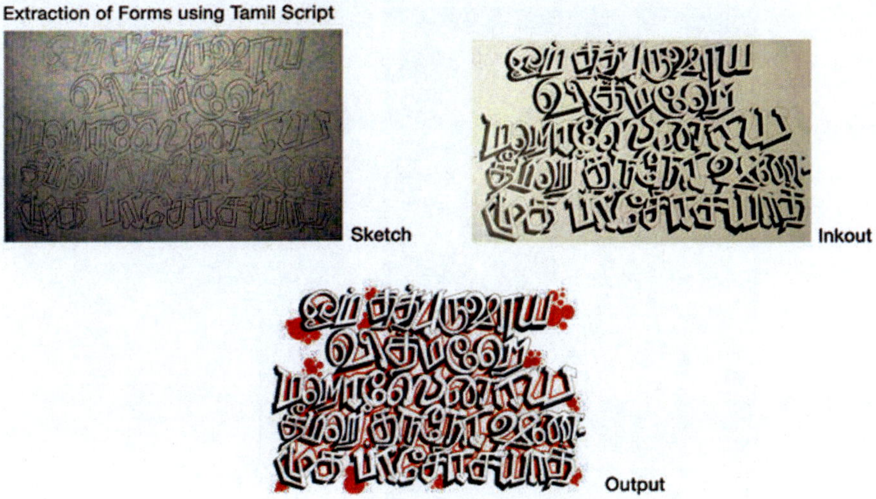

Fig. 66.7 Process to derive Tamil Graffiti done by academic typography student

66.7 Methodology

Existing graffiti is studied in terms of their writing and incorporate size, color, texture, and lettering aspect in an Indic script and learning the art of replicating it or getting close to the overall impact of it. Students were asked to capture various graffiti in vicinity as well as well-known areas. Absorbing the city from another point as the normative forms of understanding a city are differentiated zones of activity and flows of people and materials between the zones a collage was derived. Redefine the meaning of compassion and caliber in relation to graffiti using different Indic scripts. The aim was to understand the core of visual cultural display and meaning-making processes involved in this exercise (Fig. 66.8).

66.8 Observation

1. Scripts differ in appearance; the visual form is appreciated.
2. Symbol units, called *akshara*, represent graffiti at the level of both a syllable and a pictorial.
3. In contrast to the composition and script, there is visual balance with coherent to the execution.
4. Interdependency on scripts to learn individual script. (Transfer of learning)
5. As student can have different phonological awareness profiles, it is desirable to assess students in both of the languages spoken with individual system rather than replicating each other.

Bengali Script

Gujarati Script

Gujarati Script

Malayalam Script

Bengali Script

Urdu Script

Fig. 66.8 Frame from Inquilab: a documentary [3]

66.9 Conclusion

Graffiti is discourse, art, and visual culture. It is a handmade, homemade, imprint upon the built environment. Graffiti is, first and foremost, ephemeral: Its extreme and explicit temporariness is inherent in both its creation and its consumption. Graffiti brings the language of visual culture to the surfaces of built and natural environments, mediating the ways we see where we are, imprinting brick, steel, and mortar with color, script, and shade. Graffiti plays with the simultaneous gestures of representation and erasure, legibility and illegibility, authorship and anonymity. However, we define it has seeped from its own domain into other realms: From the streets into the gallery, when we take graffiti seriously as teachers, we use a range of overlapping perspectives through which we and our students can engage with popular culture. India land of masses weather street or contemporary will always deliver unexpected fallouts. Working in groups, whether intra or interdisciplinary, is above all recognized for its capacity to unlock creative potential, to promote the unexpected.

66.10 Approaching Opportunities

Further, the research paper can give possibilities to device different innovative experiments to acquire Indic scripts with different methodologies and make interesting tactile applications for the same.

Acknowledgements My sincere thanks to Dr. Prof. Santosh Kshirsagar, Dr. Sushil Shirsekar, Prof. Samip Sawant, and Mrs. Sheetal Sushil Shirsekar for helping absorb the concept and increasing the sensitivity toward the Indic scripts. Also will appreciate my former institution VIVA Institute of Applied Art elective–lettering and typography students batch 2018–2019 for the academic assignments exercised in this research paper as well as mu current institution Padma Bhushan Vasantdada Patil Pratishthan's College of Engineering and Visual Arts, India.

References

1. Mehta, D.: From within it, from outside it and from above it (2014). http://www.typoday.in/2014/spk_papers14/deshna-typo14.pdf
2. Rose, G.: In: Visual Methodologies: An Introduction to the Interpretation of Visual Materials. London: Sage Publications Ltd (2001). https://www.miguelangelmartinez.net/IMG/pdf/2001_Rose_Visual_Methodologies_book.pdf
3. Nme, G.: In: Inquilab: A Documentary (2019, April 18). Retrieved from www.youtube.com: https://www.youtube.com/watch?v=hHEhT4tUyts
4. Sammond, N., Creadick, A.: The writing on the wall: learning and teaching Graffiti. Transform. J. Inclusive Scholarship Pedagogy **24**(1–2), 138–154 (2014). Retrieved from https://www.jstor.org/stable/10.5325/trajincschped.24.1–2.0138
5. Kramer, R.: Painting with permission: legal graffiti in New York City. Ethnography **11**(2), 235–253 (2010). Retrieved from https://www.jstor.org/stable/24048062 https://journals.sagepub.com/doi/abs/10.1177/1466138109339122
6. Keni, M.: Introducing students to the concept of beauty. *Typoday.* Mumbai (2018). http://www.typoday.in/2018/spk_papers/manasi-keni-typoday-2018.pdf

Chapter 67
Study and Revival Strategies for Traditional Art Form: Case of Sindhudurg

Priyanka Mangaonkar-Vaiude and Minu Joshi

Abstract India is a country well known for its skill and handmade. As per the research data, despite having a rich heritage of traditional manufacturing skills, India's share in the international market for craft is less than 2%, while that of China stands at 30%. According to the United Nations, over the past three decades, the number of Indian artisans has decreased by 30%. These alarming numbers indicate the need to reinvest in artisans, to safeguard our history, culture, traditional knowledge and an important source of livelihood. The pressing problem faced by Indian craft sector today is to create an ecosystem that benefits all stakeholders in the value chain and help to sustain the craft and the craftsperson. Apart from the above-mentioned issue, there are various other challenges like availability of raw material, issue of mass production and time, availability of correct market value. Hence, there is an urgent need that we, as designers, understand these issues and try to create a favorable environment for the craft to sustain. This paper is an attempt to identify these issues related to survival of a craft, by taking a case study of the dying art of traditional wooden toy craft & Ganjifa cards of Sawantwadi in Konkan region of Maharashtra. This research paper is based on the primary data obtained by personal, telephonic and online interviews of artisans, people promoting handicraft and buyer. Secondary data from published research papers, journals, books, articles, and governmental data from online platform have also been analyzed.

67.1 Craft in India: Economic Statistic

Handicraft, in many ways, is an inherent part of Indian culture, history, and regional expression. Craft not only provides economic opportunities to the artists and craftsman but also satisfies the aesthetic desire of man as well as his "want" to

P. Mangaonkar-Vaiude (✉)
Practicing Architect and Visiting Faculty, Pune, India
e-mail: mangaonkar.priyanka@gmail.com

M. Joshi
Founder of Ecour Studio and Visiting Faculty, Pune, India
e-mail: ecour.ar@gmail.com

A. Chakrabarti et al. (eds.), *Design for Tomorrow—Volume 1*, Smart Innovation, Systems and Technologies 221, https://doi.org/10.1007/978-981-16-0041-8_67

express. Handicrafts of any given time and space, echo, and showcase in them the results of centuries of patient experiments of man under varying circumstances. In India, every region has a different handicraft resulting into variety of products ranging from wood work to textiles, metal work to leather, painted items to bead work, mud work, weaving, and so on. From the economic point of view, it has high capacity to generate export and foreign exchange earnings at low capital investments. According to the government statistics, India exported handicrafts worth over 128 billion Indian rupees in fiscal year 2019 [1]. Handicraft industry provides employment to more than 6 million artisans and is one of the important segments of decentralized sector in India. Statistics show that India's contribution in world toy market is less than 1% and handicraft industry's share in India's exports is just 2% [2]. Not only artisans, but various NGOs and government bodies are a vital part of this industry. These have been formed to help the growth and survival of Indian handicrafts.

67.1.1 Government Initiatives and Policies Related to Handicraft in India

Today, the craft sector in India is a fragmented and under-defined one, with policies, mandates, schemes, roles, etc., scattered across seventeen government ministries, ranging from the Ministry of Textiles to the Ministry of Labour and employment, as well as various financial institutions [3]. Although handicrafts have been one of the major sources of income (besides agriculture) in many parts of rural India [4], it is struggling to sustain due to variety of reasons discussed in the later part of paper. With most of the artists being self-employed and working on irregular basis, they lack understanding of market needs and fail to develop a competitive product [5]. To prevent this from occurring, government has already established various schemes and initiatives like "National Handicrafts Development Programme" (NHDP), comprehensive handicrafts cluster development scheme (CHCDS). These two schemes (NHDP and CHCDS) are currently being implemented by the development commissioner's office via 6 regional offices and 62 field offices pan India [6]. To make sure that the benefit of these schemes reaches the right person or organization, it has been made mandatory that NGOs or other organizations register with Niti Aayog along with an identity card called "*Pahchan*"(meaning identity) which has been issued to the genuine artisans so they can receive benefits through the "DBT" (direct benefit transfer) portal. In addition to this, various awards [7], marketing platforms, cluster developments, soft skill development platforms, and health schemes have also been introduced. Handicraft mark as a trademark is developed to maintain authenticity. Despite all these various efforts from governmental and non-governmental interventions, the craft sector is struggling to sustain. These interventions are either ineffective or fail to reach the majority of the artisans causing many traditional art forms to die.

67.2 Wooden Toy Craft, Ganjifa Cards and Sawantwadi

India is famous for its 5000-year-old traditional history of wooden toy making craft. Most of the states of India have wooden toy making as a part of their art and culture but at very few places it is practiced as a traditional craft from centuries. Sawantwadi is one such small town in Sindhudurg District of Maharashtra State, which is known for its 400-year-old history of wooden toys and Ganjifa playing cards that are traditionally made by *Chitari or Chitrakar* community. Though these communities were migrants from Karwada, they survived by supplying different handicrafts to royal families and later on adopted this craft due to its commercial success. Mostly fruits (Image 67.1), vegetables, vintage cars, cooking set (Image 67.2), bullock carts, etc., were made as wooden toys. Fruits are basically made by wood of Indian coral tree (local name: *Paangara* tree). This wood is easily available from the surrounding forest, which gets soaked in rain and dry throughout the year making it light in weight. Once the wood is dried, artisans start converting these wooden logs into different shapes of fruits with the help of simple hand tools as shown in Image 67.4. This art of carving fruits out of wood without any wastage of material is highly commendable. Along with fruits, Sawantwadi is also famous for its other traditional wooden toys.

Sawantwadi wooden toys are essentially made by joining flat-shaped solid wood pieces. Seasoned mango wood is chiseled or carved into desired shape of a toy. It further goes through various processes like sanding, painting, assembling, and finishing as shown in Image 67.4. These toys are traditionally painted with lacquer which makes them bright and vibrant and also long lasting. Generally, women artisans are involved in coloring and finishing the product. Painted articles are assembled together with the help of nails to make one piece of toy. It takes approximately two to three days to make a medium size toy. Except sanding and drilling all the steps of making a single piece of toy are done manually. Today, in spite of making different

Image 67.1 Traditional wooden fruit set

Image 67.2 Traditional wooden cooking toy set of Sawantwadi

Image 67.3 Sawantwadi Ganjifa cards

shapes on machine, final carving, touch up and painting is still done by artisans manually.

Ganjifa (Image 67.3) is a card game that is believed to have come to India from Persia and became popular during Mughal period. The mother of current heir late Rani Satwashila Devi Bhosle of Sawantwadi took personal interest in learning the craft of Ganjifa cards from *Chitrakar* community, and few crafts persons, there on were trained to safeguard and promote the art. Earlier the cards for royalty were made up of ivory, tortoise shell, mother of pearl, and cards for commoners were made up of papier mache, palm leaf, or cloth [8]. A thick sheet of paper is cut in the form of circle of diameter 5–7cms. Earlier it used to be rubbed with a stone to smoothen it for painting. The inside of the card is painted over to give a background for the main motif depending on the kind of the card as shown in Image 67.4. The main motif is first outlined using white paint and then colored. This is very intricate and time-consuming step of this craft. Once the background is laid the detailing like facial expressions, garments are done using black paint and thin brushes made up of squirrel hair. Colors used earlier were made up of natural vegetable dyes but nowadays poster colors are used. The thick borders are further detailed with polka dots or lines. There are two types of Ganjifa cards – *Bazar calam* with only lines and *Darbar calam* with line and border designs which is the specialty of Sawantwadi. After painting is done, these cards are dipped in lacquer to protect the cards and are packed in hand-painted wooden box along with playing instructions. The famous one is Dashavatara cards which consist of set of 120 cards of incarnations of lord Vishu (10 avatars with 12 cards each). The main aim of these games was to teach and tell stories from the ancient scriptures and texts.

67.2.1 Problems Faced by These Crafts

Today despite export opportunities, global market exposure and employment generation possibilities, the handicraft industry is facing a lot of challenges due to the

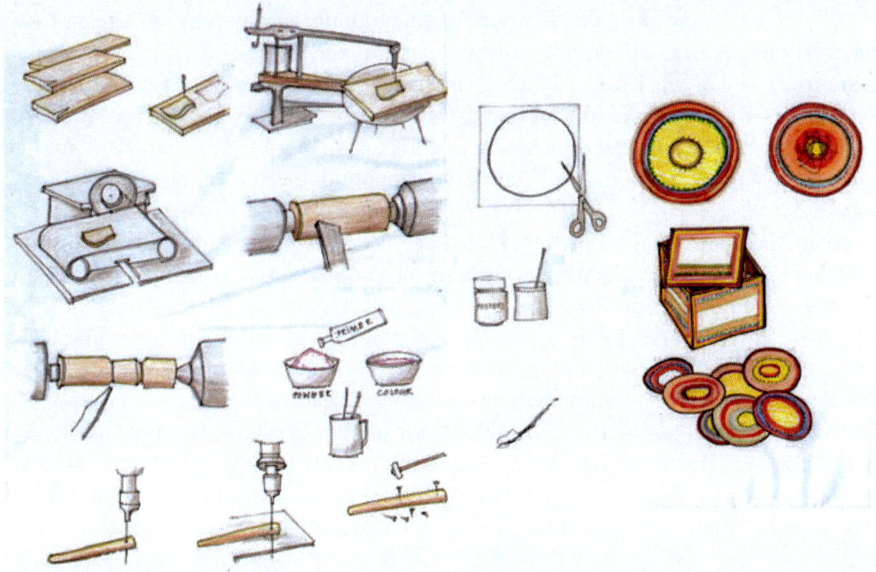

Image 67.4 Making of Sawantwadi toys and Ganjifa cards

intervention of new technology and also increased competition from different countries. As per Chitari Mohan Kulkarni, "there is very little awareness and not enough patrons to support the art and it has become financially unsustainable." From interviews and conversations with artisans, NGO, designers involved with artisans and other organizations, we have identified the following major challenges faced by artisan in this sector:

67.2.2 Competition from Outside Market and Machine-Made Items

Earlier the artisans were the ones selling the toys, so they knew the needs of the market and their designs were evolved based on the interaction with the consumer as well on what got sold. But in today's time, due to urbanization and globalization of market, artisans are facing problems in understanding the changing demands of the consumer and hence unable to create new designs at lower price and lesser time. The consumption of low-priced, machine-made products that are of higher quality is another factor affecting the demand of these handmade toys. As per Mr. Subhash Chitari, [9] Chinese toys manufacturers produce competitive machine-made toys that are less expensive than the traditional wooden toys, declining their sales tremendously. Ankita Kanekar [10] who owns a 72-year-old toy manufacturing unit in Sawantwadi says:

"China market is basically famous for manufacturing anything on cheaper rate due to various reasons like quality of raw material and GST. These Chinese toys are made from packaging wood, reducing their raw material cost but making them last for only six to twelve months. Whereas our toys have a life span of ten to twelve years. Also, the GST on handmade wooden toys is increase from 5.5 to 12% making our toys costlier. Due to all this factors we cannot compete with the cheap Chinese toys."

Today there are many wooden toys that are available in the market. Due to lack of financial support, there is no innovation in the product range of this craft. Artisans are still continuing with the age-old designs which are outdated in today's time. This leads to consumer focusing on the other available options in the market like Chinese products which are not durable but are very easily available and are cheap. The worst part is, for commercialization craftsperson themselves are keeping the Chinese products along with their products for sale. Also, there is tough competition from plastic toy manufactures which are comparatively cheaper.

If we see from the consumer point of view when discussed with a toy library owner Mrs. Snehal Bhandari from Pune, she pointed out that, "There is no innovation in the traditional Indian toys from the point of view of end used as compared to Chinese toys. There is a need of design intervention based on market needs." Even the document prepared by Maharashtra Census office published in 1968 brings out the same conclusion stating that one of the reason of static condition in the design of wooden items of Sawantwadi may be due to lack in demand from market as a result of lack in innovation [11].

67.2.3 Limited Access to Global Market

Many of the artisans are dependent only on the local market for the sale of their goods. Very few have access to global market with the help of middleman or NGO. According to a chitari Ramchandra Thakur, "When the Ganjifa art is promoted to be sold outside Sawantwadi, we receive large orders but unfortunately sometimes buyers cheat us by canceling these orders or giving us a low price for it as the order is already made and we cannot do much about it." Due to lack of market information, they are unaware of the consumer preferences globally. And hence their products are considered as traditional, old-fashioned, and not as per the modern taste. These products do not have any USP other than being traditional, which differentiate their products from other similar products in the market. There are no efforts being made for brand building or marketing for better reach of these products.

67.2.4 Availability of Good Quality Raw Materials

Due to various reasons, artisans are forced to buy substandard raw material for their toys. Urbanization resulting in reduction of forest area is one of the major reasons for non-availability of good quality wood for making these toys. Use of artificial color instead of lacquer or natural vegetable colors brings the question of safety for the children. All these factors are ultimately affecting the value, cost and authenticity of the craft. In case of Ganjifa art, the raw material is available, but the buyers do not wish to pay for premium materials, and therefore, these cards get made on paper or canvas making them less durable.

67.2.5 No Transfer of Indigenous Knowledge

The artisanal skills are passed on from generation to generation, but in present times younger generation of artisan family is not willing to take-up their family occupation; as they see their parents struggling to find the market and fair price for their products and efforts. In an interview published online, the Ganjifa card artist Ramchandra Thakur points out that he enjoys making the Ganjifa cards but would not want his children to take up the same profession as he feels that it is a loss-making business. According to him even if people show interest in the art, there is not enough demand and marketing for this art. There are hardly any families which are still working on the traditional craft of Ganjifa cards except few working for royal family of Sawantwadi. Many of these artisans are also joining the rank of casual workers [12].

67.2.6 Lack of Implementation of Government Policies

There are many government schemes and policies, as discussed earlier in the paper that are available to help the craftsperson. Each state has its own handicraft policy. Though Ganjifa cards and wooden toy making craft of Sawantwadi both are identified by Maharashtra Small Scale Industries Development Corporation (MSSIDC) but there is hardly any help that has reached to the artisans so far. It was also observed that artisans themselves are not so keen on preserving the craft, rather in earning and hence least bothered about government policies which are being introduced for them.

67.2.7 Inadequate Census Data

The lack of authentic and adequate data on artisans, their socioeconomic status, and livelihood conditions is one of the major factors affecting the planning and policy-makers working for this sector.

67.3 Cases of Surviving Crafts

Various craft forms of India have been provided with expert help from design institutes, designers, government bodies, etc., which has resulted in their revival. We will be discussing revival of Molela plaques and copper work briefly in this section.

Molela is a small town near religious village Nathdwara. This village is also known as "potters' village" which is quite popular for its Molela terracotta craft. The *Kumhars* (potters) of this village make a variety of devotional plaques. These plaques are bought by the Bhil, Gujjar, and Gajirat communities of Madhya Pradesh every year in the month of February and March. They arrive here accompanied by their priests in order to buy new votive images of their deities. The terracotta relief plaques (Image 67.5) are produced using very simple tools, motifs, and slab technique for reliefs which are quite distinct in character [13]. Previously, the main subjects depicted in these reliefs were related to gods and goddesses. Until 1981, the pottery craft of Molela remained unchanged, when Mr. Jyotindra Jain inspired them in modifying their work styles so as to create individual objects for contemporary needs.

Nowadays, they make scenes of their day-to-day life or tell a story through these reliefs. An understanding of composition, the main subject matter, and ornamentation are essential in this craft and give each item its individual character. Looking at the consumer market, changes have been made in the final product even though the technique remains the same. The ritualistic significance of the craft was slowly dying, and hence, efforts were made by giving it a decorative perspective resulting in sustaining the craft in today's time. One such example can be seen at Udaipur railway station.

Image 67.5 Traditional form of Molela terracotta relief plaques

Image 67.6 Use of Molela craft on the facade of Udaipur railway station

As featured in Udaipur times on Feb 7, 2017, the main façade (Image 67.6), interior and exterior of Udaipur railway station is decorated in Molela terracotta art by two artisans Mr. Mohanlal Kumhar and Jamnalal Kumhar with the help of local administration. It reflects how unique crafts from each region can become a great source of identity even for cities and helps travelers or visitors to remember it for a lifetime.

Copper work (Tambat art) is another such surviving craft from Pune. It came into existence during Peshwa period [14]. The Tambat community originally came to Pune from Konkan region. During the Peshwa period, this craft was majorly used for religious and military purpose for making utensils, weapons, coins, etc., for royal family. After gradual decrease in the demand of such copper artefacts, the number of artisans working on this craft also decreased [14]. Indian National Trust for Art and Cultural Heritage (INTACH) with some funding from a few of their member decided to work on the revival strategies for this craft as conveyed by Jui Tawde of INTACH a co conveyor of Pune regional chapter, in a telephonic interview. Designers and other INTACH members started spending time talking to the community, finding out what the craft is all about, possibilities in designing, market places, market solutions to come up with possible options to revive the craft. This gave rise to Studio Coppre which is a successful business model in today's time, and it runs as an individual entity. This was possible only because of design intervention. Designers worked with the craftspeople by taking their input while designing a product. As in any craft Artisans know the material and possibilities with the material in a better way than a designer. Just a designer designing something without giving it to a craftsperson to make it does not make sense. Today designers with the help of artisans have created a new product range which is best suited for current contemporary lifestyle and needs (Image 67.7). This has not only helped the craft to survive but also empowered the artisans by giving them new perspective and pride for handcrafting.

Image 67.7 Copper flower boat, water bottle and tea-light candle holder

67.4 Conclusion

Craft so far was understood as personalized occupation but nowadays is contribution to wider economy [15]. For a developing country like India, where manpower is abundant and with less capital investments, efforts can be made to sustain the existing crafts. It was mentioned in a document prepared by The Maharashtra Census office published in 1968 that wooden toy making craft of Sawantwadi may not make any great progress with respect to its expansion of business and may also vanish in distant future. But now in 2020 after 52 years also this handicraft still exists in Sawantwadi due to strong influence from nearby local market like Goa and Karwar [10] but in order to make this craft sustain and grow as a business model there is need for various interventions. A few modifications like in the purpose or added value for new application of the product used in everyday life help in revival of the craft keeping its traditional values intact. As rightly said by Jui Tawade: *"Conservation of craft go hand in hand with community aspirations and livelihood success. Collaboration between designers and craft people make the past relevant for present and the future."* Design development ideas can be achieved by having collaboration with different design institutes like NID, IDC, IIT which will help both the parties in developing new methods and reviving a craft form. Additionally, there is a need to revive the artisan–consumer linkage which existed in the olden times. This helps craftsperson to understand how to tailor their goods to meet the current demands and also consumers get to know the authenticity of handmade crafts. These linkages can be created by arranging workshops, national/international exhibitions, promoting craft tourism, creating platforms for e-commerce, collaborating with different existing brands like First cry, Fabindia, tie ups with luxury brands, hotels, corporate giants can provide a consumer platform. Soft skill training can be given to artisans that include negotiation, marketing, managing people, and ultimately managing their own business. This will reduce the dependency on middle man and will help artisans to look at their art from a business point of view. Toy market has a huge potential from the education point of view. Prime Minister Narendra Modi in his Mann Ki Baat on August 30, 2020, mentioned the need to increase local production of toys and India's share in the global toy market. Apart from this, impact of toys on various aspects of children's lives has been given significant attention in the National Educational Policy (NEP). Learning while playing, learning to make toys, visiting the places which manufacture toys all this has been included in the curriculum. He also mentioned the need for India's youth to innovate the toy industry with the help of modern technology. In this context, a toy can help children in many ways during their early stages of learning for developing different abilities. Many schools use different toys for explaining a new concept to children. While developing the design, emphasis could be given on scientific toys which can be used by different educational institutes. Ganjifa cards can be used as creative hand-painted flash cards for various subjects. The method of story-telling can be used to impart knowledge of subjects like history which are usually considered dry by the students. This will ensure the constant demand of such products. Safety and toxicity have to be kept in mind while making these toys as they will be used by children. Current international safety standards should be followed from the export point of view. The current curriculum in the schools does not include lessons regarding the importance of traditional and cultural craft of India. Identity and importance of master craftsperson is lost along with the importance of the craft. These discourages new generation from joining the family occupation and continuing the tradition as they do not see any benefit in doing so. Awareness about our cultural and traditional art and craft forms should be a part of educational system.

In this way, we can make various attempts to revive the traditional art form and sustain livelihoods of the artisans.

References

1. Keelery, S.: India-Value Of Handicrafts Exports 2011–2020. Statista. [online] Statista (2020). Available at https://www.statista.com/statistics/624202/export-value-of-handicrafts-india/ Accessed 17 October 2020
2. Gaurav A.: Indian Handicraft Industry, Jodhpur Handicrafts Exporters' Association (2020). Www.Jhea.In, www.jhea.in/Indian-handicraft-industry. Accessed 4 January 2020
3. Mubayi, Y.: Policy Gaps Study On The Crafts Sector In India. [ebook] India: European Union (2016). Available at https://www.switch-asia.eu/site/assets/files/1248/going_green_-_final_policy_gaps_study_nov_16.pdf Accessed 6 January 2020
4. Bhat, J., Yadav, P.: The sector of handicrafts and its share in Indian economy. Arabian J. Bus. Manage. Rev. (2016). [online] S3. Available at https://www.hilarispublisher.com/open-access/the-sector-of-handicrafts-and-its-share-in-indian-economy-2223-5833-1000S3_009.pdf Accessed 1 October 2020
5. Majeed, I.: Indian handicraft industry and globalization: an analysis of issues and challenges. Amer. Int. J. Res. Humanities Arts Soc. Sci. [online] **25**(1), 129–135 (2019). Available at https://www.academia.edu/38635476/Indian_Handicraft_Industry_and_Globalization_An_Analysis_of_Issues_and_Challenges Accessed 4 January 2020
6. Artisan.gov.in. (2017). [online] Available at http://artisan.gov.in/NewsPdf/Compadium2018-19.pdf Accessed 5 January 2020
7. http://www.handicrafts.nic.in/Pdf/AWARD%20GUIDELINES%202018.pdf
8. Leyden, R., Dummett, M.: Ganjifa. Victoria and Albert Museum, London (1982)
9. Jain, S.; An award-winning craftsman and the long legacy of wooden toys on his shoulders. November 2017 [online] The Better India (2020). Available at https://www.thebetterindia.com/122140/wooden-toys-sawantwadi-rural-art-maharashtra-crafts-tradition/ (Accessed 7 January 2020)
10. Kanekar, A.; Queries About Handicraft Industry. (Telephonic interview by Priyanka Mangaonkar-Vaiude) (2020)

11. Handicrafts in Maharashtra Wooden Toys Of Savantvadi and Coir Ropes of Achare. [ebook] The Maharashtra Census Office, Bombay: Government of India (2020). Available at http://lsi.gov.in:8081/jspui/bitstream/123456789/4055/1/49538_1961_RAT.pdf Accessed 11 January 2020
12. Sankaran, P.: Artisans of India. Serials Publications, New Delhi (2011)
13. Das, M. et al.:Craft documentation of terracotta relief and pottery-Molela, Pokharan & Ramgad" Unpublished: NID Ahmedabad (1994)
14. The FOLD (2018). PE_7_Sonigra—The FOLD. [Online] Available at https://thisisthefold.org/pe7-sonigra Accessed 4 January 2020
15. Saraswat, S.: sustainable craft practices and space-making in tech design contexts. In: Makers Crafters, Educators: Working for Cultural Change, Routledge, Chapter 40, Taylor & Francis group (2019)

Additional Reference

16. Ramaswamy, V.:'Through history'. In: Celebrating Craft: A Symposium on the State of Handicraft (2003)
17. Designers Meet Artisans: A Practical Guide: Craft Revival Trust, Artesanías de Colombia S A. and UNESCO (2005)
18. Kramrisch, S.: Traditions of the Indian craftsman. J. Amer. Folklore **71**(281), 224 (1958)
19. Baral, B. et al.: Wooden Toys –Sawantwadi: tradition and history of wooden toys (2020). [Online] Available at http://www.dsource.in/resource/wooden-toys-sawantwadi/introduction Accessed 7 January 2020
20. Tawade, J.: Queries About Handicraft Industry. (Telephonic interview by Priyanka Mangaonkar-Vaiude) (2020)
21. Bhandari, S.: Queries About Handmade toys. (Personal interview by Priyanka Mangaonkar-Vaiude) (2020)
22. Wood, S.: Sustaining crafts and livelihoods: handmade in India. Craft + design enquiry **03** (2012)

Other References

23. Multi Media source https://www.youtube.com/watch?v=14jqpk5ZlFs. Accessed 1 January 2020
24. Multi Media source https://www.youtube.com/watch?v=fuLJVJ6t0dQ. Accessed 1 January 2020
25. Press Information Bureau (2019) July 26 Retrieved from https://pib.gov.in/newsite/PrintRelease.aspx?relid=192314

Image References

26. Image 1 & Image 2. https://www.heartforartonline.com
27. Image 3. https://www.thehindu.com/society/history-and-culture/game-with-ganjifa-cards/article29962309.ece
28. Image 4 & Image 7 Minu Joshi
29. Image 5 Priyanka Mangaonkar Vaiude

30. Image 6. https://udaipurtimes.com/news/facelift-for-udaipur-railway-station/c74416-w2859-cid139860-s10698.htm

Part IV
Eco-design, Sustainable Manufacturing, Design for Sustainability

Chapter 68
Mapping the Topoi of Sustainable—Design—Thinking in Indian Context

Asit J. Bhatt

Abstract Here, the question is this: 'Is there an Indian way of "Sustainable—Design—Thinking"?' The paper argues that our current modes of 'sustainable—design—thinking' are limiting as they are informed by the rationalist humanist Cartesian notion of mind's autonomy over the material body whilst separating thinking from making and knowing from doing. The paper traverses across the recent developments in cognitive science and maps the paradigm shift—to view oneself not as distinct, but as an intrinsic part of all life and to remain critical to the dualist Cartesian framework—i.e. man versus environment, culture versus nature, mind versus matter, self-versus other, etc. In this 'non-dual' context, 'post-humanism' is situated.

यस्तु सर्वानि भूतान्यात्मन्येवानुपश्यति ।
सर्वभूतेषु चात्मान ततो न विजुगुप्सते ॥

The wise man, who realizes all beings as not distinct from his own self, and his own self as the self of all beings, does not, by virtue of that perception, hate anyone. (Isha Upanishada:verse 6) 'Post-humanism' calls for a 'monistic' framework, thus, the Indian ways of 'Sustainable—Design—Thinking' appear to have a great potential. Particularly, the concepts of '*advaitā*' (non-dual) and '*ahimsā*' (non-violence) that may pave a way forward for the sustainable future for all. The paper concludes not by answering the question, rather attempts to develop a ground or context through which the plausible answers may emerge.

68.1 Introduction

The idea of 'sustainable development' and the environmental concerns regarding climate change and global warming have been an integral part of the core agendas of the United Nations General Assembly (UNGA). In the year 2009, UNGA announced 22nd April as 'International Mother Earth Day' in order to propagate and also to

A. J. Bhatt (✉)
NIFT Gandhinagar, Gandhinagar, India
e-mail: asit.bhatt@nift.ac.in

A. Chakrabarti et al. (eds.), *Design for Tomorrow—Volume 1*, Smart Innovation, Systems and Technologies 221, https://doi.org/10.1007/978-981-16-0041-8_68

maintain a 'harmony with nature' while satisfying the economical, social and environmental needs in conducting our present and future life on Earth. At the 65th UNGA, in October 2010, during the second session on the 'Sustainable Development: Harmony with Nature', it was noted that many of the ancient civilisations and indigenous cultures have a rich history of understanding and maintaining a symbiotic connection that fosters a mutually beneficial relationship between human beings and nature. Doesn't this point towards the ancient Indian wisdom: '*vasudhaiva kutumbakam*'—that '*vasudhā*' (the Earth) '*iva*' (is) '*kutumbakam*' (a family)? Vandana Shiva, an eminent Indian scholar and activist, advocates such an attitude to inculcate the values of the traditional or indigenous Indian wisdom.

Indian psychologist Sudhir Kakar has noted that in India, people have a different kind of relationship to their immediate environment than the people in the west. Kakar opined that for the Indian people, outer objects do not have a separate independent existence, but these objects are intimately related to the self. In this precise sense, for the people in India, there lies almost no distinction between their environment and themselves. It is quite common among ancient aboriginal and tribal cultures to view themselves as an integral part of all living and non-living entities in their world. Amadou Hâmpaté Bâ, a prominent African intellectual has noted that as per the traditional wisdom of Bambara tribe of Mali, every actions and even gestures were considered to trigger a play of the invisible forces of life. These forces sprung from *Se*—great primitive creative power—an aspect of the supreme being known as '*Maa Ngala*' [1].

For hundreds of years, the crafts communities, in India, have manifested their traditional skills and the knowledge in maintaining their patterns of life in harmony with the patterns of nature. The idea of performing '*ayuda puja*' suggests a deep spiritual connect among the maker/artisan, matter/material and made/thing. In this context, it is quite timely, to pose the question such as this: Is there an Indian way of '(sustainable—(design)—thinking)'? This question echoes A. K. Ramanujan's question—'Is there an Indian way of thinking?' Here, Ramanujan [2], using Stanislaviski's method, poses the question in four different ways by emphasising on 'is', 'an', 'Indian' and 'thinking'; and in turn, perhaps, turning it into a kind of 'context-sensitive' playful act. Each emphasis evokes a different type of question. This performativity of the figure of speech in Stanislaviski's method does open-up new doors and windows in otherwise a closed room. Similarly, by combining 'sustainable design' + 'design thinking' = (sustainable—(design)—thinking), the question suggests that in Indian context, the idea of 'sustainable design' and 'design thinking' are closely related. Thus, the paper does not attempt to answer the question directly; rather it intends to develop a context in which the question may be posed differently so that the lines of thinking and plausible answers may emerge in the mind(s) of the reader.

68.2 Design and the 'Designerly' Thinking

There are two broadly defined categories of 'design thinking': first, 'designerly thinking', which refers to the professional practice of the designer (skills and competencies), methods of working and decision making rooted in the academic training of a designer; second, 'design thinking' imply the application of 'designerly thinking' by non-designers in the other domain such as the management [3]. 'Designerly thinking' is then further expanded into five sub-categories as follows: (1) As the creation of artefacts [4], (2) as a reflexive practice [5], (3) as a problem solving activity [6, 7], (4) as a way of reasoning/making sense of things [8–10] and (5) as a creation of meaning [11]. And 'design thinking' is then expanded into three sub-categories as follows: (1) As a design company: IDEO's way of working with design and innovation [12, 13], (2) as to approach unforeseen organisational problems [14] and (3) as a theory in the management [15]. The study points out at two distinct modes: 'designerly and design thinking' [3].

During the 1960s, 'design thinking' and 'design activism'—challenging the neo-liberal capitalist drive for utter consumerism resulting in social and environmental degradation and political subordination, appear to have emerged in tandem. The growing concerns for the environmental and social design were clearly expressed by Buckminster Fuller, Victor Papanek and others throughout the 1960s and 1970 s marking a point of departure of a kind of 'design activism' that is rooted in our current edifice under the rubric of 'sustainable design'.

68.3 Design Activism/Sustainable Design

Here, designers grapple with the irreducible complexity of the problem—to translate the existing un-sustainable production system into the preferred sustainable socio-technical system, and hence, often find themselves in distinguishing between what Herbert Simon termed as the 'well-structured' and 'ill-structured' problems [3, 6] and also between what Horst Rittel termed as the 'tame' and 'wicked' problems [6, 7]. It is therefore preferred that designers work outside their silos and team-up with the other experts in a multidisciplinary format finding solutions to the 'ill-structured' and 'wicked' problems [6]. 'Designerly and design thinking' utilises the sensibilities of the designer to satisfy needs of the people in terms of technical feasibility and economic viability [13] triggering a process towards building sustainable systems, services and products while keeping the end-user in mind.

According to Kees Dorst [16] even abduction thinking is associated with the problem solving that is also recognised in constructivist thinking as an epistemological mechanism [17]. However, the problems appear to be spectacularly contextual in the abduction thinking. The end-user centric approach is developed into the human-centred approach [18]. The human-centred approach takes human biases and heuristics into consideration, where the attitudes, beliefs, preferences and values of the end-user inform the 'designerly and design thinking'.

It is been suggested that a complex system, i.e. the climate, an environment or a society, cannot be reduced to merely a collection of its basic units, not because the system is not constituted by its basic units or constituents, but because a lot of information appears to dissolve in the process [19]. To make any prediction about such a complex system is equally ironical, in nature, as one cannot make interventions that aim to control a complex system [19] because the complex systems are very sensitive towards perturbations, even a slightest error or disturbance may develop into a major effect, and therefore, the complex system often has a short-term predictability but a long-term unpredictability. Luke Feast [19] cites an example of the weather report, where the meteorological office may analyse the data from various weather stations all over the world and forecast the weather about a week in advance. If Earth's atmosphere was a closed system, by increasing the number of stations and its computational capacity, the meteorological office could forecast a much longer period; but since the atmosphere of the Earth and its weather is a complex heterogeneous system, where intensive differences in pressure, moisture, temperature, etc., may trigger emergent effects causing errors or disturbances, and so it is unpredictable in the long-run. Thus, exact solutions cannot be attained within a complex system and so the problems have to be resolved within the real-life situation. For Feast [19], Deleuze suggests an epistemological shift in the dichotomy, similar to that of Simon and Rittel, between the 'axiomatic' (royal science) and 'problematic' (minor science).

68.4 Towards a Deleuzian Design Process

The essentialist model of design that is still been practiced and taught, originated during Renaissance; where '*disegno*' (design) implied drawing an idea from the mind into a physical world [20]. For Leon Batista Alberti [21], design was a product of thought that emerged from rational power of the mind, which implied superiority of the mental faculty over the manual faculty, the status of '*arte*' (fine art) was elevated over '*artigianato*' (handicrafts) and a genius was favoured, who was the 'mastermind' rather than the master craftsman [22]. This brought a shift in the role of a designer from the master craftsman to the 'mastermind', from a skilful expert to an intellectual genius [23].

This essentialist model may be traced back to Aristotle's ontology or the '*hylomorphic*' doctrine, where all substances are basically constituted of the form (*morphe*) as well as matter (*hyle*). Form precedes matter and carries the essence that is manifested through matter. Designer conceives the form or the essence and then represents it in a drawing. In this representational framework, the mind functions as a storehouse of passive internal representational structures and procedures [24]. Deleuze was critical to the essentialist '*hylomorphic*' doctrine à la Aristotle, where the matter (*hyle*) is merely an inert receptacle and the forms (*morphe*) are imposed on the matter (*hyle*) from without [25].

A representational view of the mind involves Cartesian 'dualism'—the mind-body dichotomy. This representational view of the mind is been challenged by the findings in cognitive science as follows: 'enactive cognition' [26, 27] 'distributed cognition' [28], 'situated cognition' [29], 'extended cognition' [30] and 'embodied cognition' [31]. And this development in the cognitive science has again shifted the focus back to what Deleuze called 'minor sciences'—such as pottery, weaving, blacksmithing, carpentry and so on, where thinking and making go hand in hand and doing and knowing are intra-active [32]. That is to say that the form (*morphe*) is developed in collaboration with the matter (*hyle*), rather than imposing the form (*morphe*) on the matter (*hyle*) from without and in doing so one meets the material halfway [25].

Luke Feast [19] calls for a design process based on the 'problematic' or 'minor sciences' model that is termed, here, as the 'Deleuzian design process' to put a multitude of approaches under an umbrella term assimilating the multi- as well as cross-disciplinary scholarship. Such a design process or 'design and designerly thinking' is indeed urgently required in order to deal with the complexity of problems that are encountered in 'sustainable design'. The '*hylomorphic*' model of 'royal sciences' views matter as a homogenous mass and an inert receptacle, while the artisanal model of 'minor science' is capable of negotiating the nonlinear, intensive and complex system and solving the 'ill-structured' or 'wicked' problems [19].

There has been a growing interest in the 'Deleuzian design process' and its 'problematic' or 'minor science' model [19, 24, 25, 33–41]. It is difficult to conceive any meaningful discourse on 'sustainable design' without taking into consideration the 'problematic' or 'minor science' model. The ideas of 'design and designerly thinking' and its 'designer-centric' approach are gradually waning due to the acknowledgement of non-human actors and their agencies and thereby decentering the designer as a primary agent in designing.

68.5 Anthropocentrism Versus Zoe-Centrism

Dr. Charles David Keeling had started 'Mauna Loa' measurements in 1958 from the Islands of Hawaii; and since then, the measurements are been registered in what is known as the 'Keeling Curve'—named after Dr. Keeling. He was perhaps the first researcher to point that the level of $CO2$ were on a constant rise and this rise of $CO2$ in the atmosphere on Earth posed a serious challenge. The first measurements, taken in 1958, registered the level of $CO2$ at 315 ppm. The most recent measurements, taken on March 25, 2019, registered the level of $CO2$ at 410 ppm. It is been accepted that since the advent of 'Industrial Revolution', human activities on Earth have geological implications and due to the exponential growth in the human activities, since nineteenth century, the Earth has moved from the '*Holocene*' into what is been proposed to be termed as the '*Anthropocene*' [42].

In 1968, NASA released the picture titled—'Earthrise', where the Earth appeared as a globe or a sphere rising over the horizon of the moon as it was viewed from Apollo 8 in orbit around the moon. Dipesh Chakrabarty [43] has noted that the epochal consciousness—in Karl Jesper's phrase—throughout the later half of the twentieth century—repudiated the idea of the Earth (*vasudhā*) as a mother, while carrying forward the history of European expansion and global capitalism. Chakrabarty has also noted that globalisation narratives are '*homo-centric*', whereas climate science literatures are '*zoe-centric*'; in other words, while humans are the central feature and occupy a core in the globalisation narratives, the climate science literatures evoke the idea of '*zoe*' that Chakrabarty articulates, after Hannah Arendt and Georgio Agamben, as the 'bare-life'. Greeks used two different terms for 'life'—'*bios*' as the form or manner in which the life is lived, and '*zoe*' as the biological fact of life. Chakrabarty [43] then reiterates the anecdote that James Lovelock recalls in writing a fiction called 'The Greening Mars' along with his friend Michael Allaby. The book speculates how humans can make the planet Mars habitable? The act of making a place habitable involves the vision of terraforming. Preparing for a new colonial expansion to a place offering distinctively novel environmental challenges. For Lovelock, such terraforming acts, are informed by the human desire to transform a place of a planetary scale by massive technological apparatus resembling gigantic bulldozers and agribusiness. However, Lovelock prefers a '*zoe-centric*' idea to make the planet Mars a home for 'life' by first making it habitable for the bacterial life.

68.6 (Sustainable—(Design)—Thinking)

Michel Foucault [44] has noted that Immanuel Kant had asked: '*Was ist der Mensch?*' (What is the human?) And in turn, rendered '*mensch*' (human) as the 'empirical-transcendental doublet'—both as the object and the subject of knowledge. Foucault further noted that the idea of 'man' as the 'episteme' appeared in the beginning of the nineteenth century. For Foucault, Nietzsche anticipated the crisis in the idea of 'man'—and had discovered the point where man and God belong to one another, and for this very reason, the death of the second (God) would also imply the disappearance of the first (man); therefore, the promise of '*übermensch*' (superman) signifies, as such 'first and foremost the imminence of the death of the man' [44].

(Sustainable—(Design)—Thinking) marks a continuum of 'sustainable design' and 'design thinking'; in order to counter the representational view of the mind and its dualistic anthropocentric framework. In this precise sense, it has an affinity with what is broadly termed as the 'post-humanism' and its conceptual foundation: a neo-Spinozist monistic ontology that assumes a radical immanence [41]. Here, '*advaitā*' (non-dual) philosophy of Adi Shankara is quite pertinent, where the plane of immanence may be understood as the '*brāhman*', where the absolute truth cannot be conceived by the isolation of the individual mind (*ego cogito*), but can only be realised by merging in its infinity. Isn't this 'unity with infinity' marked by the interdependence and oneness of all living as well as non-living entities in the cosmos?

Doesn't this oneness of all living as well as non-living entities imply the eraser of the 'man' ('*ātman*' = '*brāhman*')?

A well-known environmental activist Arne Naess' conception of 'deep-ecology' is based upon the idea of '*ahimsā*' (non-violence)—a core of the belief in the essential oneness of all life [45]. The phrase: '*ahimsā parmo dharma*' (non-violence as one's ultimate or highest duty) marks a very intimate bonding or entanglement between the '*ātman*' (individual self) and '*brāhman*' (universal self). This may also pave the way forward from what Dumont [46] called an 'individualistic society'—in which an individual defines him or herself as independent of relationships, via his or her control over the process as well as products; towards a 'holistic society'—in which an individual defines him or herself in relation to the society as a whole. The idea of an individual's duties or '*karmā*' (action or work) in the 'holistic' societies is embedded in the cultural fabric of the society and is indeed marked by one's relationship to the cosmos [47].

Artisans had acquired a pivotal place within the village communities in Indian society, as the artisans made these village communities and societies self-contained and self-sustained; which stood for 'the good of all' [1]. In the village communities, artisans work within a 'holistic' framework, while the designer mostly work within the 'individualistic' framework and perhaps, this is why Kamaladevi Chattopadhyay held her scepticism about the role of a designer in the Indian context [1].

68.7 Concluding Notes in/as Suggestions

While traversing the uneven contours of 'sustainable—(design)—thinking', the paper draws parallels and crossovers among researches in design, in cognitive science and archives of Indian wisdom to prepare a ground that may serve as a context or bases to pose the question: 'Is there an Indian way of (sustainable—(design)—thinking)?' It is an open question and it entails the spirit of A. K. Ramanujan's question—not to seek a particular answer, but to articulate a conceptual framework through which plausible answers may emerge. The paper attempts to outline the 'problematic' or 'minor science' and 'post-human' framework to think non-anthropocentrically while developing ontological as well as axiological concepts such as '*advaitā*' (non-dualism) and '*ahimsā*' (non-violence) that may pave a way forward for an Indian way of (sustainable—(design)—thinking). This may indeed help in decolonising the 'design and the designerly thinking' by appropriating the artisanal skills and knowledge; and in turn, invert the system that is currently determining the design education as well as cultural production in India.

References

1. Adamson, G.: The Craft Reader. Bloomsbury, London (2016)
2. Ramanujan, A.K.: The Collected Essays of A. K. Ramanujan. Oxford University Press, New Delhi (2013)
3. Johansson, U., Woodilla, J., Çetinkaya, M.: Design thinking: past, present and possible futures. **22**, 02 (2013). https://doi.org/10.1111/caim.12023
4. Simon, H.: The Sciences of the Artificial. MIT Press, Cambridge (1969)
5. Schön, D.: The Reflective Practitioner: How Professionals Think in Action. Basic Books, Cambridge (1983)
6. Buchanan, R.: Wicked problems in design thinking. Design Issues: vol. 08, No. 02, pp. 05–21. MIT press, Cambridge (1992)
7. Rittel, H., Webber, M.: Dilemmas in a general theory of planning. vol. 04, pp. 161–169. Elsevier Scientific Publishing Company, Amsterdam (1973)
8. Lawson, B.: How Designers Think: The Design Process Demystified. Architectural Press, Oxford (2006)
9. Cross, N.: Designerly Ways of Knowing. Berg, Oxford (2006)
10. Cross, N.: Design Thinking. Berg, Oxford (2011)
11. Krippendorff, K.: The Semantic Turn: A New Foundation for Design. Taylor and Francis, Oxford (2006)
12. Kelley, T.: The Art of Innovation: Lessons in Creativity from IDEO. America's leading design firm, Doubleday, New York (2001)
13. Brown, T.: Design thinking. Harvard Business Review, vol. 86. Harvard Business Review Press, Mumbai (2008)
14. Dunne, D., Martin, R.: Design thinking and how it will change management education. Academy of Manage. Learn. Educ. **05**(4), 512–523 (2006). ISSN 1944-9585
15. Boland, R., Collopy, F.: Managing as Designing. Stanford University Press, Redwood City (2004)
16. Dorst, K.: Design problems and design paradoxes. Des. Issues **22**(03), 04–17 (2006). https://doi.org/10.1162/desi.2006.22.3.4
17. Riegler, A.: Abductive reasoning as a way of world making. Found. Sci. **06**(04), 361–383 (2001) Springer
18. Norman, D.: In: Rethinking Design Thinking (2013). www.core77.com/posts/24579
19. Feast, L.: In: A Problematic Approach to the Science of Sustainable Design, Hong Kong, IASDR'07 (2007). ISBN 988-99101-4-4
20. Hill, J.: Building the drawing. Archit. Design **75** (2005). https://doi.org/10.1002/ad.98
21. Alberti, L.: On the art of building in ten books. In: Rykwert, J., Leach, N., Tavernor, R. (eds.) (trans) MIT Press, Cambridge (1988)
22. Vasari, G.: The lives of the artists. In: Bondonella, J., Bondonella, P., (eds.) (trans.) Oxford University Press, Oxford (1998)
23. Ingold, T.: The Perception of the Environment. Routledge, London (2000)
24. Malafouris, L.: How Things Shape the Mind. MIT Press, Cambridge (2013)
25. Herzogenrath, B.: Deleuze Guattari and Ecology. Palgrave Macmillan, New York (2009)
26. Varela, F.J., Thompson, E., Rosch, E.: The Embodied Mind. MIT Press, Cambridge (1991)
27. Noë, A.: Action in Perception. MIT Press, Cambridge (2005)
28. Hutchins, E.: Cognition in the Wild. MIT Press, Cambridge (1995)
29. Clancey, W.J.: Situated Cognition. Cambridge University Press, Cambridge (1997)
30. Clark, A.: Being There: Putting Brain, Body and World Together Again. MIT Press, Cambridge (1997)
31. Lakoff, G.: Philosophy in the Flesh: the Embodied Mind and its Challenge to Western Thought. Basic Books, New York (1999)
32. Barad, K.: Meeting the Universe Halfway: Quantum Physics and the Entanglement of Matter and Meaning. Duke University Press, Durham (2007)

33. Lather, P.: Getting Lost: Feminist Efforts Toward a Double(d) Science. University of New York Press, New York (2007)
34. Bannett, J.: Vibrant Matter: A Political Ecology of Things. Duke University Press, Durham (2010)
35. Colebrook, C.: Deleuze and the Meaning of Life. Continuum, London (2010)
36. Mazzei, L., McCoy, K.: Thinking with Deleuze in qualitative research (2010). https://doi.org/10.1080/09518398.2010.500634
37. Ingold, T.: The textility of making. Cambridge J. Econ. **34**, 01 (2010). https://doi.org/10.1093/cje/bep042
38. Ingold, T.: Making: Anthropology, Archaeology, Art and Architecture. Routledge, London (2013)
39. Parr, A., Zaretsky, M.: New Directions in Sustainable Design. Routledge, New York (2011)
40. Marenko, B., Brassett, J.: Deleuze and Design. Edinburgh University Press, Edinburgh (2015)
41. Braidotti, R.: Posthuman Ecologies, Complexity and Process After Deleuze. Polity Press, Oxford (2019)
42. Rockström, Steffen, W., Noone, K., Persson, Å., Chapin, F., Lambin, E., Lenton, T., Scheffer, M., Folke, C., Schellnhuber, H., Nykvist, B., De Wit, C., Hughes, T., van der Leeuw, S., Rodhe, H., Sörlin, S., Snyder, P., Costanza, R., Svedin, U., Falkenmark, M., Karlberg, L., Corell, R., Fabry, V., Hansen, J., Walker, B., Liverman, D., Richardson, K., Crutzen, P., Foley, J.: Planetary Boundaries: Exploring the Safe Operating Space for Humanity (2009). https://ced.agro.uba.ar/granchaco/sites/default/files/pdf/sem6/Rockstorm%20et%20al%202009.pdf
43. Chakrabarty, D.: The Human condition in the Anthropocene. The Tanner Lectures in Human Values at Yale University (2015). https://tannerlectures.utah.edu/Chakrabarty%20manuscript.pdf
44. Foucault, M.: The Order of Things. Nutech, New Delhi (2012)
45. Bhatt, A.: Mahatma Gandhi and the model of fashion education, activism and system(s) in India. Clothing Cultures **5**(1), (2018) Intellect Publication. https://doi.org/10.1386/cc.5.1.177_1
46. Dumont, L.: Homo Hierarchicus: The Caste System and its Implications. University of Chicago Press, Chicago (1980)
47. Bhatt, E., Jhabvala, R.: The idea of work. Econ. Political Weekly **39**(48), (2004) Mumbai. ISSN: 2349-8846

Chapter 69
Bio-bricks: Circular Economy and New Products

Priyabrata Rautray, Avik Roy, Deepak John Mathew, and Boris Eisenbart

Abstract Agricultural waste burning is a significant source of pollution in India, especially after the harvesting season. Bio-brick was developed as an alternative and sustainable building material that is made up of agricultural waste. And at the same time, it will also lead to the reduction of air pollution and create new jobs at the grassroots level. This research paper defines the relationship between bio-bricks and circular economy model and its benefits to the rural economy and society as a whole. It has both direct and indirect benefits such as reduction of air pollution and the potential to be reused at the EOL or as filler material. It also documents the process of initial testing and product development. Product iteration started with creating blocks of bio-bricks, followed by initial experiments to determine physical properties. These results led to the redevelopment of product in term of forms and functionalities. New iterations can also be used for industrial settings such as enclosure for machinery and equipment as a sandwich or reinforced boards. Manufacturing of bio-bricks has the potential to eliminate the problem of disposal of agricultural waste at source. It can create a new economic model for farmers and lead to the development of agriculture-based industries. Based on our research findings, we find bio-bricks can be developed as a carbon-negative, sustainable and economically viable material for construction. With the right kind of product development and incentives, it can diversify into numerous products satisfying the needs of an ecologically sensitive future.

P. Rautray (✉) · D. J. Mathew
IIT Hyderabad, Hyderabad, India
e-mail: md17resch11001@iith.ac.in

A. Roy
KIIT Bhubaneswar, Bhubaneswar, India

B. Eisenbart
Swinburne University Melbourne, Melbourne, Australia

A. Chakrabarti et al. (eds.), *Design for Tomorrow—Volume 1*, Smart Innovation, Systems and Technologies 221, https://doi.org/10.1007/978-981-16-0041-8_69

69.1 Introduction

India is home to some of the most polluted cities of the world in terms of its air quality. Data shows that around 20 out of the 30 most polluted cities of the world are in India [1]. There are numerous reasons for air pollution, but in our research, we are focussing on agricultural waste burning [2]. Crop burning results in some of the most severe air pollutions in the northern part of the country, especially in the winter months as the smoke gets trapped with the moisture in the air to create a poisonous haze called smog. This results in numerous temporary and permanent health issues of people across the region. Loss of lives due to air pollution has also surged in recent years. As architects and designers, we developed an innovative product that uses the agro-waste by converting them into a sustainable material, which can directly be used in the construction industry [3]. In this paper, we are trying to highlight our research on the relationship between circular economy and bio-bricks, initial testing and product development. We did an extensive literature review on the circular economy and how our product suits rural economic needs. Secondly, we did initial experiments on this material to find its physical properties and its appropriate uses. We have documented all the tests done on this material and its outcome, which led to new product development. Finally, based on the findings of our examination, we were able to develop appropriate products based on the strength of the material. Different compositions of bio-bricks were also tried and documented. Based on these findings, we were able to predict the socio-economical impact of bio-brick on the rural economy and how it will help in the creation of jobs at the grassroots level.

69.1.1 Bio-bricks

Bio-bricks came into existence as a solution to two significant issues, i.e. pollution, due to stubble burning and lack of sustainable building materials for construction industries in India [4, 5]. It was mainly targeted at reprocessing agricultural waste after harvest and reduce crop burning. The process of making bio-bricks is quite simple and easy to follow and was conceived, keeping in mind indigenous material and Indian rural socio-economic conditions. This sustainable building material can be made on-site or in centralised facilities and can cater to the local construction industries (Fig. 69.1 example of bio-bricks). With government initiatives like 'smart cities', 'housing for all' and development of Tier-II cities, large-scale construction has moved closer to the agriculture farmlands [6]. It is estimated to have a massive investment of around US$ 650 billion in the next 20 years [7]. Thus, sustainable development can go hand in hand with the development of the rural economy as well as reduction of air pollution due to crop burning.

Once we developed the list of primary materials and composition for bio-bricks, we went for the initial product experimentation to find out the physical properties. These tests include necessary load-bearing test, burning test, heat insulation and water

Fig. 69.1 Samples of bio-bricks

absorption test. The results from these initial tests helped and guided us in developing the next line of products. We followed the process of upcycling [8] and iterative design process, which follows a cyclic path of exploration, concept development, prototyping, experimentation and reflection [9].

69.2 Circular Economy

The circular economy is a process which turns materials which have reached their end of life (EOL) into new resources, thereby closing the loop. Thus, it will help in reducing waste, increasing productivity and jobs. According to European Commission Directive 2008/98/EC on waste; a developing economy like India should move from the current linear path of "resource-product-waste" to the three paradigms of 'reduce, reuse and recycle' as explained in Fig. 69.2.

Most of the processes of manufacturing are based on the 'take', 'make' and 'dispose' model [11], which is an unsustainable process. Even if we try to increase the efficiency of manufacturing, it will only slow down the rate of depletion of resources. Considering the limited availability of resources, production of new raw materials leads to pollution, economic exploitation and environmental disaster. Most of the building materials fall into the same category, thereby create an unsustainable process.

69.2.1 Scope of the Circular Economy in India

With India's population about to reach 1.64 billion by 2050 and the world population to hit 10 billion by 2050 [12], the need for raw materials would increase exponentially. It is expected to reach 180% of 1990′s level [13]; hence, we need to move immediately from a linear process to the model of a circular economy. India currently has 17% of the world population with a tremendous need for resources. It has one of the fastest-growing economies. India produces 62 million tons of solid waste, and according to

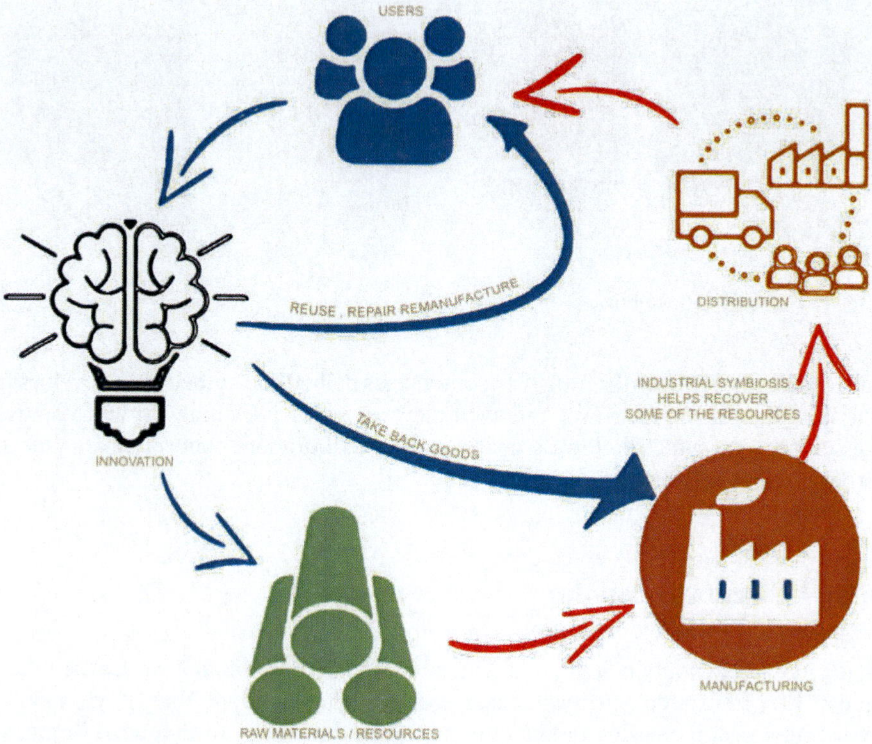

Fig. 69.2 Cyclic process of circular economy [10]

the government data, the organic waste composition will rise from 40 to 60% [14]. Only 20% of it is organic waste recycled, and the rest of it is dumped at landfill sites or burnt in case of agriculture waste. This problem is compounded by inefficient recycling processes and lack of industrial capabilities [15].

The construction industry in India is one of the largest consumers of energy and resources. It has the potential to generate up to US$ 218 billion in 2030 and US$ 624 billion in 2050 through the circular economy [16]. Industries can make a lot of savings and profits using this model. According to current data, the circular economy model can reduce greenhouse gases up to 23% by 2030 and up to 44% by 2050 [16].

The research on the manufacturing of bio-bricks is based on the circular economy model, where we are using a waste material (agro-waste) and converting it to a resource, as shown in Fig. 69.3.

Agro-waste is produced in such vast quantities in India that presently it has become difficult to dispose of and is mostly burnt. So if we can create a symbiotic relationship between agro-waste and the construction industry, then it will result in a new market for sustainable building material. Farmers can manufacture these bio-bricks at the site or can sell the agro-waste to the centralised manufacturing units. Once the EOL of the product is reached, it can be crushed again and used to make fresh bio-bricks

Fig. 69.3 Bio-bricks in the circular economic model

or as filler material, thereby completing the cycle. Thus, leading to job creations at the grassroots level and adding to the income of farmers.

69.2.2 Economics of Bio-bricks

Bio-bricks can provide both direct and indirect financial benefit to the farmers and society as a whole. The direct economic benefit to farmers can be divided into revenue generated by selling agro-waste to industrial bio-bricks manufactures and secondly by making bio-bricks themselves. The indirect benefits can be categorised into health, environmental and effective saving in building construction.

Direct benefits

Currently, due to lack of any financial interest, a large part of the stubble produced after harvest are burned on the field leading to a spike in air pollution. But by developing bio-bricks as a building material, large-scale construction industries can procure the agro-wastes after each harvest to create the sustainable building material, thereby adding value to an agricultural bi-product which is considered as a waste. Thus, now farmers not only gain from the harvest but also the agro-waste. Secondly, due to the ease of making bio-bricks, farmers can easily manufacture them and use them in local construction. This will lead to the creation of jobs at the grassroots level and keep them busy in the off-season.

Indirect benefits

Indirect or passive benefits are enormous in terms of health, environment and the construction industry. It has been estimated that crop burning leads to a loss of US $300million every year [17]. Additionally, it increases the incidences of acute respiratory infections three times. Research shows lungs capacity had dropped from a mean of force vital capacity (FVC) from 98% in 2008 to 92% in 2009 and further dipped to 88% during the post-harvest season due to crop burning [18]. Furthermore, a total annual loss due to illness, productivity loss, etc., in the state of Punjab alone is around 76 millions in Indian rupees [2].

In terms of environment, bio-bricks are compatible with the circular economy where they can be crushed and reused as raw materials when they reach their end of life (EOL) or as filler material. According to the National Policy for Management of Crop Residues (NPMCR), a substantial amount of minerals and nutrients are lost due to crop burning as well, which further increases the input cost of fertilisers [19]. Fired clay bricks use around 350 million tons of fertile topsoil leading to soil erosion [20]. India, which is the second-largest producer of fire clay bricks, consumes approximately 110 million tonnes of coal (excluding the electricity used) to produce 1000 billion blocks [21]. So if we can replace a small portion of non-load-bearing construction with bio-bricks, we can reduce a massive amount of pollution.

Bio-bricks were found to have a low average density of 423.7 kg/m3 (as calculated from prototypes) [3], this will drastically reduce the overall dead load on the superstructure of a high-rise building. Thus, reduce the consumption of concrete, steel and lowering the construction cost. Secondly, these bio-bricks are good heat and sound insulators. They can be effectively used in walls to reduce heat gain and reduce air-conditioning load, thereby making buildings more sustainable.

69.3 Product Development

Product development can be described as a continuous and dynamic process whose speed and flexibility depends on its adaptation of the needs and requirements of the ever-changing environment [22]. For our research project, we followed the process of the upcycling and iterative design process, which follows a cyclic path which includes exploration, concept development, prototyping, experimentation and reflection, as shown in Fig. 69.4. With each cycle, the process becomes more specific.

Product development starts with creating the first bio-bricks from a careful selection of dry agro-waste chopped into small particles (approximately 15 mm long). Then, this is mixed with lime slurry and stone dust [3]. Then, the mixture is poured into moulds and is compacted with a wooden dowel. Figure 69.5 shows the process of making bio-bricks graphically. Once the first batch of bio-bricks was made, we tried to explore new compositions and different forms of building material.

India has a rich cultural heritage of vernacular architecture. From the ancient time in India, lime and clay mortar have been used as a primary construction material. Inspired by the traditional building materials, natural additives were added to the mixture to improve the strength and binding of the bio-bricks, as shown in Fig. 69.6.

Bel fruit pulp (Aegle Marmelos) was the first natural binder selected for the explorative process. The fruit pulp was soaked in water for three hours after which it was crushed by hand, and the paste was added to the bio-bricks mixture before moulding. In the second composition, the equal amount of river clay slurry was added to the lime slurry and was mixed with the chopped agro-waste. And in the third alternative, the same amount of liquid molasses was added to the lime binder

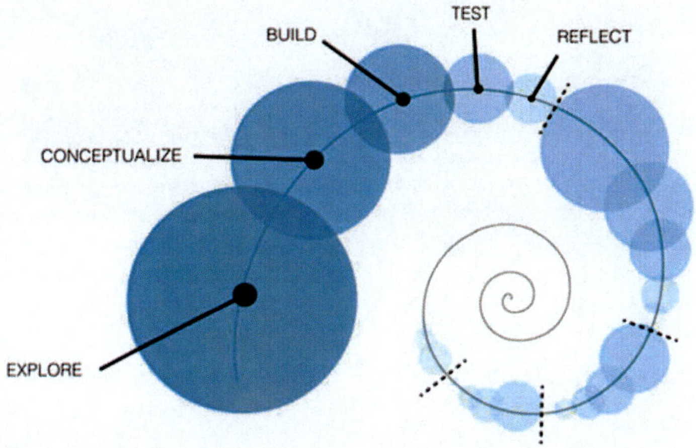

Fig. 69.4 Iterative design process with five steps [23]

Fig. 69.5 Process of making bio-bricks

Fig. 69.6 Different natural binders added to the bio-bricks mixture

before mixing. Once these compositions were ready, two different sizes of bio-bricks were made (Fig. 69.7).

Fig. 69.7 Bio-bricks sizes—(150 × 150 × 50 mm) and (100 × 100 × 100 mm)

69.3.1 Initial Testing

Once we developed a sufficient number of bio-bricks, we went for initial testing to understand the fundamental properties and direction for future product development. After a month of air drying, these bio-bricks attained their final working strength, and they were ready for the basic physical test such as load-bearing test, thermal insulation test, fire retardant test and water absorption test.

Load-bearing test

100 × 100 × 100 mm size bio-bricks were selected for the load-bearing test using a compression testing machine. Figure 69.8 highlights the load-bearing test, and the result confirmed that these bio-bricks have very low load-bearing capacities and are not suitable for load-bearing constructions. But can be used as filler walls in frame structure and as a partition wall.

Thermal insulation test

For this test, 150 × 150 × 50 mm bio-brick was used. The idea of the test was to verify if these bio-bricks were able to act as thermal insulators and help in reducing thermal gain in buildings. This test was performed using a blow torch and heating

Bio-Bricks	Load-Bearing (KN)
Sample 1	0.982
Sample 2	0.978
Sample 3	0.963
Sample 4	0.988

Fig. 69.8 Load-bearing test and result

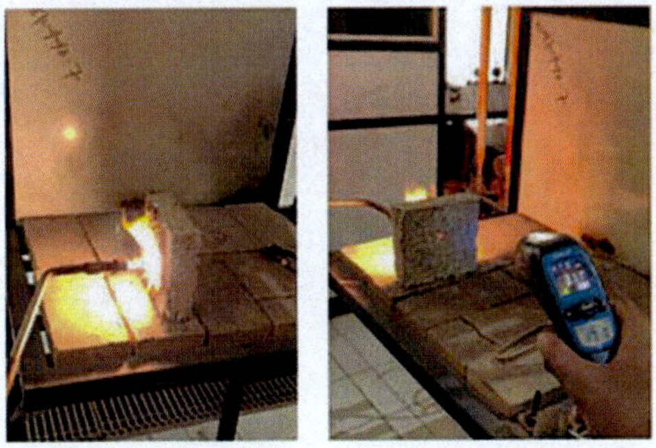

Bio-Bricks	Temperature after 3 minutes exposure to 550 °C or more
Sample 1	47.5 °C
Sample 2	49.2 °C
Sample 3	52.3 °C
Sample 4	55.1 °C

Fig. 69.9 Thermal insulation test and result

one side of the brick with flame as high as 550–600 °C for about three minutes, and the temperature was measured on the other side of the block using an optical thermometer (Fig. 69.9). The result shows that bio-bricks have excellent thermal insulation properties and can be used as insulation boards or panels.

Fire retardant test

This test was vital as we were using a highly flammable material, i.e. dry agro-waste like wheat straw or paddy straw. We had to make sure that the final building material was resistant to fire and reduce the spread of fire. For this experiment, we used a very high-temperature flame of 800 °C using a blow torch and tried to burn from the corner of the bio-brick. But to our satisfaction, once the flame was removed after three minutes, the fire died within 5 s, and it did not spread to other parts of the block. Figure 69.10 highlights the sequence of the fire retardant test.

Fig. 69.10 Sequence of fire retardant test

Bio-Bricks	Water absorption %
Sample 1	61.9
Sample 2	62.6
Sample 3	65.4
Sample 4	58.8

Fig. 69.11 Water absorption test

Water absorption test

The major component of bio-bricks being agro-waste, we wanted to check the effect of water absorption and stability of the block once soaked in water. For this experiment, we used a 150 × 150 × 50 mm block which was left in a water container for around 24 h. The bio-bricks were checked for deformation and cracks due to water absorption, and it was found to be stable after the period. For water absorption percentage, the dry weight and the weight after 24-h soaking was taken—the difference of weight divided by the original weight multiplied by100 gave the percentage (Fig. 69.11).

69.3.2 New Product Development

After completion of initial product experiment, we understood that this new building material is not suitable for load-bearing construction. Thus, we developed a new methodology to create partition wall and insulation panel along with standard bio-bricks. Figure 69.12 illustrates the laying of bio-brick material to the construction of the wall on site. Two feet high wooden boards shuttering can be used to lay two feet tall wall at a time. The bio-bricks material can be directly poured into the moulds and rammed with wooden dowels. After 24–48 h, the shuttering can be removed and placed above the previous wall, and the process can be repeated.

The inner side of the wall can be finished with wooden cladding or lime plaster, and the outer side can be completed with lime plaster or cement mortar, as shown in Fig. 69.13.

Insulation Board

From the initial thermal experiments, we found the bio-bricks material is an excellent heat insulator and has good fire retardant properties. Based on these attributes, we tried to develop insulation boards for ceiling and wall to reduce the heat gain in temporary structure or low-cost housing. Figure 69.14 shows the making of these insulation boards. These boards can be used in an industrial setting like heat and sound insulation panel or as roof panels to reduce heat gain during summer.

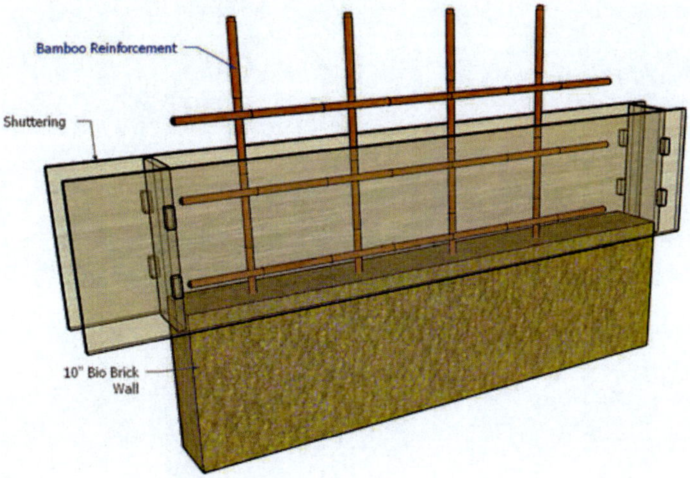

Fig. 69.12 Shuttering and bamboo reinforced for on-site wall construction

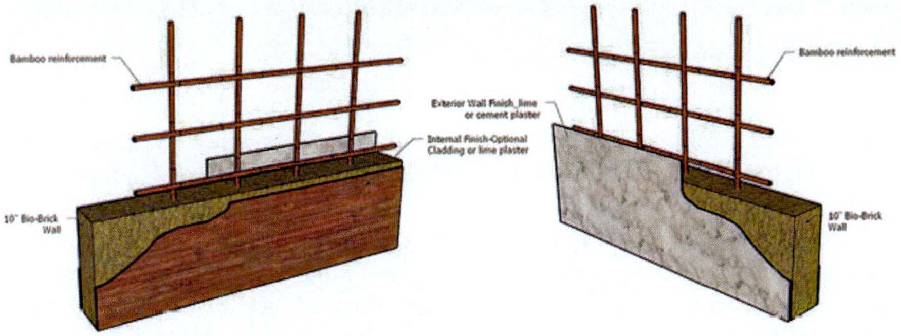

Fig. 69.13 Inner and outer wall finish

69.4 Discussion and Conclusion

Through our research, we tried to explore the circular economy and how bio-bricks as a building material in its different forms follows the ideology. We explored all the direct and indirect benefit of bio-bricks and how it is going to improve the life standard of the farmers at the grassroots level. We found bio-bricks used in construction can be reused as raw material at the end of life, thereby reducing wastage and improving recycling. This attribute of bio-bricks makes it not only sustainable but also apt for the circular economy. We postulate that large-scale adaptation of this material in construction industries will lead to a considerable reduction of air pollution due to stubble burning. Creation of jobs can boost rural economy and reduce migration of farmers in lean period or off-season.

Fig. 69.14 Making of the insulation panel

In the initial material testing, we found bio-bricks has low load-bearing strength but has high thermal insulation and fire retardant properties. These findings led to new product development of on-site casting of the wall using wooden shuttering and bamboo reinforced walls for low-cost housing. Further research is being conducted to test this construction style, and we believe that further research will bring out new novelties. Secondly, we created insulation panel boards which can be used under corrugated roofs and can be sandwiched between walls to reduce solar heat gain. This can significantly reduce the air-conditioning load of the building.

Moreover, these panels can be used in machine rooms and industrial settings to provide both heat and sound insulation. We are planning to conduct an ISO-based building material test to generate a specification sheet and get more confirmed data sets. We are also planning to study the impact of fungi, termite, rodents, etc. Thus, we can consider that bio-bricks as a sustainable building material has the potential to become material for future building construction.

References

1. IQAir.: World's Most Polluted Cities in 2019 M2.5 Ranking | AirVisual (2020). https://www.airvisual.com/world-most-polluted-cities.
2. Kumar, P., Kumar, S.: Valuing the health effects of air pollution from agricultural residue burning. *Unpublished paper* (2010)

3. Rautray, P., Roy, A., Mathew, D.J., Eisenbart, B.: Bio-Brick–development of sustainable and cost effective building material. Proc. Design Soc.: Int. Conf. Eng. Design **1**(1), 3171–3180 (2019). https://doi.org/10.1017/dsi.2019.324
4. Awasthi, A., Singh, N., Mittal, S., Gupta, P.K., Agarwal, R.: Effects of agriculture crop residue burning on children and young on PFTs in North West India. Sci. Total Environ. **408**(20), 4440–4445 (2010)
5. Madurwar, M.V., Ralegaonkar, R.V., Mandavgane, S.A.: Application of agro-waste for sustainable construction materials: a review. Construct. Build. Mater. **38**, 872–878 (2013)
6. De Pandit, S.: The role of the pradhan mantri awas yojana (urban), 2015 in financial inclusion in India. Int. J. Recent Sci. Res. **8**(8), 18959–18962 (2017)
7. Sarkar, A., Dhavalikar, U., Agrawal, V., Morris, S.: Examination of Affordable Housing Policies in India (2016)
8. Oyenuga, A.A., Bhamidimarri, R., Researcher, P.: Upcycling ideas for sustainable construction and demolition waste management: challenges, opportunities and boundaries. Int. J. Innov. Res. Sci. Eng. Technol **6**(3) (2017)
9. Swann, C.: Action research and the practice of design. Design Issues **18**(1), 49–61 (2002)
10. Stahel, W.R.: Circular economy. Nature **531** (2016)
11. Foundation, E.M.: Towards the Circular Economy: Accelerating the Scale-up Across Global Supply Chains. Ellen MacArthur Foundation Isle of Wight, UK (2014)
12. Gupta, S., Deshpande, R.: Water for India in 2050: first-order assessment of available options. Current Sci. 1216–1224 (2004)
13. Di Maio, F., Rem, P.C.: A robust indicator for promoting circular economy through recycling. J. Environ. Protect. **6**(10), 1095 (2015)
14. Joshi, R., Ahmed, S.: Status and challenges of municipal solid waste management in India: a review. Cogent Environ. Sci. **2**(1), 1139434 (2016)
15. Goyal, S., Esposito, M., Kapoor, A.: Circular economy business models in developing economies: lessons from India on reduce, recycle, and reuse paradigms. Thunderbird Int. Bus. Rev. **60**(5), 729–740 (2018)
16. Gupta, S.: Barriers and opportunities in circular economy in the construction industry in India (2019)
17. Chakrabarti, S., Khan, M.T., Kishore, A., Roy, D., Scott, S.P.: Risk of acute respiratory infection from crop burning in India: estimating disease burden and economic welfare from satellite and national health survey data for 250,000 persons. Int. J. Epidemiol. **48**(4), 1113–1124 (2019). https://doi.org/10.1093/ije/dyz022
18. Adler, T.: Respiratory health: measuring the health effects of crop burning. Environ. Health Perspect. **118**(11), A475–A475 (2010). https://doi.org/10.1289/ehp.118-a475
19. Department_of_Agriculture_&_Cooperation: National Policy for Management of Crop Residues (NPMCR) (2014)
20. Banerjee, S.: Brick kilns contribute about 9% of total black carbon emissions in India. Centre for Science and Environment (2015)
21. Skinder, B.M., Pandit, A., Sheikh, A., Ganai, B.: Brick kilns: cause of atmospheric pollution. J. Pollut. Eff. Cont. **2**(2) (2014)
22. Imai, K., Nonaka, I., Takeuchi, H.: Managing the New Product Development Process: How Japanese Companies Learn and Unlearn: Division of Research. Harvard Business School Boston, MA (1984)
23. Uğur, S.: In: Wearing Embodied Emotions: A Practice Based Design Research on Wearable Technology. Springer (2013)

Chapter 70
Sustainable Design and Development of Stubble Removing Agricultural Machine for Stopping the Burning of Paddy Stubble

Harmanpreet Singh, Prabir Sarkar, Harpreet Singh, and Fateh Singh

Abstract India is an emerging economy and has shown a great increase in the production of crops particularly after the green revolution. New kinds of seeds, fertilizers, machinery, and sowing techniques have helped a lot in meeting the needs of the country in terms of grains produced per hectare. Due to the rapid development, there is an increase in the overall pollution giving rise to greenhouse gases, and ultimately depleting the ozone layer along with the rise in the seawater level causing various problems. Air pollution is particularly increasing various respiratory diseases in small children, which is a major concern for society. Rice and wheat are the most popular cultivation in India, which produced stubble. However, wheat stubble can be used as fodder for the cattle, but rice stubble is avoided as it has high silica content. To reduce the ever-increasing pollution, the stubble removing machine (SRM) is designed and developed for fully controlling the pollution due to the burning of paddy stubble. Earlier the only method to clear the agriculture land was to burn the paddy stubble after the harvesting or use machines designed for different applications. Burning of stubble creates a lot of harmful gases like CO_x, NO_x, and their derivatives. Various other machines were developed including stubble management system (SMS) and its higher version called super SMS but were not so promising according to their application procedure. The present developed machine works on the principle of cutting, collecting, and managing the paddy stubble and clears the fields instantaneously for the next crop and also vanishing the problem of pollution due to the burning of stubble. We propose a system-level solution for the same.

70.1 Introduction

Agriculture is the basic need of any economy. India being the second largest populated country has managed easily to fulfill its food needs by rapidly improving its agriculture techniques. The major role in the country's economy is agriculture and is

H. Singh (✉) · P. Sarkar · H. Singh · F. Singh
Department of Mechanical Engineering, Indian Institute of Technology Ropar, Punjab, Rupnagar 140001, India
e-mail: harmanpreet.19mez0003@iitrpr.ac.in

A. Chakrabarti et al. (eds.), *Design for Tomorrow—Volume 1*, Smart Innovation, Systems and Technologies 221, https://doi.org/10.1007/978-981-16-0041-8_70

the only preferred job done in most of the villages. Earlier the agricultural practices including irrigation, harvesting, and maintaining of crops were done with the help of animals and labor, and the by-products from the agriculture found their efficient use within that area causing no harm to the people but with the advancement, in the agricultural tools and machinery, these practices vanished very soon [1]. The agricultural machinery though makes the work easy but also creates various problems involving excess remains of by-products after harvesting, wastage of grains, dust, pollution, and many more small problems that are often ignored [2]. Harvesting machine used for harvesting the crops like here paddy leaves the straw behind which contains the moisture and requires some time to get dry and mixed up in the soil. Earlier the harvesting was done by hands, straw was used in the fields for various activities including making ropes for tightening the other harvested crops, in construction of storage places for the cow dung cakes, animal fodder, etc. The problem related to the remains of small straw does not exist earlier and came into existence with the use of machines in the harvesting of crops.

70.2 Background

To get rid of the paddy straw fast and making the land ready for the next crop, farmers unintentionally burn the paddy straw in their fields causing pollution and greenhouse gas emission. On burning the paddy straw, the harmful gases like CO_x and NO_x along with their derivatives are released into the environment. The key concern here is the various kinds of respiratory diseases including asthma, allergy rhinitis, and many more that are degrading human life. Several machines and methods were developed to solve the problem but all have different kinds of shortcomings. The SRM was designed and developed purposely to solve the paddy straw burning problem. The machine is very efficient in collecting and conveying the paddy stubble. The paddy stubble is a naturally occurring substance and thus is a renewable resource that is also sustainable [3, 4]. The sustainable use of biomass substances like paddy straw, rice husk, bagasse, and even the corn cob for the industrial purpose is very important in terms of preventing the depletion of non-renewable resources and also the environmental degradation [5]. Though biomass is a renewable form of energy, it still produces greenhouse gases which are harmful to the environment. The reduction in greenhouse gases is observed using these biomass products when compared to the non-renewable resources [6, 7]. The details of the exhaust produced by the burning of the biomass and their combinations with other non-renewable resources are observed resulting in the reduction of CO_x, SO_x, and NO_x production [8]. There are various ways in which the biomass like paddy straw, rice husk, etc., can be used in suitable ratios with the standard fuels in the small-scale power plants resulting in better outputs in terms of money, environment, and efficient usage of agriculture wastage [9]. There is a long list of valuable products that can be made out of the paddy straw which also includes the board, plates, and bowls, etc., few of which are patented.

The agricultural waste is also used as a kind of compost that can be feed to the soil for revivifying the deficient nutrients, helpful for the health of the next crop.

Comparison with other alternatives: The machine is made to conquer the stubble burning problem in the fields, there were several kinds of method and product which were developed few of them are already in market but have few shortcomings that are discussed in Table 70.1 and alternative to which the current SRM was developed. The development of SRM costs approximately INR 3.5 lakhs and the price could be even reduced in mass production. The basis of references for the shortcomings was the personal experience of all the farmers while using these machines as they encountered the problems mentioned in Table 70.1 due to which they preferred burning the stubble rather to use the existing machines. However, there is no written scientific evidence published for the same but few interviews of farmers are commonly addressed in newspapers mentioning the problems associated with using the existing machines to manage the paddy stubble.

Table 70.1 Details of the alternative technology for paddy stubble problem

Machine	Technology	Price range	Shortcomings
Straw management system (SMS)	Unit attached behind the combine harvester	Average (1–1.25 lakhs INR approx.)	• Mileage reduction of combine • Make the machine bulky • Thrashed straw retards the growth of new seeds sown
Chopper	Addition of thrashing mechanism in reaper	High (3.25–3.50 lakhs INR approx.)	• Choking of straw in blades • Operational cost is high especially for the paddy straw
Baler	For making the knots of straw	Very high (14 lakhs INR approx.)	• Needs liner and already trimmed straw • Trimmed straw should be dry • Time-consuming process
Happy seeder	Wheat sowing machine	Average (2 lakhs INR approx.)	• Seeds sown using machine lacks nutrients, sunlight • Termite problem for the sown seed • The process used gives easy shelter to the pests and insects

70.3 Aim and Methodology

The main aim was to solve the problem of paddy stubble burning causing pollution in the air, which is also the foundation of many respiratory problems. The major problem here is the paddy stubble burning which can be simply stopped by avoiding the burning of stubble, but that initiates many other farming problems like how to manage the stubble quickly, avoiding the high cost and there should also be no delay in sowing the next crop on the same land which requires a lot of preparations. The solution to the problem must satisfy all these problems. To deal with the solution, a survey was done with the farmers.

Survey: The survey was done with the farmers, their requirements, and solutions to the problem were discussed in detail. The main discussion was how the paddy stubble can be managed efficiently with the least effort and investment along with the task should be done quickly avoiding the delay for the next crop. The survey session was like the brainstorming session and lots of solutions were suggested, which were compiled and the result was that we need to develop a machine that can manage the stubble, i.e., can cut it and take that stubble out from the field quickly so that farmers can prepare the land for next crop.

The brainstorming session was done with 7–8 active farmers from the 2 nearby villages along with the 2 professors and 10 students of IIT Ropar including the authors. The ideas that came out after the session to solve the stubble burning problem are mentioned in Table 70.2.

Table 70.2 Details of the alternative technology for paddy stubble problem

Questions	Solutions
How to get rid of the stubble from the field	• By crushing it in small parts and spreading it in fields • To make a cheaper knotter baler that can make the bundles of stubble • To make the machine that can cut and take the stubble out to some other place • To make the use of some chemical that can destroy the stubble itself in fields • Development of better techniques to sow the seeds of next crop in standing paddy stubble
Amount of money farmers can spend to clear the stubble	The majority answered it 300–500 INR per acre
Time limit to clear the stubble after harvesting of the paddy crop	Farmers were expecting the kind of machine that can clear the field within a week or less time
The machine worth 4 lakhs approximately would be preferred by farmers for managing the stubble over the burning	Nearly all of the farmers replied that the big landowners (with 20 acres or above land) can manage to buy such machine tools but the average farmers need a subsidy or other cooperative farming institutions help for purchasing the machine

The solution for making the kind of machine along with the result of various brainstorming sessions included the modification of various available farming tools like the tractor, combine harvester, or making a completely new machine that will serve the required purpose. Finally, the idea was to design and develop the machine which will perform the cutting of the paddy stubble and convey the paddy stubble to a trolley that can take the stubble away from the farming land. Then, the stubble can be used as the source of valuable products.

70.4 Development of machine

Development of the chamber: The stubble removing machine is dedicatedly built to address the problem of the stubble burning common in the northern part of India. The farming practices in the northern part of India are a bit different due to the environmental conditions, three crops are sown and harvested which includes both Kharif and rabi crops, while in other parts of the country, the Kharif crops are harvested thrice and the people even leave the land uncultivated for few months which reduces the chances for stubble burning. During the brainstorming sessions and surveys done with the farmers, it was clear that the farmer wanted to immediately clear their agricultural land for sowing the next crop after the harvest of the paddy. The main issue was studied and the SRM was designed and then simultaneously developed according to the needs of farmers.

The machine has two conveying chambers which consist of the chain drive and toothed channels that are given the motion in the anticlockwise direction for conveying the stubble. These conveying chambers are attached and are capable of conveying and transferring the stubble to the trolley, which is used to take away the stubble out from the agricultural field instantaneously.

The chamber was designed and built according to the need of conveying the paddy straw which is sticky due to the presence of moisture. Initially, the CAD drawing in 2D shown in Fig. 70.1 was made and part preparation was done.

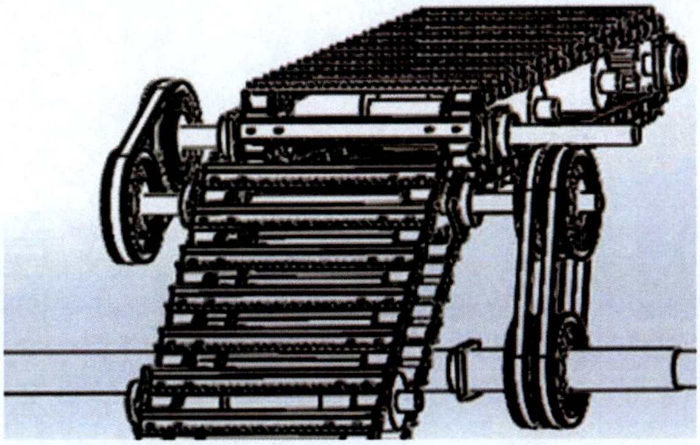

Fig. 70.1 2D CAD drawing of the conveying chamber

Fig. 70.2 Final developed machine

Cutter bar and Chassis: The machine has two large parts main chassis and cutter bar they are attached to form one machine; the final assembled machine is shown in Fig. 70.2. Cutter bar consists of cutting blade, material pushing rake, auger shaft welded with helical thread guides, and throwing fringes. The main chassis consists of a conveyor chamber which further consists of an adjustable shaft (for shock-absorbing purpose), shafts, chain drive, and material pushing bar which is adjusted on the chain drive.

The rake installed in the cutting bar pushes the stubble toward the machine. Cutting blade in cutter bar cuts the stubble from the field with approx. height of 2–3 inches from the ground. Then, the auger shaft, fringes, and helical guides come into play, they guide the trimmed stubble to the conveyor chamber. The material pushing bar sucks the material immediately and material goes through the conveyor chamber to the trolley attached at the back of the machine. The conveyor chamber has shafts, pulleys, and v-belts attached in a manner to run a mechanism of material pushing bar, which are attached on the chain drive. The whole system works with the help of a tractor, power take-off shaft is connected to the gearbox kept on the chassis of the machine and it further passes on the rotary motion through the interconnected shafts and v-belts.

70.5 Working of the Machine

The machine is tractor operated where the tractor connects to the power shaft, which is attached to the cutter bar. The cutting mechanism is coupled to the power shaft by a belt-pulley mechanism. The cutting mechanism comprising a plurality of cutting blades connected to a cutter blade strip and the cutter blade strip is connected to a bell crank. The bell crank and a cam assembly convert the rotating motion of a camshaft to the reciprocating motion of cutting blade strip. The reciprocating motion of the cutting blade strip results in the cutting of the stalks of the plants in every stroke with an approximate height of 2–3 inches from the ground. When the stalks come in contact with the two cutting blades, it gets crushed and is detached from the parent root. Further, the trimmed stalk/stubble is pushed and guided by the stubble pushing mechanism to the trolley. The machine further comprising a stubble conveying mechanism conveys the stubble to the trolley.

70.5.1 *Trials/Testing of the Machine*

After the development of the machine, it was tested for its working in nearby fields for cutting the paddy straw. Farmers were shown the working of the machine and they gave their positive response and were eager to use the machine personally. After trial 1, it was observed that the impulsive forces caused the jerks in the conveying mechanism, and rpm was also less during the cutting process in uneven fields that caused hurdles. Though the machine performed well, we made a minor change in its complete assembling like the front shaft of the chamber was made moveable for absorbing the shocks and the pulley was changed to increase the rpm.

The second trial of the machine was done after a few months with live coverage on media [10–13]. The machine was made to clear the land area at a specific time, where the machine was able to trim the stubble leaving an inch from the ground and was enough to convey that stubble to the trolley. The media, farmers evaluated the machine live and there were also few words for further development of the machine [14].

70.6 Results and Discussions

After the development of the machine, it was tested four times and shown a positive response in terms of managing the stubble. The machine is capable of cutting the stubble in one go and requires a few minutes to clean one acre of land. The stubble remains after cutting is of an inch or two and that is suitable to get mixed to the land during the plowing done for the next crop. In terms of efficiency, the machine is almost 95% efficient in cutting and conveying the stubble. As the machine is a prototype and is first of its kind, there were also a few shortcomings which can be improved with proper calculations on the next product. The machine weight management and its chassis can be improved further.

When the machine is compared with the already available solutions in the market for managing the stubble and related problems, it emerges as a better solution and is very efficient in terms of the time it takes for clearing the land according to the farmer's demand and their feedback after the trials. The straw collected by the machine is suitable to use for making valuable products like boards, plates, and bowls [15]. Moreover, these straws can be used to burn in the thermal plants alongside the standard fuel reducing the overall pollutions caused by the standard fuel and making the whole process and approach sustainable.

70.7 Conclusion

After the trials of the machine, there was a positive response from the farmers and the comparison with the other available solutions in the market made the current machine a suitable, economical and efficient way of clearing the paddy stubble from the land, which otherwise was a source of air pollution harming many lives. The paddy stubble is managed efficiently by the SRM and the stubble collected can be used in many ways including its use as a fuel with the mixture to standard fuels in thermal plants, to make the valuable products like the plates, bowls, and the boards which can be used in multiple ways which include the packaging of goods. The paddy stubble can also be used by the farmer in terms of compost that can give nutrition to the soil itself and make it fit for the next crops. Using the SRM, farmer can generate secondary income by selling his paddy straw or by manufacturing these valuable products. This machine concludes the idea of sustainability by efficiently managing the farm waste and using it as a source of good income.

IP Details of the machine

Indian patent Application No. 201911041636

Filed on: 15.10.2019

The aspect that we monopolize for our disclosure is the conveyor chamber and related parts including adjustable shaft for shock-absorbing purpose, shafts, chain drive, and material pushing bar which is adjusted on the chain drive. These combinations are specially designed to collect the stubble from the field instantaneously after harvesting the paddy crop. This instantaneous collection of stubble from the field will clear the land for the next crop without the pollution problems which use to occur when farmers have no other option to burn the stubble for clearing the agricultural land.

References

1. Ellis.E.C, Wang S.M.: Sustainable Traditional Agriculture in the Tai Lake Region of China, pp. 177–193. Agriculture, Ecosystems and Enviornment (1996)
2. Zhang, Y., Chen, D., Yin, Y., Wang, X., Wang, S.: Experimental study of feed rate related factors of combine harvester based on grey correlation, pp. 2405–8963. International Federation of Automatic Control (2018)
3. Kuti, O.: Impact of charred palm kernel shell on the calorific value of composite sawdust briquette. J. Eng. Appl. Sci. (2007)
4. Jittabut, P.: Physical and thermal properties of briquette fuels from rice straw and sugarcane leaves by mixing molasses. Energy Procedia Elsevier B.V (2015)

5. Adeleke, A.A., Odusote, J.K., Lasode, O.A., Ikubanni, P.P., Madhurai, M.: Evaluation of thermal decomposition characteristics and kinetic parameters of melina wood. Biofuels **0**(0), 1–7 (2019) Taylor & Francis
6. Eddine, B.T., Salah, M.M.: Solid waste as renewable source of energy: current and future possibility in Algeria. Waste Manage. Valorization: Alternative Technol. **3**(17), 115–141 (2017)
7. Obi, O.F., Okongwu, K.C.: Characterization of fuel briquettes made from a blend of rice husk and palm oil mill sludge. Biomass Convers. Biorefinery Biomass Convers. Biorefinery **6**(4), 449–456 (2016)
8. Adeleke, A.A., Odusote, J.K., Lasode, O.A., Ikubanni, P.P., Malathi, M., Paswan, D.: Densification of coal fines and mildly torrefied biomass into composite fuel using different organic binders. **5**(x), 0–6 (2019)
9. Sharma, S.: A comparative analysis of combustion of Pet coke and Co-firing of pet coke with rice. **5**(12), 79–84 (2015)
10. Times of India. https://timesofindia.indiatimes.com/city/shimla/now-iit-ropar-unveils-stubble-machine-calls-it-in-one-go-solution/articleshow/71946688.cms
11. Hindustan Times. https://www.hindustantimes.com/cities/iit-ropar-develops-cheaper-more-cost-effective-stubble-removing-machine/story-faYBRIb2iBqwf969g9bfTI.html
12. TheTribune. https://www.tribuneindia.com/news/archive/punjab/iit-ropar-gives-hope-develops-low-cost-machine-for-removing-stubble-857243
13. Inuth. https://www.inuth.com/india/how-iit-ropars-cost-effective-machine-can-end-stubble-burning/
14. The Tribune. https://www.youtube.com/watch?v=ke_s-Qvmaws
15. Kriya Labs IIT Delhi Initiative. https://www.kriyalabs.co.in/khaofeco.html

Chapter 71
A Three-Phase Quality Function Deployment Approach for Conceptualizing a Sustainable Product Life Cycle: Case Study of a Blower Heater

Prashant Kumar Singh and Prabir Sarkar

Abstract The increasing awareness and demand for producing environmentally sustainable products have stimulated the industries across the globe to implement eco-friendly practices in the development of various products. It is important to consider the environmental impact while conceptualizing a product, but at the same time, the choices of end users should not be overlooked. In this study, an ecodesign-based quality function deployment (QFD) approach is proposed for developing the sustainable products. The QFD approach used in this study consists of three different phases. The first phase, known as 'house of quality,' shows the relationship between the voice of customers and technical requirements of the product to be developed. In second phase, the relationship of the technical requirements with different part characteristics of the product is established. Third phase relates parts characteristics with various ecodesign practices spread across the entire life cycle of the product. The proposed methodology is applied for designing an environmentally conscious blower heater. The finding of this study shows that 'ensuring a higher durability,' 'reducing energy consumption during use' and 'ensuring easier maintenance and repair' are the most significant ecodesign practices that must be emphasized while designing a blower heater. The methodology developed in this research can assist the designers to develop eco-friendly products through the identification of the most relevant ecodesign practices for individual products.

71.1 Introduction

Consumers are the key stakeholders when it comes to developing environmentally sustainable products [1, 2]. A survey that included 9000 consumers across the globe reported that more than 50% of the consumers emphasized on the need to develop

P. K. Singh (✉) · P. Sarkar
Department of Mechanical Engineering, Indian Institute of Technology Ropar, Rupnagar, Punjab 140001, India
e-mail: pksiitrpr@gmail.com

A. Chakrabarti et al. (eds.), *Design for Tomorrow—Volume 1*, Smart Innovation, Systems and Technologies 221, https://doi.org/10.1007/978-981-16-0041-8_71

environmentally sustainable products [3]. Quality function deployment approach has continuously been used for improving the quality and functionality of products. However, it lacks the consideration of environmental criteria in the development of various products [4]. An integration of QFD with ecodesign can be an effective approach for developing eco-friendly products [5–7]. Ecodesign consists of a set of practices that can be adopted for improving the environmental performance of a product throughout its entire life cycle from cradle to grave [8, 9].

In this study, an ecodesign-based quality function deployment (EQFD) approach has been proposed for developing sustainable products. According to this approach, the concept of ecodesign takes care of the environmental needs of a product, whereas QFD focuses on the customer requirements and the functionality of the product. Rest of the article is structured as: Sect. 71.2 includes a review of the literature which is followed by the methodology of the study given in Sect. 71.3. A case study is presented in Sect. 71.4 and the conclusions drawn from this study are placed at the end under Sect. 71.5.

71.2 Literature Review

Conventional quality function deployment (QFD) method has widely been used for developing various electrical-electronic products [10]. Researchers have integrated the different environmental criteria to the QFD method for developing environmentally benign products. A green QFD methodology was developed by Cristofari et al. [11] for considering the environmental needs in the design of a product. It was further modified by Zhang [12]. QFD method has been integrated with other established methods for designing eco-friendly products. Sakao [13] proposed a method for developing a sustainable 'hair dryer' by integrating QFD, LCA and TRIZ methodologies. Zhang et al. [14] developed a method known as quality function deployment for environment (QFDE) in which the requirements of the consumers were taken care along with technical environmental measures. Younesi and Roghanian [3] integrated QFDE with a fuzzy hybrid method to prioritize the different environmental criteria for sustainable design of transformers. Some of the researchers combined ecodesign methods with QFD for better decision making while developing a new product [5, 15]. Although these studies are significant for developing eco-friendly products but are not able to identify the relevant ecodesign practices for improving the design of a product for better environmental performance. There is a need to develop a method that can identify the relevant improvement strategies for producing more environmentally benign products.

The novelty of this study lies in the fact that it integrates the relevant ecodesign practices with QFD for identifying the key ecodesign improvement strategies in early design stages with a life cycle perspective. It is achieved through the incorporation of a new phase III in QFD which is known as 'life cycle planning,' as shown in Fig. 71.1. Thus, the proposed approach takes care of consumer requirements as well as technical and environmental measures in the early design stages

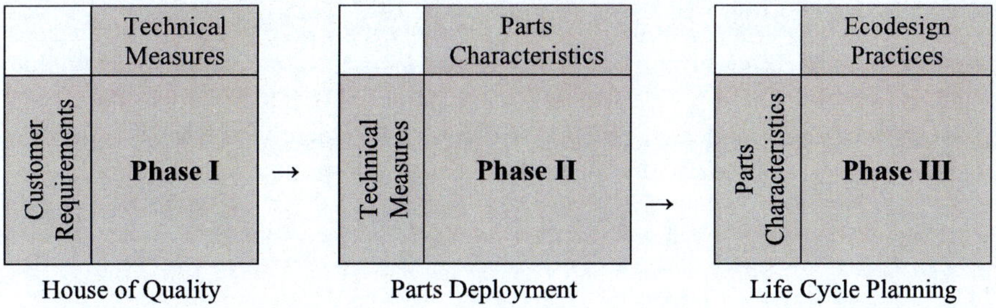

Fig. 71.1 Three different phases of QFD

while developing a product. The consumer requirements, various technical measures and relevant ecodesign practices are adopted from the literature [4].

71.3 Methodology

The methodology of this study is based on the integration of ecodesign practices with traditional quality function deployment (QFD) method. This method was introduced in Japan at the beginning of 1966 for improving the quality of the products by capturing the voice of consumers. Traditional QFD method consists of four phases, but in this study, the conventional phase III and phase IV are eliminated, and a set of ecodesign practices are accommodated in the form of a new phase III which is known as 'life cycle planning' as shown in Fig. 71.1. Thus, this study consists of three phases which are described as follows:

Phase I: This phase includes a 'house of quality' which shows the relationship among various customer requirements (CR) and technical measures (TM). Customer requirements are collected using a five-point rating scale which represents the importance of each CR. Five indicates 'very high importance' and 1 indicates 'very low importance.' The relationships among CR and TM are provided by the decision-making panel using a 1-5-9 scale where 1 indicates 'weak relationship,' 5 indicates 'medium relationship' and 9 indicates 'strong relationship.' In the end of phase I, a relative weight (RW) of each TM is computed (detail is given in Sect. 71.4.2).

Phase II: This phase is known as the 'parts deployment' phase. It shows relationships among technical measures and parts characteristics of different components of the product. The relative weights of technical measures computed in previous phase (i.e., phase I) are utilized in this phase to obtain the weight for each part of the product.

Phase III: In this study, the final phase consists of a set of ecodesign practices which are utilized for process and production planning of each part of the product. It is done for considering the various environmental requirements during the planning stage of the product. This phase is known as 'life cycle planning.' This phase represents relationships among different parts of a product and the different ecodesign practices which are implemented for developing these parts. In the end of this phase, relative weight of each ecodesign practice is computed to identify the most significant practices for sustainable development of the parts and ultimately the product.

71.4 Case Study: Blower Heater

The case study is carried out for a 'blower heater' to identify the key ecodesign practices for the development of an environmentally benign blower heater. The parameters of EQFD approach, i.e., consumer requirements, various technical measures and relevant ecodesign practices for developing an electrical appliance are obtained from the literature and a discussion with the panel of experts. The decision-making panel consists of five members which includes three designers and two production engineers each having work experience of more than eight years in consumer electrical industry. The parameters of EQFD are provided in Table 71.1.

71.4.1 Product Description

A blower heater, also called fan heater, is an electrical appliance which is generally used to warm the surrounding air in a closed space. The main components of a blower heater include a heating element, a fan, one motor, one power cord and the housing, as shown in Fig. 71.2. The weight of the blower is measured which comes out to be 3.36 kg. It has a maximum power rating of 2000 W and possesses an average lifetime of 2–3 years.

71.4.2 House of Quality

In phase I, the house of quality is constructed by assigning the 'relational strength' to each pair of customer requirements (CR) and technical measures (TM) using the inputs of decision makers (as described in Sect. 71.3). Then, the weightage of each CR is multiplied by the 'relational strength' of each TM and the resulting values in each column are added to obtain the raw score (RS) for each TM. Then, the relative weight (RW) of each TM is computed by dividing the RS of a TM by the total sum of all RS. For example, RS and RW of 'air flow' is computed as follow (see Table 71.2):

$$RS_{\text{Airflow}} = (5 \times 9) + (4 \times 1) + (4 \times 1) + (5 \times 5) = 78$$

Table 71.1 Parameters of ecodesign-based quality function deployment (EQFD)

Consumer requirements	Technical measures	Ecodesign practices
Quick heating	Air flow	Choosing non-hazardous materials
Safe operation	Overheat cutout	Selecting lightweight materials
Lightweight	Weight	Selecting recyclable materials
Portable	Volume	Ensuring recovery of components
High durability	Physical lifetime	Reducing energy consumption during use
Energy efficient	Type of materials	Ensuring easier maintenance and repair
Good aesthetics	Toxicity of materials	Ensuring recovery of materials
Non-hazardous materials	Packaging and transportation	Ensuring a safe disposal of waste
Material saving	Energy consumption	
Easy to repair	Assembly and disassembly	
Quiet operation	Noise and vibration	
Cost effective	Rate of recycled materials	

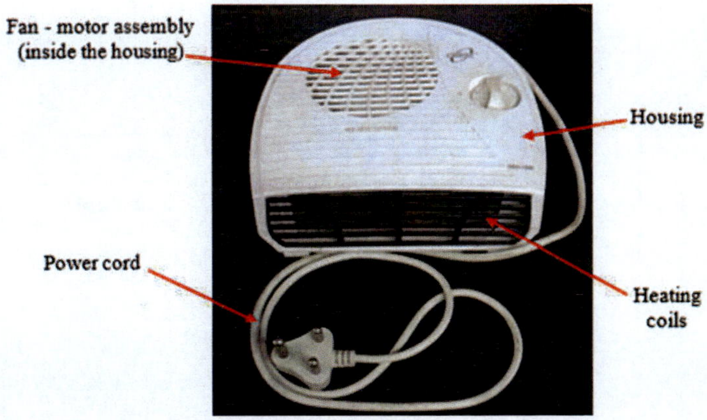

Fig. 71.2 Blower heater

Table 71.2 Phase I of QFD

		Air flow	Overheat cutout	Weight	Volume	Physical lifetime	Type of materials	Toxicity of materials	Packaging and transportation	Energy consumption	Assembly and disassembly	Noise and vibration	Rate of recycled materials
Direction for improvement		↓		↑	↑	↓	↑	↑	↑	↑		↑	↓
	W												
Quick heating	5	9	5				5			9			
Safe operation	4	1	9			1	5	9		1			
Lightweight	3			9	5		9		5	5	5	1	
Portable	2			5	5	1	1				5		
High durability	4	1		5	1	9	9	5		5	9	5	
Energy efficient	5	5	9	1	1		5			9	5	5	
Good aesthetics	3				1		5						
Non-hazardous materials	5						9	9			1		1
Material saving	4			5	5		1		5	1			5
Easy to repair	4			1	1	5	1	1			9		
Quiet operation	3						1					9	
Cost effective	4			1	1	5	5		1	1			5
Raw score (RS)		78	106	90	65	82	226	105	39	137	127	75	41
Relative weight (RW)		0.067	0.091	0.077	0.056	0.070	0.193	0.089	0.033	0.117	0.108	0.064	0.035

Note Strong relationship (9 pts), medium relationship (5 pts) and weak relationship (1 pt)

$$RW_{\text{Airflow}} = 78/(78 + 106 + 90 + 65 + 82 + 226 + 105 + 39 + 137 + 127 + 75 + 41)$$
$$= 0.067$$

It is clear from Table 71.2 that 'type of materials' is the most significant TM with an importance of 19.3% to fulfill most of the consumer requirements. 'Energy consumption' and 'assembly and disassembly' are the next two important TM with importance of 11.7% and 10.8%, respectively.

The relative weights computed in phase I are utilized in phase II (see Table 71.3). The phase II of QFD presents a relationship between technical measures and part characteristics of blower heater. The relative weight of each part of blower heater is computed in a similar manner as in phase I. 'Motor' of the blower heater stands out to be the most important part which will require the main attention in case of any change in the design. 'Housing' is the second most important part to fulfill the technical requirements of blower heater. In the final phase (i.e., phase III), a set of ecodesign practices relevant to electrical-electronic products are adopted to modify the design of the blower heater so that its environmental performance can be improved. It is given in Table 71.4.

Table 71.3 Phase II of QFD

		Heating element	Fan	Motor	Housing	Power cord
	w					
Air flow	0.067	1	5	9	1	
Overheat cutout	0.091	5			1	
Weight	0.077	3	5	9	9	0
Volume	0.056	1	5	9	9	1
Physical lifetime	0.070	9	5	9	9	5
Type of materials	0.193	5		1	9	1
Toxicity of materials	0.089	9		5	5	1
Packaging and transportation	0.033		1	1	5	
Energy consumption	0.117	9	1	9		
Assembly and disassembly	0.108	5	5	5	1	1
Noise and vibration	0.064		5	9	5	
Rate of recycled materials	0.035		1		9	1
Raw score (RS)		4.79	2.39	5.27	5.08	0.91
Relative weight (RW)		0.260	0.130	0.286	0.275	0.049

Table 71.4 Phase III of QFD

		Choosing non-hazardous materials	Selecting lightweight materials	Selecting recyclable materials	Ensuring recovery of components	Reducing energy consumption during use	Ensuring easier maintenance and repair	Optimizing the functionality	Ensuring a higher durability	Ensuring recovery of materials	Ensuring a safe disposal of waste
	w										
Heating element	0.260	5	1	5	5	9	5	5	9	5	5
Fan	0.130	1	5	1			1		5	1	
Motor	0.286	1	1	1	5	9	5	5	5	1	5
Housing	0.275	5	9	9	5		5		5	1	5
Power cord	0.049	1	1	1	1		1		1	1	
Raw score (RS)		3.14	3.72	4.24	4.15	4.91	4.28	2.73	5.84	2.04	4.11
Relative weight (RW)		0.080	0.095	0.108	0.106	0.125	0.109	0.070	0.149	0.052	0.105
Rank		8	7	4	5	2	3	9	1	10	6

In phase III of QFD, all ecodesign practices shown in Table 71.4 are analyzed against each component of the blower heater. This analysis is carried out by utilizing the relative weights of the components of blower heater (presented in the last row of Table 71.3) obtained through phase II of QFD. The phase III of the proposed QFD methodology helps to identify the most significant ecodesign practices for mitigating the negative environmental impact of a product by considering the severity of each component of the product, as shown in Table 71.4.

71.4.3 Results and Discussion

The results of three-phase QFD approach applied on a blower heater shows that 'type of materials' is the most significant technical measure to take care of most of the consumer requirements. Also, 'motor' of the blower heater is the most crucial part. 'Ensuring a higher durability' is the most important ecodesign practice followed by 'reducing energy consumption during use' and 'ensuring easier maintenance and repair' to improve the environmental performance of the blower heater. The improvements that can be achieved through the implementation of these ecodesign practices are discussed below.

i. En*suring a higher durability*: Higher durability of blower heaters can be achieved by using standard components so that they can be replaced or repaired whenever required.
ii. *Reducing energy consumption during use*: Use of lightweight materials, preferably plastics, for developing the fans. It will offer low load on the motor for moving these fans. Ensure an auto-cut system in place to cut the power supply of heating coil once it is fully heated.
iii. *Ensuring easier maintenance and repair*: Minimize the number of joining elements so that it is easy and quick to open and close the housing and the other elements for easier and periodic maintenance and repair.

The practical implication of the proposed methodology is that it will help the designers to understand that which components of a product are more prone to cause the environmental concerns. Thus, the design engineers can choose the relevant ecodesign practices to mitigate these environmental concerns during the early design stages. The main advantage of the proposed methodology is that it enables the design engineers to take decisions for improving the environmental performance of the product in early stages of a product development.

71.5 Conclusion and Direction for Future Research

In this research, an ecodesign-based three-phase quality function deployment approach is used to find the important ecodesign practices for developing an eco-friendly consumer electrical-electronic product. A blower heater has been taken as a case study for showing the applicability of the proposed method. This method is used to analyze the different consumer requirements, technical measures and ecodesign practices in three different phases. Results of this study show that 'type of materials' is a crucial technical measure, whereas 'motor' of the blower heater is the most important part which should be given more emphasis by the designers. Also, 'ensuring a higher durability,' reducing energy consumption during use' and 'ensuring easier maintenance and repair' are identified as the top three ecodesign practices for improving the environmental performance of a blower heater. This study can be used by the designers as a benchmark while developing a consumer electrical-electronic product for a better and environmentally sustainable performance.

As a part of future research, the efficacy of the proposed methodology can be further verified by conducting more case studies with a higher technical requirements and by introducing more number of relevant ecodesign practices. Also, this research is focused mainly on the category of electrical-electronic products. Future studies can be conducted to identify a set of most significant ecodesign practices relevant to some other category of products such as automotive products, pharmaceutical products and others.

References

1. Meadows, D.H., Hoffmann, E.: Consumer integration in sustainable product development. Bus. Strategy Environ. **16**, 322–338 (2007). https://doi.org/10.1002/bse.577
2. Kardos, M., Gabor, M.R., Cristache, N.: Green marketing's roles in sustainability and ecopreneurship. Case study: green packaging's impact on Romanian young consumers' environmental responsibility. Sustainability **11**, 873 (2019). https://doi.org/10.3390/su11030873
3. Younesi, M., Roghanian, E.: A framework for sustainable product design: a hybrid fuzzy approach based on quality function deployment for environment. J. Clean Prod. **108**, 385–394 (2015). https://doi.org/10.1016/j.jclepro.2015.09.028
4. Bereketli, I., Erol Genevois, M.: An integrated QFDE approach for identifying improvement strategies in sustainable product development. J. Clean Prod. **54**, 188–198 (2013). https://doi.org/10.1016/j.jclepro.2013.03.053
5. Romli, A., Prickett, P., Setchi, R., Soe, S.: Integrated eco-design decision-making for sustainable product development. Int. J. Prod. **53**, 549–571 (2015). https://doi.org/10.1080/00207543.2014.958593
6. Ocampo, L.A., Labrador, J.J.T., Jumao-as, A.M.B., Rama, A.M.O.: Integrated multiphase sustainable product design with a hybrid quality function deployment—multi attribute decision making (QFD-MADM) framework. Sustain. Prod. Consump. **24**, 62–78 (2020). https://doi.org/10.1016/j.spc.2020.06.013

7. Puglieri, F.N., Ometto, A.R., Salvador, R., Barros, M.V., Piekarski, C.M., Rodrigues, I.M., Netto, O.D.: An environmental and operational analysis of quality function deployment—based methods. Sustainability **12**, 3486 (2020). https://doi.org/10.3390/su12083486
8. Bhamra, T.A.: Ecodesign: the search for new strategies in product development. P I Mech. Eng. B J. Eng. **218**, 557–569 (2004). https://doi.org/10.1177/095440540421800509
9. Singh, P.K., Sarkar, P.: A framework based on fuzzy Delphi and DEMATEL for sustainable product development: a case of Indian automotive industry. J. Clean Prod. 118991 (2019). https://doi.org/10.1016/j.jclepro.2019.118991
10. Tsai, K.H., Yeh, C.Y., Lo, H.C., Li, C.T., Cheng, C.P., Chang, G.L.: Application of quality function deployment in design of mobile assistive devices. J. Med. Biol. Eng. **28**, 87–93 (2008)
11. Cristofari, M., Deshmukh, A., Wang, B.: Green quality function deployment. In: 4th International Conference on Environmentally Conscious Design and Manufacturing, pp. 297–304 (1996)
12. Zhang, Y.: Green QFD-II: a life cycle approach for environmentally conscious manufacturing by integrating LCA and LCC into QFD matrices. Int. J. Prod. **37**, 1075–1091 (1999). https://doi.org/10.1080/002075499191418
13. Sakao, T.: A QFD-centred design methodology for environmentally conscious product design. Int. J. Prod. **45**, 4143–4162 (2007). https://doi.org/10.1080/00207540701450179
14. Zhang, L., Zhan, Y., Liu, Z.F., Zhang, H.C., Li, B.B.: Development and analysis of design for environment oriented design parameters. J. Clean Prod. **19**, 1723–1733 (2011). https://doi.org/10.1016/j.jclepro.2011.06.012
15. Puglieri, F.N., Ometto, A., Cauchick Miguel, P.A.: Eco-design methods for developing new products based on QFD: a literature analysis. Prod. Manage. Devel. **9**, 23–29 (2011). https://doi.org/10.4322/pmd.2011.003

Chapter 72
Bio-architecture of Living Root Bridges

Srinidhi Ravishankar and Shiva Ji

Abstract Living root bridges are an excellent example of the amalgamation of human innovation and nature. The bridges are grown by methods of tree-shaping using the aerial roots of Ficus Elastica (Rubber fig tree/Indian rubber tree), which grows abundantly in the tropical rainforest regions of Khasi hills, Meghalaya. The practice has been in place for centuries to tackle the issue of crossing robust water bodies. The paper intends to understand the significance of living root bridges in the communities of Meghalaya and their potential to be used as a guiding model for contemporary design practices. It also analyses its impact and usefulness to the communities, and economic growth through domestic and international tourism.

72.1 Introduction

Meghalaya is home to over 100 living root bridges [1]. This unique phenomenon of growing the roots into bridges has evolved over centuries of effort by the people of Meghalaya. The indigenous art of tree-shaping is accelerated by the climatic factors of the state that is characterised by heavy rainfall and high humidity. These bridges are locally called "*Jingkieng Diengjri*", meaning rubber tree bridges [2]. They are mostly grown from a single tree. The living root bridges in Nongriat village and Mawlynnong village are renowned. The bridge in Umshiang, Nongriat, is especially unique for its a double-decker. The technology is not restricted to bridges alone. Ladders, platforms, walkways, and even bleachers have been created by utilising the living root methodology (Figs. 72.1 and 72.2).

S. Ravishankar · S. Ji (✉)
Indian Institute of Technology Hyderabad, Hyderabad, India
e-mail: shivaji@des.iith.ac.in

S. Ravishankar
e-mail: srinidhiravishankar@gmail.com

A. Chakrabarti et al. (eds.), *Design for Tomorrow—Volume 1*, Smart Innovation, Systems and Technologies 221, https://doi.org/10.1007/978-981-16-0041-8_72

Fig. 72.1 Living ladder (*Source* Timothy Allen)

Fig. 72.2 Roots as railing (*Source* Patrick A. Rogers)

72.2 Architecture

72.2.1 Narrative

The Khasi tribe is one of the most prominent tribal groups in Meghalaya. The forefathers of the Khasi tribe are the pioneers of living root bridges. They needed a permanent solution to cross turbulent water bodies. Bamboo or wooden bridges

Fig. 72.3 Living root bridge at Wahthyllong (*Source* Timothy Allen)

were perishable since the region receives heavy rainfall and humidity. The living root bridges were brilliantly devised by the community by leveraging the botanical advantages of Ficus Elastica [3] (Fig. 72.3). The process involves minimal material and cost of maintenance. Most important attributes required are time and patience. These bridges facilitated connectivity to isolated villages within deep jungles across rigorous waters. They can be viewed as an important socio-ecological component of the Khasi tribe wherein building these bridges is a fraternal activity. The art of growing and training the roots of the rubber tree is passed on through generations. The development and upkeep of the bridges are a communal affair. These bridges breathe life in a sense where moss, lichen, small insects, and leaves are part of the fabric. Their superior strength over concrete bridges is attributed to the fact that they are living and hence can resist ravages of termites and harsh weather. Grown over 20 years, these organic marvels can last over 500 years. Unlike modern concrete bridges, their strength only increases over time. Heavy rain that is characteristic of the region catalyses the growth of the root bridges.

72.2.2 Design of the Bridges

The living root bridges are grown, unlike concrete bridges which are built. The bridge becomes usable after 15–30 years of growth and has the capacity to hold 50 people at once. Their length may differ from 50 to 100 ft. [4]. Ficus Elastica trees are planted on either side of the river banks. The aerial root or the secondary roots of the trees are directed horizontally with the help of scaffolding across the river (Fig. 72.4).

Fig. 72.4 View from below a living root bridge (*Source* Timothy Allen)

Bamboo and wooden scaffolding are used to train the roots and make them grow in the middle. Eventually, the roots supersede the supports and traverse to the other side and anchor in the soil of the opposite bank.

In some cases, areca palm or betel nut tree trunks are used to guide the aerial roots. In this method, the roots gain added nourishment from the trunks. However, these scaffolding and supports need regular replacement since they rot due to high humidity and precipitation. In modern times, steel wire suspension bridges are used as scaffolding. This proves to be efficient since people can use the bridge while it is in the process of growth. This conventional and ecological process combined is the methodology adopted by the community in the present scenario. After the completion of the bridge skeleton, flat stones from the river bed are inserted between the threaded roots to ease walking. Eventually, as people tread and compact the surface, the stones are engulfed by the woody growth and become a part of the fabric of the bridge. Several roots are threaded together and intertwined at different intervals to form handrails and supports for people (Fig. 72.5).

72.3 Impression

Apart from being a means of commuting, the living root bridges also contribute to the surrounding ecosystem. They help mitigate soil erosion, absorb carbon from the atmosphere and help in sequestration, cause a minimal carbon footprint, and display strength even during flash floods and cyclones. They act as a great case example of socio-ecological sustainability at its best [5].

Fig. 72.5 Creating a new handrail (*Source* Timothy Allen)

In addition to this, the bridges, along with the backdrop of dense vegetation and waterbody, provide a magical experience for new visitors. They are one of the most important reasons domestic and international tourism are thriving in Meghalaya. They attract numerous tourists around the year, especially during the dry seasons (Fig. 72.6). This steady increase in notability for the region has resulted in positive economic growth and new employment opportunities. Tourism has also encouraged the communities to grow new living root bridges. To cater to the rising number of tourists, a number of homestays, cafeteria, and aiding structures are being built in and around the bridges.

The double-decker root bridge at Nongriat had weakened due to a consistent rise in the tourist footfall over the years that it had to be strengthened from the base using wooden logs. The number of total homestays has gone up from 2 to 10 in a matter of 5 years [6]. Mass tourism has also resulted in several restrictions being imposed like a two-minute limit on the bridge and five-people-at-a-time rule, which altogether diminishes the pristine character of the locale.

72.4 Results

The innovation by the local Khasi people has proven to be an object of marvel for nature-lovers, engineers, designers, and everyone alike. The superior strength of the bridges is attributed to their living character and continuous growth over hundreds of years. The living root bridges create low to no environmental damage; act as efficient carbon sequestration agents; possess exceptional qualities of durability and

Fig. 72.6 Crowd of tourist at a root bridge (*Source* Prasenjeet Yadav)

resilience; play the role of a catalyst for an entire ecosystem of moss, lichens, and other species, and most importantly, illustrate the value and potential of collective grassroots participation. The impact of the root bridges on the environment is positive and commendable, unlike concrete bridges which have a high carbon footprint. However, mass tourism can be seen as posing an imminent threat to these extraordinary structures.

72.5 Conclusion

The concept of the living root bridges reinforces a multidisciplinary approach to life that involves the perspectives of environment, botany, architecture, planning, engineering, social science, economics, and culture. The living root bridges and their underlying attributes can help to reform the contemporary building, engineering, and design practices to become more intertwined with nature for strength and sustainability. The crux of the functionalities of the living root bridges could be used as a source of inspiration to develop low-carbon impact designs. The study of the root bridges can also help in the popularisation and evolution of bio and organic architecture that revolves around symbiotic relationships. Their potential can be explored by combining the expertise of multidisciplinary experts to practically test new developments and designs with different plant species. Further investigations could be conducted to understand the structural stability of Ficus and non-Ficus species for replicability and scalability of the concept in distinct geographies and rural/urban setup.

The involvement and perseverance of the Khasi communities can be regarded as a paradigm to collective participation and subsequently, reaping of benefits. This said, there is a pressing need to address the adversity of mass tourism by educating the communities and the public at large on the concepts of minimal tourism and ecotourism for preserving the living heritage structures of Meghalaya.

Acknowledgements This work is carried out under the research project titled "Creating Digital Heritage of Representative Architectural Marvels from Each State of North East India", PI: Dr. Shiva Ji, IIT Hyderabad, India; funded by Department of Science and Technology, Ministry of Science and Technology, Govt. of India, under ICPS/IHDSR scheme.

References

1. Living Root Bridge Ecosystems of India. https://www.brokennature.org/living-root-bridge-ecosystems-india/
2. Shankar, S.: Revitalizing plant-based knowledge in northeast India. Sol. J. **8**(2) (2017)
3. Surreal Photos of India's Living Root Bridges. https://www.nationalgeographic.com/travel/destinations/asia/india/living-root-bridges-clean-village-mwalynnong-india/
4. Ludwig, F., Middleton, W., Gallenmüller, F., et al.: Living bridges using aerial roots of ficus Elastica—an interdisciplinary perspective. Sci. Rep. **9**, 12226 (2019). https://doi.org/10.1038/s41598-019-48652-w
5. Shankar, S.: Living Root Bridges: State of Knowledge, Fundamental Research and Future Application

Chapter 73
Eco-Design, Craftsmanship and Sustainability

Monikuntala Das and Amarendra Kumar Das

Abstract Eco-Design is a product design approach to assess sustainability through environment and economy. There are many aspects of Indian craftsmanship which absolutely communicate to existing Eco-Design principles pertaining to material, production, execution, wastage, environment and sustenance. This paper investigates the craftsmanship sectors of bamboo, wood, Eri silk and brass metal in Assam of North East India and compares its attributes to the successful principles of Eco-Design. The results suggest that there is a high possibility of implementing Eco-Design applications and design processes in craft sectors because there is a considerable compatibility between the nascent attributes of craftsmanship and the principles of Eco-Design, like high reliability and durability, lower energy consumption, cleaner materials and production, possibilities for recycling, etc. The segments of craftsmanship which have not been addressed yet, such as remanufacture and packaging, can undergo implementation through design thinking and execution. Technological interventions are required to bring amendments to the existing processing methods of craftsmanship to drive towards principles of Eco-Design, and therefore, sustainability. Design via craftsmanship can be effectively positioned as an Indian Design identity in the evolving modernity. The intersection can deliver an ecologically viable range of products to the urban market, through which, the design community can have its efforts calibrated towards the repercussions of waste management and climate change, along with the friction against the rapidly changing environment.

73.1 Introduction

Craftsmanship as 'cultural materialism' represents creative labour evolved into sustaining habits for problem finding and solving, using acquired skills. The Indian

M. Das (✉) · A. K. Das
Indian Institute of Technology, Guwahati, Guwahati, India
e-mail: monikuntala22@gmail.com

A. K. Das
e-mail: dasak@iitg.ac.in

A. Chakrabarti et al. (eds.), *Design for Tomorrow—Volume 1*, Smart Innovation, Systems and Technologies 221, https://doi.org/10.1007/978-981-16-0041-8_73

country has cradled craftsmanship since the time of ancient civilisations and has prolonged its survival well into the current times. Craftsmanship denotes a functioning decentralised system in a holistically sustainable environment. Its activities and reasonable consumer demands have minimal consequences on the natural environment. Indian craftsmanship has flourished and has marked its identity until Industrial Revolution, which sets its ground in the country during British Colonisation. As industrialisation gained pace, replication and fast delivery through machine made products made craftsmanship go through a major existential crisis.

Post realisation of tremendous damage of nature through industrialisation worldwide, trends of Sustainable Design and Eco Friendly Design originated to bring back Earth to its natural life giving form and revive from the mass manufacturing hysteria that the world seemed to be in. Amongst the many efforts to address sustainability in India, bringing craftsmanship back to its successful and profit-making functionality seemed to be one of the appropriate investments to apply, firstly, to stabilise the state of the Craftsmen, who constituted one-fifth portion of the country's demographics and secondly, to reduce industrialisation's impact on the natural environment.

This paper investigates few craftsmanship sectors of Assam in North East India, namely bamboo, wood, ell metal and Eri silk, and learns its attributes to find the aspects of Eco-Design that it adheres to in its natural condition. It delivers a methodology to firstly understand what Eco-Design is and what it entails, along with its relation to sustainability; secondly, to understand the institution of Indian Craftsmanship and its elements and thirdly, to understand similarities and differences between craftsmanship attributes and principles of Eco-Design by Hemel and Cramer [1] and to finally formulate a result through the determinants of craftsmanship which can be manipulated by design thinking and technology for an approach towards Eco-Design.

73.2 Relationship Between Sustainability and Eco-Design

Around 12,000 years ago, when human beings settled in agricultural settlements and gained food security, they started creating technologies and professions based out of needs for sustenance and lifestyle. Made of locally available natural materials, these produce initiated trade of the surplus, which eventually financed the governance of civilisations [2, 3]. As people's world views shifted, global legacies of nature worship were remodelled and viewed nature as a resource to be harnessed through technology.

This vision actualised in rapidly evolving Europe in around 500 AD. By 1300 AD, Europe instigated maritime research for resource-rich colonies worldwide and their conversion into production bases, which replaced the indigenous robust natural diversity to a fragile monoculture. Growing dissent between colonisers and the colonised caused numerous uprisings and Europe was unable to import labour due to bans. An alternative to this situation took to the form of Industrial Revolution. Around 1870, demands of Europe's Industrialisation created a new fetish to subjugate their colonies in Asia, Africa, Australia and America for resource extraction. Their eventual emancipation saw Europe giving away to the global civilisation age in early 1900s [2, 3].

As industrialisation spread, it caused rapid shift to urban, technocratic and industrialised systems, which increased population and economic growth, and laid the ground for the unsustainability we face today. While economic benefits of developed countries were boosted by globalisation, developing countries hosted the global production centres and also bore their ecological and social costs [4].

Sustainability began to crystallise in 1970s as an ecological concept following public dissent on Industrialisation's unprecedented development over the environment [5]. United Nations's Stockholm conference (1972) and Brundtland's Commission (1983) acknowledged the situation of accelerating degradation and its consequences [6]. Brundtland's Report represented sustainable development as 'development that meets the needs of the present without compromising the ability of future generations to meet their own needs' [7]. UN's Earth Summit and Rio Declaration (1992), Millennium Support (2000), Johannesburg Summit (2002) and Rio + 20 delegation (2012) grounded the Sustainability Development Goals which included social and economic connotations to the primarily ecological concept [8]. Various sustainability development models and diagrams were developed by Brundtland (1987), Barbier (1987), Munasinghe (1992), Daly (1996), SIGMA (2003), Runnall (2007) et al. and they presented various permutations of the four pillars of sustainability: ecological, social, economic and cultural dimensions [9], with a future inclusion of political, temporal and ethical tenets [10].

Design as a profession also crystallised when, during the Industrial Revolution, integrated artisan-based systems were divided into specialised disciplines of design, production and marketing [11]. By the late twentieth-century Western culture, designers were dissociated from craftsmen and artists as there was a distinct separation between 'having ideas' and 'making objects' [12]. The arts and crafts, Bauhaus, modernist and post-modernist movements defined evolving stages of Industrialisation. Post World War II, the depleting manufacturing power of Europe made USA the hub of production which was based on high consumption and exorbitant natural exploitation [13]. In the 1970s, the spotlight onto ecological sustainability [14] invited reflection on life cycle thinking through Green Design [15]. Eventually, the social aspect surfaced among the appropriate technology practitioners worldwide. This was the driver for Design for Environment (DfE) or Eco-Design during the 1990s, which aimed to create a profitable situation by addressing both the ecology and economy [16].

73.3 Eco-Design and Its Principles

Eco-Design is a product design approach to assess sustainability through environment and economy [17, 18]. With the help of tools and methodologies, it adopts the life cycle approach, to explore and tackle all or the greatest impacts across the product's lifecycle [19]. The challenge for eco product developers is to fulfil a need, or to provide a benefit to the customer at the lowest environment and economic 'cost' [20]. The ten most frequently suggested and successful Eco-Design principles are as follows: Recycling of Materials, High Reliability/Durability, Recycled Materials, Low Energy Consumption, Remanufacturing, Less Production Waste, Clean Production Technique, Reduction in Weight, Clean Materials and Less, Clean and Reusable Packaging [1]. Identification of data and acquired information through identified tools, make it possible to determine new flows of information to allow Eco-Design activities for product and product families [21]. This eco-effectiveness or rather, the 'right things' for mass manufacturers to do are those that lead to good growth for the future generations [22].

Meanwhile, there are various obtrusive barriers for the implementation of Eco-Design practices in an enterprise such as risks involved in the translation of an existing framework, lack of a clear environment benefit, lack of understanding and awareness amongst consumers, commercial disadvantage, lack of alternative solutions etc. [1, 19]. These barriers can be converted into promising opportunities for the application of Eco-Design in MSME's, because, unlike bigger enterprises, there is minimum risk involved in implementing changes; they are normally less bureaucratic and have quicker responses with respect to their effective internal communication channels [1, 19]. Also, for smaller enterprises, the demand for Design for Environment (DfE) exists from customer perspective and investors or corporate stakeholders as well [19].

When looked into the MSME's of India, it is noticed that one-fifth section is composed of the artist and craftsmen community [23] and when looked into the attributes of craftsmanship, it seems to be well in sync with most of the suggested and successful principles of Eco-Design.

73.4 Craftsmanship and Its Attributes

Craftsman belongs to *homo faber*, a being wholly immersed in a dimension life where all that counts are the production of things and the execution of pre-determined tasks [24]. The craftsman represents in each of us the desire to do something well. The difference between brute imitation of procedure and the larger understanding of how to use what one knows is a mark of all skill development [25].

The presence of the Indian craftsman in the midst of a simple agricultural society made possible the self-contained life of the community. The craftsman guild system existed and continued in the form of co-operative associations within the merchant community. Under British rule, the authority of trade guilds was necessarily relaxed and industrialisation imposed imports to the marked damage of handicrafts [26]. Post industrialisation design saw a call out to go back to one's roots to obscure local identities and revive the cultural diversities destroyed by the industrialist mindset [27].

What remains today is a struggling institution with the repercussions of the import of cheap machine made and replicated versions of artefacts. Its inability to adopt innovative methods of production and to adapt to the changing needs of the consumers has become one of the limiting factors in the growth of the handicraft industry [28]. Craft sector today provides livelihood and employment to 20 million people of India, which constitutes 15% of the country's population [29].

Today, craft is a family execution which has a close connection to India's caste system. It is an execution of mainly the lower classes of society. But due to the quality of the product, the craft items become the community curio in which the elite classes take pride. Craft education is transferred through generations informally. The generation next visually learns from their family members in household courtyards and eventually take part in it as they grow older. It is exceptionally a skill-based learning which focuses on muscle memory and practice via years of apprenticeship. The decentralised system allows the breakdown of the manufacturing process, and for craftsmen to find and pursue their individual expertise [30].

The market which their products cater to, fall under the category of traditional, modern, curative and functional. The craftsmen keep working on their products out of expertise, irrespective of category. They are taken forward either by middlemen, retailers and entrepreneurs for their respective markets. As time has passed and crafts have evolved, newer techniques for easier disposal have been on demand, to bring the production numbers and quality at par with the industrial market. This has called for new material processes and technological inputs [30].

The challenge for survival of crafts today is to deliver utilitarian products which conform to the urban design and lifestyle. This requires the need for diversification of products by understanding perspectives of form, function and society. Apart from utilitarian design, elite buyers appreciate the exquisiteness in contemporary, retro, ethnic, and preservative designs. To deliver such aesthetics, an adept knowledge in the section is required. The knowledge regarding crafts, its epistemology, techniques, work, ethics, etc., stays amongst the communities as a non-vocalised comprehension. Majority of this knowledge has not been documented and remains oblivious to the corresponding design communities, thus requiring recognition for future bases [30].

73.5 Examples of Craftsmanship in Assam

73.5.1 Bamboo Craftsmanship

Bhaluki village of Barpeta district in Assam has witnessed the production of many Bamboo products in its household courtyards. A venture to reduce unemployment and negative activities in the locality gave way to a bamboo craft enterprise, due to high raw material availability. From a small number of 20 craftsmen, the craft grew to around 300 craftsman families, who eventually made it their source of income. It is a decentralised system working form a central workshop, which develops it first samples of product as per requirement. The work is divided amongst craftsman based on skills, to be executed in their homes, and the finalised products are delivered back to the central workshop. They create daily utilitarian products like lamps, magazine holders, boxes, vases, coasters, beer mugs, ash trays, clocks, furniture pieces, etc. The entire family of the craftsman helps out in the process, thus making a better number for sale. The craftsmen also identify themselves as individual retailers of the products in the local markets through which they stabilise their household income.

Transportation to nearby locations is done by using local transport, distant locations are taken care of by volumetric freight systems. The products are forwarded to exhibitions, handicraft markets and emporiums. The prices are determined by material costs and transportation along with an added percentage of craftsmanship. Barpeta bamboo craftsmanship is well known in Assam for quality products and has the potential to gather better prices if the product research dimension is looked into.

73.5.2 Wood Craftsmanship

Bharalitola area of Hajo in Assam holds wooden craftsmanship where furniture making is executed and delivered to various parts of Assam. The stability of durable furniture expected by the population of semi-urban and urban societies is delivered by the material wood. This craftsmanship builds on apprenticeship in small workshops. The apprentice learns from his master till he is able to recreate his master's creations and deliver some fresh creativity. The workshops create a variety of chairs, dining sets, beds, dressing tables, etc., from a variety of wood types depending on client's speculated budgets. The raw material is bought in volumes of cubic feet from local dealers at the existing rates. The wood comes treated and after the execution by both hand and electric tools, it is delivered by local transport to the clients. Smaller and sample pieces are also taken to exhibitions for display. The pricing considers raw material, electricity, apprentice stipends, transportation and a percentage for craftsmanship.

Details developed by this craft showcases the skill of the craftsmen. Varying hues, patterns and motifs add to the beauty of the handmade wooden products. The joinery of the wooden pieces and their strength determines the sturdiness and length of the product life.

73.5.3 Brass Metal Craftsmanship

The Moriapatti settlement near Bharalitola area of Hajo is a very old Mohammedan community, originally from Assam, vested in brass metal craftsmanship. Brass Metal is an alloy of copper and zinc and is malleable in character. The craftsmen beat the brass metal discs into sheets, and those sheets into various shapes of big plates, water vessels, bowls, xorai's (vessel of Assamese identity), japi's (Assamese traditional headgear), etc. The settlement has a Mahajan who deals with the raw materials from Guwahati in the form of crude discs. He distributes the discs to the craftsmen who work with him. The roughly 300 craftsmen of the settlement and nearby villages work in their household courtyards and use various tools to bring the discs to shape with a golden skin and embellish them with patterns and motifs. The made crafts are submitted to the Mahajan and a price for craftsmanship is sought for every kilogram of brass metal. The Mahajan then further sells and distributes the products pan India, to retailers, emporiums and distributers. The price of the product includes the raw material cost, price of craftsmanship, transportation and Mahajan's benefit.

The brass metal products are famous all over Assam for their intricate detailing and aesthetic. The products though, are used as symbolic references to the Assamese culture and do not have a utilitarian aspect to it. The products are wielded on special occasions of respect and tradition and act as souvenirs. Even though the produce is so symbolic, the craftsmanship needs further ideation to bring variation in the products and put it towards a more contemporary and daily use. The craft form will slowly diminish if it is not introduced to the next generation, though means of design.

73.5.4 *Eri Silk Craftsmanship*

The indigenous forest communities of Assam's Loharghat Village Range produces hand spun and hand woven Eri silk fabric from silkworms. The communities practice sericulture as a source of food and yarn in their household courtyards. The yarn is extracted from the abandoned cocoon of the silkworm and spun with the help of drop spindles after degumification. A wide range of natural dye materials are extracted from the local woodlands, such as turmeric, lac, Indigo, Indian madder, teak which allow a wide range of colours. The dyed yarn is woven into fabric using looms.

This tradition is mostly a skill preserved by the mothers which offer an alternative income for women of all ages. The entire process of the execution does not need any input from outside community. It is self-sustained from worm to weave. Woven fabrics are collected and delivered to the nearest markets, retailers and other distributors with the help of lighter freights. The price of the produce includes sericulture-related costs, transportation and the artisan's income. The weavers take considerable pride in creating these fabrics. Ensuring a strong, sustainable economic system is paramount in ensuring a future for the people, by protecting their knowledge, land and culture.

73.6 Analysis by Comparing Craftsmanship and Eco-Design

To understand the sync of craftsmanship to Eco-Design, the ten most frequently suggested and successful Eco-Design principles [1] are put forward for an analysis

Table 73.1 Analysis of craftsmanship attributes to principles of Eco-Design

Principles of Eco-Design	Bamboo	Wood	Brass Metal	Eri Silk
Recycling of Materials	No; Degenerates into the ground	Yes; Household Recycling	Yes	Yes; Household Recycling
High Reliabilty/ Durability	Yes	Yes	Yes	Yes
Recycled Materials	No	No	Yes	No
Low Energy Consumption	Yes	No; Time for growth and Treatments	No; Thermal Process for Raw Material	Yes
Remanufacture	No	No	No	No
Less Production Waste	No	No	Yes	Yes
Clean Production	Yes	No; Chemical Treatment	No; Thermal Plant Incineration	Yes
Reduction in Weight	Yes; Due to moisture loss	No	No	No
Clean Materials	Yes	No; Treatment Chemicals	No; Thermal Plant Energy Source	Yes
Less, Clean and Reusable Packaging	No	No	No	No

of all the considered craftsmanship to see if the attributes adhere to the Eco-Design demands (Table 73.1).

73.7 Inferences

The table of comparison through colour variations indicates the existing, non-existing and probables for implementation of Eco-Design principles in each craftsmanship. The 'Yes' segments indicate those principles which are being followed by the craftsmanship in their nascent form. The 'No' segments refer to those principles which have not received any thought for interventions, though they are highly possible through means of design thinking. The 'Intermediate' segments refer to the possibility areas for amendments in the ongoing process through technological interventions which could direct the process towards a principle, currently not being practiced as one.

The four craftsmanship display a high reliability and durability of products. Bamboo products and Eri clothes remain in a household for decades altogether. Wood and brass metal products get passed along generations in the communities. Recycling of materials is done for brass metal because of the expense of raw material. Bamboo and Eri being rapidly renewable backyard raw materials do not feel the urge for recycling. Also, like wood, many a times, they are recycled within households, for different activities, till they are exhausted. Backyard raw materials also have a cleaner extraction and technique of production.

Optimisation of raw materials is always considered to reduce labour and production waste. The energy usage for bamboo and Eri Muga consumes lesser resources than wood and brass metal in terms of time of growth, treatment, and converting raw materials to workshop ready parameters. Wood and brass metal craftsmanship could benefit from amendments in the existing processing methods by switching to more environmental friendly technological interventions. The recycling and remanufacturing of the products of any of the four craftsmanship have not been taken into consideration yet but can be built by integrating a process of design thinking and execution to incorporate multiple lives of the evolving product. The ideation for packaging also has not come into effect since the products are mostly sold in local markets and long distance freights consider temporary packaging solutions before they are delivered to the clients or displayed open.

Overall, the four craftsmanship show a high compatibility with the principles of Eco-Design, and design research and technological interventions can increase the value of the craft manifold.

73.8 Implications for Practitioners

In current times of hostility towards nature, it has become imperative for designers to understand the need of interventions in the materialistic domain and its execution ability, for the planet to persist in the ever expanding user, product and system requirements. Though the legible integration of the institutions of design and craftsmanship faces challenges of socio-economic gaps, it can be reduced by designers being genuine, using basic instincts, being clear and frank with an attitude of respect for craftsman families and communicating transparent and thorough details between the buyer and the maker, to go a long way in developing a healthy business temperament. Through the craftsmen, the businesses could also cater to mass customisation for consumers as the craftsman's execution process could include alterations according to size, need and artistic expression after the basic structure of the craft is conjured. New product systems such as 3D printing could also incorporated in making scaled 3D models, joinery techniques, moulds and frameworks to enhance communication, quality and efficiency. Artisanal production might not have the finesse of industrial production but the final valuation can be based on material quality, craftsman efforts, local economy generation, social migration prevention, advantage of local knowledge for the educationally deprived and decreased social disturbances. The intention must be to integrate the craftsmen community into the urban need and aesthetics without them losing their identity [31]. Also, new identity generation helps products to sell with a new enthusiasm from consumers when delivered with a compelling story [31].

73.9 Highlights of the Study

The study suggests a high possibility of implementing Eco-Design applications and design processes in craft sectors because there is a considerable compatibility between the nascent attributes of craftsmanship and the principles of Eco-Design, like high reliability and durability, lower energy consumption, cleaner materials and production, possibilities for recycling, etc. The segments which have not been addressed yet, such as remanufacture and packaging, can undergo implementation through design thinking and execution. Technological interventions are required to bring amendments to the existing processing methods to drive towards principles of Eco-Design, and therefore, sustainability.

In the quest for an identity, design via craftsmanship can be effectively positioned as an Indian design identity in the upcoming modernity. The intersection can deliver an ecologically viable range of products to the Indian urban market or for the public infrastructure as well, and the Indian design community can have its indigenous efforts calibrated towards the repercussions of waste management and climate change, along with the friction against the rapidly changing environment.

References

1. Van Hemel, C., Cramer, J.: Barriers and stimuli for ecodesign in SMEs. J. Clean. Prod. **10**, 439–453 (2002)
2. Lloyd, C.: What on Earth Happened? In: Brief. Bloomsbury, London (2008)
3. Overy, R.: Complete history of the world. Times Books, London (2007)
4. Munasinghe, M.: Making development more sustainable: sustainomics framework and practical applications. Colombo, Munasinghe Institute for Development (MIND) (2010)
5. Hawken, P., Lovins, A., Lovins, L.H.: Natural capitalism: creating the next industrial revolution. Little Brown, Boston (1999)
6. Mann, S.: Sustainable lens: a visual guide. NewSplash Studio, Dunedin (2011)
7. Brundtland, G.H.: Our common future. Oxford University Press, Oxford (1987)
8. Barash, D.P., Webel, C.P.: Peace and conflict studies. Sage Publications, London (2002)
9. Reubens, R.: To craft, by design, for sustainability. Delft University of Technology, Delft (2006)
10. O'Connor, M.: The four spheres framework for sustainability. Ecol. Complex. **3**(2006), 285–292 (2007)
11. Walker, J.: Design history and the history of design. Pluto Press, Winchester (1989)
12. Peters, A.: The status of craft. In: Dormer, P. (ed.), The Culture of Craft, pp. 17–20. Manchester University Press, Manchester (1997)
13. Sheldon, R., Arens, E.: Consumer engineering: a new technique for prosperity. Harper and Brothers, New York (1932)
14. Papanek, V.: Design for the real world: human ecology and social change. Pantheon Books, New York (1971)
15. Whitely, N.: Design for society. Reaktion Books Ltd., London (1993)
16. Brezet, H., van Hemel, C.: Ecodesign: a promising approach to sustainable production and consumption. United Nations Environment Programme, Industry and Environment, Paris (1997)
17. Karlsson, R., Luttropp, C.: Eco-Design: what's happening? An overview of the subject area of Eco-Design and of the papers in this special issue. J. Clean. Prod. **14**(15–16), 1291–1298 (2006)

18. Bovea, M.D., Pérez-Belis, V.: A taxonomy of ecodesign tools for integrating environmental requirements into the product design process. J. Clean. Prod. **20**(1), 61–71 (2012)
19. Short, T., Lee-Mortimer, A., Luttropp, C., Johansson, G.: Manufacturing, sustainability, ecodesign and risk: lessons learned from a study of Swedish and English companies. J. Clean. Prod. **37**, 342–352 (2012)
20. Luttropp, C., Lagerstedt, J.: Eco-Design and the ten golden rules. J. Clean. Prod. **14**, 1396–1408 (2006)
21. Le Pochat, S., Bertoluci, G., Froelich, D.: Integrating ecodesign by conducting changes in SMEs. J. Clean. Prod. **15**(7), 671–680 (2007)
22. McDonough, W., Braungart, M.: Remaking the way we make things: cradle to cradle. North Point Press, New York (2002)
23. MSME's most attractive segment in commercial credit with fastest growth, Low NPAs: Report, https://economictimes.indiatimes.com/small-biz/sme-sector/sidbi-and-transunion-cibil-launch-msme-pulse/articleshow/63269240.cms?from=mdr. Accessed 12 Mar 2018
24. Fadini, U., Mari, G., Giovannini, P.: On Richard Sennett's the craftsman. Iris **2**(4), 507–521 (2010)
25. Sennett, R.: The craftsman. Yale University Press, New Haven & London (2008)
26. Coomaraswamy, A.K., Moore, A.C.: The Indian Craftsman, vol. 6. Probsthain, London (1909)
27. Rao, A.G.: Craft as a post-modern venture: experience in bamboo craft. In: PD2008 8th Brazilian International Conference (2008)
28. Ghosh, A.: Triggering innovation and creativity in traditional handicraft sectors: an Indian perspective. Manage. Insight **VIII**(1), 2–5 (2012)
29. Viswanath, B.: Enumeration of Crafts Persons in India, pp. 19–42. Madras School of Economics (2013)
30. Das, M., Das, A.K.: Craft education in design. In: DS 95: Proceedings of the 21st International Conference on Engineering and Product Design Education (E&PDE 2019), University of Strathclyde, Glasgow (2019)
31. Bheda, C., Kasliwal, A.: Design strategy and business thinking. In: Instagram Live of @designstrategistwud. World University of Design, Sonipat (2020)

Chapter 74
Identifying Criteria for the Selection of Ecodesign Methods to Integrate into SMEs Product Development

Björn Kokoschko, Laura Augustin, and Michael Schabacker

Abstract The design, development and manufacturing of products inevitably involve environmental aspects (e.g., material and energy consumption) that lead to negative environmental impacts (Herrmann in Ganzheitliches Life Cycle Management, Nachhaltigkeit und Lebenszyklusorientierung in Unternehmen. VDI-Buch. Springer, Berlin, 2010, [1]). The awareness of these issues enables an understanding of the immediate effects that design and manufacturing have on the environment and how products can affect the ecosystems around them. We face many problems, and we can only solve if the design of products is geared to ecological requirements. This requires that product development is regarded as a holistic and interlinked process. This results in the consideration of processes such as recycling, downcycling as well as maintainability and material separation in our product life cycles. Hence, production and environmental problems were treated independently of each other for a long time, and the environment had received little attention in the phases of the product development process (Huang in Design for X, Springer, Netherlands, Dordrecht, 1996, [2]). Especially small- and medium-sized enterprises will increasingly face problems in aligning their innovation activities with ecological requirements. So far there is no practical approach to holistic and interlinked product development that sufficiently takes ecological requirements into account and is easy to adopt for SMEs.

74.1 Introduction

On January 1, 2016, the 17 United Nations goals for sustainable development (SDGs) came into force [3, 4]. In addition to economic and social goals, a major focus is on

B. Kokoschko (✉) · L. Augustin · M. Schabacker
Otto-Von-Guericke University, Magdeburg, Germany
e-mail: bjoern.kokoschko@ovgu.de

L. Augustin
e-mail: laura.augustin@ovgu.de

M. Schabacker
e-mail: michael.schabacker@ovgu.de

A. Chakrabarti et al. (eds.), *Design for Tomorrow—Volume 1*, Smart Innovation, Systems and Technologies 221, https://doi.org/10.1007/978-981-16-0041-8_74

ecological goals. In order to measure its own achievement of these goals, Germany formulated a series of indicators that were adopted as part of the German Sustainability Strategy (DNS) 2016 [5], 6. A central goal of the sustainability strategy at the ecological level is the establishment of sustainable consumer and industrial products to reduce energy consumption and CO_2 emissions (goal 12) [7]. This goal can only be achieved if the design of consumer and industrial products and also the business models for their economic exploitation are geared to ecological requirements. The design and development of the consumer or industrial product must therefore be adapted to the needs of the target group in line with the corporate objectives and capabilities.

Thus, this paper analyzes literature to find criteria for the selection of methods geared to the needs of SMEs. The selection of methods for the design and development of the consumer or industrial product must therefore be adapted to the needs of the companies' circumstances, the ecological affordances, the target group in line with the corporate objectives.

74.2 Current Situation

Especially small- and medium-sized enterprises (SME), in contrast to large companies, will increasingly face problems in aligning their innovation activities with ecological requirements in terms of the DNS. In principle, many of these companies do not have a structured innovation process, as they lack research and development departments, budgeting, etc. Moreover, these companies do not have the necessary resources, capacities, innovation skills and sustainability expertise to develop ecological innovations completely independent [8]. For the federal state of Saxony-Anhalt in Germany in particular, the report on the SME offensive certifies [9] (p. 12) that there are "in many cases innovation-oriented small and medium-sized enterprises which do not have the structural and personnel prerequisites, or do not have them to the same extent, to be able to realise product innovations without support". Consequently, it is necessary to support SMEs in structuring their innovation process and aligning it with ecological requirements in the sense of the DNS—an SME-oriented step-by-step approach can make a valuable contribution here. Therefore, context-sensitive methods can be used to reduce the need to use one specific approach. This means that the selection of methods for accomplishing a task within the process model depends on the type of project, the development status, the time resources and the available data [10].

If one looks only at the area of product development, some development concepts are frequently used both in university teaching and in business practice, such as Integrated Design Engineering [11], the Design Thinking [12–14], the Agile Project Management (among others in [13, 15–17]) the VDI2221 [18, 19]. However, these concepts are not geared to the consideration of ecological requirements yet.

According to PIGOSSO et al., in contrast very specific impact categories of sustainability are considered in the eco-design literature, in which a large number of methods and tools are published [20]. In the context of this literature strand, however, the intercompatibility of the presented methods and tools is hardly or not at all questioned, and it remains unclear to what extent they can be integrated into a superordinate product development concept. Also, the tailor-made selection of methods according to the context and needs within companies is not or only insufficiently considered [20, 21]. Consequently, some researchers see precisely this variety of methods and tools as a barrier to application in industry, since it is not clear in which situations which methods with the corresponding tools are relevant [20]. There is a lack of approaches and methodologies for accelerated and optimized application by industry [20].

The approaches shown in Table 74.1 come closest to a SME-oriented approach to product development, considering sustainability requirements. On the one hand, the above-mentioned approach of Integrated Design Engineering (IDE) [11] is also

Table 74.1 Overview of a selection of concepts with strengths and weaknesses

	Strength	Weaknesses
IDE [10, 26]	• Human-centered team approach • Interdisciplinary cooperation • Step-by-step approach • Context sensitivity through a rough definition of process modules • Method building kit	• Partially insufficient consideration of sustainability aspects • So far little use by SMEs • Often only used within four phases
D4S [27, 28]	• Step-by-step approach • Practice-oriented worksheets • Team-oriented approach • Social issues such as sustainable procurement, ethical financing and ethical labour procurement are also aimed	• Different focusses complicate use for SME • Four archetypal process phases
SDS [22, 29, 30]	• Focus on product service systems • Step-by-step approach (modular)	• Focus mainly on social sustainability and service systems, relatively little consideration of environmental sustainability aspects
DfBOP [31, 32]	• Concrete method suggestions for individual process steps • Focus on social sustainability	• Limited transferability to SMEs • Low consideration of environmental sustainability aspects • Limited focus group
HCD [25, 27, 33, 34]	• Extensive practice-oriented collection of methods • Widespread and human-centered approach • Uses aspects from D4S, DfBOP, and IDEO HCD Design Kit	• Low consideration of environmental sustainability aspects • No guidance as to the procedure • More a collection of methods than a product development concept

used at the OVGU in the context of industrial projects and teaching and on the other hand further approaches such as Design 4 Sustainability (D4S) [22], System Design for Sustainability (SDS) [22], Design for Base of Pyramid (DfBOP) [23, 24] and Human Centered Design (HCD) [25]. All approaches are practical concepts, some of which provide for a concrete step-by-step approach to product development (IDE, D4S, SDS) and some of which can also cover SME-specific R&D needs (D4S, SDS, HCD). In particular, D4S and SDS seem to be easy to implement even by people from outside the industry. Compared to the other approaches, HCD is widely used in the literature and offers product developers an extensive collection of methods, but its suitability for practical application is to be rated less highly here, as the approach offers little assistance in implementing the procedure. The implemented human-centered approach (IDE, HCD) or team orientation (IDE, D4S) shall also be emphasized in some approaches.

74.3 Take Precedence and Aim of Research

With regard to the weaknesses of the individual development approaches shown in Table 74.1, it is striking that with the exception of D4S, ecological requirements are only insufficient considered. In the case of SDS and DfBOP, social aspects of sustainability are primarily targeted, with economic and ecological requirements playing a subordinate role. Since the development of products that primarily address markets in developing countries requires an effort that is likely to exceed the available resources and capacities of many SMEs.

The necessity of implementing the needs of SMEs with regard on the ecological aspects of product development into a holistic approach has been recognized. Especially for SMEs, a specific step-by-step approach seems to be a good work practice. This can practically concretize the individual phases of the interlocking product development as well as the methods and tools to be used in the individual phases. In view of the strengths and weaknesses of the above-mentioned product development approaches, it can be concluded that individual aspects of these approaches (e.g., tools) can be combined in a synergetic way for such a step-by-step approach. In conclusion, it shall be noted that previous approaches only partially take into account the underlying vision or the economic, ecological and other goals of SMEs, especially with regard to the DNS. This is necessary in order to ensure that they are willing to apply them.

A core objective of the development work is the precisely fitting applicability of the procedural concept by SMEs. Accordingly, it shall be ensured that the concept can be integrated into the companies without the need for specialized personnel. In addition, the SME-specific framework conditions, such as scarce capacities and the partial lack of data on the product life cycle, are to be considered. For the development of the approach concept, this means that a decision support for SMEs will be necessary, which explains the individual development steps as well as the methods and tools applicable in each case in detail. On the one hand, it can be avoided that the multitude of methods and tools mentioned before overburdens the non-specialized user. On the other hand, the procedural concept enables a timely implementation of the product development by means of SME practice-oriented worksheets and tools and thus counteracts the shortage of time in SMEs. Therefore, it is necessary to identify fitting methods and tools which can be used in SMEs by apply identified criteria first. Therefore, it is needed to define measurable practice-oriented and sustainability criteria, which will be used to rate the methods and then to design the process concept. In the narrower sense of sustainable products, it is necessary to question the existing product range and also the inclusion of new products in the product range [35]. BERGMANN [36] mentioned that the first two steps are the thorough needs analysis and the examination of possible immaterialization strategies. In these two steps, it is checked whether a physical product is needed at all, whether it can be immaterialised alternatively and whether it is justifiable from a sustainability point of view. Only in the third step does the actual program policy, product conception and product development begin. On the basis of the steps contained therein a sustainable product offer is created. In this paper, it is assumed that the first two steps are already finished and that the focus is on the integration of possible methods into SME processes.

For the present within this research, the definition of criteria to determine the capability to integrate a method into a SMEs product development approach and whether is mature enough is based on a research paper written at OVGU [37]. VAJNA et al. derive from case studies different criteria for comparing methods with each other (General applicability, Innovation enabling, Complexity of the method, Complexity of the supported products, Exploitation of computer support, Integration of different domains, Applicable to different kind of products, Balancing of different product goals, Analogies to nature, Inclusion of economical influences, Consideration of social sciences, Support of communication, Support of single work, Support of teamwork Systematic and predetermined procedures, Opportunistic course of action, Parallelization of activities, Dynamic reaction to changes of conditions, Ongoing comparison and evaluation of actual results, Continuous feedback, Anticipation of results, Several equivalent, but not similar solutions, Teachability, Generality, Practicability) [37].

Other researchers also defined criteria for analyzing methods but use different vocabulary to describe the criteria. Ernzer et al. focused on methods for life cycle design based on the needs of a company and give an approach to a systematic analysis of the company to select suitable methods [38]. Hereby requirements arise from relations to the product, the surrounding environment and the company. Watty et al. point out the significance of criteria selection for the choice of quality methods to achieve a high-quality product as well as product development process at reasonable costs. They correlate the benefits of larger flexibility on the market and shorter time to market, less loops in the development process, less rejection or rework, better documentation and an increasing motivation to the selection of method [39]. They refer to the importance of considering different motives of quality at the selection of a method must regard the demands of the society, companies and the customer, while even intersecting demands may differ in their motive. EDER explores "the need for a formalized engineering design methodology" and reveals a higher level of guidance in an engineering design procedure when the problem is not seen as a routine or the expertise in the product as well as the process is lacking. Especially the expertise of novice and advanced beginners needs to be considered, not in the field of the company's domain but in the field of applying new and more suitable methods of product development [40].

Within these reviews, the product, the surrounding environment, and the company itself are considered from different point of views and point out the importance of a substantial selection of criteria to select appropriate methods meeting all requirements. Within this research, the criteria have been reduced and it has also been considered to be necessary that the sustainable criteria group had to be implemented. This step needs further research and have to be validated within case studies. Baumann [41] in 2002 mentioned steps for communication are necessary for a good impact with regard to the applicability. The understanding of the role, impact and contribution of tools leads to a better learnability of methods.

A further distinction can be made between product-related and company-related criteria, both of which in turn surround criteria of legislation, standards and market-related criteria [38]. Product-related criteria include the complexity of products, the degree of ecological perfection, the number of pieces (product range) and the dominant life phases of the product. The company-related criteria are the competence in DfE (design for environment), the general method competence, the degree of innovation, resources of the DfE department, and the strategic targets of the company. Laws, standards, and the market define the surrounding criteria for methods which can be implemented into a product development process. Figure 74.1 shows an overview of the categorized criteria identified so far are structured according to priority and chronological order of the evaluation process to be carried out. Consequently, when evaluating a method for its applicability, sustainability is considered first. If this proves to be sustainable, the evaluation of the criteria can be advanced to the next category. Thus, the first evaluation step proves to be possible exclusion criteria. Furthermore, the categories of applicability and integratability have to be fulfilled in order that a method can be integrated into SME processes.

Thus, aspects of the category integratability such as learnability and practicability include a partial aspect of sustainability. These, in turn, are sometimes based

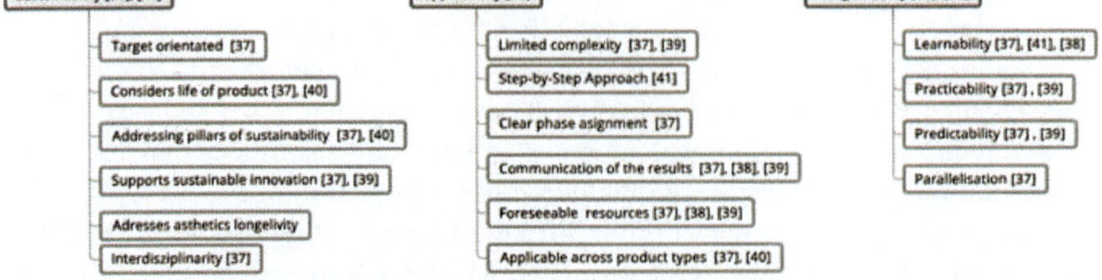

Fig. 74.1 Identified and categorized criteria

on criteria from the category of applicability, such as low complexity, step-by-step approach and communication of the results. Consequently, a low complexity increases the learnability, thus a method that is easy to learn is re-applied with a higher probability, which can lead to a higher effect in the sustainability of companies. In this phase of the work, the criteria are assigned zero to five points and summed up to the categories. This results in a maximum of 25 points for the category of sustainability, 30 points for applicability and 15 points for integratability.

For the evaluation of the identified methods with regard to their general usability in the company, a minimum number of points will have to be assigned to the categories in the process of work. In addition, the criteria and their categories must also be checked and, if necessary, adjusted. In addition, the evaluation of a method can be carried out in comparison to another conceivable method in order to compare its applicability in the company and its processes as well as its situation. A semantic differential or a comparison in a table can be used for the evaluation among themselves.

In the development of the project, it will be determined which of these concepts and methods meet the previously defined criteria. Also, it needs to become evaluated to what extend they can be used for different types of product development, e.g., generally valid or context-sensitive, and to what extent they can be effectively combined. From this analysis, a development requirement for the procedural concept is derived, which is then processed conceptually.

74.4 Summary

Up to this point, criteria have been collected for the identification of applicable methods into SMEs in order to implement sustainable methods for product development. This will lead to the carefully selection of fitting methods according to the needs of SMEs with regard to sustainable aspects. The criteria identified in this paper are summed up within the criteria tree (Fig. 74.1) which could be seen as a requirements list for methods to implement ecological requirements of a product development within SMEs. The awareness of the methods capabilities enables an understanding of the immediate effects that lead to a holistic approach. SMEs can use the additional information and support designers with methods which fit to the actual the development task, a specific phase of product development and also can be used to compare different methods.

74.5 Outlook

This approach will be improved, detailed, and verified in further research by applying it within the extension of researching methods. For the next step to create a sustainable product development approach, more methods which can be used in SMEs will be identified and rated with the developed criteria. According to the guideline VDI

2221 [18, 19], the design process is divided into different stages. Bhamra and Lofthouse [27] grouped often used tools (see Table 74.2) into five sections; Environmental Assessment, Strategic Design, Idea Generation, User-centred Design, and Information Provision. To simplify the project- and SME-specific method selection process, the identified criteria can also be used as indicators for offering fitting methods within different development tasks.

This grouping, as well as the variety of methods, is to be adopted in the following steps for the creation of the product development concept, but the group of environmental assessment is still insufficient and will be extended by some methods. These groups can be roughly assigned to the product development phases; see Fig. 74.2. What can be used to rate the criteria for phase assignment and to give the SME an overview in which phase the method can be used. After rating the methods and structure with regard to the phases, a holistic approach can be gathered. The transfer of the methods and also the criteria will be evaluated by implementing the methods within education and industry cooperations such as case studies. This leads to the opportunity to recognize not adapted characteristics of a company. What gives a view whether the methods and the criteria fit the capabilities of the company or not.

Table 74.2 Groups of a selection of methods

Groups	Methods
Environmental assessment	• Life cycle assessment • MET matrix
Strategic design	• Ecodesign web • Design Abacus • Five focal areas • Six rules of thumb • Sufficiency need assessment
Idea generation	• Information/Inspiration • Flowmaker • Attribute listing
User-centered design	• Participant observation • User trials • Product-in-use • Personae • Moodboards • Layered games • Scenario-of-use
Information provision	• Real people • Material research • Patent research

References

1. Herrmann, C.: Ganzheitliches life cycle management. Nachhaltigkeit und Lebenszyklusorientierung in Unternehmen. VDI-Buch. Springer, Berlin (2010)
2. Huang, G.Q.: Design for X. Springer, Netherlands, Dordrecht (1996)
3. United Nations publication issued by the Department of Economic and Social Affairs. In: Sustainable Development Goals Report, vol. 2019. United Nations, New York (2019)
4. Martens, J., Obenland, W.: Die agenda 2030. Globale Zukunftsziele für nachhaltige Entwicklung, 30th edn. Global Policy Forum; terre des hommes, Bonn, Osnabrück (2017)
5. BK-Amt: Deutsche Nachhaltigkeitsstrategie. Neuauflage 2016, 11044 Berlin (2016 (soweit nicht anders vermerkt), Kabinettbeschluss vom 2017)
6. BK-Amt: Deutsche Nachhaltigkeitsstrategie. Aktualisierung 2018, 11044 Berlin (2018 (soweit nicht anders vermerkt), Beschluss Bundeskabinett vom 2018)
7. Statistisches Bundesamt (Destatis): Nachhaltige Entwicklung in Deutschland - Indikatorenbericht 2018. Heike Becker, Jens Hoffmann (2018)
8. Hansen, E.G., Klewitz, J., Schaltegger, S.: Industrie Management 5/2013: Eco-Innovation. GITO mbH Verlag (2013)
9. Ministerium für Wirtschaft, Wissenschaft und Digitalisierung des Landes Sachsen-Anhalt: Mittelstandsoffensive Sachsen-Anhalt I^3—Wachstum durch Innovationen, Investitionen und Internationalisierung. Hier ist der Mittelstand (2020)
10. Wiesner, M., Kokoschko, B., Bùi Duy, L., Augustin, L.: Card Sorting basierter Ansatz zur Erarbeitung einer nutzungsgerechten Methodensammlung am Beispiel des IDE-Toolkits. In: Krzywinski, J., Linke, M., Wölfel, C. (eds.) Entwerfen Entwickeln Erleben 2019, pp. 29–40. Dresden (2019)
11. Vajna, S. (ed.): Integrated design engineering. Ein interdisziplinäres Modell für die ganzheitliche Produktentwicklung. Springer Vieweg, Berlin (2014)
12. Brenner, W., Uebernickel, F.: Design thinking for innovation. Springer International Publishing, Cham (2016)
13. Naccari, F.: Enterprise integration in the user industries—needs and current solutions. In: Kosanke, K., Nell, J.G. (eds.) Enterprise Engineering and Integration. Building International Consensus Proceedings of ICEIMT '97, International Conference on Enterprise Integration and Modeling Technology, Torino, Italy, October 28–30, 1997. Research Reports Esprit, pp. 27–33. Springer, Berlin, Heidelberg (1997)
14. Plattner, H., Meinel, C., Leifer, L.: Design thinking research. Springer International Publishing, Cham (2015)
15. Cockton, G., Lárusdóttir, M., Gregory, P., Cajander, Å.: Integrating user-centred design in agile development. Springer International Publishing, Cham (2016)
16. Kusay-Merkle, U.: Agiles Projektmanagement im Berufsalltag. Springer, Berlin Heidelberg (2018)
17. Schwaber, K., Sutherland, J.: The scrum guide. Die Spielregeln, Der gültige Leitfaden für Scrum (2017)
18. VDI: VDI 2221 Blatt 1. Entwicklung technischer Produkte und Systeme - Modell der Produktentwicklung, VDI 2221 Blatt 1. Verein Deutscher Ingenieure, Berlin (2018)
19. VDI: VDI 2221 Blatt 2. Entwicklung technischer Produkte und Systeme - Gestaltung individueller Produktentwicklungsprozesse, VDI 2221 Blatt 2. Verein Deutscher Ingenieure, Berlin (2018)
20. Pigosso, Daniela, C.A., Rozenfeld, H., McAloone, Tim, C.: Characterization of the state-of-the-art and identification of main trends for ecodesign tools and methods. Classifying three decades of research and implementation. J. Ind. Inst. Sci. **95**(4), 405–426 (2015)
21. Johansson, G.: Success factors for integration of ecodesign in product development. Environ. Manage. Health **13**(1), 98–107 (2002). https://doi.org/10.1108/09566160210417868
22. Vezzoli, C., Kohtala, C., Srinivasan, A., Xin, L., Fusakul, M., Sateesh, D., Diehl, J.C. (eds.): Product-service system design for sustainability, 1st edn. Taylor and Francis, Saltaire (2014)

23. Prahalad, C.K.: Der Reichtum der Dritten Welt. Armut bekämpfen, Wohlstand fördern, Würde bewahren, 1st edn. FinanzBuch-Verl., München (2006)
24. Prahalad, C.K.: The fortune at the bottom of the pyramid. Strategy + business **26** (2002)
25. IDEO: Human-Centered Design Toolkit: An Open-Source Toolkit To Inspire New Solutions in the Developing World (Lee, M.H. et al. Trans.), Edit the World. https://www.ideo.com/tools (2011)
26. Neutschel, B., Wiesner, M., Schabacker, M., Vajna, S.: Vorgehensweisen zum Einsatz universitärer Produktentwicklung als Innovationstreiber. In: Krzywinski, J., Linke, M., Wölfel, C. (eds.) Entwerfen Entwickeln Erleben 2016. Beiträge zum Industrial Design: Dresden 30. Juni-1. Juli 2016. Technisches Design, vol. 10. TUDpress, Dresden (2016)
27. Bhamra, T., Lofthouse, V.: Design for Sustainability. A Practical Approach, 1st edn. Design for Social Responsibility. Taylor and Francis, Abingdon, Oxon (2016)
28. Vezzoli, C., Kohtala, C., Srinivasan, A. (eds.): Product-service system design for sustainability. Greenleaf Publishing, Sheffield (2014)
29. Tukker, A., Tischner, U.: New business for old Europe. Product-service development, competitiveness and sustainability. Greenleaf Pub, Sheffield, South Yorkshire, England (2010)
30. Ceschin, F.: Sustainable product-service systems. Springer International Publishing, Cham (2014)
31. Ao, S.I. (ed.): International Multi Conference of Engineers and Computer Scientists, IMECS 2009. 18–20 March, 2009, Regal Kowloon Hotel, Kowloon, Hong Kong. Lecture Notes in Engineering and Computer Science. IAENG, Hong Kong (2009)
32. Castillo, L.G., Diehl, J.C., Brezet, J.C.: Design considerations for base of the pyramid (BoP) projects. In: Proceedings of the Nothern World Mandate: Culumus Helsinki Conference, pp. 24–26 (2012)
33. Simanis, E., Hart, S., Duke, D.: The base of the pyramid protocol: beyond "basic needs" business strategies. Innov. Technol. Govern. Global. **3**(1), 57–84 (2008)
34. Larsen, M.L.M., Flensborg, A.: Market creation toolbox. DI International Business Development (2011)
35. Hopfenbeck, W., Jasch, C.: Öko-Design. Umweltorientierte Produktpolitik. Verlag Moderne Industrie, Landsberg/Lech (1995)
36. Bergmann, G.: Umweltgerechtes Produkt-Design. Management und Marketing zwischen Ökonomie und Ökologie. Luchterhand, Neuwied, Berlin (1994)
37. Vajna, S., Kittel, K.: An approach to compare product development methods. In: Norell Bergendahl, M. (ed.) Design has never been this cool. ICED 09, the 17th International Conference on Engineering Design; 24–27 August 2009, Stanford University, Stanford, California, USA; proceedings volume. DS / Design Society, vol. 58, pp. 313–324. Design Society, Glasgow (2009)
38. Ernzer, M., Birkhofer, H.: Selecting methods for life cycle design based on the needs of a company. In: DS 30: Proceedings of DESIGN 2002, the 7th International Design Conference, pp. 1305–1310. Dubrovnik (2002)
39. Watty, R., Binz, H., Enriquez-Geppert, J.: Systematic selection of appropriate methods in quality management. In: DS 58–5: Proceedings of ICED 09, the 17th International Conference on Engineering Design, vol. 5, Design Methods and Tools (pt. 1). Palo Alto, CA, USA, 24–27.08.2009 (2009)
40. Eder, W.E.: Design engineering and needs for methodology. In: DS 58-5: Proceedings of ICED 09, the 17th International Conference on Engineering Design, vol. 5, Design Methods and Tools (pt. 1). Palo Alto, CA, USA, 24–27.08.2009 (2009)
41. Baumann, H., Boons, F., Bragd, A.: Mapping the green product development field: engineering, policy and business perspectives. J. Clean. Prod. **10**(5), 409–425 (2002)

Chapter 75
A Study on Design Concept for Comfortability of Dokhona—For Sustainability of Bodo Traditional Wear and Culture

Chaitali Brahma, Bhaskar Saha, and Debkumar Chakrabarti

Abstract Design thinking concept is a way to problem solving and also an implementation of comfortableness. Tribal women in various tribal dominated areas wear different unique style and have different taste of attire pertaining to their traditions. In this paper, the focus has been given on art and culture of Bodo inhabitant of Assam, a region in Northeast India, which appears to be a storehouse of art and culture in different way and carries a distinct identifiable design concept impression. The uniqueness of Bodo traditional attire dokhona itself is a symbol of its cultural identity. In the present scenario, it is noticed that it is getting influenced by Western outfits that creates a debate on its sustainability, with specific reference to fashion effect and discomfort arising from the point of material used in weaving of dokhona as well as the style of wearing in today's fast life. It also discusses the issues of comfortability relevant to modern lifestyle as wearing it also shows that the continuous tying up of the dokhona on chest may lead to muscle pains. This deliberation reports a survey response from various stakeholders, e.g., users, weavers, doctors and other experts, which confirms the need to go for design development looking into aspects of newness while maintaining the identity of traditional design elements with stress on ecofriendly design.

C. Brahma (✉) · B. Saha
Department of Multimedia Communication and Design, Central Institute of Technology Kokrajhar, Assam, India
e-mail: chaitali@cit.ac.in

B. Saha
e-mail: b.saha@cit.ac.in

D. Chakrabarti
Department of Design, Indian Institute of Technology Guwahati, Guwahati, India
e-mail: dc@iitg.ac.in

A. Chakrabarti et al. (eds.), *Design for Tomorrow—Volume 1*, Smart Innovation, Systems and Technologies 221, https://doi.org/10.1007/978-981-16-0041-8_75

75.1 Introduction

Indigenous and traditional attire is changing day by day, and the emerging fashion studies are gradually distancing from artifacts in the last few decades. It has been observed that there has been revival of old fashion with much better comfortable gesture through the traditional bold colors and embroidery of classic Indian attire influencing newer designs [1]. The tradition Indian attire has earned admires from people all over, even organizations like UNESCO had delivered its admiration. The craftsmanship with which artisans create traditional dresses is incredible. All forms of clothing are labor intensive and require a lot of attention to details. This has forced the textile ministry to come up with ways in order to sustain these artisans and preserve their art and talent. India is known for its myriad diversity in ethnic community clothing and distinguish one part of India from the other [2]. Northeastern India, represented with seven states, known as seven sisters, is paradise unexplored in term of tribal community-specific culture, language, dress, customs and traditions and festivals to celebrate. Both language and culture have an interface of Tibetan, Southeast Asian and East Indian culture. These states have rich cultural bonds, which give them a sense of exclusivity and belongings. The hill tribes of Northeast India are classified as Scheduled Tribes, and one of the major tribes of the Northeast is the Bodos, early settlers of the Northeast India in Assam and maintains an identifiable cultural heritage. One of their aspects of traditional glamor is the handloom sector. Weaving is being a significant part of the Bodo society [3]. The Bodos are found to have maintained their traditional artistic creativity, especially in their traditional clothing design; they have created various types of natural caricatures showcasing different design and shapes. Basically, the Bodos practice traditional methods to produce artifacts, their design and workplaces have welcomed the beneficial techniques of science and technology that has spread over the time [4]. Furthermore, types of clothing vary with types of social occasions, which indicate that the wearing of clothes is also subject to sociocultural norms. The mode of dress of the Bodo people is unique and full of colours, with style and attractiveness. The traditional attire was always handwoven, which is also a testimony of Bodo women's talent in weaving, as Fig. 75.1.

75.1.1 Bodo Ethnicity of Dokhona

Dokhona is a traditional attire of the bodo women that comprises a single piece of cloth, and serves to elegantly covered their body from chest till the ankles and tighten knots one below the armpit and the other in the waist line. Every Bodo women across all geographical regions essentially possesses a dokhona as a mark of the community's cultural and traditional identity besides using these on day-to-day basis.

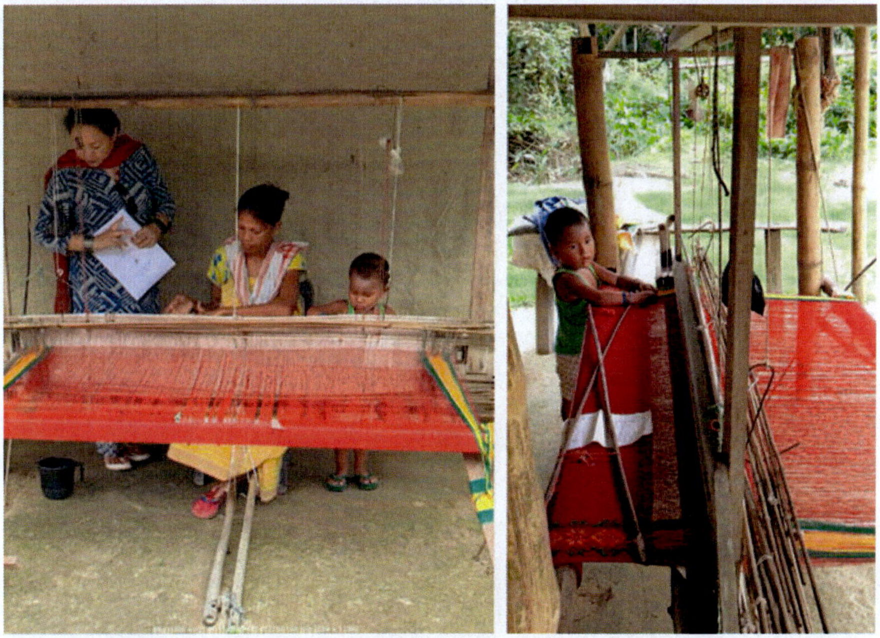

Fig. 75.1 Bodo women's talent in weaving, it is a common household activity

In order to keep pace with current changing scenario of dressing, in respect of convenience and ease of wearing as shown in Fig. 75.2, some Bodo women, particularly those who need to venture out for work and employment, preferred other dresses like salwar kameez, pants, etc. to dokhona. However, when it comes to performing household courses or holding cultural and religious functions, dokhona turns out to be indispensable. With a strong sentiment continuing across Bodo generation about clinging to and preserving the heritage, it is imperative that the modification of dokhona to enhance its functionality for meeting the needs of the time, calls for focusing on design aspects of this special attire for its continued acceptability and use [5]. Bodo women's talent in weaving through design, patterns and quality material has earned a reputation. Recently, they have started using Jacquard looms to increase production, efficiency and quality of the Bodo textiles, keeping the design elements unchanged [6]. Weaving and sericulture are integral parts of the women folk of the Bodo community. Every young woman knows the art of weaving, which they learn from early childhood within the family. Traditionally, they weave all the required cloths for the family members other than dokhona, which includes aronai, gamocha, eri chadar, shima, etc., shown in Fig. 75.3a–d.

The raw material used eri and muga (Fig. 75.4) for clothes was very popular among the non-bodos also. The endi cloth is very popular as warm wrapper till now. It is said that a Bodo girl finds it difficult to get married if she does not know the art of weaving. It is known that Bodo women are expert weavers. They are well versed in

Fig. 75.2 Bodo women wear dokhona, it is a single piece of hand woven cloth

Fig. 75.3 **a** Aronai, **b** gamocha, **c** eri chadar, **d** shima, retrieved www.google.com

weaving their dreams through their looms and if they don't then, they are castigated in the society. There are songs and folklore, which express the quality and fineness of the yarn, the design woven and the standard of cloth in the process of production [7].

The growth of ethnic consciousness at certain point of time is the sign of change to cope with the changing world orders, and this type of changes is seen in the field of social, economic, political and religion. The bringing about a change in any society is always accentuated by some vital factors—external and internal. The most vital point is the internal self-realization, and it is emanated from historical factors or past glories achieved by ancestors and aspirations for revival of the ancient glories leads

Fig. 75.4 Eri (left) and muga (right)—the raw material used for weaving, retrieved www.google.com

to redirecting the societal activities which may be termed as awakening, consciousness, movements, sometimes revolution, etc. The external factors that influence for arousing ethnic consciousness or awakening are the other civilization or achievement made by other neighboring societies or inspirations drawn from changes brought about by other revolutions around the world [8].

75.1.2 Newness and Sustainability

In the era of globalization, clothing is experiencing astonishing modernization whatever may be the competitive and creative nature of the fashion sector, some design last for a long time, some disappear quickly and some style comes back after it's gone, can be termed as mode modification [9]. The Bodo dokhona also demands certain modifications to suit today's sense of comfortable; when everyday fashion is conceived as an interactive process through which the aspiring individuals of the society consciously project their bodily self in a distinctive manner in the form of clothing style linking with belief and attitudes of the peers of the social group, the individuals tend to draw a fine line between the existing pattern of style and their appearance by dressing in a trendy manner [10]. While attempting for bringing newness in traditional attire, the intension should be able to win friends and draw appreciations by communicating their values in terms of form and appearance to suit to a social etiquette and effective form of nonverbal communication [11] of identity. We do have a social code of dressing. When it comes to casual wear, the preference could be for a mixture of values attributed to social themes and in case of formal wear, the preference stick to the conventional dressing codes and in case of sportswear, the preference is always for a comfortable and feel good factor. In case of party wear, the preference is for displaying the oomph factor and prevailing iconic values. Thus consumers are strongly influenced by their social context; they also have a creative agency in participating in fashion [12]. Therefore, it comes to an understanding that the lure of aesthetic depends on the consumers and at large on the society and are strongly influenced by their social context and do have a creative urge for participating in fashion that can be quoted as 'Consumption establishes itself a gratuitous and creative activity in a much as people reinterpret and reorganize things they have bought according to a particular style which they are continuously engaged in accomplishing' [12]. While older people prefer dokhona for all purposes, the youths do not possess the same affection, both in term of wearing as well as making practices. The youth of today allocates more time for their public appearance and they look for more trendy wear, which would lead to comfort. It is noticed that adolescents aspire for more fashion concerns than any other age group. Colleges are a place where adolescents portray their aspiration for dresses. Therefore, it is normal for an adolescence to look for trendy dress choices [13]. With the advent of fast life, it has also been seen there is a steady rise in the consumption of styles, Bodo girls are no exemption of this that have all added on a renewing feature and enhanced the existing look and feel of the customer [14, 15]. Culture is described as the control mechanism of society. There is a strong link between culture and identity elements [16, 17]; the way we dress and our identity are intimately linked to each other. Clothes display, shape and express identity while filling it with a direct material reality [18, 19] and expresses the emotions to being in time and context [20].

Clothing is a marker of class distinction, where the aspects of identity are increasingly highlighted, such as social and cultural aspects [21]. The meaning of clothing and fashion varies to people depending on different beliefs and values.

Understanding the same meaning of clothes is one condition to belong to the same cultural group, and to share the understanding of clothes and why it is worn is therefore fundamental [11].

75.1.3 Identity Elements Issues

Bodo community has maintained their clothing identity in order to mark the traditions and not much influenced by the British colonization till the present day as it is seen in many other fields [22]. The focus on local identity and authentic ethnicity has increased over the last decade in India, and the meaning of authentic Indian dressing has changed. The regional dress salwar kameez and the sari has attained a national status [23, 24], and a lot of design trials have been carried out. Indians see their traditional regional dress as 'backward' and too 'exotic,' and the more fashionable new versions of the clothes as 'progressive and modern'. Several issues, such as acceptance and comfortable usability, affect the way an Indian women dress. Styles and colors are one of the many things that are cautiously considered when an outfit is chosen.

This Bodo pride needs to be upgraded and sustain in the society with the ideas to assimilate Bodo traditions and make it known worldwide. The process of urbanization has become more prevalent across tribal areas in the post-independence India, and its impact is strongest among the northeast tribes and the Bodo tribes are no exception. With the percolation of modern education and emergence of modern employment opportunities, shift from rural to urban areas, change of occupation had led to development of district towns [25]. It is also reported that dokhona is a time-consuming wear in compared to other modern dresses as well as complained about chest pain if worn for longer time and in this fast life of constant move. So, to get over the negativity and to maintain the tradition of wearing dokhona, a new pattern can be thought of. So, studies encapsulate a brief enquiry on possible upgradation of Bodo traditional attire—Dokhona in terms of specific utility, design and production facilities with modern technology interventions and the role of technology higher learning institute situated in the region [3]. As it is evidenced that every trend has its roots somewhere in history and trends are really what fashion world go around, innovations and renovations are a big wave that is currently in vogue. With the passage of time, the Indian culture is getting blurred and we should now be careful as it would be on the verge of extinction with time. We would need to promote our cultural heritage so that superficial layers of other cultures vanish and bring out our Indian culture and trends on world platform to keep our country's treasure alive.

Various sociocultural forums confirm that dokhona has become a cultural symbol of Bodo tradition; however with the advent of external influences, this traditional heritage gradually leads to becoming endangered in course of time. This study thus analyzes the need of modifying and further upgradation of dokhona for a comfortable and safer wear to be able to sustain for Bodo identity. Safety is included here to give an understanding that the Bodo dokhona tends to slip-off once the knot is loose or pulled by mistake. The issues of uneasiness and uncomfortable are shown in Fig. 75.5. Hence, problems are visible. In view of today's globalization efforts and

Fig. 75.5 Uneasiness and uncomfort issues of wearing dokhona

practice, this study focuses to understand the current perception of stakeholders of Bodo cultural and tradition what wear in perspective of global modernization trend.

75.2 Aim and Objectives

This study looks into the retaining design aspects of cultural artifacts and sustainable expansion of Bodo traditional identity with special reference to dokhona, a household attire of Bodo women.

Objectives of the study are

1. To understand the common feel about the contemporary and extensively adopted dokhona amongst Bodo women with perceptions of communities for retaining sentimental identity and
2. To assess the possibility of motivating young minds for adopting sustainable wear by simultaneously preserving the traditional values.

75.3 Methodology

A two-stage survey on two stakeholder respondent groups, i.e., local users and practitioners, and experts were carried out, Fig. 75.6, for understanding the perception prevailed on the utility value and on the issues of possible development scope to consider for Bodo traditional attire—dokhona.

75.3.1 Respondent Groups—Sampling

The survey was conducted in Kokrajhar, Assam, meeting in-person with local people from the Bodo community who are the regular users of the product and through with the help of online questionnaire to get larger participation beyond the specific community boundary. Random sampling was considered as the locality undertaken

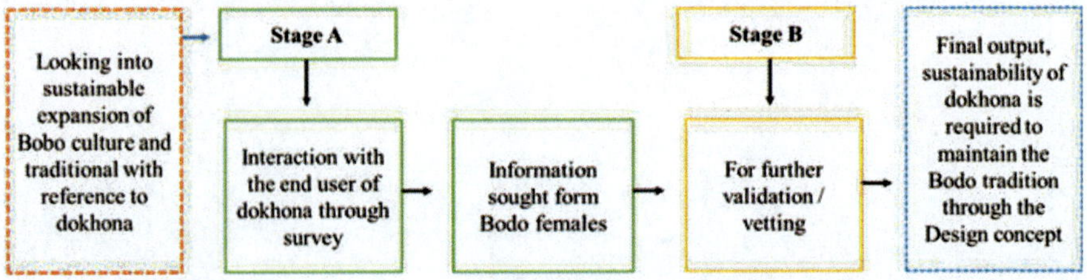

Fig. 75.6 Flowchart of study

with the help of students, expert fashion designers, doctors, ergonomists, design professors/professionals, educationist, social activists and the end users including some participants from other parts of the country who had been working in the study area and who had desire to accept and had adopted the Bodo culture and its dress. The age group between 18 and 56 years was considered so as to cover an overall response and aspects of the use of Dokhona by these end users. The survey was based on 66 responses.

75.3.2 Survey Stages

The details of survey stages are emphasized below.

75.3.2.1 Stage A: Interaction with the End Users of Dokhona Through Survey

Thirteen questions were raised to have an understanding of the prevalence of dokhona from the end users, which are summarized below.

The questionnaire was based to have an understanding if the youth of today still feels the traditional Bodo attire an integral part of the Bodo culture in spite of its diminishing use, maintaining of the identity of Bodo culture. Further, to have knowledge if users wanted to review of color combination with new designs. Some visual pictures were highlighted to understand the situation of uneasiness while working wearing dokhona and To have an understanding if long hours of wear of dokhona would lead to any health issues was also discussed. The motive to understand about overall comfort or uneasiness wearing dokhona among the females was needed. To have an opinion if any changes are required to overcome the uneasiness arising wearing dokhona was also discussed. For further understanding about weaving of dokhona could bring sustainability to the tradition was opined. Finally, to learn whether design intervention could be explored in tune with the traditional dokhona to meet its daily wear, thereby bringing in newness in dokhona.

75.3.2.2 Stage B: Consulted and Communicated with the Experts

For further vetting and validating the responses gathered from the respondents which is incorporated at stage A, communication on the information was passed on to some selected Bodo expert designers, ergonomists, professors, social activities, doctors and youth. Discussions over phone and through direct in-person meeting survey relevant information were sought from various social activities. Feedbacks were also received through individual emails. Across the table, discussions were undertaken with different design professors and ergonomists. Based on the health issued cropping on the use of dokhona on the survey reports, some physicians were consulted and information was obtained.

75.4 Result and Discussion

Research study on the sustainability of dokhona for maintaining the Bodo identity is not available and probably has not been carried out. Some scholars indeed have brought into light the making of dokhona, but the vital facts like artifacts role, its ergonomic gesture, the sustainability and its role in maintaining the Bodo identity have not been studied before this perspective work. The Bodos those who have gained awareness took the pioneer role in making efforts to rescue the tribes from being extinct by introducing socioreligious reformation movements. The combined responses against the questionnaire specifically to have the prevailed perception on the dokhona and development need as obtained from both the groups are presented in Fig. 75.4. It clearly mentions that the development would be based on its modern usability and maintaining the traditional identity with specific design element embedded in it in various ways.

According to the data received in question 1 which is based on dokhona being an integral part of the Bodo culture, 75.8% of the people strongly agreed to the criteria, while 22.7% of the participants agreed and neutral percentage was 1.5%. In question 2, which discussed woman upholding the traditional culture, 51.5% of the participants strongly agreed while 36.4% of respondents agreed and the neutral percentage was 10.6 while responding percentage in disagreement was 1.5%. In the data obtained from question 3, that asked about whether dokhona is comfortable for daily uses, 7.6% strongly agreed, 34.8% gave a neutral reply and 16.7% disagreed to the statement (Fig. 75.7).

The data showed from question 4, whether the colour can retain the traditional identity shows that 31.8% agreed to the statement, 15.2% had a neutral approach, while 50% strongly disagreed. The statement gathered 42.4% strongly agreed statement, 45.5% agreed to the statement, while 10.6% showed neutral approach according to the 5th question about new design concepts. In Question 6, which raise the problem of wearing dokhona in workplace, 21.2% agreed to the situation of the visual shown, 25.8% was neutral, 30.3% disagreed, while 19.7% strongly disagreed to the situation. In question 7, the fear of slipping of dokhona was opinion and 10.6% strongly agreed, 31.8% agreed, neutral percentage was 25.8, 21.2% disagreed, while 10.6% strongly disagreed. In question 8, discomfortable due to wearing dokhona for long hours showed strongly agreed analysis of 12.1, 22.7% agreed, 34.8% showed neutral, 22.7% disagreed and 7.6% strongly disagreed. In question 9, overall dokhona's uneasiness was discussed and 19.7% agreed to the statement, 34.8% was neutral, 27.3% disagreed while 13.6% strongly disagreed. The question 10, which asked about the requirement of changes, 9.1% strongly agreed, 34.8% agreed, 24.2% had a neutral approach while 28.8% disagreed. From question No. 11, whether dokhona relating to the Bodo identity was discussed and

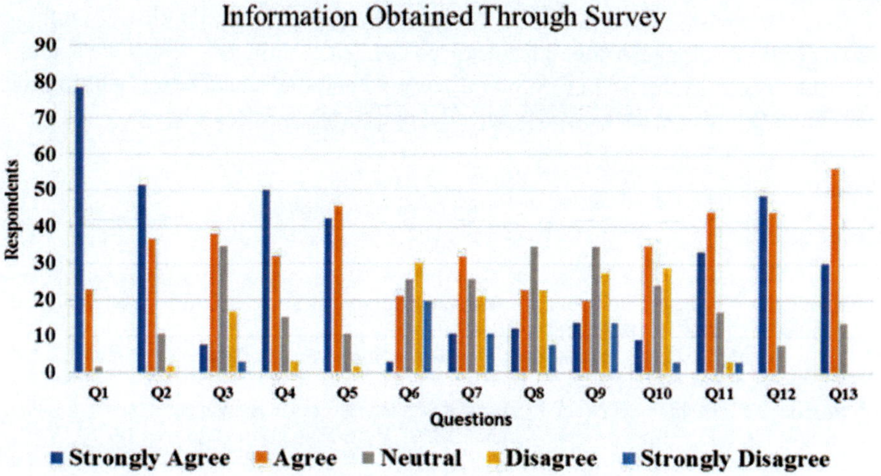

Fig. 75.7 Graphical representation of survey responses against questionnaires in general

33.3% strongly agreed to the statement, 43.9% agreed while 16.7% had a neutral approach. To have an understanding whether weaving of dokhona could lead to sustainability was discussed in question 12 and 48.5% strongly agreed to the statement, 43.9% agreed and 7.6% was neutral on the issue. Finally, in question No. 13, it was desired to understand whether design development exercise could be explored for preserving the culture and 30.3% strongly agreed, 56.1% agreed while 13.6% had neutral approach to the statement.

75.5 Conclusions

The sense of obligation of the women of the Bodo Community of the North-east India in adhering to the dokhona as a traditional full-body wear with an urge to conserve the traditional culture, heritage and identity, a change of preference in wearing other dress-forms replacing dokhona has been observed, particularly among working-ladies and school- and college-going girls, in recent times. This change has been found to be driven by the perceived need of convenience and safety, besides the influence of new fashion trends, assimilation of western cultural elements, etc. With the appreciation of the sentiment attached with the dokhona among the Bodo community, especially elderly women, the study reported herein was focused on the importance of retaining eco-design of the dokhona while incorporating subtle, yet functionally efficient, changes in respect of wear ability, safety, comfortable, etc. for sustainability. It was found that scopes of analyzing various design elements as identified by traditional practices in modern perspectives and giving a relook at their design applications for maintaining the traditionality of the Bodo dokhona have evolved. It is envisaged that the design development exercises with the above findings would give a new image to dokhona toward ensuring its continued acceptance leading to its sustenance in the Bodo Community as a valued traditional attire befitting the requirements of everyday as well as specific and purpose-suited uses.

References

1. Riello, G.: The object of fashion: methodological approaches to the history of fashion. J. Aesthetics Cult. **3**(1), 8865 (2011)
2. Swargiary, B., Boro, R.: An overview of weaving: Bodo women in the present context. Int. J. Sci. Technol. Res. **6**(11) (2017)
3. Shende, B.R.: Fashion trends and its impact on society: a case study on apparel trends of Indian society. Int. J. Res. Biosci. Agric. Technol. IJBAT **V**(3), 58–63 (2017). ISSN 2347-517X
4. Baro, N.: Tradition and change of the arts and crafts of the bodo tribes of Assam: a study. Int. Res. J. Interdiscipl. Multidiscipl. Stud. (IRJIMS) **II**(XI) (2016)
5. Pakyntein, E.H.: Census of India 1961, Assam, Vol. III, Part. V-A, pp. 284–287
6. Goswami, M.: The Bodos: culture and society. J. Bus. Manage. Social Sci. Res. **1**(1), 88–103 (2012)
7. Narzar, D.: Folklore of the Bodo Community: A Theoretical Perspective on Myths and Legends (Doctoral dissertation) (2018)
8. Singh, O.R., Guide, P.H., Basumatary, K.R.: Historical Basis of the Bodo Identity Formation
9. Choughary, A.S.: Fashion trends and its impact on Indian society. JASC J. Appl. Sci. Comput. ISSN No: 1076-5131
10. Barnard, M.: Fashion theory: an introduction. Routledge (2014)
11. Saravanan, D., Nithyaprakash, V.: Fashion trends and its impact on society. In: International Conference on Apparels and Fashion, Talmilnadu, India, vol. 30 (2016)
12. Sassatelli, R., Arfini, E.A.: Creating value, consuming Bologna: the case of DegustiBo. J. Consumer Cult. **17**(3), 542–561 (2017)
13. Erikson, E.H.: Identity youth and crisis, New York (WW Norton) (1968)
14. Crane, D.: Fashion and its social agendas: class, gender, and identity in clothing. University of Chicago Press (2012)
15. Kymlicka, W.: Multicultural citizenship: a liberal theory of minority rights. Clarendon Press (1995)
16. Geertz, C.: The interpretation of cultures: selected essays. Basic Books, New York (1973)
17. Constantin, S., Rautz, G.: Culture and identity. J. Eur. Integr. **25**(3), 189–205 (2010)
18. Ribeiro, A.: Dress and morality. Berg, Oxford (2003)
19. Davis, F.: Fashion, culture and identity Chicago (2013)
20. Barnard, M.: Fashion as communication. Routledge (2013)
21. Bhattacharjee, A.: The habit of ex-nomination: nation, woman, and the Indian immigrant bourgeoisie. Publ. Cult. **5**(1), 19–44 (1992)
22. Harcourt, E.S., Hussain, F.: Lucknow: the last phase of an oriental culture. Oxford University Press (1994)
23. Nandwani, D., Seth, S.: Salwar: Recolonizing India. Little India (2001)
24. Khakha, A.: Tribes and Urbanization in North East India and Challenges Economic & Political Weekly, ISSN(Online)2349-8846
25. Brahma, C., Saha, B., Chakrabarti, D.: Tradition and utility artifact: Bodo tradition of Assam. In: 17th International Conference on HWWE-2019

Chapter 76
Design of Bamboo Shelter Kit for Post-disaster Temporary Shelter Response

Kankana Narayan Dev and Amarendra Kumar Das

Abstract Shelter aid is an integral part of humanitarian aid in post-disaster situations. The built environment is the most affected sector in the rapidly increasing events of natural and man-made disasters. In these scenarios, sustainable design solutions need to address the loss of habitat and displacement. North-eastern part of India is very vulnerable to natural disasters primarily flood, earthquake and landslide, and annually, over a lakh of people are displaced. People in the rural areas in this region reside in bamboo houses. Recent explorations of bamboo as a material in contemporary architecture suggest an environmental stratagem towards achieving sustainability in the built environment. While, ecological considerations, social and cultural impact establishes the need for a form of architecture that is flexible, lightweight in construction. This study looks into the lifecycle of post-disaster emergency shelters with an ecological perspective to determine the critical aspects of design, material sourcing, manufacturing, delivery and disposal in an environmental friendly way possible. Further, this study attempts to design a post-disaster temporary shelter kit to establish the crucial characteristics of bamboo architecture that make it successful. The design decisions based on bamboo elements with partial prefabricated and portable components show that mobile bamboo buildings are feasible and can fulfil many different roles and are economically viable to build and operate in post-disaster situations. The study concludes with a recommendation for further iterations of the shelter kit towards sustainable reconstruction in post-disaster situations.

76.1 Introduction

The north-eastern state of Assam is one of the most vulnerable parts of India to disasters. Its geographical location in the Brahmaputra Valley has contributed to its annual

K. N. Dev (✉) · A. K. Das
Indian Institute of Technology Guwahati, Guwahati, India
e-mail: kankana.dev@gmail.com; dev176105005@iitg.ac.in

A. K. Das
e-mail: dasak@iitg.ac.in

A. Chakrabarti et al. (eds.), *Design for Tomorrow—Volume 1*, Smart Innovation, Systems and Technologies 221, https://doi.org/10.1007/978-981-16-0041-8_76

inundation by flood apart from lying in the seismic zone 5 and the socio-political turmoil where millions get displaced annually according to the Assam State Disaster Management Authority [1]. Being an Indian state that faces such catastrophes with an unqualified building stock, the design of post-disaster housing is an important subject to be studied in its various forms of post-disaster housing, viz emergency, temporary, transitional and reconstruction. "Post-disaster housing design" is one of the least explored research areas in the context of disasters in the Brahmaputra Valley. Temporary Shelters—an intermediate housing in the post-disaster housing process—is the focus of this paper. The main characteristics that distinguish a "temporary post-disaster shelter" from permanent reconstruction are; easy and fast delivery, rapid assembly, flexibility and variability, efficient and affordable technology [2].

Bamboo as a construction material in architecture has been recently a topic of research for its versatility in terms of structural durability and low carbon footprint. The literature study on bamboo suggests its usage for disaster response housing. The bamboo transitional flood shelters built in Pakistan and Indonesia aftermath the floods and tsunami are found to be highly efficient because it was found to be appropriate to the limited budget of the disaster victims. The approach of use of plastered bamboo for incremental housing projects in Indonesia is visually appealing and expresses better quality that had helped to change the poor perception of bamboo construction among the low-income community [3]. Traditionally, people in the rural areas of Assam have been residing in bamboo houses; however, the concept of delivery of a "bamboo shelter kit" in modular form for user assembly is a new design concept of temporary post-flood housing.

This paper discusses the need for emergency shelters with an ecological perspective to determine the critical aspects of design, material sourcing, manufacturing, delivery and disposal in an environmentally friendly way possible. This study aims to design a new temporary housing model applicable to the study context, using the local building resource of bamboo, which can be manufactured locally without being dependent on external humanitarian aid.

76.2 Design Criteria Development Based on Human Needs and Humanitarian Standards

From the perspective of humanitarian temporary shelter aid, the design criteria are divided into three main areas:

- Basic human necessities like the essential qualities of home and its spaces, the social environment and functions.
- Settlement requirements set by the aid organization precisely drainage, arrangement, security, water and sanitation
- Technical specifications regarding standard physical and mechanical properties of the shelter and settlement.

76.2.1 Sense of Place: Home

The process of housing is the fundamental aspect of dwelling in traditional cultures. Primitive cultures endowed individuals and households with the knowledge and materials required to coexist with the ecology of the area. The household led to the formation of the "place" of the individual and household in society [4]. The study of Maslow's hierarchy of human needs introduces a brief reflection about human needs as a source of motivation in life. A careful shelter design addressed to the creation of a welcoming and familiar environment could facilitate the recovery process of the users through its inhabitation. We find that when individuals are in control of the space, they live in, and have privacy needs met, feelings of comfort and freedom are possible. This freedom permits relaxation and the individual development of the self and the sense of a home [5]. A list of humanitarian shelter designs was studied from literature as listed in Table 76.1 to develop design criteria which are crucial to meet the housing needs of people displaced by disasters.

Table 76.1 List of design criteria based on the essential qualities of human necessities

S. No.	Design criteria	Objectives	Source
1	 Security	• Feeling of safety and security • Ensuring protection for the inhabitants and their belongings • Protection from nature and other external anti-social factors	[6]
2	 Permanence	• Experience of continuity that characterizes the home environment • Sense of belonging and connection with the place	[7]
3	 Personalization	• Exclusive control and use of the environment • Self-expression and scope for customization	[8]
4	 Privacy	• Control of space in dense and crowded environments • Sound insulation • Internal partitions	[6]
5	 Social relationships	• Relationships with family • Community meeting places • Shared spaces for cooking, washing, cleaning, etc	[5]
6	 Complexity	• Visual perception • Scope for customized assembly to avoid confusion due to repetition	[9]
7	 Identity (Form, color, light)	• Perception of physiological, environmental and cultural factors	[10]

Fig. 76.1 View of a Post-2019 Flood Relief Camp in *Chirang* District of Assam

An initial field study of a relief camp which had 75 families residing together in a community hall post-2019 floods in *Chirang*, District, Assam, was made to understand the effect on the daily lifestyle due to the displacement. The displaced community was homogenous in terms of their cultural and religious belief. Entirely agrarian, the community, was dependent on the fertile soil in the banks of the river Brahmaputra. The makeshift camp in the community hall lacked the necessary human shelter need with the improper spatial distribution with over 300 people residing in a hall space of 180 m^2. Ventilation, sanitation and lighting were not adequate for the activities of reading and stitching as observed in Fig. 76.1. Only women and children occupied the hall space, while the man slept in the outdoors, which impacted the family comfort and mutual support. From this study, we considered the design of unit shelter per family in clusters over community shelter.

76.2.2 Spatial Requirements and Functions

Space is a major constraint in post-disaster shelter and settlements [11]. A single unit is mostly used to perform multiple activities. Table 76.2 lists the design criteria which we considered while making the spatial arrangements.

76.2.3 Humanitarian Standards

The aid agencies have set up some general requirements concluded in several instruction documents during the last ten years. Among the most recognized are *The Sphere Handbook* [12], first published in 2000, communally by IFRC and several

Table 76.2 List of design criteria based on the spatial requirements and functions

S. No.	Design criteria	Objectives
1	Sleeping areas	• Indoor area with the optimum size for sleeping
2	Health and hygiene	• One latrine per 20 individuals to be considered
3	Culinary activities	• Covered space preferably shared clean and dry
4	Storage	• Storage of belongings and other relief materials
5	Washing and cleaning	• Externally provided facilities of WASH to be considered
6	Studying	• Community space among shelters one in 100 shelters
7	Outdoor playing/playing/working/meeting	• Semi-covered spaces

Table 76.3 List of design criteria based on the humanitarian standards

Dimensions	Shelter for 5
Transport	Local availability of material
Cost	Per family 150 dollars (considered as per the context of Rural Assam)
Appropriateness	Culturally and socially relevant
Assembly	Preferably made by the user community
Durability and afterlife	Contribute to reconstruction

humanitarian NGOs. It includes a specific part on minimum standards for shelters, settlements and non-food items. Table 76.3 lists the criteria given by the aid agencies.

76.2.4 Structural and Technical Requirements

The aid organizations usually set the structural and technical elements of shelter design, and we considered the following parameters. The engineering calculations are outside the scope of this study. However, we approached to verify with relevant references. This research explores bamboo material for architectural design and construction. Joinery details, involving bolted joints by employing modern tool electric drill machine and application of frame truss system, are implicit "modern bamboo construction system" [13]. Tension and compression are the two structural design principle that can be achieved by the use of bamboo. Experimental analysis of full scale joinery details is advised for use of structural bamboo due to the lack of available information and traditional joinery methods are encouraged which are in practice for a long time [14].

- Load bearing
- Thermal comfort
- Water resistance
- Vector control
- Fire resistance
- Environmental impact.

76.3 Use of Bamboo in the Built Environment of the Study Area

76.3.1 Rural Built Typology in the State of Assam

From the hazard profile of the state, we realize that the state is highly vulnerable to natural disasters, mostly floods and earthquake. The effects of floods and earthquake especially affect the human dwellings be it residential or of public use. Traditional rural housing typology, as shown in Fig. 76.2, generally consists of a homestead with a cluster of rooms arranged around an open courtyard and surrounded by betel-nut trees, plain tall trees, bamboos and vegetable gardens. A bamboo fencing surrounds the homestead. The several spaces in the homestead consist of a front yard, the living room as a block with no shared wall, the bedroom, the kitchen block and the shed. A prayer room and a granary and most of the time a tower of hay for the cow shed consist of the spatial features of the homestead. The built elements are mostly constructed using bamboo or *ikra* a variety of weed *wattle and daub* technique. Most of the dwellings do not have windows but *jalis* instead as openings. The flooring is often mud and sometimes plastered with cement. The structural columns in the new houses are made of concrete pillars with bamboo infill walls. Profile corrugated galvanized iron sheets are used for the roofing. We observed from the field study that

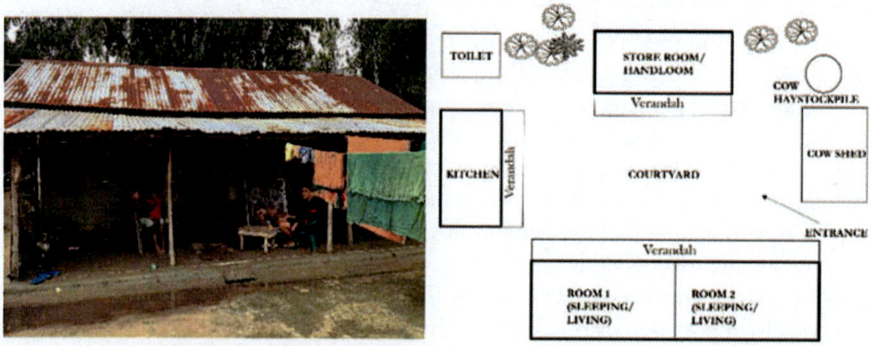

Fig. 76.2 View from the courtyard of a rural homestead and a single line plan of a rural flood-prone homestead

the flood-prone areas did consider houses on stilts as flood resilient features in some of the rural areas. However, due to the difference in the effect of flood and cultural adaptation, houses on stilts are limited to a few tribal areas only.

76.3.2 Use of Bamboo in Construction

Bamboo is the traditional building material of Assam. It is used commonly as a construction material due to its rigidity, strength and availability. With the backdrop of frequent floods in the state of Assam, and the abundant natural availability of suitable bamboo species for construction, viz *Bambusa tulda* and *Bambusa balcoa,* makes bamboo a suitable material to address the emergency housing need. During our study, we found that for the 2018 floods in *Cachar* District, Assam, ***Habitat for Humanity***—an international aid organization—distributed bamboo poles of 6 m length in a bundle of five with a tarpaulin sheet and a bucket of fasteners in the form of nails and bending wires to per affected family (Fig. 76.3).

The hot and humid weather following the floods in Assam makes it uncomfortable and prone to vector-borne diseases. The cultural and comfort levels of tents and tarpaulin cannot meet the comfort of a bamboo structure. Thus, we decided to explore design options using bamboo with wall cladding of woven bamboo mats of size 120 cm × 250 cm. Bamboo matting has a porous surface that keeps the structure cool in the hot and humid climate of Assam (Table 76.4).

76.4 Design Process of the Bamboo Shelter Kit

The contemporary DIY, "do it yourself" culture, ethic, process and tips, and its creative incorporation in academic experiences of architectural self-construction seemed to be appropriate in the said context. The design methods such as CASA

Fig. 76.3 Emergency shelters post-2019 floods Assam made with tarpaulin sheet and bamboo poles

Table 76.4 Advantages and concerns of using bamboo for post-disaster construction

Advantages	Concerns
Lightness in weight (less than 500 kg/m^3)	Variety of cross sections
Good tensile strength (30–40 N/mm^2)	Non-homogenous material
Ease of assembly	Durability in all weather conditions
Culturally relevant	Termite infestation
Local availability	Dangerous in fire
Low cost (cost of one pole of length 6 m is 150 INR)	Cost is high if chemically treated for longevity
Renewable material grows back in three years	Highly effected by humidity
Construction system can be locally derived	Skilled assistance and capacity building for advanced construction technology

(Collaborative strategies for Adaptable Architecture) and AIDA (Analysis of Interconnected Design Areas), were explored to develop several options [15]. Figure 76.4 shows the process of layout development and suitable design alternatives. The timing in the process of design is an ambiguous issue. The designer and the owner in architectural design decide together, and sometimes, it is very time consuming [16]. We considered a community consultation activity where we presented CAD drawings made using AutoCad and Sketch Up mock-up model to a representative community from the study area. The feedback obtained was considered to improve the proposal.

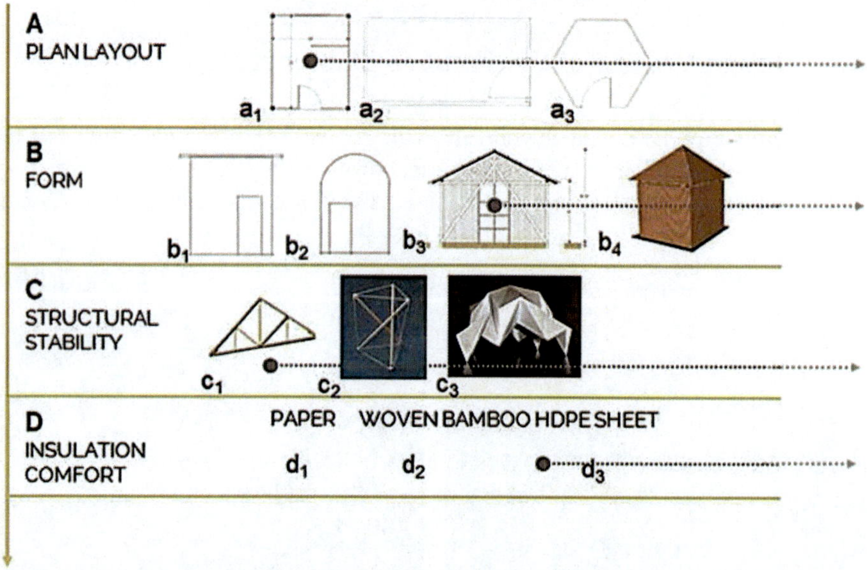

Fig. 76.4 Part of design development process using CASA collaborative strategies for adaptable architecture and AIDA analysis of interconnected design area

The layout plans were processed further for actual prototype development of the temporary shelter kit by user assembly.

76.4.1 Design Features of the Bamboo Shelter Kit

The spatial, functional, humanitarian and technical aspect of the shelter kit (Fig. 76.5).

- 9 m^2 per unit shelter.
- Maintaining the privacy of the family unit by perforated walling over openings.
- Opening of the shelter into the access pathway.
- Community cooking corner with cleaning facility is considered in the common-place.
- A shared toilet and washing facilities per every eight units.

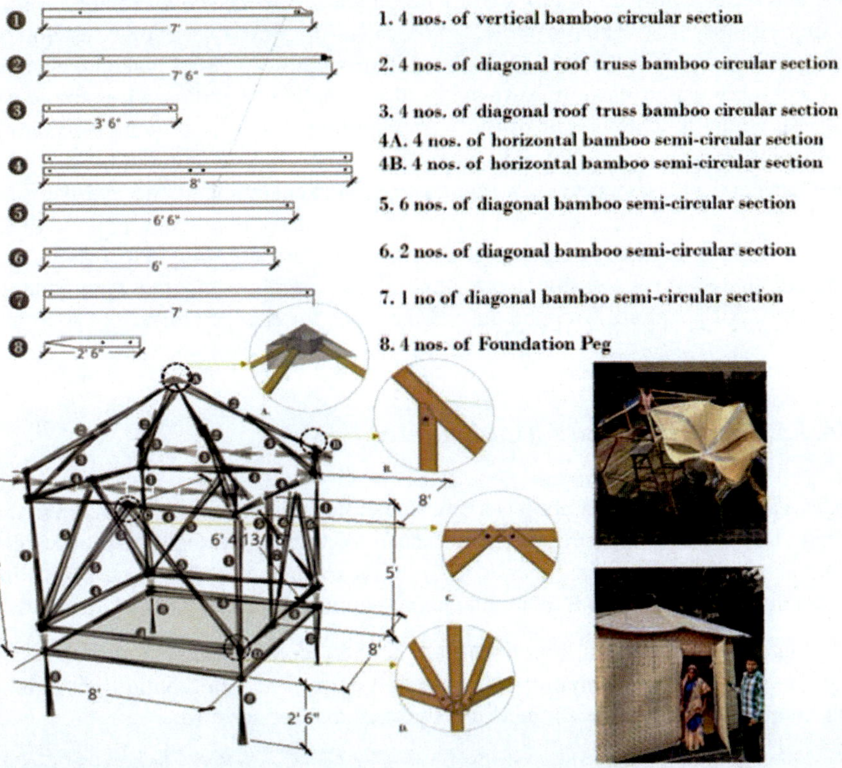

Fig. 76.5 Bamboo shelter kit: isometric view showing the skeletal frame details and to scale prototype pictures of the assembled kit

- The lowest position of the roof inside the shelter unit considered to be a minimum of 1.8 m.
- A compressed roofing layer develops a sandwiched roof with HDPE sheet and woven bamboo mats for the roofing.
- A 15 cm perforated lattice bamboo between the roof and the top of the wall cladding act as ventilation to enable the release of hot air.
- All residents were involved in setting up the shelter kit.
- User assembly thus giving scope for customization and sense of ownership.
- The affected community was involved in the process of designing. They had the skills and experience of utilizing the local resources and which makes the process fast.

76.4.2 Construction Process

The construction process which involves the transportation of the shelter kit to the site and its assembly takes just a day with only a few workmen preferably users for a single-family unit. The prefabricated details in the form of modular bamboo pole components with predefined positions in the way of drilled holes make it convenient for assembly using plastic tags/treated rods for bolts or bending wires to meet the desired strength. The weight of the Kit is optimized to be lightweight using only half bamboo splits instead of solid poles for ease in delivery without the use of any mechanical devices or cranes. The connections are kept simple for assembly by users so that if there is any shrinkage in the natural material, users can modify it to suit their needs. The delivered kit comprises of bamboo poles precut to length in different sizes and colour coded for step-by-step assembly. The prefabricated and woven bamboo wall and roof panel are wrapped around the poles for ease of delivery.

76.4.3 Protection and Maintenance

Ecological Protection: Bamboo is a renewable material which can be grown in three years. However, considering the annual need for temporary shelters, proper guidelines need to be in place to generate awareness among the local communities to find a sustainable harvest of bamboo and encourage plantations to meet future needs.

Physical Protection: Temporary shelters, unlike reconstruction, should be encouraged to be used for a temporary short period until users can rebuild their habitats. Physical protection of the shelter kit is optimized to reduce cost.

Termite Treatment: The anti-termite treatment for bamboo if chemically made is expensive. Thus, the consideration for mature culm harvest, natural water treatment and spraying of organic insecticide made using local herbs and smoking have been considered and also encouraged.

Fire Protection: Fire protection is a significant concern in bamboo houses. During the field study, we observed that in the local houses, the kitchen is detached from the homestead to avoid a fire hazard. The post-disaster temporary housing should consider centralized cooking or cooking area where a family can cook food and provisions for firewood, water and light can be made available.

Waste Management: Temporary settlements in post-disaster phase are expected to have a very short lifespan. The shelter kit having over 90% of the product as a natural material contributes minimal to the environment degradation of the shelter site.

76.5 Conclusion

The field of post-disaster housing is very multi-disciplinary uniting the principles of social, environmental and technological sustainability together with economics, logistics and politics. The design process of the post-disaster temporary shelter kit establishes the crucial characteristics of bamboo architecture in the context of disaster response. In a developing country with limited resources at our end, we must try to work with the existing resource and address its shortcomings. In the developed shelter kit, the use of bamboo has contributed to a low carbon footprint product, socially relevant and economical to the context. However, the concern of termite infestation due to non-treatment of bamboo, ecological harvesting of bamboo followed by the waterproofing of the walls gives scope for further research of exploring the material and learning the traditional treatment practices which can not only contribute to the longevity of the shelter kit but also give an aesthetic visual character due to the stains of the use of plant-based insecticides.

References

1. Assam State Disaster Management Authority, https://sdmassam.nic.in/annual_report_sub.html. Accessed 21 Apr 2020
2. Aquilino, M.J.: Beyond shelter: architecture for crisis. Thames & Hudson (2011)
3. Kristi, J.E., Widadya, M.B., Sigit, A.L.: Transitional shelter for disaster victims: bamboo core and incremental houses. Dimens. J. Archit. Built Environ. **41**, 29–36 (2014). https://doi.org/10.9744/dimensi.41.1.29-36
4. Rapoport, A.: House form and Culture. London-University Coll. New Delhi Prentice-hall India Priv. Ltd. (1969)
5. Smith, S.G.: The essential qualities of a home. J. Environ. Psychol. **14**, 31–46 (1994)
6. Saunders, G.: Transitional shelters: eight designs
7. Hayward, D.G.: Psychological concepts of home. HUD Chall. **8**, 10–13 (1977)
8. Sebba, R., Churchman, A.: The uniqueness of the home. Archit. Behav. **3**, 7–24 (1986)
9. Stamps, A.E.: Visual permeability, locomotive permeability, safety, and enclosure. Environ. Behave. **37**, 587–619 (2005)
10. United Nation Refugee agency: Core Relief Items (2012)
11. Sinclair, C.: Design like you give a damn: architectural responses to humanitarian crises. Thames & Hudson London (2006)

12. Assar: SPHERE PROJECT. United Nations High Commission for Refugees (2012)
13. Widyowijatnoko, A.: Traditional and Innovative Joints in Bamboo Construction. The Faculty of Architecture, RWTH, Aachen University, Dissertation (2012)
14. Janssen, J.J.A.: Building with Bamboo: a handbook. Practical Action Publishing, UK (2000)
15. Jones, J.C.: Design Methods. Wiley Ltd., London (1992)
16. Groat, L.N., Wang, D.: Architectural research methods. J. Wiley, New York (2002)

Chapter 77
Review of Building Energy Code and Its Implementation in Residential Sector: A Global Outlook

Kratika Piparsania and Pratul Ch. Kalita

Abstract Building energy code, standards, ratings, and labels provide minimum efficiency requirements for new, existing, and renovated buildings, ensuring reductions in energy use and emissions over the life of the building. Energy code is the subdivision of building codes, which establish baseline requirements and direct building construction. Yet, these codes will only be able to deliver outcomes when implemented. The study of building energy code needs to be considered in a consistent and comprehensive way in order to achieve low carbon development and future sustainable goals. This paper identifies the process and practices for establishing residential energy code, covering 12 countries globally, including comprehensive review for code coverage of buildings and its design, implementation context, revision schedule, penalties, incentives, materials, and certification schemes. This paper highlights a global scenario of energy code and approach followed by various countries and followed by recommended practices for India.

77.1 Introduction

Globally building floor space has expanded 65% since 2000, while energy/m^2 has improved by only 25% [1]. As per IEA, building energy codes set standards for the construction of buildings with better energy performance and are a proven method to reduce building energy consumption in buildings [2]. As of 2019, however, less than 75 countries have or are developing a mandatory or voluntary building energy code, and around 45% of those countries building codes cover just part of the buildings sector. Energy policy progress is not keeping pace with buildings sector growth []. Mandatory policies covered less than 40% of energy use and less than half of CO_2 emissions from buildings in 2017 [1]. To be in line with the sustainable development

K. Piparsania (✉) · P. Ch. Kalita
Indian Institute of Technology Guwahati, Guwahati, India
e-mail: kpiparsania@iitg.ac.in

P. Ch. Kalita
e-mail: pratukkalita@iitg.ac.in

A. Chakrabarti et al. (eds.), *Design for Tomorrow—Volume 1*, Smart Innovation, Systems and Technologies 221, https://doi.org/10.1007/978-981-16-0041-8_77

scenario by 2030: all countries need to establish mandatory building energy codes; new high-performance construction needs to rise from current 275 million m2 to more than 5 million m2 by 2030, increasing code coverage and stringency [3].

The buildings sector's share of the world's delivered energy consumption increases from about 20% in 2018 to 22% in 2050. Building energy consumption in non-Organization for Economic Co-operation and Development (OECD) countries increases at about 2% per year, about five times faster than in OECD countries, and surpasses that of OECD countries by 2025 [4]. India comes under the category of non-OECD country. Electricity remains the fundamental source of marketed energy consumption in the residential sector, and its use grows by 2.5% per year globally [5]. India and China remain the fastest-growing region in residential sector energy consumption. China adds the most residential energy consumption of any country (in absolute terms), while India experiences the fastest relative growth in residential energy consumption from year 2018 to 2050 [5]. Studies on global changes in residential energy consumption recommend on promotion of direct and indirect renewable energies to reduce energy consumption and increase in adoption of energy code and practices.

Energy policies can be successful only if they are enhanced by making them mandatory, targeting net-zero energy building, and increasing public awareness about new technologies. Stricter regulations especially in existing buildings need to be further stressed by focusing on both new technology development and more educationally-related approaches to energy saving. In a recent study on building codes and implementation states that by better understanding in implementing building energy codes, policymakers can improve the effectiveness of their code implementation systems.

77.2 Background

Integrating the element of energy efficiency with building codes is a recognized strategy which aims to reduce energy consumption in the residential and commercial sectors. Globally, countries are independently establishing building codes and implementing energy efficiency policies, programs in residential and commercial segments to decrease energy waste in the new and existing building stock.

In many countries, the central government has authority to mandate energy efficiency in buildings by forming national building codes which are often adopted and implemented by state regions and/or local municipalities. Model energy code is developed by national code development organizations for review and adoption by state and local governments and is mandatory in nature. The practice of development of code, approval, and enforcement varies considerably among nations. Stakeholders have realized that energy codes are one of the simplest, most effective tools available to reduce building energy use.

As a building's operation and environmental impact is largely determined by upfront decisions by consumers, energy codes present a unique opportunity to assure savings through efficient building design, technologies, and construction practices. Once a building is constructed, it is suggestively more expensive to achieve higher energy efficiency levels. Energy codes ensure that the building's energy use is involved as a fundamental part of the design and construction process. Socioeconomic development (improvement of human comfort levels and entertainment activities), architectural design, geography, and climate data are the main factors underpinning the energy consumption trend in residential buildings.

Building codes seek to address common barriers to energy-efficient building design. Buildings regulation set of legal and mandatory requirements for building design and their compliance provisions during the construction period aiming at promoting energy performance of building. Building envelope consists of walls, roof, and fenestration (openings including windows, doors, vents, etc.). Design of building envelope influences heat gain/loss, natural ventilation, and daylighting, which, in turn, determines indoor temperatures, thermal comfort, and sensible cooling/heating demand. Since building codes set minimum requirements for energy efficiency in buildings, several countries have developed voluntary standards, encouraging sustainability, and higher energy efficiency buildings.

India took a step forward in late 2018, developing its first national model building energy code for residential buildings. The Energy Conservation Building Code for Residential Buildings is designed to be enforced simply while also improving occupant's thermal comfort and enabling the use of passive systems [6]. Implementation of ENS will have the potential for energy savings to the tune of 125 billion units of electricity per year by 2030, which is equivalent to about 100 million ton of CO_2 emission. The code sets minimum performance standards for building envelope to limit heat gains and limit heat loss through it, also for adequate natural ventilation and adequate daylight potential [6]. The situation with ECBC code under EC ACT 2001 is that till date, it remains voluntary code with no mandates in most of the states in India. ECBC-R also has the same challenges, that it might take years, before they become mandatory in most of states.

77.3 Methodology

Understanding code and its implementation requires a clear picture on what all code constitute and what does not. Also, it is required to understand the difference between various terminologies used like code, labels, standards, and rating system. Building codes are often mistaken as building energy labels and building standards. An energy label for building constitutes consumer information on the performance of a product (the building). A building rating system is a tool that evaluates the performance of a building and its impact on the environment [7]. It comprises a predefined set of criteria relating to the design, construction, and operations of buildings. Energy standards describe how buildings should be constructed to save energy cost-effectively [7].

The diversity in residential building energy code and implementation practice among different countries poses challenges for measuring building energy code implementation and impact. This paper covers a systematic review of building energy code and their implementation systems, analyzing building energy code in 12 countries across the world and recommends on the best practices. To compile the information needed regarding the code and current practices, the reviewed literature has been accessed from data available over government websites and global networks. The key categories have been defined initially for data collection, namely code coverage, code implementation, revision and incentive structure, building material and certification and then processed further. In addition, a number of countries have implemented programs to evaluate the effectiveness of building energy code.

77.3.1 Review on Global Approaches

Building energy code tends to set minimum energy efficiency levels, but these energy savings are never realized unless states and localities implement them. This section highlights the comprehensive review of building codes and its implementation across all 12 countries. Figure 77.1 displays the timeline of year of code adoption among these countries. It showcases two versions: Current and earlier. The current versions are the revised or currently opted versions in the respective country. Earlier versions are previous versions which were implemented but not in order now.

India launched their residential energy code in late 2018, although for commercial buildings the code into picture in 2007. Developed countries have opted the code in earlier years and they have also released the revised versions.

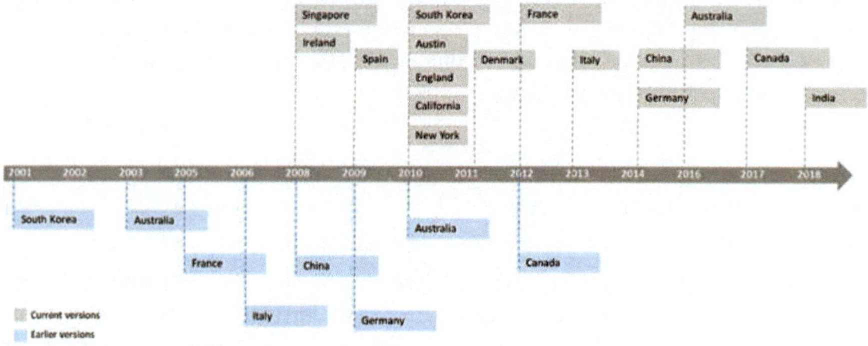

Fig. 77.1 Timeline of code adoption

77.3.2 Code Coverage

The first step in ensuring that building energy efficiency requirements apply to a significant part of buildings and have an impact on energy-intensive buildings is code coverage. Countries may have varied practices when it comes to what a code covers. Depending on the country, the code shall only apply to definite types or size of buildings; it may cover a comprehensive range of energy uses or only the building envelope. It may or may not apply to planned renovations. It is also important to note that code coverage is one way to consider the scope of implementation. Table 77.1 reflects code coverage for different countries with in depth information.

It can be observed that Italy, USA, Australia and China have not defined any building size threshold for new and existing buildings in their code. Australia, Singapore, Spain, and Turkey have threshold limits defined for renovations. Australia, China, France, Germany, Italy, South Korea, Spain, United Kingdom, and USA have included of renewable energy as a parameter in over all measures covered for building. In Turkey, buildings larger than 20,000 m^2 must use renewable energy. Most of the countries have made mandatory for their states to adopt the code.

77.3.3 Code Adoption

Implementation of energy code is generally approved by state and local bodies that are responsible for code compliance, enforcement, and training. This process ensures that new construction attains the required level of efficiency. Australia, Canada, China, India, Italy, and USA follows three-tier government system: National government develops the code, adopts national code to state requirements, and the local jurisdiction contributes in enforcing the code. For France, Germany, Singapore, South Korea, Turkey, and UK, the central government develops and adopts the code and the local jurisdiction contributes in enforcing the code. Enforcement may occur at both the design and during construction. At the design phase, an enforcement agency shall verify that plan for the building meets the specified energy efficiency requirements, while at the construction part, the code official or third party checks that construction matches with the code compliance. The built changes may also go through design review and on-site inspections. Compliance checks are divided into three phases: design, construction, and pre-occupancy checks. Australia, Canada, China, France, Germany, Italy, Singapore, USA have developed code compliance resource kits. There are training programme and tools in several countries to enable implementation of building energy code. India has launched a compliance tool for ECBC-R for evaluation and reporting. Table 77.2 reflects code adoption for different countries.

Table 77.1 Global outlook on code coverage

Country	Code	Code type	Building size threshold	Elements	Coverage
Australia	Building code of Australia	Mandatory	None for new; for renovations >2,000 m^2	Envelope, HVAC, service water heating, lighting, electric power, renewable energy, maintenance	Entire country
Canada	National energy code of Canada for buildings	Mixed	All new buildings, except residential buildings >600 m^2	Envelope, HVAC, service water heating, lighting, electric power, maintenance	Entire country
China	Design standard for energy efficiency of residential buildings	Mandatory	None	Envelope, HVAC, service water heating, electric power, renewable energy(not in all codes); lighting is in a separate code	Entire country, but rural residential code is voluntary (~45% of population is rural)
France	Energy performance of buildings directive (EPBD)	Mandatory	All new residential buildings must adhere to the standard; must have energy consumption level less than 50 kWh/m^2/pa and 80 kWh/m^2 for existing buildings	Envelope, HVAC, service water heating, lighting, electric power, renewable energy, maintenance	Entire country
Germany	Energy performance of buildings directive (EPBD)	Mandatory	All residential buildings, except for those meant for use less than four months a year	Envelope, HVAC, service water heating, lighting, electric power, renewable energy, maintenance	Entire country

(continued)

Table 77.1 (continued)

Country	Code	Code type	Building size threshold	Elements	Coverage
India	Eco-Niwas Samhita (ECBC-R)	Voluntary	All residential buildings and residential parts of 'mixed land-use projects,' both built on a plot area of ≥500 m^2	Openable window-to-floor area, visible light transmittance, thermal transmittance of roof, rETV for building envelope	Entire country (Urban India)
Italy	EPBD	Mandatory	None	Envelope, HVAC, service water heating, lighting, electric power, renewable energy	Entire country
Singapore	Code for environmental sustainability of buildings	Mandatory	None for new; all new and existing buildings that undergo major retrofit with floor area of =>2000 m^2	Envelope, HVAC, service water heating, lighting, electric power, maintenance	Entire country
South Korea	Building design criteria for energy saving	Mandatory	All residential buildings >50 households	Envelope, HVAC, service water heating, lighting, electric power, renewable energy, maintenance	Entire country
Spain	CTE	Mandatory	All residential buildings where a major renovation of more than 25% of the envelope is carried out	Envelope, HVAC, service water heating, lighting, electric power, renewable energy, maintenance	Entire Country
Turkey	TS 825	Mandatory	None for new; for existing buildings compliance with the latest version when retrofits affect at least 15% of area	Envelope; buildings larger than 20,000 m2 must use renewable energy or cogeneration	Entire Country

(continued)

Table 77.1 (continued)

Country	Code	Code type	Building size threshold	Elements	Coverage
United Kingdom	Energy performance of buildings directive (EPBD)	Mandatory	All new buildings, and existing buildings >1000 m^2 in England and wales; None for Northern Ireland	Envelope, HVAC, service water heating, lighting, electric power, renewable energy, maintenance	England and Wales have separate code, whereas Northern Ireland follow ED
United States	ASHARE and ICC	Mixed	None	Envelope, HVAC, service water heating, lighting, electric power, renewable energy	42 states adopted statewide codes; 8 rely on either national/local codes

77.3.4 Revision Schedule, Incentives and Penalties

Code revision may demand innovation that creates better products and encourages economic development. State and local building codes need to keep up with ongoing innovations in building energy, science, and technology. Some code developers push for a six-year code cycle and some for three. Incentives and penalties in the code ensure that the interests of stakeholders are aligned with the desired policy outcome, such as code implementation. Conventionally, central and state governments employ penalties, to achieve compliance. For India: Penalties for non-compliance with energy provisions in codes are decided at the state level. Currently, no state has penalties for non-compliance. Various countries offer exemptions on tax and low-interest rate on implementation of codes. India has no incentive scheme, although there are various rating programme which offer state wise incentives. Table 77.3 reflects information about code revision, incentives and penalties.

77.3.5 Building Material and Certification

Building materials with labels evidently state that they are tested for energy performance properties. Code officials may easily verify that materials are aligned with the code-compliant design or not, or if tested and labeled materials are being used. Thus, having a method for testing, rating, and labeling properties of the materials makes it easier for all stakeholders to ensure buildings are made from high-performance products. Except Australia, Italy, and India, all other countries have a mandatory provision for testing, rating and labeling building materials. Canada and India have voluntary

Table 77.2 Code adoption

Country	National	Region/State/Province	Local party
Australia	Develops performance requirements of national construction code	Adopts code; adapts national code to state requirements, including adaptations to climate zones	Enforces code
Canada	Develops code; provides tools, training and resources; issues model building codes	Adopts code; adapts national code or have their own codes	Enforces code
China	Develops and adopts code	Adopts national code to state requirements, including adaptations to climate zones	Enforces code
France	Develops and adopts code; supports accreditation; coordinates with EU	None	Enforces code
India	Develops code; provides training and resources	Adopts national code to state requirements, including adaptations to climate zones	Enforces code
Italy	Provides technical support for code development and accreditation; coordinates with EU	Adopts code and technical guidelines; supports accreditation	Enforces code
Singapore	Develops and adopts code	None	Enforces code
South Korea	Develops and adopts code; provides assistance to local governments	None	Enforces code
Spain	Develops and adopts code; coordinates with the EU	Determines forms of inspection	Enforces code
Turkey	Develops and adopts code and regulation; offers training and energy auditing; raises public awareness	None	Enforces code
United Kingdom	Develops and adopts code and regulation	None	Enforces code
United States	Develops code and regulation	Adopts code, adapts national code to state requirements	Enforces code

based certification system. Turkey initially required certification on thermal insulation which is separate from building performance certification. Australia, USA, and China have building energy code evaluation programs. Table 77.4 reflects information about building materials and certification.

Table 77.3 Code revision, incentives, and penalties

Country	Schedule	Incentives	Penalties
Australia	Irregular	Grants are offered	Denying construction permits, suspension or loss of license, refusal of permission to occupy
Canada	Every 5 years	Tax credits and low-interest loans to improve energy efficiency in buildings	Denying construction permits, suspension or loss of license, publication of names of property owners who fail to comply
China	Irregular <=5 years)	Multiple tax incentives: investment regulation tax, income tax and a value-added tax	Denying construction permits, suspension or loss of license, refusal of permission to occupy, publication of names of property owners who fail to comply
France	6 years, in coordination with EU	Exemption from property tax	Suspension or loss of license, refusal of permission to occupy
Germany	Ad hoc, in coordination with EU	Provides subsidized loans and interest rates	Denying construction permits,
India	Not mentioned	There are no incentive schemes	Not defined yet
Italy	Ad hoc, in coordination with EU	Offers a number of incentives for building retrofits	Denying construction permits,
Singapore	No clear schedule	Subsidies for building upgrade and Green Mark Incentive Scheme for New Buildings	Denying construction permits, Suspension or loss of license
South Korea	Every 4 years	Low-interest loans, tax incentives, technical support and public recognition	Denying construction permits
Spain	Every 5 years	Several types of incentives and capital grants for energy efficiency and low-interest rate loans	Not defined yet
Turkey	Irregular	None	Refusal of permission to occupy
United Kingdom	Irregular	Exemption from the stamp duty tax	Denying construction permits
United States	Every 3 years	Incentive programs and Lower interest rate loans for energy-efficient construction	Denying construction permits, suspension or loss of license, refusal of permission to occupy

Table 77.4 Building material and certification

Country	Building materials	Certification
Australia	Reference standards included in codes. There is no national labeling scheme for energy-efficient building products	NatHERS (asset rating for residential buildings) or NABERS (operational rating for new and existing non-residential buildings)
Canada	Buildings materials is tested and certified by the Canadian standards association (CSA) and voluntary certification and labeling schemes also exist, such as Energy STAR Canada (for boilers, windows, doors, heat pumps, and others)	Canada has voluntary building rating schemes, such as Super E (housing standard developed by Natural Resources Canada), Energy STAR (for commercial and new residential buildings) and EnerGuide rating system
China	Building materials are rated in China, but is not mandatory	Energy performance certificates are issued based on both design of the building and post-occupancy energy efficiency
France	Some of the standards include: HQE, EN 15,804 and EN 15,977	Article 9 of RT2012 stipulates that EPCs are mandatory at time of sale and rental since 2007 and voluntary energy endorsement label
Germany	Germany adopted CE marking under the construction products regulation (305/2011/EU-CPR), which covers energy economy and heat retention	An energy performance certificate is required in course of implementing EU's Energy Performance of buildings directive, via amendment of the energy saving ordinance
India	–	The BEE/GRIHA/ IGBC star rating system exists independently of the code
Italy	Italy does not currently have a system for testing, rating and labeling building materials, with the exception of voluntary third party certification	An EPC is required when selling or renting a property (building or building unit), when undergoing major renovation and for larger public buildings
Singapore	Singapore green labeling scheme labels bricks, tiles, insulation, windows, and many types of other materials. Singapore green building product (SGBP) certification takes into consideration the product life cycle and impact on the environment	BCA Green Mark is a green building rating system which evaluates a building for its environmental impact and performance
South Korea	Rating of building materials: windows, insulation, HVAC, lighting	Building energy performance certificate under the housing performance grading indication system, and a voluntary scheme: Korean green building certification program

(continued)

Table 77.4 (continued)

Country	Building materials	Certification
Spain	The Spanish association for standardization and certification is responsible for developing technical standards and certification programs	The energy performance certificate provides an estimate of an annual primary energy consumption and classify buildings based on CO2 emissions into classes A-G based on estimate of energy use
Turkey	Turkey follows European CE marking, which is mandatory for construction materials and cover energy economics	The thermal insulation requirements are separate from the energy identity certificates. The building performance certificates have been recently introduced
United Kingdom	The building regulations (England and Wales) 2010 and the building regulations (Northern Ireland, 2010) require rating of building materials through the various acceptable schemes. The UK accreditation service holds a schedule of testing laboratories	Energy performance certificates are not a voluntary labeling scheme, they are mandatory for all new buildings and when an existing building is sold or let
United States	Building materials are rated in the USA. Random sampling of materials is used with testing, which are conducted by certified laboratories. There are specific labels for windows, and labels are also required on some types of insulation	Energy performance certificates identified: HERS, home energy score, ASHRAE building EQ

77.4 Conclusion

There are numerous ways to assess the effectiveness of energy efficiency building codes globally, and this paper aims to give an overview of 12 countries and five elements of building energy code or review. There are many polices that were not included in this paper, such as building labeling, retrofit policies, commercial codes, and other mechanisms. However, based on the current parameters analyzed, countries that are leading the way with respect to the overall effectiveness of residential codes are USA, France, China, and Australia. This study analyzed country-specific information: Code coverage, code adoption, revision schedule, code compliance, penalties, incentives for implementation, building material, and energy certification. Most of the countries have made mandatory for their states to adopt the code. India follows voluntary code system. Penalties for non-compliance of the codes are decided at the state level in India. Currently, no state has penalties for non-compliance. Most of the countries deny on construction permits as a penalty. Australia, Canada, China, India, Italy, and USA follow three-tier government system: National government develops the code, adopts national code to state requirements, and the local jurisdiction contributes in enforcing the code. Australia, Canada, China, France, Germany, Italy, Singapore, USA have developed code compliance resource kits. India has launched a compliance tool for ECBC-R for evaluation and reporting. It provides a comprehensive summary of compliance status of all the mandatory compliance criteria of the code, for the proposed design.

This research highlights the importance of code, its components and implementation mechanism. By better understanding the range of practices in implementing building energy code, policymakers can improve the effectiveness of their code implementation systems.

References

1. IEA: Policies Database (2020). Retrieved from International Energy Agency: https://www.iea.org/policies?sector=Residential
2. IEA: Building Envelopes. International Energy Agency, Paris (2020). Retrieved from https://www.iea.org/reports/building-envelopes
3. Evans, M., Roshchanka, V., Graham, P.: An international survey of building energy code and their implementation. https://doi.org/10.1016/j.jclepro.2017.01.007
4. IEA: Global Status Report for Buildings and Construction. Paris: Global Alliance for Buildings and Construction; International Energy Agency; United Nations Environment Programme (2019). Retrieved from https://www.iea.org/reports/global-status-report-for-buildings-and-construction-2019
5. EIA: International Energy Outlook. Washington: U.S. Energy Information Administration (2019). Retrieved from https://www.eia.gov/outlooks/ieo/pdf/ieo2019.pdf
6. MoP: Standards-labelling. BEE, Ministry of Power, Government of India, New Delhi (2020). Retrieved from https://beeindia.gov.in/content/standards-labeling
7. DOE: Toolkit Definitions (2013). https://www.energycodes.gov/resource-center/ace/definitions. Accessed Sept 2020
8. EA: Policies Database (2020). Retrieved from International Energy Agency: https://www.iea.org/policies?sector=Residential
9. Energy code, Building Energy code 101—An Introduction (2010). https://www.energycodes.gov/sites/default/files/becu/BECU_Codes_101.pdf
10. Energy efficiency & renewable energy, How Are Building Codes Adopted? (2016). https://www.energy.gov/eere/buildings/articles/how-are-building-codes-adopted
11. Environmental and Energy Study Institute (EESI), The Value and Impact of Building Codes, https://www.eesi.org/papers/view/the-value-and-impact-of-building-codes
12. GBPN (Global Buildings Performance Network): Policy Tool for New Buildings (2013). https://www.gbpn.org/databases-tools/purpose-policy-comparative-tool
13. GBPN: Databases & Tools (2020). Retrieved from Global Buildings Performance Network: https://www.gbpn.org/databases-tools/bc-detail-pages/china-hot-swarm-w
14. IEA: Energy Efficiency. International Energy Agency, Paris (2019). Retrieved from https://webstore.iea.org/download/direct/2369
15. MoP: ECO Niwas Samhita (ECBC-R). BEE, Ministry of Power, Government of India, New Delhi (2018). Retrieved from https://www.beeindia.gov.in/sites/default/files/ECBC_BOOK_Web.pdf
16. SBIC: Energy Codes and Standards (2016). https://www.wbdg.org/resources/energy-codes-and-standards. Accessed Sept 2020

17. U.S. Energy Information Administration, today in energy (2018). https://www.eia.gov/todayinenergy/detail.php?id=41433
18. Young, R. Global Approaches: A Comparison of Building Energy Codes in 15 Countries. American Council for an Energy-Efficient Economy (2014). Retrieved from https://www.aceee.org/files/proceedings/2014/data/papers/3-606.pdf

Chapter 78
Exploring Two Housing Typologies in the Vernacular Architecture of Assam

Srinidhi Ravishankar and Shiva Ji

Abstract Vernacular architecture in Assam, India, is known for its seismic and weather-resistant characteristics. These underlying purposes contribute to shaping the spatial variations of traditional architecture. The paper investigates the evolution of habitat and their association in two typologies of houses in Assam—Assam-type house and Chang Ghar—to understand the distinctions of each type. Case examples are taken to analyse and compare the commonalities and distinctions over the other. It was found that one is an advanced and evolved version of the other in terms of the overall building structure, integration of materials and aspect of resilience.

78.1 Introduction

The vernacular habitat structures are built in a manner that is not only energy-efficient and sustainable but also resistant to seismic and fluvial events, wherever applicable. Vernacular architecture goes hand-in-hand with tribal art and architecture, and this makes the Northeast style of architecture stand out from the rest of India. Two major types of houses found in Assam and nearby areas are—Assam-type house and Chang Ghar. Assam-type house, otherwise called Ikra house, is an amalgamation of traditional and modern building techniques developed during the British period. Chang Ghar is an age-old tribal housing structure; it is a collective term for houses on stilts in the forest, by the river or on the hills. Mising and Karbi tribes are common dwellers of Chang Ghar.

S. Ravishankar · S. Ji (✉)
Indian Institute of Technology Hyderabad, Hyderabad, India
e-mail: shivaji@des.iith.ac.in

S. Ravishankar
e-mail: srinidhiravishankar@gmail.com

A. Chakrabarti et al. (eds.), *Design for Tomorrow—Volume 1*, Smart Innovation, Systems and Technologies 221, https://doi.org/10.1007/978-981-16-0041-8_78

78.2 Planning and Building Configuration

78.2.1 Assam-Type House

Generally, Assam-type houses are single-storeyed. However, there are a few examples of double-storeyed houses. The houses are usually rectangular in shape in the case of a single-family dwelling and L/C-shaped for a multi-family residence [1]. The spatial planning and size of the house largely depend on the number of occupants and their economic feasibility. The internal layout is planned such that the main rooms occupy the anterior spaces, followed by more private rooms. When built on slopes, the access to the house is from the hillside. There is a verandah that runs the full length of the house. The rear end of the plot houses the servicing zones—kitchen and bathroom [1] (Fig. 78.1). Assam-type houses are isolated and not constructed very adjacent to other buildings. They are known for significantly sized frontages to accommodate gardens in rural areas. However, due to space crunch, these houses in the urban setup have very minimum frontage. The basic dimensions of a typical Assam-type house are 3 × 6 m. The typical storey height is about 3.5 m [2]. An open space is incorporated as a buffer between the living space and the kitchen to prevent fire accidents. In several cases, animals are sheltered in this open space; while in other cases, they are kept separately. Presence of a rear door acts as an escape route in the case of emergencies [1]. Generally, Assam-type houses do not have elevators or fire-protected exit staircases. This type of construction is not limited to housing but used across institutional, educational and commercial typologies as well.

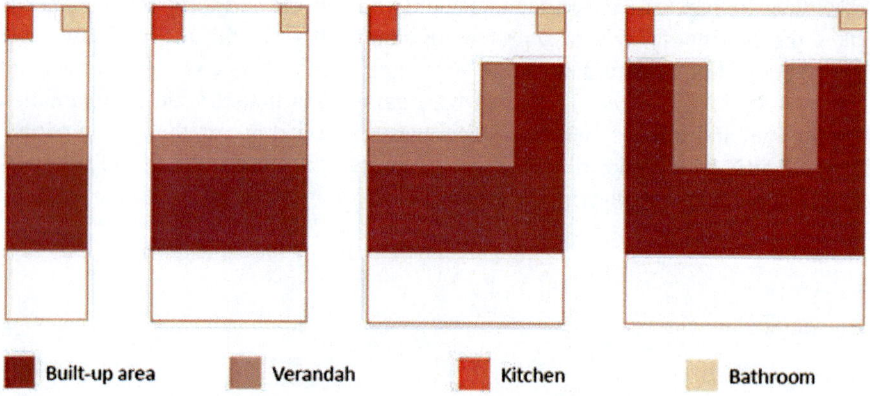

Fig. 78.1 Plan variations of a typical Assam-type house [2]

78.2.2 Chang Ghar

The house is raised on stilts from the ground and is accessed by a flight of 5–7 stairs. Usually occupied by a large family, the house is a large hall with a central kitchen. The house is in the form of a rectangle with linear planning; the rooms open into each other (Fig. 78.2). The stilted portion of the house is used as a shelter for animals in the agrarian regions. In places where flooding is a common scenario, boats are stored underneath the house to use in case of emergencies. Apart from the main house, there is also a traditional granary raised on stilts. The concept of elevation from the ground was devised to prevent animals from attacking the structure and to keep away moisture and rodents during floods.

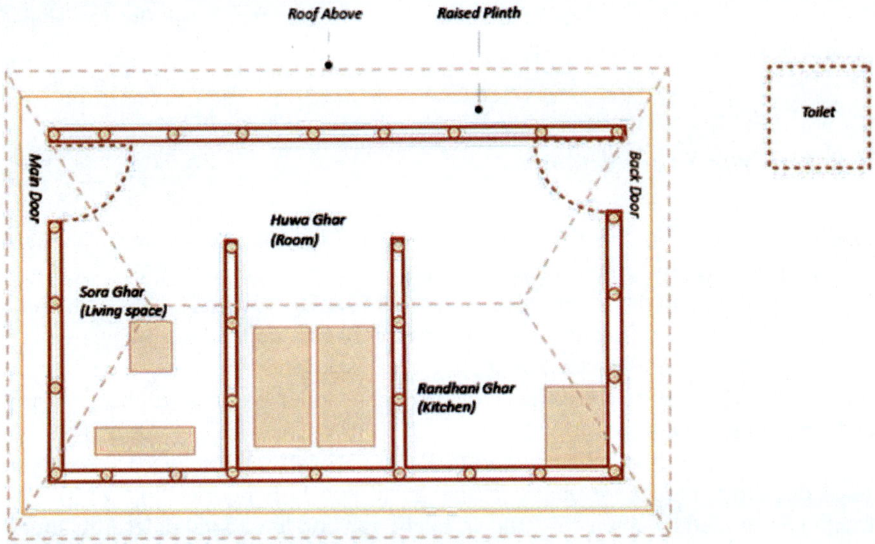

Fig. 78.2 The basic plan of a Chang Ghar unit

78.3 Building Components

This section of the paper elaborates on the different building components of a house such as the structural, flooring, walling and roofing systems and their evolution in Assam-type and Chang houses.

78.3.1 Materials

Both Assam-type houses and Chang houses use locally available materials such as bamboo, mud, cane, palm leaves, reed and timber. Bamboo being light, sturdy, available in different sizes and abundance, makes it highly suitable for construction. Their thermal properties add to the benefit of tackling harsh weather in the region. The choice of materials is one of the significant factors that make the vernacular architecture of the Northeast sustainable.

78.3.2 Structural System

Traditional Assam-type houses do not have formal foundation systems. The main vertical struts, mostly timber or sal, are driven into the ground by 600–900 mm (Fig. 78.3). The main posts have a diameter of 150–250 mm while the intermediate timber posts come in different sizes that are slightly smaller than the main posts [1]. However, in recent cases, formal plinth or burnt-brick foundation or even reinforced concrete footings may be constructed, over which wooden posts are connected using steel clamps (Fig. 78.4).

Chang Ghar is built with bamboo posts. Diagonal bracings, also made of bamboo, are used to strengthen the structure in plain as well as hilly regions (Fig. 78.5). The structure is formed by connecting the vertical posts and horizontal poles of bamboo

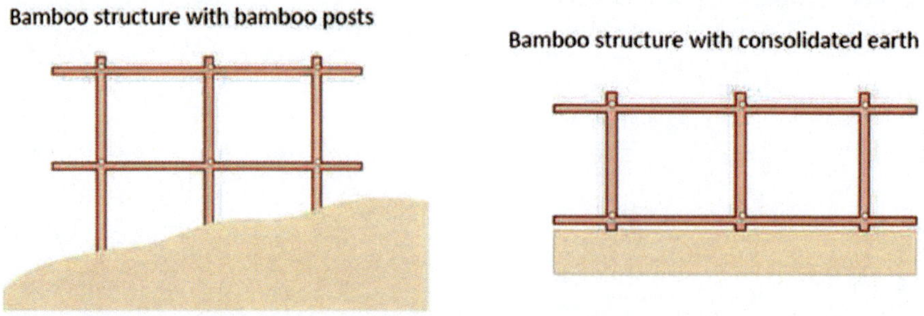

Fig. 78.3 Foundation and structural systems

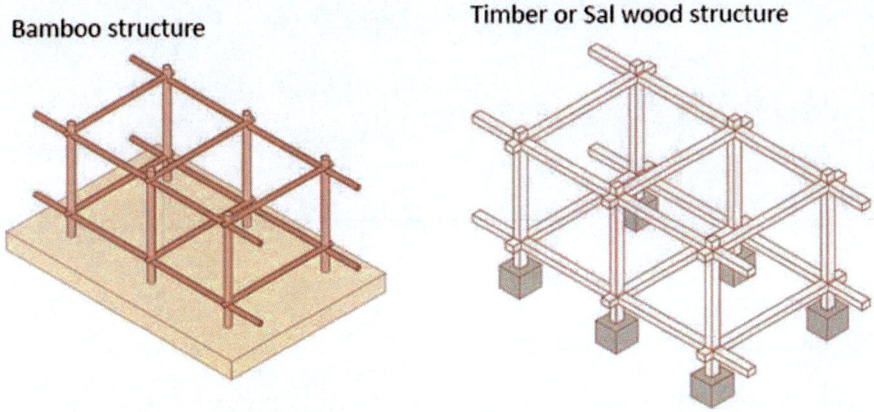

Fig. 78.4 Bamboo and timber structures

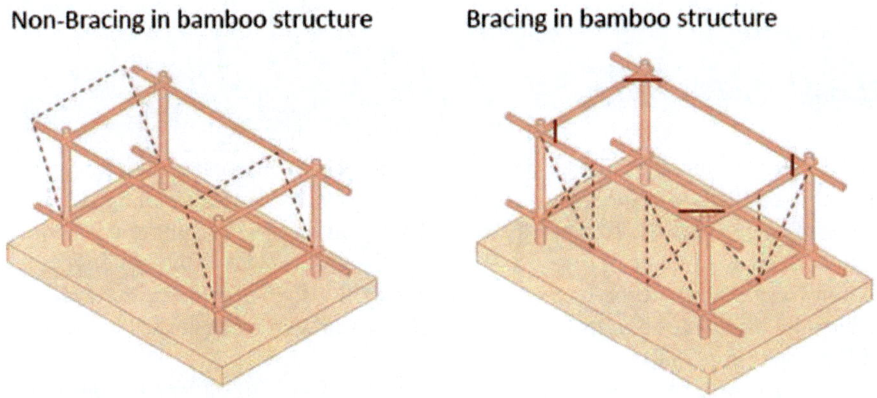

Fig. 78.5 Bracing in bamboo structure

intermediated with diagonal bracing. The stilt height of the structure is around 1.5–2.0 m from the ground. Timber poles are used as vertical struts and driven into the ground in the form of a foundation. The columns of the house are made of bamboo, and the beams are predominantly made of timber. Sometimes, mature bamboo may be used as secondary beams. In modern times, one may find slender concrete columns replacing bamboo and timber in the four corners.

78.3.3 Flooring System

The flooring in Assam-type houses is generally made up of wooden beams with wood planks on top. Often in rural areas, mud plaster flooring is followed.

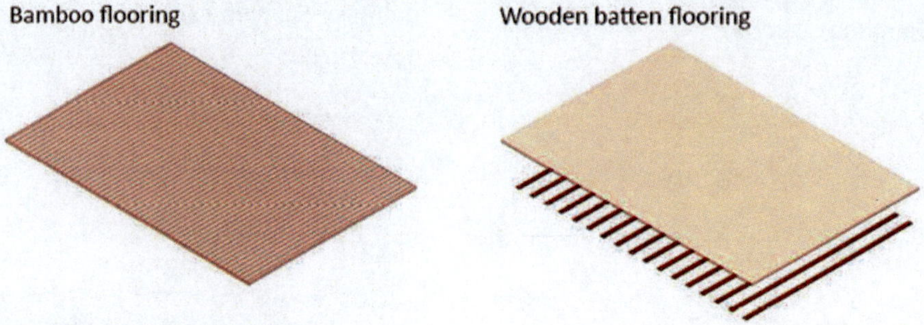

Fig. 78.6 Bamboo and wooden flooring

Chang Ghar has permeable bamboo floors that make it efficient in flood-prone regions since bamboo does not retain water. The flooring is done using split bamboo and finished with flattened bamboo (Fig. 78.6).

78.3.4 Walling System

In Assam-type houses, the wall panels are made of a bamboo framework which is then filled with Ikra reed in the vertical direction while bamboo matting is laid in the horizontal direction. To make them durable, a mud–cow dung mixture is applied onto both sides of the wall (Fig. 78.7). The usage of Ikra reed in the building is the reason for the house to be locally known as Ikra house. These days, the walls are made with brick masonry up to sill level, and chicken wire mesh is used in the place of Ikra reed (Fig. 78.8).

The walls of Chang Ghar are relatively simpler. The walling frame is generally timber or wood with woven or flattened bamboo inlays. They may be finished with a mud–dung mixture similar to Assam-type houses.

78.3.5 Roofing System

Roofing in both the examples of Assam-type and Chang Ghar is high pitched and sloped to tackle the heavy rainfall in the region. Roof trusses in Assam-type houses are made of wooden members upon which a thick layer of Ikra reeds is laid. An improved version of the roof sees a tin/GI sheet instead of reed/thatch (Fig. 78.9). The permeability of the roofing materials influences the angle of slope of the roofs. Thatched roofs are found to have steeper sloping roofs than the modern metal roofs.

The roofs of Chang Ghar are made with truss and posts method. Thatch or layers of reed are laid on top of the roof structure. Sometimes, metal roofing sheets are

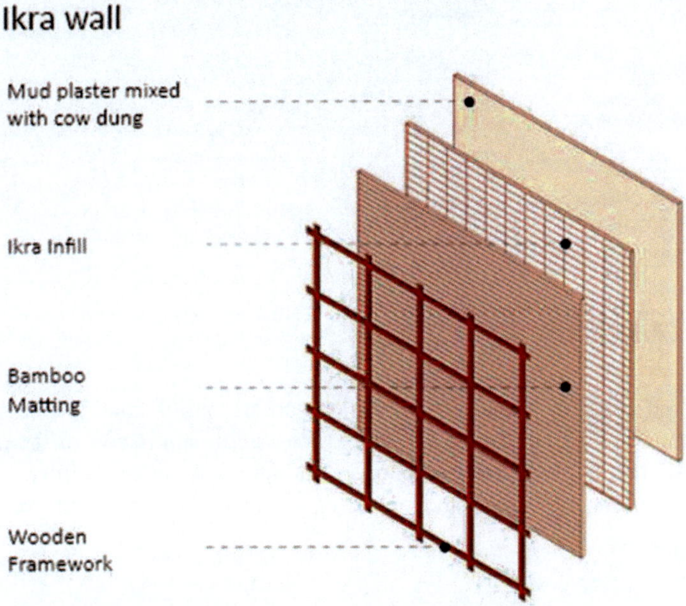

Fig. 78.7 Ikra walling system

Fig. 78.8 Modernised walling system

used, like in the case of Assam-type houses. The clear height from the floor level to the roof is approximately 3.50 m.

Both the housing types utilise the vertical spaces well by making a loft below the roof truss to store various utilities (Fig. 78.9).

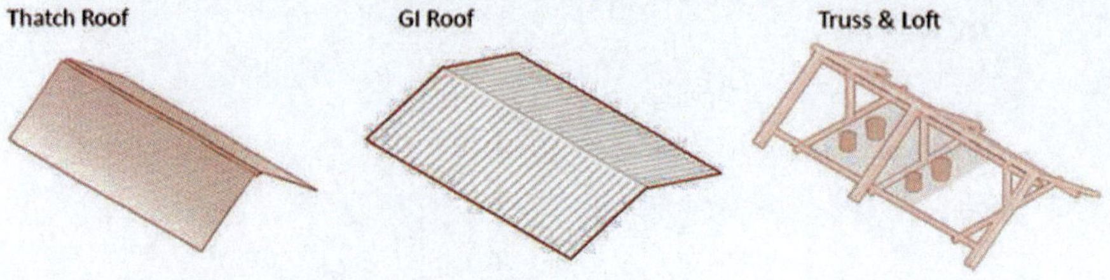

Fig. 78.9 Variations of roofing

78.3.6 Openings

The door and windows in Assam-type houses are generally small in size to cater to the high wind velocity in the region. The frames and main panels of the doors and windows are made of wood or bamboo, depending upon the availability.

In Chang Ghar, the doors and windows are made with weaved and flattened bamboo.

78.3.7 Connections and Joinery

In formal construction, nailing and bolting are the preferred method of connection. In informal cases, the connections are achieved by coir or jute ropes. However, the durability of the latter raises questions of safety. Dowel and tenon joint is the commonly seen form of wooden connections in these housing types.

78.4 Resilience

The climate of Assam is characterised by heavy downpours and frequent showers for long months. The state also lies in the seismic zone V region, which is the most seismically active zone in India. It has experienced several massive earthquakes in the past. Thus, it becomes crucial for the buildings to be able to resist damage. Vernacular architecture in Assam has evolved to ensure the safety of the inhabitants and the durability of the structures through natural disasters. Even the smallest detail or the choice of material contributes to the overall resilience of the structures.

Fig. 78.10 Example of a traditional Assam-type house (*Source* Deccan Herald)

78.4.1 Assam-Type House

The Assam-type house is characterised by good configuration and flexible connections between different elements of the building. The building materials are lightweight. The division of the wall into multiple smaller sections and frames helps in distributing lateral and shear forces during an earthquake. All these aspects of the house make it highly durable and efficient in terms of seismic activities. Their performance has been good during recent earthquakes in the region. However, since the houses are predominantly made of wood, they are prone to fires. This can be moderated by the usage of treated building materials. When these houses are built on contoured regions, the varied lengths of the vertical posts lead to unsymmetrical shaking, causing some damage [1] (Fig. 78.10).

78.4.2 Chang Ghar

Chang Ghar is designed to tackle earthquakes and moderate floods with its use of bamboo for the floor, walls, roof and the overall structure. The lightness of the material makes sure there is no heavy damage. The connections commonly made with coir or jute ropes allow for some amount of flexibility during the earthquake. Since rigidity is one of the causes for major destruction of structures during an earthquake, Chang Ghar performs well. However, during very heavy floods, the stilt houses are not able to withstand the damage (Fig. 78.11).

Fig. 78.11 Mising-tribe Chang Ghar (*Source* The Land Out There)

78.5 Results

It is evident from the case examples that the construction materials used in both Assam-type houses and Chang Ghar are alike–bamboo, wood, cane, reed, thatch, mud and coir. However, the incorporation of these materials into different building components varies when it comes to foundation, walling and flooring systems. It is also noted that several Assam-type houses are raised on stilts from the ground level when constructed on a slope similar to a Chang Ghar.

78.6 Conclusion

The Assam-type house could be seen as the advanced and evolved version of Chang Ghar in terms of the overall building structure, integration of materials and aspect of resilience. The variations in the methodology of construction of the two typologies of houses in Assam provide insight into the heterogeneous character of their location, context, culture and tribal practices, the severity of climate and disasters, and the availability of materials locally. Despite a few constraints of regular maintenance and low durability of materials, these vernacular houses perform very well with regards to weather and seismic events. The lifecycle of these structures is a closed-loop since they are all-natural materials, making the process of construction and habitation sustainable and environmentally friendly.

Acknowledgements This work is carried out under the research project titled "Creating Digital Heritage of Representative Architectural Marvels from Each State of Northeast India", PI: Dr. Shiva Ji, IIT Hyderabad, India.; funded by Department of Science and Technology, Ministry of Science and Technology, Govt. of India, under ICPS/IHDSR scheme.

References

1. Kaushik, H., Ravindra, S.: Assam-type house, World Housing Encyclopedia Report No. 154, Published by Earthquake Engineering Research Institute (EERI), USA and International Association for Earthquake Engineering (IAEE), Japan (2012)
2. Ji, S., Punekar, R.M.: Exploring attributes of vernacular Assam type house design techniques in contemporary setting. In: Reddy, B., Mani, M., Walker, P. (eds.) Earthen Dwellings and Structures. Springer Transactions in Civil and Environmental Engineering, pp 419–433. Springer, Singapore (2019)

Chapter 79
Design for Sustainable Smart Cities; An Impactful Approach Through the Role of Designers Towards Future of Mankind

Veera Venkata Atmakuri, Mohd Saim Nasim Lari, and Arun Thangaraj

Abstract According to United Nations estimates, about 70% of the world's population will live in urban areas by 2050. (https://www.un.org) Smart sustainable cities technologies and approaches, including those based on information and communications technology (ICT) solutions and Internet of things (IoT), are already helping people to overcome multiple challenges of urban development and to progress towards achieving global objectives for cities, particularly Sustainable Development Goal 11 'Make cities and human settlements inclusive, safe, resilient and sustainable' (The first United Smart Cities SMART CITY LAB opened in Vienna, supporting co-creation of solutions for smart sustainable cities worldwide 2018, [1]). The Smart Cities Mission (SCM), launched in 2015, by the Government of India, aims at building up sustainable and inclusive cities that provide core infrastructure, a clean and sustainable environment and a decent quality of life to its citizens. The strategic components of the Mission are city improvement (retrofitting), city renewal (redevelopment) and city extension (greenfield development) with smart solutions applied in service delivery and governance (SDG India Index & Dashboard 2019–2020, [2]). There are several urbanization models that incorporate digital technologies to address some of the urbanization and sustainability challenges: Digital cities feature the integration of digital technology into the city's core infrastructure systems; intelligent cities rely on the digital city infrastructure to build intelligent buildings, transportation systems, schools, enterprises, public spaces, public services, etc., and to integrate them into intelligent urban systems; and smart cities—deploy intelligent urban systems at the service of socio-economic development and improving urban

Note: All authors have made equal contribution to the paper.

V. V. Atmakuri · M. S. N. Lari (✉) · A. Thangaraj
Bangalore, India
e-mail: saim.lari@siemens.com

V. V. Atmakuri
e-mail: veeravenkata.atmakuri@siemens.com

A. Thangaraj
e-mail: arun.thangaraj@siemens.com

A. Chakrabarti et al. (eds.), *Design for Tomorrow—Volume 1*, Smart Innovation, Systems and Technologies 221, https://doi.org/10.1007/978-981-16-0041-8_79

quality of life (Estevez et al, Smart sustainable cities, reconnaissance study 2016, [3]) So, here, we are trying to deal with systems of systems in physical ecosystems powered by various ICT and digital technologies. Meanwhile, the existing technological approaches tend to latch onto the most hyped technology of the year and produce many silo-ed solutions for smart cities which, although elevates technology innovation, fails to engage its citizens. Therefore, a design intervention is needed to help bridge this and make technologies work for humans, the citizens and the primary stakeholder. This interpretation paper is an inquisition into the role of design and designers, to bring the right kind of approach when it comes to sustainable smart city design. The paper first lays down the high-level role and responsibilities of designers and its various other key stakeholders in crafting a smart city. Then, it talks about the right kind of methodologies that designers bring at all levels of sustainable smart city design—from working with official bodies in co-creation and collaboration mode to the end-users—the citizens—to understand their wants, needs and desires. The paper then details out a case study of a smart campus where the aforementioned methods are applied in order to come up with the most effective designs for the smart city that meet the official bodies' KPI requirements and at the same time, delivers best-in-class experiences to citizens.

79.1 Introduction

This is an interpretation paper which attempts to describe a design intervention approach towards design of sustainable smart cities and reports a live case study to showcase the outcomes.

79.1.1 Methodology

Designing a smart city is a collaborative process, where various users and stakeholders needs to be satisfied and promotes interaction and information exchange between various entities and groups in that ecosystem. An exploratory study was conducted to understand various standards, policies and best practices for designing a sustainable smart city in both global and Indian context. The acquired insights and knowledge were applied on an experimental basis in a live scenario and reported as a case study.

The roles and responsibilities described here in this context were defined with the understanding of the current industrial best practices across organizations and governing bodies for sustainability and smart city across globe.

79.1.2 Preamble

Currently, as there is no single official and universally accepted definition of a smart city, let us look at the smart sustainable city definition coming from United Nations Economic Commission for Europe (UNECE) and International Telecommunication Union (ITU). This definition is conceptualized after evaluating more than 100 descriptions of smart city across the globe. It is also in line with the United Nations' Sustainable Development Goal #11, 'Make cities and human settlements inclusive, safe, resilient and sustainable' which is

> A smart sustainable city is an innovative city that uses ICTs and other means to improve quality of life, the efficiency of urban operation and services and competitiveness while ensuring that it meets the needs of present and future generations concerning economic, social, and environmental aspects.

79.1.3 Global Picture

The global population is on the move. According to the United Nations, more than two-thirds of the world's citizens will live in cities by 2050, with 90% of this urban growth taking place in Asia and Africa.

Given this massive migration to urban areas, the subject of sustainability and sustainable cities is top of mind for many. And rightly so, governments, non-governmental organizations (NGOs), corporations and global leaders worldwide are all involved and grappling with a fundamental issue: How can the mega-cities of tomorrow provide essential services such as clean water and adequate waste removal? And how can they create transportation and building infrastructure that reduces the emission of pollutants into the air while taking in so many new residents?

Today, for instance, some 663 million people around the world do not have access to safe and reliable drinking water. (https://www.thirstproject.org) A mass influx of new residents into the world's largest cities will bring a surging tide of issues surrounding access to freshwater for human consumption and basic sanitation.

Mega-cities also mean 'mega' waste disposal issues. The world's cities already generate over 2 billion metric tons of solid waste each year [5]. By 2050, annual waste generation will increase another 70% [6].

And then, there are air quality issues, one of the critical challenges being carbon dioxide (CO_2) emissions. As population epicentres, cities consume over two-thirds of the world's energy and account for more than 70% of global CO_2 emissions [7]. This number will only grow as escalating urban populations demand and use more energy for heating, cooling and transportation. Smart buildings are vital for overcoming many of these challenges because they are responsible for 36% of global energy consumption and nearly 40% of direct and indirect CO_2 emissions [8].

79.1.4 Indian Context

The Smart Cities Mission is the Government of India's program launched in 2015 with an ambition to transform 100 cities in India.

The purpose of the Smart Cities Mission is to drive economic growth and improve the quality of life of people by enabling local area development and harnessing technology, especially technology that leads to smart outcomes. Area-based development will transform existing areas (retrofit and redevelop), including slums, into better-planned ones, thereby improving the liveability of the whole city. New areas (greenfield) will be developed around cities to accommodate the expanding population in urban areas. Application of smart solutions will enable cities to use technology, information and data to improve infrastructure and services. Overall development in this way will improve quality of life, create employment and enhance incomes for all, especially the poor and the disadvantaged, leading to inclusive Cities.

The implementation of the Mission at the city level will be done by a special purpose vehicle (SPV) created for the purpose. The SPV will plan, appraise, approve, release funds, implement, manage, operate, monitor and evaluate the smart city development projects.

Smart Cities Mission guidelines stipulate two strategic components for smart city development: area-based development and pan-city development initiatives.

The area-based development primarily covers three aspects:

- Retrofitting—city improvement
- Redevelopment—city renewal
- Greenfield—city extension.

While pan-city development initiatives include the application of the selected smart solution(s) to the existing city-wide infrastructure, after due consultation with citizens, a priority sector will be identified for implementation. The application of intelligent solutions will involve the use of technology, information and data to make infrastructure and services better.

79.1.5 Current Scenario

Today's cities have various challenges; some of them are listed below. These challenges not only look massive, but they also pose multiple and tricky questions, like

- What solves these challenges?
- What is the best mode of tackling these them?
- Who is the best entity to deal with them?
- Who has the right solution in addressing them?
- Who is responsible for the sustenance of deployed solutions? (Fig. 79.1)

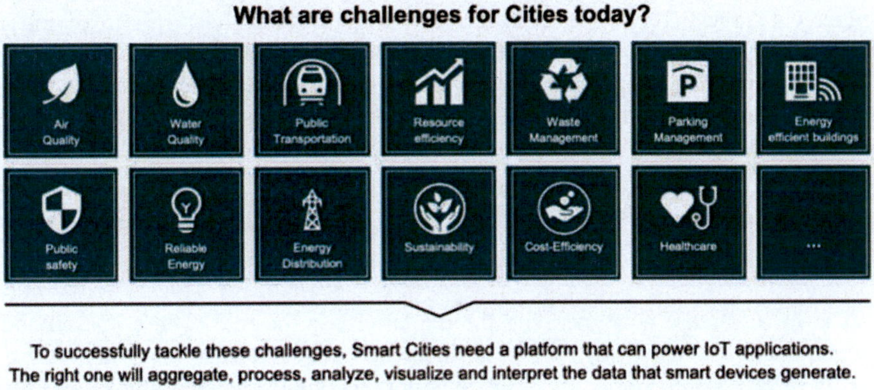

Fig. 79.1 Challenges for cities today. Image by authors

Here, as there is no single solution that can solve any of these challenges, the design team can collaborate with other stakeholders in figuring out the right solution that can be the right fit for the purpose that can ideally leverage the benefit of technologies.

Cities are changing due to new challenges emerging because of the needs of society and even because of the opportunities offered by technologies and modern instruments as in the emerging paradigm of the IoT. The awareness of social and environmental issues forced local and supranational institutions to act.

From a technology and citizen's involvement perspective, as shown in the Fig. 79.2, at a very high level, cities are classified into three:

- Digital City
- Smart City
- Sustainable Smart City.

Digital City, the primary focus is technology and innovation, where the realization of technologies is done through a set of modern deployable tools. Here, the primary goals are to digitize a system (or collection of systems) to increase the effectiveness, productivity or ease.

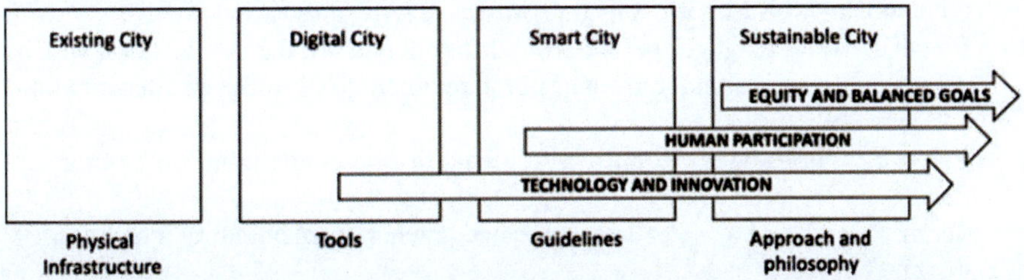

Fig. 79.2 Putting all together, a smart city is a system of systems of systems, encapsulating various in an ecosystem. *Image Source* MDPI, modified by authors

Smart City, the first change, is in terms of technology and services involving a difference in the human aspect as well. It is achieved through the social participation of the citizens. Here, guidelines about how to meet changes in services based on innovative approaches are mainly set, and effectiveness is measured on the scale of making them.

Sustainable Smart City is a new approach to services offered involving the human side of changes towards a sustainable and greener future. Here, the goals and KPIs are the primary focus, and the achievement of these goals is significant on a long-time horizon.

The main feature characterizing the digital city is the role of information and communications technology (ICT), considered a necessary tool to improve citizens' quality of life. As concerns, the smart city, this second conception, thinks the role of new technologies but focuses on citizens' roles as passive actors and active co-creators because they can contribute to the city's governance. Moreover, elements related to sustainability emerge but are intended mainly to keep, for the long term, the results obtained through the smartization process. Finally, concerning the sustainable smart city, as in the previous cases, many elements characterizing the first two conceptions described are recalled but are done so using a new approach and through the filter of modern philosophy: an equitable and balanced set of goals in line with the principles of sustainable development.

79.2 Role, Responsibilities and Methodologies of Designers in Crafting Smart Cities

In order to structure the *how* of organizational transformation, there are seven crucial roles that designers play [9].

1. **Cultural Catalyst**: Offers a safe space to try out new ideas and encourage a proactive and experimental attitude.
2. **Framework Maker**: Offers an understandable, coherent and human-centred process and structure to achieve product/service innovation.
3. **Humanizer**: Humanizing the work and tapping into the deep empathy between the teams as well as with consumers/users.
4. **Power Broker**: Engages people from different parts of the organization and, by using key user needs as a reference point, reconciles silo-induced interpretations and tensions.
5. **Friendly Challenger**: Role of a 'critical friend', challenging how things are done.
6. **Technology Enabler**: Focus on user needs when implementing a technology-driven challenge
7. **Community Builder**: Engages stakeholders and helps create a culture of participation and involvement that can survive through changing contexts.

A city or a campus, in its essence, is not much different from an organization. The major difference lies in its scale. Building upon the works of Joyce Yee [9], we can claim that at the highest level, the designer has a unique and indispensable role and responsibility in crafting a smart city. For the design of a sustainable smart city, the key roles designers play are those of a cultural catalyst, technology enabler and a community builder. Designers work with various stakeholders in understanding and coming up with the right designs and workflows that suite the local and context-sensitive needs that adhere to cultural requirements. The key stakeholders include policy makers, urban planners, standardization bodies, regulatory authorities, governing bodies, public and private enterprises and industrial players who primarily provide technological solutions to that should meet the citizens' needs in making a city smart and sustainable.

Designers and design teams work with official bodies in co-creation and collaboration mode primarily to understand their requirements and their planned key performance indicators (KPI) for the desired smart city.

Designers work with the consumers of the smart city, the citizens, to understand their wants, needs and desires. Here, the designers co-create with the citizens to fish out the most significant needs that matter to them the most.

By taking all these into account, the design teams come up with the most effective designs for the smart city that meet the official bodies' KPI requirements and at the same time deliver best-in-class experiences to citizens.

Here, the most significant challenge and opportunity are coming up with a design solution that not only makes city smart but also makes it sustainable (Fig. 79.3).

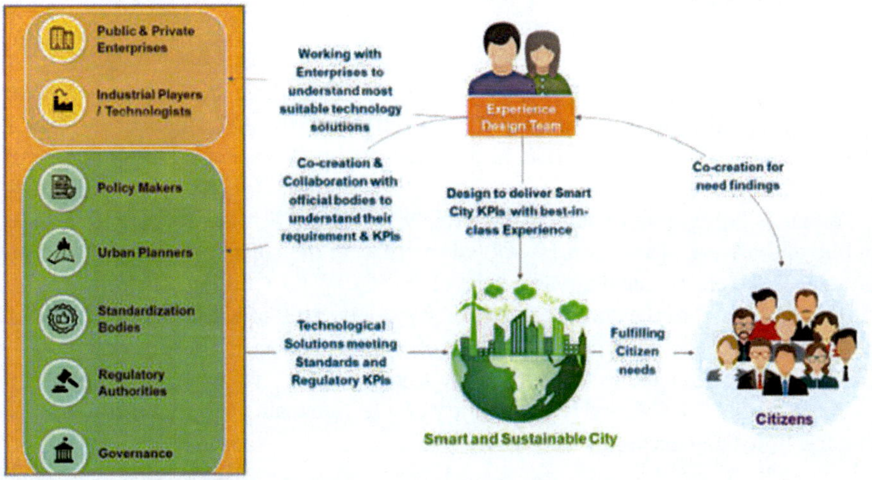

Fig. 79.3 Role of a designer and design team in crafting a smart and sustainable city. Image by authors

Fig. 79.4 Designer's circle of responsibility and influence. Image by authors

The designer and design team have a differentiating role, responsibility and influence when working with various stakeholders. As designing a smart city is a multi-competency design effort, a team of vivid expertise is required here, so a diversified group of experts covering not only the design expertise but also the team need experts from various backgrounds.

Design Team constitutes

- Various design competencies
- Local culture aspects
- Social awareness
- Human science

to name a few.

As consumers of the smart city, the design team as a whole and also each member of the group can provide their inputs as well as verify and validate the overall requirements of the smart city. It is depicted as the 'I' aspect of the influence in the Fig. 79.4.

Whereas being part of any entity (e.g. working with regulatory authorities or public and private enterprises.) involved in creating the smart city, the design team influences the engagement at various levels as the 'We' in the Fig. 79.4.

Finally, as part of the combined forces, e.g. along with officials and technologies, designers also have an excellent opportunity to influence in crafting best-in-class smart city. It is depicted as the 'Us' in the Fig. 79.4.

So, with this, the role and responsibility of a design team are vast and humongous in coming with a successful smart city.

79.3 Smart Campus Case Study

The authors applied the designers' methodologies to come up with design interventions for a smart campus. Due to the confidential nature of work, we will refer it to as 'Campus X'.

The objective of the case study was to observe the designers' role in execution of the design of Campus X and to analyse the various roles they undertook.

Campus X is an interesting case study as it brings design intervention in a highly technology-centric environment. A public event is to be held at Campus X, and the managing bodies wanted to digitalize the entire campus. In order to do this, they are making use of Internet of things which involves deploying hardware sensors at the use-case level, collecting the real-time data, aggregating the data and visualizing them as key performance indicators for stakeholders. There are two main objectives—managing day-to-day operations and strategizing sustainability activities on the campus.

Designers were able to act as 'power brokers' and 'community builders' by bringing together stakeholders (I + We), with vastly different job roles in order to define key performance indicators. For example, the designers ran an on-site collaborative brainstorming session with the stakeholders. The goal was to agree upon KPIs which were most relevant for operations and sustainability management. The designers had conducted a contextual enquiry with both teams in order to understand the KPIs of each team. They had the foresight that these KPIs were related and overlapping. However, technology focused approach hindered acceptance of such a claim given the 'then existing silo-ed culture' within both teams. In order to get the dialogue going, designers were able to bring together the operations team and the sustainability team in a common session and facilitated the session. Bringing together this mix of people from different work areas brought about an epiphany. The operations team and the sustainability team could see that there were overlaps and relationships between their KPIs. The conclusion of the session resulted in a consensus amongst both the teams and paved the way forward for design and development of the solution. With this approach, designers' role was to engage stakeholders and help create a culture of participation and involvement for consensus, thereby playing a role of community builder.

In another instance, designers were able to act as 'technology enablers' by bringing the right kind of communication methods to engage a larger audience of visitors (I + We + Us), who were going to visit Campus X. The key goal was to deliver the message of sustainability and sustainable practices through live-data storytelling in an exhibition area. The designers were able to capture the intent of sustainability messages and made them much more natural and meaningful for visitors who are outside the world of KPIs and data analysis. By bringing in the right type of communication design, designers were able to make technology and data accessible and usable for visitors. Considering that the fear of an unfamiliar user interface and user experiences is real [11], the designers wore the hat of 'technology enabler' and created solutions which eased the visitors into the new interactions as fluidly as possible and engaged them in an effective manner.

The instances from the case study of Campus X clearly illustrate designers' ever-shifting roles and subsequent methodologies to tackle the design of sustainable smart cities and shift the approach from technology-centric to human-centric. It also paves way to establish the value of design interventions, and of designers, from merely being the present interventions of a 'UI makeover', beautification or fixing workflow navigation, towards being a driving force that connects technology with the human beings at the other end; this is what makes design and designers invaluable when it comes to creating an impact towards future of mankind in the context of sustainable smart cities design.

79.4 Conclusion

The biggest challenge in the twenty-first century is knowing how to leverage the benefit of the new generation and upcoming technologies especially when technology is getting cheaper day-by-day with increased accessibility.

Deploying technology without the right mode of execution plan may not fulfil the citizens requirements to the fullest. It is where the designers/design team comes into the picture in defining the right and optimal way to cater to the citizen's needs.

Starting with proposal drafting to final solution systems deployment, designers play a considerable role, even though the technical solutions are the problem-solvers; still, the designers plan a significant role in defining, designing and mode technological solutions.

In short,

> 'Designers' prime responsibility is to make technologies work for Humans to deliver best-in-class positive experiences'.

In both the strategic components of the Smart Cities Mission, area-based development and pan-city development initiatives, designers have a role to play.

References

1. The first United Smart Cities SMART CITY LAB opened in Vienna, supporting co-creation of solutions for smart sustainable cities worldwide (2018). UNECE Press releases
2. SDG India Index & Dashboard (2019–2020). NITI Aayog
3. Estevez, et al.: Smart Sustainable Cities: Reconnaissance Study (2016)
4. United Nations: World Population Prospects 2019: Highlights (2019)
5. The World Bank: What a Waste 2.0: A Global Snapshot of Solid Waste Management to 2050 (2016)
6. The World Bank: Solid Waste Management (2019)
7. C40 Cities. https://www.c40.org
8. The International Energy Agency, IEA Efficient World Strategy
9. Yee, J. et al.: Transformations: 7 Roles to Drive Change by Design. BIS Publishers (2017)

Chapter 80
Water, a Product?

Shubham Vaidya, Aditi Singh, and Parag Anand

Abstract 'Jivanam jivinãm jivaha' is a Sanskrit shlok, cited in an Ayurveda text 'Bhavaprakash,' which translates to 'water is life to creatures.' Water has always been an important source of life on Earth. However, today's water crisis is gripping. In July 2019, Chennai witnessed an acute water crisis, after one of the main reservoirs completely dried up. A report by NITI AYOG pointed out that 21 major cities in India would run out water supply by 2020 (Aayog, "Composite water management index-A tool for water management," August 2019.). In response to this national water crisis, in 2019, the government of India urged its citizens to start a mass movement for water conservation and formed a dedicated ministry to look into water-related affairs. The paper questions this preparedness of citizens of India to counter this catastrophe and whether their systems comply with the idea of the conservation of water that do not allow for an unnecessary water wastage. The blind installation of RO filters, in urban spaces, in the fear of consuming 'impure' water, causes three times the amount of filtered water, to be drained off through reverse osmosis reject system. This has two grave consequences, one on the underground water reserves and the other on the human body. Examining these consequences, this paper will try to understand the fascination with RO filters and its hold over the masses. The paper will make an important intervention by raising the unasked question, are the RO filters necessary and if they do more harm than good? Finally, the paper will end with a feasible solution that may replace the damaging control that RO filters have over the masses. This research started with a design perspective of rethinking how RO water filters should be seen amidst this water crisis. Tests were made, for TDS and pH value, on samples collected from raw and municipal water sources with a brief survey to understand the psychological status of the market.

S. Vaidya (✉) · A. Singh · P. Anand
Department of Industrial Design, School of Planning and Architecture, New Delhi, India
e-mail: ar.id.vaidya@gmail.com

A. Singh
e-mail: ditisingh@spa.ac.in

P. Anand
e-mail: paraganand@spa.ac.in

A. Chakrabarti et al. (eds.), *Design for Tomorrow—Volume 1*, Smart Innovation, Systems and Technologies 221, https://doi.org/10.1007/978-981-16-0041-8_80

80.1 Introduction: The Problem

The widely used reverse osmosis water filtration systems that are nowadays installed in every urban household reject a certain amount of water for every fraction of purified water that they provide. Since this rejected water's volume is much higher than the volume of the purified water, these filtration systems directly contribute to the ongoing water crisis that plagues both the country and the world. The paper investigates how smart processes along with consumer awareness can provide a solution to this problem.

80.1.1 The Water Crisis: World and India

Although 70% of the earth's surface is covered with water, in this twenty-first century, there is an alarming water crisis gripping the world, where 785 million people lack even a basic drinking water service [2] and about 2.7 billion people experience water scarcity for at least one month of the year [3], according to the WHO.

India, home to the second largest population in the world, does not remain unaffected from the consequences of this crisis. In fact, Bengaluru is on the second position among the top 11 global cities that are likely to run out of drinking water by 2030, as predicted by the UN [4]. As a consequence, an exponential growth of water purification products and services has been observed in the ever growing urban areas of the country [5].

80.1.2 The Contributor: Reverse Osmosis Water Filters

Reverse osmosis water filters cause three times the amount of actual filtered water, to be drained off as a reject in the process [6]. To validate this, the authors conducted a survey with their RO water filters and found that to produce 20 L of filtered, approximately 60 L of water were drained off through the reject pipe which results in a yearly wastage of 21,900 L of semi-filtered water.

80.2 Studying Potable Water in Urban Areas

The first study tries to understand the whole process of domestic water filtration in urban areas which is followed by a detailed discussion of how the water filtration systems are contributing to the ongoing water crisis.

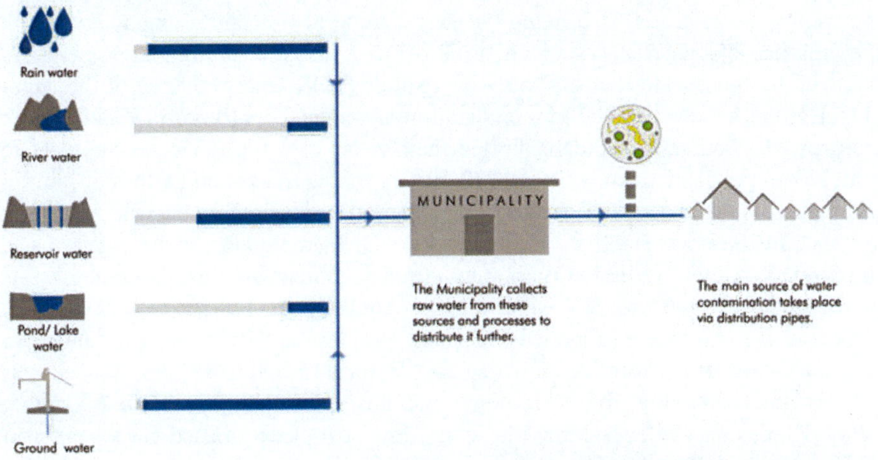

Fig. 80.1 Water supply from municipality at the urban domestic level [20]. The blue bar represents the comparative percentage, ground water is the highest. (*Source* Author's own)

80.2.1 Necessity: Potable Water

According to WHO, 'Potable water' is the water that can be used for drinking and cooking purposes [7]. Although they constitute only a small percentage, calcium, sodium, potassium, chlorine, magnesium, iron, zinc, copper, chromium, iodine, cobalt, molybdenum and selenium form essential components of this water and are present in their most stable form, and absence of these minerals would result in water to be distilled in nature [8].The pH of water should be close to being neutral (7+ on the pH scale). In concise, drinking water should be contaminant free, mineral rich, alkaline in nature and above all neutral to taste.

80.2.2 Domestic Water Supply

The water quality in urban areas mainly determined by the source of the supply and how the municipality distributes it within the available resources. The illustration (Fig. 80.1) [9] explains the water supply network and the potential point of water contamination could be in its distributions systems [10].

80.2.3 The Marketing the Water Filters

The uncertainty involved at the municipal distribution system makes the domestic households look for other filtration methods to purify the water. The filtration industry intervenes at two levels, stabilizing the excess mineral concentration and nullifying the presence of bacteria, chemicals, toxins, etc. A basic scientific assessment can be made by measuring the total dissolved solids (TDS) and pH value of the water [11]. Before the advent of commercialization of water filters (before 1980s), a large majority of population typically boiled their water or used a drip candle filter to remove dirt particles from it. Post 1980, driven by the market and a loss of faith in the municipal supplied water, the general masses started to opt for new water filtration systems. In 1983, Ion Exchange launched Zero B, a simple tap attachment for basic filtration. With an aggressive marketing strategy, Eureka Forbes also entered this market in 1980s, with its UV water purifier-Aqua Guard. Further, companies like Kent and Ion Exchange launched RO water purifiers in 1990s, but they failed to create a significant market because the cost factor attached to RO purifiers. It was not until the 2000s that these big conglomerates realized this problem, following which Hindustan Unilever came out with a low cost gravity-based carbon filter in 2004, priced at INR 2000. This phase marked the beginning of rigorous marketing by various water filtration companies to woo their customers into buying their products. By late 2000s, RO water purifiers also made a comeback, and now, a singular product would fetch them 10 times the amount of regular filters. All RO water purifiers needed maintenance, which the supplier companies cashed upon. After selling RO water purifiers, the customer got locked in for the next 4–5 years. After a stagnant market, new selling strategies were made, which totally brushed up the mass psychology of water in people's minds. Initially, it included facts and figures about the depleting water quality around the country which swiftly moved into them roping in film stars to endorse their products. Kent was the first company to use celebrities as their brand ambassadors in their marketing strategies. This marked the next phase, where brand ambassadors were projected as standards of quality lifestyle in the mind-set of people about their water filtration [12].

80.2.4 Filter Selection Method

The criteria for the selection of the best domestic filtration methods are based on total dissolved solids (TDS) and bacterial impurities. The following graphic (Fig. 80.2) explains what passes through what, in membrane filtration [13]. The important intervention here is: how can a consumer select the best filter for their water.

80.2.5 The Ultimate Solution: RO Water Filtration Process

RO water filtration system is the most commonly used water purification method. It employs multiple pre-filtration membranes driven through a pressure pump to filter out impurities of different sizes at different stages [14]. This process makes the water filtered come close to a demineralized state, Therefore, this form of filtration is very

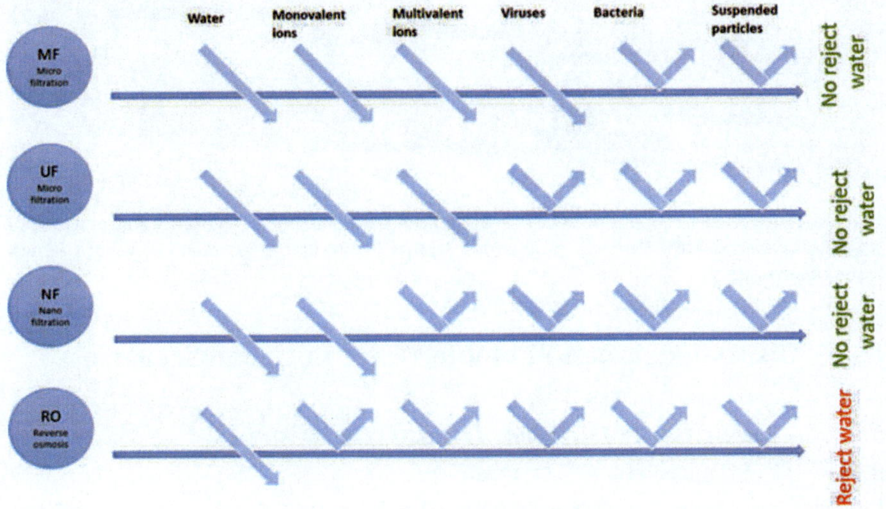

Fig. 80.2 Porosity level of different membranes. (*Source* https://www.hinesburg.org/water-project/safewaterdotorg-info-nano-and-ultrafiltration-reverse-osmosis.pdf)

effective where water has heavy metal contaminants like arsenic, fluoride, etc., in a quantity more than permissible limit. Hence, these filters can provide a solution when the quality and source of water are totally unknown.

80.2.6 Standards for Potable Drinking Water

Organizations like World Health Organization (WHO) and Bureau of Indian Standards (BIS) have set some standards (Table 80.1) for drinking water, in which TDS and pH are the initial factors to determine water [15], 16].

Table 80.1 TDS and pH standard

	WHO		BIS	
Characteristic	Upper limit	Lower limit	Upper limit	Lower limit
TDS	500 ppm	N.A	300 ppm	N.A
pH	6.5	8.5	6.5	8.5

(1) (2) (3) (4)

Fig. 80.3 Advertisements of different RO water filter brands on their e-portals. (*Source* 1-https://www.kent.co.in/; 2-https://www.livpure.com/; 3-https://www.draquaguard.co.in/; 4-https://www.bluemountro.com/)

80.3 Discussion Around Potable Water in Urban Scenario

80.3.1 Fact Check: Different Qualities of Water

Reverse osmosis (RO) filtration industry thrives on the absence of the minimum water quality standards, and their filters are inefficient if the water TDS is less than 300 ppm [17], contradicting its ubiquitous nature in India. To establish this, the authors collected domestic supplied samples of water to test its TDS. The readings ranged from 120 to 300 ppm for Delhi, 1420–1500 ppm for NOIDA and 2400–2900 ppm for Ghaziabad, which indicated that a fluctuating nature of TDS could be one of the reasons why masses are adopting these filters.

80.3.2 The Market Forces

The marketing strategy of water filtration brands relies on the assumed fact that, regardless of the source, the quality of water is not safe for consumption. On an on-ground investigation by the authors, it was noted that the facts and figures presented by RO company's representatives create and fuel this apparent truth (Fig. 80.3).

80.3.3 The Impact: Physical, Economic and Ecological Perspectives

WHO in its report published in 2004 stated that '*not only does completely demineralized water (distillate) have unsatisfactory organoleptic properties, but it also has a definite adverse influence on the animal and human organ*' [18]. However, the RO water purifiers completely demineralize the water to such an extent that the acidity level drop as low as 4.5 pH on the scale. Consumption of this water can cause serious issues in our body, as its starts extracting the deficit minerals to reach a steady state. While economically the poorer households survive on the private filtration plants supplying cans of water, the major consumer, the middle- and upper-income households, opt for these systems. It could cost about 3500 INR and can go up to 3.5 lakh INR per unit for domestic use with a minimum annual maintenance charge of 2000 INR. The only common factor which binds all the consumers together is the fear of unclean water, which is instilled by the commercial filtration companies. The major externality is borne by the soil in the form of increased concentration of total dissolved solids, as the reject water drains off and finally gets constituted with the ground water. This continues a cycle of extraction of high TDS water from the ground which further increases the concentration of reject water on RO filtration.

80.4 Water Quality Testing in a Domestic Scenario

80.4.1 The Cost of Testing

To make water potable, a certain level of filtration is required which depends on its contents. However, on a domestic level, water testing is often overlooked, and an extreme filtration method is often adopted. The authors investigated that a complete test in a government-based certified lab costs about 15,000–25,000 INR per sample, which is not an economical option. The only resort a consumer is left with is to make an involuntary decision based on the trends prevalent in his/her neighborhood and information passed through the water filtration companies.

80.4.2 Practical Solution: Apparatus

A series of basic water quality tests were conducted by the authors for TDS and pH of water with the apparatus shown in Fig. 80.4 to understand various water qualities.

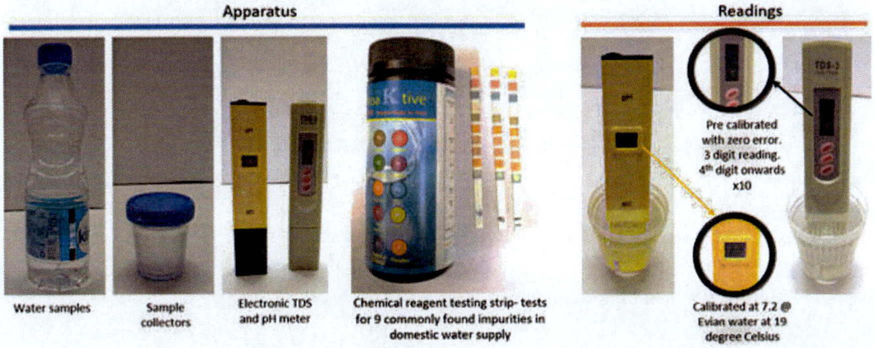

Fig. 80.4 Apparatus used to carry out water quality tests. (*Source* Author's own)

Fig. 80.5 Visual representation of results of the tested samples corresponding to their source. (*Source* Author's own)

80.4.3 Test 1: Raw and Municipal Supplied Water Samples

The samples collected were recorded for their TDS and pH value. Municipal supplied water showed adequate properties in Delhi, where the primary source is river water, but when compared to neighboring areas like Ghaziabad, the overall quality of water depleted and showed explicit levels of TDS going above 2500 ppm (Fig. 80.5).

80.4.4 Test 2: Bottled Water Samples

Samples collected from various well-known brands were tested for TDS and pH to understand the standards of water quality the industry is following (Fig. 80.6). The more expensive water bottled brands showed better TDS and pH reading compared to the local, less expensive brands. Aquafina packaged mineral water showed the properties of demineralized water.

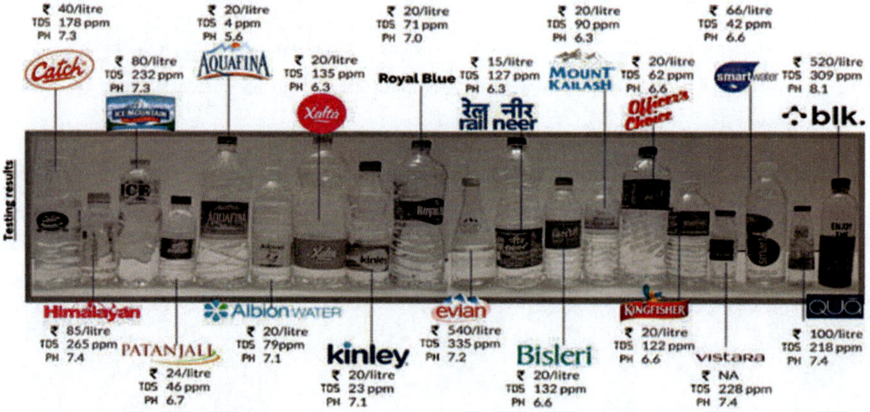

Fig. 80.6 Bottled water samples and their results. (*Source* Author's own)

80.4.5 *The Understanding of Consumer's Perspective*

An online survey was conducted across the country to understand how aware the people are about their water filtration. The survey recorded more than 400 responses in which 65.4% of the subjects were qualified to the level of postgraduate with about 69% of them belonging to or living in metropolitan cities. Most of them were aware of the source of water, its supply and its filtration processes. RO water purifiers were used by 63.8%, whereas ceramic filters were used by only 2.1% of the total subjects. The survey showed that most of the people trusted the local recommendations and 43.5% of the responses agreed that marketing had a major influence in their decision. About 95% of the subjects were uneducated about, how their water is being filtered. Around 80.4% of the subjects thought that rejected water from the RO filters cannot be reused as it might contain bacteria and other impurities. A total of 54.9% of the subjects thought of bottled water as a standard. The survey clearly pointed to the fact that RO water purifiers have portrayed a self-image of giving an ultimate resort to the consumer, as it thoroughly filters the water, whatever the quality may be. Economics does not have a major influence, as the information portrayed to the consumer is terrifying enough to push him to make a decision in favor of their sales.

80.5 Discussion of Water Quality Testing in an Urban Scenario

Our system lacks instant water testing. Although on national level, the mind-sets have started to change, and people are educating themselves about the ill effects of RO water filtration. In a recent article published in the Times of India on 31st may 2019, the National Green Tribunal (NGT) has asked the center to ban RO water purifiers where the TDS is lower than 500 ppmv [19]. NGT also suggested to raise awareness among the general masses about the ill effects of demineralized water on their health.

80.5.1 *The Correct Scale: Understanding When RO Filtration is Required*

After an extensive research and considering various international standards, a basic scale was created by the authors to understand at what TDS (in ppm) level will the RO filtration be required. An alternative widely being used is the nano filtration technology which almost eliminates the same amount of impurities as RO membrane, but without rejecting any water. It is because of the economics of setting up an RO membrane costing the same as a nano filtration membrane, that the former is given the preference, due to its extreme filtration capabilities (Fig. 80.7)..

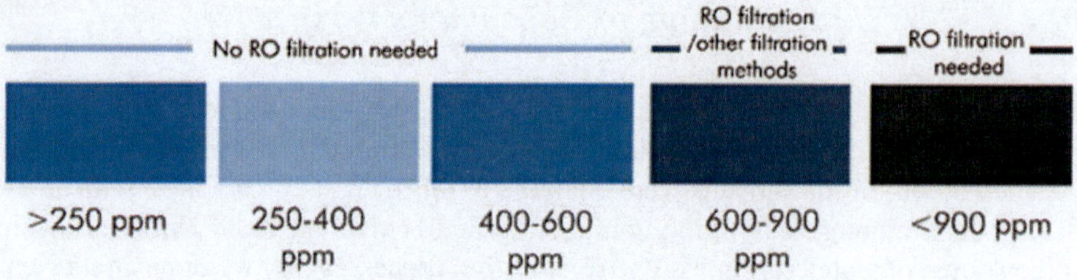

Fig. 80.7 Filtration method on basis of TDS of water. (*Source* Author's own)

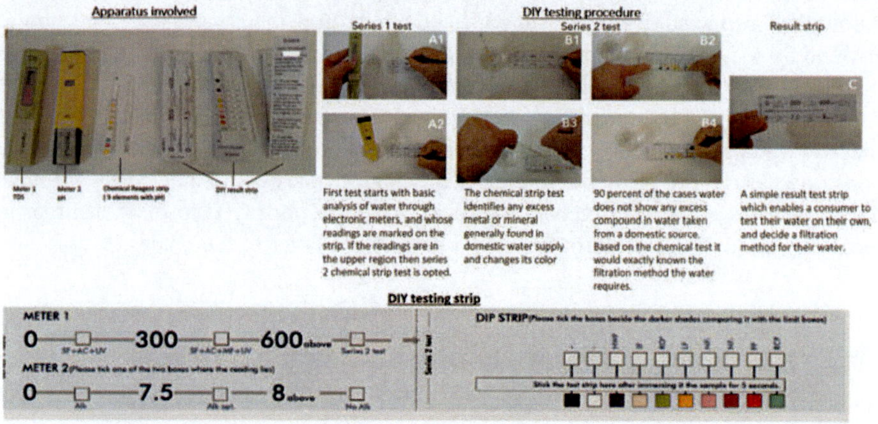

Fig. 80.8 Devised DIY strip method for water quality testing. (*Source* Author's own)

80.5.2 Instant Testing for a Custom Filter

A check box method was devised where a preliminary test suggests a filtration method, and if required, it follows up with a DIY chemical dip strip (major household water contaminants, generally 7–10 in number), which finally provides an accurate filtration solution for the given water sample (Fig. 80.8).

80.5.3 3 Stage Domestic Water Test

The domestic water testing could provide a definitive solution to the problem of blind installation of RO water filtration systems. A thorough research in understanding the systems revealed that most of the places do not need a RO water filtration. Although the water quality differs from place to place, an individual's decision of the water purification system is largely influenced by the market trends. The research was directed toward finding a solution for this problem at a domestic level. A design

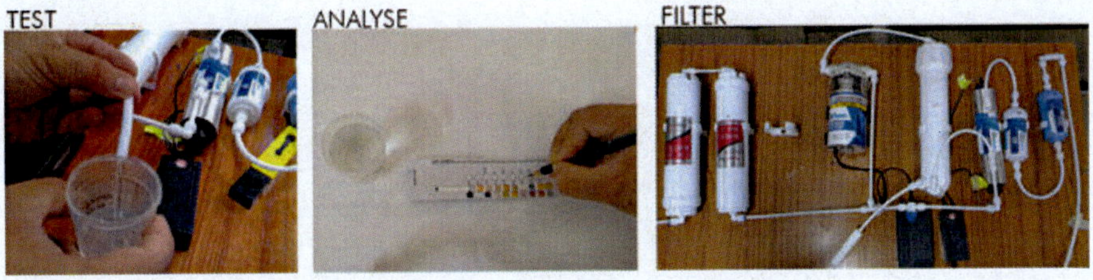

Fig. 80.9 Test, analyze and custom filtration method. (*Source* Author's own)

led exercise devised a system where existing technology could intervene to determine the water quality and propose a method to assign the most appropriate water filtration. Rethinking the whole system of testing and design implication was the main derivative of this research. A systematic and data-based solution, which could cater to the problem, resulted in a three-step procedure applicable to domestic water filtration needs. Test-Analyze-Filter (Fig. 80.9).

The first stage of the 'Test' would include a basic DIY test to understand the water quality, which would include the TDS and pH on a preliminary stage and a basic chemical strip test to evaluate the contents of water. The second stage 'Analyze' is to process this information and provide a solution based on standards set by international organizations. The third stage is based out of the first two stages which would suggest a series of filters needed by the water. According to this data, the current series of filtration membranes could be altered to provide the best filtration for the water.

80.5.4 Design-Driven Solution

The three-stage mechanism eliminates all points of human error and misinformation which influence the decision of a consumer. This concept innovates a smart system where the quality of water decides which kind of filtration it needs to achieve a potable state. This design could bridge the gap between the user and the product by providing the right information in a user-friendly way and assist the user to make an appropriate selection of filtration system. The situation demands to educate the masses about the health and environmental hazard they are progressing toward, only because of a series of uninformed decisions.

The research indicated that a majority of well-educated people lack the proper knowledge when it comes to understanding water and its filtration systems. This gap is used as an opportunity by the marketing agencies of these filtration service providers. A huge profit margin system in RO water filters and expensive service package is attracting businesses into this market. This growth has proportionally led to an increase in demand for water filters in urban areas. Considering a situation, where every urban household is relying on RO filtered water is dangerous, where millions of liters of semi-filtered water are drained off daily, to obtain a small proportion of filtered water. In this century, when all our products are approaching a smart

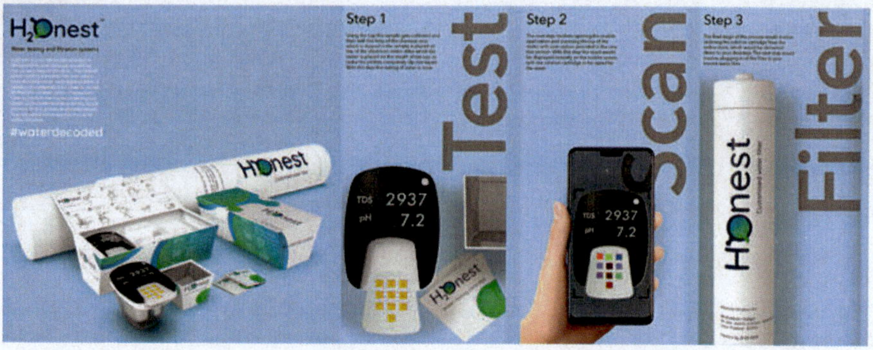

Fig. 80.10 H_2O nest, a system to instantly test, analyze and manufacture custom water filters is being developed by the author's. (*Source* Author's own)

technology to achieve the best possible outcome, the water purifiers are lagging behind. A 'Smart' water purification system, based on integration of technology with the existing filtration systems, and personalization may be a possible way forward for ensuring safe drinking water in the future. A huge scope lies in the field of water filtration, where design could possibly induce a change. The authors are investigating feasible design solutions (Fig. 80.10) which could contribute to this ongoing problem.

References

1. N. Aayog: Composite water management index—a tool for water management (2019)
2. W. H. Organisation: Fact sheet on "Drinking water, 14 June 2019. Available https://www.who.int/news-room/fact-sheets/detail/drinking-water
3. Hoekstra, A., Mekonnen, M.: Four billion people facing severe water scarcity, 12 February 2016. Available https://advances.sciencemag.org/content/advances/2/2/e1500323.full.pdf
4. The 11 cities most likely to run out of drinking water, British Broadcasting Cooperation, 11 February 2018. Available https://www.bbc.com/news/world-42982959
5. Aniket Kadam, S.B.: Water purifier Market by Technology (UV, RO, and Gravity based), Distribution Channel (Retail Sales, Direct Sales, and Online), Portability (Portable, and Non-portable), and End User (Commercial and Residential), June 2019. Available https://www.alliedmarketresearch.com/water-purifier-market
6. Krishner, F.: Reverse osmosis: need or waste, 22 March 2015. Available: https://timesofindia.indiatimes.com/city/patna/Reverse-osmosis-Need-or-waste/articleshow/46648223.cms
7. W. H. Organisation: Water sanitation hygiene. Available: https://www.who.int/water_sanitation_health/water-quality/guidelines/en/
8. Kozisek, F.: Health risk from drinking, de-mineralised water. WHO, National Institute of Public Health Czech Republic
9. W. Aid: State of Urban Water Supply in India (2018). Available https://www.wateraidindia.in/sites/g/files/jkxoof336/files/state-of-urban-water-supply.pdf
10. G. O. I. National disaster management authority, Threats to Municipal Water Supply and Water Reservoirs, July 2010

11. Anngneal, 7 Ways to Measure, Monitor, and Evaluate Water Quality, 08 December 2017. Available: https://publiclab.org/notes/anngneal/12-08-2017/7-ways-to-measure-monitor-and-evaluate-water-quality
12. Seth, B.L.: How pure are their claims? 07 June 2015. Available: https://www.downtoearth.org.in/coverage/how-pure-are-their-claims-2252
13. Safewater.org: Safe drinking water foundation- Ultra filtration, Nano filtration and Reverse osmosis. Available: https://www.hinesburg.org/water-project/safewaterdotorg-info-nano-and-ultrafiltration-reverse-osmosis.pdf
14. Dupont, R.R., Eisenberg, T.N., Middlebrooks: Reverse Osmosis in the treatment of "drinking water" (1982). Available: https://digitalcommons.usu.edu/water_rep/505
15. B.-. B. O. I. Standards, Indian Standard" drinking water"-specification (Second revision). New Delhi (2012)
16. W. H. Organisation, "Guidelines for drinking-water quality," vol. 2nd ed. Vol. 2, no. Health criteria andother supporting information, 1996.
17. WHO: Total dissolved solids in drinking-water. World Health Organistaion, Geneva (1996)
18. Rosborg, I.; Health effects of de-mineralization of drinking water. In: Drinking Water Minerals and Mineral Balance, WHO, pp. 149–160
19. Singh, P.: Problem only where borewells are being used, 30 May 2019. Available: https://timesofindia.indiatimes.com/city/delhi/problem-only-where-borewells-are-being-used/articleshow/69570087.cms
20. Das, B.: India's water resources—Availability, usage and problems, June 2009. Available: https://base.d-p-h.info/fr/fiches/dph/fiche-dph-7825.html